GEOGRAPHY
REGIONS AND CONCEPTS

To The Student: A Study Guide for the textbook is available through your college bookstore under the title *Study Guide to Accompany Geography: Regions and Concepts,* 6th edition, by H. J. de Blij and Peter O. Muller. The Study Guide can help you with course material by acting as a tutorial, review, and study aid. If the Study Guide is not in stock, ask the bookstore manager to order a copy for you.

SIXTH EDITION

GEOGRAPHY
REGIONS
AND
CONCEPTS

H. J. DE BLIJ
University of Miami

PETER O. MULLER
University of Miami

John Wiley & Sons, Inc.
New York   Chichester   Brisbane   Toronto   Singapore

ACQUISITIONS EDITOR: Barry Harmon
DEVELOPMENTAL EDITOR: Barbara Heaney
DESIGNER: Ann Marie Renzi
COPYEDITING SUPERVISOR: Gilda Stahl
PRODUCTION SUPERVISOR: Linda Muriello
PHOTO RESEARCH MANAGER: Stella Kupferberg
MANUFACTURING MANAGER: Lorraine Fumoso
MANAGING EDITOR: Joan Kalkut

Front Cover Photograph: Guido Alberto Rossi / The Image Bank
Back Cover Photograph: Steve McCorry / Magnum Photos

Library of Congress Cataloging in Publication Data:

De Blij, Harm J.
    Geography, regions and concepts / Harm J. De Blij, Peter O.
Muller.—6th ed.
        p.   cm.
    Includes bibliographical references and indexes.
    ISBN 0-471-51850-6
    1. Geography. I. Muller, Peter O. II. Title.
G128.D42 1991   90-12959
910—dc20   CIP

Printed in the United States of America

10 9 8 7 6 5 4 3

PREFACE

WORLD REGIONAL GEOGRAPHY

Our book has a twofold purpose. It discusses the world's great geographic realms and their human and physical contents, their fortunes and failures, links and barriers, potentials and prospects. It also introduces geography itself, the discipline that links human society and natural environment through a special and fascinating perspective that is first explained in the introductory chapter and then is developed as the book progresses. As such, *Geography: Regions and Concepts* constitutes an antidote against the so-called geographic illiteracy that has been diagnosed by Gallup and other recent polls, classroom tests, and public surveys. This endemic geographic illiteracy, now so commonly mentioned in the press, resulted substantially from a neglect of the very topics this book is about. Before we can usefully discuss such commonplace topics as our "shrinking world," our "global village," and our "distant linkages," we should know what the parts are, the elements that are shrinking and linking. This is not merely an academic exercise. Knowledge of the world beyond our borders is a crucial asset in consumer decision-making, international business initiatives, and the global competition that faces a nation with worldwide interests. We can gain this knowledge by studying the layout of our world—i.e., its geography—not by memorization, but by learning where people and things are

located and why, how they interact, what impels them to move or migrate, and how they prefer to shape and fashion their abodes. These are geographic themes, and every time we learn something more about a city, a region, a prevailing climate, or a cluster of resources, we also learn something more about geography itself.

OUR APPROACH

This book follows a format that has worked well for two decades, linking geography's basic concepts with an overview of the realms and major regions of our fast-changing, late twentieth-century world. We have placed more than 150 ideas and concepts in a regional perspective. Most of these concepts are decidedly geographical; others are ideas about which we believe students of geography should have some knowledge. Of course, we have not listed on the chapter-opening page every idea and concept used in that chapter (although each term listed is now indicated in boldface when it first appears in the text). Most teachers, we suspect, will want to make their own region-concept associations, and as readers will readily perceive, the book's organization is quite flexible. In fact, instructors have a great range of opportunities to shape a course to their liking, to transfer concepts, and to focus presentations mainly on conceptual matters or on regionally-oriented materials. Moreover, concepts are sometimes raised but

not pursued in depth, so that the lecturer may choose to penetrate them in greater detail and/or in a comparative regional context.

THE CURRENT EDITION

The twentieth-anniversary, sixth edition of *Geography: Regions and Concepts* contains an overall tightening of the text and an exhaustive, detailed updating of its contents. The most recent developments in the Soviet Union necessitated a complete rewriting of Chapter 2, which in addition to its previous topical and regional coverage now includes profiles of each of the 15 Soviet Socialist Republics. The Europe chapter (1), which was already scheduled for substantial revision, fully incorporates the dramatic changes of 1989–1990 in Eastern Europe and anticipates the consequences of a reunified Germany. We felt that our coverage of China could be improved as well, and most of Chapter 9 was rewritten. The chapter on Subsaharan Africa (7) also received a major overhaul, as did smaller sections of many other chapters and vignettes that are too numerous to list here. We also paid considerable attention to highlighting individual countries in the text, continuing a trend begun in the most recent editions.

As usual, we have thoroughly re-evaluated our maps, and most were revised to keep them as current and useful as possible (we even managed to include the unification of North and South Yemen, which occurred unexpectedly just as we went to press). Our maps of the world's realms and major regions received special scrutiny. The overview map of world geographic realms (Fig. I–19) was substantially modified to improve its internal regionalization. These changes were also transferred to each of our 11 chapters in the form of new, full-page chapter-opener maps that elucidate the regional framework for the coverage to come.

PRONUNCIATION GUIDES

Among the many new items introduced in the previous edition—almost all of which are retained—were the pronunciation guides located at the end of each chapter and vignette. In choosing words for inclusion (largely place names), we decided not to list words that were pronounced the way they were written unless we thought mispronunciation was likely. Although we strove for authenticity throughout, we aimed for Americanized rather than native-language pronunciations. For many place names, our initial guide was the 1988 edition of *Webster's New Geographical Dictionary*. In choosing the phonetic presentation method, we kept things as simple as possible by avoiding a formalized symbol system that would have required constant decoding. Accordingly, we employed a syllabic phonetic-spelling system with stress syllables italicized (for example, we pronounce our surnames duh-*blay* and *mull*-uh). The most frequently used vowel sounds would translate as follows: *ah* as in f*a*ther, *oh* as in t*o*ne, *au* as in *ou*t, and *uh* as in b*a*n*a*n*a*.

CONTINUING SPECIAL FEATURES OF THIS BOOK

Three continuing special features of this textbook are the geographic qualities boxes, the systematic essays, and the model boxes. Near the beginning of each regional chapter and vignette, we list, in boxed format, the 10 major "geographic qualities" that best summarize that portion of the earth's surface. Also located near the opening of each of the 10 regional chapters is a systematic essay that focuses on a major topical subfield of human or physical geography; each of these overviews was carefully selected so that its contents tie in to regional-geographic material subsequently developed in the chapter (this entire program is depicted in Fig. I–22). We also deploy three model boxes (in the introductory, first, and sixth chapters) to explore more fully the application of modeling in three fundamental areas: large-scale climate regionalization, agricultural location theory, and spatial diffusion processes and patterns.

DATA SOURCES

The population figures used in the text are our projections for 1991 (unless otherwise indicated) and are consistent with the national demographic data displayed in Appendix A. The chief source that we used as a basis for developing our projections was the 1989 *World Population Data Sheet* published by the Population Reference Bureau, Inc. The urban population figures—which entail a far greater problem in reliability and comparability—are mainly extrapolations drawn from the 1989 version of the International Data Base compiled by the U.S. Bureau of the Census, supplemented by other sources as necessary. (Previous users of this book should be aware that the [1980-based] UN data set used in earlier editions is no longer considered reliable; thus, most urban totals shown in this edition are incompatible with those used by us in 1988). At any

rate, the urban population figures used here are estimates for 1991, and they represent *metropolitan-area totals* unless otherwise specified.

ANCILLARIES

The following ancillaries were prepared to accompany this edition of the book, and they may be obtained by contacting John Wiley & Sons:

- **Student Study Guide:** Contains objectives, glossary, self-test questions, map exercises, practice exams, research guides, and outline maps to provide additional support for the student.

- **Instructor's Manual:** Contains an outline, a brief description, and a listing of key terms presented for each chapter. In addition, essay and short-answer test questions are provided with appropriate referenced page numbers where the information is first introduced in the text.

- **Test Bank:** For each chapter approximately 100 questions are keyed to the text in a multiple-choice, true-false, fill-in-the-blank, and matching format.

- **Computerized Test Bank:** Available for the IBM or the Macintosh.

- **Overhead Transparencies:** Approximately 70 four-color transparencies of maps and line drawings from the text.

- **Slide Set:** Full-color slides of approximately 100 maps and line art figures from the text.

- **Supplementary Slide Set:** Contains approximately 100 slides from the authors' personal collection of photographs not found in the text.

- **Videotape:** A collection of appearances made by H. J. de Blij as he appeared as Geography Editor on ABC's *Good Morn-*

ing America. Topics include the five-part special "Geographically Speaking" series, as well as additional informative discussions of geographical importance. Each segment runs about six minutes in length.

- **Instructor's Resource Manual:** The above ancillary items are available in a special binder to help organize the complete supplementary package.

ENVOI

Finally, to the student reader about to embark on the exploration of world geography, we leave you with the following exhortations offered by the renowned author, James Michener, in his 1970 article in *Social Education* (pp. 764–766):

The more I work in the social-studies field the more convinced I become that geography is the foundation of all. . . . When I begin work on a new area—something I have been called upon to do rather frequently in my adult life—I invariably start with the best geography I can find. This takes precedence over everything else, even history, because I need to ground myself in the fundamentals which have governed and in a sense limited human development. . . . If I were a young man with any talent for expressing myself, and if I wanted to make myself indispensable to my society, I would devote eight or ten years to the real mastery of one of the earth's major regions. I would learn languages, the religions, the customs, the value systems, the history, the nationalisms, **and above all the geography** *[emphasis added], and when that was completed I would be in a position to write about that region, and I would be invaluable to my nation, for I would be the bridge of understanding to the alien culture. We have seen how crucial such bridges can be.*

ACKNOWLEDGMENTS

In the course of this latest revision, we were fortunate to receive advice and assistance from many people.

One of the rewards associated with the publication of a book of this kind is the correspondence and other feedback that it generates. Over the years, we have heard from colleagues, students, and lay readers. Geographers, economists, political scientists, education specialists, and others have written us, almost always with helpful suggestions, often with fascinating enclosures. We have responded personally to every such letter, and our editors have communicated with many of our correspondents as well. We have, moreover, considered every suggestion made—and many who wrote or transmitted their reactions through other channels will see their recommendations in print in the current edition. The list that follows is merely representative of a group of colleagues across North America to whom we are grateful for taking the time to share their thoughts and opinions with us.

Robert J. Aalberts, Louisiana State University–Shreveport

Ian H. Ackroyd-Kelly, East Stroudsburg University (Pennsylvania)

W. Frank Ainsley, University of North Carolina-Wilmington

Chris Airriess, Ball State University (Indiana)

Joel M. Andress, Central Washington University

Joseph L. Arbena, Clemson University (South Carolina)

Saul Aronow, Lamar University (Texas)

Donna Bannasch, Webber College (Florida)

Lt. Col. C. Taylor Barnes, U.S. Air Force Academy (Colorado)

H. Gardiner Barnum, University of Vermont

Steven Bass, Paradise Valley Community College (Arizona)

Thomas F. Baucom, Jacksonville State University (Alabama)

Sanford H. Bederman, Georgia State University

John V. Bergen, Western Illinois University

Alan C. G. Best, Boston University

D.M.S. Bhatia, Austin Peay State University (Tennessee)

Mark Binkley, Mississippi State University

James H. Bower, Quincy College (Illinois)

Henry Boyce, Montgomery College (Maryland)

James A. Brooks, Central Washington University

Ayse E. Brown, Clemson University (South Carolina)

Joseph W. Brownell, SUNY Cortland (New York)

D.E. Bruyere, University of Wisconsin-Oshkosh

Fred Burbach, Seminole Community College (Florida)

Jonathan Campbell, Monroe County Community College (Michigan)

Alvar W. Carlson, Bowling Green State University (Ohio)

Gary A. Carlson, Midland Lutheran College (Nebraska)

A.K. Chakravarti, University of Saskatchewan, Canada

Sen-dou Chang, University of Hawaii at Manoa

James R. Chrisman, Black Hills State College (South Dakota)

John E. Coffman, University of Houston

Dwayne Cole, Grand Rapids Baptist College (Michigan)

Omar Conrad, Maplewoods Community College (Missouri)

Allan Cooper, St. Augustine's College (North Carolina)

John Cross, University of Wisconsin-Oshkosh

John M. Crowley, University of Montana

Brian Cullis, U.S. Air Force Academy (Colorado)

R.W. Cullison, Towson State College (Maryland)

Richard Cutler, St. Michael's College (Vermont)

Rudolph Daniels, Morningside College (Iowa)

Richard D. Dastyck, Fullerton College (California)

Claud Davidson, Texas Tech University

James Delehanty, University of Wisconsin-Madison

Dennis J. Dingemans, University of California, Davis

Clifton V. Dixon, Memphis State University

Randolph J. Dohne, Southwest Baptist University (Missouri)

John W. Donaldson, Liberty University (Virginia)

W.N. Duffett, University of Arizona

J.T. Dykstra, Menlo College (California)

Fred Edinger, Coker College (South Carolina)

Julie Elbert, University of Southern Mississippi

Jane H. Elliott, Lyndon State College (Vermont)

C.W. Fassone, Suffolk, Virginia

James S. Feng, De Anza College (California)

George A. Finchum, Milligan College (Tennessee)

Gary S. Freedom, McNeese State University (Lousiana)

James F. Fryman, University of Northern Iowa

Richard Fusch, Ohio Wesleyan University

Randy Gabrys-Alexson, University of Wisconsin-Superior

Gary L. Gaile, University of Colorado

Fred Gamble, Pensacola Junior College (Florida)

Russel Gerlach, Southwest Missouri State University

A.E. Goedeken, William Penn College (Iowa)

Daniel Good, Georgia Southern College

Marvin Gordon, George Washington University (District of Columbia)

D.E. Green, Lock Haven University (Pennsylvania)

Gary M. Green, University of North Alabama

Charles R. Gunter, Jr., East Tennessee State University

Prof. Haddad, St. Xavier College (Illinois)

Ruth F. Hale, University of Wisconsin-River Falls

Thomas J. Hannon, Slippery Rock University (Pennsylvania)

Robert Hartig, Fort Valley State College (Georgia)

Thomas L. Havill, Keene State College (New Hampshire)

John R. Healy, University of Richmond

Martha Henderson, University of Minnesota-Duluth

Roger Henrie, Central Michigan University

John Hickey, Inver Hills Community College (Minnesota)

Louise M. Hill, University of South Carolina-Spartanburg

Mario Hiraoka, Millersville University (Pennsylvania)

Deryck Holdsworth, Pennsylvania State University

Lutz Holzner, University of Wisconsin-Milwaukee

Richard F. Hough, San Francisco State University

Janice Humble, University of Kentucky

Darshan S. Kang, University of Montana

Kevin Kearns, University of Northern Colorado

Karen Kincheloe, North Harris County College (Texas)

Patricia E. Kixmiller, Miami-Dade Community College (Florida)

Thomas Klak, Ohio State University

Dorothy Lamson, Dean Junior College (Massachusetts)

Lucille Lance, Olney Central College (Illinois)

Charles W. Lasher, Lexington, Kentucky

Larry League, Dickinson State University (North Dakota)

John C. Lewis, Monroe, Louisiana

Charles Lieble, Valdosta State College (Georgia)

Robert Lind, Kearney State College (Nebraska)

Z.L. Lipchinsky, Berea College (Kentucky)

John H. Litcher, Wake Forest University (North Carolina)

Harry Loomer, University of Wisconsin Center-Barron County

Robin R. Lyons, Leeward Community College (Hawaii)

Bernard McGonigle, Philadelphia Community College

Roger McVannan, Broome Community College (New York)

Nelson Madore, Thomas College (Maine)

Robert B. Mancell, Eastern Michigan University

Christopher L. Martin, Louisiana State University-Shreveport

Kent Mathewson, Louisiana State University

Paul Meartz, Mayville State University (North Dakota)

Barry Miles, Clackamas Community College (Oregon)

Mark M. Miller, University of Southern Mississippi

Tony Misko, Cuyahoga Community College (Ohio)

Malcolm Murray, Georgia State University

Gary Nachtigall, Fresno Pacific College (California)

Godson C. Obia, Kearney State College (Nebraska)

Ray O'Brien, Bucks County Community College (Pennsylvania)

William F. O'Connor, Piedmont College (Georgia)

G.E. O'Donnell, Clinch Valley College (Virginia)

Prof. O'Malley, West Georgia College

Barry Packard, Lancaster Bible College (Pennsylvania)

Prof. Pancyzk, Bergen County Community College (New Jersey)

Linda Peake, Grand Valley State University (Michigan)

James R. Penn, Louisiana State University

Jerome M. Petry, Jamestown Community College (New York)

Paul E. Phillips, Fort Hays State University (Kansas)

Charles Pirtle, Georgetown University (District of Columbia)

John Powell, Hannibal-La Grange College (Missouri)

Lee Ann Powers, College of Southern Idaho

Vinton M. Price, Jr., Wilmington College (Ohio)

Milton Rafferty, Southwest Missouri State University

Mushtaqur Rahman, Iowa State University

L.A. Reddick, Spoon River College (Illinois)

Wolf Roder, University of Cincinnati

Rodney J. Ross, Harrisburg Area Community College (Pennsylvania)

Raj Ryali, Auburn University (Alabama)

Kurt Schroeder, University of Minnesota-Duluth

Axel Schuessler, Wartburg College (Iowa)

Earl J. Senninger, Mott Community College (Michigan)

William R. Siddall, Kansas State University

Richard G. Silvernail, University of South Carolina

Kenn Sinclair, Holyoke Community College (Massachusetts)

David R. Skeen, Muskingum College (Ohio)

Betty A. Smith, Kennesaw State College (Georgia)

Lynn R. Sprankle, Kutztown University (Pennsylvania)

George Stoops, Mankato State University (Minnesota)

J. L. Steinitz, Erie Community College (New York)

Rolf Sternberg, Montclair State College (New Jersey)

William R. Strong, University of North Alabama

George J. Suchand, California Polytechnic University

David W. Suitor, St. Mark's School (Massachusetts)

B.L. Sukhwal, University of Wisconsin-Platteville

Michael L. Talbott, Bellevue Community College (Washington)

Wayne Taylor, Athens, Texas

Ranjit Tirtha, Eastern Michigan University

J. Peter Trexler, Juniata College (Pennsylvania)

Ralph Triplette, Western Carolina University (North Carolina)

Charles C. Webb, Emporia State University (Kansas)

Jamie Whelan, McNeese State University (Lousiana)

P. Gary White, Western Carolina University (North Carolina)

Stephen E. White, Kansas State University

Morton D. Winsberg, Florida State University

Roger Winsor, Appalachian State University (North Carolina)

Charles E. Womack, Mattoon, Illinois

William Wycoff, Montana State University

Gary A. Yoggy, Corning Community College (New York)

We also wish to single out a number of people for special mention and thank them again publicly. Our colleague at the University of Miami, Ira M. Sheskin, assisted us with a number of vital tasks, not the least of which was the massive job of optically scanning the previous edition so that the manuscript could be prepared directly on the personal computer. Another faculty colleague, John (Dick) Stephens, helped us improve several maps, offered advice on many more, and oversaw the preparation of certain maps by Lisa Meday, one of our graduate students. The work of Stephen S. Birdsall (University of North Carolina), who in earlier editions contributed much of Chapters 3 and 7, continues to enhance our presentation. Gene C. Wilken (Colorado State University) and his students continued to send us detailed lists that showed us where the written text could be made more precise. Other geographers took the time to do likewise, and many of their recommendations appear in this edition; they include Canute VanderMeer (University of Vermont), Kathleen Braden (Seattle Pacific University), J.D. (Doug) Eyre (University of North Carolina), Nigel J.R. Allan (Louisiana State University), James P. Allen (California State University, Northridge), John P. Augelli (University of Kansas), Hlib S. Hayuk (Towson State University [Maryland]), and Arthur J. Krim (Salve Regina College [Rhode Island]).

We also consulted with a number of individuals on various matters. Richard L. Forstall, Chief of the Demographic Statistics Branch of the Population Division, U.S. Bureau of the Census—who has guided so many geographers so diligently for so many years—kindly gave us the benefit of his vast experience in our quest for the best international urban data; his Bureau colleague, Ed Gates, supplied us with then unpublished materials and astutely advised us on how to fill in some of the unavoidable gaps. Brock Brown (University of Colorado), one of our discipline's most gifted teachers, gave us much insight into the changing realities of presenting introductory geography to very large lecture sections. Edward J. Malecki (University of Florida) also discussed his experiences in the presentation of this material and shared with us his knowledge of high-technology activities in Europe and the Soviet Union. Richard W. Wilkie (University of Massachusetts) continued to work with us to improve our population diagrams of South America. Our valued friend, Harry Schaleman (University of South Florida), inspired us to do our best every time we encountered him, and we are also indebted to all our other colleagues who work with us on the Florida Geographic Alliance. We were delighted to receive comments on the Hypothetical Continent and Residential Preferences maps from Mr. James Michener, who graced our campus as author-in-residence during the writing of his recent work on the Caribbean. Betsy Muller again produced the gazetteer, interrupting yet another summer to do so. This new edition also owes many debts to those who helped us in the past. The pronunciation guides were originally prepared with the vital input of Melinda S. Meade (University of North Carolina), Clifton W. Pannell (University of Georgia), and Gerald G. Curtis (University of Miami, Foreign Languages). Ray Henkel (Arizona State University) made available a particularly useful manuscript entitled "Regional Analysis of the Latin American Cocaine Industry." We gratefully acknowledge, too, the comments of many anonymous reviewers, all of which were considered in the revision process. The errors that remain are, of course, ours alone.

At the University of Miami's Department of Geography, we are indebted to everyone for their support. Our faculty colleagues— Tom Boswell, Don Capone, Jan Nijman, Ira Sheskin, Dick Stephens —are all involved in the teaching of world regional geography, and their constant stream of good-natured challenges mixed with candid advice, as ever, was invaluable throughout the revision process. A wide range of critical supporting tasks were cheerfully and tirelessly performed by our office staff, headed by Susan Duncan (senior secretary).

At the National Geographic Society, where until recently H.J. de Blij served as Editor, Marta Marschalko ensured that lines of communication remained open, and we thank her for more than six years of unwavering support, assistance, and allegiance.

From start to finish, the preparation of this handsome volume has benefited immeasurably from the professionalism of the "first team" and supporting staff at John Wiley. We had never before worked with a developmental editor, but Barbara Heaney masterminded the entire production process so proficiently that we now cannot imagine working without one in the future. We were most fortunate to again have Linda Muriello as production supervisor for the manuscript, whose awesome efficiency and telephone skills not only got us through the galley and page proof stages on time, but also made us feel as if we were the only authors she was working with. Our map coordinator, and a true joy to work with, was Martin Bentz, who did a brilliant job in transforming our sometimes excessive demands into full-color cartographic reality. Another tour-de-force was turned in by Stella Kupferberg, Wiley's photo research manager, who went to incredible lengths to satisfy our every new photographic wish and turned her very sharp critical eye toward improving those pictures that were retained. Copyediting duties were most capably handled by Elizabeth Hovinen (who is also a professional geographer), and her supervisor, Gilda Stahl, who kept us on track and made many useful contributions as well. Ann Marie Renzi was the designer of this latest edition, and deserves most of the credit for the physical appearance of the final product. Joan Kalkut was instrumental in transferring the word-processed manuscript into usable form for the compositor. Barry Harmon, Wiley's new geography editor, came on board just as things were heating up; happily, he hit the ground running, and we are thankful for his enthusiastic support, attention to endless details, and exhortations to keep us going when it mattered most. Others at John Wiley who provided assistance over the past two years were Stephanie Happer, Patricia Young, Dennis Sawicki, Bonnie Lieberman, Kaye Pace, Cynthia Michelsen, Ed Starr, Catherine Faduska, Madelyn Lesure, Katharine Rubin, Lucille Buonocore, Andrea Bryant, and Hilary Newman.

Finally, we thank our wives, Bonnie and Nancy, for their constant encouragement of all our professional activities.

June 20, 1990   **H. J. de Blij**

Peter O. Muller

Coral Gables, Florida

CONTENTS

INTRODUCTION

WORLD REGIONAL GEOGRAPHY: PHYSICAL AND HUMAN FOUNDATIONS
1

PART ONE

DEVELOPED REALMS
57

PART TWO

UNDERDEVELOPED REALMS
245

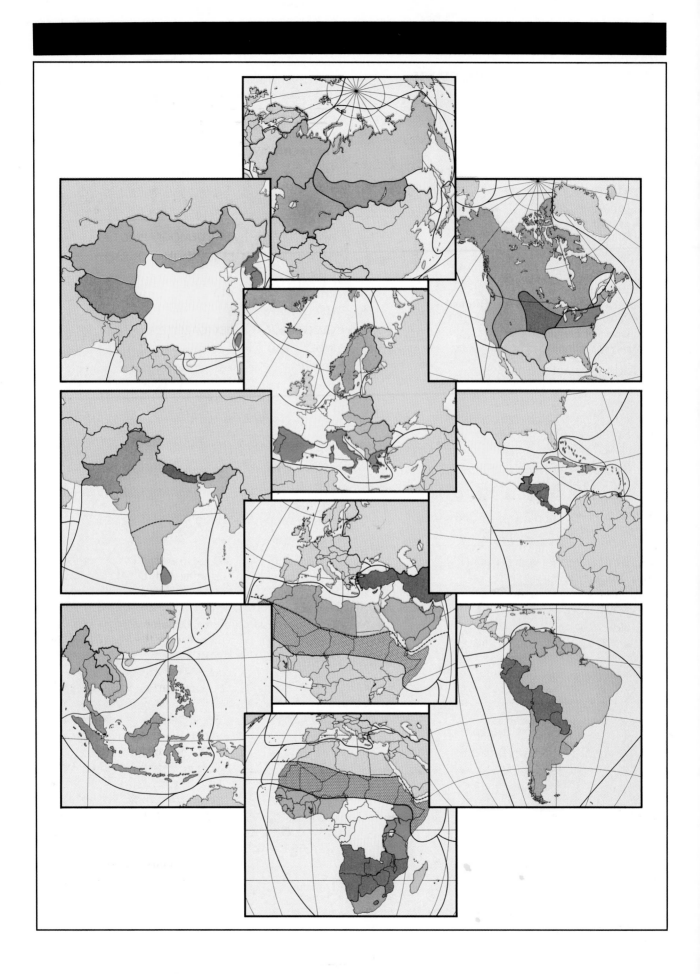

WORLD REGIONAL GEOGRAPHY: PHYSICAL AND HUMAN FOUNDATIONS

This book is about the world's great realms, surveyed and discussed in geographic perspective. Geographers study the locations and distributions of phenomena (human as well as physical) on the earth's surface. They investigate the reasons or causes behind these distributions, and in their research they try to predict how and why change will take place. The geographic perspective, therefore, is a spatial viewpoint. Just as historians focus on time and chronology, geographers concentrate on space and place. The language of geography is full of spatial terms: pattern, area, distance, direction, proximity, accessibility, isolation, clustering, and many others that reflect this spatial focus.

In this book, we use the geographic perspective and geography's spatial terminology to study the world's major geographic realms. Each of these realms of the human world (such as Europe, North America, or South Asia) possesses a special combination of cultural, environmental, historical, economic, and organizational qualities. These characteristic properties are imprinted on the landscape, giving each region its own flavor and social milieu. Geographers take a particular interest in the way people have decided to arrange and order their

---

IDEAS AND CONCEPTS

regional concepts

scale

culture

cultural landscape

natural environments

glaciation cycles

climatic regions

spatial models

population agglomerations

urbanization

development

core-periphery relationships

world geographic realms

regional geography

four basic traditions

systematic geography

map reading

map interpretation

---

living space. The street pattern of a traditional Arab town differs markedly from the layout of a Chinese settlement of similar size. The fields and farms of Subsaharan Africa look quite unlike those of the Soviet Union. Thus the study of *world realms* also pro-

vides the opportunity to examine the concepts and ideas that form the basis of the modern field of *geography*. These are our twin objectives.

CONCEPTS OF REGIONS

Modern scientific concepts can be complicated, but we use others almost without realizing it in our everyday conversation. Among the most fundamental concepts of geography are those involving the identification, classification, and analysis of regions. When we refer to some part of our country (the Midwest, for instance), or to a distant area of the world (such as the Middle East), or even to a section of the metropolitan area in which we may live (the inner central city, for example), we employ a regional concept. We reveal our perception of distant or local space, our mental image of the region to which we make reference.

Everybody has some idea of what the word *region* means, and we use the regional concept frequently in its broadest sense as a frame of reference. But **regional**

concepts are anything but simple. Take just one implication of a regional name we just used: the Midwest. If the Midwest is indeed a region of the United States, then it must have limits. Those limits, however, are open to debate. In his book *North America*, John Paterson stated that the Midwest includes the states of Ohio, Michigan, Indiana, Illinois, Wisconsin, Minnesota, Iowa, and Missouri, "but in the cultural sense it can also be said to include much of the area of heavy industry in Pennsylvania and West Virginia." Compare this definition to that of Otis Starkey and his co-authors, who defined the Midwest in their book *The Anglo-American Realm* as consisting of "most of the West North Central states [North and South Dakota, Nebraska, Kansas, Minnesota, Iowa, and Missouri] to which have been added the western parts of Wisconsin and Illinois."

These two perceptions of the Midwest as a U.S. region obviously differ. Does this invalidate the whole idea of an American Midwest? Not necessarily: the apparent conflict arises from the use of different *criteria* to give specific meaning to a regional term that has long been a part of American cultural life. Your own personal impression of the Midwest as a region is based on certain properties you have reason to consider important. When you add to your information base, you may modify your definition. Regionalization is the geographer's means of classification, and regions, like all classes, have their bases in established criteria. Classification schemes are open to change as new knowledge emerges, and so are regional definitions. A 1985 study of the Midwest underscores this. James Shortridge surveyed college students in 32 states and discovered that a sizable majority shared the perception of Starkey and his colleagues, which favored

a more westerly delimitation of the region, excluding Michigan and Ohio. This finding was consistent with other definitional variables and is probably related to the continuing overall shift of the U.S. population toward the west (and the southern-tier Sunbelt).

Regions obviously have *location*. Various means can be employed to identify a region's position on the earth, as the authors quoted previously did when they enumerated the states that form part of their conception of the American Midwest. Often a region's name reveals much about its location. During the Vietnam War (1964–1975), the name *Indochina* became familiar to Americans; it identifies an area in Southeast Asia that has received cultural infusions from India and human migrations from China. Sometimes we have a particular landscape in mind when we designate a region—for example, the Amazon Basin or the Rocky Mountains. It is also possible, of course, to denote a region's location by reference to the earth's grid system and to record its latitudinal and longitudinal position. That would give us the extent of

its *absolute location*, but such a numerical index would not have much practical value. Location is relevant only when it relates to other locations. Hence, many regional names carry references to other regions (*Middle* America, *Eastern* Europe, *Equatorial* Africa). Such names indicate a region's *relative location*, a much more meaningful and practical criterion.

Regions also have *area*. Again, this appears to be so obvious that it hardly requires emphasis, but some difficult problems are involved here. For example, certain regions are identified as the (San Francisco) Bay Area, the Greater New York Area, or Chicagoland. Everyone could probably agree that each is focused on a few internal urban concentrations, but what are the limits or boundaries of such metropolitan-centered regions? In quite another context, we use such terms as the *Corn Belt*—an agricultural region in the central United States—and the *Sunbelt*—a broad zone across the southern U.S. that has recently attracted large numbers of migrants (and employers) who wished to escape the rigors of northern win-

FIGURE I–1 Regional boundary: the edge of the Nile Valley in central Egypt.

ters. The geographical or spatial extent of a region, whether Bay Area or Corn Belt, Midwest or Middle East, cannot be established and defined without reference to its specific areal contents.

A prominent characteristic of a region's contents is *homogeneity*, or sameness. Sometimes the landscape leaves no doubt about where one region ends and another begins: in Egypt, the break between the green, irrigated, cultivated lands along the Nile River and the desert beyond is razor-sharp and all-pervading (Fig. I–1). On the map, the line representing that break is without question a regional boundary. Everything changes beyond that line—population density, vegetation, soil quality, land use. But regional distinctions are usually not so clear. The example of the U.S. Corn Belt (mapped in Fig. 3–24) provides a good contrast. Traveling northward from Kentucky into Illinois or Indiana, you would undoubtedly be struck by the increasing number of cornfields in each county. Since not all farmland is under corn, even within the Corn Belt, the difference between what you saw in Kentucky and Illinois is a matter of degree. Consequently, in order to define a Corn Belt and represent it on a map, it would be necessary to establish a criterion—for instance, 50 percent or more of all the cultivated land in a county must be devoted to growing corn. The line so drawn would delimit an agricultural region, but on the landscape it would be far less evident than the boundary enclosing Egypt's Nile Valley farmlands.

It is possible, of course, to increase the number of ingredients so that more than one condition must be satisfied before the region is delimited. To define a particular cultural region, such criteria as the use of a certain language, adherence to a specific religion, perhaps even the spatial variations of ar-

REGIONAL TERMINOLOGY

FIGURE I–2 Regional ties around the town of Garden City, Iowa.

The internal uniformity of a homogeneous region can be expressed by human (cultural, economic) as well as natural (physical) criteria. A country constitutes such a region defined in political terms, for within its boundaries certain conditions of nationality, law, government, and political tradition prevail throughout. Similarly, a natural region such as the Rocky Mountains or the Mississippi Delta is expressed by the dominance of a particular physical landscape. Quebec and the Corn Belt are uniform cultural and agricultural regions, respectively. Regions marked by this internal homogeneity are classified as *formal regions*.

Regions conceptualized as *spatial systems*—such as those centered on an urban core, an activity node, or a focus of regional interaction—are identified collectively as *functional regions*. Thus, the formal region might be viewed as static, uniform, and immobile; the functional region is dynamic, structurally active, and continuously shaped by forces that modify it.

This distinction between formal and functional regions is still debated among geographers. A formal region's evenness, some argue, is also the result of the operation of shaping forces; perhaps formal regions are less affected by change, more durable, and, therefore, more visible—but they may not be fundamentally different from functional regions. And on the landscape itself, both regional types may often be recognized simultaneously: in Fig. I–2, which shows a small Iowa town in the heart of the Corn Belt, we can observe both the functional ties (roads, storage elevators) between the town and its surrounding farms as well as the formal-region homogeneity of the agricultural spatial pattern (evenly dispersed farmsteads, repetition of field size and land uses) beyond the edge of the built-up area in the foreground. Some contributions to the continuing discussion of regional terminology in the current geographical literature are cited in the bibliography at the end of this introductory chapter.

chitectural, artistic, and other traditions might be simultaneously employed. Maps of the cultural geography of Canada, including those showing religious affiliation, dominant language spoken, land division, and settlement patterns, clearly reveal the reality of Quebec as a discrete region within the greater Canadian framework.

We can also conceptualize a region as an area consisting of places that together operate as a *system* (a system is defined as a set of objects and their mutual interactions). Thus certain regions are marked less by internal uniformity and more by a particular activity—or perhaps a set of integrated activities—that interconnects its various parts. In this way we can perceive the Bay Area and similar metropolitan entities as true regions. A large city/suburban-ring complex has a substantial surrounding area for which it supplies goods and services, from which it buys farm products, and with which it interacts in numerous ways. The metropolis's manufacturers distribute their products wholesale to regional subsidiaries. Its newspapers sell in nearby smaller towns. Maps showing the orientation of road traffic, the sources and destinations of telephone calls, the audiences of television stations, and other activities confirm the close relationship between the metropolis and its tributary region, or *hinterland*. Here again, we have a region; this time, however, it is characterized by a structured, urban-centered system of interaction that produces a *nodal* or *functional region*. This type of region is discussed further in the *box* on regional terminology.

Classifying Regions

Given the various qualities and properties of regions, their differ-

ences in dimensions and complexity, is it possible to establish a *hierarchy*, or ranking, of regions based on some combination of their characteristics? Some geographers have proposed such systems of classification, and although none of the results has gained general acceptance, it is interesting to see how they confronted this often frustrating problem. In a book entitled *Introduction to World Geography*, Robert Fuson suggested a comprehensive seven-level regional hierarchy that would divide the earth into the European and non-European worlds, and then into realms, landscapes, superregions, regions, districts, and subregions. This rank-ordered system is complicated and quite difficult to apply consistently, but it underscores the elusive nature of the problem.

Cultural geographers frequently employ a broader three-tier system. The *culture realm* (sometimes called *culture world*) identifies the largest, most complex area that can be described as being unified by common cultural traditions. For example, North or "Anglo" America (Canada and the United States) constitutes such a culture realm; Middle and South America form an additional pair of cultural realms that comprise "Latin" America. Each of these culture realms, in turn, consists of an assemblage of *culture regions*. Within the "Latin" American realms, Mexico is such a culture region, the Central American republics comprise another, Brazil a third, and so forth. Within the European culture realm, Mediterranean Europe and Eastern Europe rank as separate culture regions. These regions, in turn, can be subdivided into *subregions*. Canada is a region within the North American realm; French-speaking Quebec is a subregion within Canada. Similarly, the Iberian states (Spain and Portugal)

form a subregion within Mediterranean Europe.

Note that the regional and subregional areas tend to be identified as countries or groups of countries—which can become misleading. These country names are used because they are convenient and familiar, but national boundaries do not necessarily coincide with those of the regions or subregions in question. Mexico's regional properties spill over into the southwestern margins of the North American realm. The Sahel—a distinct subregion of West Africa—extends across parts of several of that region's countries. Thus, in a world context, realms and regions are not bounded by sharply defined limits—instead, they are separated by *transition zones* that are in many places quite broad. But if we must mark regional boundaries on maps, we should take advantage of the existing politico-geographical "grid" of countries. This has the advantage of comparative simplicity, as we will see later in this chapter when we discuss the framework of *world geographic realms* to be used in this book.

CONCEPTS OF SCALE

Regions can be conceptualized in various forms, and at different *levels of generalization* or **scale** (defined as the ratio of distance on a map to actual ground distance).

Consider the four maps in Fig. I–3. On the first map (upper left) most of the North American realm is shown, but very little spatial information can be provided, although the political boundary between Canada and the United States is evident. On the second

EFFECT OF SCALE

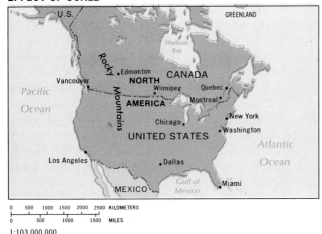

0 500 1000 1500 2000 2500 KILOMETERS
0 500 1000 1500 MILES

1:103,000,000

0 250 500 750 1000 1250 1500 KILOMETERS
0 250 500 750 MILES

1:53,200,000

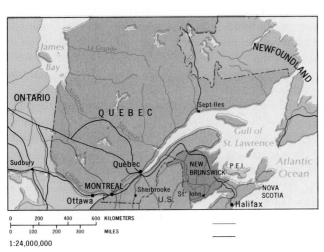

0 200 400 600 KILOMETERS
0 100 200 300 MILES

1:24,000,000

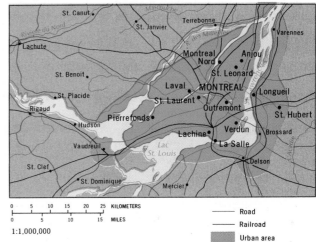

0 5 10 15 20 25 KILOMETERS
0 5 10 15 MILES

1:1,000,000

—— Road
—— Railroad
▨ Urban area

FIGURE I–3

map (upper right), eastern and central Canada are depicted in sufficient detail to permit display of the provinces, several cities, and some physical features (Manitoba's major lakes) not shown on the first map. The third map (lower left) shows the main surface communications of Quebec and immediate surroundings, the relative location of Montreal, and the St. Lawrence and Hudson/James Bay drainage systems. The fourth map (lower right) reveals the metropolitan layout of Montreal and its adjacent hinterland in considerable detail.

Each of the four maps has a scale designation, which can be shown as a *bar graph* (in miles and kilometers in this case) and as a fraction—1:103,000,000 on the first map. The fraction is a ratio indicating that one unit of distance on the map (one inch or one centimeter) represents 103 million of the *same* units on the ground. The smaller the fraction (in other words, the larger the number in the denominator), the smaller the scale of the map. Clearly, this *representative fraction (RF)* on the first map (1:103,000,000) is the smallest, and that of the fourth map (1:1,000,000) is the largest. Comparing maps number 1 and number 3, we find that on the *linear* scale, number 3 has a repre-

sentative fraction that is more than four times larger than number 1. When it comes to *areal* representation, however, 1:24,000,000 is more than 16 times larger than 1:103,000,000 because the linear difference prevails in both dimensions (the length *and* the breadth of the map).

In a book that surveys world realms, it is obviously necessary to operate at relatively small scales. When studying regions or subregions in greater detail, our ability to specify criteria and to "filter" the factors we employ increases as we work at larger scales. On occasion that method will be used, for example when urban centers are the topic of concern (as suggested by Montreal in Fig. I–3). But most of the time our view will be the more macroscopic and general—the small-scale view of the world's geographic realms.

Besides scale, maps exhibit a number of other basic properties. Familiarity with them simplifies the task of reading and interpreting the maps in this book; these properties are discussed in the *box* on pp. 53–55.

CONCEPTS OF CULTURE

The realms and regions to be discussed in the chapters that follow are, in part, defined by humankind's *cultures*. Geographers approach the study of culture from several vantage points, and one of these, the analysis of cultural landscape, is central to our regional concern. Therefore, we should carefully consider the concept of culture. The word **culture** is not always used consistently in the English language, which can lead to some difficulties in establishing its scientific meaning. When we speak of a "cultured" individual, we tend to mean someone with refined tastes in music and the arts, a highly educated, well-read person who knows and appreciates the "best" attributes of his or her society. But among social scientists, the term culture refers not only to the music, literature, and arts of a society, but also to all the other features of its way of life—prevailing modes of dress, routine living habits, food preferences, the architecture of houses as well as public buildings, the layout of fields and farms, and systems of education, government, and law. Thus, culture is an all-encom-

passing concept that identifies not only the mosaic of lifestyles of a people, but also their overriding values and beliefs.

This is not to suggest that anthropologists and other social scientists have found defining the concept of culture easy. If you read some of their basic literature, you will find that anthropologists have had as much difficulty with definitions of the culture concept as geographers have had with the regional concept. A culture may be the total way of life of a people—but is it their *actual* way of life ("the way the game is played") or the standards by which they give evidence of *wanting* to live through their statements of beliefs and values ("the rules of the game")? There are strong differences of opinion on this, and as a result the various definitions have become quite complicated.

Anthropologist E. Adamson Hoebel says in his *Anthropology: The Study of Man* that culture is

the integrated system of learned behavior patterns which are characteristic of the members of a society and which are not the result of biological inheritance . . . culture is not genetically predetermined; it is noninstinctive. . . . [Culture] is wholly the result of social invention and is transmitted and maintained solely through communication and learning.

This definition raises still another question: how is culture carried over from the one generation to the next? Is this entirely a matter of learning, as Hoebel insists, or are certain aspects of a culture indeed instinctive and, in fact, a matter of genetics? This larger question is the concern of sociobiologists and not cultural geographers, although some of its side issues such as *territoriality* (an allegedly human instinct for territorial possessiveness) and *proxemics*

(individual and collective preferences for nearness or distance in different societies) clearly have important spatial dimensions.

But even without these theoretical concerns, the culture concept remains difficult to define satisfactorily. In 1952, anthropologists Alfred Kroeber and Clyde Kluckhohn identified no fewer than 160 definitions—all of them different—and from these they distilled their own:

Culture consists of patterns, explicit and implicit, of and for behavior and transmitted by symbols, constituting the distinctive achievements of human groups, including their embodiments in artifacts . . . the essential core of culture consists of traditional (that is, historically derived and selected) ideas and especially their attached values; culture systems may, on the one hand, be considered products of action, and on the other as conditioning elements of further action.

Some of the definitions from which this one was synthesized, together with a few more recent ones, appear in the *box* on culture concepts.

For our purposes, it is sufficient to stipulate that culture consists of a people's *beliefs* (religious, political), *institutions* (legal, governmental), and *technology* (skills, equipment). This notion is broader than that adopted by many contemporary anthropologists, who now prefer to restrict the concept to the interpretation of human experience and behavior as products of systems of symbolic meaning. It is also important to keep in mind that definitions of this kind are never final and absolute; rather, they are arbitrary and designed for a particular theoretical purpose.

The culture concept is defined to facilitate the explanation of human behavior. Each discipline has

THE CULTURE CONCEPT

Below are several definitions of the concept of culture as developed by some prominent social scientists:

That complex whole which includes knowledge, belief, art, morals, law, custom, and any other capabilities and habits acquired by man as a member of society.
Edward B. Tylor (1871)

The sum total of the knowledge, attitudes, and habitual behavior patterns shared and transmitted by the members of a particular society.
Ralph Linton (1940)

The mass of learned and transmitted motor reactions, habits, techniques, ideas, and values—and the behavior they induce.
Alfred L. Kroeber (1948)

The man-made part of the environment.
Melville J. Herskovits (1955)

A way of life which members of a group learn, live by, and pass on to future generations.
Ann E. Larimore et al. (1963)

The learned patterns of thought and behavior characteristics of a population or society.
Marvin Harris (1971)

The acquired knowledge that people use to interpret experience and to generate social behavior.
James P. Spradley and David W. McCurdy (1975)

The sum of the morally forceful understandings acquired by learning and shared with the members of the group to which the learner belongs.
Marc J. Swartz and David K. Jordan (1976)

different requirements, and would construct contrasting "operational" definitions. Of the definitions in the *box*, cultural geographers would be particularly attracted to the Herskovits definition, because they are especially interested in the way that members of a society perceive and exploit their resources, the way they maximize the opportunities and adapt to the limitations of their environment, and the way they organize the portion of the earth that is theirs.

This last aspect, the way human societies organize their portions of the earth's surface, goes to the heart of our approach in this book. Human works carve long-lasting if not permanent imprints into the land: Roman structures still mark some European countrysides and many Roman routes of travel have now evolved into Europe's major highways. Over time, regions take on certain dominant qualities that collectively create a regional character, a personality, a distinct atmosphere. This, in large part, is the basis for our division of the human world into major geographic realms.

The Cultural Landscape

Culture is expressed in many ways as it gives visible character to a region. Aesthetics play an important role in all cultures, and often a single scene in a photograph or a picture can reveal to us, in general terms, in what part of the world it was made. The architecture, the clothing of the people, the means of transportation, and perhaps even the goods being carried reveal enough to permit a good guess. This is because the people of any culture are active agents of change; when they occupy their part of the earth's available space, they transform the land by building structures on it, creating lines of transport and communication, parceling out the fields, and tilling the soil.

This composite of human imprints on the surface of the earth is called the **cultural landscape**, a term that came into general use in geography during the 1920s. Carl Ortwin Sauer, for several decades professor of geography at the University of California, Berkeley, developed a school of cultural geography that was focused around the concept of cultural landscape. In a paper written in 1927, entitled "Recent Developments in Cultural Geography," Sauer proposed his most straightforward definition of the cultural landscape: *the forms superimposed on the physical landscape by the activities of man.* He stressed that such forms result from the operation of cultural *processes*—causal forces that shape cultural patterns—which unfold over a long time and involve the cumulative influences of successive occupants.

Sometimes these successive groups are not of the same culture. Settlements built by European colonizers a century ago are now occupied by Africans; minarets of Islam rise above the buildings of certain Eastern European cities, evincing an earlier period of hegemony by the Muslim Ottoman Empire. In 1929, Derwent Whittlesey introduced the term *sequent occupance* to categorize these successive stages in the evolution of a region's cultural landscape, a concept explored further on page 427.

The durability of the concept of cultural landscape is underscored by its redefinition in 1984 by the contemporary scholar J.B. Jackson. His statement closely parallels Sauer's: *a composition of man-made or man-modified spaces to serve as infrastructure or background for our collective existence.* Thus, the cultural landscape consists of buildings and roads and fields—and more. But it also possesses an intangible quality, an atmosphere or flavor, a sense of place that is often easy to perceive and yet difficult to define. The smells and sights and sounds of a traditional African market are unmistakable, but try recording those qualities on maps or in some other objective way for comparative study!

Geographers have long grappled with this problem of recording the less tangible characteristics of the cultural landscape, which are often so significant in producing the total regional personality. Jean Gottmann, a Franco-British geographer, put it this way in his book, *A Geography of Europe:*

To be distinct from its surroundings, a region needs much more

than a mountain or a valley, a given language or certain skills; it needs essentially a strong belief based on some religious creed, some social viewpoint, or some pattern of political memories, and often a combination of all three. Thus regionalism has what might be called iconography *as its foundation: each community has found*

for itself or was given an icon, a symbol slightly different from those cherished by its neighbors. For centuries the icon was cared for, adorned with whatever riches and jewels the community could supply.

Gottmann is trying here to define some of the abstract, intangible

FIGURE I–4 Africa's rural landscape in central Kenya.

qualities that also go into the make-up of a region's cultural landscape.

The more concrete properties are easier to observe and record. Take, for instance, the urban "townscape"—a prominent element of the overall cultural landscape—and compare a major U.S. city with, say, one in Japan. Visual representations of these two metropolitan scenes would reveal the differences quickly, of course, but so would maps of the two urban places. The American city with its rectangular layout of the *central business district (CBD)* and its widely dispersed suburbs contrasts sharply with the clustered, space-conserving Japanese city. Using a rural example, the spatially lavish subdivision and ownership patterns of American farmland (Fig. I–2) look unmistakably different from the traditional African countryside, with its irregular, often tiny patches of land surrounding a village (Fig. I–4). Still, the totality of a cultural landscape can never be captured on a photograph or map because the personality of a region involves far more than its prevailing spatial organization: one must also include its visual appearance, its noises and odors, the shared experiences of its inhabitants, and even their pace of life.

CHANGING NATURAL ENVIRONMENTS

Sauer defined the cultural landscape as the forms superimposed on the natural (physical) landscape by human activity, and Whittlesey introduced the idea of cultural landscape evolution as sequent occupance. Both concepts contain the element of *change*—changing cultural traditions, changing societies, changing regional contents and personalities. But the earth itself, the physical landscape, has been presumed to be a comparatively static stage on which the cultural scenes were played out. Changes are brought to this natural landscape largely through humanity's works: building irrigation canals and dams in river valleys, terracing hillsides, substituting cultivated crops for natural vegetation. The physical world has been viewed as passive and stationary, while the human world has produced the dynamics of activity and change.

In recent decades, this assumption has been challenged. Of course, the earth has always been more receptive to human occupance in certain areas than in others, a variability that is reflected on every map of world population distribution. However, the physical world varies not only in space, but also in time. Today, archaeologists excavate ancient cities located in deserts, but when those cities were built, no arid climate prevailed there. In Roman times, farmlands near North Africa's Mediterranean coast produced large quantities of produce from fields amply watered by rain and irrigation, but today those Roman aqueducts lie in ruins and the fields are abandoned. Icy tundra conditions once prevailed where the heart of the Soviet Union lies today.

The earth, therefore, is a variable stage too, both in time and space. This is true even of the surface itself, with its component plains, plateaus, hills, and mountains (Fig. I–5). The planet may be over 4.5 billion years old, and its solid crust began to form about 4 billion years ago. Eventually, the earth differentiated into a number of concentric shells, with the surface *crust* underlain by a thicker *mantle*. Physical geographers have known for more than a century

that the surface of the crust has undergone momentous change: mountain ranges have arisen only to be eroded away again, continental regions (such as the U.S. Interior Lowlands) were invaded by the ocean and lay inundated for millions of years as sediments accumulated. Then, in 1915, Alfred Wegener published a book in which he presented evidence that the continents themselves are mobile and were once united in a gigantic landmass he called *Pangaea* (meaning "all-earth"). The breakup of this supercontinent, he reasoned, has taken place over the last 80 to 100 million years, and the continents are still moving.

Quite recently it was discovered that the earth's outer shell, the crust and the uppermost layer of the mantle (together called the *lithosphere*), consists of a set of rigid geologic *plates*. The exact number and location of these 15 or so tectonic plates are still not certain, but it is known that they average some 60 miles (100 kilometers) in thickness and that the largest plates are of continental dimensions (Fig. I–6). Moreover, these plates are in motion, emerging along great fissure zones in the ocean basins and colliding elsewhere. Where they meet, one plate usually descends under the other, and there is great deformation and crumpling of the crust that often produces seismic (earthquake) and volcanic activity. Human communities located in such hazardous zones are accustomed to sudden and sometimes violent changes in their physical environment. What thousands of years of quiet weathering and erosion cannot accomplish, an earthquake (or volcanic eruption) can achieve in seconds as the residents of Soviet Armenia discovered after violent tremors struck without warning in late 1988 (Fig. I–7).

A comparison of Figs. I–5 and I–6 helps to explain the spatial distribution of the earth's present landscapes. Great mountain ranges, complete with volcanic zones and earthquake-prone belts, extend from western South America through Middle and North America to East Asia and the Pacific islands, including New Zealand east of Australia. Here the Pacific Plate and its neighbors, the Philippine, Cocos, and Nazca plates, collide against the Indian-Australian, the Eurasian, and the two American plates; no wonder this contact zone is often called ''The Pacific Ring of Fire.'' Africa and Australia, on the other hand, lie at the heart of the African and Indian-Australian plates, respectively, and are much less subject to such stresses and deformation. Other relationships between Figs. I–5 and I–6 are readily apparent, but it is important to remember that the map of the world tectonic plates is being modified as new evidence and reinterpretations come to light. For example, some physical geographers believe that the African Plate—shown here as an unbroken cohesive unit of the lithosphere—in fact consists of several segments.

Mobile tectonic plates carry the earth's landmasses along as they move, and Wegener's hypothesis of *continental drift* now appears substantially correct (see pp. 400–402). But it is important to view the process of continental drift in the perspective of human evolution on this planet. Current research in East Africa indicates that the human family began to emerge between 12 and 14 million years ago, and that the use of stone tools developed between 2 and 3 million years ago. A fossil site in Tanzania has yielded evidence of a stone structure, probably the foundation of a simple hut, dated as 1.8 million years old. But the emergence of larger human communities, made possible by the domestication of plants and animals and attended by the development of

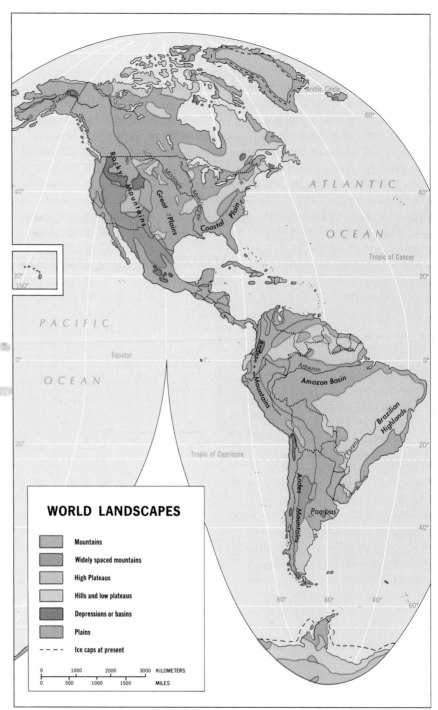

WORLD LANDSCAPES

Mountains

Widely spaced mountains

High Plateaus

Hills and low plateaus

Depressions or basins

Plains

- - -   Ice caps at present

FIGURE I–5

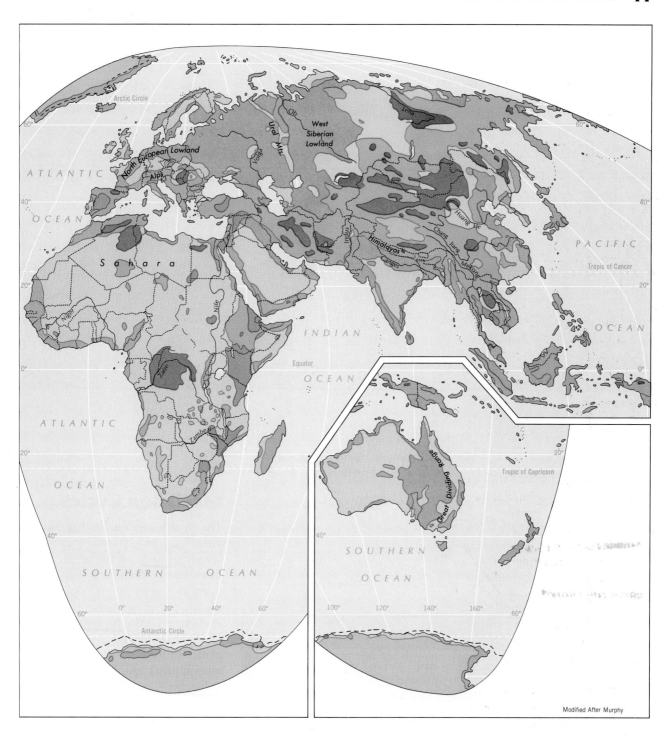

Modified After Murphy

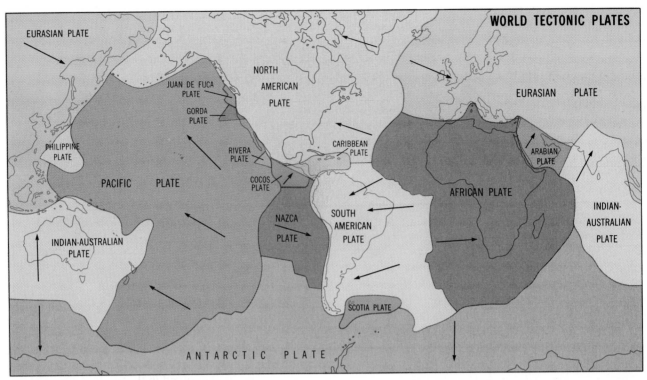

FIGURE I—6

FIGURE I—7 A panoramic view of the devastation in the city of Leninakan, following the 1988 earthquake that struck Armenia in the south-central Soviet Union.

urban centers, is the product of only the past 10,000 years—a tiny fraction of time in the context of earth history (and even evolutionary human history). Thus, recorded time has not been nearly long enough to observe all the momentous changes we know the earth's landscapes have undergone.

Glaciation Cycles

The earth's lifetime has been divided by geologists into four major stages, or _eras_, on the basis of information derived from the study of rocks and fossils. Obviously, the least is known about the oldest of these eras, the Precambrian, covering the period from our planet's formation until about 570 million years ago. Next comes the Paleozoic Era, from 570 to 225 million years ago, followed by the Mesozoic Era that lasted until about 65 million years before the present (B.P.). The latest era, the one during which we live, is the Cenozoic. Each of these lengthy eras is subdivided into _epochs_, and the Cenozoic is marked by seven of these. The two most recent Cenozoic epochs, the Pleistocene (ca. 2 million to 10,000 years ago) and the Recent or Holocene (10,000 years B.P. to the present), witnessed the emergence of humankind and the eventual attainment of civilization.

Geologic eras and epochs are identified through the study of rock layering, fossil assemblages, mountain building, and other evidence yielded by the earth's crust. Many epochs begin with the deposition of huge thicknesses of sedimentary rocks and end when those rocks are bent and broken by tectonic activity (internal earth forces that deform the crust). But the Pleistocene is no ordinary epoch. It witnessed the height of the *Late Cenozoic Ice Age* that began about 3.5 million years ago in the preceding epoch. During the Ice Age, large icesheets intermittently covered all of Canada, most of the north-central and northeastern United States, and a substantial portion of Europe (Fig. I–8).

The earth's climatic and vegetative environments changed drastically when these continental-scale glaciers advanced toward the mid-latitudes. There were several such outbreaks of the icesheets during the more than 3 million years of the Late Cenozoic glaciation; these icesheets would expand and then contract, only to push outward again. The newest evidence suggests that there were as many as 24 of these advances and retreats, and that the Late Cenozoic Ice Age was but the latest in a series of such events in the earth's environmental history.

Most geologic time charts indicate that the Pleistocene Epoch has ended and that we are now in a new epoch called the Holocene (or sometimes simply ''Recent''). It is likely, however, that the human world of the past 10,000 years

has evolved during an *interglaciation*, a pause between glacial advances when milder temperatures have enormously expanded the earth's living space—temporarily. In this view, the possibility is strong that the continental icesheets will again overspread those large areas previously affected by Late Cenozoic glaciations (Fig. I–8). If this should happen, a renewed ice age would have an impact far beyond those regions directly affected. Climates throughout the world would change: moist areas would dry up; areas now dry would begin to receive ample moisture; temperatures, even in tropical zones, would decline.

Whatever the future, there can be no doubt that the modern world's human cultures emerged and evolved in the wake of the

FIGURE I–8

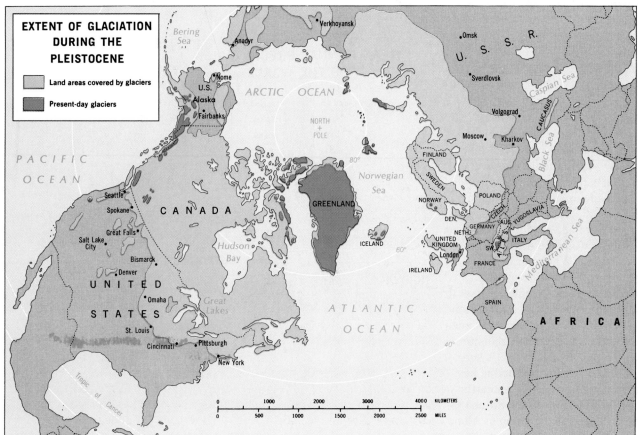

EXTENT OF GLACIATION DURING THE PLEISTOCENE

Land areas covered by glaciers

Present-day glaciers

withdrawal of the last of the Pleistocene glaciers. Certainly there were active human communities during earlier phases, but the great transformations—plant and animal domestication, urbanization and civilization, mass migration, the agricultural and industrial revolutions—occurred during the past 10,000 years. We cannot know as yet why these momentous environmental changes took place when they did, but they have been accompanied, especially during the past century, by an unprecedented increase in human numbers. This population explosion (about which we will say more in Chapters 1 and 8) is taking place right now while the earth's available living space is at a maximum. The implications of a return of the icesheets and their impact on the earth's climates and habitable areas stagger the imagination. Nevertheless, many environmental scientists warn that such a sequence of events may well lie ahead.

Water—Essence of Life

The French geographer Jean Brunhes, writing in his book *Human Geography*, remarked that "every state, and indeed, every human establishment, is an amalgam made up of a little humanity, a little soil, and a little water." He might have added that without water there would be neither humanity nor soil. When the United States in 1976 sent two space probes to Mars, scientists on earth waited for the crucial information to be relayed back from the landed vehicles: was there any moisture on our neighboring planet's surface? Moisture would be the key to life on Mars as here on earth; but the Martian surface proved to be as barren as the moon's. Alone

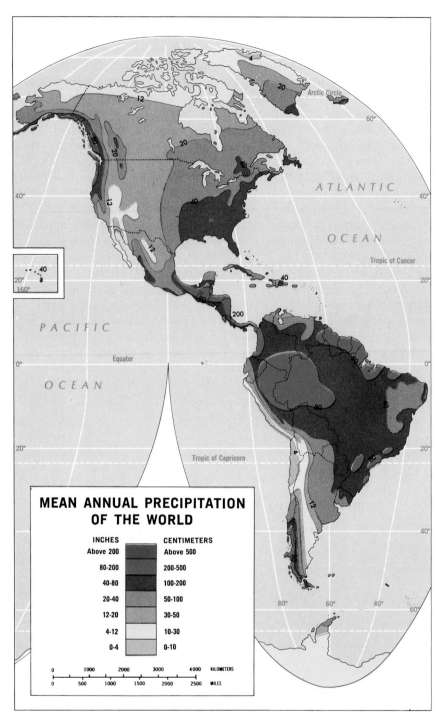

MEAN ANNUAL PRECIPITATION OF THE WORLD

INCHES		CENTIMETERS
Above 200		Above 500
80-200		200-500
40-80		100-200
20-40		50-100
12-20		30-50
4-12		10-30
0-4		0-10

FIGURE I–9

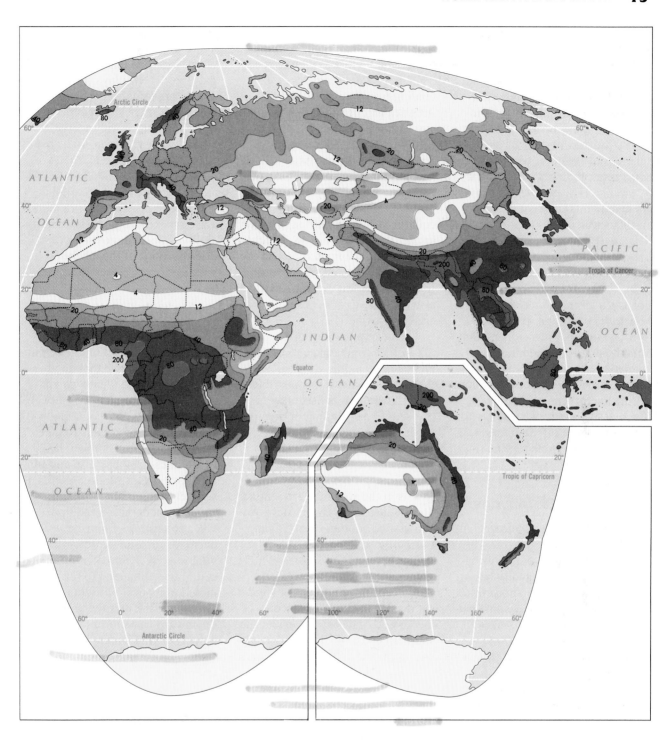

among the planets of our solar system, the earth possesses a *hydrosphere*—a cover of water in the form of a vast ocean, frozen polar icesheets, and a moisture-laden atmosphere. Technically, all the earth's water, even that in lakes and streams, is part of the hydrosphere, but the great world ocean constitutes about 97 percent of it by volume. This water body covers just over 70 percent of the earth's surface, but it would do little good if a mechanism did not exist to bring moisture from that ocean to the land. This mechanism, the *hydrologic cycle,* functions as a circulation system. Moisture evaporates into the air from the oceanic surface, and this humid mass of air then moves over land where, by various atmospheric processes, condensation occurs and precipitation falls. Much of the water then returns to the ocean as runoff via rivers, and this continuous cycle repeats itself.

Precipitation Distribution

The earth's landmasses do not share equally in this provision of moisture, and much of the historical geography of humanity has involved the search and competition for well-watered areas. Global patterns of precipitation are mapped in Fig. I–9, which should be viewed as the latest frame from a piece of Late Cenozoic film, a still picture of continually changing conditions. It represents the distribution of moisture conditions as they prevail today, but it differs from the world precipitation map as it would have looked when the Middle East's Fertile Crescent witnessed the domestication of crops or when North Africa, thousands of years later, was a granary of the Roman Empire. Moreover, the map will look quite different a thousand years from

now. Today, the map reveals an equatorial zone of heavy rainfall, where annual totals often exceed 80 inches (200 centimeters), extending from mainland Middle America through the Amazon Basin, across smaller areas of West and Equatorial Africa, and into South and Southeast Asia. This low-latitude zone of high precipitation gives way to dry conditions in both poleward directions. In equator-straddling Africa, for example, the Sahara lies to the north of the tropical humid zone and the Kalahari Desert lies to the south. Interior Asia and central Australia also are very dry, as is southwestern North America.

The general pattern of equatorial wetness and adjacent dryness is broken along the coasts of all the continents, and it is possible to discern a certain consistency in this spatial distribution of precipitation. Eastern coasts of continents and islands in tropical as well as mid-latitude locations receive comparatively heavy rainfall, as in the southeastern United States, eastern Brazil, eastern Australia, and southeastern China. Furthermore, a narrow zone of higher precipitation exists at higher latitudes on the western margins of the continents, including the Pacific Northwest coast of the United States and Canada, the southern coast of Chile, the southern tip of Africa, the southwestern corner of Australia, and, most important, the western exposure of the great Eurasian landmass—Europe.

The distribution of world precipitation as reflected in Fig. I–9 results from an intricate interaction of global systems of atmospheric and oceanic circulation as well as heat and moisture transfer. Whereas the analysis of these systems is a part of the subject of physical geography, we should remind ourselves that even a slight

change in one of them can have a major impact on a region's habitability. It is also important to know that Fig. I–9 displays the *average* annual precipitation on the continents, but no place on earth has a guarantee that it will receive, in any given year, precisely its average rainfall. In general, the variability of precipitation increases as the recorded average total decreases. In other words, rainfall is least dependable just where reliability is needed most—in the drier portions of the inhabited world.

Precipitation and Human Habitation

Even heavy, year-round precipitation is no guarantee that an area can sustain large, dense populations. In equatorial areas, the heavy precipitation combined with high temperatures leads to the faster destruction of fallen leaves and branches by bacteria and fungi (see Evapotranspiration *box*). This severely inhibits the formation of *humus,* the dark-colored upper soil layer consisting of nutrient-rich decaying organic matter, which is vital to soil fertility. The drenched soil is also subjected to the removal of its best nutrients through *leaching*—the dissolving and downward transport of nutrients by percolating water—so that only oxides of iron and aluminum remain at the surface to give the tropical soil its characteristic reddish color. Such tropical soils (called oxisols) support the dense rainforest, but they cannot carry crops without massive fertilization. The rainforest thrives on its own decaying vegetative matter, but when the land is cleared, the leached soil proves to be quite infertile. Not surprisingly, the Amazon and Zaïre (Congo) Basins are not among the world's most populous regions.

EVAPOTRANSPIRATION

The map of world precipitation distribution (Fig. I–9) should be viewed in the context of temperature distribution. From the map one might conclude that all the areas that receive over 40 inches (100 centimeters) of rainfall are thereby equally well supplied with moisture. But there are equatorial areas where temperatures average over 77°F (25°C) that receive 40 inches, and much cooler places—for example, parts of New Zealand and Western Europe—where 40 inches are also recorded. Obviously, evaporation from the ground goes on much more rapidly in the tropics than in the mid-latitude zones.

Similarly, evaporation from vegetation also speeds up in equatorial regions. This evaporation from leaf surfaces is actually a three-stage process. Roots of plants absorb water from the soil. This water is then transmitted through the organism and reaches the leafy parts, which transpire in warm weather much as we humans perspire. From the surface of the leaves, the moisture evaporates. Thus, a plant acts like a pump, and the process of evaporation from vegetation is actually a process of transpiration plus evaporation—or *evapotranspiration*.

Thus 40 inches of rainfall in a tropical area may very well be *in*adequate, because if the amount lost by evaporation and evapotranspiration is calculated, it could exceed 40 inches—which means that the plants would use even more moisture if it were available. Some of those "moist" tropical areas, even those with over *60* inches (150 centimeters) of rainfall annually, can be shown to be moisture-deficient. In other areas, the seasonality of precipitation is so pronounced that there is a deficiency during part of the year. On the other hand, in cooler parts of the world, just 30 inches (75 centimeters) of rainfall may be enough to keep the soil moist and the vegetation adequately supplied. A map like Fig. I–9 is of necessity a generalization, and it is important to know what it fails to reveal.

Climatic Regions

It is not difficult to discern the significance of precipitation distribution on the map of world climates (Fig. I–10). Determining **climatic regions**, however, has always presented problems for geographers. In the first place, climatic records are still scarce, short-term, or otherwise inadequate in many parts of the world. Second, weather and climate tend to change gradually from place to place, but the transitions must be presented as authoritative-looking lines on the map. In addition, there is always room for argument concerning the criteria to be used and how these criteria should be weighed. Vegetation is a response to prevailing climatic conditions. Should boundary lines between climate regions therefore be based on vegetative changes observed in the landscape no matter what precipitation and temperature records show? The debate still goes on.

Figure I–10 displays a regionalization system devised by Wladimir Köppen and modified by Rudolf Geiger, which we later generalize further in our discussion of Fig. I–11. This scheme has the advantage of comparative simplicity and is based on a triad of letter symbols. The first (capital) letter is the critical one: the *A* climates are humid and tropical; the *B* climates are dominated by dryness; the *C* climates are humid and comparatively mild; the *D* climates reflect increasing coldness; and the *E* climates mark the frigid polar and near-polar areas.

Humid Equatorial (*A*) Climates

The humid equatorial, or tropical, climates are marked by high temperatures all year and by heavy precipitation. In the *Af* subtype, the rainfall comes in substantial amounts every month; but in the *Am* areas, there is a sudden enormous increase due to the arrival of the annual wet *monsoon* (the Arabic word for "season" [see p. 456]). The *Af* subtype is named after the vegetation association that develops there—the tropical rainforest. The *Am* subtype, prevailing in part of peninsular India, in a coastal area of West Africa, and in sections of Southeast Asia, is appropriately referred to as the monsoon climate. A third tropical climate, the savanna (*Aw*), has a wider daily and annual temperature range and a more strongly seasonal distribution of rainfall. As Fig. I–9 indicates, savanna rainfall totals tend to be lower than those in the rainforest zone, and the associated seasonality is often expressed in a "double maximum." This means that each year produces two periods of increased rainfall separated by pronounced dry spells. In many savanna zones, inhabitants refer to the "long rains" and the "short rains" to identify those seasons; a persistent problem in these regions is the unpredictability of the rain's arrival. Savanna soils are not among the most fertile, and when the rains fail, the specter of hunger arises on these humid grasslands. Savanna regions are far more densely peopled than rainforest areas, and millions of residents of the savanna subsist on what they manage to cultivate. Rainfall variability under the savanna regime is their principal environmental problem.

Dry (*B*) Climates

Dry climates occur in lower as well as higher latitudes. The difference between the *BW* (true desert) and moister *BS* (semiarid steppe) varies, but may be taken to lie at about 10 inches (25 centimeters) of annual precipitation. Parts of the central Sahara receive less than 4 inches (10 centimeters) of rainfall. A pervasive characteristic of the world's arid areas is an enormous daily temperature range, especially in subtropical deserts. In the Sahara, there are recorded instances where the maximum daytime shade temperature was over 120°F (49°C) followed by a nighttime low of 48°F (9°C).

Humid Temperate (*C*) Climates

These mid-latitude climate areas, as the map shows, almost all lie just beyond the Tropics of Cancer and Capricorn. This is the prevailing climate in the southeastern United States from Kentucky to central Florida, on North America's west coast, in most of Europe and the Mediterranean, in southern Brazil and northern Argentina, in coastal South Africa and Australia, and in eastern China and southern Japan. None of these areas suffers climatic extremes or severity, but the winters can be fairly cold, especially away

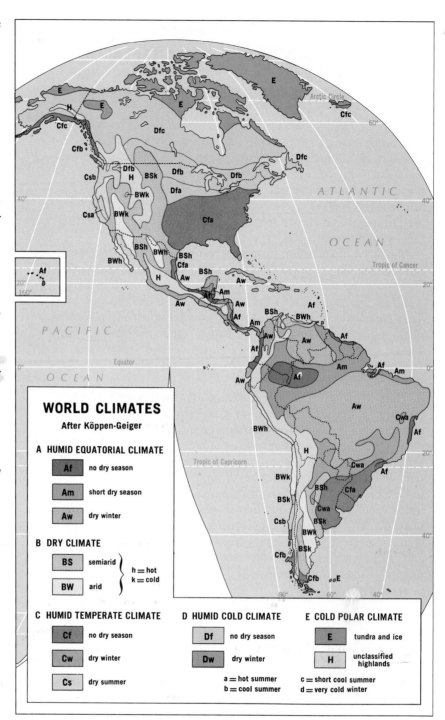

FIGURE I–10

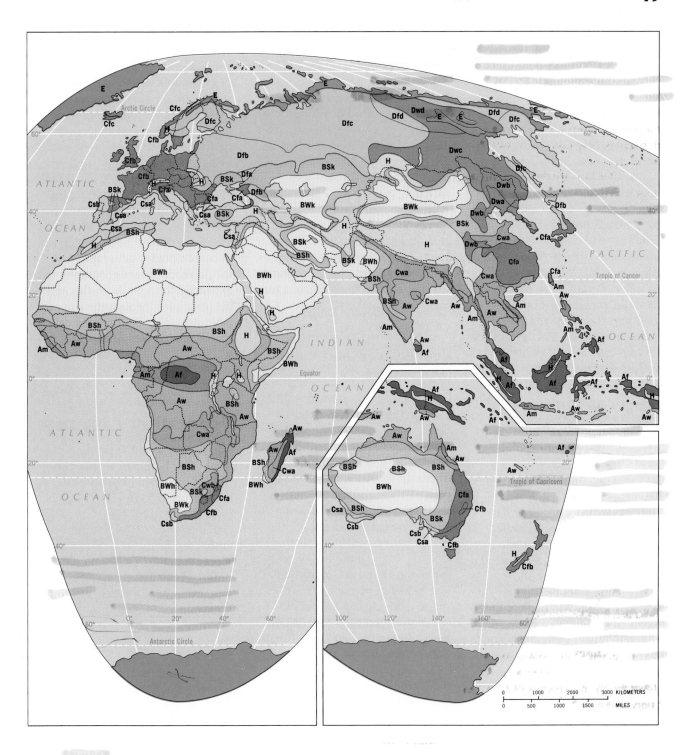

from moderating water bodies. These areas lie about midway between the winterless equatorial climates and the summerless polar zones.

The humid temperate climates range from quite moist, as along the densely forested coasts of Oregon, Washington, and British Columbia, to relatively dry, as in the so-called Mediterranean (dry-summer) areas that include not only coastal southern Europe and northwest Africa, but also the southwestern tips of Australia and Africa, central Chile, and Southern California. In these Mediterranean areas, the scrubby, moisture-preserving vegetation creates a natural landscape very different from that of richly green Western Europe.

Humid Cold (*D*) Climates

The humid cold (or "snow") climates may be called the continental climates, for they seem to develop in the interior of large landmasses, as in the heart of Eurasia or North America. No equivalent land areas at similar latitudes exist in the Southern Hemisphere; consequently, no *D* climate occurs there at all.

Great annual temperature ranges mark these humid continental climates, and very cold winters and relatively cool summers are the rule. In a *Dfa* climate, for instance, the warmest summer month (July) may average as high as 70°F (21°C), but the coldest month (January) might average only 12°F (−11°C). Total precipitation, a substantial part of which comes as snow, is not very high, ranging from over 30 inches (75 centimeters) to a steppe-like 10 inches (25 centimeters). Compensating for this paucity of precipitation are cool temperatures that inhibit the loss of moisture via evaporation and evapotranspiration.

Some of the world's most productive soils lie in areas under humid cold climates, including the U.S. Midwest, parts of the Soviet Union's Ukraine, and Northeast China. The period of winter dormancy, when all water is frozen, and the accumulation of plant debris during the fall combine to balance the soil-forming and enriching processes. The soil differentiates into well-defined nutrient-rich layers, and a substantial store of organic humus accumulates. Even where the annual precipitation is light, this environment sustains extensive coniferous forests.

Cold Polar (*E*) and Highland (*H*) Climates

Cold polar (*E*) climates are differentiated into true icecap conditions, where permanent ice and snow keep vegetation from gaining a foothold, and the "tundra," where up to four months of the year may have average temperatures above freezing. Like *rainforest*, *savanna*, and *steppe*, the term *tundra* is vegetative as well as climatic, and the boundary between the *D* and *E* climates on Fig. I–10 can be seen to correspond quite closely to that between the needleleaf forests and arctic tundra on Fig. I–12. Finally, the *H* climates—unclassified highlands mapped in gray—resemble the *E* climates in a number of ways. High elevations and complex topography of major mountain systems often produce near-arctic climates above the tree line, even in the lowest latitudes (such as the Andes in western equatorial South America).

Values of Classification Schemes

We have already pointed out that mapping world climates in Fig.

I–10 (and global precipitation in the preceding map) involves a good deal of generalizing the much more complicated and detailed environmental patterns that actually exist across the surface of the earth. Köppen and Geiger, who performed their work in the early decades of this century, realized the value of generalization, which allowed them to concentrate on the big picture unencumbered by less essential local complexities. This kind of methodology persists in contemporary geography. In fact, in recent years geography has expanded the search for theoretical principles through the use of laboratory-like abstractions called **spatial models** (see *Model Box 1*, p. 21).

Vegetation Patterns

The map of world vegetation patterns (Fig. I–12) shows the closeness of the spatial relationship between climatic regions and plant associations. This map depicts the global distribution of natural vegetation, regionalized according to *biomes*—macroscale plant communities occupying large geographical areas, marked by similarities in vegetation structure and/or appearance. But because so much of that plant life has been destroyed or modified by humans, the regions shown in Fig. I–12 represent the natural vegetation that exists or would exist as a result of long-term plant succession and adaptation to prevailing climatic conditions. This latter *climax vegetation* may be a rainforest, as in the Amazon Basin, or it may be the scrub and bush that just manages to take hold in the driest steppe.

We should not, however, lose sight of the changeable character of Late Cenozoic environments. Like the maps of precipitation and

M O D E L   B O X   1

THE HYPOTHETICAL CONTINENT AND MODELS IN GEOGRAPHY

A modern approach to generalization in both physical and human regional geography is through the development of **models**. Peter Haggett, in his book *Locational Analysis in Human Geography*, offers an especially lucid definition: *in model-building we create an idealized representation of reality in order to demonstrate its most important properties*. He further points out that using models is necessitated by the complexity of reality—to understand how things work, we must first filter out the main processes and their responses from the myriad details with which they are embedded in a highly complicated world. Models, therefore, provide a simplified picture of reality to convey, if not the entire truth, a useful, essential part of it.

The effectiveness of modeling is demonstrated in Fig. I–11, which shows the generalized distribution of climates on a *Hypothetical Continent* that tapers southward from the upper latitudes of the Northern Hemisphere to the mid-latitudes of the Southern Hemisphere. Its shape represents the amount of land at each latitude on the world map: Eurasia and Canada are widest at about 50° north, whereas at 35° south only the relatively narrow southern extremities of South America, Africa, and Australia interrupt the vast Southern Ocean that girdles the globe. Within the Hypothetical Continent, the distribution of climate regions—using the identical color scheme of Fig. I–10—represents a considerable smoothing out of the Köppen/Geiger system. This tightly controlled simplification makes the overall

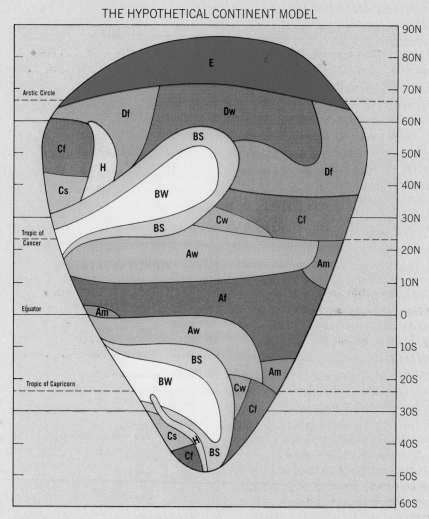

THE HYPOTHETICAL CONTINENT MODEL

FIGURE I–11

geographical pattern of climate types much easier to comprehend and think about, and, with the exception of place references, is an equally effective spatial expression of the text discussion of climate regions. Once the generalities are understood, one can easily shift one's attention back and forth between the idealized and *empirical* (actual) maps—Figs. I–11 and I–10—to see how and why certain distortions occur in reality.

A similar Hypothetical Continent could be constructed to model the regionalization (and interrelationships) of precipitation or vegetation or soils or even population. Models play an increasingly vital role in our understanding of the world's regional spatial structure. We will be introducing additional Model Boxes (in Chapters 1 and 6) to elaborate on a pair of the most important of such geographic abstractions.

climates, this map of world vegetation is a still from a motion picture, not a permanent end product. There are places where one plant association can be observed gaining on another, signaling change. In West Africa, the steppe is encroaching southward on the savanna, and the Sahara to the north, in turn, is gaining on the steppe. Inexorable climatic change may be involved in this, although human communities in the region have played a major role in modifying the natural environment and thereby greatly accelerating this *desertification*. In any case, large parts of the regions shown in Fig. I–12 do not actually support the climax vegetation communities shown in the map's legend.

The vegetative cover of the earth consists of trees, shrubs, grasses, mosses, and an enormous variety of other plants. At the global scale, as we know, plant geographers group this mass of vegetation into the five biomes mapped in Fig. I–12: forest, savanna, grassland, desert, and tundra. Note that these biomes are not confined to particular latitudes; there is forest in the equatorial zone, in the subtropics poleward of the Tropics of Cancer and Capricorn (23½° North and South latitude, respectively), in mid-latitude areas such as the eastern United States, and in higher latitudes on the margins of the tundra. The adaptation of the forest species is what differs. Equatorial and temperate forests have leafy evergreen trees, whereas cold-climate, high-latitude forests have coniferous trees with thin needles. In the mid-latitudes, trees are deciduous and shed their leaves each autumn.

Another prominent impression gained from Fig. I–12 relates to the vastness of the savanna grasslands. The bulk of Africa south of the Sahara is savanna country, as

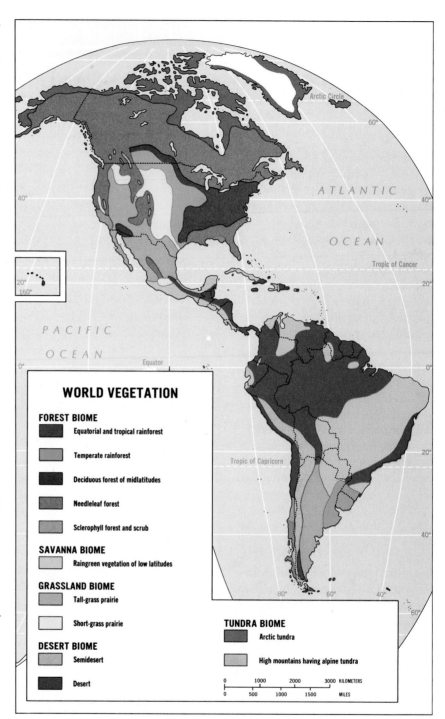

WORLD VEGETATION

FOREST BIOME

Equatorial and tropical rainforest

Temperate rainforest

Deciduous forest of midlatitudes

Needleleaf forest

Sclerophyll forest and scrub

SAVANNA BIOME

Raingreen vegetation of low latitudes

GRASSLAND BIOME

Tall-grass prairie

Short-grass prairie

DESERT BIOME

Semidesert

Desert

TUNDRA BIOME

Arctic tundra

High mountains having alpine tundra

FIGURE I–12

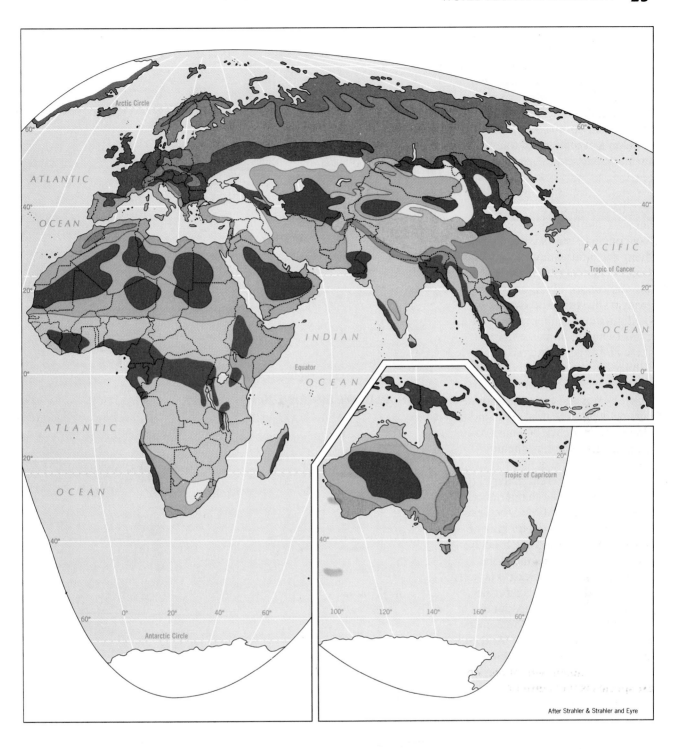

After Strahler & Strahler and Eyre

are most of eastern Brazil and India. The savanna also prevails in interior Southeast Asia and in northern Australia. In later chapters, we will have frequent occasion to refer to the vagaries of the savanna environment with which hundreds of millions of the world's farmers must cope.

Soil Distribution

We conclude this overview of the human habitat with an examination of another vital ingredient for the sustenance of life on this planet—the soil. The earth's soils have proven even more difficult to classify and regionalize than climate and vegetation, and Fig. I–13 is only one possible alternative among several. Research into the processes of soil formation continues to produce new data, and as this latest evidence becomes available, regionalization schemes change.

Because the parent material (the rocks beneath), the temperature, the moisture conditions, the vegetation, and the terrain (degree of slope steepness) all vary from place to place across the globe, there is enormous diversity of soils as well. Some soils are infertile and cannot sustain crops; others can carry two or even three crops annually, year after year. And still today, liberating technologies notwithstanding, the great majority of the world's people depend directly on the local soil for their food. Any map of global population distribution, to a considerable degree, reflects the productiveness of the soils of particular areas—and their infertility elsewhere.

Figure I–13 also displays areas of correspondence with world distributions of climate types and vegetation, reminding us that the soil is a responsive component of the total environmental complex.

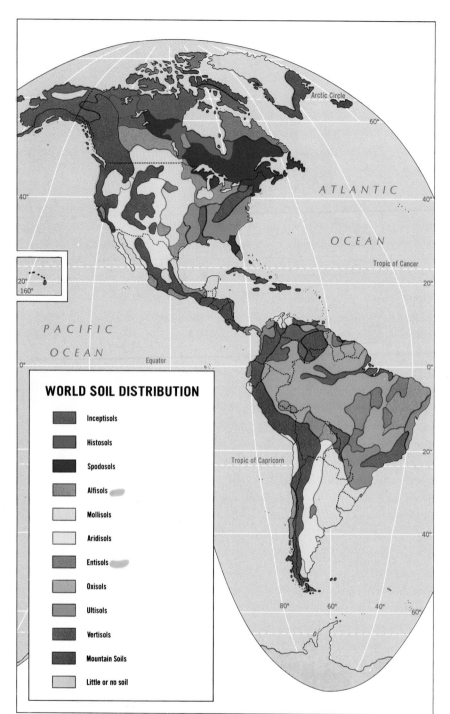

FIGURE I–13

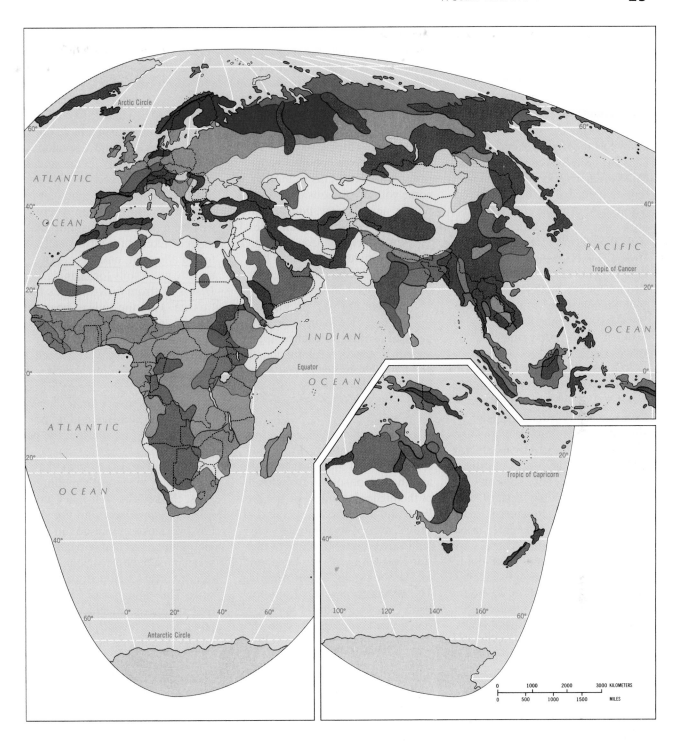

It is important, again, to remember that this map of the world distribution of soils represents a generalization of reality. After many years of experimentation, pedologists (soil scientists) produced an all-encompassing classification system known as the Comprehensive Soil Classification System (*CSCS*); because so many less successful attempts preceded it, they call it the Seventh Approximation. According to this classification, the world's soils are grouped into 10 Orders, which in turn are divided into 47 Suborders, 185 Great Groups, about 1000 Subgroups, 5000 Families, and 10,000 Series. No small-scale map like Fig. I–13 could even begin to communicate all this detail.

Some of the new soil names are self-explanatory and easy to understand. *Oxisols*, for example, are the excessively leached soils of the tropics, the familiar reddish-colored soils marked by high concentrations of the *oxi*des of iron and aluminum. These soils are what the subsistence farmers of the savannas and equatorial areas of Africa and South America must confront and cultivate. Oxisols are often thick and deeply weathered, but their needed nutrients have been heavily leached, or washed downward. Tropical vegetation that grows in this soil sustains itself by absorbing nutrients directly from fallen leaf matter. When the farmer clears the land, the oxisols may carry a crop for a year or two, but then they become exhausted and fail.

Another soil order with an appropriate name comprises the *aridisols*. These are salt-rich, infertile soils of rainfall-deficient areas and, as Fig. I–13 shows, they are widely distributed across the earth, occurring in the southwestern United States, in western South America, and in vast areas of Africa, Asia, and Australia.

In the high latitudes, where soil-forming conditions are inhibited by periods of extreme cold, limited warmth from the sun, the thin covering of tundra mosses and lichens, and poor drainage, the incompletely developed soils are called *inceptisols* (which also occur on coastal and river floodplains in lower latitudes). This is another name with an obvious derivation, but the neighboring *spodosols* are not. The spodosols, which extend across northern Canada and much of high-latitude Eurasia, are better developed and support great stands of pine and spruce (the "Needleleaf Forests" in Fig. I–12). They owe their name to the peculiar ash-like, bleached appearance that results from the constant and intense removal of matter from the upper soil layer; Russian farmers refer to this as *podzol* (ashy soil).

Between the oxisols (with the related *ultisols* and *vertisols*) and the aridisols of the lower latitudes and the inceptisols and spodosols of the higher latitudes, lie the *mollisols* and the *alfisols*. As Fig. I–13 shows, these two soil orders extend across vast stretches of interior North America, Eurasia, and, to a lesser extent, South America. The mollisols (*mollis* is Latin for "soft") possess a dark, humus-rich upper layer, and they never become hard as the aridisols do, even under comparatively dry conditions. As the map indicates, mollisols are located in intermediate positions between dry and humid climates (see Fig. I–9). These are the grassland soils that support large livestock herds or, when farmed, sustain vast expanses of grain crops. Alfisols occur in a broad zone in central Canada, in the U.S. Midwest, and in Europe and the Soviet Union. These soils evolve under a wide range of environmental conditions and are generally quite fertile, often sup-

porting highly intensive agriculture.

Figure I–13 shows limited areas of *entisols* (rec*ent*) in North America and Eurasia but larger zones in Africa and Australia. Entisols, like inceptisols, have not existed long enough to develop maturity. They may lie on sand accumulations, on recently deposited river *alluvium* (silt), or in areas subject to strong erosion where surface material is constantly removed. Except for intrinsically fertile alluvium, entisols are not usually good soils for farming. *Histosols* (from *histos*, Greek for "tissue") are the soils of bogs and moors, and they primarily consist of organic rather than mineral matter. Frequently waterlogged, histosols may develop as peat accumulations. The largest areas of histosol exposure lie in Canada, south of Hudson Bay and in the far northwest just south of the permafrost limit.

Finally, Fig. I–13 shows the distribution of the world's mountain soils, the soils of high-relief terrain. A comparison with Fig. I–5, however, proves that not all mountainous or high-elevation areas must necessarily contain thin, poorly developed, or stony soils. The highlands of Ethiopia, for example, form an area of comparatively fertile alfisols.

It is also interesting to compare Fig. I–13 with Fig. I–14 (showing the world distribution of population). The valley and delta of the lower Nile River in northeastern Africa, the basin of the Ganges River in India and Bangladesh, and the plain of the Huang He (Yellow River) in China all contain recent alluvial soils, and through their nearly legendary fertility they sustain many millions of people. Today, fully 95 percent of Egypt's 58 million people live within 12 miles (20 kilometers) of the Nile waterway. Hundreds of millions of

Indians and Chinese depend directly on alluvial soils in the river basins of the Ganges and Huang He, where crops are grown that range from corn to cotton, wheat to jute, rice to soybeans. These are only the most prominent examples of such alluvium-based agglomerations; in Pakistan, Bangladesh, Vietnam, and many other countries, the alluvial soils of river valleys provide in abundance what the generally infertile upland soils do not.

These riverine population clusters also provide a link with humanity's past. In the fertile valleys of Southwest Asia's rivers, the art and science of irrigation may first have been learned. And two of the world's oldest continuous cultures—Egypt and China—still retain heartlands near their ancient geographic hearths of thousands of years ago.

SPACE AND POPULATION

Less than 30 percent of the earth's surface is land area. But, as we observed in the course of our survey of global environments, large parts of that land surface cannot support any substantial number of inhabitants. Arid deserts, rugged mountain ranges, and frigid tundras constitute just a few of the less habitable parts of the world, but together they cover nearly half the landmasses—and we have not even begun to account for places where unproductive soils, persistent disease, frequent drought, and other more localized conditions keep population numbers comparatively low.

As a result, the map of world population distribution (Fig. I–14) prominently displays the clustering of huge numbers of people in areas of fertile and productive soils. For thousands of years, following the beginning of crop and animal domestication, communities remained dependent on soils and pastures in their immediate vicinity; the large-scale, worldwide food transportation of today is essentially a phenomenon of the past 200 years, a product of industrial and technological revolutions. True, the ancient Romans imported grains from North Africa and colonial powers even centuries ago brought shiploads of spices and sugar across the oceans. But the overwhelming majority of the world's peoples continued to subsist on what they could cultivate locally. When groups migrated, they did so in search of new lands to be opened up, new pastures to be exploited. If they succeeded, as the Chinese of North China did when they moved southward, they thrived. If they failed, they often starved.

Today's map of world population is another one of those stills from that Late Cenozoic film, for it shows the current stage of humankind's expansion and dispersal across the habitable space on the earth's surface. And the picture is changing rapidly. After thousands of years of relatively slow growth, world population during the past two centuries has been expanding at an increasing rate. It took about 17 centuries from the time of the birth of Christ for the world to add 250 million people; now, that same number will be added over just the next two and a half years. This is the subject of discussions in Chapters 1 and 8, but it is important to realize that this modern explosive growth is *not* filling in the remaining available living space on the globe. Rather, the already-crowded areas are becoming even more so. Thus the food produced in the ricefields of Asia, plentiful though it is, must be shared by even more and hungrier people.

At the moment, we will confine ourselves to an overview of the world population as it is distributed today. The 1991 world population was estimated to be some 5.4 billion, and it is broken down by realm and country in *Appendix A* (pp. 589–592). More than one-fifth of this total (1.1 billion) resides in one country, China, and the next most populous single state, India, has over 870 million inhabitants. Yet, both India and China still contain extensive areas that are nearly devoid of permanent population, as Fig. I–14 reveals. China's habitable, agriculturally productive environments lie concentrated in that country's east, and comparing Fig. I–13 with Fig. I–14 leaves no doubt about the association between fertile land and the clustering of population.

Population Agglomerations

Figure I–14 shows that the earth presently contains four major **population agglomerations**. The three largest are all located on a single landmass—Eurasia—and include East Asia, South Asia, and Europe; the fourth of these world-scale clusters is found in eastern North America. The spatial prominence of these population concentrations on the world map of humankind is even more apparent in Fig. I–15. This specially transformed map, or *cartogram*, is not based on the traditional scale representation of distance or area. Instead, countries are drawn proportional to their population, so that those containing large numbers of people are "blown up" in *population-space* while those containing smaller numbers are "shrunk" in size accordingly.

East Asia

The greatest of the four population agglomerations is the *East Asia* cluster, adjoining the Pacific Ocean from Korea to Vietnam and centering, of course, on China itself. The map indicates that the number of people per unit area tends to decline from the coastal zone inland, but two ribbon-like extensions can be seen to penetrate the interior (e.g., **A** and **B** on Fig. I–14). Reference to the map of world landscapes (Fig. I–5) proves that these extensions represent populations concentrated in the valleys of China's major rivers, especially the Huang He (Yellow) and the Chang Jiang (Yangzi). This serves to remind us that the great majority of the people of East Asia are farmers, not city dwellers. True, there are great cities in China, and some of them such as Shanghai and Beijing (the capital) rank among the largest in the world. But the total population of these and the other cities is far outnumbered by the farmers—those who need the river valleys' soils, the life-giving rains, and the moderate temperatures to produce crops of <u>wheat and rice</u> to feed not only themselves but also the residents of the cities and towns.

South Asia

The second major concentration of world population also lies in Asia and displays many similarities to that of East Asia. At the heart of this *South Asia* cluster lies India, but it extends also into neighboring Pakistan, Bangladesh, and island Sri Lanka. Again, note the coastal orientation of the most densely inhabited zones, and the finger-like extension of high-den-

WORLD POPULATION DISTRIBUTION
One dot represents 100,000 people.

INDIVIDUAL COUNTRY TOTALS SHOWN ON TABLE IN APPENDIX A.

FIGURE I–14

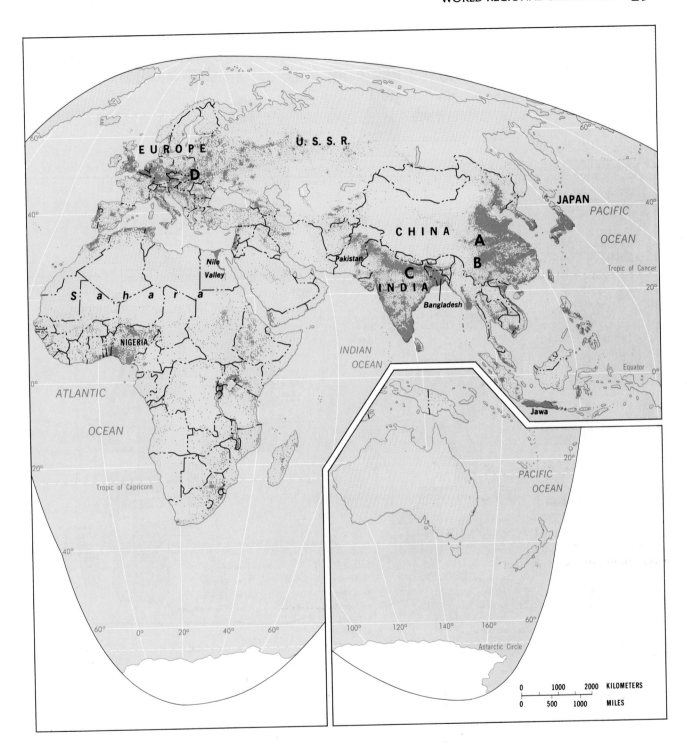

sity population into northern India (**C** on Fig. I–14). This is one of the great agglomerations of people on earth, focusing on the broad plain of the central and lower Ganges River Basin. The South Asia population cluster in 1991 numbers over 1.1 billion people; at present rates of growth, it will be almost 1.4 billion in 2000. Our map also shows how this region is sharply marked off by physical barriers: the Himalaya Mountains rise to the north above the Ganges Lowland, and the desert takes over west of the Indus River Valley in Pakistan. In this confined region population is growing more rapidly than almost anywhere else on earth, and its capacity to support those people has, by every estimate, already been exceeded. As in East Asia, the overwhelming majority of the people are farmers, but here in South Asia the pressure on the land is even greater. In Bangladesh, 121 million people are crowded into an area about the size of Iowa, and nearly all of these people are farmers. Over large parts of Bangladesh, the rural population density is more than 2500 per square mile (1000 per square kilometer). By comparison, the 1991 population of Iowa was about 2.8 million, with a rural density of just 18 per square mile (10 per square kilometer).

Europe

Further inspection of Fig. I–14 reveals that the third-ranking population cluster—*Europe*—also lies on the biggest landmass, but at the opposite end of Eurasia from China. An axis of very dense population extends from the British Isles eastward into Soviet Russia. It includes large parts of West and East Germany, Poland, and the western Soviet Union; it also incorporates the Netherlands and Belgium, parts of France, and

northern Italy. This European cluster (including the contiguous U.S.S.R.) counts over 700 million inhabitants, which puts it in a class with the two larger concentrations—but there the similarity ends. A comparison of the population and physical geography maps indicates that, in Europe, terrain and environment appear to have less to do with population distribution than in the two Asian cases. See, for instance, that lengthy extension marked **D** in Fig. I–14 that protrudes into the interior of the Soviet Union. Unlike the Asian inland extensions, which reflect fertile river valleys, the European population axis is associated with the orientation of Europe's coalfields, the power resources that fired the Industrial Revolution. If you look more closely at the landform map (Fig. I–5), you will note that comparatively dense population occurs even in rather mountainous, rugged country—for example, along the boundary zone between Czechoslovakia and Poland. Another contrast lies in the number of Europeans who live in cities and towns. Far more than in Asia, Europe's population agglomeration is composed of numerous cities and towns, many of them products of the Industrial Revolution. In the United Kingdom, fully 90 percent of the people live in such urban places; in West Germany, nearly 95 percent; and in Belgium, over 95 percent. With so many people concentrated in urban areas, the rural countryside is more open and sparsely populated than in East and South Asia, where fewer than 30 percent of the people reside in cities and towns.

North America

The three world population concentrations just discussed (East Asia, South Asia, and Europe) account for about 3 of the world's

more than 5 billion people. Nowhere else on the globe is there any population cluster with a total even half of any of these. Look at the dimensions of the landmasses on Fig. I–14 and consider that the populations of South America, Subsaharan Africa, and Australia together total *less* than that of India alone. In fact, the next ranking cluster is *Eastern North America*, comprising the east-central United States and southeastern Canada; however, it is only about one-quarter the size of the smallest of the Eurasian concentrations. As Fig. I–14 clearly shows, this region does not possess the large, contiguous high-density zones of Europe or East and South Asia. The North American population cluster displays European characteristics, and it even outdoes Europe in some respects. Like the European region, much of the population is concentrated in several major metropolitan centers, whereas the rural areas remain relatively sparsely settled. The leading focus of the North American cluster lies in the urban complex that lines the U.S. northeastern seaboard from Boston to Washington, and includes New York, Philadelphia, and Baltimore—the great multimetropolitan agglomeration that urban geographers call *Megalopolis.*

There are other large urban concentrations within North America as well. In fact, three more megalopolis-like regions can be identified, one linking Chicago–Detroit–Cleveland–Pittsburgh, the second Montreal–Toronto–Windsor in the Canadian provinces of Quebec and Ontario, and the third San Francisco–Los Angeles–San Diego in California. If you study Fig. I–14 carefully, you will note other prominent North American metropolises standing out as small clusters of high-density population, among them St. Louis, Kansas City, Denver, Dallas, and Seattle.

Smaller Clusters

Still further examination of Fig. I–14 leads us to recognize substantial local population clusters in Southeast Asia. It is appropriate to describe these as discrete (discontinuous) clusters, for the map confirms that they are actually a set of nuclei rather than a contiguous population concentration. Largest among these nuclei is the Indonesian island of Jawa (Java), with about 120 million inhabitants. Elsewhere in this corner of Asia, populations cluster in the lowlands and deltas of major rivers such as the Mekong in Indochina and the Chao Phraya in Thailand. Neither these river valleys nor the rural surroundings of the cities have population concentrations comparable to those of either China to the north or India to the west, and under normal circumstances Southeast Asia is able to export rice to its hungrier neighbors. However, decades of (still-ongoing) strife have disrupted the region to such a degree that its productive potential has not been attained.

The remaining continents do not have population concentrations comparable to those we have already considered. Africa's 683 million inhabitants cluster in above-average densities in West Africa (where Nigeria has a population larger than its official census, probably approaching 150 million) and in a zone in the east extending south from Ethiopia to South Africa. Only in North Africa, however, is there an agglomeration comparable to the crowded riverine plains of Asia: Egypt's Nile Valley and Delta with 58 million residents. Significantly, it is the pattern here—not the numbers—that resembles Asia. As in

East and South Asia, the Nile's valley and delta teem with farmers who cultivate every foot of the rich, fertile alluvial (riverine) soil. But the Nile's gift (see Fig. I–1) is a miniature compared to its Asian equivalents: the lowlands of the Ganges, Chang Jiang, and Huang He contain many times the number of inhabitants who manage to eke out a living along the Egyptian Nile.

The wide-open spaces in South America and Australia and the peripheral distribution of the modest populations of these continents suggest that here we might find considerable space for the world's huge numbers. And indeed, South America could probably sustain more than its present 302 million, as Australia can undoubtedly accommodate more than 17 million. But the population growth rate of South America's countries is higher than the world average, and Australia's severe environmental limitations hardly qualify the smallest continent as a safety valve for Asia's farming millions.

Urbanization

Urbanization increasingly marks the current stage of humanity's expansion throughout the world, although not everywhere with the same intensity. Hope Tisdale has defined urbanization as

a process of population concentration. It proceeds in two ways: the multiplication of the points of concentration and the increasing in size of individual concentrations. . . . Urbanization is a process of becoming, [implying movement] *from a state of less concentration to a state of more concentration.*

Over the past four decades, the population clustered in cities worldwide has more than doubled to a total exceeding 2 billion, thereby making urbanites of more than 4 out of every 10 people on earth. The significance of this latest increase in global urbanization can be seen in Fig. I–16. Throughout this century urban population

FIGURE I–16

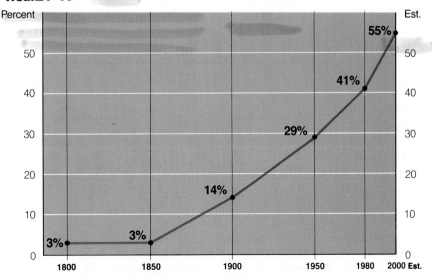

PERCENT OF WORLD POPULATION RESIDING IN SETTLEMENTS OF 5000 OR MORE, 1800–2000

has been increasing at an accelerating rate, particularly since 1980; in fact, if projections hold, the urbanization pace of the second half of the twentieth century will just about double all world urban growth that occurred before 1950. Table I–1 breaks down this recent urbanization trend by major world region for each decade since 1950. Observe that urban population growth has steadily increased in every region, with rapid recent advances occurring in the less developed parts of the world. Keep in mind that although the overall *percentage* of urban population is lower in less developed regions, the much greater *absolute* number of people makes even a 1 percent cityward shift a major movement. Moreover, the impact of such migration is felt even more deeply because these cities are invariably unable to house or employ the newcomers, as we will shortly see in the section on underdeveloped countries.

Urbanization as a force of change is now so pervasive in contemporary world regional geography that we will be considering the various manifestations of this process and its responses throughout this book. As a prelude, we offer here some basic ideas about the nature of cities—the spatial focal points containing sizable concentrations of people and activities that are produced by the operation of world urbanization processes. Truman Hartshorn defines cities as "centers of power—economic, political, and social . . . they are where the action is in terms of innovations and control." He further points out that cities are agglomerations of people "with a distinctive way of life, in terms of employment patterns and organization" and that they contain "a high degree of specialized and segregated land uses and a wide variety of social, economic, and political institutions that coordinate the use of [urban] facilities and resources." We are also reminded that cities today have expanded into vast *metropolitan* complexes, which are composed of the older core or central city and a surrounding outer suburban city (at least in the developed world) that is rapidly enlarging in size and urban function. Although this definition by an urban geographer captures a wide range of functions and properties, we should also be aware of additional uses of the term on the world scene. Viewed specifically, as Jack Williams and his co-authors tell us, "the term *city* is essentially a political designation, referring to a place governed by some kind of administrative body or organization." Seen more broadly, on the other hand, the largest cities (especially when they serve as capitals) are nothing less than the foci—indeed complete microcosms—of their national cultures. In succeeding chapters, we will introduce additional urban terminology and concepts, including a capsule survey of urban geography itself near the beginning of Chapter 3.

POLITICAL GEOGRAPHY

We return in later chapters to additional issues involving world population (problems of density, growth patterns, migration, well-being), beginning with an overview of the subfield of population geography at the outset of the next chapter. Our survey of world realms, however, must also be preceded by an introduction to the politico-geographical complexities of the late-twentieth-century world.

States

The limited land area of the globe is presently divided into approximately 170 national states and 50 dependent territories. The states range in population from China's 1.136 billion to Liechtenstein's 31,000. In terms of territory, the largest is the Soviet Union with 8,600,400 square miles (22,275,000 square kilometers), and the smallest is Monaco, containing less than 1 square mile. But even when the smallest *microstates* are left out of such comparisons (as is done in the table displaying national data in Appendix A), the range remains enormous. In fact, some full-fledged members of the United

TABLE I–1 Urban Population as a Percentage of Total Population by Major World Region, 1950–90

Region	1990 (est.)	1980	1970	1960	1950
North America	77	74	70	67	64
Europe	73	69	64	58	54
Soviet Union	71	65	57	49	39
Latin America	71	65	57	49	41
Asia	34	29	25	22	17
Africa	36	28	23	18	15
World Total	46	41	38	34	29

Source: *Patterns of Urban and Rural Population Growth* (New York: United Nations, Dept. of Intl. Economic and Social Affairs, Population Studies, No. 68, 1980).

Nations such as Iceland and Brunei have populations of under 350,000.

Inequalities among the world's national states are not confined to size and population. Large territorial size is no guarantee of resource wealth: some of the world's larger states are among the poorest. Zaïre (in Equatorial Africa), Sudan (much of which is Sahara), India (with its huge population), and Brazil (better off but not rich despite its huge area) are among countries where size is not matched by comparable shares of the earth's natural resources. Neither do large numbers of people ensure national strength: Indonesia has nearly 200 million inhabitants, and Nigeria contains about twice as many people as West Germany. Some states with small populations (Sweden, Switzerland) enjoy high standards of living, while many populous countries are poor.

Boundaries

The familiar world political map (Fig. I–17) is a recent product of national territorial competition and adjustment. Just a little over a century ago, large areas of Africa and Asia remained beyond the jurisdiction of the encroaching imperialist powers. Finally, when colonial spheres of influence collided, boundaries were often drawn to assign dependencies and to facilitate European administration. The map of modern Africa was largely created or confirmed at the Berlin Conference held in the mid-1880s; most likely, no one present realized that those hastily defined colonial boundaries would, within a century, constitute the borders of independent states.

The world's boundary framework is always under pressure in certain areas, and recent years have witnessed several changes. The border between North and South Vietnam was eliminated. The boundary between Morocco and former Spanish Sahara was also erased, and the territory of the latter divided between Morocco and Mauritania along a new border. Moreover, following a major conflict on the island of Cyprus, a *de facto* boundary between Greeks and Turks appeared and functioned in some respects as a recognized international border. In

FIGURE I–18 Hadrian's Wall: a relict linear boundary between England and Scotland, demarcated by the Romans in the second century A.D.

Lebanon, too, demarcated boundaries appeared between Christian and Muslim zones during the conflicts of the 1970s and 1980s, but they remain in doubt amid the unrelenting upheaval in that country. In fact, the entire question of Israel's boundaries with its neighbors continues to plague Middle East political affairs. And, most recently, following the demise of Soviet control over Eastern Europe, the reunification of Germany eradicates the border that separated its eastern and western components between 1945 and 1991.

These actual and prospective changes involve comparatively minor modifications of a boundary framework whose major outlines have proved quite durable. When African and Asian colonies attained independence in the post-World War II era, some observers predicted that the ''boundaries of imperialism'' would soon be abandoned in favor of more realistic dividing lines. But despite some significant steps in this direction (the India–Pakistan border, for example, created when the British colonial era came to an end in 1947), the chief elements of the pre-independence boundary framework have remained intact—as they are today.

The concept of the linear political boundary is quite old. The Romans and Chinese built walls for this purpose 2000 years ago, and relics remain on the cultural landscape; an outstanding example is Hadrian's Wall (Fig. I–18), built by the Romans to demarcate the edge of their empire along a line near the border between modern-day England and Scotland. In post-Roman Europe, rivers often served as trespass lines, and territorial competition and boundary delimitation were part of Europe's political gestation. European colonial expansion during the seventeenth and eighteenth centuries,

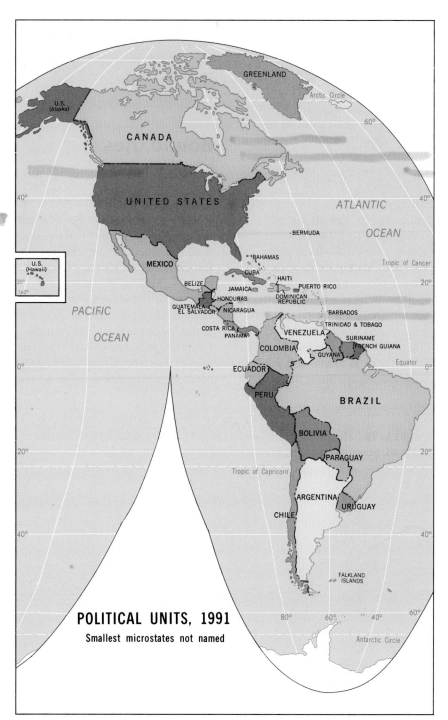

POLITICAL UNITS, 1991
Smallest microstates not named

FIGURE I–17

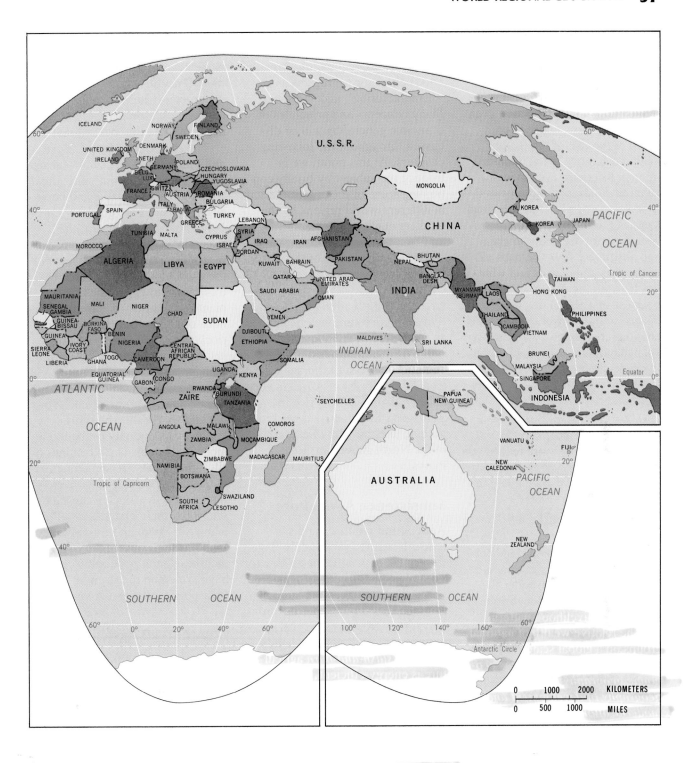

and the new power arising from the Industrial Revolution, hastened a process that had begun early but had progressed slowly. European concepts of territorial acquisition and delimitation, even at sea, were now imposed on much of the rest of the world. Within a century, what remained of open frontiers, even in ice-covered Antarctica, disappeared. Today, we are witnessing the final stage of this territorialization process as states are about to claim and absorb the bulk (perhaps all) of the seas and oceans and what lies beneath them. These and other political spatial realities will be discussed in many of the regional chapters that follow, with Chapter 10 highlighting the subfield of political geography.

THE GEOGRAPHY OF ECONOMIC DEVELOPMENT

As a final prelude to our regional survey of the world's geographic realms, we need to consider two important economic dimensions of humankind: livelihood patterns and the relative wealth of national groups on a global scale. The first of these dimensions involves *economic geography*; the second, the *geography of development and modernization*.

Economic Geography

This subfield of geography is concerned with both the various ways in which people earn a living and how the goods and services they produce in order to earn that in-

come are spatially expressed and organized. Geographers tend to group various occupations broadly into such major aggregates as "manufacturing" and "services." Many concepts and principles of economic geography will be discussed throughout the following chapters. At this point, we introduce the subject by briefly contrasting some of the practices used in different world regions by the workers engaged in humanity's leading occupation—*agriculture*.

The great majority of the earth's peoples, industrial and technological progress notwithstanding, still farm the soil for a living. This activity of agriculture may be defined as the deliberate tending of crops and livestock in order to produce food (or fiber). But farming in the tropical rainforest of Africa is very different from growing rice in Asia—and the Asian paddies look nothing like the vast wheat farms of the North American Great Plains.

Subsistence Agriculture

Agricultural practices and systems, therefore, vary widely. In the tropics, it is often necessary to cut down and burn the original forest vegetation to clear a patch of land that will support a crop (probably a root crop) for one year, perhaps two. It is all done by hand, and such *slash-and-burn* agriculture is energy-efficient. Machines mean very little in areas where cleared land must soon be abandoned—because of the leaching of soil nutrients—in favor of a new patch nearby. But neither can such shifting cultivation support a very dense population, even on the forest margins where the rainfall decreases a bit and the soils are somewhat less leached. Figure I–14 shows that huge areas of lowland Equatorial Africa, South America, and Southeast Asia have

a sparse population. Extensive (nonintensive) subsistence farming is the rule here, even where the rainfall declines to 40 inches (100 centimeters) annually (Fig. I–9), and a dry season permits harvesting corn and other hardy grains.

The ricefields of South, East, and Southeast Asia sustain subsistence farmers, too, but cultivation here is highly intensive and population densities are great (as Fig. I–14 shows). Once again, the Asian rice culture is highly efficient in terms of energy inputs: most of the paddies are still prepared by ox-drawn plow, and the rice is planted by hand—hundreds of millions of hands. But these Asian ricelands would largely comprise just another subtropical zone of meager subsistence agriculture were it not for the alluvial soils in the great river basins. Highly fertile and replenished by rains—which bring not only needed moisture, but also new coatings of silt—these soils are sometimes capable of sustaining two or even three crops in a single year, *one after the other*. The masses of humanity in South and East Asia depend on such soils for survival, and these great population clusters have grown on the strength of their persistent productivity.

Commercial Agriculture

Commercial agriculture in the wheatlands of the American Great Plains presents quite another picture. Vast, almost unbroken fields of grain cover the flat countryside; the soil was fertilized by machine, the sowing was done by machine, and so was the harvesting. This type of modern commercial agriculture is also quite labor-efficient, requiring few hands, but it is less efficient in terms of energy requirements.

Between these extremes lie other agricultural systems. The

plantation, a specialized commercial enterprise in tropical and subtropical environments, is capable of supporting a population of medium density. Elsewhere, the complex, integrated exchange type of farming dominates in the hinterlands of the large urban areas of Western Europe and the eastern United States as well as in Japan, a country in which space is at a premium, distance to markets is often crucial, and the soil is heavily fertilized and tended so that it will produce as much as possible.

Development and Modernization: Advantaged Versus Disadvantaged Countries

The world geographic realms to be discussed in this book are grouped into two main categories: *developed* realms and *underdeveloped* (or *developing*) realms. These terms are used because they have become commonplace, as have such related terms as rich and poor, haves and have-nots, and the like. But perhaps the best designation of degree of **development** would be *advantaged* and *disadvantaged*. In United Nations parlance, the disadvantaged countries are primarily the countries of the *Third World*, the capitalist and socialist systems, respectively, constituting the first two (see Fig. 9–1). Generally speaking, Third World countries find themselves in a disadvantageous economic position in relation to the developed, economically powerful states of the world, whether capitalist or socialist.

Grouping the advantaged countries together, they include most of the states of Europe, the Soviet Union, Canada and the United States, Australia and New Zealand, Japan, Singapore, Israel, and, by some definitions, perhaps South Africa. Against this comparatively short list of countries (which contain a little less than 25 percent of the world's population) are the rest of the countries at various levels of underdevelopment: those of Middle and South America, Africa, and southern and eastern Asia. Not all these less developed countries are at the same level of underdevelopment (just as some developed countries are ahead of others). Few would argue that Haiti, Bolivia, Ethiopia, Bangladesh, and Indonesia are not underdeveloped countries (*UDCs*). But to place Argentina, Uruguay, Chile, Venezuela, Mexico, Brazil, Turkey, China, and South Korea in the same category creates an issue. Some economic geographers suggest that it is appropriate to recognize, between the developed countries (*DCs*) and the UDCs, a middle tier of countries that, although still possessing many characteristics of UDCs, are now "emerging" or at a "takeoff" stage.

Although we routinely (and arbitrarily) divide our world into developed and underdeveloped portions, no universally applicable criteria exist to measure development accurately. Leading indexes are grouped under seven headings (see *box*), but they remain arbitrary and subject to debate. However, what these numerical indexes fail to convey is the time factor. The DCs are the advantaged countries—but why cannot the UDCs and "emerging" countries catch up? It is not, as is sometimes suggested, simply a matter of environment, resource distribution, or cultural heritage (a resistance to innovation, for example). The sequence of events that led to the present division of our world began long before the Industrial Revolution: Europe, even by the middle of the eighteenth century, had laid the foundations for its colonial expansion. The Industrial Revolution then magnified Europe's demands for raw materials, and its manufactures increased the efficiency of its imperial control. Western countries thereby gained an enormous head start, while colonial dependencies remained suppliers of resources and consumers of the products of the Western industries. Thus was born a system of international exchange and capital flow that really changed very little when the colonial period came to an end. Underdeveloped countries, well aware of their predicament, accused the developed world of perpetuating its long-term advantage through *neocolonialism*—the entrenchment of the old system under a new guise.

Symptoms of Underdevelopment

The disadvantaged countries suffer from numerous demographic (population-related), economic, and social ills. Their populations tend to have high birth rates and moderate to high death rates; life expectancy at birth is comparatively low (see Chapter 8 and Appendix A). A large percentage of the population (as much as half) is 15 years of age or younger. Infant mortality is high. Nutrition is inadequate; diets are not well balanced; protein deficiency is a common problem. The incidence of disease is high; health-care facilities are inadequate; there is an excessively high number of persons per available doctor; and hospital beds are too few in number. Sanitation is poor. Substantial numbers of school-age children do not go to school; illiteracy rates are high.

Rural areas are overcrowded

MEASURES OF DEVELOPMENT

What distinguishes a developed economy from an underdeveloped one? Obviously, it is necessary to compare countries on the basis of certain measures—the question cannot be answered simply by subjective judgment. No country is totally developed; no economy is completely underdeveloped. We are comparing *degrees* of development when we identify DCs and UDCs. Our division into developed and underdeveloped economies is arbitrary, and the dividing line is always a topic of controversy. There is also the problem of data. Statistics for many countries are inadequate, unreliable, incompatible with those of others, or simply unavailable.

The following measures are frequently used to gauge levels of economic development:

1. *National product per person.* This is determined by taking the sum of all incomes achieved in a year by a country's citizens and dividing it by the total population. Figures for all countries are then converted to a single currency index for purposes of comparison. In DCs, the index can exceed (U.S.) $10,000; in some UDCs, it can be below $100.

2. *Occupational structure of the labor force.* This is given as the percentages of workers employed in various sectors of the economy. A high percentage of laborers engaged in the production of food staples, for instance, signals a low overall development level.

3. *Productivity per worker.* This is the sum of production over the period of a year divided by the total number of persons in the labor force.

4. *Consumption of energy per person.* The greater the use of electricity and other forms of power, the higher the level of national development. These data, however, must be viewed to some extent in the context of climate.

5. *Transportation and communication facilities per person.* This measure reduces railway, road, airline connections, telephone, radio, television, and so forth, to a per capita index. The higher the index, the higher the development level.

6. *Consumption of manufactured metals per person.* A strong indicator of development levels is the quantity of iron, steel, copper, aluminum, and other basic metals used by a population during a given year.

7. *Rates.* A number of additional measures are employed, including literacy rates, caloric intake per person, percentage of family income spent on food, and amount of savings per capita.

low because its demand on the available land is higher. There is little production for the local market because distribution systems are poorly organized and demand is weak. On the farms, yields per unit area are low, subsistence modes of life prevail, and the specter of debt hangs constantly over the peasant family. These circumstances preclude investment in such luxuries as fertilizers and soil conservation techniques. As a result, soil erosion and land denudation scar the rural landscapes of many UDCs. Where areas of larger-scale modernized agriculture have developed, they produce for foreign markets with little resultant improvement of domestic conditions.

In the urban areas, appalling overcrowding, poor housing, inadequate sanitation, and a general lack of services prevail. Job opportunities are insufficient, and unemployment is always high. Per capita income is low, savings per person are minimal, and credit facilities are poor. Families spend a very large proportion of their income on food and basic necessities. The middle class remains small; often a substantial segment of that middle-income population consists of foreign immigrants.

These are some of the criteria that signal underdevelopment, and the list is not complete. For example, one of the geographic properties that mark UDCs is the problem of internal regional imbalance. But even in UDCs, there are local exceptions to the general economic situation. The capital city may appear as a skyscrapered model of urban modernization, with thriving farms in the immediate surroundings and factories on the outskirts. Road and rail may lead to a bustling port where luxury automobiles are unloaded for use by the privileged elite. Here, in the country's core, the rush of "progress" is evident—but travel a few miles into the countryside

and suffer from poor surface communications. Men and women do not share fairly in the work that must be done; women's workloads are much heavier, and children are pressed into the labor force at a

tender age. Landholdings are often excessively fragmented, and the small plots are farmed with outdated, inefficient tools and equipment. The main crops tend to be cereals and roots; protein output is

CORE-PERIPHERY REGIONAL RELATIONSHIPS

As we have just seen, ours is a world of uneven development in which the advantages lie with the DCs. But how has this situation become so extreme? And what perpetuates it? Regional contrasts in development have been attributed to many factors. Climate has been blamed for it (as a conditioner of human capacity), as has cultural heritage (e.g., resistance to innovation), colonial exploitation, and, more recently, neocolonialism. Obviously, the distribution of accessible natural resources and the territorial location of countries play a major part. Opportunities for interaction and exchange have always been greater in certain areas than in others. Isolation from the mainstreams of change has been a disadvantage. These factors, to a greater or lesser degree depending on location, certainly have played their roles.

One of the hallmarks of national spatial organization is the evolution of *core areas*, foci of human activity that function as the leading regions of control and change. Most commonly today, we tend to associate this regional concept with a country's heartland—its largest population cluster, most productive and influential region, the area possessing the greatest centrality and accessibility that usually contains a powerful capital city. These ideas will be developed and applied to the world's major realms in ensuing chapters, and they will also be shown to be of historic importance in the emergence of ancient city-states, culture hearths, seats of empires, and modern technological revolutions. Consider this: such cores would not have developed without contributions from their surrounding

areas. One of the earliest developments in ancient cities was the creation of organized armed forces to help rulers secure taxes and tribute from the people living in the countryside. Thus, core and periphery (margin) became functionally tied together: the core had requirements, and the periphery gave.

Have modern times ended this core-periphery relationship? The answer not only is negative—the situation has actually worsened. You can see core (*have*) and periphery (*have-not*) relationships in every country in the world. In many African countries (Kenya, Senegal, and Zimbabwe, to name just three examples in different regions), the former colonial capital now lies at the focus of the national core area. Its tall buildings and government centers symbolize the concentration of privilege, power, and control. Here are the trappings of development, the paved roads and streetlights, hospitals and schools, corporate headquarters and businesses, industries and markets. Surrounding the capital are farms producing high-priced goods, especially perishables, for the city dwellers who can afford to buy them. But a little farther out into the countryside, life swiftly changes. In the periphery, the core area is distant, even ominous—always a concern. What will be paid for crops on the markets in the core? What will the cost be of goods made there and needed here, in the countryside? How high will taxes be? This core-periphery arrangement is a system that keeps regional development contrasts high. The core may give the impression of being a thriving metropolis and may remind one of

a city in a developed country. But the periphery is the land where underdevelopment reigns.

Looking at the world as a whole from this perspective, we see that core-periphery systems now encompass not just individual countries but the entire globe. Today, the whole of Western Europe functions as a core area; North America forms a second core area, and Japan is the third within the Western capitalist world. Moreover, the western Soviet Union constitutes a fourth global core area, the dominant focus of the communist world. Therefore, what was said about that powerful city-region in an otherwise underdeveloped country applies, at the macroscale, to these global core areas as well. Whether in Europe, North America, Japan, or the western U.S.S.R., power, money, influence, and decision-making organizations are concentrated here.

Imagine that you, as a geographer, are asked to locate a factory (say a textile plant) to sell products in core-area markets. Salaries and wages in the core area are high, so you look for an alternative in the periphery, where wages are much lower. Raw materials also can be bought locally. So you advise the corporation to establish its factory in a certain country in the periphery, thereby initiating a relationship between core-based business and periphery-based producers. Now, however, the corporate management expresses concern over political stability in that periphery country. Soon you are talking to government representatives about guarantees, and the economic connection also becomes a political one.

Is all this necessarily disadvantageous to the countries in the periphery, the mainly underdeveloped countries, the countries of what we have called the Third World? After all, the core-area corporations and businesses make investments, stimulate economies, and put people to work in the countries of the periphery. The problem is that a dependency relationship develops that soon tends to diminish any such advantage. The profits earned return mostly to the core area and benefit the periphery only minimally. When you see that ultramodern high-rise hotel towering above the sandy beaches of a Caribbean island and watch its staff at work, you observe both sides of it. True, an investment was made here and jobs were created. But the hotel's profits go back to the multinational corporation's bank account—in New York, Tokyo, or London.

Another problem has to do with the sociopolitical effects. That textile factory we mentioned needs a manager and other administrative personnel. These people tend to be (or become) part of the *elite*, the "upper class" in that periphery country, a kind of cadre of representatives of the world core area in the underdeveloped realms. Because they represent core-area interests, these people have divided loyalties: what is best for core-area enterprises is not always in the best interest of their home country. Such examples of core-periphery interactions abound, and the entire world is now interconnected to a far greater degree than many of us realize. Superimposed on the traditional cultural landscapes of the Third World is a growing global network of interaction that reaches into even the

smallest village store in the remotest part of a UDC. Thus, the numerous geographic realms and regions we will come to recognize in this book are, above all, enmeshed in a worldwide economic-geographic system—a system tightly oriented to the world core area. Moreover, this web now includes the Soviet Union. Researchers sometimes blame the core-periphery dilemma on capitalist economics, but, in fact, the relationships between the U.S.S.R. and its communist "Second World" client states have not been that different. Like the three great capitalist core areas, state-directed economic enterprise aims to lower costs, strengthen the core, and influence events in the periphery.

Thus, the countries of the periphery find themselves locked within a global economic system over which they have no control. Even countries that find themselves possessing commodities in high demand in the core (such as the oil of the OPEC states) have difficulty converting their temporary advantage into longer-term parity with core-area powers. Countries that depend heavily for their income on the export of raw materials such as strategic minerals (or single crops such as bananas or sugar) are at the mercy of those who do the buying. But there are other reasons for the depressing condition of many periphery countries. Like the contemporary city, the world core area beckons constantly, siphoning off the skilled and professional people from the periphery. How many doctors, engineers, and teachers from India are working in England and the United States? And how badly India now needs such trained people! Every loss is

magnified—but oppressive political circumstances in periphery countries contribute as "push" factors in this "brain drain."

Another side to the continuing underdevelopment of periphery countries relates to traditional cultural attitudes and values. The introduction of core-area tentacles into outlying regions where traditional culture is strong may lead to a reaction, perhaps a resurgence of fundamentalism. This, in turn, restricts further economic change. The invasion of modern ways can be unsettling. Do not forget that the long traditions of societies in the periphery have generated large bureaucracies whose interests are threatened by the kind of change development brings with it. Occasionally, the tangible presence of the external (core) installation such as an office building, airline agency, or retail establishment is the target of violent attack. Such incidents symbolize ideological opposition to the intrusion they represent.

The countries of the periphery, therefore, confront severe problems of many kinds. They are pawns in a global economic game whose rules they cannot touch, let alone change. Their internal problems are intensified by the aggressive involvement of core-area interests. Their resource use (such as allocation of soils to producing local food versus export crops) is strongly affected by foreign interference. They suffer far more than core-area countries do from environmental degradation, overpopulation, and mismanagement. They possess inherited disadvantages that have grown, not lessened, over time. The widening gaps that result between enriching cores and persistently impoverished peripheries clearly are a threat to the future of the world.

(or even to the squatter "shack-towns" at the edge of the city), and you will probably find that almost nothing has changed. And just as the rich countries become richer and leave the poorer ones farther behind, so the gap between progressing and stagnant regions *within* UDCs grows larger. It is a problem of global dimensions, because similar **core-periphery relationships** exist at the international scale—a topic explored in the *Opening Essay* on pages 41–42.

There can be no doubt that the world economic system works to the disadvantage of the UDCs, but sadly it is not the only obstacle the less advantaged countries face. Political instability, corruptible leaderships and elites, misdirected priorities, misuses of aid, and traditionalism are among the conditions that commonly inhibit development. External interference by interests representing powerful DCs have also had negative impacts on the economic as well as the political progress of UDCs. Underdeveloped countries even get caught in the squeeze when other UDCs try to assert their limited strength: for example, when the Organization of Petroleum Exporting Countries (OPEC)—mostly underdeveloped states themselves—raise the price of oil, then energy and fertilizers slip still farther from the reach of the poorer UDCs not fortunate enough to belong to this favored group. As the DCs get stronger and wealthier, they leave the underdeveloped world ever farther behind: the gap is still widening, and the prospects for the UDCs are not bright. The theme of development and modernization runs throughout the chapters (4–10) on the Underdeveloped Realms; an overview of the role of geography in this interdisciplinary inquiry is presented at the outset of Chapter 9.

GEOGRAPHIC REALMS OF THE WORLD

Earlier in the chapter, we discussed the classification of world regions and introduced the three-tier hierarchy of culture realm, region, and subregion. The world regional system to be used in this book, however, is based not on culture realms but on broader **world geographic realms**. Thus, our global spatial framework (Fig. I–19) represents much more than a regionalization of cultural-geographical phenomena and cultural landscapes. It also reflects the leading features of economic, urban, political, physical, and historical geography, and is a synthesis of human geography as a whole, not just cultural geography. Cultural geographers, for example, might combine Middle and South America under the single rubric of "Latin" America because of the strength of the Latin cultural imprint. But that imprint is far stronger in South America than in Middle America, as will be noted later. The contrast is sufficiently strong to separate Middle America as a *geographic* realm from South America. Differences in physical, historical, political, and economic geography confirm the boundary between realms ④ and ⑤ on the map.

Thirteen geographic realms form the structural framework for our global regional survey. Five of them comprise the developed world (Europe, Australia–New Zealand, the Soviet Union, North America, Japan), and the remaining eight consist of assemblages of underdeveloped countries. We now introduce these realms in a series of profiles (to be read in conjunction with the map in Fig. I–19) that preview the coming 10 chapters and 5 vignettes.

Europe ①

Europe merits designation as a world realm despite the fact that it occupies only a small portion of the Eurasian landmass—a fraction that, moreover, is largely made up of Eurasia's western peninsular extremities. Certainly Europe's size is no measure of its global significance: no other part of the world is or ever has been so packed full of the products of human achievement and the source of so many innovations and revolutions that transformed areas far beyond its own borders. Over the past several centuries, the evolution of world interaction consistently focused on European states and their capitals. Time and again, despite internal wars, despite the losses of colonial empires, despite the impacts of external competition, Europe has proved to contain the human and natural resources needed to rebound and renew its progress.

Among Europe's greatest assets is its internal natural and human diversity. From the warm shores of the Mediterranean to the frigid Scandinavian Arctic and from the flat coastlands of the North Sea to the grandeur of the Alps, Europe presents an almost infinite range of natural environments. An insular and peninsular west contrasts against a more continental east. A resource-laden backbone extends across the center of Europe from England toward the east. Excellent soils produce harvests of enormous quantity and variety. And the population includes people of many different stocks, peoples grouped under such familiar names as Latin, Germanic, and Slavic. Europe has its cultural minorities as well—for example, the Hungarians and the Finns. Immigrants continue to stream into Europe, contributing further to a diversity that has been an advantage to Europe in uncountable ways. Today's resilient Europe, especially in its west, is a realm dominated by great cities, intensive transport networks and mobility, enormous productivity, a large and often very dense population, and a persistently sophisticated technology.

Australia–New Zealand (1A)

Just as Europe merits recognition as a continental realm, although it is merely a peninsula of Eurasia, Australia has achieved similar identity as the anchor of the island realm of Australasia. Australia and New Zealand are European outposts in an Asian–Pacific world, as unlike Indonesia as Britain is unlike India. Although it was spawned by Europe and its people and economy are Western in every way, Australia as a continental realm is a far cry from the crowded European world. The image of Australia is one of impressive modern cities, wide rural expanses, huge herds of livestock (perhaps also the pests, rabbits and kangaroos), deserts, and beautiful coastal scenery. There is much truth in such a picture: 15 of Australia's slightly more than 17 million people are concentrated in the country's urban areas. In this as well as other respects, Australia is more like the United States than other realms. With its British heritage, single language (except for that of a small indigenous minority), and advanced economy, Australia–New Zealand's identification as a realm rests on its remoteness and spatial isolation.

The Soviet Union (2)

The Soviet Union—the world's largest country in areal extent—

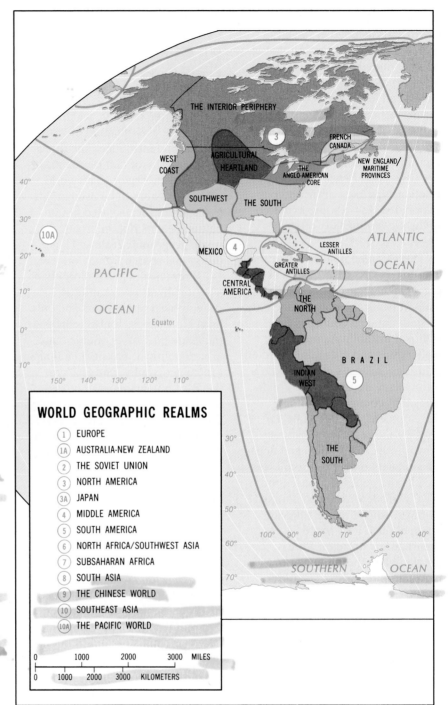

WORLD GEOGRAPHIC REALMS

1. EUROPE
1A. AUSTRALIA-NEW ZEALAND
2. THE SOVIET UNION
3. NORTH AMERICA
3A. JAPAN
4. MIDDLE AMERICA
5. SOUTH AMERICA
6. NORTH AFRICA/SOUTHWEST ASIA
7. SUBSAHARAN AFRICA
8. SOUTH ASIA
9. THE CHINESE WORLD
10. SOUTHEAST ASIA
10A. THE PACIFIC WORLD

FIGURE I–19

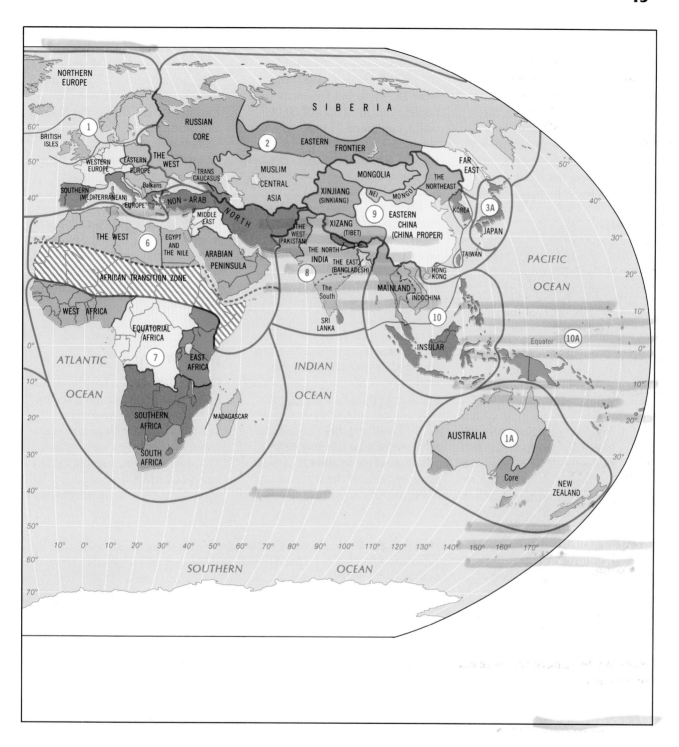

constitutes a geographic realm not just because of its size. To its south lie the Japanese, Chinese, South and Southwest Asian realms, all clearly different culturally and economically from the realm that Russia forged. The Soviet realm's western boundary is always subject to debate, but neither Finland nor Eastern Europe have been areas of permanent Russian or Soviet domination.

The events of the twentieth century have greatly strengthened the bases for the recognition of the Soviet Union as a geographic realm. Tsarist Russian expansionism was halted, the Communist revolution came, and a whole new order was created, transforming the old Russian Empire into a strongly centralized socialist state whose hallmark was economic planning. The new political system awarded the status of Soviet Socialist Republic (S.S.R.) to areas incorporated into the former empire, including those inhabited by Muslims and other minorities within the eastern U.S.S.R. In the economic sphere, the state took control of all industry and production, and agriculture was collectivized.

The disastrous dislocation of World War II notwithstanding, the past half-century has seen much progress in the Soviet Union. The state has risen from a backward, divided, near-feudal country to the position of a world superpower with a strong individualistic (and also centrally directed) culture, an advanced technology, and a set of economic and social policies that have attracted world attention and, in some instances, emulation.

North America ③

The North American geographic realm consists of Canada and the United States, two of the most heavily urbanized and industrialized countries in the world. In the United States in 1987, there were 282 metropolitan areas with populations in excess of 60,000, which contained 77 percent of all the people in the country; a substantial proportion of the remainder, moreover, lived in urban places larger than 10,000.

The North American realm is characterized by its large-scale sophisticated technology, its massive consumption of the world's resources and commodities, and its unprecedented mobility and fast-paced lifestyles. Suburbs citify and blend into one another as metropolises coalesce, surface and air transport networks intensify, and skylines change; the landscape of Southern California's San Bernardino County exquisitely captures these dynamics (Fig. I–20). Overcrowded shopping centers, traffic jams, airport delays, pollution of air and water—these are also some of the by-products of North American technocracy.

Both the United States and Canada are pluralistic societies, with Canada's cultural sources being in Britain and France and those of the United States in Europe, Africa, Asia, and Middle and South America. In both Canada and the United States, minorities remain separate from the dominant culture, thereby giving rise to a number of persistent social problems. Quebec is Canada's French-dominated province, and in the United States patterns of racial segregation endure with black Americans overwhelmingly concentrated in particular urban areas.

Japan ③A

On Fig. I–19 Japan hardly seems to qualify as a geographic realm. Can a group of comparatively small islands qualify as a separate realm? Indeed they can—when 124 million people inhabit those islands and when they produce there an industrial giant, an ultramodern

FIGURE I–20 Freeway corridors form the heart of today's burgeoning outer suburban cities. San Bernardino County east of Los Angeles, with the central city skyline visible on the horizon.

urban society, and a vigorous national power. Japan is unlike any other non-Western country, and it exceeds many Western-developed countries in a number of spheres. Like its European rivals, Japan built and later lost a colonial empire in Asia. During its tenure as a colonial power, Japan learned to import raw materials and export finished products. Even the calamity of World War II—when Japan became the only country ever to suffer nuclear attack—failed to extinguish Japan's forward push. In the postwar era, Japan's overall economic growth rate was the highest in the world. Today, raw materials no longer come from colonies, but are purchased all around the world. Its high-quality products are sold practically everywhere: almost no city on earth is without Japanese cars in its streets, few photography stores lack Japanese cameras, and many scientific laboratories use Japanese precision equipment. From stereos to VCRs to oceangoing ships, Japan's manufactures flood the world's markets.

Japan, therefore, constitutes a geographic realm by virtue of its role in the world economy today, the transformation—in a single century—of its national life and its industrial power. Yet Japan has not turned its back on its old culture; modernization is not all that matters to the Japanese, who still adhere to ancient customs, including the rites of the Shinto religion, tea ceremonies, *Kabuki* drama, arranged marriages, and traditional song and dance. There is also still a reverence for older people, and despite all the factories, sprawling cities, high-speed trains, and jet aircraft, Japan is still a land of shopkeepers, handicraft industries, home workshops, carefully tended fields, and space-conserving settlements. Only by holding on to their culture's old

traditions could the Japanese have tolerated the onrush of their new age.

Middle America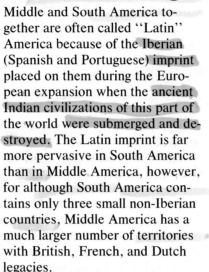

Middle and South America together are often called "Latin" America because of the Iberian (Spanish and Portuguese) imprint placed on them during the European expansion when the ancient Indian civilizations of this part of the world were submerged and destroyed. The Latin imprint is far more pervasive in South America than in Middle America, however, for although South America contains only three small non-Iberian countries, Middle America has a much larger number of territories with British, French, and Dutch legacies.

Middle America consists of Mexico, the seven Central American republics from Guatemala to Panama, and the island chains of the Caribbean Sea to the east. In pre-European times, this was a significant hearth of human development: great cities arose a thousand years ago, crops and livestock were domesticated, progress was made in the sciences and the arts, and empires were forged. This realm, too, was the scene of the first European arrival in the Americas just 500 years ago; for some time, it remained the New World core from which the white invaders' influences radiated outward. The early importance of the slave trade is still reflected in the high percentages of black people among Middle America's island (and certain mainland) populations.

South America

The continent of South America is also a geographic realm—a region

shared by Portuguese-influenced Brazil and a Spanish colonial domain now divided among nine seprate countries. As in mainland Middle America, the Roman Catholic religion dominated life and culture, and systems of landownership, tenancy, taxation, and tribute were transferred from the Iberian Old World to the New. Today, South America still carries the impress of its source region in the architecture of its cities, in the visual arts, and in its music.

South America remains one of the world's more sparsely peopled realms, and much of the continent's interior is virtually empty terrain. As Fig. I–14 shows, the distribution of population is not only coastally oriented but is also heavily clustered, and to a considerable degree those clusters exist in isolation from one another. Internal communication and interaction have not been leading qualities in South America, whose individual countries often possess stronger ties with Europe than they do with their neighbors.

Whereas South America does not yet suffer from the population pressures that bedevil other parts of the underdeveloped and emerging world (including several countries in Middle America to the north), the realm nevertheless confronts some serious liabilities. These include the nature of landownership, control of the means of production, and the resistance of those in authority against the pressures of change. Under such circumstances, South American experiments in democratic government have failed, and authoritarian rule repeatedly arouses violent response.

Brazil is treated as a separate vignette because of its vast size: it contains 51 percent of the South American population and 48 percent of the continental land area. In many ways, this giant country epitomizes the potentials and problems of South America as a whole. Its development and modernization plans continue to be the realm's most ambitious.

North Africa/Southwest Asia ⑥

The North Africa/Southwest Asia realm is known by several names, none of them satisfactory: the Islamic realm, the Arab world, the dry world. Undoubtedly, the powerful influence of the Islamic (Muslim) religion is this realm's overriding cultural quality, for Islam is more than a faith: it is a way of life, one that is vividly expressed in the human landscape (Fig. I–21). The contrasts within this realm are underscored when we note its contents, extending as it does from Morocco and Mauritania in the west to Iran and Afghanistan in the east and from Turkey in the north to the African Horn in the south. Huge desert areas separate highly clustered populations (see Fig. I–14); this isolation perpetuates cultural discreteness, and we can distinguish regional contrasts within the realm quite clearly. The term *Middle East* refers to one of those regions (Fig. I–19), the five countries that lie between the eastern Mediterranean and the Persian Gulf. The oil-rich Arabian Peninsula is another region, whose holy city of Mecca is the international focus of Islam.

Rural poverty, strongly conservative traditionalism, political instability, and conflict have marked the realm in recent times, but this is also the source area of many of the world's great religions and the site of ancient culture hearths and early civilizations. Had we drawn Fig. I–19 several centuries ago, we would have shown Arab-Turkish penetrations of Iberian and southeastern Europe and streams of trade and contact reaching the length of Africa's east coast and into Southeast Asia. So vigorously was Islam propagated beyond this realm, that to this day there are more adherents to the Muslim faith *outside* the so-called Arab world than within it.

Subsaharan Africa ⑦

Between the southern margins of the Sahara Desert and South Africa's southernmost Cape Province lies the geographic realm of Subsaharan Africa. Its boundary with the North Africa/Southwest Asia realm to the north is a broad zone of transition across the empty expanses of the Sahara; here coincidence with political boundaries cannot be employed. In Sudan, for instance, the north is part of the Arab world, but the south is quite distinctly African. So strong are these regional contrasts within Sudan that it would be inappropriate to include the entire country in either realm. Therefore, Sudan— as well as Ethiopia, Chad, Nigeria, and Mali—is bisected (Fig. I–19).

The Subsaharan Africa realm is defined by its mosaic of hundreds of languages belonging to specific African linguistic families and by its huge variety of traditional local religions that, unlike world religions such as Islam or Christianity, remain essentially "community" religions. With few exceptions, Africa is also a realm of farmers. The subsistence way of life was changed very little by the European colonialists; tens of thousands of villages all across Subsaharan Africa were never brought into the economic orbits of the Europeans (Fig. I–4). Root crops and grains are still grown in ancient, time-honored ways with the hand hoe and the digging stick. It is a backbreaking, low-yield proposition, but the African agriculturalist does not have much choice.

This realm consists of four regions: West Africa, East Africa, Equatorial Africa, and Southern Africa. The last of these is dominated by the country of South Africa, which is treated as a separate

FIGURE I–21 The unity of faith and culture is a hallmark of the Islamic townscape, in this case the holy city of Qum in northern Iran.

vignette because it is engaged in a monumental power struggle between the white minority and African peoples. In the early 1990s, that conflict has intensified South Africa's continuing racial upheaval, a tragedy with sharp geographical overtones whose background we will explore in detail.

South Asia ⑧

The familiar triangular shape of India outlines a subcontinent in itself—a clearly discernible physical region bounded by mountain range and ocean, inhabited by a population that constitutes one of the greatest human concentrations on earth. The scene of one of the oldest civilizations, it became the cornerstone of the vast British colonial empire. Out of the colonial period and its aftermath emerged six states: India, Pakistan, Bangladesh, Sri Lanka, Nepal, and Bhutan.

Europe was a recipient of Middle Eastern achievements in many spheres, and so was India. From Arabia and the Persian Gulf came traders across the sea. The tidal wave of Islam came from the northwest over land, across the Indus River, through the Punjab, and into the Ganges Plain. Along this route had also come the realm's modern inhabitants, who drove the older Dravidian peoples southward toward the tip of the subcontinent.

Long before Islam reached India, major faiths had already arisen here, religions that still shape countless lives and attitudes. Hinduism and Buddhism emerged before Christianity and Islam, and the postulates of the Hindu religion have dominated life in India for several thousand years. They include beliefs in reincarnation, in the goodness of holy men and their rejection of material things, and in the inevitability of a hierarchical

structure in life and afterlife. This last quality of the Hindu faith had expression in India's *caste system*, the castes being rigid social strata and steps of the universal ladder. It had the effect of locking people into social classes, the lowest of which—the untouchables—suffered a miserable existence from which there was no escape. Buddhism was partly a reaction against this system, and the invasion of Islam was facilitated by the alternatives it provided. Eventually, the conflicts between Hinduism (India's primary religion today) and Islam led to the partitioning of the realm into India and Pakistan, an Islamic republic that subdivided again with the birth of Bangladesh in 1971.

If this realm is one of contrasts and diversity, there are overriding qualities nevertheless. It is a region of intense adherence to the various faiths; a realm of thousands of villages and several teeming, overpopulated large cities; and an area of poverty and frequent hunger, where the difficulties of life are viewed with an acquiescence that comes from a belief in a better new life—later.

The Chinese World ⑨

China is a national state as well as the heart of a geographic realm. The Chinese embody what may be the oldest of the world's continuous civilizations. It was born in the middle basin of the Huang He (Yellow River) and now envelops an area of nearly 3.7 million square miles with a population of more than 1.1 billion. Alone among the ancient culture hearths, China's spawned a major modern state of world stature, with the strands to the source still intact. The Chinese people still refer to

themselves as the "People of Han," the great dynasty (207 B.C. to A.D. 220) that was a crucial formative period in the country's evolution. But the cultural individuality and continuity of China were already established 2000 years before that time, perhaps even earlier.

In the lengthy process of its evolution as a regionally distinct culture and a great national state, China has had an ally in its isolation. Mountains, deserts, and sheer distance protected China's "Middle Kingdom" and afforded the luxury of stability and comparative homogeneity. Not surprisingly, Chinese self-images were those of superiority and security—the growing Chinese realm was not about to be overrun. There were invasions from the steppes of inner Asia, but the intruders were repulsed or absorbed; there would always be a China. The European impact of the nineteenth century finally ended China's invincibility, but it was held off far longer than in India, it lasted a much shorter time, and it had less permanent effect.

Unlike India, China's belief system was always concerned more with the here and now, the state and authority, than with the hereafter and reincarnations. The century of convulsion that ended with the Communist victory in 1949 witnessed a breakdown in Chinese life and traditions, but there is much in the present system of government that resembles China under its dynastic rulers. China may have undergone a transformation into a communist society, but Chinese attitudes toward authority, the primacy of the state, and the demands of regimentation and organization are not new.

Today, China is moving toward world-power status once again, and there are imposing new industrial developments and growing cities. But for all its modernization, China remains a realm of crowded farmlands, carefully diked floodplains, intricately terraced hillsides, and cluttered small villages. Crops of rice and wheat are still meticulously cultivated and harvested, and the majority of the people still bend to the soil.

Southeast Asia ⑩

Southeast Asia is nowhere nearly as well defined a geographic realm as either South Asia or China. It is a mosaic of ethnic and linguistic groups, and this corner of the world has been the scene of countess contests for power and primacy. Spatially, the realm's discontinuity is quite obvious: it consists of a peninsular mainland—where populations tend to be clustered in river basins—and thousands of islands forming the archipelagoes (island chains) of Indonesia and the Philippines.

During the colonial period, the term *Indochina* came into use to denote the eastern rim of mainland Southeast Asia. The term is a good one, for it reflects the major sources of cultural influence that have affected the entire realm. The great majority of Southeast Asia's inhabitants have ethnic affinities with the people of China, but it was from India that the realm received its first strong, widely disseminated cultural imprints— Hindu and Buddhist faiths, architecture, and key aspects of social structure. The Muslim faith also arrived via India. From China came not only ethnic ties, but also elements of culture; Chinese modes of dress, plastic arts, boat construction, and other qualities were widely adopted throughout Southeast Asia. In more recent times, a major migration from China to the cities of Southeast Asia further strengthened the impact of Chinese culture on this realm.

WHAT DO GEOGRAPHERS DO?

A systematic spatial perspective and an interest in regional study are the unifying themes and enthusiasms of geography. Geography's practitioners include physical geographers, whose principal interests are in the study of geomorphology (land surfaces), in research on climate and weather, in vegetation and soils, and in the management of water and other natural resources. There are also geographers who concentrate their research and teaching on the ecological interrelationships between the physical and human worlds; they study the impact of humankind on our globe's natural environments and the influences of the environment (including such artificial contents as air and water pollution) on human individuals and societies. Other geographers are regional specialists, often concentrating their work for governments, planning agencies, and multinational corporations on a particular region of the world. Still other geographers—who now comprise the largest group of practitioners—are devoted to certain topical or systematic subfields such as urban geography, economic geography, cultural geography, and many others (see Fig. I–22); they perform numerous tasks associated with the identification and resolution (through policy-making and planning) of spatial problems in their specialized areas. And, as in the past, there are still many geographers who combine their fascination for spatial questions with technical knowhow. Cartography, geographic information systems, remote sensing, and even environmental engineering are among specializations listed by the 10,000-plus professional geographers of North America. In Appendix B (pp. 593–599), entitled *Opportunities in Geography*, you will find considerable information on the field, how one trains to become a geographer, and the many exciting career options that are open to the young professional.

The Pacific World ⑩ᴀ

Between Asia–Australia and the Americas lies the vast Pacific Ocean, larger than all the land areas of the earth combined. In this great ocean lie tens of thousands of islands, large and small. This fragmented, culturally complex realm defies effective generalization. Population contrasts are reflected to some extent in the regional diversification of this sprawling Pacific world. Accordingly, the islands from New Guinea eastward to the Fiji group are called *Melanesia* (*mela* means black). The people here are black or very dark brown and have black hair and dark eyes. North of Melanesia, and east of the Philippines, lies the island region known as *Micronesia* (*micro* means small [for the islands]), and the people here show a mixture between Melanesians and Southeast Asians.

In the vast central Pacific, east of Melanesia and Micronesia and extending from the Hawaiian Islands to the latitude of New Zealand, is *Polynesia* (*poly* means many). Polynesians are widely known for their strong physique; they have somewhat lighter skin than other Pacific peoples, wavier hair, and rather high noses. Their ancestry is complex, including

Indian, Melanesian, and other elements. Anthropologists also recognize a second Polynesian group, the neo-Hawaiians, a blend of Polynesian, European, and Asian ancestries. Yet, despite the complexity of human occupance within the dispersed Pacific realm, the Polynesians, in their songs and dances, their philosophies regarding the nature of the world, their religious concepts and practices, their distinctive building styles, their work in stone and cloth, and in numerous other ways have built a culture of strong identity and distinction.

We have defined the 13 realms depicted in Fig. I–19 on the basis of a set of criteria that include not only cultural elements, but also political and economic circumstances, relative location, and modern developments that appear to have lasting qualities. Undoubtedly our scheme possesses areas that are open to argument, but such a debate can by itself be quite instructive. We have indicated some locations where doubts may exist; there are more, as further reading and comparisons of our framework to others would underscore.

GEOGRAPHY AND INTERNATIONAL UNDERSTANDING

A number of governmental and private-sector reports in the 1980s criticized the quality of education in the United States. One of the greatest deficiencies that they identified was the appalling lack of international knowledge possessed by American citizens at precisely the time when their nation was entering a new era dominated by global economic and social interrelationships.

Responding to this problem of world ignorance within the United States, a special committee convened by the Association of American Geographers demonstrated that geography contributes heavily to international understanding and the development of global perspectives by emphasizing:

1. The relationships of societies, cultures, and economies around the world to specific combinations of natural resources and of the physical and biological environment.
2. The importance of location of places with respect to one another, as depicted on appropriate maps.
3. The diversity of the regions of the world.
4. The significance of ties of one country with another through the flow of commodities, capital, ideas, and political influence.
5. The world context of individual countries, regions, and problems.

As was pointed out in the Preface, the need for greater international awareness is a central aim of this book. Accordingly, all five emphases listed above are stressed throughout the regional chapters and vignettes that follow.

THE PLACE OF REGIONAL STUDIES IN CONTEMPORARY GEOGRAPHY

Geography, which is often called the parent of all the sciences, is one of the very oldest fields of learning, and **regional** study has been a focus of the discipline throughout its evolution over the past 3000 years. During the third century B.C., the ancient Greek scholar Eratosthenes first began to use the term **geography** (*geo*, earth + *graphy*, writing). By Roman times, the search for meaningful physical and human regions had become well established, and maps frequently incorporated such concepts as climatic zones. This valuable tool of regionalization greatly helped geographers to interpret the ever-expanding mass of environmental and cultural information they were accumulating.

This approach also assisted them in developing a broad spatial perspective of the world and its contents, which was being assembled as regional study was integrated with three other maturing geographical traditions: *earth science* (physical geography), *culture-environmental interaction* (human ecology), and *spatial organization* (the locational structure of spatial distributions). In combination, these **four basic traditions** established geography as nothing less than a formal and comprehensive system of thinking that could be applied to any aspect of human affairs involving space, a *holistic* (integrating) discipline in the same sense that history is with respect to answering questions involving a time element. Besides its obvious academic applications, contemporary geography is also showing strength in a more practical way:

increasing numbers of geographers are being employed these days in the private and public sectors to help solve spatial problems of every variety (see *box*, p. 50).

In modern practice, the four traditions have usually taken turns in overshadowing each other rather than always operating as co-equals. For example, American geography opened this century with the earth science tradition at the forefront, but by 1925 the culture-environment approach was dominant. By 1940, regionalism had begun a period of ascendancy that lasted until the early 1960s, when the spatial organization approach quickly eclipsed the position of area studies. The picture in the 1990s is somewhat more complex:

although the spatial organization tradition still holds center stage, there are signs that the pendulum is swinging back toward regional studies. This trend may well strengthen in the immediate future as the United States begins to cope with its citizens' inadequate understanding of the rest of the world at a time when the nation is entering a new era dominated by international relationships (see *box*, p. 51).

As contemporary geography continues to evolve as both a social science and an environmental science, the spatial concept of region is firmly rooted as a unifier of the two. Reaffirming the role of regional study today, Richard Morrill noted that this approach is one

of the two very justifications for having a distinct discipline of geography:

[Its major contribution is] *its capacity to describe and analyze how diverse physical and human processes* [play out and] *interact to produce particular regional landscapes, cultures and places. . . .* [Regions] *are the manifest taxonomy of geography, the empirical, interrelated composites of phenomena which geographers, as both scientists and humanists, strive to explain and understand.*

Morrill's other justification is geography's concern with developing ''[theories and] principles of human-environmental interaction

FIGURE I–22

THE RELATIONSHIP BETWEEN
REGIONAL AND SYSTEMATIC GEOGRAPHY

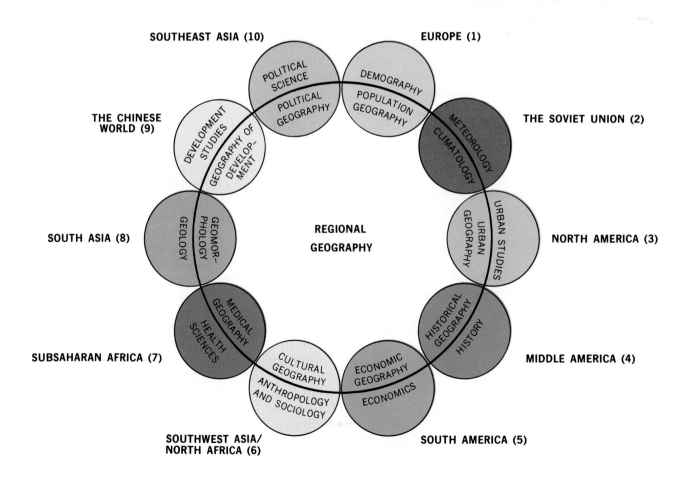

and spatial organization [that] hold across broad classes of physical and human processes.'' Most often, this involves specializing in a topical or **systematic** subfield of geography. A political geographer, for instance, would study politico-spatial concepts and principles as they occur across all the world's regions; he or she might be searching for a useful classification of boundary types, which can be developed by studying boundary evolution and perception in each of the geographic realms. Of course, by studying every major systematic branch of geography across the entire set of realms, the same total world picture obtained from a complete regional survey would emerge. A number of geographers have conceptualized this close linkage between regional and systematic geography. A most appropriate model was developed by Nevin Fenneman in 1919. He saw regional geography as a unifying theme at the core, integrating the various systematic subfields that were distributed around the discipline's circumference, where they also spilled across geography's limits to connect with their own neighboring parent disciplines (for example, political geography's link to political science). Because systematic geography is so closely tied to the regional geography of the 1990s, we will highlight one of these topical subfields in an essay at the beginning of each of the succeeding chapters. This program is illustrated in Fig. I–22, where the ten leading subfields we have chosen to discuss are presented in the framework of the Fenneman scheme.

Finally, before we commence our survey of the world's geographic realms, we call your attention to the *box* on pages 53–55, which will assist you in reading, interpreting, and understanding the many maps employed in the coming chapters.

MAP READING AND INTERPRETATION

As we have seen throughout this chapter, maps are basic tools that are used to gain an understanding of patterns in geographic space. In fact, they constitute an important visual or *graphic communications* medium whereby encoded spatial messages are transmitted from the cartographer (mapmaker) to the map reader. This shorthand is necessary because the real world is so complex that a great deal of geographical information must be compressed into the small confines of maps that can fit onto the pages of this book. At the same time, cartographers must carefully choose which information to include; these decisions force them to omit many things in order to prevent cluttering a map with less relevant information. For example, Fig. I–24 shows several city blocks in central London, but avoids mapping individual buildings because they would interfere with the main information being presented—the spatial distribution of cholera deaths.

Deciphering the coded messages contained in the maps of this book—**map reading**—is not difficult, and it becomes quite easy with a little experience in using this ''language'' of geography. The need to miniaturize portions of the world onto small maps has already been discussed in the section on *map scale* (pp. 4–6), and two additional contrasting examples are provided in Figs. I–23 and I–24. *Orientation*, or direction, on maps can usually be discerned by reference to the geographic grid of latitude and longitude.

FIGURE I–23

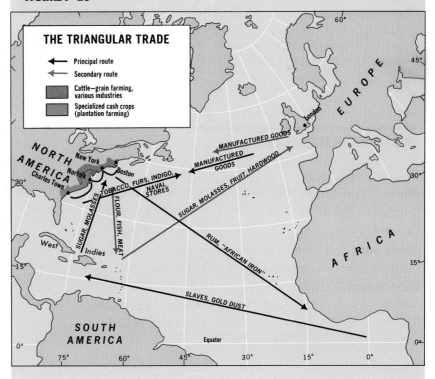

Latitude is measured from 0° to 90° north and south of the equator (parallels of latitude are always drawn in an east–west direction), with the equator being 0° and the North and South Poles being 90°N and 90°S, respectively. Meridians of *longitude* (always drawn north–south) are measured 180° east and west of the *prime meridian* (0°), which passes through the Greenwich Observatory in London, England; the 180th meridian, for the most part, serves as the *international date line* that lies in the middle of the Pacific Ocean. Inspection of Fig. I–23 shows that north is not automatically at the top of a map; instead, the direction of north curves along every meridian, with all such lines of longitude converging at the North Pole. The many minor directional distortions in this map are unavoidable—it is geometrically impossible to transfer the grid of a three-dimensional sphere (globe) onto a two-dimensional flat map. Therefore, compromises in the form of *map projections* must be devised in which properties such as areal size and distance are preserved, but directional constancy is sacrificed.

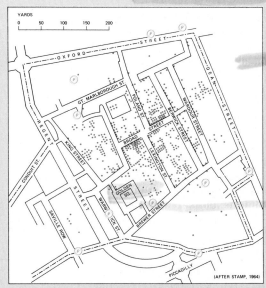

FIGURE I–24

Once the background mechanics of scale and orientation are understood, the main task of decoding the map's content can proceed. The content of most maps in this book is organized within the framework of point, line, and area symbols, which are made even clearer through the use of color. These symbols are usually identified in the map's *legend*, as in Fig. I–23. Occasionally, the map designer omits the legend, but must tell the reader verbally in a caption or within the text what the map is about. Figure I–24, for instance, is a map of cholera deaths in the London neighborhood of Soho during the outbreak of 1854, with each Ⓟ symbol representing a municipal water pump and each red dot the location of a cholera fatality. *Point symbols* are shown as dots on the map and can tell us two things: the location of each phenomenon and, sometimes, its quantity. The cities of New York and London on Fig.

PRONUNCIATION GUIDE

Beijing   (bay-*zhing*)
Biome   (*bye*-ohm)
Brunhes   (*broon*)
Cenozoic   (sen-oh-*zoh*-ick)
Chang Jiang   (chahng-jee-*ahng*)
Chao Phraya   (*chow*-pruh-yah)
Eratosthenes   (errah-*tos*-thuh-neez)
Gottmann   (*got*-mun)

Huang He   (*hwahng-huh*)
Humus   (*hue*-muss)
Islam   (iss-*lahm*)
Jawa   (*jah*-vuh)
Köppen   (*ker*-pun)
Leninakan   (len-uh-nah-*kahn*)
Liechtenstein   (*lish*-ten-shtine)
Monaco   (*mon*-uh-koh)
Montreal   (mun-tree-*awl*)

Pangaea   (pan-*jee*-uh)
Pleistocene   (*ply*-sto-seen)
Qum   (*koom*)
Sahel   (suh-*hell*)
Sauer   (*sour*)
Sri Lanka   (sree-*lahng*-kuh)
Ukraine   (yoo-*crane*)
Wegener   (*vay*-ghenner)
Zaïre   (zah-*ear*)

I–23 and the dot pattern of cholera fatalities on Fig. I–24 (with each red dot symbolizing one death) are examples. *Line symbols* connect places between which some sort of movement or flow is occurring. The "triangular trade" among Britain and its seventeenth-century Atlantic colonies (Fig. I–23) is a good example—each leg of these trading routes is clearly mapped, the goods moving along are identified, and the more heavily traveled principal routes are differentiated from the secondary ones. *Area symbols* are used to classify two-dimensional spaces and thus provide the cartographic basis for regionalization schemes (as we saw in Figs. I–10 through I–13). Such classifications can be developed at many levels of generalization. In Fig. I–23, blue and beige areas broadly offset land from ocean surfaces; more specifically, red and green area symbols along the eastern North American coast delimit a pair of regions specializing in different types of commercial agriculture. Area symbols may also be used to communicate quantitative information: for example, the gray zones in Fig. I–9 delimit semiarid regions within which annual precipitation averages from 12 to 20 inches/30 to 50 centimeters.

Map interpretation—the explanation of cartographic patterns—is one of the geographer's most important tasks. Although that task is performed for you throughout this book, readers should be aware that today's practitioners use many sophisticated techniques and machines to analyze vast quantities of areal data (for some insights into computer mapping and remote sensing see pp. 331 and 362). These modern methods notwithstanding, geographic inquiry still focuses on the search for meaningful *spatial relationships*. This longstanding concern of the discipline is nicely demonstrated in Fig. I–24. By showing on his map that cholera fatalities clustered around municipal water pumps, Dr. John Snow was able to persuade city authorities to shut them off; almost immediately, the number of new disease victims dwindled to zero, thereby confirming Snow's theory that contaminated drinking water was crucial in the spread of cholera.

REFERENCES AND FURTHER READINGS

(BULLETS [●] DENOTE BASIC INTRODUCTORY WORKS)

- Allen, James P. "Core-Periphery Systems as a Framework for Cross-Cultural Perspectives in World Regional Geography," in Baskauskas, Liucija, ed., *Unmasking Culture: Cross-Cultural Perspectives in the Social and Behavioral Sciences* (Novato, Calif.: Chandler & Sharp, 1986), pp. 119–141.

Berry, Brian J. L. *Comparative Urbanization: Divergent Paths in the Twentieth Century* (New York: St. Martin's Press, 2 rev. ed., 1981).

Brunhes, Jean. *Human Geography* (London: George Harrap, trans. Ernest F. Row, 1952). Quotation taken from p. 40.

- Brunn, Stanley D. & Williams, Jack F., eds. *Cities of the World: World Regional Urban Development* (New York: Harper & Row, 1983).

Chorley, Richard J. & Haggett, Peter, eds. *Models in Geography* (London: Methuen, 1967).

- de Blij, Harm J. & Muller, Peter O. *Human Geography: Culture, Society, and Space* (New York: John Wiley & Sons, 3 rev. ed., 1986).

- de Souza, Anthony & Porter, Philip W. *The Underdevelopment and Modernization of the Third World* (Washington, D.C.: Association of American Geographers, Commission on College Geography, Resource Paper No. 28, 1974).

Dickinson, Robert E. *The Regional Concept* (London: Routledge & Kegan Paul, 1976).

Dogan, Mattei & Kasarda, John D., eds. *The Metropolis Era: Vol. I, A World Of Giant Cities; Vol. II, Mega-Cities* (Newbury Park, Calif.: Sage Publications, 1988).

Fenneman, Nevin M. "The Circumference of Geography," *Annals of the Association of American Geographers*, 9 (1919): 3–11.

Fuson, Robert. *Introduction to World Geography: Regions and Cultures* (Dubuque, Iowa: Kendall/Hunt, 1977). Regional ranking scheme on pp. 2–5.

Geography and International Knowledge (Washington, D.C.: Association of American Geographers, Committee on Geography and International Studies, 1982).

- Glassner, Martin I. & de Blij, Harm J. *Systematic Political Geography* (New York: John Wiley & Sons, 4 rev. ed., 1989).

- *Goode's World Atlas* (Chicago: Rand McNally, 18 rev. ed., 1990).

Gottmann, Jean. *A Geography of Europe* (New York: Holt, Rinehart & Winston, 4 rev. ed., 1969). Quotation taken from p. 76.

Haggett, Peter. *Locational Analysis in Human Geography* (London: Edward Arnold, 1965). Definition taken from p. 19.

Haggett, Peter. *The Geographer's Art* (New York: Basil Blackwell, 1990).

- Harris, Chauncy D., chief ed. *A Geographical Bibliography for American Libraries* (Washington, D.C.: Association of American Geographers/National Geographic Society, 1985).

- Hartshorn, Truman A. *Interpreting the City: An Urban Geography* (New York: John Wiley & Sons, 1980). Quotation taken from pp. 1–2.

Hoebel, E. Adamson. *Anthropology: The Study of Man* (New York: McGraw-Hill, 4 rev. ed., 1972).

Jackson, John Brinckerhoff. *Discovering the Vernacular Landscape* (New Haven: Yale University Press, 1984).

James, Preston E. & Martin, Geoffrey J. *All Possible Worlds: A History of Geographical Ideas* (New York: John Wiley & Sons, 2 rev. ed., 1981).

Kroeber, Alfred L. & Kluckhohn, Clyde. "Culture: A Critical Review of Concepts and Definitions," *Papers of the Peabody Museum of American Archaeology and Ethnology*, 47 (1952), entire issue. Definition taken from p. 181.

Lanegran, David A. & Palm, Risa, eds. *An Invitation to Geography* (New York: McGraw-Hill, 2 rev. ed., 1978).

Massey, Doreen B. & Allen, John, eds. *Geography Matters! A Reader* (New York: Cambridge University Press, 1985).

Miller, G. Tyler, Jr. *Living in The Environment: An Introduction to Environmental Science* (Belmont, Calif.: Wadsworth, 6 rev. ed., 1990).

Morrill, Richard L. "The Nature, Unity and Value of Geography," *Professional Geographer*, 35 (1983).

Quotations taken from pp. 5–6.

• Muehrcke, Phillip C. *Map Use: Reading-Analysis-Interpretation* (Madison, Wisc.: JP Publications, 2 rev. ed., 1986).

Paddison, Ronan & Morris, Arthur S. *Regionalism and the Regional Question* (New York: Basil Blackwell, 1988).

Paterson, John H. *North America* (New York: Oxford University Press, 8 rev. ed., 1989). Quotation taken from p. 181 of the 5th ed.

Pattison, William D. "The Four Traditions of Geography," *Journal of Geography*, 63 (1964): 211–216.

Sauer, Carl O. "Cultural Geography," *Encyclopedia of the Social Sciences* (New York: Macmillan, Vol. 6, 1931), pp. 621–623.

Shortridge, James R. *The Middle West: Its Meaning in American Culture* (Lawrence, Kan.: University Press of Kansas, 1989).

Small, John & Witherick, Michael, eds. *A Modern Dictionary of Geography* (London & New York: Routledge, 2 rev. ed., 1989).

Starkey, Otis P. et al. *The Anglo-American Realm* (New York: McGraw-Hill, 2 rev. ed., 1975).

Quotation taken from p. 138.

Strahler, Arthur N. & Strahler, Alan H. *Modern Physical Geography* (New York: John Wiley & Sons, 3 rev. ed., 1987).

Tisdale, Hope. "The Process of Urbanization," *Social Forces*, 20 (1942): 311–316. As quoted in Berry, 1981, p. 27.

Wegener, Alfred. *The Origin of Continents and Oceans* (New York: Dover Publications, trans. John Biram, 1966).

• Wheeler, James O. & Muller, Peter O. *Economic Geography* (New York: John Wiley & Sons, 2 rev. ed., 1986).

Whittlesey, Derwent S. et al. "The Regional Concept and the Regional Method," in James, Preston E. & Jones, Clarence F., eds., *American Geography: Inventory and Prospect* (Syracuse, N.Y.: Syracuse University Press, 1954), pp. 19–68.

Williams, Jack F. et al. "World Urban Development," in Brunn, Stanley D. & Williams, Jack F. eds., *Cities of the World: World Regional Urban Development* (New York: Harper & Row, 1983), pp. 2–41. Quotation taken from p. 6.

PART ONE

DEVELOPED
REALMS

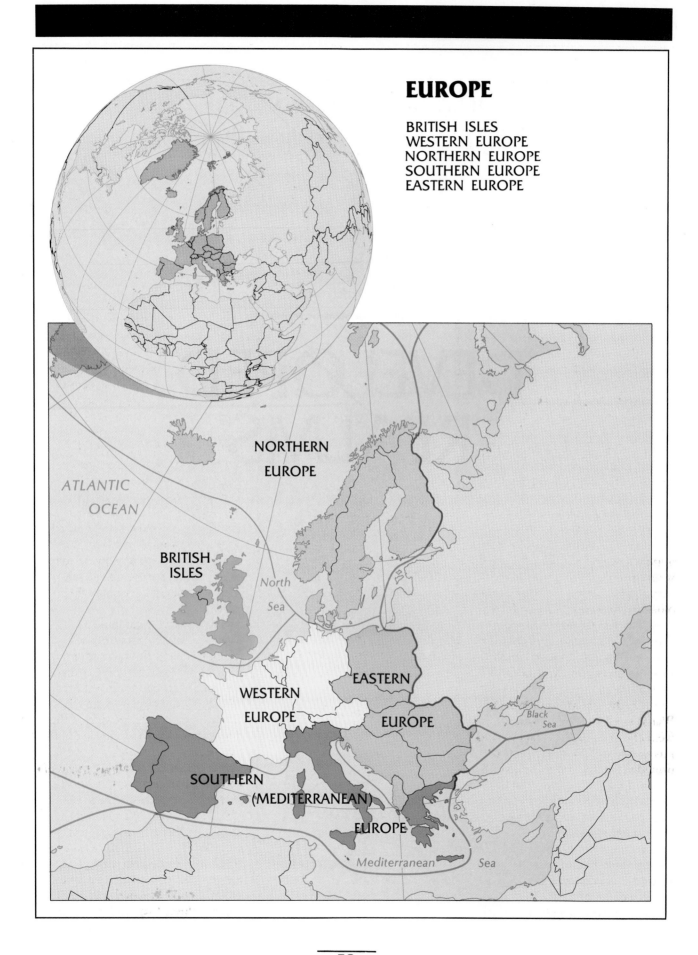

EUROPE

BRITISH ISLES
WESTERN EUROPE
NORTHERN EUROPE
SOUTHERN EUROPE
EASTERN EUROPE

ATLANTIC
OCEAN

NORTHERN
EUROPE

BRITISH
ISLES

North
Sea

EASTERN

WESTERN
EUROPE

EUROPE

Black
Sea

SOUTHERN

(MEDITERRANEAN)

EUROPE

Mediterranean   Sea

RESILIENT EUROPE: CONFRONTING NEW CHALLENGES

For centuries, Europe has been the heart of the world. European empires spanned the globe and transformed societies far and near. European capitals were the focal points of trade networks that controlled distant resources. Millions of Europeans migrated from their homelands to the New World as well as to newly settled parts of the Old, creating new societies from North America to Australia. While Europe's colonial powers competed abroad, they fought each other at home. In agriculture, in industry, and in politics, Europe went through revolutions—and then promptly exported those revolutions throughout the world, which served to consolidate the European advantage. Europe's decline (which may well be temporary) began during the second of two disastrous twentieth-century wars. In the aftermath of World War II (1939–1945), Europe's weakened powers lost the colonial possessions that for so long had provided wealth and influence. But a measure of this realm's fundamental strength is that resilient European states, even without their empires, continued to thrive and today search for ever more effective ways to integrate their considerable resources.

As noted in the introductory chapter, Europe decidedly consti-

IDEAS AND CONCEPTS

population geography

demographic transition model

migration

relative location

areal functional specialization

nation-state

von thünen's isolated state

industrial location

organic theory of state evolution

spatial interaction principles
complementarity
transferability
intervening opportunity

primate city

central business district (cbd)

european urbanization trends

conurbation

devolution

site/situation

acid rain

shatter belt

balkanization

irredentism

supranationalism

tutes a world geographic realm despite its comparatively small size on the peninsular margin of western Eurasia. For more than 2000 years, Europe has been a focus of human achievement, a hearth of innovation and invention. And Europe's human resources were matched by its large and varied raw material base. When the opportunity or the need arose, Europe's physical geography proved to contain what was required; most recently, the seafloor adjacent to Europe's North Sea coasts has yielded enough oil to make several surrounding countries exporters of energy in the 1990s.

For so limited an area (slightly more than half the size of the United States), Europe's internal natural diversity is probably unmatched. From the warm shores of the Mediterranean to the frigid Scandinavian Arctic and from the flat coastlands of the North Sea to the grandeur of the Alps, Europe presents an almost infinite range of natural environments. The insular and peninsular west contrasts strongly against a more interior, continental east. A raw-material-laden backbone extends across the middle of Europe from England eastward to Poland (and beyond), yielding coal, iron ore, and other valuable minerals. And this diver-

POPULATION GEOGRAPHY

Population geography is concerned with the *distribution*, *composition*, *growth*, and *movement* of people as related to spatial variations in living conditions on the earth's surface. The subject differs from *demography*—the interdisciplinary study of population characteristics—in that it focuses on the geographic expression through which population data are linked to specific areas.

Population distribution is perhaps the most essential of all geographical expressions because the way in which people have arranged themselves in space at any given time represents the sum total

FIGURE 1–1

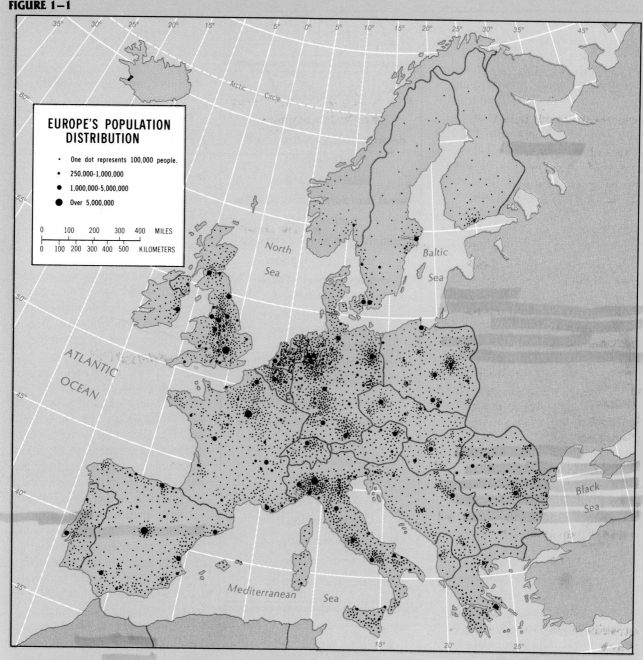

EUROPE'S POPULATION DISTRIBUTION

· One dot represents 100,000 people.
· 250,000-1,000,000
· 1,000,000-5,000,000
· Over 5,000,000

0   100   200   300   400   MILES
0   100 200 300 400 500   KILOMETERS

of adjustments they have made to their overall environment. Worldwide, as we saw in the previous chapter (Fig. I–14), humanity has sorted itself out to the extent that we can identify four primary concentrations of population: East Asia, South Asia, Europe, and eastern North America. Since this chapter treats the geography of the third cluster, we examine Europe in more detail here.

The distribution of Europe's population is shown in Fig. 1–1 and reflects a cumulative response to the forces that have progressively shaped this continent's *ecumene* (habitable areas) for at least 20 centuries. Physical extremes of terrain and climate inhibited settlement in mountainous zones and across much of the Arctic margin to the north, whereas favorable environmental conditions encouraged the massive peopling of fertile valleys and coastal lowlands. Variable economic conditions related to perceived modern opportunities spawned several migrations toward prospering areas, especially industrial cities. Descriptively, we may analyze the map according to three spatial criteria: pattern, dispersion, and density. By *pattern* we mean the geometric arrangement of a distribution. Europe's pattern is quite complex, but certain features do stand out such as overall unevenness and linearity of settlement within major river valleys and transportation corridors. *Dispersion* describes the extent of spread of a distribution. Although people are widely dispersed across the realm as a whole, the clustering or localization of population in large metropolitan areas is a dominant feature of the map. *Density* measures the frequency of occurrence of a phenomenon within a

given area. Europe's population concentration of 500 million is one of the most densely settled parts of the world; realmwide, the figure is 266 persons per square mile (103 per square kilometer). In certain countries, however, the people/land ratio is much higher; in the crowded Netherlands, this *arithmetic density* figure is 942 people per square mile (364 per square kilometer), but the *physiologic density*—the number of persons per unit area of arable land—jumps to over 3200 per square mile.

Population composition is usually grouped by age and sex as shown in Fig. 1–2, which displays Sweden's population profile. Since growth is a function of the birth rate, the relatively modest proportion of females in the childbearing years (ages 15 to 44) indicates that this nation has a very low growth rate (which, in fact, averaged just above zero for the 1980s). At the same time, an outstanding health-care system promotes longevity

FIGURE 1–2
SWEDEN: AGE–SEX STRUCTURE

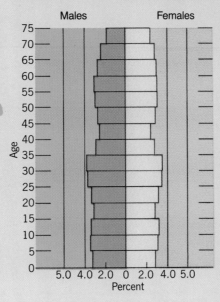

and fairly low death rates. Thus, overall, the population is an *aging* one, which is quite typical of European and most other developed countries in the late twentieth century.

In general, the degree of demographic change affecting a country can be represented in this simple formula:

$$\text{Total Population} = \text{Starting Population} + \text{Births} - \text{Deaths} + \text{Immigration} - \text{Emigration}$$

The population of the world as a whole is now growing older: by 2025, there will be more people over 60 than under 25. Third World countries, however, are not expected to share this characteristic to any great degree: younger people will continue to dominate age-sex structures there for many decades to come, and these nations will continue to account for about 90 percent of humankind's population growth until 2000, when 6.3 billion persons (versus 5.4 billion in 1991) are expected in the world.

Population growth, therefore, continues to be a major world crisis. As famine, disease, and other so-called growth "checks and balances" have been reduced in many parts of the world, population totals have accelerated; from 1650 to 1991 alone, the global total has surged more than tenfold, from 500 million to over 5 billion. In a period of such rapid growth, population *doubling times* become compressed: whereas 170 years were required after 1650 for a population of 500 million to double, only about 39 years are now needed to double today's population of 5.4 billion.

Counteracting this discouraging

outlook are the population growth changes experienced by countries that have undergone modern economic development in the form of the Industrial Revolution. These may be summarized in the three-stage **Demographic Transition Model**, a generalization derived from the historic case of the United Kingdom (Fig. 1–3). Stage 1 is the pre-industrial era before 1800 (shown in light green), in which a rural agricultural population predominated, with rather high birth and death rates but an overall low growth rate. Following the onset of industrialization and the huge cityward migration that it spawned, things began to change. Better health standards and new medical advances began to eradicate disease, and the effect on lowering death rates was dramatically demonstrated during the nineteenth century in Stage 2 (dark orange). At the same time, however, birth rates stayed high, and the steady drop in mortality—particularly infant mortality—resulted in an enormous net population increase. Unchanged attitudes toward family formation supported persistently high birth rates because the former rural residents of the new industrial cities had always viewed children as additional laborers who could augment a family's income. Eventually, overcrowded living conditions and the enactment of child-labor laws altered the social and economic values that spawned large families. By the middle of this century, Stage 3 (light orange) was achieved, marked again by a convergence of birth and death rates but at a far lower overall level.

Population movements are another major concern of this subdiscipline because the redistribution of people constantly changes the world's patterns of resource use. Indeed, the process of **migration**—involving the purposeful relocation of one's permanent residence—is so pervasive a topic in contemporary world regional geography that it can only be introduced here (it is developed further on pp. 120–121). Economic factors strongly propel many of the world's leading population flows. The search for a higher-paying job and a better life is universal, and perceived economic opportunity—the belief that the grass is greener in another locale—annually convinces millions to relocate their homes and families. Other forces also enter the picture and act as catalysts of migration, among them cultural pressures, changing political conditions, and the need to escape from hazardous natural environments.

FIGURE 1–3

THE DEMOGRAPHIC TRANSITION

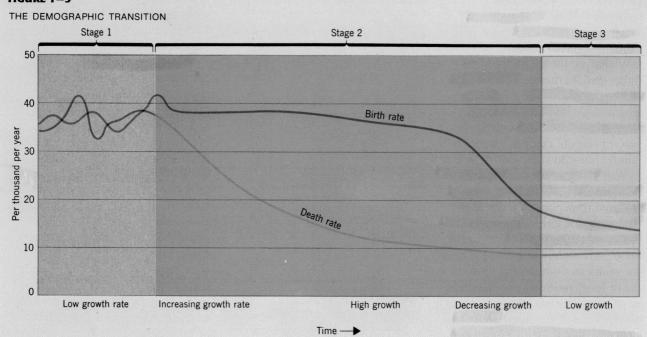

sity is not confined to the physical makeup of the continent. The European realm contains peoples of many different cultural-linguistic stocks, including not only Latins, Germanics, and Slavs, but also numerous minorities such as the Finns, the Hungarians, and various Celtic-speaking groups.

This diversity of physical and human content alone, of course, does not constitute such a great asset for Europe. If differences in habitat and culture automatically led to rapid human progress, Europe would have had many more competitors for its world position than it did. But in Europe, there were virtually unmatched advantages of scale and proximity. Globally, Europe's **relative location**—at the heart of the land hemisphere—is one of maximum efficiency for contact with the rest of the world (Fig. 1–4).

Regionally, Europe is also far more than a mere western extremity of the Eurasian landmass. Almost nowhere is Europe far from that essential ingredient of European growth—the sea—and the water interdigitates with the land here as it does nowhere else on earth. Southern and Western Europe consist almost entirely of peninsulas and islands, from Greece, Italy, France, and the Iberian Peninsula (Spain and Portugal) to the British Isles, Denmark, Norway, and Sweden. Southern Europe faces the Mediterranean, and Western Europe virtually surrounds the North Sea as it looks out over the Atlantic Ocean. Beyond the Mediterranean lies Africa, and across the Atlantic are the Americas. Europe has long been a place of contact between peoples and cultures, of circulation of goods and ideas. The hundreds of miles of navigable waterways; the easily traversed bays, straits, and channels be-

RELATIVE LOCATION: EUROPE IN THE LAND HEMISPHERE

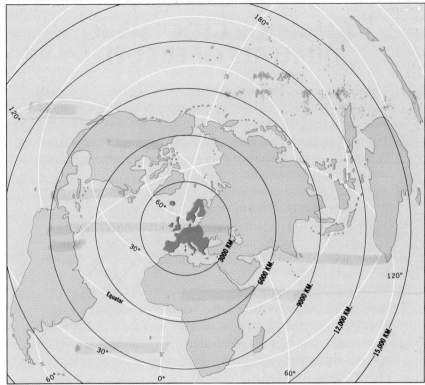

Azimuthal equidistant projection centered on Hamburg

FIGURE 1–4

tween numerous islands and peninsulas and the mainland; and the highly accessible Mediterranean, North, and Baltic seas all provided the avenues for these exchanges. Later, even the oceans were tamed to provide for similar long-distance spatial interaction.

This historic advantage of moderate distances on European waters applies on the mainland as well. Europe's Alps seem to form a transcontinental divider, but what they separate still lies in close juxtaposition (Alpine passes have for centuries provided several corridors for contact). Consider Rome and Paris: the distance between these long-time control points of Mediterranean and north-western Europe is less than that between New York and Chicago or Miami and Atlanta. No place in Europe is very far from anyplace else on the continent, and nearby

places are often sharply different from each other in terms of economy and outlook. Short distances and large differences make for much interaction—and that has marked the geography of Europe for over a millennium.

LANDSCAPES AND OPPORTUNITIES

Europe may be small in areal size, but its physical landscapes are varied and complex. It would be easy to establish a large number of physiographic regions, but in doing so we might lose sight of the broader regional pattern—a pattern that has much to do with the

TEN MAJOR GEOGRAPHIC QUALITIES OF EUROPE

1. The European realm consists of the western extremity of the Eurasian landmass, a locale of maximum efficiency for contact with the rest of the world.

2. Europe's lingering and resurgent world influence results largely from advantages accrued over centuries of global political and economic domination.

3. Europe's nation-states emerged from durable power cores that formed the headquarters of world colonial empires.

4. Europe is marked by strong internal regional differentiation (cultural as well as physical), exhibits a high degree of functional specialization, and provides multiple exchange opportunities.

5. The European natural environment displays a wide range of topographic, climatic, vegetative, and soil conditions and is endowed with many industrial resources.

6. European economies are dominated by manufacturing, and the level of productivity has been high. Levels of development decline from west to east.

7. Europe's population is generally healthy and well fed, exhibits low birth and low death rates, enjoys long life expectancies, and constitutes one of the world's three largest population clusters.

8. Europe's population is highly urbanized, highly skilled, and well educated.

9. Europe is served by efficient transport and communications networks that promote extensive trade and other forms of international spatial interaction.

10. Europe is making important progress toward economic integration and political unification, and the push toward still stronger coordination continues.

way the European human drama unfolded. Accordingly, Europe's landforms can be grouped regionally into four units: the Central Uplands, the Alpine Mountains in the south, the Western Uplands, and the great North European Plain (Fig. 1–5).

The very heart of Europe is occupied by an area of hills and small plateaus, with forest-clad slopes and fertile valleys. These *Central Uplands* also contain the majority of Europe's productive coalfields. When the region emerged from its long medieval quiescence and stirred with the stimuli of the Industrial Revolution, towns on the Uplands' flanks grew into cities, and farms gave way to mines and factories.

The Central Uplands are flanked on the south by the much higher Alpine Mountains (and to the west and north by the North European Lowland). The *Alpine Mountains* include not only the famous Alps themselves, but also other ranges that belong to this great mountain system. The Pyrenees between Spain and France (one of Europe's few true barriers), Italy's Appennines, Yugoslavia's Dinaric Ranges, and the Carpathians of Eastern Europe are all part of this Alpine system that extends even into North Africa (as the Atlas Mountains) and eastward into Turkey and beyond. Although the Alps are rugged and imposing, they have not been a serious obstacle to communication; traders have operated along their pass routes for many centuries.

Europe's western margins are also quite rugged, but the *Western Uplands* of Scandinavia, Scotland, Ireland, France's Brittany, Portugal, and Spain are not part of the Alpine system (maximum elevations are markedly lower than those in the Alps). This western arc of highlands represents older geologic mountain building, contrasting sharply with the relatively young, still active, earthquake-prone Alpine mountains. Scandinavia's uplands form part of an ancient geologic shield underlain by old crystalline rocks now bearing the marks of the Pleistocene glaciation. Spain's central plateau or *Meseta* is also supported by comparatively old rocks, now worn down to a tableland.

The last of Europe's landform regions is also its most densely populated. The *North European Lowland* (also known as the Great European Plain) extends in a gigantic arc from southwestern France through the Paris Basin and the Low Countries across northern Germany and eastward through Poland. Southeastern England, Denmark, and the southern tip of Sweden also are part of this region, which forms a continuous belt on the mainland from southern France to the eastern Baltic (Fig. 1–5). Most of the North European Lowland lies below 500 feet (150 meters) in elevation, and local relief rarely exceeds 100 feet

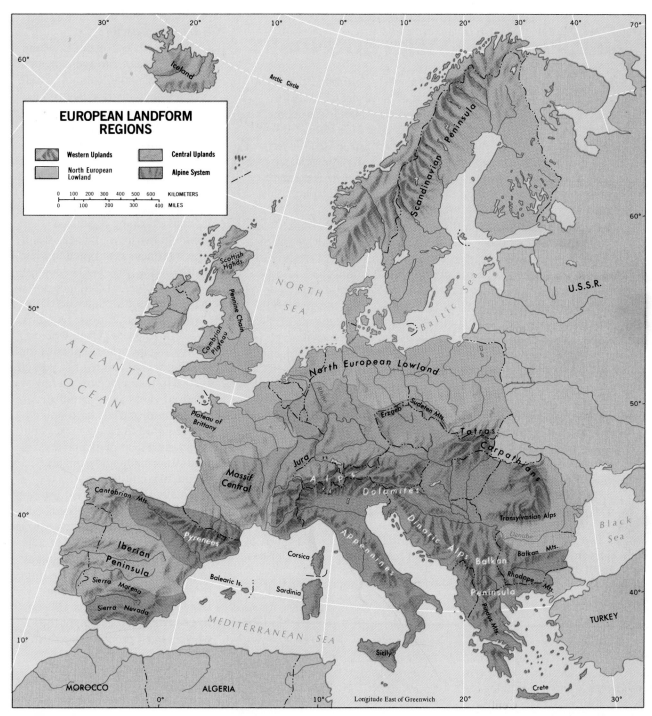

FIGURE 1–5

(30 meters). Make no mistake; this region may be topographically low-lying and flat or gently rolling, but there its uniformity ends. Beyond this single topographic factor there is much to differentiate it internally. In France, it includes the basins of three major rivers—the Garonne, Loire, and Seine. In the Netherlands, for a good part, it is made up of land reclaimed from the sea, enclosed by dikes and lying below sea level. In southeastern England, the higher areas of the Netherlands, northern Germany and Denmark, southern Sweden, and farther eastward, it bears the marks of the Pleistocene glaciation that withdrew only a few thousand years ago (see Fig. I–8). Each of these particular areas affords its own opportunities as soils and climates vary, giving rise to some of the world's most

EUROPE: THE EASTERN BOUNDARY

The European realm is bounded on the west, north, and south by Atlantic, Arctic, and Mediterranean waters, respectively. But Europe's eastern boundary has always been a matter for debate. Some scholars place this boundary at the Ural Mountains (deep inside the U.S.S.R.), thereby recognizing a ''European'' Soviet Union and, presumably, an ''Asian'' one as well. Others argue that because there is a continuous transition from west to east (that continues into the Soviet Union), there is no point in trying to define any boundary.

Still, the boundary used in this chapter—marking Europe's eastern boundary as the border with the Soviet Union—has geographic justifications. Here diversity changes into uniformity, variety into sameness. Eastern Europe shares with Western Europe its fragmentation into several states possessing distinct cultural geographies, a political condition that sets it apart from the giant to the east. Historical and cultural contrasts mark large segments of this boundary, notably in Finland, Hungary, and Romania (see Fig. 1–12). The Soviet Union's postwar hegemony in Eastern Europe did not wipe out certain ideological and iconographic differences between the two regions. Eastern Europe retains its discrete nationalisms, latent conflicts, pressures, and vicissitudes, qualities most vividly demonstrated by the anti-communist upheaval that swept through the region in 1989. But the Soviet Union remained the great monolith for the seven decades following the 1917 revolution—a transforming structure now bedeviled by its own internal cultural forces seeking to break free.

FIGURE 1–6 Village and farmlands on the North European Lowland, here in West Germany near the physiographic contact zone with the Central Uplands.

productive (and prestigious) agricultural pursuits.

The North European Lowland has been one of Europe's major avenues of human contact. Entire peoples have migrated across it; armies have repeatedly marched through it. At an early stage, there were trading ties with southern parts of the continent. As settlement took place, agricultural diversity became a hallmark, and land use came to be dominated by intensive farming organized around a myriad of villages from which farmers commuted to their nearby fields (Fig. 1–6). Today, centuries later, it is still not possible to speak of a ''dairy belt'' or ''wheat belt'' in Europe like those in North America. Even where one particular crop dominates the farming scene, some other pursuit is sure to take place only a few hundred yards away. And because agricultural space is both limited and required to feed a huge population, every bit of available cropland is normally under highly productive cultivation.

Finally, Europe's great lowland possesses yet another crucial advantage: its multitude of navigable rivers, emerging from higher adjacent areas and wending their way to the sea. In addition to the three rivers of France already mentioned, the Rhine-Meuse (Maas) river system serves one of the world's most productive industrial and agricultural areas, reaching the sea via the Netherlands; the Weser, Elbe, and Oder penetrate northern Germany; the Vistula traverses Poland. In southeastern Europe, the Danube rivals the Rhine in regularity of flow and navigability. Thus, north of the Alpine Mountain System, Europe's major rivers create a radial pattern outward from the continent's interior highlands. In this way, the natural waterways as well as the land surface of the North European Lowland favor traffic and trade. Over many centuries, the Europeans have improved the situation still further by connecting navigable stretches of rivers with artificial canals, developing a complete network of water transport routes. These waterways, and the roads and railroads that followed, combined to bring tens of thousands of localities into contact with one another. New techniques and innovations could spread rapidly; trade connections and activity intensified continuously. No region in Europe provided greater opportunities for all this human interaction than the Lowland.

HERITAGE OF ORDER

Modern Europe was peopled in the wake of the Pleistocene's most recent glacial retreat—a gradual withdrawal that saw cold tundra turn into deciduous forest and ice-filled valleys into grassy vales. On Mediterranean shores Europe witnessed the rise of its first great civilizations—on the islands and peninsulas of Greece, and later in Italy. Greece lay exposed to the influences radiating from the advanced civilizations of the Middle East (see Chapter 6), and the intervening eastern Mediterranean was crisscrossed by maritime trade routes.

Ancient Greece

As the ancient Greeks forged their city-states and intercity leagues, they made intellectual achievements as well (which peaked during the fourth century B.C.). Their political philosophy and political science became important products of Greek culture, and the

writings they left behind have influenced politics and government ever since. But there was more to ancient Greece than politics, and great accomplishments were also recorded in such fields as architecture, sculpture, literature, and education. Because of the fragmentation of their habitat, there was local experimentation and success, followed by active exchanges of ideas and innovations. Individualism and localism were elements that the Greeks turned to their advantage, but internal discord was always present and in the end it got the better of them. The constant struggle between the two major cities, Athens and Sparta, spearheaded their decline, and by 147 B.C. the Romans had defeated the last sovereign Greek intercity league.

The Roman Empire

Nevertheless, what the ancient Greeks had accomplished was not undone: they had transformed the eastern Mediterranean into one of the cultural cores of the world, and Greek culture became a major component of Roman civilization. The Roman successors, however, made their own essential contributions. The Greeks never achieved politico-territorial organization on the scale accomplished by Imperial Rome, and much progress was also made in such spheres as land and sea communications, military organization, law, and governmental administration. The Roman Empire during its greatest expansion (in the second century A.D.) extended from Britain to the Persian Gulf and from the Black Sea to Egypt (Fig. 1–7). Facing little opposition, the vast empire could organize internally without interference, and in Europe it evolved into the continent's first truly interregional political unit.

Given the variety of cultures

that had been brought under Roman control and the resulting exchange of ideas and innovations, there were many possibilities for economic interaction. This process of economic development (for such it really was) had a profound impact on the whole structure of Mediterranean and Western Europe. Areas that had hitherto supported only subsistence modes of life were drawn into the greater economic framework of the state, and suddenly there were distant markets for products that had never found even local markets before. In turn, these areas received the farming know-how of the heart of the Roman state so that they could increase their yields and benefit even further. Foodstuffs came to Rome from across the Mediterranean as well as from southeastern and southwestern Europe. With an ever more diversified population of perhaps a quarter of a million, the city itself was the greatest single marketplace of the empire and the first real metropolitan urban center of Europe.

This urban tradition came to characterize Roman culture throughout the empire, and many cities and towns founded by the Romans continue to function and grow today. Whereas the economic base of many of these places has changed, others, notably the ports, still perform the same functions as they did in Roman times. Roman urban centers were also connected by an unparalleled network of highway and water routes, and many European roads today still follow exactly the corridors laid out by Roman engineers (Fig. 1–8). But more than anything else, the Roman Empire left Europe a legacy of ideas, of concepts that long lay dormant but eventually played their part when Europe again discovered the path of progress. In political and military organization, effective administra-

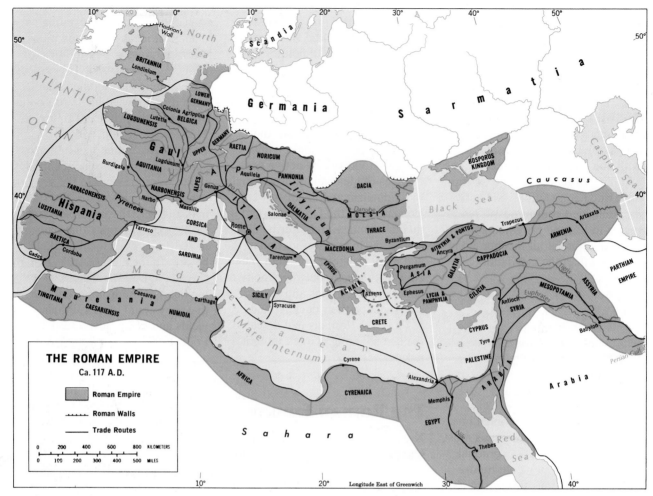

FIGURE 1—7

FIGURE 1—8 The enduring imprint of a transport artery on the cultural landscape: this road and bridge near the southwestern Spanish town of Mèrida was originally built by the Romans.

tion, and long-term stability, the Empire was centuries ahead of its time. Moreover, never was a larger part of Europe unified by acquiescence than it was under the Romans. And at no time did Europe come closer to obtaining a *lingua franca* (common language) than it did during the age of Rome.

Finally, Europe's transformation under Roman rule heavily involved the geographic principle of **areal functional specialization**. Before the Romans brought order and connectivity to their vast domain, much of Europe was inhabited by tribal peoples whose livelihoods were on a subsistence level. Many of these groups lived in virtual isolation, traded little, and fought over territory when encroachment occurred. Peoples under Rome's sway, however, were brought into Roman economic as well as political spheres, and farmlands, irrigation systems, mines, and workshops appeared. Thus, Roman-dominated areas began to take on a characteristic that has marked Europe ever since: *particular peoples and particular places concentrated on the production of particular goods.* Parts of North Africa became granaries for urbanizing (European) Rome; Elba, a Mediterranean island, produced iron ore; the area near Cartagena in southeastern Spain mined and exported silver and lead. Many other locales in the Roman Empire specialized in the production of certain farm commodities, manufactured goods, or minerals. The Romans knew how to exploit their natural resources; at the same time, they also learned to use the varied productive talents of their subjects.

DECLINE AND REBIRTH

The eventual breakdown and collapse of the Empire in the fifth century A.D. could not undo what the Romans had forged in the spreading of their language, in the dissemination of Christianity (in some ways the sole strand of permanence through the ensuing Dark Ages), in education, in the arts, and in countless other spheres. But ancient Rome's decline was attended by a momentous stirring of Europe's peoples as Germanic and Slavic populations moved to their present positions on the European stage. The Anglo-Saxons invaded Britain from Danish shores, the Franks moved into France, the Allemanni traversed the North

NATION-STATE AS CONCEPT

As Europe went through its periods of rebirth and revolutionary change, the politico-geographical map was transformed. Smaller entities were absorbed into larger units, conflicts resolved (by force as well as negotiation), boundaries defined, and internal divisions reorganized. European **nation-states** were in the making.

But what is a nation-state and what is not? The question centers in part on the definition of the term *nation*. The definition usually involves measures of homogeneity: a nation should comprise a group of tightly knit people who speak a single language, have a common history, share the same cultural background, and are united by common political institutions. Accepted definitions of the term suggest that many states are not nation-states because their populations are divided in one or more important ways. But cultural homogeneity may not be as important as a more intangible "national spirit" or emotional commitment to the state and what it stands for. One of Europe's oldest states, Switzerland, has a population that is divided along linguistic, religious, and historical lines, but Switzerland has proven to be a highly durable nation-state nonetheless.

A nation-state, therefore, may be defined as a political unit comprising a clearly defined territory and inhabited by a substantial population, sufficiently well organized to possess a certain measure of power, with the people considering themselves to be a nation having certain emotional and other ties that are expressed in their most tangible form in the state's legal institutions, political system, and ideological strength. This definition essentially identifies the European model that emerged in the course of the realm's long period of evolution and revolutionary change. France is often cited as the best example among Europe's nation-states, but Italy, Britain, Spain, Poland, Hungary, and Sweden are also among countries that satisfy the terms of the definition to a great extent. Yugoslavia is an example of a European state that cannot at present be designated as a true nation-state.

European Lowland and settled in Germany. Capitalizing on the disintegration of Roman power, numerous kings, dukes, barons, and counts established themselves as local rulers. Europe was in turmoil, and it was invaded in its weakness from Africa and the Middle East. In Iberia, the Arab-Berber Moors conquered a large region; in Eastern Europe, the Ottoman Turks extended their Islamic empire. The townscapes of southern Spain and the Balkans

still carry the cultural imprints of these Muslim invasions.

After nearly a thousand years of feudal fragmentation during the Dark and Middle Ages, modern Europe began its emergence in the second half of the fifteenth century—some date this rebirth from 1492, the year of Columbus's first arrival in the New World. At home, monarchies strengthened at the expense of feudal lords and landed aristocracies, and in the process forged the beginnings of **nation-**

states (see *box*). Abroad, Western Europe's developing states were on the threshold of discovery—the discovery of continents and riches across the oceans. Europe's emerging powers were fired by a new national consciousness and pride, and there was renewed interest in Greek and Roman achievements in science and government. Appropriately, this period is referred to as Europe's *Renaissance*.

The new age of progress and rising prosperity was centered in Western Europe, whose countries lay open to the new pathways to wealth—the oceans. While Columbus ventured to the Americas, Eastern Europe was under attack from the Ottoman Turks, who pushed across the Balkan Peninsula and penetrated as far as modern Austria. Had the Ottomans not been preoccupied with ongoing conflicts in Asia, they might have copied the Romans in making the Mediterranean Sea an interior lake of their empire. In that case, Western Europe would have been concerned with self-defense instead of scrambling for the wealth of the distant lands of which it had become newly aware.

Thus protected, the highly competitive monarchies of Western Europe could afford to engage in economic nationalism that operated in the form of *mercantilism*. The objectives of this policy were the accumulation of as large a quantity of gold and silver as possible, and the use of foreign trade and colonial acquisition to achieve that end. Mercantilism was promoted and sustained by the state; precious metals could be obtained either by the conquest of peoples in possession of them or indirectly by achieving a favorable balance of international trade. Thus, there was stimulus not only to seek new territories where such metals might lie, but also to produce goods at home that could be sold profitably abroad. The matrix of nation-states in Western Europe

was now taking shape ever more rapidly, and the spiral had been entered that was to lead to great empires and a period of world domination.

THE REVOLUTIONS

Strife and dislocation punctuated Europe's march to world domination. Much of what was achieved during the Renaissance was destroyed again as powerful monarchies struggled for primacy, religious conflicts dealt death and misery, and the beginnings of parliamentary government fell under new tyrannies. Nevertheless, revolutions—in several spheres—were in the making. Economic developments in Western Europe ultimately proved to be the undoing of absolute monarchs and their privileged, land-owning nobilities. The city-based merchant was gaining wealth and prestige, and the traditional measure of affluence—land—began to lose its status in these changing times. The merchants and businesspeople of Europe were soon able to demand political recognition on grounds the nobles could not match. Urban industries were thriving; Europe's population, more or less stable at about 100 million since the mid-sixteenth century, was on the increase.

The Agrarian Revolution

This transformation was heightened by an ongoing *agrarian revolution*—the significant metamorphosis of European farming that preceded the Industrial Revolution and helped make possible

sustained population increase during the seventeenth and eighteenth centuries. The Netherlands, Belgium, and northern Italy paved the way with their successes in commerce and manufacturing. The stimulus provided by expanding urbanization and markets led to improved organization of land-ownership and agriculture. Some of the new practices spread to England and France, where traditional communal landownership began to give way to individual landholding by small farmers. Land parcels were marked off by fences and hedges, and the new owners readily adopted innovations to improve crop yields and increase profits.

At the same time, methods of soil preparation, crop rotation, cultivation, harvesting, and livestock feeding improved. More effective farm equipment was adopted; there was better planning and experimentation; storage and distribution systems became more efficient. In the growing cities and towns, farm products fetched higher prices. New crops were introduced, especially from the Americas; the potato now became a European staple. More and more of Europe's farmers were drawn from subsistence into profit-driven market economies. Later, the manufactured products of the Industrial Revolution further stimulated the transformation of the realm's agriculture.

The Industrial Revolution

Europe also had some industries before the *Industrial Revolution* began. In Flanders and England, specialization had been achieved in the manufacturing of woolen and linen textiles. In what is now southwestern East Germany, iron ore was mined and smelted. European manufacturers produced a

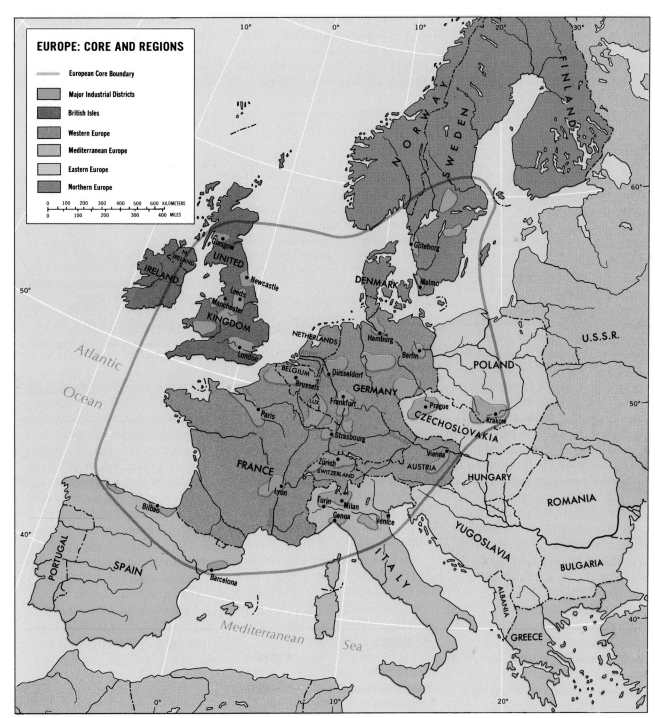

EUROPE: CORE AND REGIONS

- European Core Boundary
- Major Industrial Districts
- British Isles
- Western Europe
- Mediterranean Europe
- Eastern Europe
- Northern Europe

FIGURE 1–9

wide range of goods for local markets, but their quality was often surpassed by textiles and other wares from India and China. This provided European entrepreneurs with the incentive to refine and mass-produce their products. Since raw materials could be shipped home in virtually unlimited quan-

tity, if they could find ways to mass-produce these commodities into finished goods, they could bury the Asian industries under growing volumes and declining prices.

Now the search for better machinery was on, especially improved spinning and weaving

equipment. The first steps in the Industrial Revolution were not all that revolutionary, for the larger spinning and weaving machines that were built were still driven by the old source of power—water running downslope. But then, in the 1780s, James Watt and others succeeded in devising a steam-driven engine, and soon this invention was adapted for various uses. About the same time, it was realized that coal (converted into carbon-rich coke) was a greatly superior substitute for charcoal in smelting iron. These momentous innovations had rapid effect. The power loom revolutionized the weaving industry. Iron smelters, long dependent on Europe's dwindling forests for fuel, could now be concentrated near coalfields. Engines could move locomotives as well as power looms. Ocean shipping entered a new age.

England had an enormous advantage, for the Industrial Revolution occurred when British influence was worldwide and the significant innovations were achieved in Britain itself. The British controlled the flow of raw materials, they held a monopoly over products that were in global demand, and they alone possessed the skills necessary to make the machines that manufactured the products. Soon the fruits of the Industrial Revolution were being exported, and the spatial pattern of modern industrial Europe began to take shape (Fig. 1–9). In Britain, industrial regions, densely populated and heavily urbanized, developed near coalfields in the English Midlands, Newcastle to the northeast, southern Wales, and along the Clyde River in Scotland.

In mainland Europe, a belt of major coalfields extends from west to east, roughly along the southern margins of the North European Lowland, due eastward from southern England across northern France and southern Belgium, the Netherlands, Germany (the Ruhr), western Bohemia in Czechoslovakia, and Silesia in southern Poland. Iron ore is found in a broadly similar belt, and the industrial map of Europe reflects the resulting concentrations of economic activity (Fig. 1–9). Another set of manufacturing regions emerged in and near the growing urban centers of Europe, as the same map demonstrates. London—already Europe's leading urban focus and Britain's richest domestic market—was typical of these developments. Many local industries were established here, taking advantage of the large supply of labor, the ready availability of capital, and the proximity of so great a number of potential buyers. Although the Industrial Revolution thrust other places into prominence, London's primacy was sustained, and industries in and around the British capital multiplied.

Political Revolutions

Europe had had long experience with experiments in democratic government, but the *political revolution* that swept the realm after 1780 brought transformation on an unprecedented scale. Overshadowing these events was the French Revolution (1789–1795), but France's—and Europe's—political catharsis lasted into the twentieth century as the rising tide of nationalism eventually affected every monarchy on the continent.

In France, the popular revolution had plunged the country into years of chaos and destruction. Only when Napoleon took control in 1799 was stability restored. Napoleon personified the new French republic, and he reorganized France so completely that he laid the foundations of the modern nation-state. He also built an empire that extended from German Prussia to Spain and from the Netherlands to Italy. Although his armies were ultimately repelled, Napoleon had forever changed the politico-geographical map of Europe. His French forces had been joined by nationalist revolutionaries from all over the continent; one monarchy after another had been toppled. Even after his final defeat at Waterloo in 1815, there were popular uprisings in Spain, Portugal, Italy, and Greece. In France itself, the monarchy briefly resurfaced, but France was to be a republic, as it still is today. If the monarchy survived elsewhere, its powers were short-lived or sharply curtailed. Europe had had its first real taste of democracy and nationalist power, and it would not revert to its old ways.

GEOGRAPHIC DIMENSIONS OF MODERNIZATION

Modern Europe has emerged from an age of revolutionary change—in livelihoods and commerce, industrialization and technology, government and war. During the nineteenth century, certain geographers tried to interpret the forces and processes that were shaping the new Europe, and some of their work still holds interest today.

The agrarian revolution that accompanied the Industrial Revolution was observed attentively by an economist-farmer named Johann Heinrich Von Thünen (1783–1850), who in 1826 fashioned one of the world's first geographical models. Von Thünen owned a large farming estate in northeastern Germany, and for four decades he kept meticulous records of his landholding's transactions. He became most interested in a subject that still fascinates economic geographers today—the effects of distance and transportation costs on the location of productive activity. Using all the data he gathered, Von Thünen began to write about the spatial structure of agriculture. His studies were published under the title *The Isolated State*, and his methods in many ways constitute the foundations of modern location theory. In fact, Von Thünen's conclusions are still being discussed in the scholarly literature, and his main ideas are presented in *Model Box 2* (pp. 74–76).

Industrial and Urban Intensification

It is not surprising that the industrialization of Europe—or, rather, its industrial *intensification* following the Industrial Revolution—also became the topic of geographic research. What influences affected **industrial location**? How was Europe's industrialization channeled? Again, the first important studies were conducted by German geographers, mostly during the second half of the nineteenth century. Much of this work was incorporated in a volume by Alfred Weber (1868–1958), published in 1909 and entitled *Concerning the Location of Industries*. Like Von Thünen, Weber began with a set of limiting assumptions in order to minimize the complexities of the real Europe. But unlike Von Thünen, Weber dealt with activities that take place at particular *points* rather than across large areas. Manufacturing plants, mines, and markets are located at specific places, and so Weber created a model region marked by sets of points where these activities would occur. He eliminated labor mobility and varying wage rates, and thereby could calculate the ''pulls'' exerted on each point in his theoretical region.

In the process, Weber discerned various factors that affect industrial location. He recognized what he called ''general'' factors that would affect all industries—dominated by transport costs for raw materials and finished products—and ''special'' factors (such as perishability of foods). He also differentiated between ''regional'' factors (transport and labor costs) and ''local'' factors. The latter, Weber argued, involve agglomerative (concentrating) and deglomerative (deconcentrating) forces. Take the case of London discussed earlier: industries located there, in large part, because of the advantages of locating together. The availability of specialized equipment, a technologically sophisticated labor force, and a large-scale market made London (as well as Paris and other big cities not positioned on rich resources) an attractive site for many manufacturing plants that could benefit from agglomeration. On the other hand, such concentration may, over time, create strong disadvantages, chiefly competition for space, rising land prices, and environmental pollution. Eventually, an industry might move away and deglomerative forces would set in.

Europe's industrialization also speeded the growth of many of its cities and towns. In Britain in the year 1800 only about 9 percent of the population lived in urban areas, but by 1900 some 62 percent lived in cities and towns (today the figure has surpassed 90 percent). All this was happening as the total population skyrocketed as well (see Fig. 1–3). As industrial modernization came to Belgium and Germany, to France and the Netherlands, and to other parts of Western Europe, the entire urban pattern changed. The nature of this process—the growth and strengthening of towns and cities—and related questions also became topics of geographic study, just as agriculture and industrialization had (the principles of urban geography are covered in the Systematic Essay that opens Chapter 3).

Politico-Geographic Order

Europe's convulsive political revolution was also the object of geographic study, and thereby was born the field of modern political geography. Still another German geographer, Friedrich Ratzel (1844–1904), observed the fortunes and failings of European states and sought to identify underlying spatial principles. He suggested that states are like biological organisms passing through stages of growth and decay. Just as an organism requires food, Ratzel reasoned, so the state needs space. Any state, to retain its vigor and to continue to thrive, must have access to more space, the essential life-giving force. Only by acquiring territory and through the infusion of newly absorbed cultures could a state sustain its strength. From his study of European historical geography, he formally proposed an **organic theory of state evolution**, linking the struggles among Europe's nation-states to biological competition. Invariably, he claimed, states in ascendancy were gaining territory; those in decline were losing it. In an article entitled ''Law of the Spatial Growth of States,'' published in 1896, he attempted to discern laws that govern the growth of states; essentially, each of his seven laws forms an element of the organic theory—growth is vitality.

Certainly, Europe appeared to provide ample confirmation of Ratzel's theory. The British Empire spanned the world, and British power was nearing its zenith. France, having lost its European empire, expanded its sphere of influence in Africa and Asia. Other European states thrived on their overseas domains. Ratzel's writings appeared in scholarly journals and were couched in theoretical language; but some of his readers were less theoretical and sophis-

M O D E L B O X 2

THE VON THÜNEN MODEL

Von Thünen's Isolated State model was so named because he wanted to establish, for purposes of analysis, a self-contained country devoid of outside influences that would disturb the internal workings of the economy. Thus, he created a sort of regional laboratory within which he could identify the factors that influence the locational distribution of farms around a single urban center. In order to do this, he made a number of limiting assumptions. First, he stipulated that the soil and climate would be uniform throughout the region. Second, no river valleys or mountains would interrupt a completely flat land surface. Third, there would be a single centrally positioned city in the Isolated State, the latter surrounded by an empty, unoccupied wilderness. Fourth, the farmers in the Isolated State would transport their own products to market by oxcart, directly overland, straight to the central city. This, of course, is the same as assuming a system of radially converging roads of equal and constant quality; with such a system, transport costs would be directly proportional to distance.

Von Thünen integrated these assumptions with what he had learned from the actual data collected while running his estate, and he now asked himself: What would be the ideal spatial arrangement of agricultural activities within his Isolated State? He concluded that farm products would be grown in a series of concentric zones outward from the central market city. Nearest to the city would be grown those crops that perished easily and/or yielded the highest returns (such as vegeta-

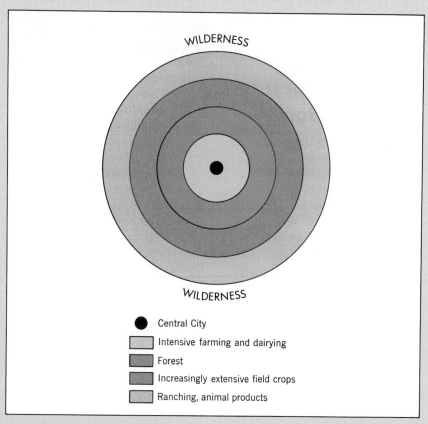

FIGURE 1–10

bles), because this readily accessible farmland was in great demand and, therefore, quite expensive; dairying would also be carried on in this innermost zone. Farther away would be potatoes and grains. And eventually, since transport costs to the city increased with distance, there would come a line beyond which it would be uneconomical to produce crops. There the wilderness would begin.

Von Thünen's model incorporated four zones or rings of agricultural land use surrounding the market center (Fig. 1–10). The first and innermost belt would be a zone of intensive farming and

dairying. The second zone, Von Thünen said, would be an area of forest, used for timber and firewood. Next, there would be a third ring of increasingly extensive field crops. The fourth and outermost zone would be occupied by ranching and animal products, beyond which began the wilderness that isolated the region from the rest of the world.

If the location of the second zone, the forest, seems inappropriate, it should be noted that the forest was still of great importance in Von Thünen's time as a source of building materials and fuel. All that was about to change with the onslaught of the Industrial Revolu-

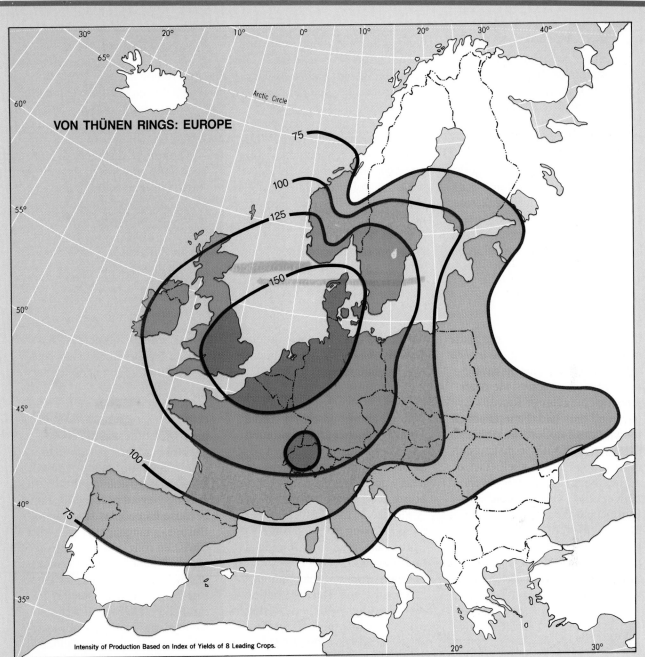

VON THÜNEN RINGS: EUROPE

Intensity of Production Based on Index of Yields of 8 Leading Crops.

Source: Adapted from Van Valkenburg and Held, *Europe,* Wiley, New York, 1952.

FIGURE 1–11

tion, and modern applications of the Von Thünen model to the developed world no longer contain a forestry ring. But there are lots of essentially pre-industrial towns and cities left in the world. When Ronald Horvath studied Addis Ababa, the capital of Ethiopia, in this context (see Fig. 6–30), he found a wide continuous belt of eucalyptus forest surrounding that city, positioned more or less where Von Thünen would have predicted it would be, and serving functions similar to those attributed to the forest belt of the Isolated State.

Von Thünen knew, of course, that the real Europe (or world) did not present idealized situations exactly as he had postulated them.

Transport routes serve certain areas more efficiently than others. Physical barriers can impede the most modern of surface communications. External economic influences invade every area. But Von Thünen wanted to eliminate these disruptive conditions in order to discern the fundamental processes at work shaping the spatial layout of the agricultural economy. Later, the distorting factors could be introduced one by one and their influence measured. But first he developed his model in theoretical isolation, basing it on total regional uniformity.

It is a great tribute to Von Thünen that his work still commands the attention of geographers. The economic-geographic landscape of Europe has changed enormously since his time, but geographers still compare present-day patterns of economic activity to the Thünian model. Such a comparison was made by Samuel van Valkenburg and Colbert Held (Fig. 1–11), whose map of Europe's agricultural intensity reveals a striking ring-like concentricity. The overriding spatial change since Von Thünen's time is the improvement in transportation technology,

which permitted the Isolated State to expand from *micro-* to *macroscale*. Thus, the model is no longer centered on a single city, but on the vast urbanized area lining the southern coasts of the North Sea, which now commands a continent-wide Thünian agricultural system.

The role of distance (to markets, from raw materials, and so forth) in the development of the spatial pattern of the economy remains the subject of many other modern studies, and geographers acknowledge that Von Thünen first formulated the crucial questions.

ticated, and ultimately there arose in Germany a school of *Geopolitik* (geopolitics). This group gave practical expression to the deterministic character of Ratzel's writings, and they laid the intellectual foundations for German military expansionism—not in distant colonies but in Europe itself. Twice during the first half of the twentieth century, Germany plunged Europe and the world into war. The aftermath of the second of these conflicts, the now-ended postwar era (1945–1990), has been the formative period of contemporary Europe.

THE EUROPEAN REALM TODAY

The nation-states of Europe are among the world's oldest, and the colonial empires of European powers were among the most durable. Wars and revolutions notwithstanding, European nations survived, and out of their long-term stability they forged a confidence that is a hallmark of European culture. Centuries of exploitation of overseas domains amassed fortunes at home and established a foreign influence that continued when the colonial era ended. But all of those advantages were nearly destroyed forever in the horror of World War II.

However, like the proverbial phoenix rising from its ashes, Europe emerged from that devastation to enter a new postwar era of reconstruction, realignment, and resurgence. Reconstruction was aided by the billions of dollars made available through the Marshall Plan (1948–1952) by the United States. Realignment came in the form of the *Iron Curtain* separating the Soviet-dominated East from Western Europe; it also involved the emergence of several international "blocs" consisting of states seeking to promote multinational cooperation. The resurgence of Europe continued despite the loss of colonial empires, recurrent political crises, energy shortages, labor problems, and other obstacles. Europe's momentum has carried the day once again, and a restructured realm is on the move in the 1990s.

Although Europe quite clearly constitutes a geographic realm, it has little of the overall homogeneity that such world-scale regional identity might imply. It is sometimes postulated that Europe may be viewed as a regional unit because its peoples share Indo-European languages (Fig. 1–12), Christian religious traditions (see Fig. 6–2), and common Caucasian (European) racial ancestry (see Fig. 7–2). But these human cultural and physical traits extend well beyond Europe, and, in any case, are not strong unifying elements within the European mosaic. As Fig. 1–12 shows, Hungarians and Finns are among European groups who do not speak Indo-European languages; indeed, for so small a realm, Europe is a veritable Tower of Babel. As for common religious traditions, Europe has a history of intense and destructive conflict over issues involving religion; shared Christian principles, for ex-

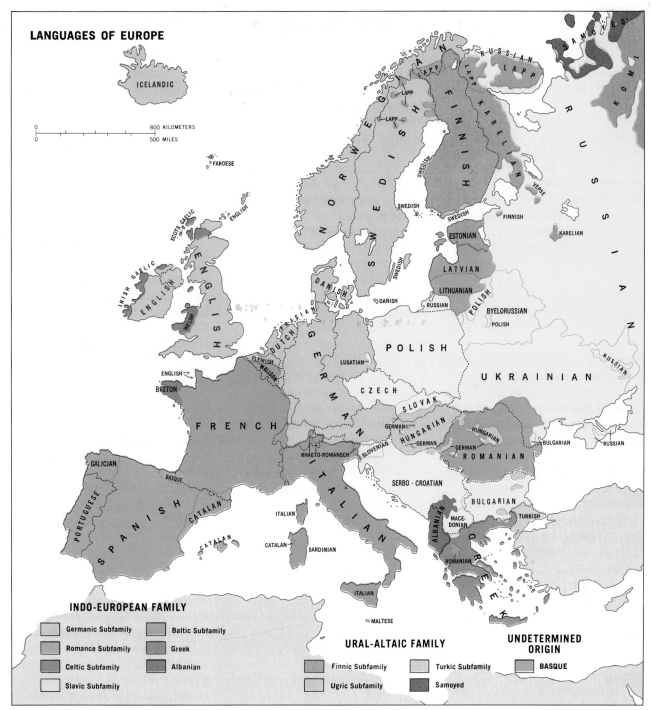

FIGURE 1–12

ample, have done little to bring unity to Northern Ireland. And again, Europe's purported common racial ancestry masks strong internal differences in observable physical characteristics between Spaniard and Swede, Scot and Sicilian.

Overcoming Problems of Political Fragmentation

Not surprisingly, the persistence of internal cultural variations has produced political and economic fragmentation across the European realm. Ever since the unifying period of the Roman Empire, Europe has been divided into numerous political entities. Not counting the smallest of these (Monaco, Andorra, Liechtenstein, and the lesser microstates), there are now

27 states listed in the table in Appendix A. Many of them are true nation-states with strong identities and traditions.

During the past half-century, however, Europeans have fully come to recognize the disadvantages of continuing to operate within such a highly fragmented mosaic. Even before the end of World War II, three countries (the Netherlands, Belgium, and Luxembourg) finalized plans to create a cooperative economic organization to be called *Benelux*. Later, the terms of the Marshall Plan mandated greater international cooperation, and 16 countries of Western Europe formed the Organization for European Economic Cooperation (OEEC) in 1948. Ever since, the states of Western Europe have pursued the prospect of European unification. Today, the greatest success is the 12-member European (Economic) Community (EC)—the so-called *Common Market*—incorporating almost all of Europe's core area. The EC's ambitious plan to lift all remaining barriers to economic unity after 1992 will be another major step forward, perhaps signaling the opening of a new era of progress toward eventual political unification.

Intensifying Spatial Interaction

Greater economic integration is a logical outcome because Europe's environments and resources—probably to a larger degree than in any world area of similar dimensions—present outstanding opportunities for human contact and interaction. Conceptually, **spatial interaction** is best organized around a triad of **principles** developed by the American geographer Edward Ullman (1912–1976): (1) complementarity, (2) transferability, and (3) intervening opportu-

nity. **Complementarity** occurs when one area has a surplus of an item demanded by a second area. The mere existence of a resource in a locality is no guarantee that trade will develop—that resource must specifically be needed elsewhere. Thus, complementarity arises from regional variations in both the supply and demand of human and natural resources. **Transferability** refers to the ease with which a commodity may be transported between two places. Sheer distance, in terms of both the cost and time of movement, may be the major obstacle to the transferability of a good; therefore, even though complementarity may exist between a pair of areas, the problems of economically overcoming the distance separating them may be so great that trade cannot begin. The third interaction principle, **intervening opportunity**, maintains that potential trade between two places, even if they satisfy the necessary conditions of complementarity and transferability, will only develop in the absence of a closer, intervening source of supply.

European economic geography has always been stimulated by internal complementarities. A specific example involves Italy, which ranks foremost among Mediterranean countries in economic development but lacks adequate coal supplies. For a long time its industries have depended on coal imports from the rich deposits of Western Europe. At the same time Italian farmers grow crops that cannot be raised in the cooler climate north of the Alps. Citrus fruits, olives, grapes, and early vegetables are in high demand on Western Europe's markets. Hence, Italy needs northwestern Europe's coal; the latter region imports Italian fruits and wines. This case of double complementarity is not counteracted by transferability restric-

tions: the physical barrier of the Alps has long been breached by rail and highway routes, and the two-way trade flow is most attractive to shippers because freight-carrying vehicles will not have to return home empty. Moreover, because northwestern Europe and Italy are the closest sources of coal and fruits, respectively, there are no intervening opportunities to disrupt spatial interaction, and a thriving mutual exchange of these surplus commodities is nurtured.

Historically, there are countless examples of this kind throughout Europe, all dependent on efficient transportation linkages. Fortunately, a good circulation system has characterized Europe since Roman times, and the steady improvement of transport technology spawned trading relationships involving ever more numerous and distant places. Today's excellent network of railroads, highways, and air routes is fully integrated thanks to the efforts of the postwar planners of a united Europe. Currently, two major projects are under construction that will significantly enhance northwestern Europe's already splendid passenger-rail operations. In 1993, the long awaited tunnel beneath the English Channel is scheduled to open, directly connecting Britain with the continent and halving the surface travel time between London and Paris. And by the end of the 1990s, a new high-speed rail network will link the major cities of Western Europe, modeled on the French TGV (*train à grande vitesse*) that already connects Paris with both Lyon and Le Mans at average speeds of 186 miles/300 kilometers per hour (Fig. 1–13).

Problems of Aging

The European population, in general, enjoys a good standard of liv-

FIGURE 1–13   The high-speed TGV on its dash across the French countryside between Paris and Lyon, a trip of more than 350 miles (550 kilometers) that takes a mere two hours.

ing as well as high levels of health, education, and housing. But the industrial base (still one of the world's great manufacturing complexes along with those in the United States, Japan, and the Soviet Union) that brought these comforts is today aging and in decline. These problems are perhaps most evident at the very source of the Industrial Revolution in Britain, where manufacturing operations have now become increasingly inefficient. Output per worker is low and declining, and overall productivity lags far behind that of the other industrial countries. British manufacturing plants, once among the world's most modern, are now outmoded; the investments needed to keep them current and competitive have not been made, and factory closings have become commonplace since 1980. Moreover, there is already considerable evidence that this lat-

est British manufacturing trend is spreading into mainland Europe as the Industrial Revolution itself did nearly two centuries ago. This decline has much to do with the historic economic changes accompanying the shift from the industrial to the "postindustrial" era (a matter treated at length in Chapter 3). Characteristically, European economies are responding to these newest challenges and making the transition—something they must do if they are to keep pace with their competitors in the other developed realms.

Europe in the 1990s is also affected by another problem of aging—the rapid aging of its population. As the table in Appendix A indicates, Europe exhibits the lowest natural-increase rate of any world realm (0.3 percent annually). Moreover, the populations of many of the realm's industrialized countries are currently growing at

even lower rates (the United Kingdom, West Germany, Italy, Belgium, and Sweden); in fact, the increase rate for West Germany, Italy, Denmark, and Hungary is actually zero or *negative*. This demographic revolution is spearheaded by younger Europeans, who are deciding to marry later and have fewer children. The consequences of this intensifying population decline are enormous. Structurally, European population profiles will be dominated by middle-aged and elderly people, a "graying" hastened by the continuing rise in life expectancies. In the economic arena, labor shortages (particularly of skilled workers) will intensify while pension payments to retirees will skyrocket. To fill the growing employment void, immigration (mainly from the Third World) is almost certain to increase substantially, a steady inflow already changing the social complexion of many European cities (a topic discussed later in the *box* on pp. 92–93). Let us now consider Europe's urban scene in broad perspective.

Contemporary European Urbanization

As we noted in the introductory chapter, cities, especially the largest metropolitan centers, are the crucibles of their nation's culture. In his 1939 study of the pivotal role of great cities in the development of national cultures, Mark Jefferson postulated the law of the **primate city,** which stated that "a country's leading city is always disproportionately large and exceptionally expressive of national capacity and feeling." Although rather imprecise, this "law" can readily be demonstrated using European examples. Certainly Paris personifies France in countless ways, and there is nothing in England to rival London. In both of these primate cities, the culture and history of a nation and empire are indelibly etched in the urban landscape. Similarly, Vienna is a microcosm of Austria, Warsaw is the heart of Poland, Stockholm is Sweden, and Athens is Greece. Today, each of these (together with the other primate cities of Europe) sits atop a hierarchy of urban centers that has captured the lion's share of national population growth since World War II.

Europe at 73 percent urban ranks among the most highly urbanized realms in the world (see Table I–1, p. 34), though more so in its west and north than in the east and south. In both Northern and Western Europe, the regional mean is now 84 percent urban, with the United Kingdom, West Germany, and Belgium exceeding

FIGURE 1–14 Juxtaposition of tradition and ultramodernity: the cityscape of the CBD in Frankfurt, West Germany.

90 percent. And even where the average percentage of urbanization is lower in Eastern (63 percent) and Southern Europe (68 percent), the towns and cities contain a much larger share of the population than in Asian or African countries.

Patterns Within Cities

At the intraurban scale, although the trend of the past three decades has slowly been toward United

States–style suburbanization, the overall appearance of the cityscape in Western Europe is still quite different from the North American urban scene. Lawrence Sommers underscores the reasons for this contrast:

Age is a principal factor, but ethnic and environmental differences also play major roles in the appearance of the European city. Politics, war, fire, religion, culture, and economics also have

played a role. Land is expensive due to its scarcity, and capital for private enterprise development has been insufficient, so government-built housing is quite common. Land ownership has been fragmented over the years due to inheritance systems that often split land among sons. Prices for real estate and rent have been government controlled in many countries. Planning and zoning codes, as well as the development of utilities, are determined by government policies. These are characteristics of a region with a long history, dense population, scarce land, and strong government control of urban land development.

The internal spatial structure of the European *metropolis*—which consists of both the central city and its suburban ring—is very much typified by the urban pattern of the London region (see map p. 85). The whole metropolitan area is still focused on the large city at its center, especially the downtown **central business district (CBD)**, which is the oldest part of the urban agglomeration and contains the region's main concentration of business, government, and shopping facilities as well as its wealthiest and most prestigious residences. Although the CBD continues to be revered as the political, economic, and cultural core, a growing number of skyscrapers increasingly transform its richly historical landscape (Fig. 1–14). Wide residential sectors radiate outward from the CBD across the rest of the central city, each one home to a particular income group and thereby perpetuating geographically the strong class system in European society. Lower-income groups are usually located in crowded neighborhoods adjacent to, and downwind of, the inner-city industrial district, whereas middle-income residents congregate at lower densities well away from the factories and their effluents.

The Metropolitan Periphery

Beyond the central city is a sizable suburban ring, but residential densities here are much higher than in the United States—because the European tradition is one of living in apartments rather than detached single-family houses and because there is a far greater reliance on public transportation, which necessitates a compact development pattern. Although automobile-based suburbanization has increased with the proliferation of high-speed freeways since 1970 (the Germans pioneered these superhighways with the first *autobahns* in the 1930s), it has not yet produced a significant amount of low-density residential sprawl (quite possibly because home heat-

ing fuels and gasoline are priced up to three times current U.S. levels). Therefore, the typical European suburb is still a high-density satellite town or village surrounded by open countryside that is heavily used for recreational purposes.

The preservation of open space and the limitation of sprawl development in the metropolitan periphery is the legacy of an urban and regional planning movement that has been a popular tradition (and source of government growth policy) throughout Europe for much of this century. Modern planning originated in Britain shortly before 1900—as a reaction to the horrors of early industrial urbanization—and soon spread across the continent. The centerpiece of these efforts was the New Towns Movement, which involved the controlled dispersal of certain people and economic activities (including light manufacturing) to new self-

FIGURE 1–15 A nighttime view of La Défense, a U.S.-style suburban business center located 12 miles from the center of Paris.

sufficient, lower-density towns located in the outer suburbs within an hour's train ride of the CBD. Although the most vigorous period of new-town construction occurred between 1945 and 1960, it is still expected that over 5 percent of the population of England and Scotland will reside in more than 30 of these settlements by 2000. Postwar continental Europe has adopted this innovation even more widely, and such showcase new towns as Tapiola (Finland), Vallingby (Sweden), and Lelystad (the Netherlands) rank among the most advanced state-of-the-art urban communities in the world. Another major planning feature of the contemporary European metropolis is the encircling band of open space or "greenbelt" that separates most large cities from their suburbs (see map on p. 85). This was a response to the rapid urbanization of the pre–World War II period that threatened to overrun the entire countryside, and its implementation guaranteed most city dwellers the same access to nearby open space enjoyed by their suburban counterparts.

As this century draws to a close, the European urban landscape will continue to change. The latest **European urbanization trend** is the accelerating suburbanization of population, most evident in Britain, France, West Germany, Switzerland, and the Low Countries. But is the ring road around Paris (*Boulevard Péripherique*) or London's new circumferential M-25 motorway really about to become the "Main Street" of its metropolis in the manner that the Capital Beltway, for instance, has transformed the urban geography of the Washington, D.C. area? Despite the rise of major new suburban business centers (Fig. 1–15), the answer for now is almost surely no, given the apparently enduring commitment of Europeans toward preserving their rich urban

heritage. But we must remind you that we began this overview of the spatial patterning of Europe's metropolitan areas by observing that the slow but steady trend since mid-century has been toward Americanization—and that might ultimately prove to be a force too powerful to resist.

REGIONS OF EUROPE

Earlier in this chapter, on the regional map of Europe (Fig. 1–9, p. 71), the realm's core area was delineated. This boundary encloses Europe's heartland, historically a large-scale functional region of dynamic growth, dense population, burgeoning urbanization, and great industrial productivity. We now turn to a more traditional overview of the regions shown on that map, which were derived by grouping sets of European countries together. Each regional grouping reflects those countries' proximity, historical-geographical association, cultural similarities, common habitats, and shared economic pursuits. On these bases, identify five regions of Europe: (1) the British Isles, (2) Western Europe, (3) Northern or Nordic Europe, (4) Mediterranean Europe, and (5) Eastern Europe.

The British Isles

Off the coast of mainland Western Europe lie two major islands, surrounded by a constellation of tiny ones, that comprise the British Isles (Fig. 1–16). The larger of the two major islands, which also lies nearest to the mainland (a mere 21 miles or 34 kilometers at the closest point), is the island of *Britain*;

its smaller neighbor to the west is *Ireland*. Although Britain and Ireland continue to be known as the British Isles, the British government no longer rules over the state that occupies most of Ireland, the Republic of Ireland or *Eire*. After a very unhappy period of domination, the London government in 1921 gave up control over that part of the island, which remains overwhelmingly Roman Catholic; but the northeastern corner of Ireland, where English and Scottish Protestants had settled, was retained by the British and is called *Northern Ireland*.

Although all of Britain is part of the United Kingdom, political divisions exist here too. *England* is the largest of these units, the center of power from which the rest of the region was originally brought under unified control. The English conquered *Wales* in the Middle Ages, and *Scotland* was tied to England in the seventeenth century when a Scottish king ascended the English throne. Thus, England, Wales, Scotland, and Northern Ireland became the *United Kingdom (UK)*.

Highland and Lowland Britain

Britain's natural environment is varied, and the relative location of different subareas matters much. The most simple and meaningful division of Britain is into a *lowland* and a *highland* region. Most of England, with the exception of the Pennine Mountains and the peninsular southwest, lies in Lowland Britain; Highland Britain consists mainly of Scotland and Wales. Lowland Britain, therefore, lies opposite the western periphery of the North European Lowland on the continent, and it is essentially a geological continuation of it (Fig. 1–5). In this part of England, the relief is low, soils are generally good, rainfall is ample, and agri-

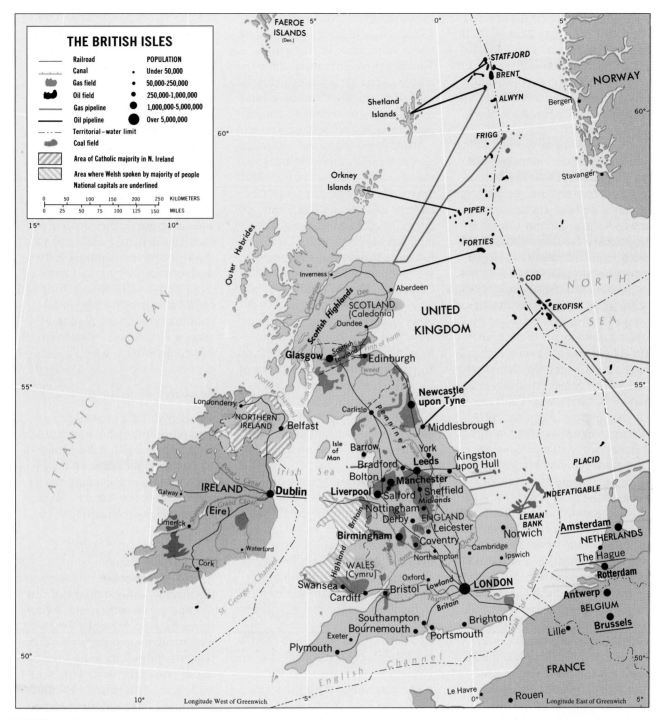

FIGURE 1–16

cultural productivity is relatively high.

Lowland Britain, particularly the accessible Thames Basin around London in southeastern England, has long been the Isles' power center. Before the birth of the British nation and empire fol-lowing the eleventh century, nu-merous peoples had penetrated and fought over this subregion, among them the Celts, the Romans (who founded London and in-troduced Christianity), the Ger-manic Angles and Saxons, the Vikings, and the Normans (who had invaded from France under William the Conqueror in A.D. 1066). In its political development, Britain benefited from its insular separation from Europe, and En-gland became the focus of regional power and economic organization. Centered on London, England steadily evolved into the core area of the British Isles. Ultimately, the British achieved a system of parliamentary government that had no peer in the Western world, and England rose from a regional core area into the headquarters of a global empire.

The economic development that to a considerable extent precipitated this political progress came in stages that could be accommodated without disrupting the whole society. Britain's original subsistence economy gave way to the commercial era of mercantilism, and the Industrial Revolution was foreshadowed by the early development of manufacturing based on the hydropower provided by streams flowing off the Pennines, Lowland Britain's mountain backbone. Then, when the Industrial Revolution came, coalfields in the Midlands (the crescent surrounding the southern Pennines) provided the needed energy source. The eastern Midlands deposits supported the industries of Sheffield, Leeds, and Nottingham; those to the west sustained Manchester, Salford, and Bolton (Fig. 1–16). Smaller coalfields were also exploited south of the Pennines near Birmingham, and in the northeast near Newcastle.

The impact of the Industrial Revolution in England was felt most strongly in these central and northeastern industrial regions. Despite poor working and living conditions in these manufacturing centers, population totals soared and Britain prospered. Nowhere was the principle of areal functional specialization better illustrated than here. Woolen-producing cities lay east of the Pennines, centered on Leeds and Bradford; the cotton textile industries of Manchester clustered on the western side of the range. Birmingham and its nearby satellite cities concentrated on the production of steel and

metal products; Nottingham came to specialize in hosiery; boots and shoes were made in Leicester and Northampton. In northeastern England, shipbuilding and the manufacture of chemicals were the leading industries along with coal-mining and iron and steel production.

Today, the industrial specialization of British cities continues, but the resource situation has changed drastically as local raw materials became exhausted. A one-time exporter of coal and iron ore, England has in recent years been forced to import these commodities as domestic supplies dwindle. Of course, Britain's industries always did use raw materials from other countries—wool from Australia and New Zealand; cotton from the United States, India, and Egypt. But with the breakdown of the British Empire and the unrelenting competition from rival industrial powers, Britain is today hard-pressed to maintain its position as a major Western industrial power.

The overall energy situation, on the other hand, has been improving, and Britain is for the moment self-sufficient in oil and natural gas, thanks to recent discoveries beneath the North Sea. European countries facing the North Sea have allocated sectors of seafloor to themselves (see Fig. 1–16). The British sector has proven to contain rich fuel reserves, and the UK has also purchased concessions in Norway's productive Ekofisk field, from which a pipeline was opened in 1974. With several offshore oilfields now operating, Britain's northeastern coast has particularly benefited. But while British holdings on the North Sea floor continue to yield new discoveries of petroleum and natural gas reserves, it is estimated that today's self-sufficiency will not last far into the next century.

Lowland Britain, in addition to its industrial development, also

has the vast majority of Britain's good agricultural land. However, there is not nearly enough land to feed the British people (57 million of them) at present standards of caloric intake. The good arable land is mainly concentrated in southeastern England, with a smaller area of good soils near the port city of Liverpool in west-central Britain. Where the soils are adequate, there is intensive cultivation of grain crops, potatoes, and sugar beets. But much of Lowland Britain, because of soil quality, coolness, and excessive moisture, cannot support field crops and is suitable only for pasture. Therefore, Britain must import a large quantity of food, enough to feed up to two-thirds of its people.

Regions of Britain

The British Isles constitute a region of the European realm, and both Britain and Ireland can be divided into subregions. In the case of Britain, four such units can be identified: (1) the Affluent South; (2) the Stagnant North; (3) Scotland; and (4) Wales.

The Affluent South
As industrial decline accelerates in the manufacturing regions of central and northern England, a dual economy is emerging in the 1990s, with its major geographic division lying roughly along a line connecting Bristol in the west with Norwich in the east (Fig. 1–16). South of that line, high-technology and service industries are thriving, dominated by financial, banking, engineering, communications, and energy-related activities. Nowhere is this boom greater than in the Southeast, the area anchored by London and its immediate hinterland, which is home to more than 20 million of the UK's 57 million inhabitants.

Metropolitan London itself contains 9.1 million people, and its classic European-city layout of built-up zones and greenbelt is mapped in Fig. 1–17. At the same time, London centers one of the world's great **conurbations**, the general term used by geographers to describe vast multimetropolitan complexes (such as Megalopolis, discussed on p. 32) that formed from the coalescence of two or more major urban areas. The central city, of course, is the country's historic focus, the seat of government, the headquarters of numerous industrial and commercial enterprises, the leading port, the most concentrated and richest market, and the national transportation hub—in short, the primate city.

The remainder of southern Britain contains two other areas worth noting. One is the tapering peninsula that extends southwestward from Bristol, a world apart from the crowded Southeast. This area belongs to Highland Britain, offering scenic country that attracts many tourists, but otherwise is an economic backwater that has long been awaiting development. And to the northeast of metropolitan London lies a round bulge of land that constitutes the third component of the British South—East Anglia. Here is one of England's oldest settlement areas, which attracted Germanic invaders and eventually developed a major fishing industry. Today, the sea is still important (offshore gasfields and still-productive fishing)—but so is agriculture, because East Anglia contains some of Britain's best soils.

The Stagnant North

To the north of the Bristol–Norwich dividing line is a very different England, one dominated by the social and economic decline associated with the aging and disintegration of the once-prodigious regional industrial complex (Fig.

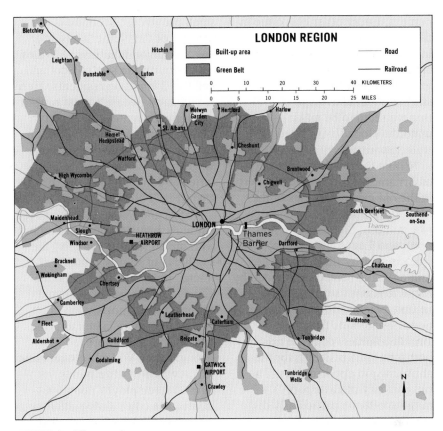

FIGURE 1–17

1–18). Unemployment is one of the most significant measures of this north-south cleavage in Britain's economic geography: at the end of the 1980s, the unemployment rate for all of England stood slightly above 6 percent, but the rate for the North was just over 10

FIGURE 1–18 The harsh impact of rapid industrial decline in northern England: a stilled factory surrounded by recent inner-city decay in Liverpool.

percent. Every other economic indicator tells the same story. Even in car ownership rates, the North lags so far behind the rest of England (214 per 1000 population versus the national average of 319) that we seem to be making an underdeveloped/developed country comparison.

Within the northern subregion of Britain, conditions are bleakest in the old industrial cities of the Midlands, including Manchester (4.0 million), Sheffield (1.1 million), and Leeds (1.6 million). The woes of nearby Birmingham (2.1 million) have been compounded by racial tensions as its large low-income black and Asian populations (immigrants from former British colonies) compete grimly with long-time residents for the shrinking job opportunities. The old port city of Liverpool (1.4 million) offers a particularly distressing outlook to its young generation: the docks have almost totally shut down; over 50,000 jobs vanished during the 1980s; and the unemployment rate hovers around 50 percent for youths, many of whom will never find a job and will be forced to live out their lives on welfare. Seeking to alleviate these severe problems of the North, the British government and the private sector recently launched some new revitalization programs in the Tyneside area surrounding Newcastle (1.4 million). New facilities, especially industrial parks, have been quite successful, but local unemployment has not dropped markedly and the longer-term impact of these investments remains uncertain.

Scotland and Wales

Scotland and Wales are distinct subregions of Britain because they have political as well as cultural, economic, and physiographic identity. Moreover, their relationships with long-dominant England are changing. By European standards of scale, these are by no means small provinces: Scotland is twice the size of the Netherlands and has 5.2 million inhabitants; Wales is one-fourth the size of Scotland and has 2.9 million residents. Here in Highland Britain's most rugged and remote territories ancient Celtic peoples found refuge from encroaching invaders, and their languages still survive after centuries of domination by the English (Fig. 1–12).

Wales and Scotland shared in Britain's Industrial Revolution but with mixed results. Initially, the coalfields of southern Wales attracted industries, but when the better deposits were exhausted and production costs went up, competition from manufacturing centers in England proved too strong. Soon, Wales's countryside was ravaged by strip mines, its now-underemployed people dislocated, its cities slum-ridden—and many Welsh citizens left their country for new opportunities elsewhere. Scotland was more fortunate. Along the narrow ''waist'' of the Clyde Valley (Clydeside) and Firth of Forth, Scotland possessed an extensive coalfield and nearby iron ore deposits, which formed the basis for long-term manufacturing development (notably in shipbuilding). This lowland corridor became the region's core area, with industrial Glasgow (1.9 million) at one end and the cultural focus of Edinburgh (700,000) at the other. Although Clydeside shares with the English North much of the industrial decline caused by obsolescence, Glasgow has revived and is now a budding center of industrial research, especially in electronics. In addition, Scotland now fronts a North Sea that produces lucrative oil and natural gas.

Wales and Scotland also share a resurgent nationalism. The government in London (where Scottish and Welsh representatives are far outnumbered by their English counterparts) has always pursued economic policies that favor England, often to the clear disadvantage of Scotland and Wales. When a lingering recession in the 1970s brought wide new support for longstanding Scottish and Welsh nationalist movements, strong demands for regional autonomy began to be heard. Given Scotland's importance to the successful new energy industries of the North Sea, particularly in the wake of the crisis in Northern Ireland, London swiftly learned that it must adjust at an early stage to regional pressures. Thus, Parliament has in recent years undertaken formal discussions under the theme of **devolution** (see *box*)—the redistribution of authority in the United Kingdom and the restructuring of the country's political framework. Although separatist proponents had only limited successes in recent elections, the issue is likely to remain current in British politics for decades to come.

Fractured Ireland

Northern Ireland

The smallest of the UK's four units continues to be London's most serious domestic problem. Northern Ireland, with a population of 1.6 million, occupies the northeastern one-sixth of the island of Ireland, a legacy of England's colonial control. Despite a poor resource base, a diversified economy has emerged here as a result of infusions of British capital and raw materials. In Belfast (650,000), the shipbuilding industry grew to significant proportions; so did textile (mostly linen) manufacturing in Londonderry (190,000). But unemployment has been an erosive reality, and Northern Ireland has not fared well since 1970—economically or socially.

Northern Ireland, severed from the Republic of Ireland when the latter became independent from London in 1921, has always been

DEVOLUTION—GLOBAL SYMPTOM?

The possibility that regional self-government may come to Wales and Scotland is unthinkable to many Britons; numerous Canadians view the prospect of an independent Quebec as inconceivable; and Nigeria went to war in the 1960s to prevent the secession of its southeastern region. Has the evolution of the state system come to an end? Will a reverse process—*devolution*—become common-place?

Indications are that many states will confront the specter of fragmentation. Yugoslavia's fragile federal structure is under increasing pressure from Slovenian and other separatists. Canada faces the possible secession of Quebec if its French-speaking population is not recognized as a "distinct society." Eritrean guerrillas continue to fight for separation from Ethiopia. Kurds in Iran, Pakhtuns in Pakistan, Muslims in the Philippines, Tamils in Sri Lanka, Turks in Cyprus—the number of regional secessionist movements appears to be growing, and the list of potential problem areas expands as well. Such movements are vulnerable to involvement in the greater ideological struggle in the modern world as the major powers try to exploit them to their own advantage.

different from the rest of Ireland because of its substantial Protestant population. About two-thirds of its people trace their ancestry to Scotland or England, from which their predecessors came across the Irish Sea to settle. Only about 35 percent of Northern Ireland's people are Catholics, in contrast to the Republic, which is overwhelmingly Roman Catholic. No spatial separation marks the religious geography of Northern Ireland: Protestants and Catholics live in clusters throughout the area.

For the past two decades, Northern Ireland has remained on the brink of civil war as Protestants and Catholics attacked each other in terrorist campaigns. Catholics have long protested what they perceive to be discrimination by the Protestant-dominated local administration, especially in housing and employment; Protestants accuse Catholics of seeking union with the Republic. Since 1972, the territory has been ruled directly from London, but even a large British

armed force sent to maintain order has been unable to end the continuing sectarian violence.

Republic of Ireland

The republic to the south (Eire) is one of Europe's youngest independent states, having fought itself free from Britain only 70 years ago. Without protective mountains and without the resources that might have swept their country ahead in the great forward rush of the Industrial Revolution, the Irish—a nation of peasants—faced their English adversaries across the narrowest of seas. During the seventeenth century the British conquered the surprisingly resistant Irish, expropriated their good farmland, and turned it over to (mostly absentee) landlords; those peasants who remained faced exorbitant rents and excessive taxes. The English also placed restrictions on every facet of life, especially in Catholic southern Ireland, though somewhat less in the Protestant north.

The island on which Eire is located is shaped like a saucer with a wide rim, which is broken toward the east where lowlands open to the sea. A large plainland is thus enclosed, but this topographic advantage does not result in better agricultural opportunities than in Scotland or Wales. As Fig. I–9 shows, Ireland is wetter than England, and excessive moisture is the great inhibiting element in farming here. Hence, pastoralism is once again the dominant agricultural pursuit, and a good deal of potential cropland is turned over to fodder. But how can 60 inches (150 centimeters) of rainfall, Eire's annual average, be too much for cultivation when there are parts of the world where 60 inches is barely enough? Here, the *evapotranspiration* factor comes into play (see p. 17), and this is a good illustration of the deceptive nature of such climatological maps as Fig. I–9—deceptive unless more is known than the map tells us. Ireland's rain comes in almost endless soft showers that drench the ground and keep the air cool and damp. Rarely is the atmosphere really dry enough to permit the ripening of grains in the fields. Yet, without such a dry period, the crops are damaged or destroyed, and so there are severe limitations on what a farmer can plant without too great a risk.

In Ireland, the potato quickly took the place of other crops as a staple when it was introduced from America in the 1600s; it does particularly well in such an environment. But even the potato could not withstand the effects of the heightened precipitation that fell during several successive years in the 1840s. Having long been the nutritive basis for Ireland's population growth, which by 1830 had reached 8 million, the potato crop failed repeatedly; ravaged by blight and soaking, potatoes by the millions simply rotted in the ground. Now the

Irish faced famine: over a million people died, and nearly twice that number left the country within ten years of this calamity. It started a pattern, as emigration offset and even exceeded natural increase. Together, Ireland and Northern Ireland now have a population of 5.2 million (3.6 million in Ireland proper), barely two-thirds of what it was when the famine of the 1840s struck. Not many places in the world can point to a decline in population since the days of the Industrial Revolution. Moreover, the exodus still continues in the 1990s, particularly among the educated younger Irish who are fleeing at a rate of 250,000 per year (*7 percent* of Eire's total population!). Economic causes are the main driving force in this emigration (Eire's unemployment rate is more than double the European norm), but also cited frequently is the conservatism of Irish cultural and religious life.

This outflow of young talent notwithstanding, efforts are being made to diversify Ireland's economy and speed the country's development. Dublin (1.0 million), the capital, has become the center for a growing number of light industries. In the trading arena, long-dominant farm exports are being balanced by manufactures that today account for about half the annual foreign sales. Ireland's decision to join the Common Market in 1973 is further evidence of the country's desire to modernize. But this task is a difficult one, and Ireland today is still more reminiscent of agrarian Eastern Europe than the Western European heartland it has recently joined.

Western Europe

The essential criteria for identifying a Western European region are those that give Western Europe

GERMANY'S REUNIFICATION

FIGURE 1–19 The parting of the Berlin Wall in November 1989. A joyous throng of West Berliners greets a steady flow of their eastern neighbors just after the opening of this hated barrier, which for the previous 28 years stood as an ugly reminder of the city's bifurcation by this segment of the Iron Curtain.

The prospect for the unification of West and East Germany became reality in 1990 (Fig. 1–19). On both sides of the partition line (superimposed at the end of World War II), there is much en-

the characteristics of a continental core: this is the Europe of industry and commerce, of great cities and bustling interaction, of functional specialization and areal interdependence (Fig. 1–9). This is dynamic Europe, whose countries founded empires while forging democratic governments, and whose economies staged an astounding recovery from the devastation of World War II five decades ago.

But if it is possible to recognize some semblance of historical and cultural unity in British, Scandinavian, and Mediterranean Europe, that quality is absent here in the melting pot of Western Europe. Europe's western coreland may be a contiguous area, but in every other way it is as divided as any

part of the continent is or ever was. Unlike the British Isles, there is no *lingua franca* here; unlike Scandinavia, there is considerable religious fragmentation, even within individual countries; unlike Mediterranean Europe, there is no common cultural heritage. Indeed, the core region we think we can recognize today based on economic realities could be shattered at almost any time by political developments. It has happened before.

The two leading states of Western Europe are France and West Germany. In addition, there are the Low Countries of Belgium, the Netherlands, and Luxembourg (collectively called *Benelux*), and the Alpine states of Switzerland

thusiasm for reunification, but some politico-geographical questions enter into the picture:

1. *European fears.* Germany during the present century twice plunged the world into war. From the United Kingdom to the Soviet Union, there are millions who fear that "the third time never fails." Thus, there is pressure for a demilitarized Germany if there is to be a united Germany, just as Japan was demilitarized in 1945.
2. *Economic domination.* A united Germany would dominate the economic geography of Europe; already, the economy of West Germany is the world's fourth largest. The UK and France would be disadvantaged by the consolidation of German economic power.
3. *Boundaries.* The Oder-Neisse Line became East Germany's eastern border with Poland in 1945; the Germans lost a wide zone of territory to the Poles in the reconstruction of the postwar European map. Unless East and West Germany unequivocally state that there will be no challenge to this superimposed boundary, non-German Europeans can envision the rise of a new German nationalism based on the issue of "lost" territories.
4. *The capital.* Bonn has served as West Germany's capital in postwar times, but Berlin was Germany's primate city before 1945 and would be likely to regain its former position as the national headquarters. However, Berlin lies close to a united Germany's eastern border, a peripheral location it did not have before Poland acquired the area east of the Oder and Neisse rivers. Therefore, the relocation of the capital to Berlin could also spur territorial demands.

and Austria. Northern Italy, by virtue of its industrialization and its interconnections with Western Europe, was identified as part of the continental core (Fig. 1–9); but for the purposes of the present regionalization of Europe, all of Italy will be treated as part of Mediterranean Europe. Finally, the partitioning of Germany into West and East at the end of World War II poses a special regional problem, for East Germany's cultural-geographic and historical ties are with Western rather than Eastern Europe. But postwar East Germany was incorporated into the Soviet sphere of influence and reorganized along socialist lines. Now the collapse of the commu-

nist system in Eastern Europe in 1989 will result in the restoration of a single Germany in the 1990s (see *box*). In this chapter, East Germany is mentioned under both Western and Eastern Europe, and use of the label *Germany* is intended to refer to both East and West Germany.

France and West Germany

Territorially, Western Europe is dominated by France; economically, West Germany is paramount. In many other ways, these two mainland powers present interesting and sometimes enlightening contrasts. France is an old state, by most measures the oldest

in Western Europe. Germany is a young country, born in 1871 after a loose association of German-speaking states had fought a successful war against . . . the French! Modern France bears the imprint of Napoleon, who died just a few years after the political architect of Germany, Bismarck, was born.

On the map of Western Europe, it would seem that Germany, smaller than France in area, also has a less advantageous geographic position than its neighbor (Fig. 1–20). France has a window on the Mediterranean, the North Atlantic Ocean, the English Channel, and, at Calais, even a corner on the North Sea. Germany, on the other hand, has only short coastlines on the North Sea and the icy Baltic, and the rest of its territory seems tightly landlocked by the Netherlands and Belgium to the west, by the Alps to the south, and by Poland in the east. But such appearances can be deceptive. Actually, France is at a disadvantage when it comes to foreign trade, because none of its natural harbors are particularly good; moreover, most of France's rivers and inland waterways are not navigable by large oceangoing ships. Therefore, much waterborne trade destined for France goes through other European ports such as London, Rotterdam, and Antwerp. Germany is much better off. Although the mouth of the important Rhine River lies in the Netherlands, West Germany contains most of its course, which runs past Europe's leading heavy industrial complex, the *Ruhr*. Here, the Rhine is almost as good a connection with the North Sea and the Atlantic as a domestic coast would be. The Dutch port of Rotterdam, located at the mouth of the Rhine, is Europe's largest in terms of tonnage handled and serves as a more effective outlet for Germany than any French port is for France.

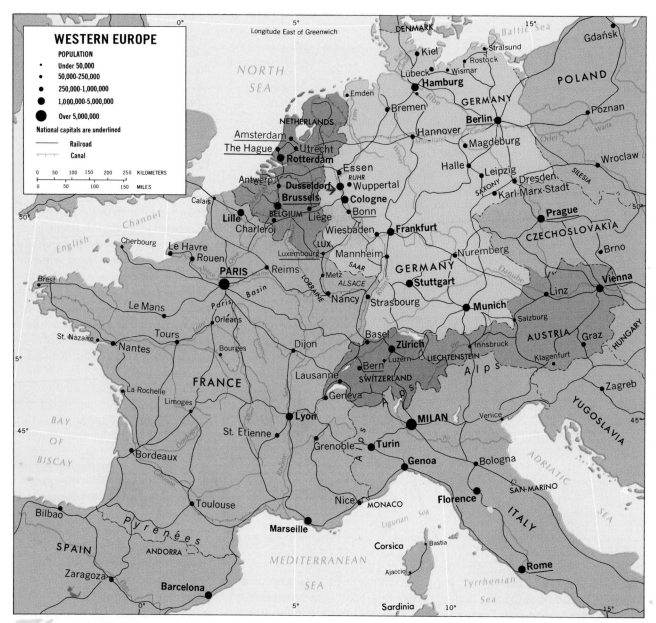

Fig. 2.0

France

Another sharp contrast between France and West Germany lies in the degree of urbanization in the two countries. Unlike overwhelmingly urbanized (94 percent) West Germany, only 73 percent of the French population reside in towns and cities. True, Paris is without rival in France (and perhaps mainland Europe), but there is a huge gap between this cultural focus and primate city of 8.7 million and the second-ranking French city,

Lyon (1.3 million). Why should Paris, without major raw materials in its immediate vicinity, be so dominantly large? Paris owes its origins to advantages of **site** and its later development to a fortuitous **situation** (Fig. 1–21). Whenever an urban center is studied, these are two highly significant locational qualities to consider. A city's *site* refers to the actual physical attributes of the place it occupies—whether the land is flat or hilly, whether it lies on a river or coast,

whether there are any obstacles to future expansion such as ridges or marshes. By *situation* is meant the geographic position of the city with reference to surrounding areas of productive capacity, the size of its hinterland, the location of nearby competing towns—in other words, the greater regional framework within which the city finds itself.

Paris was founded on an island in the Seine River, a place of easy defense where the waterway was often crossed. Exactly when pre-Roman settlement on this site, the *Ile de la Cité*, actually began is not known, but it functioned as a Roman outpost some 2000 years ago. For many centuries, this defensibility of the site continued to be important, but the island soon proved to be too small and the city began to expand along the banks of the river (Fig. 1–21). Soon, complementary advantages of Paris's situation were revealed. The city lay near the center of a large and prosperous agricultural area, and as a growing market its focality increased steadily. Significantly, the Seine is joined near Paris by several navigable tributaries, providing access to various parts of the Paris Basin. Via these rivers (the Oise, Marne, and Yonne) converging from the northeast, east, and southeast, and via the canals that later extended them even farther, Paris can be reached from the Loire Valley, the Rhône-Saône Basin, the Lorraine industrial area, and from the Franco-Belgian border zone of coal-based manufacturing. Of course, there are land connections as well. The political reorganization that Napoleon brought to France included the construction of a radial system of roads—followed later by railroads—that focused directly on Paris (Fig. 1–21, inset map).

If it is not difficult to account for the greatness of Paris, it is quite another matter to account for the relatively limited development of French industrial centers. In France, there is no Birmingham, no Glasgow—certainly no Ruhr. And yet, there is coal as well as good quality iron ore. What is lacking? For one thing, large supplies of high-quality coal; for another, the juxtaposition of such coal with cheap transport facilities and sizable population concentrations. So French manufacturers did what Europeans have done almost everywhere: unable to com-

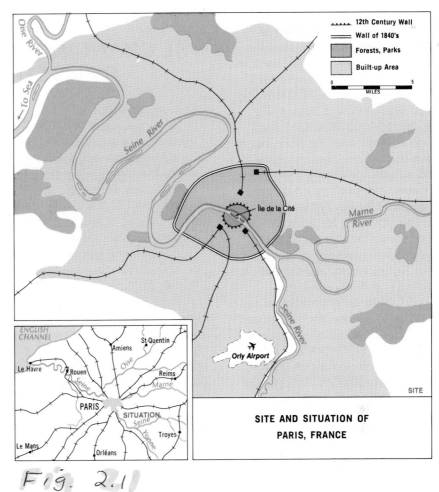

SITE AND SITUATION OF PARIS, FRANCE

Fig. 2.1

pete in volume, they specialized—in high-quality textiles, precision equipment, automobiles, and, of course, wines and cheese.

France achieved greater strength in agriculture (it is still Europe's leading producer) with a much larger area of arable land than Britain or Germany, and with the benefits of a moist temperate climate free of extremes. Even so, there exists a wide variety of local conditions, each with its special opportunities and limitations. Wheat is grown on the best soils and is France's leading crop; oats and barley are raised on soils of lesser quality. Spatially, French agriculture is marked by an enormous diversity of production, so much so that it is impossible to delineate land-use regions—except in the vineyards of the Rhône-

Saône, Garonne, and Loire valleys.

Germany

Germany—West or East—has no Paris, and West Germany is far from self-sufficient in farm production. But despite the absence of a primate city, West Germany has more than twice the number of metropolitan areas containing over a half million population: 13 versus 6 in France. And, as might be expected, West Germany also is more heavily industrialized than its southwestern neighbor.

The majority of Germany's cities lie along the zone of contact between the North European Lowland and the Central Uplands (see Fig. 1–5). This, as we know, is also the leading zone of coal deposits, and they gave rise to pre-war Germany's three major industrial areas: the *Ruhr* near the Dutch border in the west; the *Saxony* area along the Czechoslovakian border (now in East Germany); and *Silesia*, now in Poland (Fig. 1–20). Postwar West Germany was left with only the Ruhr, but it has become the greatest heavy manufacturing complex in Europe. Today, it remains an advantaged industrial region, but many of its rapidly aging factories are experiencing the same economic-geographic problems that bedevil the British North, and the Ruhr's future is now uncertain (Fig. 1–22). With the restoration of ties between East and West Germany, Saxony—a manufacturing complex more oriented to skill and quality—will play a growing role, particularly after it is brought up to Western standards of efficiency and modernization.

West and East Germany count a total of 78 million inhabitants, and a great deal of good agricultural land would be required to feed so large a population. But farming is limited by hilly country in the south and by sandy soils in the north. Nevertheless, with staples including the ever-present potato, rye, and wheat, Germany manages to produce about three-quarters of the annual caloric intake of its population. This is not quite as good as France, but far surpasses what Britain manages to do—all in all a remarkable achievement.

Benelux

Three political entities are crowded into the northwestern corner of Western Europe: Belgium, the Netherlands, and tiny Luxembourg, collectively referred to by their first syllables ("Be-Ne-Lux"). These states are also frequently called the Low Countries, a very appropriate label because most of the land is extremely flat

THE CHANGING SOCIAL GEOGRAPHY OF WESTERN EUROPE

Europe in modern times has sent millions of its inhabitants to populate the Americas, Australia, and other overseas realms. Where they overwhelmed local communities, the white settlers from Europe created new societies in the European mold. Even where they remained in the minority, as in South Africa and Algeria, they drastically changed their new homelands.

But now Europe is experiencing an immigrant invasion itself. Since the end of World War II, immigrants have come to Europe—especially to Western Europe—in large numbers. By 1990, about 13 million immigrants were living in Western Europe, more than 8 percent of the region's population. Almost all came from the Third World, with many arriving from comparatively nearby countries (such as Turkey and Algeria) to take jobs that became available during Europe's industrial boom of the 1960s and 1970s. Others came from more distant one-time colonies, such as Indonesia, Angola, and Suriname, exercising their right to do so as subjects of the colonial power. The great majority, whatever their origin, settled in Europe's great urban areas, where jobs and other opportunities were located.

When Europe boomed, these immigrants found employment and were generally welcomed. But when Western Europe's economies slowed in the 1980s, the newcomers were less welcome. Opposition to further immigration arose. Competition between the immigrants and Europeans for scarcer jobs increased. Social problems in the cities intensified. For the Third World immigrants, Europe did

FIGURE 1–22 Aging factories and other suggestions of manufacturing decline along the waterfront of Duisburg, near the confluence of the Ruhr and Rhine rivers.

not prove to be the melting pot that Europeans found in America. Charges of discrimination were made against European governments, and many European cities with ethnic neighborhoods (where the immigrants mostly cluster) found themselves with problems for which they were not prepared.

Major European cities, including capitals such as Paris and Amsterdam, contain large cohesive neighborhoods where the immigrants have implanted some of their own culture. France's leading cities (not only Paris, but also Lyon and Marseille) have suburbs where street signs are in Arabic, where Islam rules, and where the atmosphere is that of urban Morocco, Tunisia, or Algeria. The Turkish imprint on West German cities is similarly strong, though of 1.8 million Turks only a few thousand have been permitted to become German citizens. In Amsterdam, the 300,000-plus Surinamese immigrants are changing the face of the city—nearly half of that former colony's population now lives in the Netherlands. Today, over half of inner-city Amsterdam's schoolchildren are non-Dutch, and the city's social geography has altered rapidly as many white residents moved to outer suburbs. Black (mainly Surinamese) inner-city numbers grow and soon will constitute more than one-third of the population. Such changes cannot occur without difficulty, and Amsterdam suffers from increased crime and related problems.

Thus, Western Europe, itself a patchwork of societies and traditions, faces the need for new adjustments and the reality of a new social map. Europe has been described as the most international of realms; it is also becoming the most intercultural, as the rapid aging of its population virtually guarantees the influx of more and more immigrants.

and lies near (and in the Netherlands even below) sea level. Only toward the southeast, in Luxembourg and eastern Belgium's Ardennes, is there a hill-and-plateau landscape with elevations in excess of 1000 feet (300 meters). As with France and Germany, there are major differences between Belgium and the Netherlands. Indeed, there are such contrasts that the two countries find themselves in a position of complementarity.

Belgium is marked by two industrial corridors. One is the coal-based east-west axis through Charleroi and Liège, where there are heavy industries. The second corridor of lighter and more varied manufacturing extends north from Charleroi through Brussels to Antwerp. The diversified industrial products of these areas include metals, textiles, chemicals, and specialties such as pianos, soaps, and cutlery. The Netherlands, on the other hand, has a large agricultural base (along with its vitally important transport functions); it can export dairy products, meats, vegetables, and other foods. Hence, there was mutual advantage to the Benelux economic union of the 1940s, because it facilitated the reciprocal flow of needed imports to both countries and it doubled the domestic market.

The Benelux countries are among the most densely populated on earth, and space is truly at a premium. Some 25 million people inhabit an area about the size of Maine (which contains 1.2 million). For centuries, the Dutch have been expanding their living space—not at the expense of their neighbors, but by wresting it from the sea. The greatest project so far is the draining of almost the entire Zuider Zee, begun in 1932 and scheduled for completion after the turn of the century. In the southwest, the islands of Zeeland are being connected by dikes and the water pumped out, creating additional *polders* (reclaimed lands). Another future project involves the islands that curve around northern Holland, which may be connected by dikes to dry and reclaim the intervening Wadden Sea.

Three cities—Amsterdam, Rotterdam, and The Hague—anchor the Netherlands' triangular core area. Amsterdam (2.0 million), the constitutional capital, remains very much the focus of the Netherlands, with a bustling commercial center, a busy port, a variety of light manufactures—and an increasingly heterogeneous social complexion (see *box*). Rotterdam (1.1 million), Europe's busiest port, is the shipping gateway to Western Europe, commanding the entries to both the Rhine and Meuse (Maas) rivers. Its modern development mirrors that of the German Ruhr-Rhineland; thus, ongoing manufacturing decline in the Rhine hinterland, as well as increased competition from other European ports, is eroding Rotterdam's historic situational advantage. The third city in the triangle, The Hague (750,000), is the seat of the Dutch government and the home of the United Nations' World Court. Collectively, these three cities of the triangular core have spawned a conurbation called *Randstad*, with their coalescence creating a ring-shaped multi-metropolitan complex that surrounds a still rural center (the literal translation of "rand" is edge or margin). A more precise labeling of the conurbation would be Randstad-Holland, named for the country's western provinces in which it is located; "Holland" is also used interchangeably with "Netherlands" by many of the Dutch (see *box*).

In contrast to Belgium, the Dutch resource base has always been heavily agricultural. With a premium on space, rural population densities are very high, and practically every square foot of available soil is in some form of productive use. Only in the southeast does the Netherlands share the Campine belt of good coal deposits that extends across northern Belgium. Newly discovered natural gas reserves have been opened up in the northeast, further adding to Dutch energy supplies, and future development of this fuel is promising.

With the existing limitations of space and raw materials, the Dutch and Belgians have also turned to managing international trade as a leading economic activity. It is hard to think of any other countries that could be better positioned for this. Not only do the Rhine and Meuse form primary arteries that begin in the interior and terminate in the Low Countries, but Benelux itself is also surrounded by the most productive countries in Europe. Belgium's capital city, Brussels (2.2 million), is also a global city of consequence. This historic royal headquarters, positioned awkwardly astride the Flemish-French (Walloon) linguistic dividing line across Belgium (see Fig. 1–12), has become an international administrative center. Hundreds of multinational corporations with European interests have their central

THE NETHERLANDS OR HOLLAND?

The Netherlands is among countries that have two names (the United Kingdom/England; the Soviet Union/Russia) of which only one is technically correct. The Kingdom of the Netherlands has 11 provinces, of which two are named North Holland and South Holland. These two provinces also are the country's most important, containing the three leading cities, the largest population cluster, most of the heavy industry, and much of the agricultural capacity. North and South Holland form the historic heartland of the Netherlands, and many citizens simply call themselves "Hollanders."

Both names reflect the country's dominant physical property. *Nederland* (the Dutch spelling) means "low" country; *Holland* means "hollow" country—which, given the fact that more than one-third of it lies below sea level, is an appropriate name.

But not all Netherlanders approve of the other designation. Even in this small country, there is regionalism. Frieslanders or Groningers of the two northernmost provinces are not likely to call themselves Hollanders. Thus, the correct geographic name, the Netherlands, should be employed when reference is made to the country as a whole; Holland may be used to identify the Dutch heartland that faces the North Sea.

offices here, enhancing the city's role as a financial center and commercial-industrial complex. And Brussels has also become the administrative headquarters for international economic and military organizations such as the Common Market (EC) and its many subsidiaries as well as NATO (North Atlantic Treaty Organization).

Switzerland and Austria

Switzerland and Austria share a landlocked, increasingly peripheral location and the mountainous topography of the Alps—and little else. On the face of it, Austria would seem to have the advantage over its western neighbor; it is twice as large in area and has a bigger population than Switzerland. A sizable portion of the upper Danube Valley lies in Austria, and the Danube is to Eastern Europe what the Rhine is to West-

ern Europe. Moreover, no Swiss metropolis can boast of a population even half as large as that of the famous Austrian capital, Vienna (2.3 million). Austria also has considerably more land that is relatively flat and cultivable, a prize possession in this part of Europe. And that is not all. From what is known of the resources buried within the Alpine topography, Austria is again the winner, with deposits of iron ore, coal, bauxite, and even some petroleum; Switzerland has hardly any exploitable mineral deposits.

From all this, we might infer that Austria should be the leading country in this part of Europe, and that impression would probably be strengthened by a look at some cultural geography. Take the map of European languages (Fig. 1–12). Austria is a unilingual state in which only one language (German) is spoken throughout the country.

But in Switzerland no less than four languages are in use: German is spoken in the largest (northern) part; French over the western quarter; Italian in the extreme southeast; and in the mountains of the central southeast lies a small remnant of Romansch usage. Similarly, and adding to the picture of disunity, Swiss religious preferences are evenly split between Protestantism and Roman Catholicism whereas Austria is 85 percent Catholic.

Switzerland

These hindrances notwithstanding, it is the Swiss people who have forged for themselves a superior standard of living, and it is the Swiss state, not Austria, that has achieved greater stability, security, and progress. The world map (Fig. I–5) sometimes seems to suggest that mountainous countries share certain limitations on development. It is therefore tempting to generalize about the impact of mountainous terrain—and its frequent corollary, *landlocked location*—as preventing productive agriculture, obstructing the flow of raw materials, and hampering the dissemination of new ideas and innovations. Tibet, Afghanistan, and the Andean portions of South American states seem to prove the point. That is why Switzerland is such an important lesson in human geography: all the tangible evidence suggests that here is a European area that will be economically deprived and lack internal cohesion—but the actual situation is exactly the opposite!

The Swiss, through their skills and abilities, have overcome a seemingly restrictive environment; they have made it into an asset that has permitted them to keep pace with industrializing Europe. First they took advantage of their Alpine passes to act as middlemen in interregional trade; then they

FIGURE 1–23   The alpine topography of Switzerland, showing the importance of the forest as a safeguard against landslides and avalanches where human occupance clings to the steep slopes.

used the water cascading from those mountains to produce the hydroelectric power that spawned a high-quality, specialized industrial base; and finally they learned to accommodate with professional excellence the tourists who came to visit their beautiful mountain country (Fig. 1–23). Simultaneously, Swiss farmers were learning how to get the most out of their efforts on the limited non-highland rural space available, and the country today successfully specializes in dairy farming and the worldwide export of cheeses and chocolate products.

The majority of the country's population is concentrated in the central plateau, where the land is at lower elevations, and here lie the major cities. Zürich (850,000) is the largest and functions as a banking center of global importance. Geneva (400,000) is the famous city of international organizations and conferences, and between them lies Bern (225,000), the Swiss capital. Here, too, are found many of the industrial centers, which must import virtually all their raw materials, but their products (precision machinery, instruments, tools, fine watches) have so prestigious a reputation that these Swiss exports are guaranteed a prominent place on the world market. Thus, a strong case can be made that Switzerland is part and parcel of the European core, whereas Austria is not.

Austria

Austria is a much younger state than Switzerland, and its history is far less stable. Modern Austria is a remnant of the Austro-Hungarian Empire that fell during World War I; subsequently, the country suffered through convulsions far more reminiscent of Eastern Europe than Switzerland's peaceful Alpine isolation. Stability finally returned in 1955 when the last foreign occupiers withdrew, but ever since then Austria's most difficult problem has been its reorientation to Western Europe. Even Austria's physical geography seems to demand that the country look eastward: it is at its widest, lowest, and most productive in the east; the Danube flows eastward; and Vienna also lies closer to the eastern perimeter. Perhaps the new era dawning in Eastern Europe in the aftermath of Soviet domination will offer major oppor-

tunities in that direction, but with the interruptions and setbacks of the twentieth century, Austria is not well equipped to catch up with competitive Western Europe.

Nordic Europe (Norden)

In three directions from its core area, Europe changes quite drastically. To the south, Mediterranean Europe is dominated by Greek and Latin influences, and by the special habitat produced from the combination of Alpine topography and a Mediterranean climatic regime. To the east lies truly continental Europe, which is less industrialized and urbanized than the coreland. And to the north lies the Nordic Europe of Scandinavia, Finland, and Iceland (Fig. 1–24), almost all of it separated by water from what we have defined as the European core.

Despite its peripheral location, Nordic Europe is not "underdeveloped" Europe. To the contrary, a great deal has been achieved here. But in terms of resources and in realmwide context, the northern region of Europe is not particularly rich (although the exploitation of the North Sea oilfields has lately boosted Norway's economy). Northern Europe possesses the realm's most difficult environments: cold climates, poorly developed soils, steep slopes, and sparse mineral resources mark much of the region, with only small Denmark offering better opportunities thanks to fertile soils and lower relief. This region's conditions are reflected by its comparatively small population—its five countries contain just 23.1 million, a total less than that of Benelux. The overall land area, on the other hand, is almost the size of the entire European

core. People go where there is a living to be made, but the living in most of Scandinavia is not easy.

Several aspects of Nordic Europe's location have much to do with this. First, this is the world's northernmost group of states; although the Soviet Union, Canada, and the United States possess territory at similar latitudes, each of these much larger countries has its national core area in a more southerly position. The Northern Europeans themselves call their region *Norden*, an appropriate term indeed. Second, Norden, as viewed from Western Europe, is on the way to nowhere. How different would the relative location of the Norwegian coast be if important world shipping routes paralleled the shoreline on their way to and from the European core? Third, except for Denmark, all of Norden is separated by water from the rest of Europe. As we know, water has often been an ally rather than an enemy in the development of Europe—but mostly where it could be used for the interchange of goods. Norden, however, lies separated from Europe *and* relatively isolated in its northwestern corner. The only major exception to the bleak Scandinavian rule is found in Denmark and southern Sweden, which are really extensions of the North European Lowland.

Nordic Europe's relative isolation did have some positive consequences as well. Its countries have a great deal in common. They were not repeatedly overrun by different European peoples. The three major languages—Danish, Swedish, and Norwegian—are mutually intelligible, allowing people to converse without requiring an interpreter. Icelandic belongs to the same Germanic language subfamily (see Fig. 1–12); only Finnish is of totally different origins, but Finland's long period of con-

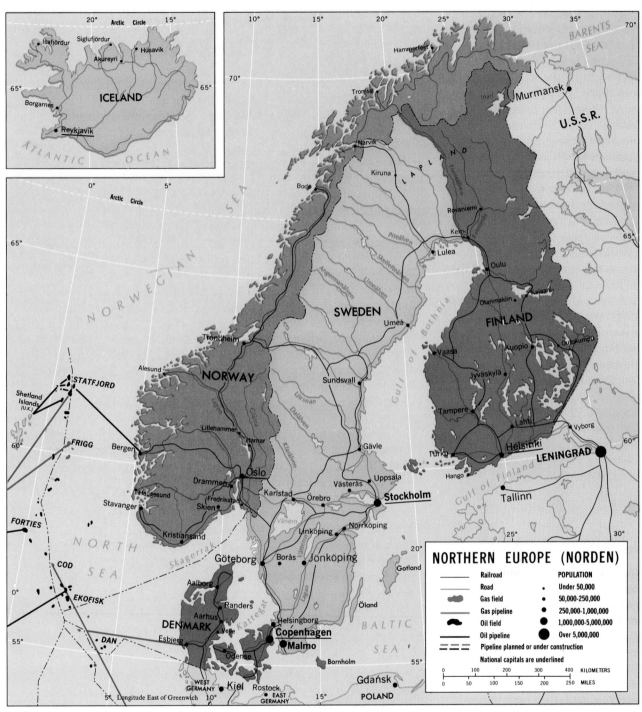

FIGURE 1—24

tact with Sweden has helped to overcome the linguistic barrier. Furthermore, in each of the Scandinavian countries, there is overwhelming adherence to the same Lutheran church, and in each it is the recognized state religion. Finally, there is considerable similarity in the political evolution of the Scandinavian states: democratic representative parliaments emerged early, and individual rights have long been carefully protected.

Because of their higher-latitude location, the five Nordic states also share generally common environmental conditions. Understandably, populations tend to be concentrated in the more southerly parts of each country. Although the warm waters of the Atlantic Ocean temper the Arctic cold and keep Norway's ports open year-round, their effect diminishes both northward and landward. Most important is Scandinavia's high mountain backbone that blocks air moving eastward across Norway and into Sweden. Not only does this highland limit the milder maritime belt to a narrow strip along the Norwegian coast, but by its elevation it also allows Arctic conditions to penetrate southward into the heart of the Scandinavian Peninsula. One cannot help speculating on what Western Europe's environment might have been if that Scandinavian backbone had continued southward through the Low Countries into France and Spain. At the same time, the lack of a physiographic barrier between Scandinavia and Western Europe has permitted airborne industrial pollution to become a major environmental hazard (see *box*).

Norway

Norway, with its long Atlantic coast, consists almost entirely of mountains whose valley soils have been stripped away by glaciation (Fig. 1–26). Only in the southeast around the capital of Oslo (800,000), in the southwest below Bergen, and on the west coast near Trondheim are there limited areas of good soil and agriculture (apart from tiny bottomland patches in fjorded valleys near the coast). But even though Norway came off second best in the division of land resources on the Scandinavian Peninsula, it has turned to the sea, which more than makes up for this deficiency. The Norwegian fishing industry is one of the world's largest: its fleets ply all the

ACID RAIN

One of the most discussed environmental perils of recent years is **acid rain**. It is caused by the considerable quantities of sulfur dioxide and nitrogen oxides released into the atmosphere as fossil fuels (coal, oil, natural gas) are burned. These pollutants combine with water vapor contained in the air to form dilute solutions of sulfuric and nitric acids, which are subsequently washed out of the atmosphere by rain or other types of precipitation such as fog and snow.

Although acid rain usually consists of relatively mild acids, they are sufficiently caustic to do great harm to certain natural *ecosystems* (the mutual interactions between groups of plant or animal organisms and their habitat). There is much evidence that this deposition of acid is causing lakes and streams to acidify (resulting in fish kills), forests to become stunted in their growth, and acid-sensitive crops to die in affected areas. In cities, the corrosion of buildings and monuments is both exacerbated and accelerated. To some extent, acid rain has always been present in certain humid environments, originating from such natural events as volcanic eruptions, forest fires, and even the bacterial decomposition of dead organisms. However, during the past century as the global Industrial Revolution has spread ever more widely, the destructive capabilities of natural acid rain have been greatly enhanced by human actions.

The geography of acid rain is most closely associated with patterns of industrial concentration and middle- to long-distance wind flows. The highest densities of coal- and oil-burning are associated with large concentrations of heavy manufacturing, such as those already discussed for Britain and Western Europe. As these industrial areas began to experience increasingly severe air pollution problems in the second half of the twentieth century, many nations (including the United States in 1970) enacted environmental legislation to establish minimal clean-air standards for the first time. For industry, the easiest solution has often been the construction of very high smokestacks (1000 feet [300 meters] or higher is now quite common) in order to disperse pollutants away from source areas via higher-level winds. These longer-distance winds have been effective as transporters, with the result, of course, that more distant areas have become the dumping grounds for sulfur and nitrogen-oxide wastes. Regional windflows all too frequently steer these acid rain ingredients to wilderness areas, where livelihoods depend heavily on tourism, agriculture, fishing, and forestry. Where such airborne pollution crosses international borders, political problems develop (notably between the United States and Canada) and can be expected to intensify in the future.

The spatial distribution of acid rain within Europe offers a classical demonstration of all this. As Fig. 1–25 shows, high-sulfur-emission sources are located in the major manufacturing complexes of England, France, Belgium, the Germanys, Poland, and Czechoslovakia. The map also indicates that prevailing winds from these areas converge northward toward Scandinavia, with a pariculanly severe acid rain crisis occurring in southern Norway. Lake acidity

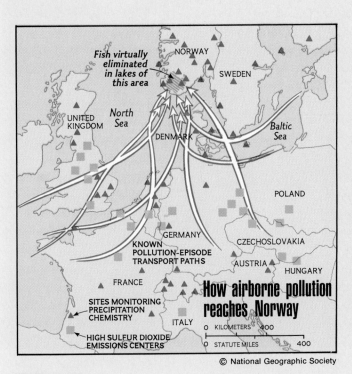

Fish virtually
eliminated
in lakes of
this area

NORWAY

SWEDEN

North
Sea

Baltic
Sea

UNITED
KINGDOM

DENMARK

POLAND

KNOWN
POLLUTION-EPISODE
TRANSPORT PATHS

GERMANY

CZECHOSLOVAKIA

AUSTRIA

HUNGARY

FRANCE

**How airborne pollution
reaches Norway**

SITES MONITORING
PRECIPITATION
CHEMISTRY

ITALY

0   KILOMETERS   400

0   STATUTE MILES   400

HIGH SULFUR DIOXIDE
EMISSIONS CENTERS

© National Geographic Society

FIGURE 1–25

here is already in the moderately caustic 4-to-5 range on the pH scale of 0 to 14 (7 is neutral), and most fish species, the phytoplankton they feed on, and numerous aquatic plants have been obliterated.

Scandinavia is not alone in these environmental problems. Central Europe is also severely affected, and its forest ecosystems are now so badly damaged by acid rain that vast woodland areas are diseased and dying. More than 25 percent of Austria's forests are threatened; in Switzerland, the proportion nears 40 percent. The worst devastation, however, has occurred in West Germany, where over 50 percent of the woodlands are afflicted and entire forests (such as the once-lovely Black Forest in the country's southwestern corner) are all but decimated. Government agencies have belatedly recognized this ecological disaster, but their proposed remedies may well be a case of too little too late. In the Alps, deforestation has caused particular alarm: besides the visual blight, the destruction of trees—nature's own barriers against landslides and avalanches—jeopardizes hundreds of villages located on the steeper slopes (see Fig. 1–23).

oceans, especially the highly productive fishing grounds that lie close to Norway's own waters. Another maritime pursuit is the Norwegian merchant marine (again one of the largest anywhere), which is mostly hired to transport goods between other trading countries. But what is ultimately likely to prove to be the most important benefit of the sea remained unrealized until the 1970s, when Norway's share of the North Sea floor began to yield undreamed-of wealth in petroleum and natural gas, heralding the dawn of a bright new economic future.

Sweden

Norway's eastern neighbor on the Scandinavian Peninsula is also favored in many respects. Of Sweden's two agricultural zones, the leading one lies at the southernmost end of the country, just across the Kattegat Straits from Denmark. This area resembles Denmark agriculturally, except that more grain crops, especially wheat, are raised here; Malmö (760,000) is the area's chief metropolis and agricultural service center. Sweden's other farming zone lies astride a line drawn southwest from the capital, Stockholm (1.6 million), to Göteborg (950,000). Here dairying is the main activity, but farming is overshadowed by industry. Swedish manufacturing, unlike that of certain Western European countries, is scattered through dozens of small and medium-sized towns.

Unlike Denmark and Norway, Sweden has the resources to sustain such industries. For a long time, Sweden served as an exporter of raw or semi-finished materials to industrializing countries; but increasingly, the Swedes are making finished products themselves, honing skills and specializing much the way the Swiss have

FIGURE 1–26 A glacial trough containing a deep inlet of the North Sea on Norway's fjorded coast.

done. The list of its well-known products is now quite long, and includes furniture, stainless steel, automobiles, electronic goods, and glassware. And there is a great deal more, much of it based on relatively minor local resources. Apart from iron, of which there is a great deal in the area around Kiruna north of the Arctic Circle, the Swedes possess copper, lead, zinc, manganese, and even some silver and gold.

Finland

In Finland, the alternatives are fewer. Most of the country is too cold and its glacial soils too thin to sustain permanent agriculture. Where farming is possible, mostly along the warmer coasts of the south and southwest, the objective is self-sufficiency rather than export. Known mineral deposits are few, but the Finns still have suc-

ceeded in translating their limited opportunities into a healthy economic situation; once again, the skills of the productive population play a major role. The domestic market sustains a textile industry centered at Tampere, and metal industries for shipbuilding and locally needed machinery at Turku and the capital, Helsinki (975,000). Elsewhere, Finland's extensive forests have long supported pulp, paper, and timber production (an activity also widespread in Norway and Sweden); wood and wood products annually account for over half of the country's exports.

Iceland

The westernmost of Norden's (and Europe's) countries is Iceland, a hunk of active volcanic rock that emerges above the surface of the frigid waters of the North Atlantic just south of the Arctic Circle. Its

population has the dimensions of a microstate (263,000), and about half the people are concentrated in the capital, Reykjavik. Iceland shares with Scandinavia its difficulties of terrain and climate (even more severe here), and it also has ethnic affinities with continental Norden (its settlers came from Norway and Denmark). Not until 1918 did Iceland gain its modern-era autonomy; only in 1944 were all remaining political ties to Denmark renounced and a republic finally established. Icelandic economic geography is almost entirely oriented to the sea, with fish products yearly accounting for over 80 percent of the country's exports.

Denmark

Denmark is territorially the smallest of the countries of Northern Europe, but its population of 5.1 million ranks it second in the region after Sweden. Consisting of the Jutland Peninsula and adjacent islands between Scandinavia and Western Europe, Denmark has a comparatively mild, moist climate. It also has level land and soils good enough to sustain intensive agriculture over 75 percent of its area. Denmark exports dairy products, meats, poultry, and eggs, mainly to its chief trading partners, the United Kingdom and West Germany. It has been estimated that Denmark's farm production could feed nearly 20 million people annually.

Denmark's capital, Copenhagen (1.7 million), is also Norden's largest metropolitan center. It has long been the place where large quantities of goods are collected, stored, and transshipped, because it lies at the *break-of-bulk* point where many oceangoing vessels are prevented from entering the shallow Baltic Sea. Conversely, ships with smaller tonnages ply the Baltic and bring their cargoes to these collecting stations. Thus, Copenhagen is an *entrepôt* whose transfer functions maintain the city's position as the lower Baltic's leading port.

Mediterranean Europe

From Northern Europe, we turn to the four Mediterranean countries of the south: Greece, Italy, Spain, and Portugal (Fig. 1–27). With this shift from near-polar Europe into near-tropical Europe, it is reasonable to expect strong contrasts; indeed, there are many—but there also are similarities. Once again, we are dealing with peninsulas: two are occupied singly by Greece and Italy, and the third—the Iberian Peninsula—jointly by Spain and Portugal. Once again, there is effective separation from the Western European core. Greece lies at the southern end of Eastern

Europe and has the sea and the Balkans between it and the rest of the realm; Iberia lies separated from France by the Pyrenees, which through history has proven to be a major mountain barrier; and then there is Italy. Southern Italy lies far removed from Western Europe, but the north is situated very close to it, pressing against France, Switzerland, and Austria. For many centuries, northern Italy was in close contact with Western Europe and developed less as a Mediterranean area than as a part of the core area of Western Europe.

The Scandinavian countries share a common cultural heritage, and so do the countries of Mediterranean Europe. Firm interconnections were established early by the Greeks and Romans. This unity, as we know, did not last. New political arrangements replaced the old, and the Romanic language differentiated into Portuguese, Spanish, and Italian. In Greece, the Roman tide was resisted to a surprising extent, but the underlying shared legacy remains strong to this day.

Like Norden, Mediterranean Europe lies largely within a single climatic region, and from one end to the other the environmental opportunities and problems are similar. The opportunities lie in the warmth of the near-tropical location, and the problems are largely related to the moisture supply, especially its limited quantity and the dryness of the warmest months. In topography and relief, too, Mediterranean countries share similar conditions—conditions that would look quite familiar to a Norwegian or a Swede. Much of Mediterranean Europe is upland country with steep slopes and poor, thin, rocky soils. For its agricultural productivity, the region largely depends on river valleys and coastal lowlands. As with most rules there are exceptions, as

in northern Italy and northern and interior Spain, but generally the typical Mediterranean environment prevails.

Neither is Mediterranean Europe better endowed with natural resources than Scandinavia. Both Greece and Italy are deficient in coal and iron ore (Italian industry relies on massive imports); only in northwestern Iberia are there sizable deposits of these commodities positioned close to each other. But Spain, best endowed of all Mediterranean countries, has only lightly industrialized, exporting its minerals to Europe's coreland on the scale of an underdeveloped, raw-material-supplying country. Woodland resources in Southern Europe also offer a discouraging picture: unlike Scandinavia, the Mediterranean region stands largely denuded, its once extensive forests obliterated to supply fuel, building materials, and additional agricultural space. With regional fuels in generally short supply, hydroelectric power opportunities have been developed, but to a lesser extent than in Nordic Europe. Only Italy—particularly in the north near the Alps and their dependable precipitation—has had much success. Elsewhere, given the seasonal and rather low rainfall, there are frequent water shortages when streams run dry and water levels behind dams go down.

From what we know of Mediterranean dependence on agriculture, the historical geography of trade routes and urban growth, and the general topography of Southern Europe, we can fairly well predict the distribution of the region's 118 million inhabitants. In Italy, no fewer than 25 million of the country's 57 million people are concentrated in and near the basin of the Po River in the north. Elsewhere, large population clusters also exist in coastal lowlands and river valleys, on both flanks of the

Appennine Mountain spine, all the way from Genoa to Sicily. In Greece, today as in ancient times, densely settled coastal plains are separated from one another by relatively empty highlands; the largest agglomerations are in the lowland dominated by Athens and on the western side of the Peloponnesus Peninsula. Both Spain and Portugal also contain high-density settlement in their coastal lowlands, although the interior Meseta Plateau is more hospitable than Greece's rocky uplands.

Thus, Mediterranean population distribution is dominated by peripheral location, by heavy concentrations in productive areas (usually coastal and riverine lowlands), and by a varying degree of isolation on the part of these clusters. Spatial interaction between central and southern Greece, between the east and west coasts of Italy, and between Atlantic and Mediterranean Iberia is, therefore, not always effective; terrain and distance are the obstacles. Although it is difficult to say exactly what constitutes overpopulation, there has long been excessive population pressure on the land and resources in many parts of the Mediterranean Basin. Perhaps the sharpest contrast between Nordic and Mediterranean Europe lies in the living standards of the people: while economic specialization, limited population growth, and government policies to distribute wealth equitably have produced standards of living in most of Scandinavia that are at least equal to those of the European core, much of Mediterranean Europe has lagged well behind.

Greece

This is perhaps the least favorably endowed of the Mediterranean countries: less than a third of its area is presently capable of supporting cultivation, under an acre

FIGURE 1–27

per person when Greece's extremely high rural population density is factored in. Thus, many Greek farmers are forced to engage in near-subsistence agriculture, because their incomes are insufficient to buy modern farm machinery, fertilizer, and essential irrigation equipment. Yet, agriculture is Greece's mainstay, for industrial opportunities are few (Greeks have long emigrated to seek work in the industrial centers of Western Europe). Greek farmers raise wheat and corn for the home market, and tobacco, cotton, and such typical Mediterranean produce as olives, grapes, citrus fruits, and figs for export.

Six out of ten Greeks live in cities and towns, with the urban hierarchy dominated by primate Athens (3.5 million). This capital city is also the commercial, cultural, and historic focus of Greece, and despite its limited industry,

Athens has grown to a size far beyond what would be expected for such a relatively poor, agrarian country. With its outport of Piraeus, it stands at the head of the Aegean Sea; Athens also has a major international airport. Moreover, thanks to its heritage of ancient structures, the Greek capital is one of the Mediterranean's major tourist attractions—despite a serious air pollution problem that threatens the fragile monuments of antiquity. Yet another source of national income is Greece's large merchant marine fleet that competes for cargoes wherever and whenever they need to be hauled.

Spain and Portugal

At the other end of the Mediterranean, the Iberian Peninsula is less restrictive in the opportunities it presents for development. Iberia is much larger than Greece, and raw

materials are in far more plentiful supply. But the countryside is overpopulated, and one price Spain has paid for its slow industrial development was that its population explosion occurred mainly in rural areas where pressures already were high. Land, especially in the north, was heavily divided; farms grew ever smaller and less efficient; agriculture expanded onto poorer soils even though their productivity was bound to be low. These problems were also compounded by the slow breakup of the entrenched *latifundia* land-tenure system, under which massive numbers of tenant farmers labored on huge estates controlled by wealthy (often absentee) landowners. These, then, are some of the reasons why Southern Europe's crop yields are frequently more than 50 percent lower than those of Western Europe and why so many Mediterranean

farmers are caught up in long-term cycles of poverty.

Spain's major industrial area is located in the far northeast in Catalonia, centered on metropolitan Barcelona (4.2 million). Here, it is neither a favorable location nor a rich local resource base that has stimulated industrialization; rather, the vigorous and progressive outlook of the Catalans has produced this development, aided by a strong regional identity that has enabled these people to forge ahead of the rest of Spain. The Basque provinces around the north-central city of Bilbao (1.0 million) constitute another manufacturing zone, particularly for metal and machine industries. To the west, people have agglomerated in Galicia near the coal-mining areas of the Cantabrian Mountains. Two other sizable population clusters are found in Andalusia in the south, where the

lowland of the Guadalquivir River opens into the Atlantic, and in the center of the country, around the capital of Madrid (4.5 million). In Portugal, which possesses an Atlantic but no Mediterranean coast, the majority of the population is distributed along the coastal lowlands rather than on the Iberian Plateau; Lisbon (2.4 million) and Porto (1.1 million) are the leading urban centers.

Although Spain in recent decades has lagged behind Western Europe, its economy is now taking a spectacular new direction. The triggering event occurred in 1986 when Spain (along with Portugal) joined the Common Market, thereby signaling a new determination to close the development gap between itself and the European core. The response from north of the Pyrenees was immediate: by 1988 the Spanish were being inundated with so much foreign investment that they recorded Europe's fastest economic growth rate, launching a boom that persists into the early 1990s. Among the major attractions to American and Japanese as well as European investors are Spain's stable democracy, market-oriented government policies, tax incentives, low wage levels, and an increasingly affluent domestic marketplace. Spain's climatic amenities are also playing a role, not only in the rush to build new factories, but also in the steady upsurge of tourism and the construction of recreational facilities, condominiums, and retirement communities. This, of course, sounds very much like recent growth in the U.S. Sunbelt, and the analogy is appropriate. Spaniards already talk about their country as the "Florida" and "California" of Europe, as resorts, high-technology industry, and ultramodern fruit and vegetable farms swiftly multiply, especially along the Mediterranean coast. However, as in the United States, this kind of growth has brought problems of traffic congestion, infrastructure overload, and skyrocketing land prices. A more serious by-product is the spottiness of new development. Many places are booming, but vast areas are untouched, and Spain's unemployment rate still hovers above 15 percent. Thus, dark shadows remain interspersed with the bright sunshine as Spain continues down the long road toward real economic parity with Western Europe.

Italy

Much of what has been said about Mediterranean habitats and economies applies to southern Italy and the islands of Sicily and Sardinia. But in many ways Italy is two countries, not one. While the north has had the opportunities—and the advantages of proximity to the European core—to sustain development in the style of Western Europe, the south (or Mezzogiorno) has for centuries been a laggard and stagnant region. Together, north and south count over 57 million inhabitants (nearly equal to Spain, Portugal, and Greece combined), bound by the capital city of Rome (3.0 million), which lies in the transition zone between the two contrasting Italian halves. By every measure, Italy is Mediterranean Europe's leading state—in the permanence of its contributions to Western culture, in the productivity of its industry and agriculture, and in its overall living standard.

Today, the core area of Italy has shifted northward from the vicinity of historic Rome to the area known as Lombardy in the central Po Basin. This broad lowland of the Po and its tributary rivers, lying between the Appennines and

FIGURE 1–28 Part of outer Milan's highly-diversified, ultramodern industrial landscape. Although small plants predominate, their extraordinarily efficient manufacturing operations have earned them a prestigious international reputation.

the Alps, is the largest in Mediterranean Europe. As the climatic map (Fig. I–10) shows, the Po Plain has an almost wholly non-Mediterranean regime, with a much more even rainfall distribution throughout the year. Certainly, this area possesses superior advantages for agriculture, but what marks it today is the greatest development of manufacturing in Southern Europe. It is all a legacy of the early period of contact with northwestern Europe via nearby transalpine trading routes, an old exchange vigorously renewed when the stimulus of the Industrial Revolution came. Hydroelectric power from Alpine and Appennine slopes is the only local resource, other than a large and skilled labor force. But northern Italy imports huge quantities of iron ore and coal, and it now ranks as Europe's largest steel producer after West Germany. Italian industry seeks to create precision equipment, in which a minimum of metal and a maximum of skill produce the desired products and revenues. The metal industries are led by the manufacturing of automobiles, for which Turin (1.7 million) is the chief center. At nearby Genoa (950,000), Italy's leading port, there is a major shipbuilding industry.

The principal metropolis of northern Italy is Milan (4.8 million), located in the heart of Lombardy. This is not only Italy's financial, banking, and service-industry headquarters, but also the leading manufacturing center in all of Mediterranean Europe. No Italian city rivals Milan's range of industries—from farm equipment to television sets, from fine silk to pharmaceuticals, from chinaware to shoes (Fig. 1–28). Factories here tend to be small operations, but they are so well managed and so automated that they have acquired a global reputation for efficiency, adaptability, and speed.

In fact, the enterprising Milanese are so productive that they account for fully 30 percent of Italy's national income with only 8 percent of its total population. This performance has not only made Milan one of Europe's wealthiest cities, but lately has helped propel the Italian economy past that of the United Kingdom; by century's end, Italy will have passed France as well, and is likely to be challenging West Germany for the continent's top position.

Eastern Europe

Between the might of the Soviet Union and the wealth of industrialized Western Europe lies a region of transition and fragmentation—Eastern Europe (Fig. 1–29). Its position with reference to the major cultural influences in this part of the world at once explains a great deal. To to the west lie Germanic and Latin cultures, politically represented by Germany and Italy; to the east looms the Slavic culture realm, dominated by Russia; and to the south lie Greece and Turkey, whose impact has also been felt strongly when the Byzantine and Ottoman empires took in sizable portions of this ever-unstable region. Throughout Eastern Europe's modern history, these strong external forces have collided here, resulting in endless invasions, territorial give-and-take, and political fragmentation. Such a fractured, constantly stressed area is often called a **shatter belt**, and is further reflected by Eastern Europe's especially complex human mosaic of linguistic, religious, and ethnic groups.

As defined here, Eastern Europe consists of seven states: Poland, the largest in every way; Czechoslovakia and Hungary, both landlocked; Romania and Bulgaria, facing the Black Sea;

and Yugoslavia and Albania along the Adriatic Sea. These countries form the fifth and easternmost regional unit of Europe, whose eastern boundary also forms the eastern limit of the European geographic realm. This is Europe at its most continental, most agrarian, and most underdeveloped. Overall, it was not as well endowed as Western Europe or the U.S.S.R. with the essentials for industrialization, and it was the most remote of all parts of Europe from the sources of those innovations that brought about the Industrial Revolution. For four and a half decades following World War II, Eastern Europe looked eastward for directions in its political and economic development. The Soviet Union gained hegemony here in the mid-1940s—thanks to the presence of the Red Army as well as the cooperation of local communist parties—and communist forms of totalitarian control, resource use, and political organization prevailed. Although the countries of Eastern Europe were not remade into Soviet Socialist Republics (as were the former Baltic states of Estonia, Latvia, and Lithuania), they did become satellites of the Soviet Union and were drawn completely into the Soviet economic and military spheres. All that, of course, changed dramatically at the end of the 1980s.

The collapse of the Soviet-dominated communist system in the autumn of 1989 is but the latest reminder that external power has never succeeded in unifying the shatter belt that is Eastern Europe. Within the region, even the framework of political boundaries (which largely evolved from the 1919 Paris Peace Conference that concluded World War I) has not eliminated its internal ethnic problems. In fact, it really was impossible to arrive at any set of boundaries that would totally

FIGURE 1–29

satisfy all the peoples involved. So intricate is the ethnic patchwork that different social groups simply had to be joined together. At the same time, people who had affinities with each other were separated by the new international borders, and every country in Eastern Europe as constituted in 1919 found itself with sizable minorities to govern. Such fragmentation and division was especially pronounced in the southern portion of the region, from which emanated a new term—**balkanization** (see *box*).

The internal distribution of minorities was important too, particularly when those groups were located near a border. In such cases, adjacent states often began to call for a transfer of authority over these populations based on ethnic, historical, or some other grounds. To cite just one example among many, Transylvania, the eastern part of the Hungarian Basin, was severed from Hungary and attached to Romania—but Hungary openly laid claim to it (and still does today). When a certain state, through appeals to a regionally concentrated minority in an adjacent state, seeks to acquire the people and territory on the other side of the boundary, its action is termed **irredentism**. The term derives from the name of a political pressure group in northernmost Italy whose principal objective was the incorporation of *Italia Irredenta* (Unredeemed Italy), a neighboring part of the South Tyrol in Austria's Alps.

Eastern Europe between the world wars was plagued by problems of irredentism, but only after 1945 were some of them eliminated by further boundary revision (Fig. 1–29). Poland was literally shoved westward for well over 100 miles when the Soviet Union annexed much of the eastern half of that country. Then-Communist Poland, in turn, was compensated by attaching to it a large chunk of defeated Germany in the west. The U.S.S.R. also took for itself the easternmost portions of Czechoslovakia and Romania. The Soviets then further simplified the regional ethnic situation through their postwar control over East Germany by initiating the migration, voluntary as well as forced, of ethnic Germans from several East European countries (where they formed sizable prewar minorities) to this part of their homeland. Elsewhere, similar migrations have carried Hungarians from Czechoslovakia and Romania to Hungary, Ukrainians and Belorussians from Poland to the Soviet Union, and Bulgars from Romania to Bulgaria. This is not to suggest that Eastern Europe's minority problems are solved, and with the umbrella of Soviet dominance removed, irredentist prob-

BALKANIZATION

The southern half of Eastern Europe is sometimes referred to as the Balkans, or the Balkan Peninsula. This refers to the triangular landmass whose points are at the tip of the Greek Peloponnesus, the head of the Adriatic Sea, and the northwestern corner of the Black Sea. The name comes from a mountain range in Bulgaria, but it has also become a concept, born of the reputation for division and fragmentation in this Eastern European subregion. Any dictionary carries a definition of the term *balkanize*—for example, "to break up (as a region) into smaller and often hostile units," to quote Webster.

Certainly, division and hostility have been outstanding characteristics of Eastern Europe. Its peoples—Poles, Czechs, Slovaks, Magyars, Bulgars, Romanians, Slovenes, Croats, Serbs, and Albanians—have often fought each other; at other times, they have been in conflict with forces from outside the region. Various empires have tried to incorporate all or parts of Eastern Europe in their domains. In the end, however, none was completely successful—as the most recent failure of the 45-year-long Soviet effort underscores. Countless boundary shifts have taken place: time and again, local minorities have risen in revolt against their rulers. Culturally, Eastern Europe is divided by strong internal differences, expressed in language and religion, and sustained by intense nationalisms. With the removal of Soviet hegemony over the region in 1989, these forces can be expected to come to the forefront in the 1990s.

lems are likely to surface again in the 1990s.

Poland and Czechoslovakia

Several geographic bases exist to separate Eastern Europe's two northernmost countries from the others. Poland, facing the Baltic Sea, is the region's largest state territorially and demographically (39 million); Czechoslovakia, mountainous and surrounded, is less than half as large (16 million), but it is ahead in economic development. Poland's traditionally agrarian economy began to yield to industrialization only after World War II, when centralized communist planning decreed the exploitation of the country's industrial opportunities. Czechoslovakia, on the other hand, has been more directly exposed to in-

fluences from Western Europe; it lay more directly in the paths of commercial exchange and industrial development along Europe's east-west axis, and the Czechs had longstanding ties with the West.

Poland

This country lies largely in the North European Lowland, so that there are few internal barriers to effective communication. In the south, along the Czech border, Poland shares the foothills of the Sudeten and Carpathian mountains where the raw materials necessary for heavy industry are located. Warsaw, the capital and primate city (2.1 million), lies close to the center of Poland in a productive agricultural area near the head of navigation on the Vistula River and has made itself the focus of a radiating network of transportation routes that reaches all parts of the country. The leading manufacturing complex is found in Silesia in southern Poland, an industrial crescent anchored by the three cities of Krakow, Katowice, and Wroclaw (combined population: 5.8 million).

The Polish south also has a highly fertile black soil belt—broadening eastward into the Soviet Ukraine—that sustains intensive farming, with wheat as the major crop. In central Poland, thin and rocky glacial soils support only rye and potato crops; the Baltic rim possesses such poor soils that cultivation must give way to pastureland and moors. Nearly a half-century of communism, however, has taken its toll on Polish farming, largely because of repeated failures to collectivize agriculture. The Soviet pattern of great investment and support for industry, usually at the expense of badly needed agricultural progress, is evident throughout Eastern Europe, and it will take the region years to undo this economic-geographic imbalance.

Czechoslovakia

This southern neighbor shares with Poland the Silesian industrial region, with local sources of raw materials lying astride the gap between the Sudeten Mountains and the Tatra extension of the Carpathians. This area is referred to as Moravia, and the gap as the Moravian Gate, because of its vital importance as a low-lying passageway between the Danube Valley and the North European Plain. Located midway between Bohemia in the west and Slovakia in the east, this growing industrial area is of increasing importance, emphasizing the production of steel, metals fabrication, and chemicals.

The western sector of Czechoslovakia, focused on the mountain-enclosed Bohemian Basin, has always been an important core

area in Eastern Europe, cosmopolitan in character and Western in its exposure and development. With its Elbe River outlet, this westward orientation was maintained for centuries, but in 1919 the Slovaks were attached to Czech-dominated Bohemia. The Slovak eastern portion of Czechoslovakia, mountainous and rugged, is much more representative of Eastern Europe than is the Czech portion of the state. Unlike the more advanced Bohemian-Moravian west, Slovakia is mostly rural and underdeveloped, a situation that persisted under postwar Soviet dominance. Prague, the capital (1.4 million), is not a centrally positioned headquarters, but lies in the far west in the heart of Bohemia. Nonetheless, it is in every way Czechoslovakia's primate city. Prague, itself a major manufacturing center, lies near the upper Elbe River at the middle of the country's greatest concentration of wealth in the Bohemian Basin. The surrounding mountains contain many valleys in which lie small industrial towns that specialize, Swiss-style, in numerous high-quality manufactures. In Eastern Europe, the Czechs have always led in technology and engineering skills, and their products, from automobiles to textiles, find their way to Western as well as communist markets.

Hungary and the Balkan Peninsula

The five countries that make up the southern tier of Eastern Europe seem almost to have been laid out at random, with little regard for the potential unifying features of this part of the region. The comparative orderliness of Poland and Czechoslovakia, one a land of uniform plains and the other a rather well-defined mountain state, is lost here, especially in the Bal-

kans, where imposing highlands abound, large and small basins are sharply differentiated and separated, and the ethnic situation is even more confused than elsewhere in Eastern Europe. At least the Slovaks had this in common with the Czechs: they were neither Poles nor Hungarians and their attachment to the Czech state seemed a reasonable solution. But in the Balkans, such solutions have been much harder to come by (as we saw in the *box* on p. 107).

The great Balkan unifier that might have been is the Danube River, which originates in southern Germany, traverses Austria, and then crosses central Eastern Europe, forming first the Czechoslovakia-Hungary border, then the Yugoslavia-Romania boundary, and finally the Romania-Bulgaria border. It is indeed anomalous that such a great transport artery, which could form the focus for a large region, is instead a constant dividing line. After it emerges from the Austrian Alps, the Danube flows across the Hungarian Basin, which, although largely occupied by Hungary, is shared also by Yugoslavia, Romania, and Czechoslovakia. Then it squeezes through the Transylvanian Alps at the Iron Gate (near Orsova), and into the basin that forms its lower course, which is shared by Romania and Bulgaria. No other river in the world touches so many countries, but the Danube has not been a regional bond.

In a very general way, it can be argued that progress and development in Eastern Europe decline from west to east and also from north to south. In both Poland and Czechoslovakia, the western parts of the country are the most productive. Hungary contains the largest and the most productive share of the Hungarian Basin, and is better off than Yugoslavia immediately to the south. On the

FIGURE 1–30 The graceful arc of the Danube flowing amid the architectural splendors of Hungary's capital city, Budapest.

whole, Yugoslavia is far ahead of its small southern neighbor, Albania. And on the eastern side of the Balkan Peninsula, Romania has the advantage over Bulgaria to the south. Thus, Hungary is southeastern Europe's leading state in several respects; given its pivotal position in Eastern Europe, this is not surprising.

Hungary

The Hungarians (or Magyars) themselves form a minority in the Balkan subregion, since they are neither of Slavic nor of Germanic stock. They are a people of Asian origin, who arrived in Hungarian Basin in the ninth century A.D. Ever since, they have held on to their fertile lowland, retaining their cultural and linguistic identity, al-

though at times losing their political sovereignty. The capital, Budapest (2.3 million), is today more than ten times as large as the next-ranking Hungarian city, a reflection of the rural character of the country. With its Danube port and its extensive industrial development, its cultural distinctiveness, and its nodal location within the state, Budapest epitomizes the general situation in Eastern Europe, where urbanization has been slow and where the capital city is normally the only urban center of any magnitude (Fig. 1–30).

Hungary's rural economy has not been without problems. After the country was delimited in 1919, large estates began to be carved into small holdings. Productivity rose, but soon World War II and

its attendant destruction, especially of livestock, set the rural economy back. In the postwar era, the Soviets coveted fertile Hungary as a potential breadbasket for Eastern Europe, but a long struggle to improve food output significantly by collectivized farming made little headway. Through it all, the country remained agriculturally self-sufficient (and the region's only food exporter), with its harvests of wheat, corn, barley, oats, and rye. Today, with the constraints of communism removed, Hungary may finally be on its way to fully realizing its agricultural opportunities. Industrially, too, Hungary has not yet been able to take full advantage of its potential. There is some coal, and iron ore can be brought in via the Danube for steel production; this has been done in quantity only recently. For a long time, Hungary, like so many countries in the developing world, was exporting millions of tons of raw materials—particularly bauxite (aluminum ore), with which it is abundantly endowed—to other areas.

Despite these limitations, Hungary (population: 10.5 million) has become Eastern Europe's most successful country, enjoying a living standard well above the region's norm. Ironically, the economic growth sustaining this trend originated in the aftermath of the second Soviet invasion (the first occurred in 1944) that crushed the short-lived revolution of 1956. By convincing fellow citizens to think of the advantages of accommodating to renewed Soviet domination, Hungary's leaders were quickly able to stabilize the country; the grateful Soviets soon quietly began to grant Hungary permission to experiment in free markets, private initiatives, and greater personal freedoms. This balancing of communism and capitalism provided incentives for the resourceful Hungarians to make substantial progress—and began to set the stage for the historic breakthrough of mid-1989 that saw Hungary pave the way for Eastern Europe's remarkable exodus from the Soviet sphere.

Yugoslavia

Yugoslavia, south of Hungary, shares a part of the Danube Basin, and this northeastern area has become the country's heartland. Yugoslavia was also created after World War I, when Serbs, Croats, Slovenes, Macedonians, Montenegrons, and lesser ethnic groups were combined in an uneasy union under the royal house of Serbia. The monarchy collapsed in the chaos of World War II, but the country was resurrected as a federative socialist republic under the leadership of Marshal Tito in 1945. Tito, who had led the fight against the Germans, took Yugoslavia on a communist course, but when the Soviets tried to impose their directives on him, he asserted his country's independence from Moscow. Seeking a more balanced approach to its economic problems, Yugoslavia went slowly in its farm collectivization efforts while industry was guided by national planning.

Yugoslavia today is a federal state consisting of the six Socialist Republics of Slovenia, Croatia, Bosnia-Hercegovina, Montenegro, Macedonia, and Serbia (Serbia also contains two Autonomous Provinces, Vojvodina in the north and Kosovo in the south). But these republics, although created along ethnic lines, contain many minorities: Yugoslavia has Hungarians, Albanians, Slovaks, Bulgarians, Romanians, Italians, and even Turks within its borders. Approximately 36 percent of Yugoslavia's 24 million people are Serbian, and about 18 percent are Croat (or Croatian); less than 8 percent of the population is Slovene. Yugoslavia's politicogeographical problems can be discerned from Fig. 1–29: the complicated mixture of peoples, their unequal locational access to the country's productive opportunities, and regional inequalities in progress and development. Add to these residual nationalism and latent hostilities, and it is something of a miracle that Yugoslavia has survived so long with stability. That status quo, however, will undergo change in the 1990s. As the new decade opened, ethnic Albanians and Serbs were in open conflict in Kosovo; Slovenes and Croats were demanding greater autonomy to pursue their own cultural and political affairs; and fear of an uprising, particularly like the one that overthrew the Romanian government in 1989, was prompting serious talk about introducing a multiparty system that would terminate the dominance of communism.

Albania

If our generalization concerning declining development and southerly location in the Balkans holds true, then Albania should be even less developed than mountain-and-plateau Yugoslavia. And it is. With only 3.3 million people, Albania ranks last in Europe (excluding Luxembourg and Iceland) in both population and territorial size. Most Albanians eke a subsistence out of livestock herding and farming the one-seventh of this mountainous country that can be cultivated at all. Perhaps because of its poverty and the limited opportunities for progress (consisting of some petroleum exports, tobacco cultivation, and chrome ore extraction), Albania turned to Maoist China for ideological as well as material support. Certainly Albania was worth more to China in this context than it was to the Soviet Union, which already controlled much of Eastern Europe. Today, while the region transforms all around it, Albania has cautiously introduced a limited program of political reform that could gradually end decades of repression and regional isolation.

Romania

The region's two Black Sea states, Romania and Bulgaria, also confirm the southward lag in progress and development. Romania is both richer and larger—more than twice the size of its southern neighbor in terms of territory as well as population (23 million versus 9 million). The potential advantages of Romania's compact shape are to some extent negated by the giant arch of the Carpathian and Transylvanian mountains (Fig. 1–5). To the south and east of this arch lie the plains and low hills of the Danube Basin, and to the west lies the Romanian share of the Hungarian Basin—with a sizable Hungarian population that forms one of Eastern Europe's ubiquitous minorities. As we have noted before, a mountain-plain association often provides mineral wealth, and Romania is well endowed with raw materials. Most notably, in the Carpathian foothills around Ploesti lies Europe's major continental oilfield, together with large deposits of natural gas; these have long enabled Romania to be a net exporter of fuels.

After World War II, the Romanian state found itself under Soviet domination, and thus began the familiar sequence of centralized planning, further agrarian reform involving collectivization, an emphasis on industrialization, and accelerated urbanization. But progress was not that spectacular, and despite Romania's considerable domestic resources, industrial development has largely been geared to local markets. Increasingly, between 1967 and 1989, Romania came under the dictatorial grip of its brutal Communist Party chief, Nicolae Ceausescu, who squandered the country's meager wealth, nurtured a bizarre personality cult, and forced most citizens to suffer miserable living conditions. Not surprisingly, when the revolution came at the end of 1989, it was Eastern Europe's most violent by far (Fig. 1–31).

Bulgaria

A Bulgarian state did not appear until 1878, emerging from a long history of Turkish domination. Bulgaria is a mountain country, except for the northern Danube lowland and the plains of the Maritsa River, which to the south forms the boundary between Greece and Turkey. Even the capital, Sofia (1.2 million), is located in the far interior, away from the plains, the rivers, and the Black Sea; the highlands were used as protection for the headquarters of the weak embryo state, and it has remained here. The east-west range that forms Bulgaria's backbone and separates the Danube and Maritsa basins carries the subregion's name—the Balkan Mountains. Much of the rest of Bulgaria is mountainous as well, and its large rural population lives clustered in basins and valleys separated by rough terrain. The Turkish period eliminated the wealthy landowners, so that Bulgarian farms tend to be small gardens, carefully cultivated and quite productive. These garden plots survive today, although collectivization has been carried further here than in any other Eastern European country. On the bigger collective farms, the domestic staples of wheat, corn, barley, and rye are grown; but the smaller gardens produce vegetables, olives, plums, grapes, and other fruits (reminding us that we are again approaching the Mediterranean).

At the time of transition to Soviet domination in the mid-1940s, Bulgaria was in a different position from its Eastern European neighbors, Romania and Yugoslavia. It was poorer than both, and it had still less in terms of agricultural and industrial resources, but there was a lengthy history of positive association with the Russians. The Bulgars are also more Slavicized than the Romanians, and in the postwar era they proved to be the U.S.S.R.'s most loyal East European satellite. Typically, Bulgaria's response was minimal when political and economic reforms began to sweep across the region in 1989. Nevertheless, change has begun and is likely to continue now that it is clear the Soviets will not stand in the way of the transformation of their once tightly controlled Eastern European bloc.

FIGURE 1–31   The climax of Eastern Europe's recent emergence from the communist era: fires rage at the presidential palace in the center of Romania's capital, Bucharest, on December 22, 1989.

EUROPEAN UNIFICATION

Individually, no European country constitutes a power of world stature comparable to the United States or the Soviet Union. But in combination, Europe's component states would indeed have world-power status. Although Europeans have long recognized the potential advantages of unification, prior to the twentieth century only the Romans had succeeded in unifying Italy and Iberia, Greece and Britain. In the aftermath of World War I, the states of Europe met to consider the possibility of Pan-European unification, but their aspirations soon were dashed by the events of the 1930s and World War II. Since 1945, however, Europe has made unprecedented progress toward economic integration and political unification (see *box*). There is, as yet, no United States of Europe, but critical first steps have been taken, and the push toward still stronger coordination continues.

It all began during World War II with Benelux, as we noted earlier, and this first move was followed by the infusion of U.S. Marshall Plan aid (1948–1952) and the creation of the Organization for European Economic Cooperation (OEEC). An important principle of **supranationalism** (international cooperation involving the voluntary participation of three or more nations in an economic, political, or cultural association) is that political integration, if it is ever to occur, follows rather than precedes economic cooperation. In this way, the infusion of Marshall Plan funds, by generating the need for an international European economic-administrative structure, also promoted coordination in the political sphere. As early as 1949, the Council of Europe convened for the first time; it met in a city that was proposed as the future capital for a united Europe—Strasbourg, located on the French bank of the Rhine River. For 30 years, the Council was little more than a forum for the exchange of

SUPRANATIONALISM IN EUROPE

1944, September	Benelux agreement signed.
1947, June	Marshall Plan proposed.
1948, January	Organization for European Economic Cooperation (OEEC) established.
1949, January	Council for Mutual Economic Assistance (COMECON) formed.
1949, May	Council of Europe created.
1951, April	European Coal and Steel Community (ECSC) Agreement signed (effective July 1952).
1957, March	Treaty of Rome signed, establishing European Economic Community (EEC) (effective January 1958).
	European Atomic Energy Community (Euratom) Treaty signed (effective January 1958).
1959, November	European Free Trade Association (EFTA) Treaty signed (effective May 1960).
1965, April	EEC-ECSC-Euratom Merger Treaty signed (effective July 1967).
1968, July	All customs duties removed for intra-EEC trade; a common external tariff is established.
1973, January	United Kingdom, Denmark, and Ireland become members of the EEC, thereby creating "The Nine."
1979, March	European Monetary Agreement linking the currencies of EC (EEC has been shortened) members goes into effect.
1979, June	First general elections to a European Parliament are held. First session of the 410-member legislature is held the following month in Strasbourg, France.
1981, January	Greece becomes member of the EC, thereby creating "The Ten."
1986, January	Spain and Portugal become members of the EC, thereby creating "The Twelve."
1986, February	Single European Act ratified, targeting the end of 1992 as the date for eliminating all barriers to the free movement of people, goods, services, and capital among the 12 EC member states.

ideas and opinions; representatives of its 18 member states had no executive authority, but their views did have an impact on national governments.

The year 1979 brought a momentous development when, for the first time, (410) representatives to a *European Parliament* were elected rather than appointed. European political parties played a role in the formation of this regional legislature, and its composition began to reflect more accurately the political mainstreams of non-Communist Europe. It was another step in the direction of a united Europe, and monthly meetings began to take place in Strasbourg. Subsequent elections in 1984 and 1989 enlarged this organization to 518 members, and the Parliament may yet evolve into a true European government.

Achievements in economic cooperation, though often difficult to come by, have been more consequential. Soon after the OEEC was established, the creation of the European Coal and Steel Community (ECSC) was proposed, with the principal objective of lifting the restrictions that impeded the flows of coal, iron ore, and steel among the mainland's six prime producers—France, West Germany, Italy, and the Benelux countries. By 1951 the ECSC became reality, and the mutual advantages of this arrangement soon encouraged the six participants to go further. Gradually, through negotiation and agreement, they enlarged their sphere of cooperation to include reductions of tariffs and a free flow of labor, capital, and non-steel commodities. Ultimately, in 1958 they launched the Common Market, the *European Economic Community (EEC)*, which is today known by its shortened name—*European Community (EC)*. This organization linked virtually all mainland Europe's core area, and its combined assets in terms of resources, skilled labor, and markets were enormous.

One very significant development related to the creation of the EC was the decision of the United Kingdom not to join it. This was itself a move based on supranational considerations: there was fear in Britain that participation

would damage evolving relationships with its (formerly colonial) Commonwealth countries, many of whom relied on the British as their chief trading partner. Thus, Britain stayed out of the EC, but it also made its own effort to create closer economic bonds within Europe. In 1959 it took the lead in establishing the *European Free Trade Association (EFTA)*, comprising, in addition to the UK, three Scandinavian countries (Sweden, Norway, and Denmark), the two Alpine states of Switzerland and Austria, and Portugal. This scattered group of countries, with their relatively small populations, limited resource bases, and restricted purchasing power, added up to something much less than the Common Market. They soon became known as the "Outer Seven," while the contiguous EC states of the mainland core were called the "Inner Six."

Within a few years of the creation of the EFTA, the United Kingdom changed its position on EC membership and decided to seek entry. Now, however, the political attitude on the continent had changed; France took an inflexible position during most of the 1960s, obstructing British participation despite the willingness of the other EC members to ratify admission. Finally, by the early 1970s, the path was cleared for Britain's entry, and the UK became a member of the European Community (together with Denmark and Ireland) in 1973. By 1978, the nine member states of the EC had voted in favor of the future admission of Greece, Spain, and Portugal, a process completed by 1986 (Fig. 1–32A). Alongside the EC, several other multinational associations function to foster European unity, including the EFTA (now with six member countries), Euratom (to develop peaceful uses of nuclear energy), the European Space Research Or-

ganization (to coordinate space programs), and, of course, the North Atlantic Treaty Organization (NATO), the cooperative military association that links many states west of the former Iron Curtain (Fig. 1–32B). Nonetheless, as the 1990s open, the spotlight is on the pacesetting European Community, which is attempting to achieve a major new breakthrough in economic unification by January 1, 1993.

When the EC reached its current level of 12 member states in the mid-1980s, Europe was in a slow-growth phase marked by high unemployment and rising international competition from the United States and Japan. Moreover, several countries were reintroducing border tariffs and trade quotas. To counteract those trends, it was decided in 1986 to give economic integration another big push by ratifying the Single European Act. This action adopted a report entitled "Completing the Internal Market," which called for nothing less than the creation of a single EC market for goods and financial services by the end of 1992. Specifically, the Act mandates the drafting and implementation of 279 directives toward achieving that ambitious goal. Accepting the challenge, the member countries resolved to turn this checklist into reality. However, despite their enthusiasm, it is unlikely that remaining trade barriers and financial restrictions can be eliminated before the deadline. By 1990, progress was evident on more than half of the Act's directives, but a number of critical elements in the plan had not come to fruition, including the centralized European Bank, a single EC currency, and a uniform tax system. Still, the accomplishment of all 279 directives is given a good chance for later in the 1990s, and with it the realization of the world's biggest single trading bloc (account-

EUROPEAN SUPRANATIONALISM: 1990

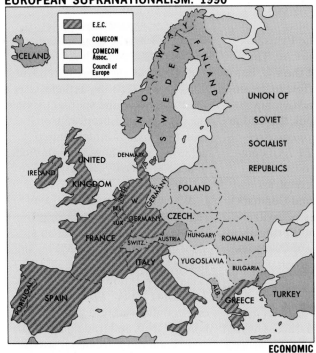

FIGURE 1–32A

FIGURE 1–32B

ing for about 40 percent of all global commerce) and marketplace (with 327 million consumers).

Europe is currently coming together in ways that could not have been foreseen even a few years ago. Mobility of people, goods, and capital is at an all-time high, and forces have been set into motion that will further intensify this interaction. International borders function less to divide than ever

before, and license plates bearing the exhortation "Europa!" can be seen on cars and trucks from Rotterdam to Rome. The swift demise of the Soviet satellite system in Eastern Europe removed major ideological divisions with the rest of the realm, raising the tantalizing possibility of wider regional integration between the EC and COMECON (the supranational Council for Mutual Economic Assistance

[Fig. 1–32A], established by the Soviets in their eastern bloc in 1949 to answer U.S. efforts at promoting postwar economic cooperation among the countries of non-Communist Europe). Europe in the past has transformed the world; today, by vigorously restructuring itself, Europe is loudly signaling its intention to again be one of the most influential of the developed geographic realms.

PRONUNCIATION GUIDE

Adriatic (ay-dree-*attic*)
Aegean (uh-*jee*-un)
Andalusia (unda-looh-*see*-uh)
Appennines (*app*-uh-nynz)
Ardennes (ar-*den*)
Barcelona (bar-suh-*loh*-nuh)
Basque (*bask*)
Bern (*bairn*)
Bohemia (boe-*hee*-mee-uh)
Bosnia-Hercegovina (*boz*-nee-uh hert-suh-go-*vee*-nuh)
Bucharest (*boo*-caressed)
Budapest (*booda*-pest)
Byzantine (*bizz*-un-teen)
Calais (kah-*lay*)
Campine (kam-*peen*)
Carpathians (kar-*pay*-theons)
Cartagena (karta-*hay*-nuh)
Ceausescu, Nicolae (chow-*shess*-koo, *nick*-oh-lye)
Celt(ic) (*kelt* [ick])
Charleroi (*sharl*-rwah)
Conurbation (konner-*bay*-shun)
Croat (*kroh*-ut)
Croatian (kroh-*ay*-shun)
Czechoslovakia (check-oh-slow-*vah*-kee-uh)
Danube (*dan*-yoob)
Défense, La (day-*fawss*, lah)
Devolution (*dee*-voh-looh-shun)
Duisburg (*dyoose*-boorg)
Ecumene (*eck*-yooh-meen)
Edinburgh (*eddin*-burruh)
Eire (*air*)
Elbe (*elb*)
Entrepôt (*ahntra*-poh)
Erzgebirge (*airtss*-guh-beer-guh)
Frankfurt (*frunk*-foort)
Gaelic (*gale*-ick)
Galicia (guh-*lee*-see-uh)
Garonne (guh-*ron*)

Geopolitik (gayo-polly-*teek*)
Göteborg (*goat*-uh-borg)
Guadalquivir (gwahddle-kee-*veer*)
Hegemony (heh-*jeh*-muh-nee)
Iberia (eye-*beery*-uh)
Ile de la Cité (*eel* duh-lah-see-*tay*)
Jutland (*jut*-lund)
Katowice (kott-uh-*veet*-suh)
Kattegat (*kat*-ih-gat)
Kiruna (kih-*roona*)
Kosovo (*kaw*-suh-voh)
Krakow (*krah*-kow)
Leicester (*less*-ter)
Lelystad (lelly-*shtahd*)
Le Mans (luh-*maw*)
Liège (lee-*ezh*)
Lingua franca (*leen*-gwuh *frunk*-uh)
Loire (luh-*wahr*)
Luxembourg (*lux*-em-borg)
Lyon (lee-*aw*)
Macedonia (massa-*doh*-nee-uh)
Madrid (muh-*drid*)
Malmö (*mal*-muh)
Maritsa (muh-*reet*-suh)
Marseille (mar-*say*)
Mercantilism (*merr*-ken-teel-ism)
Mèrida (*mare*-uh-duh)
Meseta (meh-*say*-tuh)
Meuse/Maas (*merz*/*mahss*)
Mezzogiorno (met-so-*jorr*-no)
Milan (muh-*lahn*)
Moravia (more-*ray*-vee-uh)
Norwich (*norritch*)
Oder-Neisse (oh-der *nice*)
Oise (*wahz*)
Peloponnesus (pelloh-puh-*nee*-zus)
Piraeus (puh-*ray*-uss)

Ploesti (ploh-*ess*-tee)
Prague (*prahg*)
Pyrenees (*peer*-unease)
Randstad (*rund*-stud)
Ratzel (*raht*-sull)
Reykjavik (*rake*-yah-veek)
Rhône-Saône (*roan* say-*oan*)
Romania (roh-*main*-yuh)
Romansch (roh-*mahn*-sh)
Ruhr (*roor*)
Seine (*senn*)
Sicily (*sih*-suh-lee)
Silesia (sye-*lee*-zhuh)
Slav(ic) (*slahv* [ick])
Slovakia (slow-*vah*-kee-uh)
Slovenia (slow-*vee*-nee-uh)
Sofia (*so*-fee-uh)
Strasbourg (*strahss*-boorg)
Sudeten (soo-*dayten*)
Suriname (soor-uh-*nahm*-uh)
Tatra (*taht*-truh)
Thames (*temz*)
Train à grand vitesse (*tran* ah-grawnd vee-*tess*)
Turku (*toor*-koo)
Ukraine (yoo-*crane*)
Ural (*yoor*-ull)
Vienna (vee-*enna*)
Vistula (*vist*-yulluh)
Vojvodina (*voy*-vuh-deena)
Von Thünen (fon-*too*-nun)
Walloon (wah-*loon*)
Weber (*vay*-buh)
Weser (*vay*-zuh)
Wroclaw (*vraught*-slahv)
Yugoslavia (yoo-goh-*slah*-vee-uh)
Zeeland (*zay*-lund)
Zuider Zee (*zyder* zee)
Zürich (*zoor*-ick)

REFERENCES AND FURTHER READINGS

(BULLETS [●] DENOTE BASIC INTRO-
DUCTORY WORKS ON THE REALM OR
SYSTEMATIC ESSAY TOPIC)

Beckinsale, Monica & Beckinsale, Robert P. *Southern Europe: A Systematic Geographical Study* (New York: Holmes & Meier, 1975).

Burtenshaw, David et al. *The City in West Europe* (New York: John Wiley & Sons, 1981).

Chapman, Keith. *North Sea Oil and Gas: A Geographical Perspective* (London: David & Charles, 1976).

Chisholm, Michael. *Rural Settlement and Land Use: An Essay in Location* (London: Hutchinson University Library, 3 rev. ed., 1979).

● Clayton, Keith M. & Kormoss, I.B.F., eds. *Oxford Regional Economic Atlas of Western Europe* (London: Oxford University Press, 1971).

Clout, Hugh D., ed. *Regional Development in Western Europe* (New York: John Wiley & Sons, 3 rev. ed., 1987).

● Clout, Hugh D. et al. *Western Europe: Geographical Perspectives* (London & New York: Longman, 2 rev. ed., 1989).

Demko, George J., ed. *Regional Development Problems and Policies in Eastern and Western Europe* (New York: St. Martin's Press, 1984).

● Diem, Aubrey. *Western Europe: A Geographical Analysis* (New York: John Wiley & Sons, 1979).

East, W. Gordon. *An Historical Geography of Europe* (London: Methuen, 1962).

Embleton, Clifford, ed. *Geomorphology of Europe* (New York: Wiley-Interscience, 1984).

Glebe, Günther & O'Loughlin, John, eds. *Foreign Minorities in Continental European Cities* (Wiesbaden, West Germany: Franz Steiner Verlag, 1987).

● Gottmann, Jean. *A Geography of Europe* (New York: Holt, Rinehart & Winston, 4 rev. ed., 1969).

Greenhouse, Steven. "With Spain in Common Market, New Prosperity and Employment," *New York Times*, January 15, 1989, pp. 1, 9.

Hall, Ray, & Ogden, Philip. *Europe's Population in the 1970s and 1980s* (Cambridge, U.K.: Cambridge University Press, 1985).

● Hoffman, George W., ed. *Europe in the 1990's: A Geographic Analysis* (New York: John Wiley & Sons, [6 rev. ed. of *A Geography of Europe: Problems and Prospects*], 1989).

Hoffman, George W. & Dienes, Leslie. *The European Energy Challenge: East and West* (Durham, N.C.: Duke University Press, 1985).

Houston, James M. *A Social Geography of Europe* (London: Gerald Duckworth, 1963).

● Ilbery, Brian W. *Western Europe: A Systematic Human Geography* (New York: Oxford University Press, 2 rev. ed., 1986).

Jefferson, Mark. "The Law of the Primate City," *Geographical Review*, 29 (1939). Quotation taken from p. 226.

John, Brian S. *Scandinavia: A New Geography* (London & New York: Longman, 1984).

● Jones, Huw R. *A Population Geography* (Savage, Md.: Rowman & Littlefield, 2 rev. ed., 1990).

● Jordan, Terry G. *The European Culture Area: A Systematic Geography* (New York: Harper & Row, 2 rev. ed., 1989).

● Knox, Paul L. *The Geography of Western Europe: A Socio-Economic Survey* (Totowa, N.J.: Barnes & Noble, 1984).

Lagerfeld, Steven & Joffe, Josef. "Europe 1992," *The Wilson Quarterly*, 14 (Winter 1990): 56–81.

Lawday, David. "Europe's [Mediterranean] Sun Belt Also Rises," *Newsweek*, July 18, 1988, pp. 27–29.

Lewis, Flora. *Europe: A Tapestry of Nations* (New York: Simon & Schuster, 1987).

● Mellor, Roy E. H. & Smith, E. Alistair. *Europe: A Geographical Survey of the Continent* (New York: Columbia University Press, 1979).

Nelan, Bruce W. "Resurrecting Ghostly Rivalries: Eastern Europe Discovers That National Hatreds and Prejudices Increasingly Haunt the Land," *Time*, January 29, 1990, pp. 50–54.

Newman, James L. & Matzke, Gordon E. *Population: Patterns, Dynamics, and Prospects* (Englewood Cliffs, N.J.: Prentice-Hall, 1984).

Park, Chris. *Acid Rain: Rhetoric and Reality* (London & New York: Routledge, 1989).

Parker, Geoffrey. *The Logic of Unity: A Geography of the European Economic Community* (London & New York: Longman, 3 rev. ed., 1981).

Pounds, Norman J.G. *An Historical Geography of Europe, 1800–1914* (New York: Cambridge University Press, 1985).

Pringle, Dennis G. *One Island, Two Nations?: A Political Geographical Analysis of the National Conflict in Ireland* (New York: John Wiley & Sons, 1985).

Redman, Christopher. "Charging Ahead . . . Western Europe is Marching to Its Own Drummer," *Time*, September 18, 1989, pp. 40–45.

"Rising Racism on the Continent: Immigrants Face Economic Hardship and Increasing Prejudice," *Time*, February 6, 1984, pp. 40–45.

● Rugg, Dean S. *Eastern Europe* (London & New York: Longman, 1986).

Sommers, Lawrence M. "Cities of Western Europe," in Brunn, Stanley D. & Williams, Jack F., eds., *Cities of the World: World Regional Urban Development* (New York: Harper & Row, 1983), pp. 84-121. Quotation taken from p. 97.

Turnock, David. *Eastern Europe: An Economic and Political Geography* (London & New York: Routledge, 1989).

Turnock, David. *The Human Geography of Eastern Europe* (London & New York: Routledge, 1989).

Wheeler, James O. & Muller, Peter O. *Economic Geography* (New York: John Wiley & Sons, 2 rev. ed., 1986), Chap. 13.

White, Paul. *The West European City: A Social Geography* (London & New York: Longman, 1984).

Williams, Allan M. *The West European Economy: A Geography of Post-War Development* (Savage, Md.: Rowman & Littlefield, 1988).

BIBLIOGRAPHICAL FOOTNOTE

Individual countries (or groups of countries such as the Low Countries and the Alpine states) have also been the subject of many geographies, both for Europe and the remaining 12 world realms. A good beginning is a continuing series of sketches in *Focus*, published four times a year by the American Geographical Society in New York City. The Van Nostrand *Searchlight Series* of paperbacks, many dating back to the 1960s, includes discussions of many of the world's nations and areas; unfortunately, these are out of print in the 1990s (and are therefore not cited in this book), but they can still be found in many libraries. Aldine's *World Landscapes Series* contains many incisive works. Also refer to Praeger's *Country Profiles Series* (many now published by its branch, Westview Press of Boulder, Colo.) and Methuen's *Advanced Geographies Series*.

AUSTRALIA AND NEW ZEALAND:
European Outpost

Australia evokes images of faraway isolation, enormous expanses and uninhabited spaces, vast livestock ranches, beautiful coastlines, and modern cities (Fig. A–1). Australia is a continent and a nation-state—the center of a discrete realm of the world because its population and culture are European. If Australia were populated by peoples of Malayan stock with ways and standards of living resembling those of island Indonesia and mainland Southeast

FIGURE A–1   Sydney Cove, the heart of Australia's largest metropolis and port, and site of the first European settlement just over 200 years ago.

TEN MAJOR GEOGRAPHIC QUALITIES OF AUSTRALIA-NEW ZEALAND

1. Australia and New Zealand lie remote from the places with which they have the strongest cultural and economic ties.

2. Australia and New Zealand constitute a geographic realm by virtue of territorial dimensions, relative location, and cultural distinctiveness—not population size.

3. Australia has the lowest average elevation and the lowest overall relief of all the continental landmasses.

4. Australia is marked by a vast arid and semiarid interior, extensive open plainlands, and marginal moister zones.

5. Australia, with a large and diverse natural resource base, is a continent of substantial untapped potential.

6. Australia's population has a very low arithmetic and a low physiologic density.

7. Australia's population distribution is decidedly peripheral as well as highly clustered.

8. A very high percentage of Australia's population is concentrated in a small number of major urban areas.

9. Australia's indigenous (black) population was almost completely submerged by the European invasion, remains numerically small, and participates only slightly in modern society.

10. Australian agriculture is highly mechanized and produces large surpluses for sale on foreign markets. Huge livestock herds feed on vast pasturelands.

Asia (and had a comparable history), then it is quite possible that it would be viewed today only as an exceptionally large island sector of the Asian continent. But just as Europe merits recognition as a continental realm, despite the fact that it is merely a peninsula of "Eurasia," so Australia has achieved identity as the island realm of "Australasia." Australia and New Zealand are European outposts in an Asian Pacific world, as unlike Indonesia as Britain and America are unlike India.

Although Australia was spawned by Europe and its people and economy are Western in every way, Australia as a geographic realm is a far cry from the crowded, productive, complex European world. Australia's entire population (just over 17 million) is only slightly larger than that of the Netherlands, but Australia is about 200 times as large territorially. This gives an average population density of 5.6 per square mile (2.2 per square kilometer) and suggests that Australia is a virtual population vacuum on the very edge of overpopulated Asia. But so much of Australia is arid or semiarid (Fig. I–10) that only about 8 percent of its total area is agriculturally productive, and much of this moister part of the continent is too rugged for farming. By some calculations, only 1 percent of Australia's total area of nearly 3 million square miles (7.7 million square kilometers) is prime land for intensive cultivation. Certainly Australia could support many more people than it does today, but it is no feasible outlet for Asia's millions. Australia's *total* population is less than India's *annual increase*.

MIGRATION AND TRANSFER

Australia's indigenous peoples probably arrived on the island continent from Southeast Asia by way of New Guinea across a land bridge that existed when enlarged Pleistocene glaciers lowered global sea levels. The original Australians were black peoples with ancestral roots in East Asia, and they never numbered more than 300,000 to 350,000. They were hunters and gatherers who lived in small groups, using different and mutually unintelligible languages; geographically, they were more numerous in eastern than in western Australia. These earliest Australians skillfully adapted themselves to a difficult natural environment, and almost every narrative about them remarks on their detailed knowledge of the terrain, its opportunities and limitations.

Today, the first Australians are a nearly forgotten people. They could not withstand the impact of the European invasion, which was delayed because of Australia's **remoteness** and its initial unattractiveness. Portuguese, Spanish, and Dutch explorers briefly landed on Australian coasts between 1550 and 1750, but not until the journeys of James Cook, the British sea captain, did Australia finally enter the European orbit. When Cook visited the east coast in 1770, he was the first European to see this part of Australia. At the behest of the British government, he returned in 1772 and again in

1776; by that time, it was determined that *terra australis incognita* should become a British settlement. In 1788, white settlement in Australia began—and with it the demise of the aboriginal societies.

The Europeanization of Australia involved the movement of hundreds of thousands of emigrants, most of them British, from Western Europe to the shores of a little-known landmass in another hemisphere. This type of external migration has affected human communities from the earliest times, but it reached an unprecedented level during the nineteenth century. Before the 1830s, fewer than 3 million Europeans had left their homelands to settle in the colonies. But between 1835 and 1935, perhaps as many as 75 million departed for other lands—for the Americas, for Africa, and for Australia (Fig. A–2).

If so many people had not left Europe during the great population expansion that accompanied the Industrial Revolution, Europe's population problems would have been far more serious than they were. Even so, Europe during the nineteenth century was, for many people, an unpleasant place to be. Famine swept Ireland; war and oppression overtook much of the mainland; and the cities of industrializing Britain, Belgium, and Germany contained some awful living conditions. Many Europeans abandoned their homes to seek a better life across the ocean, even though they were uncertain of their fate in their new environments.

Many more Europeans came to the Americas than went to Africa or Australia. What led to their selection of the New World? Studies of the **migration** decision indicate that the intensity of a

migration flow varies with such factors as: (1) the perceived degree of difference between one's home and the planned destination, (2) the effectiveness of the information flow—the news sent back by those who migrated to those who stayed behind awaiting details, and (3) distance. A century ago, a British social scientist, Ernst Georg Ravenstein, studied migration in England and concluded that there is an inverse relationship between the incidence of migration and the distance between source and destination. Subsequent studies have modified Ravenstein's conclusions. The concept of **intervening opportunity**, for example, holds that people's perception of a faraway destination's comparative advantages is changed when there are closer opportunities. In the days when British people emigrated to lands other than North America, many chose South Africa

FIGURE A–2

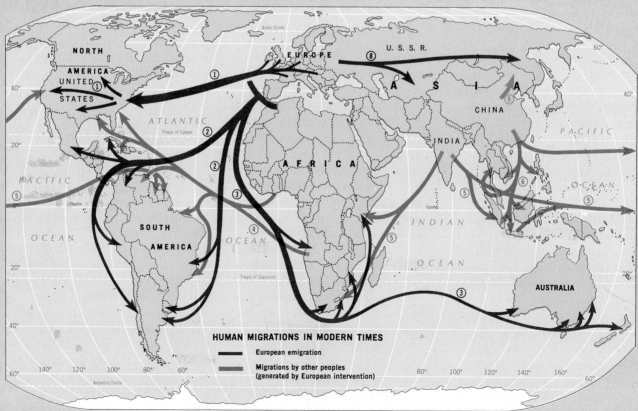

HUMAN MIGRATIONS IN MODERN TIMES

— European emigration

— Migrations by other peoples (generated by European intervention)

rather than cross yet another ocean to reach Australia.

Studies of migration also refer to *push factors* (conditions that tend to motivate people to move away) and *pull factors* (circumstances that attract people to a new destination). Australia was a new frontier, a place where one might acquire a piece of land, a herd of livestock, or where a piece of property might prove to hold valuable minerals; the skies (it was said) were clear, the air fresh, the climate much better than that of England.

Many of the first Europeans to arrive in Australia, however, were not free but in bondage. We have just described *voluntary migration*, in which movers made their own decisions to relocate. But thousands of the first European Australians did not reach their new homeland this way. European countries used their overseas domains as dumping grounds for people who had been convicted of some offense and sentenced to deportation—or "transportation" as the practice was called. This is a form of *forced migration*, in which voluntary push-pull factors play no role. "Transportation" had long been bringing British convicts to American shores, but in the late 1770s this traffic was impeded by the Revolutionary War. British jails soon overflowed, but judges continued to sentence violators to be deported; at this time Britain was undergoing its own economic and social revolution, and offenders were numerous. An alternative to America now had to be found, and it was not long before Australia was suggested. From the British point of view, the far side of Australia (the east coast) was an ideal place for such a penal colony. It lay several thousand miles away from the nearest British colony and would hardly be a threat; its environment appeared to be such that the convicted deportees

might be able to farm and hunt. In 1786, therefore, an order was signed making the southeastern corner of Australia, known as New South Wales, a penal colony. Within two years, the first party of convicts arrived at what is today the harbor of Sydney (Fig. A–1) and began to try to make a living from scratch.

For those of us who have learned of the horrible treatment of black slaves transported to the Americas, it is revealing to see that European prisoners sentenced to deportation were no better off despite the fact that the offense of many of them was simply their inability to pay their debts. The story of the second group of deportees to Australia gives an idea: more than 1000 prisoners were crammed aboard a small boat; 270 died on the way and were thrown overboard. Of those who arrived at Sydney alive, nearly 500 were sick; another 50 died within a few days. What the colony needed was equipment and healthy workers; what it got was hardly any supplies and ill and weakened people. The first white Australians hardly seemed the vanguard of a strong and prosperous nation. But by the middle of the nineteenth century, as many as 165,000 deported offenders had reached Australian shores, and London was spending a great deal of money on the penal stations. Meanwhile, the numbers of free colonists were increasing, convicts were entering society after serving their sentences, and many new settlements were founded. Australia's reputation in England began to change. As its penal image faded, the continent's spacious beauty and economic opportunities beckoned. The last convict ship arrived in 1849. The flow of voluntary immigrants grew, and along Australia's coastline several thriving colonies emerged whose tentacles soon reached into the interior.

LIVELIHOODS AND URBANIZATION

Australia's physiographic regions are identified in Fig. A–3, but another regionalization of Australia is more meaningful. Australia is a coastal rimland with cities, towns, farms, and forested slopes giving way to a dry, often desert-like interior that Australians themselves call the "Outback" or the "Inland" (Fig. A–4). The habitable coastal rimland is not continuous, for both in the south and in the northwest the desert reaches the sea (Fig. I–10). In the east lies the Great Dividing Range, extending from Cape York in the north to the island of Tasmania in the south. Between these highlands and the sea lie the fertile, well-watered foothills where Europeans in Australia had their start. Across the Great Dividing Range to the west lie the extensive grassland pastures that catapulted Australia into its first commercial age—and on which still range one of the greatest sheep herds on earth (over 150 million sheep annually produce more than one-fourth of all the wool sold in the world).

But Australia is not just a nation of stockbreeders. The steppe and near-desert country where sheep are raised, and the moister northern and eastern regions where cattle by the millions are ranched, these are Australia's sparsely inhabited areas (Fig. I–14) where sheep stations lie separated by dozens of miles of parched countryside, where towns are small and dusty, where a true frontier still exists. The great majority of modern Australians (86 percent) live in the cities and towns on or near the coast. In this respect Australia is quite similar to Europe, for in Britain, too, about nine out of every ten people live in urban areas.

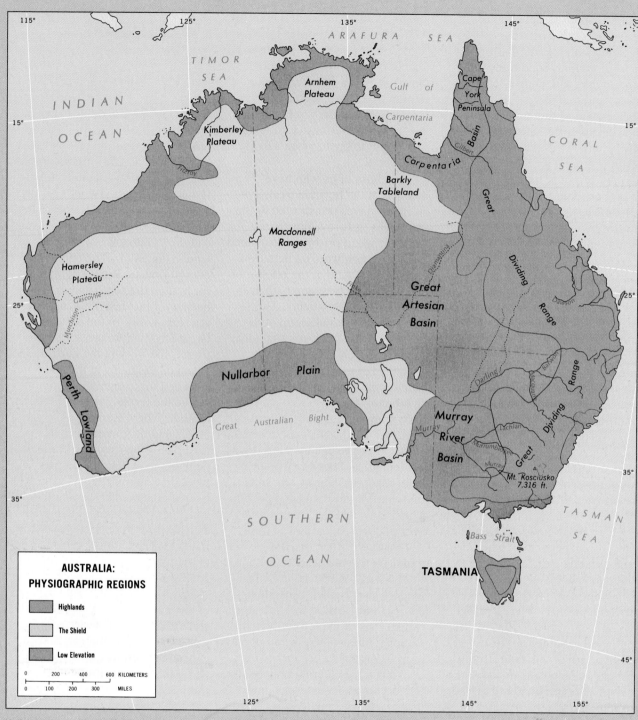

Modified After: McGraw Hill Book Co. <u>Man's Domain</u>

FIGURE A–3

Australia's cities began as penal colonies (including Sydney and Brisbane), as strategic settlements to protect trade and confirm spheres of influence (Perth in the southwest), or as centers of local commerce and regional markets (Melbourne and Adelaide). They became the foci of self-governing colonies; for a long time, Australia and Tasmania were seven ocean ports with separate hinterlands. These political regions were delimited by Australia's now-familiar pattern of straight-line boundaries, which were completed by 1861 (Fig. A–5). Notwithstanding their shared cultural heritage, the Australian colonies found themselves at odds not only with London over colonial policies, but also with

FIGURE A—4 The vastness of the arid Outback as seen from atop Ayers Rock, the famous geologic outcrop that lies close to Australia's geographic center about 250 miles southwest of Alice Springs.

each other over economic and political issues. National integration was a slow and difficult process, even on the ground. Each of the Australian colonies laid down its own railroad lines, not with a view toward coordinated continent-wide transport, but with local objectives. In the process, three different railroad gauges came into use (see red printing on Fig. A–5). The unification of these separate systems was costly and difficult, and not until 1970 was it possible to go by rail from Sydney to Perth without changing trains.

Australians also share with Europeans their increasing mobility. Australia now counts more automobiles per capita than even the United States, and Australians, like North Americans, are quite accustomed to long-distance travel. Australia is nearly as large as the contiguous United States, but its population of 17 million is what the U.S. white population was in the 1840s. In Australia, neighbors are often far removed: Perth in Western Australia is as far from Sydney as Los Angeles is from Atlanta, and Brisbane and Adelaide are as far apart as Washington, D.C., and Miami. Not surprisingly, there are today more road miles per person in Australia than in the United States.

Australia's cities retain some of the competitive atmosphere of the colonial period, but they are as modern, efficient, spacious, and suburbanized as most cities of the Western world. Sydney (3.5 million), capital of New South Wales and Australia's largest city and leading port, has a long history of rivalry with Melbourne (3.1 million), which at one time overtook Sydney in size and was Australia's temporary capital. Sydney lies on a magnificent site, its skyscrapered central city overlooking the famed Sydney Harbour Bridge (Fig. A–1). However, its architecture is no match for less spectacular but historically more interesting Melbourne, which is still the capital of the state of Victoria. Between them, metropolitan Sydney and Melbourne are home to almost 40 percent of Australia's total population, and the 2.4 million residents of Brisbane (Queensland) and Adelaide (South Australia) account for an additional 14 percent. Perth, Western Australia's capital, has 1.4 million inhabitants, whereas Hobart on Tasmania has under 200,000; the smallest of the regional capitals is Darwin (Northern Territory), with about 75,000 residents.

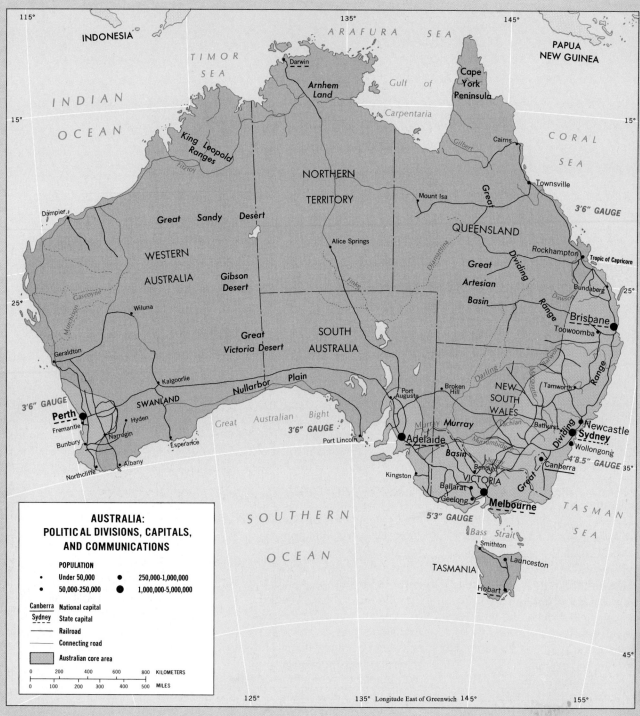

FIGURE A–5

ECONOMIC GEOGRAPHY

Agriculture

Sheep-raising thrust Australia into the commercial age, and the technology of refrigeration brought European markets within reach of Australian beef producers, who also constitute an important sector of the national economy. But there is more to Australian rural land use than herding livestock. Commercial crop farming concentrates on the production of wheat, and Australia is one of the world's leading exporters of that grain, which brings in about two-thirds as much foreign income derived from wool.

Wheat production is concentrated in a broad crescent from the vicinity of Adelaide into Victoria and New South Wales along the rim of the Murray Basin (Fig.

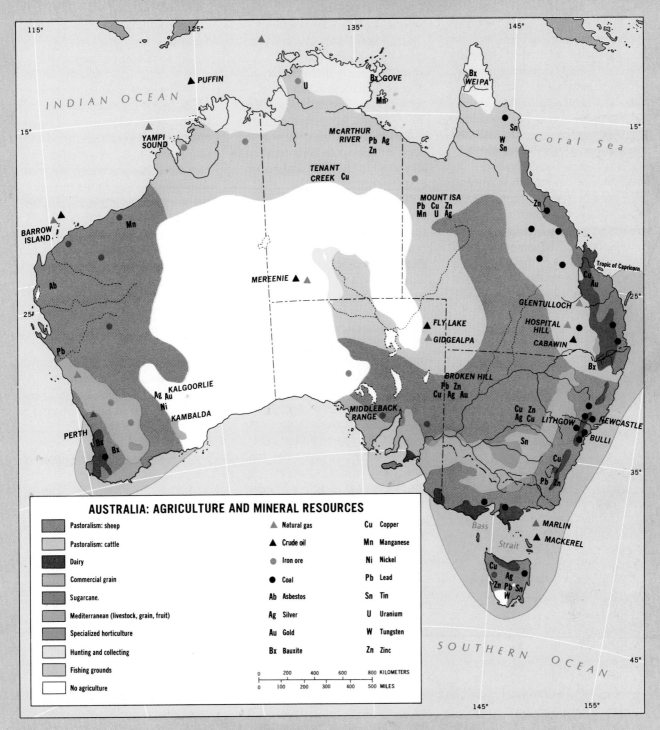

AUSTRALIA: AGRICULTURE AND MINERAL RESOURCES

Pastoralism: sheep	▲ Natural gas	Cu	Copper
Pastoralism: cattle	▲ Crude oil	Mn	Manganese
Dairy	● Iron ore	Ni	Nickel
Commercial grain	● Coal	Pb	Lead
Sugarcane.	Ab Asbestos	Sn	Tin
Mediterranean (livestock, grain, fruit)	Ag Silver	U	Uranium
Specialized horticulture	Au Gold	W	Tungsten
Hunting and collecting	Bx Bauxite	Zn	Zinc
Fishing grounds			
No agriculture			

FIGURE A–6

A–6). It is also raised in a wide belt behind Perth in Western Australia. Although these zones are mapped as ''commercial grain'' farming, sheep and wheat share the land in a unique rotation system. Under this mixed crop-and-livestock farming, the sheep use the cultivated pasture for several

years; it is then plowed for wheat sowing, and after the wheat harvest, the soil is rested again. This system, aided by the Australians' innovations in mechanized equipment, has created a highly lucrative industry. Australian wheat yields per acre are still low by North American standards, but

the output per worker is about twice that of the United States.

As Fig. A–6 shows, dairying with cultivated pastures takes place in what Fig. I–9 shows to be areas of high precipitation. The location of the dairy zones also suggests their development in response to Australia's comparatively large urban markets. Even these humid areas of Australia are occasionally afflicted by damaging droughts, but the industry is normally capable of providing more than the local market demands. Dairying is, in fact, the most important rural industry as measured by numbers of people employed.

It is logical that irrigation should be attempted in Australia, the world's driest continent. Unfortunately, the opportunities for irrigation-assisted agriculture are quite limited. The major potential lies in the basin of the Murray River, shared by the states of Victoria (which leads in irrigated acreage) and New South Wales; here, rice, grapes, and citrus fruits are among the leading irrigated crops. Moreover, on the east coast, in Queensland and northern New South Wales, the cultivation of sugarcane is partly under irrigation.

Mineral Resources

Nothing shook the Australian economy as much as the discovery of gold in 1851 in the territories of New South Wales and Victoria. In no way—politically, socially, or materially—was Australia prepared for what followed the news of the gold finds. True, gold had been known to exist prior to 1851, but only when some Australian diggers returned from California's fields to prove the lucrative character of the Australian ores did the rush begin. During the decade from 1851 to 1861, the population of the colony of Victoria increased more than sevenfold to well over a half million. Yet Melbourne, the colony's leading town, was just 16 years old when the rush began and had none of the amenities needed to serve so many people. New South Wales, which in those days still included all of what is now

Queensland and the Northern Territory, saw its population rise to over 350,000, and its pastoral economy was rudely disturbed by the rush of diggers to Bathurst in the foothills west of Sydney. Overall, Australia's population nearly tripled in the 1850s; gone was the need for subsidized immigration in all areas but the west and northeast.

The new wealth of Australia brought problems of accommodation for the new settlers, but it also brought great advantages. A new prosperity came to every Australian colony, and by the 1850s the continent produced 40 percent of the world's gold. It inspired successful searches for other minerals, and discoveries are still being made today. Long believed poor in petroleum and natural gas supplies, Australia now seems to have sizable reserves of both fuels after all, with recent finds in the Great Artesian Basin and on the floor of the Bass Strait between the mainland and Tasmania. Recently, too, nickel deposits were found near the center of Western Australia's gold-mining area, Kalgoorlie; in particular, the nickel ore discovered at nearby Kambalda may constitute the world's largest known reserve of this important ferroalloy. Even today, the Australian frontier still holds its secrets and surprises.

As Fig. A–6 suggests, Australia's mineral deposits are spatially dispersed and quite varied. The country was fortunate in being so well endowed with coal deposits, because its waterpower prospects are minimal (except in Tasmania and the southeast) and its petroleum discoveries were long delayed (the Bass Strait fields were not found until the 1960s). The chief coalfields, as the map indicates, lie in the east, notably around Sydney, in the hinterland of Brisbane, and in the zone to

the north lying inland from Rockhampton. From here, this energy resource is sent by coastal shipping to areas that are deficient in coal. But coal is widely distributed—even Tasmania and Western Australia mine some deposits and can thus keep their import requirements down. Overall production has lately risen steadily, and Australia has joined the ranks of the world's leading coal exporters (about one-sixth of the country's export sales derive from this raw material). However, problems may loom because world prices have remained fairly low as other coal producers such as South Africa, Indonesia, and China are increasing their output as well.

The most famous Australian mining district undoubtedly is Broken Hill (in northwestern New South Wales), which has neither gold, coal, nor iron. Discovered in the aftermath of the gold rush, Broken Hill became one of the world's leading lead- and zinc-producing areas, and the enormous income derived from the export sales of these minerals supplied much of the capital for Australia's industrialization. Japan has become one of the principal buyers, and Australia's "Japanese connection" has proven quite lucrative; in fact, during the 1980s Japan regularly purchased between 25 and 30 percent of all Australian exports. In northwestern Queensland, the Mount Isa ore body yields a similar mineral association, and in both areas important uranium deposits have also been developed. But these products are only a small part of the total Australian inventory. Tasmania's copper, Queensland's tungsten and bauxite deposits on Cape York Peninsula, and Western Australia's asbestos represent the vast range of the continent's minerals. There are few non-metallic resources in which minerally rich

Australia is truly deficient—and the real search has just begun.

Manufacturing

For a long time, Australia was to Britain what colonies always were: a supplier of needed resources and a ready market for British manufactured products. But World War I, at a crucial stage, cut these trade connections—and Australia for the first time was left largely on its own. Now it had to find ways to make some of the consumer goods it had been getting from Britain. By the time the war was over, Australian industries had made a great deal of progress. Wartime also pushed the development of the food-processing industries, and today Australian manufacturing is quite diversified, producing not only machinery and equipment made of locally manufactured steel, but also textiles and clothing, chemicals, foods, tobacco, wines, and paper—among many other items. As might be expected, these industries are located where the facilities for manufacturing are best and where markets exist—in the large metropolitan areas.

Australia is not yet a major exporter of finished industrial goods. Manufacturing of steel products, automobiles, textiles, chemicals, and electrical equipment are oriented to the domestic market. Although this market is small numerically, it is a considerable factor: the gross domestic product per capita in Australia for 1987 was just under (U.S.) $11,000. The success of domestic industry is a matter of prime concern to the government, because Australian manufactures have not taken their place on world markets; foreign products therefore pose a challenge, both in competing against locally-produced goods as well as

threatening an overly unfavorable import-export balance. Thus, plentiful raw materials notwithstanding, Australian manufacturing faces high production costs and long distances to foreign markets, and it cannot expect to compete overseas. Indeed, Australia's list of imports still includes machinery, equipment, chemicals, foods, and beverages that could, in part at least, be produced at home. In overall context, therefore, Australia's manufacturing industries still exhibit some symptoms of a developing country.

POPULATION POLICIES

Not surprisingly, the state capitals of Australia became the country's major industrial centers. These urban complexes offered the best locational advantages for industry—concentrations of labor and capital, transport nodality, access to government—and they also constituted the major markets. In addition, these metropolitan areas have absorbed the vast majority of the country's recent immigrants. The advantages of immigration to Australia are obvious: a strengthening of the country's still small population, an expansion of the skills and numbers of the labor force, an increase in the size of the domestic market, and a growing consolidation of the Australian presence on its large island continent.

Before the 1980s, Australia had been widely known for its selective immigration policies, virtually excluding persons who are not white from the opportunity to settle there. Since 1901, successive Australian governments adhered to an unofficial "White Australia"

policy. After World War II international criticism mounted, so in the 1950s the practice began to soften. First, the government approved the admission of family members of nonwhite Australians. Then in 1966, it became possible for Asians to apply for Australian residence—but only those with urgently needed skills. Thus, the lion's share of immigration continued to come from Europe, a flow dominated by Britishers but also including numerous Italians, Germans, Greeks, and Eastern Europeans.

In the meantime, the indigenous (aboriginal) Australian population, which numbered as many as 350,000 two centuries ago, has been reduced to a mere remnant. It is the familiar story of contact between indigenous peoples living a life of subsistence in small groups and the onslaught of an advanced society: these black Australian communities were driven away, broken up, often destroyed. Today, fewer than 50,000 Australians still have a totally aboriginal ancestry; another 175,000 or so have mixed origins. Only a few communities still pursue the original life of hunting and gathering in the remotest corners of the Northern Territory, Western Australia, and northern Queensland. The remainder are scattered across the continent, working on cattle stations, subsisting on reserves set aside for them by the government, or performing menial jobs in towns and cities. Lately, the Australian conscience has been aroused, and an effort has begun to undo centuries of persecution and neglect—but these attempts are met, understandably, with suspicion and distrust. In affluent Australia, the aboriginal Australians remain victims of poverty, disease, insufficient education, and even malnutrition.

Australian immigration policies

have continued to evolve, with a significant modification in 1979 that raised immigration quotas and liberalized admissions requirements. These changes met opposition, particularly from labor leaders who charged that the immigration of a large number of workers without guaranteed jobs would create a docile work force and heighten unemployment. Against this background there also arose the issue of Asian refugees, most of whom were ''boat people'' desperately fleeing Vietnam and other parts of Communist Indochina. They came by way of Malaysia and Indonesia, most of them arriving at or near the northern port of Darwin. Their numbers and problems of adjustment mounted, and a national debate arose. The Australian labor movement, always the focus for the ''White Australia'' immigration policy, traditionally opposed Asian immigration; furthermore, the Australian Labour party (then the opposition party) had opposed the Vietnam War. Thus, refugees escaping the new order in Vietnam were not welcome on several grounds in addition to their circumvention of established immigration policy.

As the debate ran its course, attitudes regarding Asians evolved further during the 1980s. Convinced that Australia's future economic growth depends on closer ties with its Western Pacific neighbors to the north, the government led a careful campaign to admit more immigrants, especially from Southeast Asia. This effort met with less resistance than ever before, and since 1984 Asians and Pacific Islanders have accounted for over 40 percent of the annual immigrant inflow. By no means have refugees dominated this influx (even though Australia has accepted more Vietnamese boat people than any other country). A large proportion of the latest Asian immigration consists of skilled people and originates in such relatively affluent places as Hong Kong, Singapore, and the more developed areas of Thailand, Malaysia, Indonesia, and the Philippines. The appeal of Australia as a land of opportunity exerts a powerful hold on nearby Asians (not unlike the lure of U.S. soil in Middle America), and that pull can be expected to intensify in the future. Today, Asians still constitute only 4 percent of Australia's population and their rate of influx remains modest; but these Asians are likely to make increasingly important contributions to Australia's development, and that could win them wider acceptance and lead to further liberalization of immigration policies.

AUSTRALIAN FEDERALISM

Australia may be a European offshoot, but in at least one respect it differs significantly from most European countries. That difference is **federalism**: Australia is a federal state, whereas all but a few European countries are unitary states. The term *unitary* derives from the Latin *unitas* (unity), which in turn comes from *unus* (one). A *unitary state*, then, is unified and centralized. It is not surprising that many European states, long ruled with absolute authority from a royal headquarters, developed into unitary states. The distant colonies had no such tradition, whether in Canada, the United States, or Australia. In Australia, six colonies for most of the nineteenth century went their separate ways, competing for hinterlands and arguing over trade policies. Eventually, however, the idea that they might combine into some greater national framework began to take hold. But this would be no unitary state: each colony wanted to retain certain rights. Thus, Australia became a *federal state*.

The federal concept also has ancient Greek and Roman roots, and the term originates from the Latin *foederis*, meaning league or association. In practice, the federal system is one of alliance and coexistence, and the recognition of differences and diversity. Of course, there must be an underlying foundation for a federal union: in Australia, this was primarily the colonies' common cultural heritage. But there were other bases as well. The colonists worried about a foreign invasion from Asia or some mainland European power. They particularly feared Germany, because in the 1880s the Germans embarked on a vigorous colonization campaign in Africa and the Pacific. In 1883, the colonial government of Queensland took control of the eastern part of the island of New Guinea because it feared a German takeover there. Also, the colonies shared the objectives of selective immigration; a coordinated ''White Australia'' policy would be more easily administered on a continent-wide basis. Finally, the colonies realized that their internal trade competition was damaging and precluded a coordinated commercial policy on world markets.

With such positive prospects, it would seem that federation could have been accomplished quickly and without difficulties—but this was not the case. Victoria's manufacturers wanted protective tariffs; public opinion in New South Wales favored freer trade. There was wrangling over the site of the future federal capital, and no colony was willing to yield to another the advantage of becoming the home of the new federal headquarters. But in 1900, lengthy conferences succeeded in creating a federal constitution that was subsequently approved by the voters of the colonies and confirmed by London. From January 1, 1901, the colonies became the *Commonwealth of Australia* consisting of six states and the federal territories of the future capital as well as the Northern Territory, the sparsely peopled rectangular area that extends from Arnhem Land south into the heart of the Australian desert (Fig. A–5).

Canberra: The Federal Capital

More than one federal government has found the selection of a federal capital a difficult task requiring compromise and consent. Australia experienced this problem, which arose mainly from the intense rivalry between the adjacent southeastern states of Victoria and New South Wales. Victoria's capital, Melbourne, was the chief competitor of Sydney, headquarters of New South Wales. Shortly before the implementation of federation, the premier of New South Wales managed to secure an agreement that the federal capital would lie in that state in return for two concessions. First, the new capital would *not* be Sydney, but a totally new city to be built at a site to be jointly selected. Second, Melbourne would function as the temporary capital until the federal headquarters was ready for occupation.

The site of Canberra was chosen in 1908, and the buildings necessary for the government functions were ready by 1927. Canberra, unlike other major Australian cities, lies away from the coast; it is situated at the foot of the Great Dividing Range about halfway between Sydney and Melbourne (Fig. A–5). Once the site

FIGURE A–7 Canberra, Australia's planned capital city, which was completed in 1927.

was selected, a worldwide architectural design competition was held in order to acquire the best city plan; the U.S. architect Walter Burley Griffin was chosen for the task. Today, Canberra (325,000) is a modern, widely dispersed urban area that comprises the Australian Capital Territory, a federal district separated from the state of New South Wales (as

Washington, D.C. is from the state of Maryland). Nonetheless, Canberra shares with other comparatively new federal capital cities a sense of remoteness and isolation, and the city itself, endowed with impressive public architecture, ceremonial places, and parks (Fig. A–7), still retains an atmosphere of coldness and distance. Canberra lacks Sydney's drive and bustle,

Melbourne's style and comfort, and Adelaide's sense of history—and is thus a far cry from being the pulse of Australia.

A Youthful State

If states do indeed experience life cycles, Australia exhibits the energy and vigor of youth. Many observers have likened Australia today to the United States of a century and a half ago: still discovering major resources, still in the early stages of the penetration of national territory, settlement still essentially peripheral, the interior still a frontier—albeit a modern frontier with airstrips rather than covered wagons.

Australia's cities, where the overwhelming majority of Australians live, reflect the country's youthfulness in several ways. The largest, Sydney, contains over 20 percent of the entire Australian population. Sydney also anchors a metropolitan area of over 4600 square miles/12,000 square kilometers (larger than New York or London), but internal circulation is hampered by the lack of adequate highways as available facilities have been unable to keep pace with the still-expanding population and its automobiles. The central city is also severely congested: traffic jams on the approaches to the picturesque but outmoded Harbour Bridge (opened in 1932), linking Sydney's north and south (Fig. A–I), are among the world's worst. Major arteries, optimistically called the Great Western Highway and the Pacific Highway, begin at best as wide city streets whose traffic is bedeviled by numerous stoplights, lane inconsistencies, and inadequate posted directions. And the only access to the airport is through still more of those crowded streets.

Australia has never had a space problem, and the cities reflect this as suburbs sprawl far and wide. Downtown Sydney lies at the heart of a conurbation consisting of five incorporated cities and more than 30 municipalities, a built-up region that extends more than 50 miles (80 kilometers) from north to south. Greater Melbourne's dimensions are only slightly smaller. Few remnants in the urban landscape reveal the cities' ages: modern towering skyscrapers dominate the central business districts. Melbourne, capital of Victoria, retains more architectural interest and is a more cultured city than Sydney. Perth, the capital of enormous Western Australia, is the youngest of the state capitals, a thriving urban center that is still the most frontier-like among Australia's large metropolises.

NEW ZEALAND

Like Australia, New Zealand is a product of European expansion. In another era, New Zealand, located over 1000 miles (1700 kilometers) southeast of Australia, would have been included in the Pacific realm because prior to the Europeans' arrival it was occupied by the Maoris, a people with Polynesian roots. Today, New Zealand's population of 3.4 million is more than 85 percent European, and the Maoris form a minority of about 450,000—with many of mixed ancestry.

New Zealand consists of two large mountainous islands and numerous scattered smaller islands (Fig. A–8). The two large islands, with South Island somewhat larger than North Island, look diminutive in the great Pacific Ocean, but yet

together they are larger than Britain. In sharp contrast to Australia, the two islands are mainly mountainous or hilly, with several peaks rising far higher than any on the Australian landmass. South Island has a spectacular snowcapped range appropriately called the Southern Alps, with peaks reaching beyond 11,700 feet (3500 meters). Smaller North Island has proportionately more land under low relief, but it also has an area of central highlands along whose lower slopes lie the pastures of New Zealand's chief dairying district. Hence, while Australia's land lies relatively low in elevation and has much low relief, New Zealand's is on the average quite high and exhibits mostly rugged relief.

Thus, the most promising areas must be the lower-lying slopes and lowland fringes on both islands. On North Island, the largest urban area, Auckland (1.0 million), occupies a comparatively low-lying peninsula. On South Island, the largest lowland is the agricultural Canterbury Plain, centered on Christchurch (360,000). What makes these lower areas so attractive, apart from their availability as cropland, is their magnificent pastures. Such is the range of soils and pasture plants that both summer and winter grazing can be carried on. Moreover, a wide variety of vegetables, cereals, and fruits can be produced in the Canterbury Plain (Fig. A–9), the chief farming region. About half of all New Zealand is pastureland, and much of the farming is done to supplement the pastoral industry in order to provide fodder when needed. Sheep and cattle dominate these livestock-raising activities, with wool and meat providing over 50 percent of the islands' export revenues.

Despite their contrasts in size, shape, physiography, and history, New Zealand and Australia have a

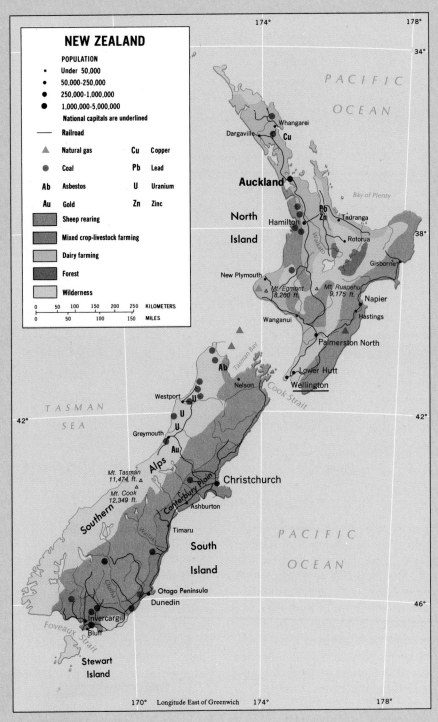

FIGURE A–8

though New Zealand lags behind Australia in industrial development, the country's manufactures are becoming more diverse all the time; textiles and clothing, wood products, fertilizers, steel and aluminum products, and some machinery are now made in New Zealand.

Spatially, New Zealand shares with Australia its pattern of peripheral development (Fig. I–14), imposed not by desert but by high rugged mountains. The country's major cities—Auckland and the capital of Wellington (together with its satellite of Hutt, 335,000) on North Island, and Christchurch and Dunedin (140,000) on South Island—are all located on the coast, and the entire railway and road system is peripheral in its configuration (Fig. A–8). This is more pronounced on South Island than in the north, for the Southern Alps are New Zealand's most formidable barrier to surface communication.

Compared to brash, progressive, and modernizing Australia, New Zealand seems quieter. A slight regional contrast might be discerned perhaps between more forward-looking North Island and conservative South Island, but the distinction fades in the light of Australia's historic internal urban and regional rivalries. Nothing in New Zealand compares to the variety of the urban experience in Australia; a sameness pervades New Zealand, and probably contributes to the high rate of emigration, mainly to Australia. If New Zealand fails to satisfy many of its younger citizens, it is not because of a lack of personal security. The government has developed an elaborate cradle-to-grave system of welfare programs that are affordable, given the the country's high incomes and standards of living (taxes also are high). Whether this has suppressed entrepreneurship and initiative is a

great deal in common. Apart from their joint British heritage, they share a sizable pastoral economy, a small local market, the problem of great distances to world markets, and a desire to stimulate and develop domestic manufacturing.

The high degree of urbanization in New Zealand (85 percent of the total population) indicates another similarity to Australia: substantial employment in city-based industries (particularly processing and packing of livestock products). Al-

FIGURE A–9 New Zealand's Canterbury Plain—South Island's leading agricultural area—with the Southern Alps in the background to the west.

matter for debate, but New Zealand is a country with few excesses and much stability. In any case, the government in the mid-1980s began a major restructuring of the economy into a free market business environment—with wide approval by the electorate.

Other significant change can be discerned as well. Auckland not only is New Zealand's largest city, it also may be the largest Polynesian city of all, with about 100,000 Maoris, Samoans, Cook Islanders, Tongans, and others making up more than 10 percent of the metropolitan population. The twentieth century has witnessed a revival of Maori culture—and a

persistently slow pace of integration of the Maori minority into New Zealand society (nearly all the Maoris reside on North Island, where three-quarters of all New Zealanders live). In the 1990s, however, the Maori presence has become the leading national issue, because the Maoris have launched a vigorous effort to get the courts to enforce the terms of the 1840 Treaty of Waitangi that granted the British sovereignty over New Zealand. In particular, the Maoris have laid claim to at least half of the national territory, asserting that their tribal lands were illegally wrested away by British settlers. Judicial rulings in the late 1980s

began to support the Maori position, and the government may soon be required to return land that Maori tribes can prove once belonged to them. The Waitangi treaty was intended to be the framework for a partnership with the Maoris. If that promise can finally be fulfilled, New Zealand's future could be shaped by a harmonious bicultural society. If it fails, the specter of serious racial polarization looms because the more-rapidly-growing Maori population could double to 25 percent of New Zealand's total within the next two decades.

PRONUNCIATION GUIDE

Adelaide (*addle*-ade)	Canberra (*kan*-burruh)	Melbourne (*mel*-bun)
Auckland (*awk*-lund)	Dunedin (duh-*need*-nn)	Tasmania (tazz-*may*-nee-uh)
Brisbane (*brizz*-bun)	Maori (*mah*-aw-ree/*mau*-ree)	Waitangi (*wye*-tonggy)

REFERENCES AND FURTHER READINGS

(BULLETS [●] DENOTE BASIC INTRODUCTORY WORKS ON THE REALM)

Atlas of Australian Resources, 3rd Series (Canberra: Division of National Mapping, 6 vols., 1980–1987).

● Barrett, Rees D. & Ford, Roslyn A. *Patterns in the Human Geography of Australia* (South Melbourne: Macmillan of Australia, 1987).

Bolton, Geoffrey C. *Spoils and Spoilers: Australians Make Their Environment, 1788–1980* (Winchester, Mass.: Allen & Unwin, 1981).

● Burnley, Ian H. *Population, Society and Environment in Australia* (Melbourne: Shillington House, 1982).

Burnley, Ian H. & Forrest, James, eds. *Living in Cities: Urbanism and Society in Metropolitan Australia* (Winchester, Mass.: Allen & Unwin, 1985).

Cameron, Roderick W. *Australia: History and Horizons* (New York: Columbia University Press, 1971).

Courtenay, Percy P. *Northern Australia: Patterns and Problems of Tropical Development in an Advanced Country* (London & New York: Longman, 1983).

● Cumberland, Kenneth B. & Whitelaw, James S. *New Zealand* (Chicago: Aldine, 1970).

"Environment and Development in Australia," *Australian Geographer,* 19 (May 1988): 3–220.

Hamnett, Stephen & Bunker, Raymond, eds. *Urban Australia: Planning Issues and Policies* (Melbourne: Nelson Wadsworth, 1987).

Hanley, Wayne & Cooper, Malcolm, eds. *Man and the Australian Environment: Current Issues and Viewpoints* (Sydney: McGraw-Hill, 1982).

● Heathcote, Ronald L. *Australia* (London & New York: Longman, 1975).

● Heathcote, Ronald L., ed. *The Australian Experience: Essays in Australian Land Settlement and Resource Management* (Melbourne: Longman Cheshire, 1988).

Heathcote, Ronald L. & Mabbutt, J. A. *Land, Water and People: Essays in Australian Resource Management* (Winchester, Mass.: Allen & Unwin, 1988).

Hofmeister, Burkhard. *Australia and Its Urban Centers* (Berlin: Gebrüder Borntraeger, 1988).

Hughes, Robert. *The Fatal Shore* (New York: Alfred A. Knopf, 1987).

● Jeans, Dennis N., ed. *Australia: A Geography. Volume 1: The Natural Environment* (Sydney: Sydney University Press, 1986); *Volume 2: Space and Society* (Sydney: Sydney University Press, 1987).

Jeans, Dennis N., ed. *Australian Historical Landscapes* (Winchester, Mass.: Allen & Unwin, 1984).

Johnston, Ron J., ed. *Society and Environment in New Zealand* (Christchurch, N.Z.: Whitcombe & Tombs, 1974).

● Lewis, George J. *Human Migration: A Geographical Perspective* (New York: St. Martin's Press, 1982).

Linge, Godfrey & Frazer, R. *Atlas of New Zealand Geography* (Wellington, N.Z.: Reed, 1966).

Logan, Malcolm I. et al. *Urbanization: The Australian Experience* (Melbourne: Shillington House, 1981).

● McKnight, Tom L. *Australia's Corner of the World* (Englewood Cliffs, N.J.: Prentice-Hall, 1970).

Meinig, Donald W. *On the Margins of the Good Earth: The South Australian Wheat Frontier, 1869–1884* (Chicago: Rand McNally, 1962).

● Powell, Joseph M. *An Historical Geography of Modern Australia: The Restive Fringe* (New York: Cambridge University Press, 1988).

Powell, Joseph M., ed. *Urban and Industrial Australia* (Melbourne: Sorrett, 1974).

Powell, Joseph M. & Williams, Michael, eds. *Australian Space, Australian Time: Geographical Perspectives* (New York: Oxford University Press, 1975).

Rich, David C. *The Industrial Geography of Australia* (Sydney: Methuen, 1987).

Roberts, John E., ed. *Bold Atlas of Australia* (Sydney: Ashton Scholastic, 1983).

Spate, Oskar H.K. *Australia* (New York: Praeger, 1968).

Statham, Pamela, ed. *The Origins of Australia's Capital Cities* (Sydney: Cambridge University Press, 1989).

Terrill, Ross. *The Australians* (New York: Simon & Schuster, 1987).

White, Paul E. & Woods, Robert I., eds. *The Geographical Impact of Migration* (London & New York: Longman, 1980).

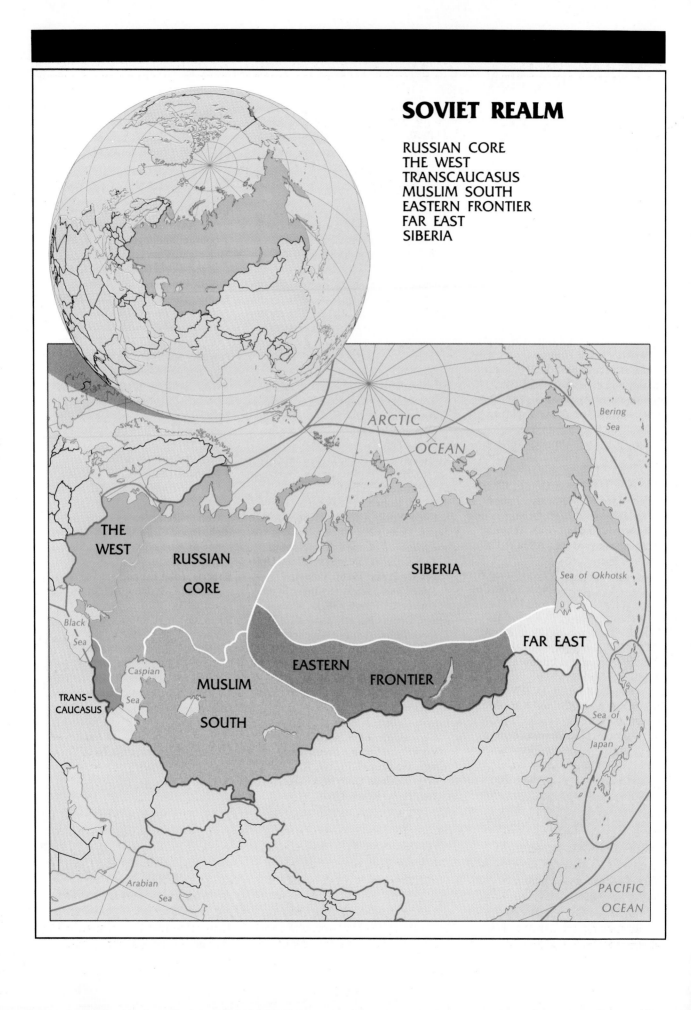

SOVIET REALM

RUSSIAN CORE
THE WEST
TRANSCAUCASUS
MUSLIM SOUTH
EASTERN FRONTIER
FAR EAST
SIBERIA

THE
WEST

RUSSIAN

CORE

SIBERIA

*Black
Sea*

*Caspian
Sea*

TRANS-
CAUCASUS

MUSLIM

SOUTH

EASTERN

FRONTIER

FAR EAST

ARCTIC

OCEAN

*Bering
Sea*

Sea of Okhotsk

*Sea of
Japan*

*Arabian
Sea*

PACIFIC
OCEAN

THE SOVIET UNION IN TRANSITION

From the Baltic Sea in Europe to the Asian shore of the Pacific Ocean and from the Arctic coast in the north to the borders of Iran in the south lies the Soviet Union— political region, culture area, historic empire, ideological hearth, and world superpower. Its 8.6 million square miles (22.3 million square kilometers) constitute nearly one-sixth of the earth's land surface and half of the vast landmass of Eurasia. Territorially, the Soviet Union is about two and a half times as large as the United States or China. A country of massive continental proportions, the Soviet Union stretches across 11 time zones. When people awake in the morning in Vladivostok on the Sea of Japan, it is still early the previous evening in Leningrad on the Gulf of Finland. The ground is permanently frozen across the northern tundra that borders the Arctic Ocean, but subtropical crops grow on the balmy Georgian shore of the Black Sea.

The Soviet Union does not, however, have a population to match its huge territory. Its 295 million people are far outnumbered by 1.1 billion Chinese and 871 million Indians, neighbors on the Asian continent. And, as Fig. I–14 reveals, the Soviet population is still strongly concentrated in what is sometimes called "Euro-

---

IDEAS AND CONCEPTS

climatology

imperialism

colonialism

irredentism (2)

federation

acculturation

economic planning

migration (2)

core area

exclave

centrality

inaccessibility

heartland theory

frontier

---

pean" Russia—the Soviet Union west of the Ural Mountains, the north-south trending range that crosses Russia from the Kara Sea to Kazakhstan (see Fig. 2–5). Much of the remainder consists of vast Siberia (the name means "sleeping land") and the Soviet Far East; both regions are virtually empty, an unchanged, little-known frontier. In those huge areas beyond the Urals, the Soviets are discovering additional mineral riches, but the cost of

building transport systems to carry them to the industrial centers of the western U.S.S.R. may be prohibitive for a long time to come.

EVOLUTION OF A MULTINATIONAL STATE

The Soviet Union is a state of many nations. Over centuries of expansion, the Russian state became an imperial power that annexed and incorporated peoples of many different nationalities and cultures. Until the first decade of this century, the Russians carried on the campaign of a succession of tsars: they absorbed frontiers, pushed their boundaries outward, and overthrew uncooperative rulers. By the time the ruthless Russian regime faced revolution among its own people, the tsars had bequeathed the winners in that struggle—the communists who forged the Soviet Union—with the largest empire on earth. Virtually all the colonized peoples were incorporated into the new communist state, including Christian Armenians and Georgians, Shi'ite Muslim Azerbaydzhanis, and

S Y S T E M A T I C E S S A Y

CLIMATOLOGY

Climatology is one of the several branches of physical geography. Climatologists analyze the distribution of climatic conditions over the earth's surface. They study the causes behind this distribution and the processes that change it. Such topics as the earth's long-term cooling (glaciation) and short-term warming (the so-called "greenhouse effect"), desertification, and hurricane tracking are part of this field.

The term *climate* implies an average, a long-term record of *weather* conditions at a certain place or across a region. Weather conditions, therefore, are recorded in specifics for any given moment in time: the temperature, the amount of rainfall, percentage of humidity, wind speed and direction, and other data. Climate, on the other hand, is described in more general terms, as shown in Fig. I–10 (pp. 18–19). *Humid Equatorial* climates, *Dry* climates, and *Cold Polar* climates are marked by certain prevailing characteristics that can be mapped, resulting in the regional framework in Fig. I–10.

It is important to remember that Fig. I–10 is a kind of cartographic snapshot of the earth's changing climatic environment. Just 10,000 years ago, the map would have looked quite different; the most recently developed icesheets were just receding. Even only one century ago, we would, at a larger scale, be able to detect differences. In West Africa, for example, the Sahara Desert has spread southward far enough that we would be able to see the evidence on a hundred-year-old map.

To interpret the global climatic map, we must understand the processes and conditions that underlie both weather and climate. Our planet orbits the sun in such a way that the tropics receive the maximum radiation (heat) and the polar areas the minimum. But warmth is redistributed by the atmosphere and by the oceans. The whole planet is enveloped by a layer of atmosphere; and about 70 percent of the earth's crust lies under water. The earth's rotation, coupled with the differential heating between equatorial and polar areas, sets up a system of redistribution of warmth from the tropics to the higher latitudes. Without the Gulf Stream current and its offshoot, the North Atlantic Drift, Europe would be as cold and barren as Labrador. Without the moisture brought from the oceans by humid air masses, the U.S. Midwest would be a desert rather than the nation's breadbasket.

In the earth's great ocean basins, water circulates in giant cells or *gyres* which, because of the earth's rotation, move slowly in a clockwise direction in the Northern Hemisphere and in a counterclockwise flow south of the equator. This places warm water, moving from low toward high latitudes, along east coasts, and cool water, flowing from polar latitudes toward the tropics, along west coasts. The Europe-warming North Atlantic Drift is actually a deviation from this pattern, caused by the configuration of North Atlantic coastlines.

Flowing over land and sea are currents of air, or *winds*. Again, the earth's rotation and the sun's heating generate the pattern, but atmospheric circulation is further complicated by the sizes, shapes, and topographies of the landmasses. We know some of these wind belts by familiar names: the easterly *trade winds* that form persistent belts in the northern and southern tropics, and which propelled the earliest transoceanic sailing ships; and the *westerlies*, the mid-latitude wind belt that carries weather across much of the United States from west to east. Other wind and air-pressure belts may be less familiar, but we have learned to pay attention to the behavior of high-speed, upper-atmosphere winds called *jet streams*. These tube-like streams of air snake around the globe in the middle latitudes, alternately forming northward and southward loops. By doing so, they interrupt average conditions, sometimes steering frigid polar air into lower latitudes, and at other times wafting tropical air poleward to soften a winter's cold.

The complex climatic pattern in Fig. I–10 results from these and other variables. Just as the ocean circulates in gyres, so the atmosphere is patterned into circulating cells called *highs* or *lows*, depending on the pressure or weight of the air they contain. High-pressure cells are associated with cold, heavy, "stable" air. Low-pressure cells tend to contain warmer, moister, less stable air which, if caused to rise, quickly condenses and precipitates its moisture.

But the capacity of an air mass to carry moisture is limited. When warm moist air is carried by the westerlies across Europe, it may rain—because the air rises against a cool mountain side; because it rides up against a colder, stabler air mass; or because it crosses warm terrain and rises through convection. By the time that air

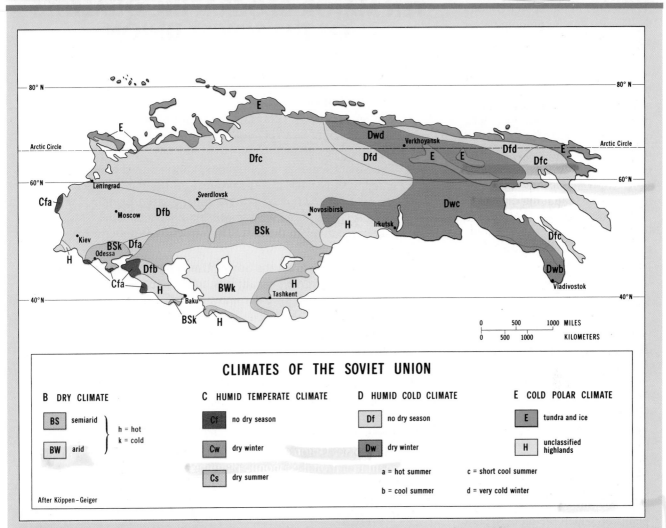

FIGURE 2–1

mass reaches Poland, it may already have lost much of its moisture; thus, when it reaches the Soviet Union, it may be dry. Countries that lie in the deep interiors of continents, perhaps separated from the ocean by mountains, are at a disadvantage when it comes to the essence of life—water.

The Soviet Union lies remote from the warm North Atlantic Drift current, and moisture-laden air masses must cross Europe before they reach its farmlands. As Fig. 2–1 shows, not only is the U.S.S.R. a high-latitude country, but it also lies exposed to polar coldness. No protective mountain barriers stretch across the north, and Arctic air has free reign from the Kola to the Kamchatka Peninsula. But in the south, from where warm moist air might have come, the Soviet Union is virtually ringed by high mountains. As a result, the entire southern region, from the Caspian Sea to the mountains on the border with China, is desert or semiarid steppe, climatic regions that do not yield until conditions become those of a cool-summer continental climate.

This climatic map (Fig. 2–1) goes far to explain why the majority of the Soviet population today remains concentrated in the western one-sixth of this gigantic country, why agriculture remains a critical problem, and why areas that are known to contain valuable raw materials remain virtually uninhabited. Russian tsars sought warm-water ports because the country's higher-latitude harbors froze up for several months each year; Soviet planners have diverted entire rivers to bring cultivation to desert lands. The Soviet Union is a storehouse of mineral wealth, but it faces a host of environmental problems that can be summarized under one rubric: climate.

Sunni Muslim Kazakhs. Today, more than 100 nationalities are represented in the Soviet Union; the Russians, who predominate, constitute just 51 percent of the total of nearly 300 million. As we will see later in this chapter, many of these nationalities were assigned a homeland on the complicated map the Soviet planners created. But by far the largest unit on the Soviet map is Russia itself, officially the Russian Soviet Federative Socialist Republic, the hearth of Russian **imperialism** and Soviet communism. Therefore, it is not surprising that throughout the period of Soviet organization, people have continued to call the U.S S.R. by its old name—Russia.

A Eurasian Heritage

The historical geography of the Soviet realm focuses on its western region: on Russia west of the Urals, on the Ukraine, and on neighboring areas. Although our knowledge of Russia before the Middle Ages is only fragmentary, it is clear that peoples moved in great migratory waves across the plains on which the modern state eventually was to emerge. The dominant direction seems to have been from east to west; many groups came from central Asia and left their imprints on the makeup of the population. Scythians, Sarmatians, Goths, and Huns came, settled, fought, and were absorbed or driven off. Eventually, the Slavs emerged as the dominant people in what is today the Ukraine; they were peasants, farming the good soils of the plains north of the Black Sea. Their first leadership came not from their own midst, but from a Scandinavian people to the northwest they called the Varangians (better known to us as the Vikings), who had for some time played an im-

---

TEN MAJOR GEOGRAPHIC QUALITIES OF THE SOVIET UNION

1. The Soviet Union is by far the largest territorial state in the world. Its area is more than twice as large as the next ranking country (Canada).

2. The Soviet Union's enormous area lies mainly at high latitudes; much of it is very cold or very dry. Large rugged mountain zones accentuate harsh environments.

3. The Soviet Union is a contiguous, multinational empire dominated by Russians. It is the product of Russian colonialism and imperial subjugation of nations, ranging from Estonia, Latvia, and Lithuania in the west to Muslim khanates in the south and Siberian natives in the east. This structure is under great stress in the 1990s.

4. For so large an area, the Soviet Union's population of 295 million is comparatively small; it is, moreover, overwhelmingly concentrated in the westernmost one-quarter of the country.

5. Development in the Soviet Union remains focused in "European" Russia west of the Ural Mountains. Here lie the major cities, industrial regions, densest transport networks, and most productive farming areas.

6. National integration and economic development east of the Ural Mountains in Soviet Asia extend mainly along a narrow corridor that connects Omsk, Novosibirsk, Irkutsk, Khabarovsk, and Vladivostok.

7. The Soviet Union constitutes the world's largest-scale experiment in national economic planning.

8. The Soviet Union is the heart of what is left of the socialist world (as opposed to the capitalist world and the Third World).

9. Its large territorial size notwithstanding, the Soviet Union suffers from land encirclement within Eurasia; it has few good and suitably located ports.

10. The Soviet Union has more neighbors than any other state in the world. Boundary conflicts and territorial disputes involve several of these neighboring states, ranging from capitalist Japan and hard-line China to theocratic Iran, former satellite Romania, and nonaligned Finland.

---

portant role in the fortified trading towns (*gorods*) of the area.

The first Slavic state (ninth century A.D.) came about for reasons that will be familiar to anyone who remembers the importance of the transalpine routes across the middle of Europe. The objective was

to render stable and secure an eastern crossing of the continent, from the Baltic Sea and Scandinavia in the northwest to Byzantine Europe and Constantinople in the southeast. This route ran southward from the Gulf of Finland to Novgorod (positioned on

Lake Ilmen), and then crossed what is today Belorussia to Kiev. From there, it proceeded along the Dnieper River to the shores of the Black Sea.

Novgorod, near Scandinavia and close to the Baltic coast, was the European center on this long route; it had a cosmopolitan population and had trade connections with the Hanseatic League. It benefited from its position to become the capital of its area, a distinctive princedom known as a *Rus*. Kiev, on the other hand, lay in the heart of the land of the Slavs. Its situation was near an important confluence of the Dnieper, and it was located near the zone of contact between the forests of middle Russia and the grassland steppes of the south. Kiev also had centrality: it served as a meeting place of Scandinavian with Mediterranean Europe. Though distinctly different from northern Novgorod, Kiev became a Europeanized urban center as well. It too was the center of a Rus, the Kievan Rus. During the eleventh and twelfth centuries, Kiev was the political and cultural focus for a large region. Briefly, Novgorod and Kiev even united, and regional stability brought still greater prosperity.

Mongols

Prosperity, however, attracts competition, and the Kievan Rus suffered internal division and external invasion. The external threat came from the Mongol Empire far to the east in inner Asia, which had been building under Genghis Khan. The Tatar hordes rode into the Kievan Rus, and the city as well as the state fell. Many Russians fled into the forests, where the horsemen of the open steppes were far less threatening. What remained of the western Rus, then, lay in the forest between the Baltic and the

SOVIET UNION

The Soviet Union is the product of the Revolution of 1917 when more than a decade of rebellion against the rule of Nicholas II led to the tsar's abdication. Russian revolutionary groups were called *soviets* (meaning ''councils''), and these soviets had been engaged in revolutionary activity since the first Russian workers' revolution in 1905. In that year, thousands of workers marched on the palace in protest, and the tsar's soldiers opened fire on the marchers, killing and wounding hundreds. Throughout the years that followed, Russia was engaged in costly armed conflicts, government became more corrupt and less effective, and social disarray prevailed. In 1917, a coalition of military and professional men forced the tsar to abdicate, and Russia was ruled by a Provisional Government until November (by the Gregorian calendar), when the country experienced its first and only truly democratic election.

In the meantime, however, the Provisional Government had allowed the return to Russia of exiled Bolshevik activists—Lenin from Switzerland, Trotsky from New York, and Stalin from Siberia. In the political struggle that ensued, the Bolsheviks gained control over the revolutionary soviets, and this ushered in the age of communism. Since 1924, the country has officially been known as the Union of Soviet Socialist Republics (U.S.S.R.), for which *Soviet Union* is shorthand. Sometimes the country is still referred to by its pre-revolutionary name, Russia. But Russia is only one (though by far the largest and dominant) of the 15 republics that constitute the U.S.S.R.

steppes, and in that area a number of weak feudal states arose. Many of these were ruled by princes who paid tribute to the Tatars in order to be left in relative peace.

From among these feudal states, the one centered on Moscow (Muscovy) possessed superior locational advantages. It was positioned on a river that formed a route to Novgorod, and it was centrally situated with respect to other Russian trading sites. While the Russians managed to hold off their Asian enemies, and the westward drive of the Tatars' ''Golden Horde'' slowed, a major state emerged where the old Kievan Rus had lain. This was the Grand Duchy of Lithuania, which for a time extended from near the Baltic Sea to the vicinity of the Black

Sea. When Moscow became the dominant Russian center of power, the Lithuanians, allied with the Poles, controlled a region that included present-day Poland, Belorussia, the Ukraine, and adjacent areas.

Soon Moscow's geographic advantages began to play their role. During the fifteenth century, Moscow's ruler was able to take control of Novgorod. But Muscovy became a real tsardom under Ivan (IV) the Terrible (1530–1584), who began Russia's advances into non-Russian territory—a campaign that did not end until well after the tsarist regime was terminated by the Communist revolution of 1917. Ivan first established control over the entire basin of the important Volga River to the east, inflicted heavy defeats on the Tatars, and then pushed eastward across the Urals into Siberia.

Cossacks

This eastward expansion of Russia was carried out by a relatively small group of semi-nomadic peoples who came to be known as Cossacks, and whose original home was in the present-day Ukraine. Opportunists and pioneers, they sought the riches of the eastern frontier, chiefly fur-bearing animals, as early as the sixteenth century. By the middle of the seventeenth century they reached the Pacific Ocean, defeating Tatars in their path and consolidating their gains by constructing *ostrogs*, strategic fortified way-stations along river courses. Before the eastward expansion halted in 1812

(Fig. 2–2), the Russians had moved across the Bering Strait to Alaska and down the western coast of North America into what is now northern California (see *box*).

Tsar Peter the Great

By the time Peter the Great took over the leadership of Russia (he reigned from 1682 to 1725), Moscow lay at the center of a great empire—great at least in terms of the territories under its hegemony. As such, emergent Russia had many enemies. The Mongols were no longer a menace, but the Swedes, no longer allies in trade, threatened from the northwest as did the Lithuanians. To the west there was a continuing conflict with the Germans and the Poles. And to the southwest, the Otto-

man Turks were heirs to the Byzantine Empire and also posed a threat. Peter consolidated Russia's gains and did much to make a modern European-style state out of the loosely knit country. He wanted to reorient the empire to the Baltic, to give it a window on the sea, and to make it a maritime as well as a land power. In 1703, following his orders, the building of Petersburg (later St. Petersburg) began. It was built by Italian architects at the tsar's behest, and they designed numerous ornate buildings that were arranged around the grandiose city's many waterways in a manner reminiscent of Venice (Fig. 2–3).

St. Petersburg, known today as Leningrad, is positioned at the head of the Gulf of Finland which opens onto the Baltic Sea. Not only did it provide Russia with an

FIGURE 2—2

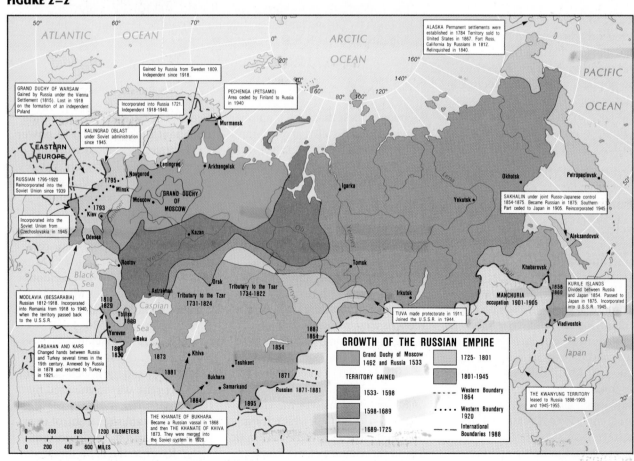

GROWTH OF THE RUSSIAN EMPIRE

FIGURE 2—3   A view across part of the center of stately Leningrad, which as St. Petersburg was built by some of southern Europe's leading architects.

important maritime outlet, but the city was also designed to function as a *forward capital*: it lay on the doorstep of Finland, which at that time was a Swedish possession, and thus represented the Russian determination to maintain its presence in this strategic area. In 1709, Peter's armies defeated the Swedes, confirming Russian power on the Baltic coast. Four years later, Peter took the momentous step of moving the Russian capital from Moscow to the new Baltic headquarters, where it remained until 1918.

Peter was an extraordinary leader, in many ways the founder of modern Russia. In his desire to remake Russia—to pull it from the forests of the interior to the western coast, to open it to outside influences, to end its comparative isolation—he left no stone unturned. Not only did he move the capital, but he himself, aware that the future of Russia as a major force lay in strength at sea as well as power on land, went to Holland to work as a laborer in the famed Dutch shipyards to learn how ships were most efficiently built. Peter wanted a European Russia, a maritime Russia, a cosmopolitan Russia. He developed St. Petersburg into a leading seat of power as well as one of the most magnificent cities in the world, and it remains to this day a European-style city apart from most others in the Soviet Union.

Tsarina Catherine the Great

During the eighteenth century, Tsarina Catherine the Great, who ruled from 1762 to 1796, continued to build Russian power, but on another coast and in another area: the Black Sea in the south. Here the Russians confronted the Turks, who had taken the initiative from the Greeks; the Byzantine Empire had been succeeded by the Turkish Ottoman Empire. But the Turks were no match for the Russians. The Crimean Peninsula soon fell, as did the old and important trading city of Odessa. Before long, the whole northern coast of the Black Sea was in Russian hands. Soon afterward, the Russians penetrated the area of the Caucasus to the southeast, and in due course they took Tbilisi, Baku, and Yerevan. But as they pushed farther into the corridor between the Black and the Caspian seas, they faced growing opposition from the British—who held sway in Persia (modern-day

RUSSIANS IN NORTH AMERICA

The first white settlers in Alaska were Russians, not Western Europeans, and they came across Siberia and the Bering Strait, not across the Atlantic and North America. Russian hunters of the sea otter, valued for its high-priced pelt, established their first Alaskan settlement at Kodiak Island in 1784. Moving southward along the North American coast, the Russians founded additional villages and forts to protect their tenuous holdings, until they reached as far as the area just north of San Francisco Bay, where they built Fort Ross in 1812.

But the Russian settlements were isolated and vulnerable. European fur traders began to put pressure on their Russian competitors, and Moscow found the distant settlements a burden and a risk. In any case, American, British, and Canadian hunters were decimating the sea otter population and profits declined. When U.S. Secretary of State William Seward offered to purchase Russia's holdings in 1867, Moscow quickly agreed—for $7.2 million. Thus, Alaska, including its lengthy southward coastal extension, became American territory. Although Seward was ridiculed for his decision—Alaska was called ''Seward's Folly'' and ''Seward's Icebox''—his reputation was redeemed when gold was discovered there in the 1890s. The twentieth century has proved Seward's action one of great wisdom, strategically as well as economically. At Prudhoe Bay, off Alaska's northern Arctic slope, large oil reserves are being exploited. And like Siberia, Alaska probably contains other yet unknown riches.

pansion and consolidation, to the first setback in the Russian drive for territory. In 1892, the Russians began building the Trans-Siberian Railroad in an effort to connect the distant frontier more effectively to the western core. As the map shows (Fig. 2–2), the most direct route to Vladivostok was across northeastern China (Manchuria). The Russians wanted China to permit the construction of the last link of the railway across Manchurian territory, but the Chinese resisted. Taking advantage of the 1900 Boxer Rebellion in China, Russia responded by annexing Manchuria and occupying it. This brought on the Russo-Japanese War of 1905, in which the Russians were disastrously defeated; Japan even took possession of southern Sakhalin Island (which they called Karafuto). For the first time in nearly five centuries, Russia sustained a setback that resulted in a territorial loss.

Colonial Legacy

Thus Russia—recipient of British and European innovations in common with Germany, France, and Italy—expanded by colonialism too. But where other European powers traveled by sea, Russian influence traveled overland into Central Asia, Siberia, China, and the Pacific coastlands of the Far East. What emerged was not the greatest empire, but the largest *territorially contiguous* empire in the world. It is tempting to speculate what would have happened to this sprawling realm had European Russia (for such it still was) developed politically and economically in the manner of the other European power cores. At the time of the Japanese war, the Russian tsar ruled over more than 8.5 million square miles (22 million square kilometers), just a tiny fraction less than the area of the Soviet Union today. Thus the modern

Iran)—as well as the Turks, and their advance was halted short of its probable ultimate goal: controlling a coast on the Indian Ocean.

The Nineteenth Century

Russian expansionism was not yet satisfied. While extending the empire southward, the Russians also took on the Poles, old enemies to the west, and succeeded in taking most of what is today the Polish state, including the capital of Warsaw. To the northwest, Russia took over Finland from the Swedes (1809). During most of the nineteenth century, however, the Russian preoccupation was with Central Asia, where Tashkent and Samarkand came under St. Petersburg's control. The Russians here were still bothered by raids of

nomadic Mongol horsemen, and therefore sought to establish their authority over the Central Asian steppe country as far as the edges of the high mountains that lay to the south. Thus Russia gained a considerable number of Muslim subjects, for this was Islamic Asia they were penetrating; but under tsarist rule, these people acquired a sort of ill-defined protectorate status while retaining some autonomy. Much farther to the east, a combination of Japanese expansionism and a decline of Chinese influence led Russia to annex from China several provinces to the east of the Amur River. Soon thereafter in 1860, the port of Vladivostok on the Pacific was founded.

Now began the course of events that was to lead, after five centuries of almost uninterrupted ex-communist empire, to a very large extent, is the legacy of St. Petersburg and European Russia, not the product of Moscow and the socialist revolution.

THE PHYSICAL STAGE

The Soviet Union is not only the world's largest country—it also is the world's northernmost large country. The Black Sea coast, the U.S.S.R.'s balmiest area, lies at the approximate latitude of Canada's southeastern Ontario; Edmonton, Alberta, the highest-latitude large Canadian city, lies farther south than Moscow.

Climatic Patterns

Study of the spatial aspects of weather and climate comprises the field of *climatology* (see pp. 136–137). In the Introduction we noted the distribution of global climatic regions (see Fig. I–10), based on a system of classification that involved measurements of temperature, moisture, vegetation response, and other factors. A look at the climatic map can often reveal much about a country's problems, and so it is with the Soviet Union. With high latitude and with remoteness from sources of warm water and moist air come climates dominated by drought and by wide temperature ranges. The term *continentality* is used to describe a climatic environment that is remote from moderating maritime influences. The Humid Cold D climates (Fig. 2–1) are also frequently called "Humid Continental" climates. Even if there is sufficient moisture, the growing season may be too short for many crops, and soils may be poorly developed. Compare Fig. 2–1 with Fig. I–12, and note how much of the northern U.S.S.R. is covered by needleleaf-forest and tundra vegetation, reflecting the frigid conditions that prevail there.

Where temperatures become warmer, drought is a problem. The south-central zone of the Soviet Union is a region of desert and steppe climates. Where water can be made available through irrigation, crops grow well. In the areas shown as BWk and BSk (k for cold winters—even here), the Soviet planners began huge irrigation projects. Some were successful, but others led to environmental disasters as surface streams were diverted, groundwater supplies dwindled, chemical pollution worsened, and public health was threatened. The Aral Sea, in the approximate center of the BWk region, has lost about half its water surface over the past three decades because streams feeding it have been diverted for irrigation (Fig. 2–4). Only in the Ukraine (see the small area of Dfa—f for

FIGURE 2–4   One of the world's greatest environmental disasters is the Aral Sea in south-central U.S.S.R., which has shrunk by nearly 50 percent since the 1950s. As the streams that fed it were diverted to irrigate the surrounding desert, the Aral's edges receded by several miles. These fishing boats strewn along the remains of a canal were left high and dry on a newly emerged desert of sand and salt—a dramatic reminder of the losing struggle of fishermen to maintain access to the rapidly retreating waters.

year-round moisture and *a* for warm summers) and in areas adjoining the Black Sea does the Soviet Union have what much of Europe and lower-latitude North America enjoy: moist moderate climates and good soils for intensive farming. Soviet dependence on annual grain imports to satisfy domestic demand is one of Moscow's greatest economic (and political) problems.

Physiographic Regions

Some of the forces shaping the Soviet Union's harsh environments can be gathered from the map of its physiography (Fig. 2–5). Mountains ring this giant

country from the Caucasus in the southwest to the highland complexes of the Pacific east. Warm tropical air has little opportunity to penetrate here, while cold Arctic air masses can sweep southward without interruption. The Arctic fringe is a lowland sloping gently toward the Arctic Ocean. The simplest physiographic subdivision of the Soviet Union produces nine regions (Fig. 2–5). The *Russian Plain* ① is the eastward continuation of the North European Lowland, the theater within which arose the Russian state. At its heart lies the Moscow Basin. To the north, the countryside is covered by needleleaf forests similar to those of Canada, whereas to the south lie the grainfields of the grassland Ukraine. The Russian Plain is bounded on the east by the

Ural Mountains ②, not a high range but prominent because it separates two extensive plains. The Ural range forms no barrier to transportation, and its southern end is quite densely populated; this region has yielded a variety of minerals, including oil.

South of the Urals lies the *Caspian-Aral Basin* ③, a region of windswept plains of steppe and desert. Its northern section is the Kirghiz Steppe. In the south, as we know, the Soviets are trying to conserve what little water there is through irrigation schemes along rivers that drain into the Aral Sea. East of the Urals lies Siberia. The *West Siberian Plain* ④ has been described as the world's largest true unbroken lowland; it is the basin of the Ob and Irtysh rivers. Over the last 1000 miles (1600

FIGURE 2–5

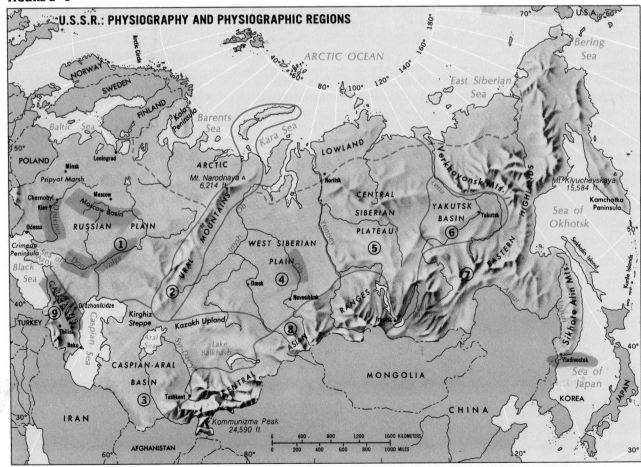

U.S.S.R.: PHYSIOGRAPHY AND PHYSIOGRAPHIC REGIONS

FIGURE 2-6   A passenger train under way on the double-tracked Trans-Siberian Railroad in south-central Siberia.

the aging Trans-Siberian line to the south (see Fig. 2–11).

The southern margins of the Soviet Union are also marked by mountains: the *Central Asian Ranges* ⑧ from the vicinity of Tashkent in the west to Lake Baykal in the east, and the *Caucasus* ⑨ in the corridor between the Black and Caspian seas. The Central Asian Ranges rise above the snow line and contain extensive mountain glaciers. The annual melting brings down alluvium to enrich the soils on the lower slopes and water to irrigate them. The Caucasus form an extension of Europe's Alpine mountain system and exhibit a similar topography, but they do not provide convenient passes. The trip by road from Tbilisi to Ordzhonikidze (via the Georgian Military Highway) is still a circuitous adventure in surface travel.

THE SOVIET UNION IN THE TWENTIETH CENTURY

Russia's enormous territorial expansion produced an empire, but nineteenth-century Russia was infamous for the wretched serfdom of its landless peasants, the exploitation of its workers, the excesses of its nobility, and the sumptuous palaces and riches of its rulers. The arrival of the Industrial Revolution had introduced a new age of misery for those laboring in factories. There were ugly strikes in the cities, and from time to time the peasants rebelled against the nobility. But retribution was always severe, and the tsars remained in tight control. And then, in 1905, Russian forces fighting

kilometers) of its course, the Ob falls less than 300 feet (90 meters). The northern area of the West Siberian Plain is permafrost-ridden, and the central area is marshy. But the south carries significant settlement, including the cities of Omsk and Novosibirsk, within the corridor of the Trans-Siberian Railroad (Fig. 2–6).

East of the West Siberian Plain the country begins to rise, first into the *Central Siberian Plateau* ⑤, a sparsely settled, remote, permafrost-afflicted region. Here winters are long and extremely cold, and summers are short; the area still remains barely touched by human activity. Beyond the *Yakutsk Basin* ⑥, the terrain becomes mountainous and the relief high. The *Eastern Highlands* ⑦

are a jumbled mass of ranges and ridges, precipitous valleys, and volcanic mountains. Lake Baykal lies in a trough that is over 5000 feet (1500 meters) deep. On the Kamchatka Peninsula, volcanic Mount Klyuchevskaya reaches nearly 15,600 feet (4750 meters). The northern area is the Soviet Union's most inhospitable zone, but southward along the Pacific coast the climate is less severe. Nonetheless, this is a true frontier region. The forests provide opportunities for lumbering, a fur trade exists, and there are gold and diamond deposits. To help develop this promising area, the Soviets have constructed the new Baykal-Amur Mainline (*BAM*) Railroad, a 2200-mile-long (3540 kilometer-long) route that roughly parallels

achieve with its own resources and labor force the goals that had for so long eluded it. The chief political and economic architect in this effort was also a revolutionary leader—V.I. Lenin (born Vladimir Ilyich Ulyanov). Among his solutions to the country's problems are the present-day political framework of the Soviet Union and its centrally planned economy.

The Political Framework

Russia's great expansion had brought a large number of nationalities under tsarist control, and now it was the turn of the revolutionary government to seek the best method of organizing this heterogeneous ethnic mosaic into a smoothly functioning state. The tsars had conquered, but they had done little to bring Russian culture to the peoples they ruled. The Georgians, Armenians, Tatars, and residents of the Muslim khanates of Central Asia were among dozens of individual cultural, linguistic, and religious groups that had not been "Russified." The Russians themselves in 1917, however, constituted only about one-half of the population of the entire country (the proportion today remains about the same, but the ethnic Russian population has slipped from 54 to 51 percent since 1970). Thus, it was impossible to establish a Russian state instantly over the whole of this vast political region, and these diverse national groups had to be accommodated.

The question of the nationalities became a major issue in the young Soviet state after 1917. Lenin, who brought the philosophies of Karl Marx to Russia, talked from the beginning about the "right of self-determination for the nationalities." The first response by many of Russia's subject peoples was to proclaim independent republics, as was done in the Ukraine, Georgia, Armenia, Azerbaydzhan, and even in Central Asia. But Lenin had no intention of permitting any breakup of the state. In 1923, when his blueprint for the new Soviet Union went into effect, the last of these briefly independent units was fully absorbed into the sphere of the

FIGURE 2–8

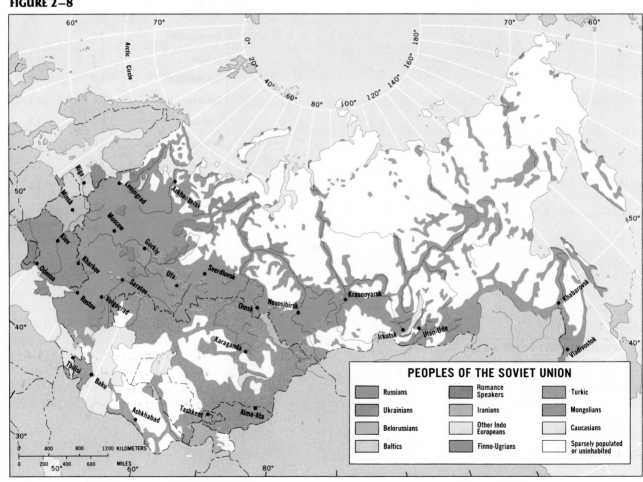

PEOPLES OF THE SOVIET UNION

Russians	Romance Speakers	Turkic
Ukrainians	Iranians	Mongolians
Belorussians	Other Indo Europeans	Caucasians
Baltics	Finno-Ugrians	Sparsely populated or uninhabited

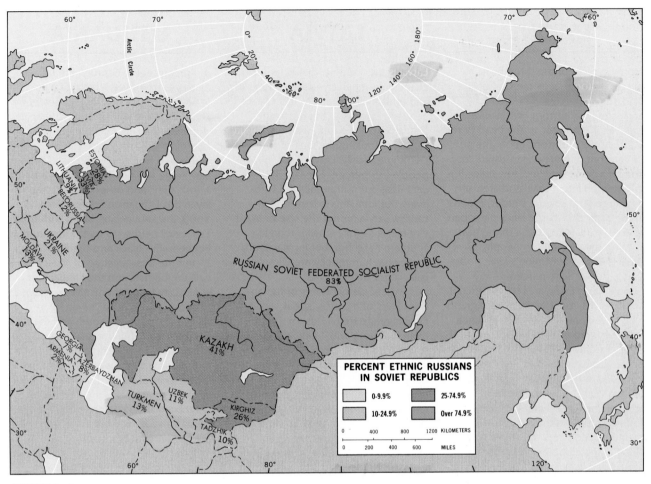

PERCENT ETHNIC RUSSIANS
IN SOVIET REPUBLICS

0-9.9%	25-74.9%
10-24.9%	Over 74.9%

FIGURE 2–9

Moscow government. Nonetheless, the political framework of the Soviet Union remains essentially based on the cultural identities of its incorporated peoples. Although there have been modifications since the days of Lenin, the major elements of the original system are still there. The country is divided into 15 Soviet Socialist Republics (or S.S.R.s for short), each of which broadly corresponds to one of the major nationalities (Fig. 2–7). As the map shows, the largest of these S.S.R.s (in fact, three times as large as all the others combined) is the Russian Soviet Federative Socialist Republic or R.S.F.S.R.; the remaining 14 republics lie in an arc around the western core of Russia, extending from the Gulf of Finland to Central Asia. However, the map of

nationalities, even a simplified representation such as Fig. 2–8, is much more complicated than the political map (Fig. 2–7). Within the S.S.R.s, territories of lesser rank were assigned to smaller minorities. "Autonomous" republics or autonomous oblasts (republics-within-republics), autonomous regions, and other units in the Soviet political framework may be nationality-based. As the profiles of the 15 S.S.R.s on the next five pages reveal, this framework has not always withstood the demands or satisfied the needs of the people involved.

The Russian S.F.S.R.

Theoretically, the 15 republics of the Soviet Union possess equal

standing. In practice, the Russian republic leads—and dominates. With half the country's population, its core area, the capital, and about three-quarters of the U.S.S.R.'s territory, Russia remains the nucleus of the Soviet Union. Russians migrated into all the non-Russian republics (Fig. 2–9), often to implement planning decisions made for those republics in Moscow and also to play leading roles in the political life (i.e., the Communist Party) there. The Russian language has been taught in the schools of all the other republics, but none of the other languages is required to be taught in the R.S.F.S.R.

As the dominant Soviet republic, the R.S.F.S.R. extends territorially from the Arctic Ocean to shorelines on both the Black and

ESTONIA

Northernmost of the three Baltic republics, Estonia faces the Gulf of Finland. Its capital, Tallinn, lies directly across the water from Helsinki. Leningrad lies to the east, at the head of the gulf. Estonia's territory, about the size of Vermont plus New Hampshire, consists of a coastal plain of low relief. It is fragmented into a mainland area and two large (and several smaller) islands in the Baltic Sea. Large lakes mark part of the border with Russia to the east.

With only 1.6 million people, Estonia is the Soviet Union's least populous republic. Sixty percent of the population is Estonian, but 28 percent are Russians, most of them immigrants who have come to the country since its incorporation into the U.S.S.R. in 1940. Estonian nationalism, however, never was extinguished. Germanic rule prevailed in the thirteenth century, and Estonia was endowed with its Lutheran religious heritage. Later the Swedes took control, and they in turn yielded their Baltic holdings to Tsar Peter the Great in the eighteenth century. Estonia freed itself from Russian rule during the Revolution, and it was an independent country between the two World Wars.

Estonia is an industrialized country that benefits from ample electric power derived from large deposits of oil shale; electricity from Estonia supplies consumers as far away as Leningrad. Economic links with the Russian hinterland are strong, but political ties have been weakening. Estonian nationalists call their country an occupied nation, and demands for independence arose soon after *perestroika* became Moscow's policy. In 1990, Estonian leaders announced plans to introduce a new currency to replace the Russian ruble.

LATVIA

Latvia is the "middle" Baltic republic in every way—spatially, demographically, and territorially. Spatially, it is situated between Estonia to the north and Lithuania to the south, with the capital, Riga, overlooking the Gulf of Riga. Demographically, Latvia's 2.8 million people (55 percent Latvian, 33 percent Russian) outnumber the Estonians, but Lithuania is larger. Territorially, Latvia (about the size of West Virginia) is slightly smaller than its southern neighbor, but its glaciated, somewhat hummocky area is substantially larger than Estonia.

Latvians, descendants of the ancient Balts of the area, speak a distinctive language derived from that of their ancestors. Most Latvians adhere to the Lutheran church, although there is a substantial minority of Roman Catholics.

Latvians have seen their country fall to Germans, Poles, Swedes, and Russians. By the end of the eighteenth century, Latvia had become a victim of the Russian tsars' drive to gain wider access to the sea, but independence was restored during World War I. From then until 1939, Latvia was a fully independent country. But in that year, Hitler's Germany and Stalin's Soviet Union signed a bilateral agreement that assigned all three of the Baltic states to the Russian "sphere of influence." The following year, the Soviet Union annexed Latvia and its neighbors, although the United States never has recognized this forced incorporation.

Latvia has made considerable economic progress and is an urbanized and industrialized country. Scientific equipment, machinery, and a wide range of consumer goods are made and exported, and the economy has attracted many Russian immigrants. In 1990, Latvia's Popular Front was calling for total independence from Moscow.

LITHUANIA

In the fourteenth century, the Lithuanians were a powerful regional force in what is today the western Soviet Union. The Grand Duchy of Lithuania dominated an area from the Baltic coast in the north to the Ukraine in the south, and from the environs of Warsaw in the west to Tula in the east. Minsk, Kiev, and other major centers lay in this Lithuanian sphere. Around 1600, the Lithuanians, allied with the Poles, still held sway over a vast region between the Black and the Baltic seas.

But the Russian state of Muscovy was on the rise, and the Lithuanians eventually found themselves, wedged between the Poles (allies no longer) and Germans to the south and the Latvians to the north, in a small coastal Baltic state. Today, Lithuania has just a tiny window to the sea, wide enough for the port of Klaipeda but little else. Unlike its two coast-oriented Baltic neighbors, Lithuania has positioned its capital, Vilnius, in the interior, very near the border with Belorussia.

Lithuania's nearly flat territory is slightly larger than that of Latvia. It contains a notable resource in the form of amber, a golden-hued, translucent fossil resin that is hard enough to be fashioned into ornamental objects such as beads and rings.

Lithuanians make up 80 percent of the total population of 3.6 million; only 9 percent are Russians, reflecting the country's slower pace of modernization. The Lithuanian language, with its ancient roots, and the Lithuanian Roman Catholic and Lutheran churches, serve as ingredients for a strong nationalism that has survived Soviet annexation (1940) and its aftermath. By 1990, Lithuania had taken the lead among the Baltic states in moving toward secession from the Soviet Union.

BELORUSSIA

Belorussia lies in the shatter belt that took the brunt of impact during World War II. Its compact territory, about the size of Kansas, lies centered on the capital and largest city, Minsk. Belorussia has five neighbors: Poland to the west, Russia to the east, Lithuania and Latvia to the north, and the Ukraine to the south. Political and economic directions in these neighbors vary, and Belorussia cannot remain unaffected by those changes.

Belorussia's population of 10.5 million is 80 percent Belorussian (a Slavic people) and only 12 percent Russian. Belorussians (or "White" Russians) do not have a history of national sovereignty, and in their landlocked republic nationalism has not been a potent force.

Although Belorussia was one of the four union republics when the U.S.S.R. was created in 1922, its role in the Soviet Union's economy was overshadowed by the others, notably the Ukraine. Its major resource was peat; forests covered as much as one-third of the territory; marshes extended across the south. Already lagging, the republic's limited agricultural and industrial economies were shattered by World War II. One-quarter of the population was killed, and recovery was slow.

Minsk, almost totally rebuilt after the war, has become an important industrial center because it was linked to the Soviet Union's expanding network of oil and gas pipelines, which provide energy and raw materials for a petrochemical complex. Some oil has been discovered in the republic itself; a potash deposit is being developed for fertilizer production; and agriculture, though still inefficient, is expanding. Nevertheless, Belorussia remains comparatively isolated and stagnant in a changing realm.

THE UKRAINE

The Ukraine is critical not only to the future of a Soviet Union, but to the prospects of Russia itself. Facing the Black Sea, endowed with a wide range of resources, blessed with a comparatively moderate climate, and served by good transport networks, the Ukraine is a cornerstone of the Soviet Union.

Of all the non-Russian republics, the Ukraine has the strongest ingredients of nationhood. Its population of 53 million is three-quarters Ukrainian (and 21 percent Russian). Its territory, nearly the size of Texas, is larger than that of any European country, even a reunified Germany. Its capital and primate city, Kiev, is one of the realm's leading historic, cultural, and economic centers. In the late 1980s, the Ukraine was producing nearly half of the entire U.S.S.R.'s total farm output by value, in addition to a wide array of raw materials ranging from coal and oil to iron ore and manganese.

In the Ukraine, *perestroika* will meet what may become its most consequential challenge. Stirrings of nationalism, grounded in religious issues and encouraged by Ukrainian groups living outside the U.S.S.R., already have occurred. A millennium ago, Byzantine Christianity came to Kiev, eventually to give rise to the Russian Orthodox Church. Today, the Orthodox Greek and Roman Catholic churches have large numbers of active and nominal adherents whose intentions are not exclusively religious.

As one of the four union republics, the Ukraine owes its relationship with Russia in part to the Cossacks, a group of powerful warriors based in the eastern part of its area. These Cossacks helped secure the Ukraine for Russia. After its founding as a Soviet republic, the Ukraine was expanded repeatedly; in 1940 it acquired a part of Romanian Bessarabia, and in 1954 the Crimean Peninsula was incorporated.

MOLDAVIA

On the map, Moldavia looks like an extension of Romania—and it is. Two-thirds of its population of 4.4 million is Moldavian, which is to say Romanian. This landlocked territory between the Prut and Dnestr rivers, about twice the size of New Jersey, was taken from Romania by the Soviets in 1940.

But that was only the latest in a centuries-long series of transfers involving the land and people of Bessarabia, as this area was formerly called. Along with Romania, Bessarabia was part of the Ottoman Empire; when the Turks withdrew, they ceded it to the Russians in 1812. After World War I, Moldavia rejoined Romania, but in the aftermath of World War II the Soviets annexed the area and established a Soviet Socialist Republic there. The Russians imposed their culture, including the use of their Cyrillic alphabet to spell the Romanian (Romance) dialect spoken in Moldavia. In 1989, the Moldavians were allowed to reintroduce Roman characters to write their language, the result of a long campaign that stoked fires of nationalism.

Moldavia lies landlocked by a narrow strip of the Ukraine from the Black Sea, and is situated in the hinterland of the port of Odessa. The capital, Kishinev, reflects the modest industrial development that has taken place, but Moldavia remains primarily an agricultural and pastoral area, one of the Soviet Union's breadbaskets. Best known for its viticulture and very good wines, Moldavia also produces corn and wheat as well as vegetables and fruits. Soils, like those of the Ukraine, are fertile, and on extensive northern pastures cattle and pigs are raised for dairy products and meat. Given this agricultural base, the Russian presence is comparatively small, amounting to about 13 percent of the population. In fact, Ukrainians in Moldavia slightly outnumber the Russians.

GEORGIA

Soviet Transcaucasia consists of three S.S.R.s, of which only one, Georgia, has a Black Sea coast. Georgia, somewhat smaller than South Carolina, is a country of high mountains and fertile valleys and a complicated political geography. Its population of 5.5 million is nearly 70 percent Georgian, but it also includes Armenians (9 percent), Russians (7 percent), Azerbaydzhanis or *Azeris* (5 percent), and smaller numbers of Ossetians and Abkhazians. Within the republic lie three minority-based autonomous entities, the Abkhazian and Adjarian Autonomous Soviet Socialist Republics and the South Ossetian Autonomous Region.

Sakartvelo, as the Georgians call their country, has a long and turbulent history. Tbilisi, the capital for 15 centuries, lay at the core of a major empire around the turn of the thirteenth century, but the Mongol invasion ended this era. Next, the Christian Georgians found themselves in the path of Islamic wars between Turks and Persians. Turning northward for protection, the Georgians soon were annexed by the Russians, who were looking for warm-water exits. Like other peoples overpowered by the tsars, the Georgians seized on the Russian Revolution to reassert their independence; but the Soviets reincorporated Georgia in 1921 and proclaimed a Georgian Soviet Socialist Republic in 1936. Josef Stalin was a Georgian.

Georgia is renowned for its scenic beauty, warm and favorable climates, varied agricultural production (especially tea), timber, manganese, and other products. Georgian wines, tobacco, and citrus fruits are much in demand. The diversified economy is healthy.

Anti-Russian sentiment was worsened in 1989 when Soviet troops suppressed protest demonstrations with heavy loss of life. Strikes and railroad blockades have accompanied calls for independence.

ARMENIA

Armenia occupies some of the most rugged and earthquake-prone terrain in the Soviet Transcaucasus. Territorially the smallest of all the Soviet republics, Maryland-sized Armenia has a prorupt shape (see Chapter 10) that has the effect of fragmenting its eastern neighbor, Azerbaydzhan, into two parts.

More than 90 percent of Armenia's predominantly Christian population of 3.5 million is Armenian, and less than 3 percent is Russian. But Armenians also form a majority of 75 percent in the Nagorno-Karabakh Autonomous Region, an Armenian exclave encircled (and administered) by Muslim Azerbaydzhan.

This spatial recipe for trouble, engineered by Soviet sociopolitical planners in the 1920s, now destabilizes the region. Armenians, who adopted Christianity 17 centuries ago, for more than a millennium have sought to establish a secure homeland here on the margins of the Muslim world. Their ancient domain was overrun by Turks, Persians, and later by Russians. During World War I, the Ottoman Turks decimated their Armenian adversaries on the empire's eastern flank. At the end of the war an independent Armenia briefly arose, but it lasted just two years. In 1920 it became a Soviet republic, and in 1936 it was proclaimed one of the 15 constituent republics of the Soviet Union.

Yerevan, the capital, lies within sight of the Turkish border. Mineral-rich and producing a variety of subtropical fruits, Armenia has opportunities for development; hydroelectric power is plentiful. But severe setbacks have afflicted the republic. A disastrous earthquake struck in 1988, killing tens of thousands (see Fig. I–7). Conflict between Armenians and Muslim Azeris, initially over conditions in Nagorno-Karabakh, spread to other areas shared by these peoples. Soviet intervention suppressed the violence, but the root causes remain.

AZERBAYDZHAN

The 7 million people of Azerbaydzhan—the *Azeris*—occupy a corner of Transcaucasia that lies separated from Russia by mountains, from its western neighbors by religion and ethnicity, and from the Muslim republics to the east by the waters of the Caspian Sea. Small wonder that the Muslim Shi'ite Azeris look southward across the border to Iran for support. About 4 million Azeris live in northern Iran; they speak the language and practice the faith of their Soviet counterparts. Late in 1989, Azerbaydzhani nationalists tore down once-impregnable boundary fences and crossed into Iran in large numbers, many reportedly returning with weapons in hand—weapons for the struggle against Christian Armenians in their midst and against Russian occupiers from the north.

Azerbaydzhan marks the northward limit of the Islamic march into the Black-Caspian corridor. The Tatars declared independence here in 1918, but the Red Army conquered them and made the area a Soviet republic in 1920. The complex political geography of Maine-sized Azerbaydzhan includes the Nagorno-Karabakh Autonomous Region (population 175,000; three-quarters Armenian) and the Nakhichevan Autonomous S.S.R. (200,000), an Azeri exclave created by a corridor of Armenian territory.

Baku, the coastal capital, has grown into the U.S.S.R.'s fifth largest city, based on nearby oil reserves, associated industries, and an expanding port function. Many Armenians came to work here, and friction developed. In 1989, Azeris attacked the Armenian communities, killing many; of more than 200,000, only 10 percent remained by the time Soviet forces had gained control. Demands for independence and threats of an Afghanistan-style guerrilla war arose. The errors of the 1920s haunt Transcaucasia today.

KAZAKHSTAN

Kazakhstan is not only territorially the largest non-Russian Soviet republic: it is larger than all the others combined. Its more than 1 million square miles (about 30 percent of the land area of the United States) extend eastward from the Caspian Sea to the Chinese border.

In a Soviet Union of many nationalities, Kazakhstan may have the most complex cultural mosaic of all. Although the total population is only 17.9 million, more than 100 ethnic groups are represented. Kazakhs now constitute just 36 percent of the total; they are outnumbered by Russians (41 percent). Ukrainians (6 percent) also form a significant minority.

The Kazakhs are a people of Mongol descent, who adopted Sunni Islam and speak a Turkic language. A loosely knit nation of nomadic herders, they were overpowered by the Russians during the nineteenth century. The present S.S.R. is much larger than the original Kazakh domain, since the Soviets welded territory to it before and after its proclamation as a union republic in 1936. The capital, Alma-Ata, lies in the far southeast, but development has been concentrated in the Russified north and center of Kazakhstan.

Kazakhstan is the scene of grand Soviet planning projects—and dismal failures. In the 1950s, Moscow launched the Virgin and Idle Lands Program to cultivate and irrigate vast pasturelands. Millions of acres of land were opened up and production expanded, but climatic variability, excessive groundwater use, and indiscriminate pesticide application created an environmental disaster. In the 1980s, the Aral Sea was drying up and health statistics showed the devastating impact of chemical pollution.

Kazakhstan's political and cultural mixes are explosive, touching off both anti-Russian and interethnic violence sporadically. In addition, Kazakhstan has laid claim to territory in neighboring Uzbekistan and Turkmenistan.

UZBEKISTAN

To a greater degree than even Kazakhstan, Uzbekistan has been blighted by Soviet agricultural development policy. Cotton, grown under irrigation in the desert, has become the economic mainstay at the cost of severe environmental degradation. To make Uzbekistan the world's third largest cotton producer (it harvests two-thirds of the entire Soviet output), water has been diverted from rivers; streams have dried up, and the Aral Sea in 1990 had shrunk to only about half its 1960 size. Pesticides pollute drinking water; locally grown foods are unsafe. Infant mortality is soaring, and cancer rates are high.

With 19.9 million people (70 percent Uzbek), this is the third most populous Soviet republic, and Uzbeks form the third largest Soviet national group. Adherents of Sunni Islam, they ruled Central Asia until their Turkestan khanates of Khiva and Bokhara were absorbed into the Soviet Union in 1924. The Russian presence is just over 11 percent, in sharp contrast to neighboring Kazakhstan. Industrial development is minor. Uzbekistan produces half of the Soviet Union's rice.

The capital of California-sized Uzbekistan, Tashkent (the U.S.S.R.'s fourth largest city), carries prominently the imprint of Islam; some modern industry has developed here. The city also has been the scene of anti-Russian demonstrations reflecting anger over environmental issues and inadequate Muslim representation in government. Past Soviet practices of exiling perceived enemies of the state to Central Asia now result in strife as well: the local Meshketian minority, evicted by Stalin from their home in Georgia for "likely" collaboration with the enemy during World War II, has been attacked by Uzbeks who resent their comparative well-being in the fertile Fergana Valley. Relations between Uzbekistan and Kazakhstan are strained over territorial issues.

TURKMENISTAN

Ashkhabad is the Soviet Union's southernmost capital, the focus of a desert republic inhabited by just 3.5 million people. Extending from the shores of the Caspian Sea to the boundary with Afghanistan, this area was part of Muslim Turkestan before it became a Soviet republic in 1925. As large as Nevada and Utah combined, Turkmenistan was the frontier domain of many nomadic peoples when Soviet efforts to modernize the area began. Although 70 percent of the republic's inhabitants are Turkmen, many ethnic divisions still fragment the predominantly Muslim population. In addition, there are Russians (13 percent, mostly in the towns), Ukrainians, Uzbeks, Kazakhs, Tatars, and Armenians.

Soviet efforts to stabilize the population and to make it more sedentary center on a massive project: the Kara Kum Canal, begun in the 1950s to bring water from mountains to the east into the heart of the desert. By 1990, the canal, which eventually will reach the Caspian Sea, was over 700 miles (1100 km) long, and more than 2.7 million acres had been brought under cultivation. Cotton, corn, fruits, and vegetables are now farmed. But many Turkmen people still herd sheep, and Astrakhan fur remains a valuable export.

The Soviet central government has accused Iran of **irredentism** toward the people of Turkmenistan, but joint plans for better surface communications between Tehran and the republic were going ahead in the early 1990s. In late 1989 and early 1990, many Armenians who fled the Azeri pogroms in Baku crossed the Caspian Sea to seek initial safe haven in Turkmenistan. Remoteness and ribbon development have helped keep Turkmenistan calm in turbulent times, but it does border Iran and Afghanistan.

TADZHIKISTAN

In the rugged terrain of southeastern Muslim Central Asia, the political geography is as complex as it is in Transcaucasia. Tadzhikistan (also spelled *Tajikistan*), Kirghizia, and eastern Uzbekistan look like pieces of a jigsaw puzzle, their territories interlocked and, in Tadzhikistan's case, fragmented.

Iowa-sized Tadzhikistan borders China as well as Afghanistan. Its east is dominated by the gigantic Pamirs, where high glacier-sustaining mountains feed rivers such as the Amu-Darya, source of irrigation water for the neighboring desert republics of Turkmenistan and Uzbekistan.

Tadzhiks constitute 60 percent of the population of 5.4 million; they are of Persian (Iranian) origin and speak a Persian language, but are Sunni (rather than Shi'ite) Muslims. Many minorities share the republic's land, including Uzbeks, concentrated in the northwest, who comprise nearly one-quarter of the total; Russians (10 percent); Ukrainians; Armenians (some recently arrived from strife-torn Transcaucasia); and Ismaili Muslims (for whom the Gorno-Badakhshan Autonomous Region was created in mountain valleys in the Pamirs).

The capital, Dushanbe, lies in the west, not far from lands claimed by the republic from Uzbekistan. When Tadzhikistan was established, this was an area of handicraft industries and semi-nomadic herding. Soviet rule brought large industrial enterprises, mining, and irrigated agriculture. But local traditions are durable (Muslim fundamentalism is on the rise), and most Tadzhiks today remain farmers, herders, and producers of subtropical fruits and grains. In 1990 unrest broke out, and Dushanbe was the scene of anti-minority and anti-Russian riots. Established as a constituent republic of the Soviet Union in 1929, Tadzhikistan still is a remote frontier.

KIRGHIZIA

The great ranges of the Tianshan straddle the center of Kirghizia, a republic whose outline on the map seems to relate to neither topography nor culture. Once part of Turkestan, this territory fell to Russian expansionism during the nineteenth century, when it was a remote frontier land. Once under Soviet control, it became an Autonomous Republic before it was elevated to the status of constituent republic of the Soviet Union in 1936.

Kirghizia's boundaries were defined on the basis of nationality as interpreted by Soviet planners in the 1920s, but today the indigenous inhabitants, the Kirghiz, represent just under 50 percent of the population of 4.2 million. Russians number more than 25 percent, Uzbeks about 13 percent. Ukrainians, Tatars, and other minorities also inhabit Nebraska-sized Kirghizia. Like their neighbors to the west and south, the Kirghiz are Sunni Muslims.

The capital, Frunze, lies on the north slope very near the border with Kazakhstan. Surface communications between northern Kirghizia and neighboring Kazakhstan are better than those within Kirghizia itself; the Tianshan remains a formidable barrier, although a tunnel is under construction. Towns in southwestern Kirghizia are linked by road and rail to Uzbekistan, from where many immigrants have come to work in factories and to farm.

Pastoralism is the republic's mainstay. In addition to sheep and cattle, the Kirghiz raise yaks for meat and milk; the yak can survive on high-altitude pastures unsuitable for other livestock. Irrigated lowlands adjacent to the mountains yield wheat, fruits, and vegetables. Most industries constitute processing and packaging plants for locally produced textiles and foods, as well as other consumer goods. Remote and disjointed, Kirghizia has little national identity. Neighboring China has at times laid claim to Kirghizian territory.

RUSSIA

The Russian Soviet Federative Socialist Republic (R.S.F.S.R.) contains just over three-quarters of the total area of the Soviet Union. Ethnic Russians comprise about 80 percent of its population of 150 million. Russia is the motherland, the heart of the old empire and the new Soviet Union, the source of its ideology and the core of its strength. An estimated 70 percent of the Soviet Union's annual industrial and agricultural production comes from Russia.

But Russia is no homogeneous region, either in cultural or in political terms. Included in its huge area are 16 Autonomous Soviet Republics, ranging in size from the vast Yakut A.S.R. in eastern Siberia with nearly 1.2 million square miles (3.1 million sq. km.) down to the tiny North Ossetian A.S.R. with about 3,100 square miles (8,000 sq. km.) in the Caucasus. In some of these republics, which have a lesser status than Soviet Socialist Republics (S.S.R.s), Russians are in a plurality but do not constitute a majority. In others, Russians are a minority. Ethnic patterns and political expediencies combine to generate the complicated Russian map.

The Russian Communist Party's dominance over political and economic affairs appears now to be threatened. Disputes and defections are eroding it, and an unprecedented surge of alternate preferences is confronting it. Russian nationalism also is emerging as a force of consequence. The Soviet-style Russian map may become another victim of *perestroika*.

Caspian seas, and from the Gulf of Finland to the Pacific Ocean. The entire Far East and all of Siberia form part of the Russian republic although, as Fig. 2–8 shows, Mongolians and Turkic-speaking minorities do live there. Indeed, even in the west, where the Russian population remains concentrated, numerous minorities are incorporated within the R.S.F.-S.R. In accordance with Soviet administrative principles, many of these national groups have local self-government in designated areas. The R.S.F.S.R includes 6 Territories, 49 Regions, 16 Autonomous Soviet Republics, 5 Autonomous Regions, and 10 Autonomous Areas. This system is unwieldy and cumbersome, and today fails in many instances to solve the problems it was designed, 70 years ago, to overcome.

An example is the Tatar Autonomous Soviet Republic, a ''republic'' of 4 million located in the hinterland of Kazan, a city east of Moscow on the Volga River. Kazan was an important Tatar center, but when Ivan the Terrible defeated the Tatars, he destroyed 400 mosques (at the same time, this tsar built the great St. Basil's Cathedral in Moscow). Many Tatars continued to live in the area even after the Russian conquest, and they participated in the Revolution of 1917. When the new Soviet Union was designed, the Tatars were rewarded with their own Autonomous Soviet Socialist Republic, an area of about 26,000 square miles (68,000 square kilometers), including Kazan, in which they were the ethnic majority. Today, however, the Tatars no longer are the majority in their own republic. About 55 percent of the population is Russian and only 40 percent Tatar. Kazan and the republic lay in the path of Soviet industrial expansion, and development schemes for the Tatar republic included truck fac-

tories, chemical industries, and engineering plants. These, in turn, attracted Russian workers, and the population balance shifted. The Tatars began to object, their anger already aroused by cultural Russification and the discouragement of their Muslim faith. Now the Tatars are trying to revive their remaining mosques, rekindle their Arabic language, and stem the tide of Slavic culture. The political framework of the past is a Soviet liability, but boundaries, once delimited, are not easily relocated.

Compared to what has occurred in other ethnic-designated areas in the Soviet Union, developments in the Tatar Autonomous Soviet Republic have been benign. Reaction

to Russification elsewhere has been much more violent (see the republic profiles on pp. 150–154), and even the remote Muslim republics in Central Asia have been affected. Russification and Sovietization are efforts at what cultural geographers call **acculturation**, the transfer of cultural traditions and practices from a dominant community to a weaker one. In perhaps no other part of the world was a more comprehensive effort at acculturation made. Russian practices and Soviet programs were systematically imposed from Moscow onto a patchwork of peoples ranging from fellow Slavs to Mongols and Muslims. Despite huge investments and elaborate systems

FIGURE 2–10 Harvesting wheat on a giant state farm (*sovkhoz*) in the eastern Ukraine. Despite mechanization, agricultural productivity here falls well below the level of the rest of the developed world.

of control, the program ultimately failed—emphasizing the strength and durability of older cultures and religions.

The Economic Framework

The geopolitical changes that accompanied the founding of the Soviet Union were accompanied by a gigantic economic experiment: the conversion of the empire from a capitalist system to communism. From the early 1920s onward, the country's economy would be *centrally planned*—that is, all decisions regarding **economic planning** and development were made by the communist leadership in Moscow. Soviet planners had

two principal objectives: (1) to speed industrialization, and (2) to collectivize agriculture. In order to accomplish this, the entire country was mobilized, with a national planning commission (*Gosplan*) at the helm. For the first time ever on such a scale, and for the first time in accordance with Marxist-Leninist principles, a whole country was organized to work toward national goals prescribed by a central government.

The Soviet planners believed that agriculture could be made more productive by organizing it into huge state-run enterprises. Landowners were expropriated, private farms taken away from the farmers, and land was consolidated into collective farms. Initially, it was intended that all such land would be part of a *sovkhoz*,

literally a grain-and-meat factory in which agricultural efficiency, through maximum mechanization and minimum labor requirements, would be at its peak (Fig. 2–10). But the Soviets ran into opposition from many farmers, who tried to sabotage the program in various ways, hoping to retain their land. These opponents were ruthlessly dealt with, but the planners realized that a smaller collective farm known as a *kolkhoz* would be more easily established where acceptance of the reform program had been achieved. After 15 years of reform, about 90 percent of all farms, ranging from large estates to peasant holdings, had been collectivized.

Productivity, however, did not rise as the Soviet planners had hoped. Farmers like to tend their

FIGURE 2–11

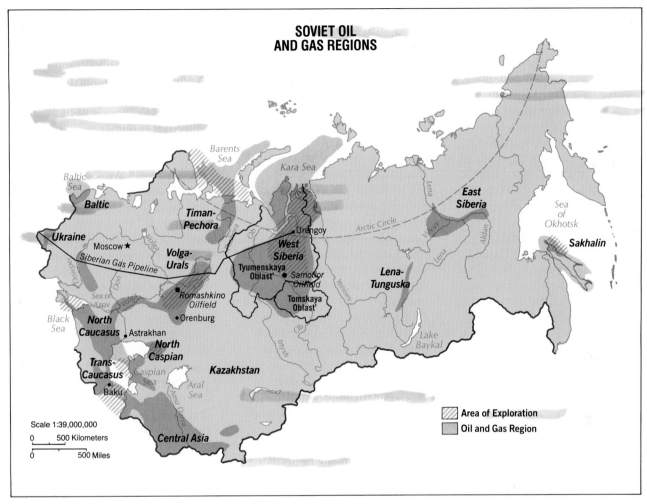

SOVIET OIL AND GAS REGIONS

Area of Exploration
Oil and Gas Region

Scale 1:39,000,000

0   500 Kilometers

0   500 Miles

FIGURE 2—12

own land, and they take better care of it than they do the land of the state. Resentment over the harsh imposition of the program undoubtedly played its role, too, but ultimately the two persistent problems were poor management and weak incentives for farmers to do their best. Add to this the recurrent weather problems caused by the country's disadvantageous relative location, and it is not surprising that Soviet farm yields were below expectations—and below requirements. Throughout the three generations of communist rule, the Soviets have had to turn to the outside world for food supplies.

Even as collectivization proceeded, the Soviets brought mil-

lions of acres of new land into cultivation through ambitious irrigation schemes. During the 1950s, they launched the Virgin and Idle Lands Program in Kazakhstan, turning pasturelands into cotton fields. The program transformed parts of the Kazakh republic (see p. 153), but success was linked to environmental disaster as diverted streams dried up, groundwater was poisoned by pesticides, and people in large numbers became ill. Again, the grand design created new problems as it solved old ones.

The Soviet planners hoped that collectivized and mechanized agriculture would free hundreds of thousands of workers to labor in the industries they wanted to es-

tablish. Enormous amounts of money were allocated to the development of manufacturing; the Soviets knew that national power would be based on the country's factories. The transport networks were extended; a second rail route to the Far East (the BAM) was completed, and such remote places as Alma-Ata and Karaganda were connected to the system (Fig. 2–11). Energy development was given priority, and compared to agriculture, the U.S.S.R.'s industrialization program had good results. Productivity rose rapidly, and when World War II engulfed the country, its industries were able to generate the equipment needed to repel the German invaders. But even in this context the

Soviet grand design held liabilities for the future. Not needing to respond to market pressures or certain cost factors, the Soviet planners assigned the production of certain goods to particular places, sometimes disregarding locational considerations. For example, the production of office filing cabinets might be assigned to a factory in Minsk. No other factory anywhere else would be permitted to produce this equipment—even if supplies of needed raw materials ran out near Minsk and were available, say, near Kuybyshev about a thousand miles away. Thus, the industrial map of the U.S.S.R. locked certain places into particular production tasks. When the political system began to crack, that proved to be a major impediment to reform.

Its vast size and varied geology endow the Soviet Union with a wide range of raw materials for industry, as we will observe in greater detail in the regional discussion that follows. In the all-important field of fossil fuels (oil, natural gas, coal), the U.S.S.R. is the world's leading producer of the first two and ranks third in coal output. Nevertheless, the country faces problems in sustaining its position as the world's top (overall) energy producer and one of its leading consumers and exporters. The problem is geographic: energy resources near existing industrial areas and population centers are rapidly being depleted, whereas vast fuel deposits lie in remote parts of the country in environments that deter exploitation (Fig. 2–12). Already, the Soviet Union west of the Urals relies for more than half of its annual supply of energy on imports from Siberia and Central Asia. The future may prove some of these sources to be undependable.

Soviet foreign-trade income depends heavily on oil and gas exports, which in the late 1980s were accounting for about 60 percent of all export revenues. European countries are the U.S.S.R.'s chief customers, but Soviet oil and liquefied gas also go to Japan, India, and Brazil. To sustain their energy output, Soviet planners have invested heavily in nuclear-power production, a program that came to international attention following the disastrous accident at the Ukraine's Chernobyl nuclear power plant in 1986. At the time, nuclear reactors were providing more than 10 percent of all electricity consumed; today, the Soviet nuclear-reactor program is going forward again, undeterred by the concerns raised in the aftermath of Chernobyl. The provision of plentiful electric power is an indispensable ingredient of Soviet development programs, and everything possible must be done to overcome distance and isolation. Where coal is running out, where hydroelectric power is not available, and where other fuels are too far away, the nuclear alternative is an imperative.

SOVIET POPULATION AND MIGRATION

Little more than a century ago, Russia had twice as many inhabitants as the United States; even as late as 1900, the Russian population was about 125 million whereas the U.S. population was 76 million. Yet today, the Soviet population is only about 18 percent larger than that of the United States. What suppressed Soviet population growth? The unhappy answer is that the twentieth century brought repeated wartime disaster to the Soviet people, in spite of the fact that their country was rising to the position of world superpower. Moreover, the Soviet regime during the dictatorial rule of Stalin (1929–1953) killed between 5 and 10 million, either by execution or in famines caused by the collectivization movement. But the major destruction, staggering in its magnitude, was caused by the two World Wars.

World War I, the Revolution, and associated dislocation and famine resulted in about 17 million deaths and about 8 million deficit births (i.e., births that would have taken place had the population not been reduced by such large numbers). World War II cost the Soviet Union approximately 27 million deaths and 13 million deficit births. Therefore, during the twentieth century, the Soviet population lost over 70 million in destroyed lives and unborn children; American losses in all its twentieth-century wars stand well below 1 million. In addition, there was the effect of **migration**: the Soviet Union was affected by substantial emigration, while the United States received millions of immigrants. Hence, the American population grew rapidly, whereas the Soviet increase was severely limited; in the half-century from the beginning of World War I to 1964, the United States gained some 90 million people (from a base of 100 million), while the Soviet Union gained about 65 million. Thus, during that period, the U.S. population nearly doubled as the Soviet population grew by only one-third.

Naturally, the demographic results of these calamities have concerned Soviet governments and planners. The deficit of births is still reflected today by a shortage of younger people in the labor force; at higher ages, women still outnumber men to a greater extent than perhaps in any other country in the world. For a period after

FIGURE 2–13   To attract people from west of the Urals, towns of the Eastern Frontier must offer better housing. This is a large residential complex (exhibiting a lavishly low density by Soviet standards) under construction in Bratsk, a rapidly growing city at the site of a huge dam on the Angara River, 400 miles downstream from its Lake Baykal outlet.

World War II, the government pursued policies whereby families were encouraged to have a large number of children, and mothers who had 10 children received the Order of Mother Heroine from the Presidium of the Supreme Soviet. This vigorous pursuit of expansionist population policies has now been slackened, and Soviet planners push other objectives: to accommodate the strong movement from rural areas to the cities and, especially, to lure labor to population centers in the remote trans-Ural areas of the country (Fig. 2–13). Nevertheless, concern over low birth rates has emerged again in recent years, and incentives for larger families may be reinstituted.

Just 70 years ago, less than one-fifth of the Soviet population was urbanized, a reflection of the country's state of development shortly after the takeover of the Communist administration. In 1940, the urban proportion reached one-third of the population; by 1960, just under half of the population (then 210 million) resided in cities and towns. Today, about 70 percent live in urbanized areas, a figure that still lags behind North American and Western European standards. The continuing migration toward the cities has produced housing problems and serious overcrowding in the many miles of medium-sized apartment buildings that line the streets of Soviet cities from Leningrad to Vladivostok (Fig. 2–14).

As Fig. I–14 shows, the bulk of the Soviet Union's population remains heavily concentrated in the Russian West and in the Ukraine to the south. But that distribution has changed, because the western concentration was even greater a century ago. On the map of world migrations (Route ⑧ in Fig. A–2, p. 120), note that the Soviet realm has been the scene of a major *internal* migration stream: an eastward and southeastward movement that has begun to fill distant frontiers. Part of this migration flow is reflected by the map showing the percentage of ethnic Russians now living in the 14 non-Russian republics (Fig. 2–9). Even republics as remote from the Soviet population core as

FIGURE 2–14 Soviet-style urban overcrowding: a neighborhood in the Belorussian capital, Minsk.

Kirghizia have substantial Russian populations. This is part of the Russification program that accompanied the Soviet development program (note that Kazakhstan is more than 40 percent Russian), but the map does not tell the whole story.

For many years, Soviet planners have encouraged western residents to move to the frontier—to the cities and towns beyond the Urals, to Siberia and the Far East. Incentives such as better living space and significantly higher wages and fringe benefits (decidedly un-communist attractions) were offered to families willing to move to such places as Komsomolsk and Khabarovsk. Over time, this program has had the effect of creating a ribbon of eastward frontier development (visible on Fig. I–14), and poles of Soviet activity and growth in the Far East. As the exploitation of east-

ern raw materials becomes necessary, and as eastern areas become politically sensitive, a growing Soviet (especially Russian) presence will be essential to the strength of the state. Fifty years ago, only about 25 percent of the Soviet population resided east of the Urals; today, that proportion exceeds 40 percent. Compared to the vast size of the trans-Ural east, that may seem a small cohort, but the continuing eastward migration is absolutely vital to the future of the country.

REGIONS OF THE SOVIET REALM

The Soviet Union is a realm in transition. In 1990, constituent republics of the U.S.S.R. plotted

secession; others demanded political, economic, and other social change. During the l980s, astonishing developments took place in the country, and the words *glasnost* (openness) and *perestroika* (restructuring) became part of the international vocabulary. A visionary president, Mikhail Gorbachev, led the Soviet Union in new directions.

In a previous edition of this book, we described the Soviet Union as a complex of contradictions:

The Soviet Union . . . is a country of large, impressive cities, efficient communications, modern industries, and mechanized farms. It also is a country of vast empty spaces, remote settlements, inadequate transport facilities, and badly outmoded rural dwellings. Against the confidence and accomplishments of Soviet planners, the treatment of many Jewish citizens and dissident authors, scientists, and artists stands in stark contrast. The Soviet Union has powerful armed forces, great universities, magnificent orchestras, spectacular space programs—but it still deploys the sinister State Security Committee (KGB) to spy on its citizens, sends nonconformists to ''mental hospitals,'' and nurtures the intimidating specter of Siberian exile. Soviet world influence works to liberate faraway colonies from foreign domination, but Moscow finds itself unable to accommodate expressions of nationalism in the satellite states of Czechoslovakia, Hungary, or Poland.

This paragraph effectively summarizes the reasons why *glasnost* and *perestroika* found fertile soil in the Soviet Union; it also underscores how much has changed in a very short time. Moscow's domination of Eastern Europe is at an end; Czechoslovakia, Hungary, and Poland have new political systems without the threat of Soviet tanks in the streets of their capitals. The Soviets' mighty military machine is withdrawing from its former satellites. Soviet involvement in foreign campaigns elsewhere also is declining: its armies were withdrawn from Afghanistan, and Ethiopia is receiving less Soviet help in its war against Eritrean (and other) rebels.

But the most dramatic developments have occurred within the U.S.S.R. itself. The murderous excesses of Stalin's regime have been confronted. Wartime misdeeds have been acknowledged. Human rights conditions have improved. The mass media now report the news with greater candor; the Chernobyl disaster was the first major incident about which

news was given directly to Soviet citizens by Soviet radio, television, and newspapers. Dead revolutionaries, whose names had been sullied, were rehabilitated and honored. Living dissidents, once exiled in the Russian tradition, were welcomed into a newly structured parliament. The most famous nonconformist, Dr. Andrei Sakharov (1921–1989), returned to Moscow to participate in open, sometimes bitter debate over the legacies of Lenin. In the early 1980s, all this was unthinkable. Barely ten years later it had become routine.

Legacies of the past have always weighed heavily on Russian society, and the present period of *perestroika* is no exception. The Gorbachev regime wanted to restructure an inefficient economy and an unwieldy political system, but moving the Russian monolith was no easy task. The political system strained under demands for independence, demands for greater autonomy, demands for territorial adjustments, demands that ethnic Russians cease playing

dominant roles in the Communist parties of non-Russian republics. *Glasnost* empowered a host of interest groups to voice their wishes: environmentalists, Muslim fundamentalists, Russian nationalists, and many others. Expectations were raised, but solutions could not be produced as rapidly as these many constituencies demanded.

In the meantime, social conditions deteriorated: consumer goods were fewer, waiting lines were longer than ever (Soviets are accustomed to standing for hours in lines, whether for a restaurant table or a bag of potatoes [Fig. 2–15]). The economic straitjacket created by 70 years of planning programs could not be loosened overnight. Gorbachev wanted to move the Soviet economy from absolute control through central planning toward a more mixed system, in which capitalist-style incentives and market forces would coexist with planned production. But three generations of communist control had left the Soviet Union without experience in cru-

FIGURE 2–15 The most famous Soviet queue of 1990: the waiting line to get into the U.S.S.R.'s first McDonald's, newly opened in the heart of Moscow just a short distance from Red Square.

KALININGRAD: A RUSSIAN EXCLAVE

A careful look at Fig. 2–7 reveals that the possible secession of Lithuania will have an important geographic effect. A small part of the Russian Republic, centered on the Baltic coastal city of Kaliningrad, would be isolated from the R.S.F.S.R. Overland, this small oblast can only be reached through Lithuania or through Poland.

If Kaliningrad were merely a fishing town, this might not matter. But Kaliningrad, the former German port of Königsberg, is a major Soviet naval base. When territorial claims were settled after World War II, the Soviets were careful to claim Königsberg and its naval facilities; but they did not attach this area to Lithuania (which they had annexed in 1940 as the war started).

Like the Russian tsars before them, the Soviets always have been sensitive about losing warm-water outlets. The independence and secession of Lithuania would create a Soviet *exclave*, and a military exclave to boot. Small wonder that President Gorbachev went personally to Vilnius to plead for restraint on the part of Lithuanian nationalists.

cial areas and without traditions of high productivity and quality control. At a time when the U.S.S.R. most needed foreign revenues for its exports, there were few takers for Soviet manufactures. A major liability was the Soviet currency, the ruble. Artificially valued against foreign currencies, the ruble had worth inside the Soviet Union, but none outside. While the Soviet Union was run as a giant corporation by national planners, this made sense—or so it seemed. For *perestroika*-minded reformers, however, it was another obstacle.

When, therefore, we undertake our survey of the regional map of the Soviet realm, we should keep in mind that the U.S.S.R. is being transformed—but that long-entrenched patterns do not change instantly. If, for example, one or more of the Baltic states secedes from the Soviet Union, that will not change the essentials of geography (save for one notable exception, the R.S.F.S.R. **exclave** of Kaliningrad [see *box*]). Their hinterlands will still be Russian; their economic futures are unavoidably bound up with Russian markets. Political secession is feasible. Economic disentanglement is another matter.

The Russian Core

The heartland of a state, as noted in the Opening Essay in the introductory chapter, is its **core area**. Here, a large part of the population is concentrated, and large cities, major industries, dense transport networks, intensively cultivated lands, and other essentials of the country cluster within a relatively small region. Core areas of long standing carry the imprints of culture and history especially strongly. The Russian core of the Soviet Union, broadly defined, is the region that extends from the western border of the Russian S.F.S.R. to the Ural Mountains in the east (Fig. 2–16). This is the Russia of Moscow and Leningrad, of the Volga River and its industrial cities, of farms and forests. Here, the Muscovy Russians asserted their power and began the formation of the Russian Empire, forerunner to the Soviet Union.

Central Industrial Region

At the heart of the Russian core lies the *Central Industrial Region* (Fig. 2–17). The precise definition of this subregion varies, as all regional definitions are subject to debate. Some geographers prefer to call this the *Moscow Region*, thereby emphasizing that for over 250 miles (400 kilometers) in all directions from the capital, regional orientations are toward this historic focus of the state. As the map shows, Moscow has maintained its decisive **centrality**: roads and railroads radiate in all directions to the Ukraine in the south; to Minsk, Belorussia, and Europe in the west; to Leningrad and the Baltic coast in the northwest; to Gorkiy and the Urals in the east; to the cities and waterways of the Volga Basin in the southeast (a canal links Moscow to the Volga, the Soviet Union's most important navigable river); and even to the subarctic northland that faces the Barents Sea (see *box*).

Moscow itself is a transforming metropolis (1991 population: 10.5 million) as high-rise apartment complexes increasingly dominate the residential landscape (Fig. 2–18). Although this helps to alleviate the Soviet capital's severe housing shortage, nearly all Russian cities are horribly overcrowded, with most people forced to accept unreasonably cramped personal living spaces while they endlessly wait to move up on long

FIGURE 2—16

FIGURE 2—18 The fringes of Moscow being pushed ever farther into the countryside as new high-rise residential suburbs transform the metropolitan periphery. This recently completed "micro-center" now houses over 100,000 people.

waiting lists for roomier quarters. Moscow is also the focus of an area that includes some 50 million inhabitants (over one-sixth of the country's total population), many of them concentrated in such major cities as Gorkiy (1.5 million)—the automobile-producing "Soviet Detroit"; Yaroslavl (675,000)—the tire-producing center; Ivanovo (525,000)—the heart of the textile industry; and Tula (565,000)—the mining and metallurgical center where lignite (brown coal) deposits are worked.

Leningrad, as the Soviets renamed Petrograd in 1924 after Lenin's death, remains the Soviet Union's second city, with 4.7 million people. Leningrad has none of Moscow's locational advantages,

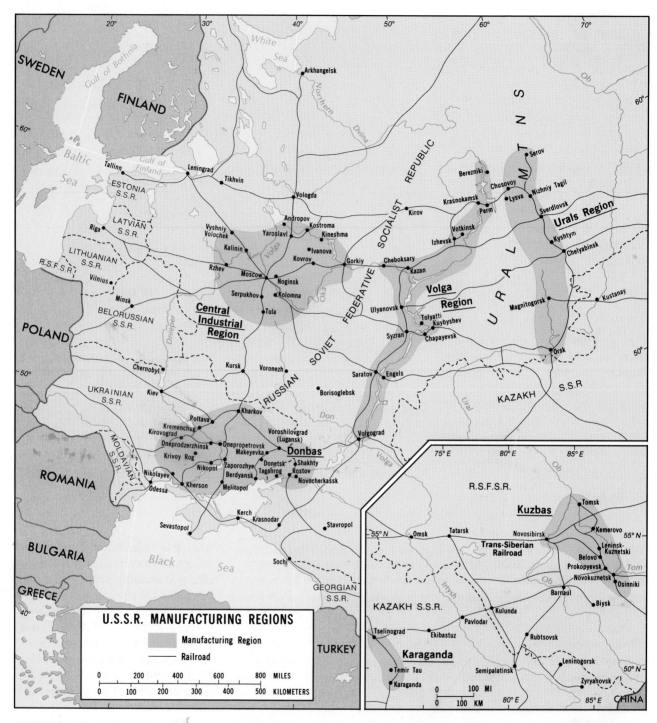

FIGURE 2-17

at least not with respect to the domestic market. It lies well outside the Central Industrial Region and, in effect, in the northwestern corner of the country, 400 miles (650 kilometers) from Moscow. Neither is it better off than Moscow in terms of resources: fuels, metals, and foodstuffs must all be brought in, mostly from far away. The Soviet emphasis on self-sufficiency has even reduced Leningrad's asset of coastal location, because some raw materials might be imported much more cheaply across the Baltic Sea from foreign sources than from domestic sites in distant Central Asia (only bauxite deposits lie nearby, at Tikhvin). But Leningrad was at the vanguard of the Industrial Revolution in Russia, and its specialization and skills have remained. Today, the city and its immediate environs contribute about 5 percent of the country's manufacturing, much of it through the building of high-quality machinery. In addition to the usual association of industries (metals, chemicals, textiles, and food processing), Leningrad has major shipbuilding plants and, of course, its port and naval station. This productive complex, although not enough to maintain its temporary advantage over Moscow, has kept the city in the forefront of modern Soviet development.

Povolzhye: The Volga

A second region lying within the Russian Core is the *Povolzhye*, the Russian name for an area that extends along the middle and lower valley of the Volga River. It would be appropriate to call this the Volga Region, for that greatest of Soviet rivers is its lifeline, and most of the cities that lie in the *Povolzhye* are situated on its banks (Fig. 2–17). In the 1950s, a canal was completed to link the lower Volga with the lower Don River (and thereby the Black Sea), extending this region's waterway system still farther.

The Volga River was an important historic route in old Russia, but for a long time neighboring regions overshadowed it. The Moscow area and the Ukraine were far ahead in industry and agriculture. The Industrial Revolution that came late in the nineteenth century to the Moscow area left the *Povolzhye* little affected. Its major function remained the transit of foodstuffs and raw materials to and from other regions.

This transport function is still important, but things have changed in the *Povolzhye*. First, World War II brought a time of furious development because the

---

FACING THE BARENTS SEA

North of the latitude of Leningrad, the region we have named the Russian Core takes on Siberian properties. But the Soviet presence in this remote northland is much stronger than it is in Siberia proper. Two substantial cities, Murmansk (475,000) and Arkhangelsk (450,000), are road-and-rail-connected outposts in the shadow of the Arctic Circle.

Murmansk lies on the Kola Peninsula, not far from the border with Finland (Figs. 2–7 and 2–5). In its hinterland lie a variety of mineral deposits, but Murmansk is particularly important as a naval base. During World War II, allied ships brought supplies to Murmansk; the city's remoteness shielded it from German occupation. After the war, it became a base for nuclear submarines. This city is also an important fishing port and a container facility for cargo ships.

Arkhangelsk is located near the mouth of the Northern Dvina River where it reaches an arm of the White Sea (Fig. 2–17). Its site was chosen by Ivan the Terrible during Muscovy's early expansion; Ivan wanted to make this the key port on a route to maritime Europe. But Arkhangelsk, mainly a port for lumber shipments, suffers from a more restricted ice-free season than Murmansk, whose port can be kept open with the help of the North Atlantic Drift ocean current (and by icebreakers when the need arises).

Nothing in Siberia east of the Urals rivals either of these cities—yet. But their very existence and growth prove that Siberian barriers to settlement can be overcome.

Volga region, located east of the Ukraine, was protected by distance from the German armies that invaded from the west. Second, in the postwar era, the Volga-Urals region proved to be the greatest source of petroleum and natural gas in the entire Soviet Union. From near Volgograd (formerly Stalingrad) in the southwest to Perm (1.2 million) on the Urals' flank in the northeast lies a belt of major oilfields (Fig. 2–12). Once these deposits were believed to comprise the Soviet Union's largest reserve but, as the oil and gas region map shows, later discoveries in western Siberia indicate that even more extensive oil- and gasfields lie beyond the Urals. Still, because of their location and

size, the Volga region's fossil-fuel reserves retain their importance. Third, the transport system has been greatly expanded. The Volga-Don Canal directly connects the Volga waterway to the Black Sea; the Moscow Canal extends the northern navigability of this river system into the very heart of the Central Industrial Region; and the Mariinsk Canals provide a link to the Baltic Sea. Today, the Volga region's population exceeds 25 million, and the cities of Kuybyshev, Volgograd, Kazan, and Saratov are all in the 1.0-to-1.3-million range. Manufacturing has also expanded into the middle Volga Basin, emphasizing more specialized engineering industries. The huge Fiat auto assembly plant in Tol-

FIGURE 2–19 Beneath one of the worker-exhortation signs that are so characteristic of Soviet factories, newly-assembled Fiats are lined up inside the massive autoworks at Tolyatti in the middle Volga valley near the city of Kuybyshev.

yatti (720,000), for example, is one of the world's largest of its kind (Fig. 2–19).

The Urals

The Ural Mountains form the eastern limit of the Russian Core. These mountains are not particularly high; in the north they consist of a single range, but southward they broaden into a zone of hilly topography. Nowhere do they form an obstacle to east-west transportation. An enormous storehouse of metallic mineral resources, located in and near the Urals, has made this area a natural place for industrial development. Today, the Urals Region, well connected to the Volga Region and to the Russian industrial heartland, extends from Serov in the north to Orsk in the south (Fig. 2–17).

The Urals Region rose to prominence during World War II and its aftermath, when it benefited from remoteness from the German invaders, allowing its factories to support the war effort without threat of destruction. Coal had to be imported, but the problem of

energy supply here has been considerably relieved by oilfields discovered in the zone lying between the Urals and the Volga Region (Fig. 2–12). More recent West Siberian oil and gas exploitation (from fields lying northeast of the Urals) increasingly supports development in the Urals as well.

The Central Industrial, Volga, and Urals regions form the anchors of the Russian core area. For decades they have been spatially expanding toward one another, their interactions ever more intensive. These regions of the Russian Core stand in sharp contrast against the comparatively less-developed, forested, Arctic north and the remote mountainous south between the Black and the Caspian seas; thus, even within this Russian coreland, there still are frontiers awaiting growth and development.

The West

The Soviet Union's Western Region (Fig. 2–16) consists of the six republics of Estonia, Latvia,

Lithuania, Belorussia, Moldavia, and the Ukraine (see profiles on pp. 150–151). As the manufacturing map (Fig. 2–17) shows, the dominant subregion in this western tier of S.S.R.s is the Ukraine.

Soviet strength has been built on a mineral wealth of great variety and volume. With its immense area, the country possesses an almost limitless range of raw materials for industry, although the best remaining deposits are increasingly located in remote areas and require heavy investments in transportation. In the Ukraine, the Soviet Union has one of those regions where major deposits of industrial resources lie in relatively close proximity. The Ukraine began to emerge as a leading region of heavy industry toward the end of the nineteenth century, and one major reason for this was the Donets Basin, one of the world's greatest coalfields. This area, known as the *Donbas* for short, lies north of the city of Rostov, and in the early decades this field produced over 90 percent of all the coal mined in the country. Much of the *Donbas* coal is of the high grade needed to manufacture steel. Today, although supplies are declining, the Donets Basin still accounts for over 25 percent of the total Soviet coal output.

What makes the Ukraine unique in the Soviet Union is the location, less than 200 miles (320 kilometers) from all this *Donbas* coal, of the Krivoy Rog iron ores (Fig. 2–17). Again, the quality of the deposits is high, although the better ores are being worked out and the industry is now turning to the concentrating of poorer-grade deposits. Major metallurgical industries arose on both the Donets coal and the Krivoy Rog iron: Donetsk (1.2 million) and its satellite Makeyevka (515,000) dominate the *Donbas* group (constituting what might be called the "Soviet Pittsburgh"), whereas Dnepropetrovsk (1.3 million) is the chief center of the Krivoy Rog cluster.

One way or another, all the major cities located nearby have benefited from this fortuitous juxtaposition of minerals in the east-central Ukraine: Rostov (1.1 million), Volgograd (1.0 million), and Kharkov (1.7 million) near the cluster of *Donbas* cities, the leading Black Sea port of Odessa (1.3 million), and even Kiev (2.8 million) not far from the Krivoy Rog (740,000) agglomeration. And like West Germany's Ruhr, the Ukraine industrial region lies in an area of dense population (offering readily available labor), good agricultural productivity, and adequate transportation systems; it also lies near large markets. Moreover, this region has provided alternatives when exhaustion of the better ores began to threaten: not only are extensive lower-grade deposits capable of sustaining production through the foreseeable future, but additional iron reserves are located near Kerch on the eastern tip of the Crimean Peninsula, close enough for use in the established plants of the Ukraine. However, the biggest development in recent years is the opening of the Kursk Magnetic Anomaly, halfway between the *Donbas* and Moscow. This area may technically lie outside the Ukraine (by 50 miles or so), but its rich iron ores are increasingly critical to keeping the heavy industrial complexes of the Ukraine going. In fact, if the latest estimates are correct, the Kursk deposits may be among the largest anywhere on earth.

And this is not all. The Ukraine and neighboring areas yield several other essential raw materials and energy resources for heavy industry. Chief among these is manganese, a ferroalloy vital in the manufacture of steel; about 12 pounds (5.4 kilograms) of manganese ore, on the average, go into

making a ton of steel. The Soviet Union has ample supplies of this vital commodity; deposits just east of Krivoy Rog at Nikopol and those lying between the greater and lesser ranges of the Caucasus Mountains are the two leading sources of manganese in the world. Finally, there is the heavily used pipeline connection from the still-important Baku oilfield complex, located not far to the southeast (at least by Soviet standards of distance) along the southwestern shore of the Caspian Sea.

The Ukraine provided the Soviets with their opportunity to gain rapidly in strength and power; when World War II broke out, this was the center of heavy industry in the country. But favorable as its concentration of resources and manufacturing was, it also contributed to Soviet vulnerability. The industries of the Ukraine and the Transcaucasian oilfields were prime objectives of Germany's invading armies. As the Soviets were forced to withdraw, they dismantled and even destroyed more than 1000 manufacturing plants to avoid their falling into the hands of the enemy. Thus began a series of developments that has pulled the Soviet center of economic gravity steadily eastward, to the Urals and beyond. Soviet planners were impressed with the need for greater regional dispersal of industrial production, not just because the Ukrainian and other western mines might eventually become exhausted, but also for strategic reasons.

The Ukraine (loosely translated, the name means "frontier"), for all the Soviets' encouragement of new economic development and the eastward march of population, remains a cornerstone of the country. With 53 million people, the Ukrainian S.S.R. contains about one-fifth of the entire Soviet population (though less than 3 percent of the land), and its industrial and

agricultural production is enormous. With all that spectacular industrial growth, it is easy to forget that much of the Ukraine was pioneer country because of its agricultural possibilities, not its known minerals. Even today, half of its people live on the land rather than in those industrial cities; in the central and western parts of this republic, rural densities exceed 400 per square mile (150 per square kilometer). Wheat and sugar beets cover the landscape where the soils are best—in the heart of the Ukraine—and where moisture conditions are optimal. Yet water supplies here do not exist in abundance; Fig. 2–1 shows that much of the Ukraine lies in a semiarid steppe zone, and therein looms a future problem as continued high productivity exerts ever-greater pressures on the republic's limited water resources.

The Transcaucasus Region

Smallest of the Soviet Union's geographic regions, the Transcaucasus consists of three non-Russian republics (Georgia, Armenia, and Azerbaydzhan) at odds with each other—and with their Russian comrades. Here Russian forces and Muslim armies met, and Ottomans and Turks contested for space with Georgians and Armenians. As the profiles of these S.S.R.s (p. 152) reveal, neither Soviet political nor economic planning has succeeded in stabilizing this geographically divided region.

In the Transcaucasus Region, the Soviet Union borders Iran and Turkey and confronts Christian religions as well as Muslim fundamentalism. The political framework designed to accommodate ethnic and religious identities is complicated and poorly planned; the Armenian Republic bisects the

Muslim Azerbaydzhan Republic, and an exclave of Armenian Christians is surrounded by (and was designated to be administered by) Muslim Azeris. Strife among Georgians and incorporated minorities has also invalidated the political design there.

Economically, the Transcaucasus has considerable importance. Tea, subtropical fruits, and other agricultural products are exported northward. Oil deposits along the Caspian shore made Baku (1.7 million) a major port and the Caspian Sea a significant link between energy supply and demand.

The Muslim South

East of the Caspian Sea lies a region that has been given various geographical names, including *Soviet Central Asia* and the *Transcaspian Region*. We name it the Muslim South because this region's dominantly Islamic population constitutes the paramount cultural influence here, 70 years of Sovietization notwithstanding (Fig. 2–20). This is a land of vast desert expanses and oases, scarce surface water, irrigated agriculture, and local modernization grafted onto older Muslim ways. In the west and center of the region, the land is mainly flat or slightly undulating, a typical desert landscape broken by the shrinking Aral Sea (see Fig. 2–4). In the southeast, the relief increases, and in the republics of Kirghizia and Tadzhikistan lie snowcapped peaks that give rise to waters used for irrigation hundreds of miles away.

Although this region is divided into only five Soviet republics, more than 100 distinct nationalities occupy areas of the Muslim South—and not all of them are Muslim. The politico-geographical map has been adjusted repeatedly since it was first delimited in the 1920s, and further changes undoubtedly lie ahead. In the mountainous southeast, three republics—Tadzhikistan, Kirghizia, and Uzbekistan—lie territorially interlocked in such a way that it is easier to travel from one republic to the other than from one part to another of the same republic. To their north, Kazakhstan, largest of all the non-Russian republics, has its capital, Alma-Ata (1.2 million), situated in a remote southeastern corner whereas its economic development has focused on areas hundreds of miles away.

Soviet planners invested heavily

FIGURE 2–20   Muslim-dominated Khiva, Uzbekistan—the quintessential urban landscape of Islam.

in these subregions, especially in enormous irrigation projects in central Kazakhstan. This republic became one of the world's leading cotton exporters, and the region yields a variety of warm-climate crops for Soviet consumers. The northern zone of Kazakhstan, however, experienced quite different development. This area lay in the path of the Russian expansion eastward, and, from the Urals Region in the west to the Kuznetsk Basin in the east, transport routes were laid across it. Northern Kazakhstan, politically still a part of the Kazakh S.S.R., has geographically become a sector of the Eastern Frontier region. This helps explain why Kazakhstan, a republic established for the Kazakhs and other Muslim minorities, today has more ethnic Russians among its population of 18 million than Kazakhs (41 to 36 percent). The distribution of population, however, is such that Russians dominate in the modernized north, and Kazakhs and others in the rest of the republic.

FIGURE 2–21   Coal is still a vital resource in the manufacturing geography of southern Siberia, and most of it is consumed within the region. This colliery is part of the fuel and power-generating complex located in the Kansk-Achinsk Basin adjacent to the Kuzbas.

The Eastern Frontier

The eastward march of Soviet development had made the Karaganda-Tselinograd area one of its hubs of growth. Karaganda, once a small Kazakh town, has become a center for iron and steel production along with chemicals and other manufactures. With a population of more than 700,000, Karaganda is a Russified city that exchanges raw materials and finished products with other Soviet production zones, especially the Urals Region. Tselinograd (300,000) emerged in the 1950s an administrative and collection center for the troubled Virgin and Idle Lands Program; together, the two cities are linked by rail to the major developing areas of the Eastern Frontier (inset, Fig. 2–17).

From the beginning of Soviet rule,

the penetration and consolidation of the Eastern Frontier were priority planning goals. Concerns over effective national control along sensitive eastern boundaries was part of the reason; security in the event of renewed German invasion from the west was another. The exploitation of mineral resources was a further incentive.

Northeast of the Karaganda area, and fully 1200 miles (2000 kilometers) east of the Urals, lies the third-ranking region of heavy manufacturing in the Soviet Union—the Kuznetsk Basin or Kuzbas (inset, Fig. 2–17). In the 1930s, this area was first opened up as a supplier of raw materials (especially coal) to the Urals, but that function steadily diminished in importance as local industrialization accelerated (Fig. 2–21). The

original plan was to move coal from the Kuzbas west to the Urals and to let the returning trains carry iron ore east to the coalfields, but good-quality iron ore deposits were subsequently discovered in the vicinity of the Kuznetsk Basin itself. As the new resource-based Kuzbas industries grew, so did its urban centers. The leading city, Novosibirsk (with 1.6 million inhabitants) stands at the intersection of the Trans-Siberian Railroad and the Ob River as the very symbol of Soviet enterprise in the vast eastern interior. To the northeast lies Tomsk (530,000), one of the oldest Russian towns in all of Siberia, founded three centuries before the Bolshevik uprising and now caught up in the modern development of the Kuzbas region. Southeast of Novosibirsk lies

Novokuznetsk, a city of nearly 700,000 people that specializes in the fabrication of such heavy engineering products as rolling stock for the railroads; aluminum products using Urals bauxite are also manufactured here.

Impressive as the concentration of coal, iron, and other resources may be in the Kuznetsk Basin, the industrial and urban development that has taken place here must in large measure be attributed, once again, to the ability of the state and its planners to promote this kind of expansion, notwithstanding what capitalists would see as excessive investments. In return, they were able to push the country vigorously ahead on the road toward industrialization, with the hope that certain areas would successfully reach "takeoff" levels. Then they would require progressively fewer direct investments as growth would, to an ever-greater extent, become self-perpetuating. The *Kuzbas*, for instance, was expected to grow into one of the Soviet Union's major industrial agglomerations, with its own important market and with a location, if not favorable to the Urals and points west, then at least fortuitous with reference to the developing markets of the Soviet Far East.

Between the *Kuzbas* and Lake Baykal lies one of those areas that has not yet reached the takeoff point. This 1000-mile-long (1600-kilometer-long) area has impressive resources, including coal, but Soviet planners have been uncertain exactly where to apply the necessary investments to make this a significant industrial district. In addition to major hydroelectric power plants already operating at Bratsk (275,000) and elsewhere along the Angara River, there are further possibilities of harnessing water power from the Yenisey River. Timber is in plentiful supply, brown coal is used to generate

electricity, and oil is piped all the way from the Volga-Urals fields to be refined at Irkutsk. Thus, Krasnoyarsk (950,000) and Irkutsk (680,000) lie several hundred miles farther again from the established core of the country. Although the distant future may look bright—especially if Soviet plans for its Pacific margins become reality—the situation right now presents difficulties that are not lessening.

The Far East

The Soviet Union has about 5000 miles (8000 kilometers) of Pacific coastline—more than the United States (including Alaska). But most of its coastal margin lies north of the latitude of the state of Washington, and from the point of view of coldness, the Soviet coastline lies on the "wrong" side of the Pacific. The port of Vladivostok (680,000), the eastern terminus of the Trans-Siberian Railroad, lies at a latitude about midway between San Francisco and Seattle, but must be kept open throughout the winter by icebreakers—something unheard of in these American ports. The climate, to say the least, is harsh. Winters are long and bitterly cold; summers are cool. Nonetheless, the Soviets are determined to develop their Far East region as intensively as possible, and their resolve has recently been spurred by the ideological differences between the Soviet Union and China. In the Asian interior, high mountains and empty deserts mark most of the Chinese-Soviet borderland zone; south of the Kuznetsk Basin and Lake Baykal, the Republic of Mongolia functions as a sort of buffer between Chinese and Soviet interests. But in the Pacific area, the Soviet Far East and Northeast China confront each other directly across an uneasy river boundary (along the Amur and its tributary,

the Ussuri). Hence, the Soviets want to consolidate their distant outpost (which also is not far from the extremities of Alaska), and are offering special inducements to their western residents to "go east."

Despite its severe climate and difficult terrain, the Far East region is not totally without assets. Although most of it still consists of very sparsely populated wilderness, the endless expanses of forest (see Fig. I–12) have slowly begun to be exploited by a timber industry whose major problems are the small size of the local market, the bulkiness of the product, and—as always in Soviet Asia—the enormous distances to major consuming areas. Lumbering centers and fishing villages break the emptiness of the countryside. Fish is still the region's leading product; from the Kamchatka Peninsula to Vladivostok, Soviet fishing fleets sail the Sea of Okhotsk and the Pacific in search of salmon, herring, cod, and mackerel to be frozen, canned, and shipped via rail to far-off western markets.

In terms of minerals, too, the Soviet Far East has possibilities. A deposit of high-quality coal lies in the Bureya River Valley (a northern tributary of the Amur), and a lignite (brown coal) field near Vladivostok is being exploited. On Sakhalin Island there are locally important quantities of bituminous coal and oil. Not far from Komsomolsk (350,000) lies a body of inferior iron ore, and this city has become the first steel producer in the region. In addition to lead and zinc, the tin deposits north of Komsomolsk are important because they constitute the U.S.S.R.'s leading source.

Thus, the major axis of industrial and urban development within the Pacific coastal margin has emerged along the Amur-Ussuri river system, from Vladivostok in the south—with its military instal-

FIGURE 2–22 A recent view of Nakhodka's docks and part of its excellent natural harbor. This port on the Sea of Japan is expanding rapidly and could become the U.S.S.R.'s leading maritime trade center by the late 1990s.

(approximately 10 million) and the available farmlands of the Amur-Ussuri valleys. But Soviet planners have not been sure just how to direct industrialization here. Large investments in energy supply are needed to take advantage of some magnificent opportunities for hydroelectric development in the Amur and tributary valleys. But the choice is difficult: to spend here—in an area of long-term isolation, modest resources, labor shortage, and particularly harsh weather—or to spend money where conditions are less bleak. Future relationships with China may have much to do with the pace of development in the Soviet Far East.

Siberia

East of the Urals and north of the ribbon-like Eastern Frontier lies vast, bleak, frigid, forbidding Siberia (Fig. 2–16). Larger than the conterminous United States but inhabited by only an estimated 15 million people, Siberia quintessentially symbolizes the Russian environmental plight: vast distances, cold temperatures worsened by strong Arctic winds, difficult terrain, poor soils, and limited options for survival. But Siberia is also a region of resources. From the days of the first Russian explorers and Cossack adventurers, word of Siberia's riches filtered back to the west. Gold, diamonds, and other precious minerals were found. Later, metallic ores including iron and bauxite were discovered. Still more recently, the Siberian interior proved to contain sizable quantities of oil and natural gas (Fig. 2–12) and began to contribute importantly to the energy supply of the Soviet Union.

As the physiographic map (Fig. 2–5) shows, several major rivers—the Ob, Yenisey, and Lena—flow

lations, shipbuilding, and fish-processing plants—to Komsomolsk, the iron and steel center in the north. Near the confluence of the Amur and Ussuri Rivers lies Khabarovsk (725,000), the city with the greatest advantages of centrality and the region's leading urban center. Here, metals manufacturing plants process the iron and steel of Komsomolsk, chemical industries benefit from Sakhalin oil, and timber is used for the burgeoning furniture industry. During the 1980s, Soviet planners also encouraged the rapid growth of another Far East city, the port of Nakhodka 55 miles (90 kilometers) up the coast from Vladivostok (Fig. 2–22). This city of 300,000 is now well on its way to becoming the region's leading trade center, with intensifying linkages to nearby Japan. If all goes according to plan, Nakhodka's modern port (built with Japanese assistance) will become the Soviet Union's largest by the year 2000.

The farther eastward the Soviet Union's developing areas lie, the greater is their **inaccessibility** and the larger their need for self-sufficiency in view of the ever-increasing cost of transportation. Therefore, the Pacific region exchanges far fewer commodities with points in the Soviet west than does the *Kuzbas*, although some raw materials from the Far East (such as tin) do travel to the west in return for relatively small tonnages of *Kuzbas* resources. In terms of food supply, the Far East cannot come close to feeding itself despite its rather small population

gently northward across Siberia and the Arctic Lowland into the Arctic Ocean. Hydroelectric power development in the basins of these rivers has generated electricity used in the extraction and refinement of local ores, and in the lumber mills that have been set up to exploit the vast Siberian forests. Soviet planners have even given serious consideration to grandiose schemes to reverse the flow of Siberian streams—to divert Siberian water onto parched farmlands far to the southwest.

The human geography of Siberia is fragmented, and much of the region is, for all intents and purposes, uninhabited (Fig. 2–8). Ribbons of Russian settlement have developed—note that the Yenisey River can be traced on this map of Soviet peoples (a series of small settlements north of Krasnoyarsk), and that the upper Lena Valley is similarly fringed by ethnic Russian settlement. But these ribbons and other islands of habitation are separated by hundreds of miles of empty territory. During the worst excesses of the Stalinist period (and long afterwards as well), dissidents and criminals were exiled to Siberian mining and lumbering camps to serve their terms at hard labor under the harshest of conditions; many never returned. The Siberian region also has indigenous inhabitants, and Fig. 2–8 shows that they are concentrated in the east.

The principal ethnic group here is the Yakut people, who are of Mongol ancestry and have settled in the central basin of the Lena River. The Yakut were awarded their own Autonomous Soviet Republic, which covers much of eastern Siberia and is larger even than the Kazakh S.S.R. (but does not hold the same rank in the political framework). Traditionally, the Yakut are reindeer herders; fishing also is an important activity. But the future may hold a different direction for these people. Their

capital, Yakutsk (200,000), appears to lie near major oil and gas deposits, and associated development is likely to transform the heart of this remote area.

Siberia, Russia's freezer, is stocked with goods that may become mainstays of national development. Already, precious metals and mineral fuels contribute importantly to the Soviet economy. As time goes on, we may expect Siberian resources to sustain the development of the Eastern Frontier; and as this happens, this southern zone of development will expand northward as well as eastward, bringing more of Siberia into the national sphere. One significant step in that direction has already been accomplished with the recent completion of the BAM (Baykal-Amur Mainline) Railroad. This high-speed line, north of and parallel to the Trans-Siberian Railroad, extends 2200 miles (3540 kilometers) eastward from Tayshet (near Krasnoyarsk) directly to Komsomolsk (Fig. 2–11); it is not only a faster link to the Pacific coast, but is also expected to become the axis of a major new industrial corridor early in the twenty-first century.

HEARTLANDS AND RIMLANDS

The emergence of Russia (as the dominant force of the Soviet Union) onto the world political stage has been a subject of debate and analysis for nearly a century. In the early part of this century, Russian forces were defeated by the Japanese, and the country was in chaos on the eve of the 1917 Revolution. Who would predict that a national state centered on decaying Russia would become a world power before the century's close? The British geog-

rapher Sir Halford Mackinder (1861–1947) did just that. In an article published in 1904, he analyzed the potential strengths and weaknesses of Eurasian regions, concluding that one such region, which he called the *pivot area*, enjoyed a combination of natural protection and resource wealth that would propel its occupants to world power. This pivot area (Fig. 2–23, upper left) consisted of the Moscow region, the Volga Valley, the Urals, Central Asia, and western Siberia.

Mackinder's article sparked a heated debate. In 1919, he published a revision of his idea, incorporating Eastern Europe into the framework (Fig. 2–23, lower left). He now called the crucial (and expanded) strategic area the *heartland*, and argued that control over Eastern Europe would mean global power. This became known as Mackinder's **Heartland Theory:**

Who rules East Europe commands the Heartland;
Who rules the Heartland commands the World Island;
Who rules the World Island commands the World.

Mackinder's theory seemed to become a prescription for national policy: the Germans during World War II sought to control Eastern Europe and pushed into the heartland, and then the Soviets tried to subjugate postwar Eastern Europe. The World Island, in Mackinder's parlance, consisted of Eurasia and Africa; this huge territory would fall to the heartland power, leading ultimately to world domination.

Other geographers took a different view of the capacities and limitations of the heartland, and some, notably Nicholas Spykman, suggested that the future of Eurasia might lie in a *rimland* rather than in Mackinder's heartland (Fig. 2–23, upper right). But no one could fail to note the Soviets' own

EURASIAN HEARTLANDS

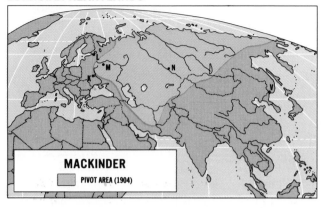

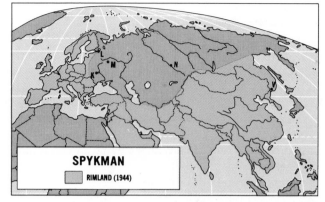

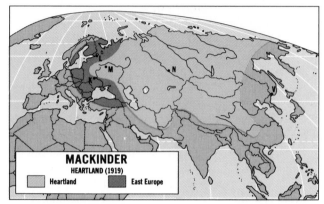

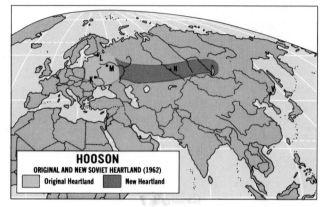

FIGURE 2–23

reaction: David Hooson sketched the Soviet planners' attempt to move the Soviet core area eastward, away from the vulnerable west (Fig. 2–23, lower right). In an age of intercontinental missiles and nuclear weapons, a dispersed core area confers a certain degree of security upon a threatened state. Until very near the time of his death in 1947, Mackinder participated in the continuing discussion, convinced that he was as correct near mid-century as he had been in 1904.

SOVIET NEIGHBORS

The Soviet Union has borders with 12 neighbors. As the inheritor of a Russian empire, the U.S.S.R. has unfinished boundary business with

several of these adjoining states. When Soviet boundaries were defined, some affected countries, such as China, were in a weakened state; other countries, like Poland and Romania, lost territory to the Soviet Union in the aftermath of World War II. When the Russian tsars sent their armies far afield to conquer distant territories, large areas on the margins of the empire were under uncertain authority. Just a century ago, such ill-defined **frontiers** still separated countries and colonizers. This enabled the stronger powers to impose borders on weaker ones. In a relatively short time, most frontiers disappeared from the world map, replaced by boundaries defined according to power relationships or by treaties that were not always fair.

Territorial issues, as we have noted, arise in connection with the demands of some of the non-Russian republics in the west. The

annexation of Moldavia may be reversed, which would reposition the Ukrainian (not the Russian) boundary. But the most serious territorial issues lie in central and eastern Asia, all of them involving China.

Areas and Boundaries

When tsarist forces occupied and secured the Far East, China was in disarray. Repeatedly since the Soviets consolidated the tsarist empire, Chinese governments have demanded the "return" of areas on the Sino-Soviet border. These include virtually all of the Soviet Far East, a part of the Chita area northeast of Mongolia, a portion of the Altay territory northwest of Mongolia, and parts of the Kazakh and Kirghiz S.S.R.s (Fig. 2–24).

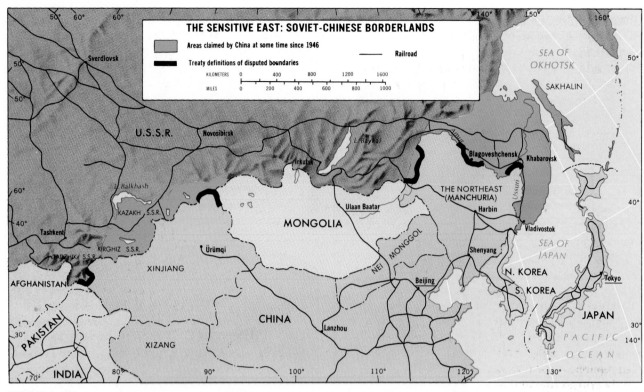

FIGURE 2–24

More specific border issues also have arisen in several places. The tsars managed to impose boundary treaties upon the Chinese during the period from 1858 to 1864 (see Fig. 2–2), but China has now repudiated those treaties. As Fig. 2–24 shows, five boundary segments (shown by heavy black lines) have caused particular trouble: three sections of the China-Soviet border in the Far East and the Eastern Frontier, one section affecting the Kazakh Republic, and one involving the Tadzhik Republic. The most serious of these has been the boundary along the Amur and Ussuri Rivers in the Far East. In 1969, there even was fighting between Soviet and Chinese troops at Chen Pao (Damansky) Island in the Ussuri River; soon thereafter came skirmishes along the Central Asian boundary. Both the territorial and boundary issues have been dormant since the 1970s, but they constitute latent points of friction. The changes affecting the Muslim

Soviet republics could rekindle the Central Asian border disputes, especially since these boundaries separate peoples of close affinities. Thus, the ripple effects of instability in the Soviet system could reach the sensitive borderland zone at a time when China is looking for a unifying political cause.

GEOGRAPHY OF THE SOVIET UNION IN THE 1990s

So far-reaching are the changes affecting the Soviet Union during its current transition that it is tempting to focus on these current events as we assess the geography of this important realm. But we should not forget the more perma-

nent, durable geographic characteristics of the U.S.S.R., because these will still prevail even after the present period of change has ended. The list of ''Ten Major Geographic Qualities of the Soviet Union'' (p. 138) contains the realm's key properties: its enormous size—and inherent problems of distance and accessibility; its high-latitude position and consequent coldness; its nearly landlocked position and associated dryness; its Russian-dominated politico-geographical framework (with liabilities we have discussed throughout this chapter); its regionally unbalanced development; and the effects of its long-term planned economic evolution, which will require considerable time to modify. Add to this the properties of the Soviet population—its relatively small size, ethnic and cultural division, demographic imbalance—and we can see why real change will be difficult to achieve in the short term. For all its official encouragement, the government of the U.S.S.R. has been unable to induce the kind of eastward migration it would really like; the Russian Core is still the heartland of the Soviet Union. A legacy of centuries is not erased in a few generations.

But the geography of the Soviet Union *is* changing, and maps of the future may well show territorial losses: secession is a right guaranteed in the original Soviet constitution, and some of the U.S.S.R.'s non-Russian republics could separate from Moscow. But even this development should not be exaggerated. The "loss" of the three Baltic republics, all or part of the Transcaucasus, or parts of the Muslim South would not threaten the survival of the Russian Republic. Only one non-Russian republic—the Ukraine—plays a truly critical role in the overall Soviet economic system. Any serious threat of Ukrainian secession would have a lasting, perhaps destructive impact on Russia. Otherwise, we should remember that just as France survived without its colonies in Algeria, West Africa, and Indochina, and just as the United Kingdom thrived after it

lost its "indispensable" British Empire, so the Russian state would manage without its contiguous colonies, the Ukraine excepted.

It is also well to remember that President Gorbachev is not the first Soviet leader to seek liberalization. Premier Nikita Khrushchev, three decades ago, tried to usher in an early version of *glasnost*. He exposed Stalin's excesses, and in the early 1960s statues of Stalin lay in pieces from Leningrad to Yerevan. Cultural exchanges were encouraged and tourists invited. But the Khrushchev period ended abruptly in 1964 with his ouster, and the Soviet Union quickly returned to its darker days. Undoubtedly, the initiatives of President Gorbachev have gone much further, but in a society accustomed to the kind of democracy that communism brings, reversals are always possible.

Perhaps it is a quieter geographic change that will have more durable consequences. It is often said that Russians now make up just over half of the total Soviet population, and that they will soon find themselves in the minority in their own

country. In fact, the *Slavic* majority in the country is quite large: Russians, Ukrainians, and Belorussians make up about two-thirds of the population. Nevertheless, net population growth in the Slavic U.S.S.R. (not counting a growing emigration) is quite low, whereas population growth in the Muslim republics is high. (between 1979 and 1989, the Muslim increase exceeded 25 percent). Thus, the Soviet government will be faced not only with the problems arising from an intensifying Muslim fundamentalism in a state that has long espoused atheism, but it also will confront problems of economic development associated with rapid population growth and limited opportunity. These problems are all the more serious because they are regionally concentrated; moreover, the Islamic areas in question lie against sensitive borders that face neighbors with irredentist histories. After the present period of *perestroika* has led to a new order, the Soviets'—and Russia's—old geographic problems will still challenge the state.

PRONUNCIATION GUIDE

Abkhazian (ab-*kahz*-zee-un)

Adjarian (uh-*jar*-ree-un)

Alma-Ata (ahl-muh-uh-*tah*)

Altay (*al*-tye)

Amu-Darya (uh-mooh-*dahr*-yuh)

Amur (uh-*moor*)

Angara (ahng-guh-*rah*)

Aral (*arrel*)

Arkhangelsk (ahr-*kan*-jelsk)

Armenia (ar-*meeny*-uh)

Ashkhabad (*ash*-kuh-bahd)

Astrakhan (*astra*-kahn)

Azeri (ah-*zerry*)

Azerbaydzhan (ah-zer-bye-*jahn*)

Baku (bah-*kooh*)

Barents (*barrens*)

Baykal (bye-*kahl*)

Belorussia (bell-oh-*russia*)

Bering (*beh*-ring)

Bessarabia (bess-uh-*ray*-bee-uh)

Bokhara (boh-*kahr*-ruh)

Bolshevik (*boal*-shuh-vick)

Bratsk (*brahtsk*)

Bureya (buh-*ray*-yuh)

Caucasus (*kaw*-kuh-zuss)

Chen Pao (chen-*bau*)

Chernobyl (*chair*-nuh-beel)

Chita (chih-*tah*)

Crimean (cry-*mee*-un)

Cyrillic (suh-*rill*-ick)

Czechoslovakia (check-oh-slow-*vah*-key-uh)

Dnepropetrovsk (duh-nep-roh-puh-*trawffsk*)

Dnieper (duh-*n'yepper*)

Dnestr (duh-*nyess*-truh)

Donbas (*dahn*-bass)

Donetsk (duh-*netssk*)

Dushanbe (dooh-*shahm*-buh)

Dvina (duh-vee-*nah*)

Eritrea (erra-*tray*-uh)

Fergana (fare-*gahn*-uh)

Frunze (*froon*-zuh)

Georgia (*george*-uh)

Glasnost (*gluzz*-nost)

Gorbachev, Mikhail (gor-buh-choff, meek-*hyle*)

Gorkiy (*gore*-kee)

Gorno-Badakhshan (gore-noh-bah-dahk-*shahn*)

Gyre (*jyer*)

Irkutsk (ear-*kootsk*)

Irtysh (ear-*tish*)

Ismaili (izz-*mye*-lee)

Ivanovo (ee-*vah*-nuh-voh)

Kaliningrad (kuh-*leen*-un-grahd)

Kamchatka (komm-*chut*-kuh)

Kansk-Achinsk (kahnsk-uh-*chintsk*)

Kara (*kahr*-ruh)

Karaganda (kahra-gun-*dah*)

Kara Kum (kahr-ruh-*koom*)

Kazakhstan (kuzz-uck-*stahn*)

Kazan (kuh-*zahn*)

Kerch (*kertch*)

Khabarovsk (kuh-*bahr*-uffsk)

Khan, Genghis (*kahn, jing*-gus)

Khanate (*kahn*-ate)

Kharkov (*kahr*-koff)

Khiva (*kee*-vuh)

Khrushchev, Nikita (*kroos*-choff, nih-*kee*-tuh)

Kiev (*kee*-yeff)

Kirghiz(ia) (keer-*geeze* [ee-uh])

Kishinev (*kish*-uh-neff)

Klaipeda (*klye*-puh-duh)

Klyuchevskaya (klee-ooh-*cheff*-skuh-yuh)

Kolkhoz (*koll*-koze)

Komsomolsk (kom-suh-*mawlsk*)

Königsberg (*kay*-nix-bairk)

Krasnoyarsk (kruzz-noh-*yarsk*)

Krivoy Rog (krih-voi-*roag*)

Kursk (*koorsk*)

Kuybyshev (*kwee*-buh-sheff)

Kuzbas (kooz-*bass*)

Kuznetsk (kooz-*netsk*)

Lena (*lay*-nuh)

Lenin (*lennin*)

Mackinder, Halford (muh-*kinn*-duh, *hal*-ferd)

Makeyevka (muh-*kay*-uff-kuh)

Mariinsk (muh-rih-*eentsk*)

Meshketian (mesh-*ketty*-un)

Minsk (*meentsk*)

Moldavia (mol-*day*-vee-uh)

Moscow (*moss*-cau)

Murmansk (moor-*mahntsk*)

Muscovy (muh-*skoe*-vee)

Muslim (*muzz*-lim)

Nagorno-Karabakh (nuh-*gor*-noh kah-ruh-bahk)

Nakhichevan (nah-kee-chuh-*vahn*)

Nakhodka (nuh-*kaught*-kuh)

Nikopol (nih-*kaw*-pul)

Novgorod (*nahv*-guh-rahd)

Novokuznetsk (noh-voh-kooz-*netsk*)

Novosibirsk (noh-voh-suh-*beersk*)

Oblast (*ob*-blast)

Odessa (oh-*dessa*)

Okhotsk (oh-*kahtsk*)

Ordzhonikidze (or-johnny-*kidd*-zuh)

Ossetian (ah-*see*-shee-un)

Pamir (pah-*meer*)

Perestroika (perra-*stroy*-kuh)

Perm (*pairm*)

Povolzhye (puh-*voll*-zhuh)

Prut (*prroot*)

Riga (*ree*-guh)

Ruble (*rooble*)

Ruhr (*roor*)

Rus (*roos*)

Sakartvelo (sah-*kart*-vuh-loh)

Sakhalin (*sock*-uh-leen)

Sakharov, Andrei (*sock*-uh-roff, *ahn*-dray)

Saratov (suh-*raht*-uff)

Sarmatian (sahr-*may*-shee-un)

Scythian (*sith*-ee-un)

Serov (*sair*-roff)

Shi'ite (*shee*-ite)

Siberia (sye-*beery*-uh)

Slavic (*slah*-vick)

Sovkhoz (*sov*-koze)

Spykman (*spike*-mun)

Stalin (*stah*-lin)

Tadzhik (tah-*jeek*)

Tadzhikistan (tah-jeek-ih-*stahn*)

Tallinn (*tah*-lin)

Tashkent (tahsh-*kent*)

Tatar (*taht*-uh)

Tayshet (tye-*shet*)

Tbilisi (tuh-*bill*-uh-see)

Tianshan (tyahn-*shahn*)

Tikhvin (*tik*-vun)

Tolyatti (tawl-*yah*-tee)

Transcaucasia (tranz-kaw-*kay*-zhuh)

Tselinograd (seh-*linn*-uh-grahd)

Tula (*too*-luh)

Turkmenistan (terk-*men*-uh-stan)

Ukraine (yoo-*crane*)

Ulyanov, Vladimir Ilyich (ool-*yah*-nof, *vlah*-duh-mir *ill*-yitch)

Ural (*yoor*-ull)

Ussuri (ooh-*soor*-ree)

Uzbek (*ooze*-beck)

Uzbekistan (ooze-beck-ih-*stahn*)

Varangian (vuh-*range*-ee-un)

Vilnius (*vill*-nee-uss)

Vladivostok (vlad-uh-vuh-*stahk*)

Volgograd (*voll*-guh-grahd)

Yakut(sk) (yuh-*koot*-[sk])

Yaroslavl (yar-uh-*slahv*-ull)

Yenisey (yen-uh-*say*)

Yerevan (yair-uh-*vahn*)

REFERENCES AND FURTHER READINGS

(BULLETS [●] DENOTE BASIC INTRO-
DUCTORY WORKS ON THE REALM
OR SYSTEMATIC ESSAY TOPIC)

Allworth, Edward, ed. *Ethnic Russia in the U.S.S.R.: The Dilemma of Dominance* (Elmsford, N.Y.: Pergamon Press, 1980).

● Bater, James H. *The Soviet Scene: A Geographical Perspective* (London & New York: Routledge, 1989).

Bater, James H. & French, Richard A., eds. *Studies in Russian Historical Geography* (New York: Academic Press, 2 vols., 1983).

● Brown, Archie et al., eds. *Cambridge Encyclopedia of Russia and the Soviet Union* (New York: Cambridge University Press, 1982).

Buck, Trevor & Cole, John P. *Modern Soviet Economic Performance* (New York: Basil Blackwell, 1987).

Chew, Allen F. *Atlas of Russian History: Eleven Centuries of Changing Borders* (New Haven, Conn.: Yale University Press, 1967).

Clem, Ralph S. "Russians and Others: Ethnic Tensions in the Soviet Union," *Focus*, September–October, 1980.

● Cole, John P. *Geography of the Soviet Union* (Stoneham, Mass.: Butterworth, 1984).

● Critchfield, Howard J. *General Climatology* (Englewood Cliffs, N.J.: Prentice-Hall, 4 rev. ed., 1983).

Dellenbrant, Jan A. *The Soviet Regional Dilemma: People, Planning, and Natural Resources* (Armonk, N.Y.: M.E. Sharpe, 1986).

Demko, George J. & Fuchs, Roland J., eds. *Geographical Studies on the Soviet Union: Essays in Honor of Chauncy D. Harris* (Chicago: University of Chicago, Department of Geography, Research Paper No. 211, 1984).

Dewdney, John C. *U.S.S.R. in Maps* (New York: Holmes & Meier, 1982).

Dewdney, John C. *The U.S.S.R.: Studies in Industrial Geography* (Boulder, Colo.: Westview Press, 1976).

Dienes, Leslie. *Soviet Asia: Economic Development and National Policy Choices* (Boulder, Colo.: Westview Press, 1987).

Dienes, Leslie & Shabad, Theodore. *The Soviet Energy System: Resource Use and Policies* (New York: Halsted Press/V. H. Winston, 1979).

French, Richard A. & Hamilton, F. E. Ian, eds. *The Socialist City: Spatial Structure and Urban Policy* (New York: John Wiley & Sons, 1979).

● Gregory, James S. *Russian Land, Soviet People: A Geographical Approach to the U.S.S.R.* (London: George G. Harrap, 1968).

Hooson, David J. M. *The Soviet Union: Peoples and Regions* (Belmont, Calif.: Wadsworth, 1966).

Hosking, Geoffrey. *The Awakening of the Soviet Union* (Cambridge, Mass.: Harvard University Press, 1989).

● Howe, G. Melvyn. *The Soviet Union: A Geographical Study* (London & New York: Longman, 2 rev. ed., 1986).

Jensen, Robert G. et al., eds. *Soviet Natural Resources in the World Economy* (Chicago: University of Chicago Press, 1983).

Karklins, Rasma. *Ethnic Relations in the U.S.S.R.: The Perspective from Below* (Winchester, Mass.: Allen & Unwin, 1986).

Lewin, Moshe. *The Gorbachev Phenomenon: A Historical Interpretation* (Berkeley: University of California Press, 1988).

Lydolph, Paul E. *Climates of the Soviet Union* (Amsterdam: Elsevier Scientific Publishing Co., 1977).

● Lydolph, Paul E. *Geography of the U.S.S.R.* (Elkhart Lake, Wisc.: Misty Valley Publishing, 1990).

Mackinder, Halford J. *Democratic Ideals and Reality* (New York: Holt, 1919).

● Mather, John R. *Climatology: Fundamentals and Applications* (New York: McGraw Hill, 1974).

Mellor, Roy E. H. *The Soviet Union and Its Geographical Problems* (London: Macmillan, 1982).

"Muslims in the U.S.S.R.," *Aramco World*, January–February, 1990, pp. 1–43.

Nove, Alec. *Glasnost in Action: Cultural Renaissance in Russia* (Winchester, Mass.: Unwin Hyman, 1989).

Nove, Alec. *The Soviet Economic System* (Winchester, Mass.: Allen & Unwin, 2 rev. ed., 1980).

● Parker, William H. *The Soviet Union* (London & New York: Longman, 2 rev. ed., 1983).

Sagers, Matthew J. & Green, Milford B. *The Transportation of Soviet Energy Resources* (Totowa, N.J.: Rowman & Littlefield, 1986).

Sanchez, James J. *A Bibliography for Soviet Geography* (Chicago: Council of Planning Librarians, 1985).

Scherer, John L., ed. *U.S.S.R.: Facts and Figures Annual* (Gulf Breeze, Fla.: Academic International Press, annual).

Smith, Hedrick. *The New Russians* (New York: Random House, 1990).

"Soviet Disunion: Growing Cries for Independence Bring Gorbachev's Empire to the Breaking Point," *Time*, Special Section, March 12, 1990, pp. 26–52.

● Symons, Leslie, ed. *The Soviet Union: A Systematic Geography* (London & New York; Routledge, 2 rev. ed., 1990).

Treadgold, Donald W. *Twentieth Century Russia* (Boulder, Colo.: Westview Press, 7 rev. ed., 1990).

U.S.S.R. Energy Atlas (Washington, D.C.: U.S. Central Intelligence Agency, 1985).

Wixman, Ronald. *The Peoples of the U.S.S.R.: An Ethnographic Handbook* (Armonk, N.Y.: M.E. Sharpe, 1984).

Wood, Alan, ed. *Siberia: Problems and Prospects for Regional Development* (London & New York: Methuen, 1987).

Zum Brunnen, Craig & Osleeb, Jeffrey P. *The Soviet Iron and Steel Industry* (Totowa, N.J.: Rowman & Littlefield, 1986).

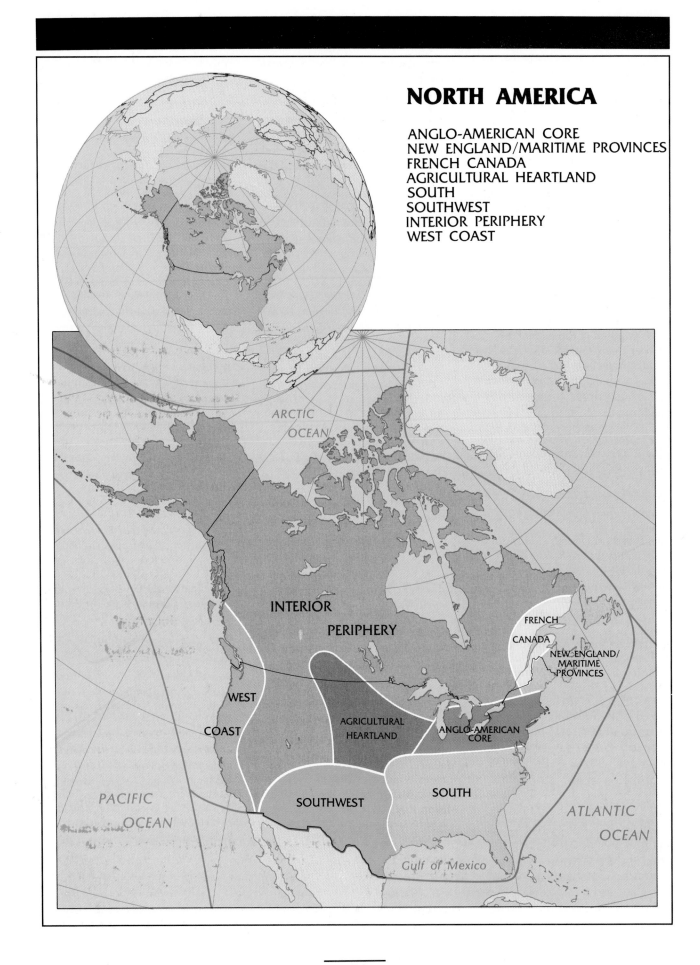

NORTH AMERICA

ANGLO-AMERICAN CORE
NEW ENGLAND/MARITIME PROVINCES
FRENCH CANADA
AGRICULTURAL HEARTLAND
SOUTH
SOUTHWEST
INTERIOR PERIPHERY
WEST COAST

CHAPTER 3

NORTH AMERICA: THE POSTINDUSTRIAL TRANSFORMATION

The North American realm consists of two countries that are alike in many ways. In the United States and in Canada, European cultural imprints dominate. Indeed, the realm is often called *Anglo-America*: English is the official language of the United States, and shares equal status with French in Canada. The overwhelming majority of churchgoers adhere to Christian faiths. Most (but not all) of the people trace their ancestries to various European countries. In the arts, architecture, and other spheres of cultural expression, European norms prevail.

North American society is the most highly urbanized of the world's realms; nothing symbolizes the New World quite as strongly as the skyscraper panorama of New York, Toronto, Chicago, or San Francisco. North Americans are also hypermobile, with networks of superhighways, air routes, and railroads efficiently interconnecting the realm's far-flung cities and regions. Commuters stream into and out of suburban activity centers and central-city downtowns by the millions each working day. A U.S. family relocates, on the average, once every five and a half years.

In the 1990s, North America has entered an entirely new age,

IDEAS AND CONCEPTS

urban geography

cultural pluralism

time-space convergence

physiographic province

rain shadow effect

pollution

culture hearth

epochs of metropolitan evolution

megalopolitan growth

eras of intraurban structural evolution

suburban downtown

urban realms model

mental maps

economies of scale

postindustrial revolution

nine-nations hypothesis

---

the third since the discovery of the New World by Columbus 500 years ago. The first four centuries were dominated by agriculture and rural life; the second age—industrial urbanization—has endured over the past century but is now nearing an end. In its place, the United States and Canada are to-

day experiencing the maturation of a *postindustrial* society and economy, which is dominated by the production and manipulation of information, skilled services, and high-technology manufactures, and operates within an increasingly global-scale framework of business interactions. As blue-collar rapidly yields to white-collar employment and as the automated office becomes the dominant workplace, fundamental dislocations are felt throughout North America. The aging Manufacturing Belt of the northeastern quadrant of the United States (and southeastern Canada) is often called the "Rust Bowl," whereas glamour and prestige are attached to such Sunbelt locales as suburban San Francisco's Silicon Valley and its high-technology counterparts in Texas, North Carolina, Florida, Arizona, Colorado, and Southern California.

Not surprisingly, the human geography of the United States and Canada is undergoing a parallel transformation as dynamic new locational forces surface. In fact, the realm's durable internal regional structure is being reshaped to the extent that a new mosaic of *nine nations* can (and will) be discussed at the conclusion of this chapter. Whatever the outcome, the winners and losers in the current

manding nothing less than outright separation from the rest of Canada.

Internal regionalism also affects Canada to the west of *Franco-phone* (French-speaking) Quebec. The country's core area lies mainly in Ontario, Canada's most populous province, centered on metropolitan Toronto (3.2 million); Ontario's French-speaking minority, constituting only about 5 percent of the population, is clustered mostly in the east near the Quebec border. Farther west lie the interior prairie provinces of Manitoba, Saskatchewan, and Alberta, where the French join the Germans and Ukrainians to form small minorities, ranging from 4 to 7 percent of each provincial population. The French cause is weakest in Canada's westernmost province, British Columbia, which is third in population and focuses on Greater Vancouver (1.9 million); here, over 80 percent of the population claims English ancestry, whereas only about 1.5 percent are ethnic French, well behind the area's rapidly growing Asian community.

No multilingual divisions affect the unity of the North American realm's other federation, but **cultural pluralism** of another kind prevails south of the Canada–U.S. border. More persistent in the U.S. cultural mosaic than language or ethnicity is the division between peoples of European descent (about 85 percent of the population) and those of African origin (roughly 12 percent). Despite the significant progress of the modern civil rights movement, which decidedly weakened *de jure* racial segregation in public life, whites still refuse to share their immediate living space with blacks; thus, *de facto* residential segregation is all but universal. Although separatist regional thinking on the provincial/state scale of Quebec does not exist, a strong case can be made that persisting local racial segregation has, in effect, produced two societies—one white and one black—that are both separate and unequal.

Economic Affluence and Influence

The United States and Canada rank among the most highly advanced countries of the world by every measure of national development, possessing the highest living standards on earth. But the good life is not shared equally by all of North America's residents. Deprivation is surprisingly widespread, with notable spatial concentrations in the United States inside the inner-ring slums of big cities and on rural reservations containing Indians or Eskimos (Native Americans). In the worst of these poverty pockets malnutrition is commonplace, although its severity pales by comparison to the daily misery experienced by the Third World poor.

North America's highly developed societies have clearly achieved a global leadership role, which arose from a combination of history and geography. Presented with a rich abundance of natural and human resources over the past 200 years, Americans and Canadians have brilliantly converted these productive opportunities into continent-wide affluence and worldwide influence as their booming Industrial Revolution surpassed even Europe's by the early decades of this century.

Perhaps the greatest triumph accomplished by human ingenuity in the North American realm was the hard-won victory over a difficult, large-scale environment: by persistently improving transportation and communications, the United States and Canada were finally able to organize their entire countries spatially across great distances (exceeding 2500 miles/4000 kilometers) in the needed east-west direction across a terrain in which the grain of the land—particularly mountain barriers—is consistently oriented north-south. As breakthroughs in railroad, highway, and air transport technology have succeeded each other during the past 150 years, geographic space-shortening—or **time-space convergence**—allowed distant places to become ever nearer to one another and constantly enabled people and activities to disperse more widely.

Although they continue to have their share of differences, the United States and Canada maintain close and cordial relations, and the border between them is by far the longest open international boundary on earth (over 75 million people cross it each year). Clearly, the two countries are firmly locked together in a mission of free-world leadership. What is now happening in the North American realm—and why—matters enormously to every other country. The United States and Canada constitute the most advanced region in the world; wherever it goes in the new postindustrial age, many of the rest will eventually follow.

NORTH AMERICA'S PHYSICAL GEOGRAPHY

Before we examine North America's human geography more closely, it is important to consider the physical setting in which this key geographic realm is rooted. The North American continent extends from the Arctic Ocean to

Panama, but we will confine ourselves here to the region north of Mexico—a region that still stretches from the near-tropical latitudes of southern Florida and Texas to the subpolar lands of Alaska and Canada's far-flung Northwest Territories. The remainder of the North American continent, which constitutes a separate realm (together with the island chains of the Caribbean Sea), comprises *Middle America* and will be treated in Chapter 4.

Physiography

North America's physiography is characterized by its clear, well-defined division into physically homogeneous regions called **physiographic provinces**. Such a region is marked by a certain degree of uniformity in relief, climate, vegetation, soils, and other environmental conditions, resulting in a scenic sameness that comes readily to mind. For example, we identify such regions when we refer to the Rocky Mountains, the Great Plains, the Appalachian Highlands. However, not all the physiographic provinces of North America are so easily delineated.

A complete picture of the continent's physiography is seen in Fig. 3–4. The most obvious aspect of this map of North America's physiographic provinces is the north-south alignment of the continent's great mountain backbone, the Rocky Mountains, whose rugged, often-snow-covered topography dominates the western segment of the continent from Alaska to New Mexico. The major feature of eastern North America is another, much lower chain of mountain ranges called the Appalachian Highlands. These eastern uplands also trend approximately north-south and extend from Canada's Atlantic Maritime Provinces to Alabama. The orientation of the

Rockies and Appalachians is important because, unlike Europe's Alps, they do not form a topographic barrier to polar or tropical air masses flowing southward or northward, respectively, across the continent's interior.

Between the Rocky Mountains and the Appalachians lie North America's vast interior plains, which extend from the shores of Hudson Bay to the coast of the Gulf of Mexico. These flatlands can be subdivided into several provinces. The major regions are: (1) the great Canadian Shield, which is the geologic core area containing North America's oldest rocks; (2) the Interior Plains or Lowlands, covered mainly by glacial debris laid down by meltwater and wind during the Late Cenozoic glaciation; and (3) the Great Plains, the extensive sedimentary surface that slowly rises westward toward the Rocky Mountains. Along the southern margin, these interior plainlands merge into the Gulf–Atlantic Coastal Plain, which extends from southern Texas along the seaward margin of the Appalachian Highlands and the neighboring Piedmont until it ends at New York's Long Island.

On the western side of the Rocky Mountains lies the zone of Intermontane Basins and Plateaus. This physiographic province (within the contiguous United States) includes: (1) the Colorado Plateau in the south, with its thick sediments and spectacular Grand Canyon; (2) the lava-covered Columbia Plateau in the north, which forms the watershed of the Columbia River; and (3) the central Basin-and-Range country (Great Basin) of Nevada and Utah, which contains several extinct lakes from the glacial period as well as the surviving Great Salt Lake.

The reason this province is called *intermontane* has to do with its position between the Rocky

Mountains to the east and the Pacific Coast mountain system to the west. From the Alaskan Peninsula to Southern California, the west coast of North America is dominated by an almost unbroken corridor of high mountain ranges whose origins stem from the contact between the North American and Pacific Plates (Fig. I–6). The major components of this coastal mountain belt include California's Sierra Nevada, the Cascades of Oregon and Washington, and the long chain of highland massifs that line the British Columbian and southern Alaska coasts. Two broad valleys in the conterminous U.S. portion of the province, which contain dense populations, are the only noteworthy interruptions: California's Central (San Joaquin–Sacramento) Valley, and the Puget Lowland of Washington State that extends southward into western Oregon's Willamette Valley.

Climate

The various climatic regimes and regions of North America are clearly depicted on the world climate map (Fig. I–10) and model (Fig. I–11). In general, temperature varies latitudinally—the farther north one goes, the cooler it gets. Local land-and-water-heating differentials, however, distort this broad pattern. Because land surfaces heat and cool far more rapidly than water bodies, yearly temperature ranges are much larger where continentality (see p. 137) is greatest.

Precipitation generally tends to decline toward the west—except for the Pacific coastal strip itself—as a result of the **rain shadow effect**, whereby most Pacific Ocean moisture is effectively screened from the continental interior (see *box*). This broad division into Arid (western) and Humid (eastern)

America, however, is marked by a very fuzzy boundary that is best viewed as a wide transitional zone. Although the separating criterion of 20 inches (50 centimeters) of annual precipitation is easily mapped (see Fig. I–9), that generally north-south *isohyet* (the line connecting all places receiving ex-

actly 20 inches per year) can and does swing widely across the drought-prone Great Plains from year to year because highly variable rains from the Gulf of Mexico come and go in unpredictable fashion.

On the other hand, in Humid America, precipitation is far more

regular. The prevailing westerly winds (blowing from west to east—winds are always named for the direction *from which they come*) that normally come up dry for the large zone west of the 100th meridian, pick up considerable moisture over the Interior Lowlands and distribute it

FIGURE 3–4

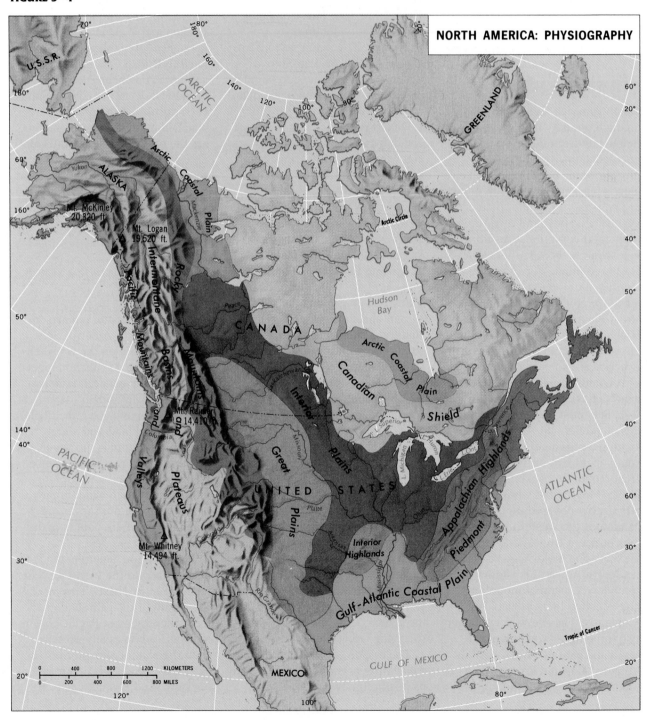

NORTH AMERICA: PHYSIOGRAPHY

throughout eastern North America. A large number of storms develop here on the highly active weather front between tropical Gulf air to the south and polar air to the north. Even if major storms do not materialize, local weather disturbances created by sharply contrasting temperature differences are always a danger. There are more tornadoes (nature's most violent weather) in the central United States each year than anywhere else on earth. And in winter, the northern half of this region receives large amounts of snow, especially just downwind from the Great Lakes, whose moisture often reinforces the precipitative potential of moving cyclonic storms.

Fig. I–10 shows the absence of humid temperate (*C*) climates from Canada (except along the narrow Pacific coastal zone) and the prevalence of cold in Canadian environments. East of the Rocky Mountains, Canada's most *moderate* climates correspond to the *coldest* of the United States. Nevertheless, Southern Canada does share the environmental conditions that mark the Upper Midwest and Great Lakes areas of the United States, so that agricultural productivity in the prairie provinces and in Ontario is substantial. Canada is a leading food exporter (chiefly wheat), as is the United States, in spite of its comparatively short growing season.

Soils and Vegetation

As the world soil distribution map shows (Fig. I–13), the wide variety of climates in North America has helped spawn an even more complex pattern of soil regions. In general, the realm's soils also reflect the broad environmental partitioning into Humid and Arid America. Where precipitation annually exceeds 20 inches (50 centimeters), humidland soils tend to be leached and *acidic* in chemical content.

THE RAIN SHADOW EFFECT

The dryness of most of western North America is a result of the relationship between climate and physiography. At these latitudes, the prevailing wind direction is from west to east, but the moisture-laden air moving onshore from the Pacific is unable to penetrate the continent because high mountains stand in the way. When eastward-moving humid Pacific air reaches the foot of the Cascade Mountains and other north-south ranges, it is forced to rise up the western (windward) slope in order to surmount the topographic barrier. As the air rises toward the 8000- to 12,000-foot (2400- to 3600-meter) summit level, it steadily cools. Since cooler air is less able to hold moisture, falling temperatures produce substantial rainfall (and snowfall), in effect "squeezing out" the Pacific moisture as if the air were a sponge filled with water. This altitude-induced precipitation is called *orographic* (mountain) rainfall; the two other major types of precipitation are associated with *fronts* (boundaries along which air masses of considerably different temperatures come in contact, sometimes spawning spiraling storm systems called cyclones) and with *convection* (the localized rising and cooling of warm moist air).

Although precipitation totals are quite high on the upper windward slopes of the Sierra Nevada–Cascade Mountain chain, the major impact of the orographic phenomenon is felt to the east. Robbed of most of its moisture content by the time it is pushed across the summit ridge, the eastward-moving air now rushes down the leeward (downwind) slopes of the mountain barrier. As this air warms, its capacity to hold moisture greatly increases, and the result is a warm dry wind that can blow strongly for hundreds of miles inland. This widespread existence of semiarid (and in places even truly arid) environmental conditions is known as the *rain shadow effect*—with mountains quite literally creating a leeward "shadow" of dryness. In fact, as air masses move west to east across the Intermontane and Rocky mountain provinces, they may again be subjected to orographic uplift (and even further drying out) as they pass over additional highland barriers. This reinforces the spreading of dryness, in most years, well to the east of the Rockies. Here in the Great Plains, Pacific moisture sources are usually blocked, and the area must depend on fickle south-to-north winds from the Gulf of Mexico for much of its warm-season precipitation; no wonder this region is susceptible to recurring droughts and was the scene of dust-bowl environmental disasters in the 1930s and 1950s. In all, the vast dry area extending from the West Coast mountains halfway across the continent to the eastern margin of the Great Plains (in Southern Canada as well as the United States) comprises what may be called *Arid America*; alternatively, the eastern half of the realm and the narrow strip between the Pacific shore and its nearby parallel ranges—where annual precipitation exceeds 20 inches (50 centimeters)—can be viewed as *Humid America*.

Elsewhere in the world, many examples of rain shadows can be observed downwind from where mountains block prevailing moist windflows. Sweden (p. 98) is a good case in point.

Since crops do best in soils that are neither acidic nor *alkalinic* (higher in salt content), fertilization is necessary to achieve the desired level of neutrality between the two. Arid America's soils are typically alkalinic and must be fertilized back toward neutrality by adding acid compounds. Although many of these dryland soils, particularly in the Great Plains, are quite fertile, settlers learned over a century ago that water is the main missing ingredient in achieving their agricultural potential. In the 1970s, the *center-pivot irrigation* method (Fig. 3–5) was perfected and finally provided a real opportunity to expand more intensive farming west from the Corn Belt into the drier center and western portions of the Plains. Glaciation has also enhanced the rich legacy of fertile soils in the central United States, both from the deposition of mineral-rich glacial debris left by meltwater and from thick layers of fine wind-blown glacial material (called *loess*) in and around the middle Mississippi Valley.

Vegetation patterns are displayed on the world map (Fig. I–12), but the enormous human modification of the North American environment in modern times has all but reduced this regionalization scheme to the level of the hypothetical. Nonetheless, the Humid/Arid America dichotomy is again a valid generalization: the natural vegetation of areas receiving more than 20 inches (50 centimeters) of water yearly is *forest*, whereas the drier climates give rise to a *grassland* cover. The forests of North America east of 100°W longitude—plus those of the Pacific rim and the higher elevations of inland mountain ranges—tend to make a broad transition by latitude. In the Canadian North needleleaf forests dominate, but these coniferous trees become mixed with broadleaf deciduous trees as one crosses the border into the U.S. Northeast. As one proceeds toward the Southeast, broadleaf vegetation becomes dominant, except for large stands of pine forests along the drier sandy soils of the Gulf–Atlantic Coastal Plain. In South Florida, the realm's only tropical climate begets a savanna-like grassland in combination with large coastal swamps such as the Everglades. The only noteworthy departure in the limited western forest zone is the brushy tree-and-shrub landscape (*chaparral*) of Mediterranean Southern California. Arid America mostly consists of short-grass prairies or *steppes*. The only area of true *desert* (less than 10 inches/25 centimeters of annual rainfall) is found in southeastern California and adjoining southwestern Arizona, with cactus and other leathery shrubs predominating.

Hydrography (Water)

Surface water patterns in North America are dominated by the two major drainage systems that lie between the Rockies and the Appalachians: (1) the five Great Lakes that drain into the St. Lawrence River, and (2) the mighty Mississippi–Missouri river network, supported by such major tributaries as the Ohio, Tennessee, Platte, and Arkansas rivers. Both are products of the last episode of Late Cenozoic glaciation, and together they amount to nothing less than the best natural inland waterway system in the world. Human intervention has further enhanced this network of navigability, mainly through the building of canals that link the two systems as well as the St. Lawrence Seaway (opened in 1959, but increasingly obsolete because it cannot accommodate today's larger ships).

Elsewhere, the northern east coast of the continent is well served by a number of short rivers leading inland from the Atlantic. In fact, many of the big northeastern seaboard cities of the United States (such as Richmond, Baltimore, and Philadelphia) are located at the waterfalls that marked the limit to tidewater navigation.

FIGURE 3–5 Circular fields, the signature of center-pivot irrigation on the rural landscape.

Rivers in the Southeast and west of the Rockies at first offered little practical value owing to orientation and navigability problems. In the Far West, however, the Colorado and Columbia rivers have now become critically important as suppliers of drinking and irrigation water as well as hydroelectric power as the Pacific coast continues its regional development.

Groundwater is also a crucial resource in many parts of the United States. *Aquifers*, or underground reservoirs contained within porous water-bearing rock layers, underlie many states and have become essential to the maintenance of people and activities on the surface above. In dry areas especially, they are the cornerstones of development—such as in the central Great Plains where the Ogal-

lala Aquifer is the only steady source of water for the heavily agricultural economy. Problems quickly arise, however, when increasing usage begins to exceed the leisurely rate of natural recharge. Human intervention can also produce more serious consequences, as when the dumping of sewage and chemical wastes contaminates groundwater supplies. Such issues raise a whole new dimension in North American physical geography, and merit a closer look.

Human Environmental Impacts

More than a century of advanced industrial technology has taken its

toll on the natural environment of North America. For decades, the growing problems of air and water **pollution** were ignored until a public outcry in the United States during the 1960s forced the federal government to establish the Environmental Protection Agency (EPA) and take a leading role in enforcing new pro-environmental legislation. Although substantial progress was made in the 1970s, administrations since 1980 have been criticized for weakening the federal enforcement apparatus. Also during the 1970s, cancer surpassed heart disease as the leading killer in the United States. Since many cancers, especially lung cancer, are environmentally related, medical geographers have increased their study of the subject (this subdiscipline is reviewed in the Systematic Essay in Chapter

FIGURE 3–6

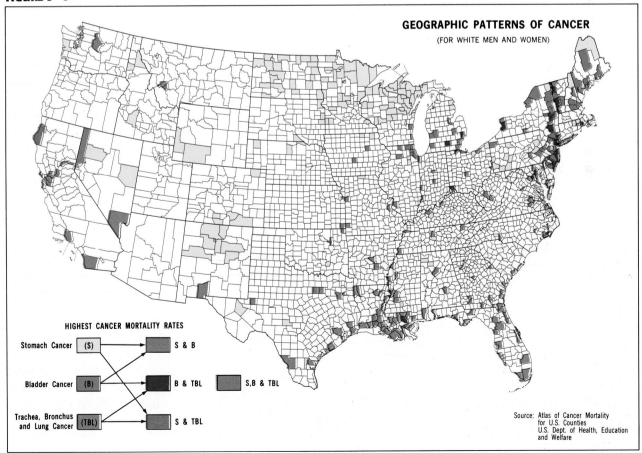

GEOGRAPHIC PATTERNS OF CANCER
(FOR WHITE MEN AND WOMEN)

HIGHEST CANCER MORTALITY RATES

Stomach Cancer (S) → S & B
Bladder Cancer (B) → B & TBL
Trachea, Bronchus and Lung Cancer (TBL) → S & TBL

S,B & TBL

Source: Atlas of Cancer Mortality for U.S. Counties U.S. Dept. of Health, Education and Welfare

7). The spatial distribution of cancer mortality (the newest map available) is shown in Fig. 3–6, which reveals that respiratory-system cancer coincides with a number of major manufacturing and refining centers.

One of the most severe *air pollution* problems of large metropolitan complexes is smog. Dozens of major cities experience this annoying hazard, with Los Angeles and Denver among the worst offenders on the continent (Fig. 3–7). Smog is usually created by a temperature inversion in which a warm dry layer of air, hundreds of feet above the ground, prevents cooler underlying air from rising; this causes the surface air to become stagnant, thereby trapping automo-

bile and industrial emissions that intensify and rapidly turn into chemical smog (the word is a contraction of "smoke" and "fog"). One of the realm's most serious *water pollution* problems is acid rain (see pp. 98–99). Although at least 75 percent of North America's sulfur and nitrogen emissions emanate from the U.S. Manufacturing Belt (particularly Illinois, Indiana, Ohio, and Michigan), the lion's share of the resultant acid rain appears to fall in southeastern Canada as prevailing winds blow from the Midwest toward eastern Ontario and southern Quebec. The issue, the thorniest in the relationship between the United States and Canada, has yet to be resolved.

POPULATION IN TIME AND SPACE

The current distribution of North America's population is shown in Fig. 3–8. This map is once again the latest still in a motion picture, one that has been unreeling for nearly four centuries since the founding of the first permanent European settlements on the northeastern coast of the United States. Slowly at first, then with accelerating speed after 1800 as one major technological breakthrough followed another, Americans and Canadians took charge of their remarkable continent and pushed the settlement frontier

FIGURE 3–7 Smog is an ever-present hazard in the Los Angeles Basin, and was even reported by its Spanish discoverers who observed the local atmospheric effects of cooking fires above Indian villages. This is central Los Angeles on a typical day, looking northeast toward the CBD and the San Gabriel Mountains (whose ridge top emerges into the clean air above the murk).

westward to the Pacific. Undoubtedly, the population distribution of North America has undergone the swiftest, most dramatic changes of any realm over the past 150 years. North Americans are still the most mobile people in the world—nearly 19 percent of the U.S. population changes residence each year—a trait deeply rooted in the American national character and nativist culture. Migrations in the 1990s continue to reshape the United States, perhaps the most important being the persistent interregional shift of people and livelihoods toward the south and west (the so-called Sunbelt), away

FIGURE 3–8

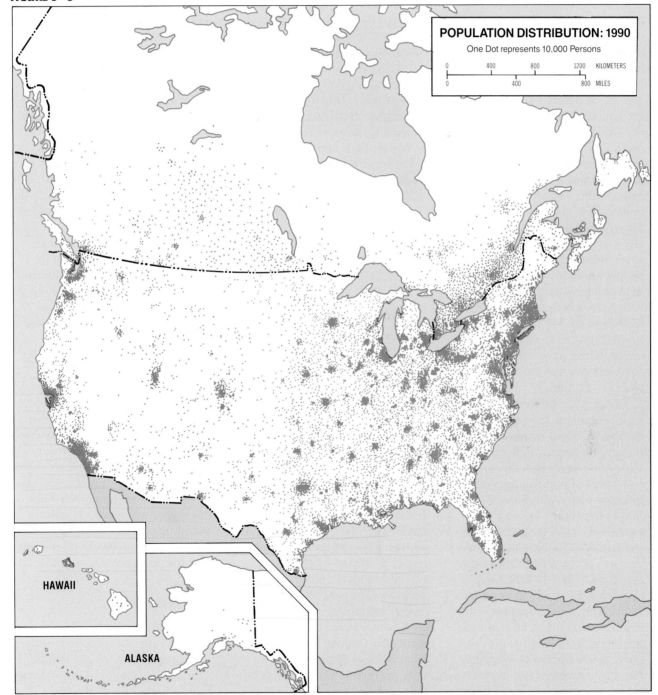

POPULATION DISTRIBUTION: 1990

One Dot represents 10,000 Persons

Source: Adapted from U.S. Census Bureau and Census Canada.

from the north and east. One significant finding of the 1980 census was that, for the first time, the geographic center of the American population had crossed west of the Mississippi River, capping a 200-year movement in which every decennial census reported a westward shift.

In order to understand the contemporary population map, we need to review the more important forces that have shaped, and continue to shape, the distribution of North Americans and their activities. Even before independence, the United States was perceived as the world's premier "land of opportunity" and attracted a steady influx of immigrants who were rapidly assimilated into the societal mainstream. Despite tight quotas, immigration continues, with well over 1 million new arrivals annually—at least 2 out of 3 of whom enter illegally. Within the realm, people have sorted themselves out to maximize their proximity to existing economic opportunities, and they have shown little resistance to relocating as the nation's changing economic geography has successively favored different sets of places over time.

During the past century, these transformations have spawned a number of major migrations. Although the American frontier closed 100 years ago, the westward shift of population continues, but now with a pronounced deflection to the south. The explosive growth of cities, triggered by the Industrial Revolution in the final decades of the nineteenth century, launched a rural-to-urban migratory stream that lasted into the 1960s. The middle decades of this century also witnessed significant migration from south to north—particularly by blacks—but that, too, has ended and is even reversing as sizable numbers of affluent blacks today return south. This is

part of the much larger north-to-south movement toward the Sunbelt mentioned above, a migratory stream that remains strong for the following reasons: (1) the U.S. economy and its better jobs are clearly shifting in that direction; (2) the historically dominant westward push is increasingly inseparable from the shift toward the southern tier; (3) the substantial retirement migration of the affluent elderly to such states as Florida and Arizona, though slowing, shows little sign of ending; and (4) the new tide of immigration from Middle America is overwhelmingly directed toward the zone adjacent to the southern border.

Let us now look more closely at the historical geography of these changing population patterns, viewing first the initial rural influence, and then the decisive impacts of industrial urbanization and its postindustrial aftermath.

Populating Rural America Before 1900

The geographical distribution of North America's population is rooted in the colonial era of the seventeenth and eighteenth centuries. That era began just after 1600 with a number of discoveries and land claims made by such European powers as England, France, Spain, and the Netherlands. The French were concerned with penetrating the continental interior, propelled by their desire to establish a lucrative fur-trading network. The other major (and larger) colonizing force, the English, concentrated their settlement efforts along the coast of what is today the northeastern U.S. seaboard. These British colonies quickly became differentiated in their local economies, a

diversity that was to endure and eventually shape American cultural geography. The northern colony of New England (Massachusetts Bay and environs) specialized in commerce; the southern Chesapeake Bay colony (Tidewater Virginia and Maryland) emphasized the large-scale plantation farming of tobacco; the Middle Atlantic area in between (southeastern New York, New Jersey, eastern Pennsylvania) was home to a number of smaller independent-farmer colonies. As these three adjacent colonial regions thrived and yearned to expand after 1750, the British government kept the inland frontier closed and exerted ever tighter economic controls, thereby unifying the three groups of colonies in their growing dislike for the now heavy-handed mother country.

By 1775, escalating tensions produced open rebellion on the American side of the Atlantic and the onset of the eight-year-long Revolutionary War, which resulted in English defeat and independence for the newly formed United States of America. In the aftermath of this upheaval, the United Empire Loyalists, favoring Britain, fled north to a safe refuge in Ontario and the maritime provinces, where they later (in 1867) formed the Canadian confederation together with both the descendants of the early French settlers of the lower St. Lawrence Valley of Quebec and an older Scottish maritime colony (Nova Scotia) located on a peninsula extending into the Atlantic east of Maine.

The U.S. West

Meanwhile, in the fledgling nation to the south, the western frontier swung open and the old British Northwest Territory (Ohio-Michigan-Indiana-Illinois-Wisconsin) was settled rapidly, as much of its soil proved to be highly favorable for agriculture—a remarkable improvement over the relatively infertile seaboard soils. This added yet another element to the increasingly varied American environment; but diversity had by then been converted into a positive force through the establishment of widespread trading ties, including cross-Appalachian commodity flows, that were based on interregional complementarities and the emergence of an economy whose spatial organization was assuming national-scale proportions.

By the time the westward-moving frontier swept across the Mississippi Valley in the 1820s (Fig. 3–9), it was clear that the three former seaboard colonies had become separate **culture hearths**—source areas and innovation centers (*A*, *B*, and *C* on Fig.

3–9) from which migrants carried cultural traditions into the interior. The New England region (*A*) influenced the southern and western margins of the Great Lakes, with settlers here creating a cultural landscape that reflected New England's architecture and village patterns. Immediately to the south in the Trans-Appalachian West lay a much larger area that focused on the Ohio Valley but stretched as far south as southern Tennessee, which similarly resembled Middle-Atlantic Pennsylvania (*B*); and along the Atlantic and Gulf Coastal Plain, from Delaware to newly purchased Louisiana, lay an emerging culture area dominated by the tobacco- (and now cotton-) plantation tradition that had originated in the Tidewater Maryland/Virginia hearth (*C*).

The northern half of this vast

interior space soon became well unified as transport linkages improved. By 1860 the railroad had replaced earlier plank roads and canals, providing a more efficient network that offered fast and cheap long-distance transportation, thereby producing significant time-space convergence among once-distant regions within the northeastern quadrant of the United States. The American South, however, did not wish to integrate itself economically with the North to any great degree, preferring instead to export cotton and tobacco to overseas markets. This divergent regionalism, together with its insistence on preserving slavery, soon led the South into secession and the ruinous Civil War (1861–1865). In the aftermath, the South took decades to rebuild.

FIGURE 3–9

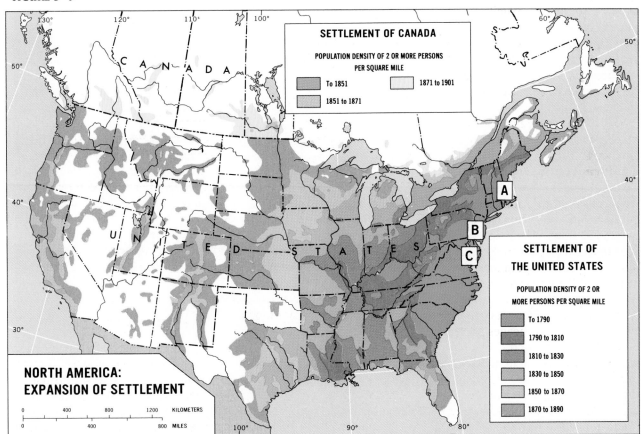

SETTLEMENT OF CANADA

POPULATION DENSITY OF 2 OR MORE PERSONS PER SQUARE MILE

To 1851
1851 to 1871
1871 to 1901

SETTLEMENT OF THE UNITED STATES

POPULATION DENSITY OF 2 OR MORE PERSONS PER SQUARE MILE

To 1790
1790 to 1810
1810 to 1830
1830 to 1850
1850 to 1870
1870 to 1890

NORTH AMERICA: EXPANSION OF SETTLEMENT

The second half of the nineteenth century also saw the frontier cross the western United States, although at first it was a "hollow" frontier: parts of the Pacific coast were settled before the dry steppelands of the Great Plains region far to the east (Fig. 3–9). As early as 1869, agriculturally booming California was linked to the rest of the nation by transcontinental railroad, and these same steel tracks also began to open up the bypassed Great Plains. The semiarid Plains, it turned out, were quite fertile after all, particularly for raising wheat on a large scale. This region was deficient only in water supplies, which could be stretched through careful farming methods or enhanced by irrigation near river and aquifer sources. Two new inventions—automated farm machinery for large wheatfields, and barbed-wire fences to protect crops from grazing cattle—further supported the development of grain farming and the regional economy.

The Canadian West

The Canadian experience in westward regional development, though it involved a far smaller national population, was similar. A strong bicultural division prevailed in the east here as well (British versus French settlers), but these differences resulted in the 1867 British-led confederation agreement rather than civil war. The initial push to the west also encountered physical obstacles—not the Appalachians but the barren Canadian Shield, a heavily glaciated and swampy flatland totally devoid of productive soils, that stretched from just north of Lake Ontario almost all the way to the 100th meridian (Fig. 3–4). The Great Plains on the western side of the Shield—known here as the prairie provinces—offered an environment for dryland wheat farming identical to the one south of the U.S. border (except for a shorter growing season) and provided Canada's only opportunity to develop a large-scale agricultural region. To the west of the Plains, there again arose the barrier of the Rockies–Intermontane

FIGURE 3–10 The rectangular symmetry of the Township and Range system covers most of the U.S. landscape west of the Appalachians, here near Lompoc in California's central coastal zone.

Plateaus–Pacific Highlands, narrower in Canada but lacking the convenient mountain passes that were available in the U.S. Nevertheless, a transcontinental railway was heroically pushed across to Vancouver in the 1880s, and Canada, too, had connected its Atlantic and Pacific coasts by rail. This effort was, at least in part, urged forward by the fear that U.S. westward expansion would surge north across the U.S.–Canada boundary if Canadians did not organize, use, and populate their westernmost lands.

The U.S. Land Survey System

By the time the U.S. frontier closed in the 1890s, today's rural settlement pattern was firmly in place, anchored to a set of enduring national agricultural regions (discussed later in this chapter). At the local level, except for the Thirteen Original States and Texas, rural population was distributed within the square geometry of the *Township-and-Range* land-division system. This nationwide checkerboard-like scheme (Fig. 3–10) was designed by Thomas Jefferson and his associates before 1800 to survey properties easily and disperse settlers evenly across newly opened farmlands. To this day, it still encompasses the largest area of planned rural settlement anywhere on earth.

The closure of the frontier also coincided with an accelerating shift in the distribution of the U.S. population: the census of 1900 revealed that more people lived in newly developed metropolitan areas than in the countryside. An urban revolution was following in the wake of the Industrial Revolution, which had taken hold in North America after the Civil War a generation earlier. To be sure, cities had been vitally important

since colonial times. But the degree and extent of urbanization by 1900 was approaching massive proportions, and it was clear, as the twentieth century opened, that the tilt from rural to urban America was just beginning.

North American Industrial Urbanization, 1900–1970

In the United States, the Industrial Revolution occurred almost a century later than in Europe, but when it finally did cross the North Atlantic in the 1870s, it took hold so successfully and advanced so robustly that only 50 years later America was surpassing Europe as the world's mightiest industrial power. Thus, the far-reaching societal changes and demographic transition (see Fig. 1–3) that Europe's industrializing countries experienced were greatly accelerated in the United States, fueled further by the arrival of more than 25 million European immigrants—who overwhelmingly headed for jobs in the major manufacturing centers—between 1865 and 1914.

The impact of industrial urbanization occurred simultaneously at two levels of generalization. At the national scale or *macroscale*, a network or system of new cities rapidly emerged, specializing in the collection, processing, and distribution of raw materials and manufactured goods, linked together by an efficient web of long-distance and local railroad lines. Within that urban system, at the *microscale*, individual cities prospered in their new roles as manufacturing centers, generating a wholly new internal structure that still forms the spatial framework of most of the central cities of

America's large metropolitan areas. We now examine the urban trend at both of these scales.

Macroscale Urbanization

The rise of the national urban system in the late nineteenth century was based on the traditional external role of cities: providing goods and services for their hinterlands in return for raw materials. This function had been present since colonial times, with preindustrial North American cities in the early nineteenth century (which were modest in size and spatial extent) located at the most accessible points on the existing transportation network so that movement costs could be minimized. Since handicrafts and commercial activities were already clustered in cities, the emerging industrialization movement gravitated toward them. These urban centers also contained concentrations of labor and capital, provided a large market for finished goods, and possessed favorable transport and communications connections. Furthermore, they could readily absorb the hordes of newcomers who would cluster by the thousands around new factories built within, and just outside, these compact cities. Their constantly growing incomes, in turn, permitted the newly industrialized cities to invest in a bigger local infrastructure of private and public services as well as housing, and thereby convert each round of industrial expansion into a new stage of urban development. Once generated, this whole process unfolded so quickly that planning was impossible; quite literally, America awoke one morning near the turn of the twentieth century to discover it had unexpectedly built a number of large cities.

The rise of the national urban system, unintended though it may have been, was a necessary by-product of industrialization, without which rapid U.S. economic development could not have taken place. This far-flung hierarchy of cities and towns now blanketed North America and came to serve its local populations with all the conveniences of modern life. Even though it first emerged during the Industrial Revolution of the 1870–1910 period, the American urban system was in the process of formation for several decades preceding the Civil War. The evolutionary framework of the system from 1790 to 1970 is best summarized within the model developed by John Borchert, consisting of four **epochs of metropolitan evolution** based on transportation technology and industrial energy.

The preindustrial *Sail-Wagon Epoch* (1790–1830) was the first stage of development, marked by slow, primitive overland and waterway movements. The leading cities were such northeastern ports as Boston, New York, and Philadelphia—none of which had yet emerged as the primate city—which were at least as oriented to the European overseas trade as they were to their still rather inaccessible western hinterlands (though the Erie Canal was completed as the epoch came to a close). Next came the *Iron Horse Epoch* (1830–1870), dominated by the arrival and spread of the steam-powered railroad, which steadily expanded its network from east to west until the transcontinental line was completed as the epoch ended. Accordingly, a nationwide transport system had been forged, coal-mining centers boomed (to keep locomotives running), and—aided by the easier movement of raw materials— small-scale urban manufacturing began to disperse outward from its New England hearth. The national urban system started to take shape:

New York advanced to become the primate city by 1850, and the next level in the hierarchy was occupied by burgeoning new industrial cities such as Pittsburgh, Detroit, and Chicago. This economic/urban development process crystallized during the third stage—the *Steel-Rail Epoch* (1870–1920)—which encompassed the American Industrial Revolution. Among the massive forces now shaping the growth and full establishment of the national metropolitan system were: (1) the rise and swift dominance of the all-important steel industry along the Chicago–Detroit–Pittsburgh axis (as well as its coal and iron ore supply areas in the northern Appalachians and Lake Superior district, respectively); (2) the increasing scale of manufacturing that necessitated greater agglomeration in the most favored raw-material and market locations for industry; and (3) the use of steel in railroad construction, which permitted significantly higher speeds, longer hauls of bulk commodities, and the time-space convergence of hitherto distant rail nodes. The *Auto-Air-Amenity Epoch* (1920–1970) comprised the final stage of North American industrial urbanization and maturation of the national urban hierarchy. The key innovation was the gasoline-powered internal combustion engine, which underwrote ever-greater automobile- and truck-based regional and metropolitan dispersal. And as technological advances in manufacturing spawned the increasing automation of blue-collar jobs, the U.S. labor force steadily tilted toward white-collar personal and professional services to manage the industrial economy. This kind of productive activity responded less to traditional cost- and distance-based location forces and ever more strongly to the amenities (pleasant environments)

available in suburbia and the outlying Sunbelt states in a nation now fully interconnected by jet travel and long-distance communication networks.

The growth of the American urban system and the national integration of its industrial-based economy produced dramatic spatial changes as population relocated to keep pace with shifting employment opportunities. The most notable regional transformation was the early twentieth-century emergence of the continental *core area* or American Manufacturing Belt, which contained the lion's share of industrial activity in both the United States and Canada. As Fig. 3–11 shows, the geographic form of the core region—which includes Canada's southernmost Ontario— was a great rectangle whose four corners were Boston, Milwaukee, St. Louis, and Washington, D.C. However, because manufacturing is such a spatially concentrated activity, the core area should not be thought of as a continuous factory-dominated landscape. In truth, less than 1 percent of the territory of the Manufacturing Belt is actually devoted to industrial land use. Moreover, most of its mills and foundries are clustered tightly into a dozen districts centering on metropolitan Boston, Hartford–New Haven, New York–northern New Jersey, Philadelphia, Baltimore, Buffalo, Toronto–Hamilton– Windsor, Pittsburgh–Cleveland, Detroit–Toledo, Chicago–Milwaukee, Dayton–Cincinnati, and St. Louis.

At the subregional scale, as transportation breakthroughs permitted progressive urban decentralization and **megalopolitan growth**, the expanding peripheries of major cities soon coalesced to form a number of conurbations (see pp. 84–85). The most important of these by far was the *At-*

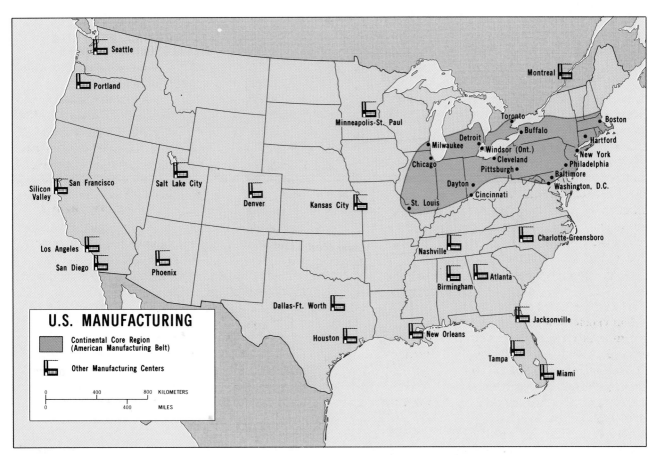

FIGURE 3–11

lantic Seaboard Megalopolis (Fig. 3–12), the 600-mile (975-kilometer)-long urbanized northeastern coastal strip extending from southern Maine to Virginia, that contains metropolitan Boston, New York, Philadelphia, Baltimore, and Washington. This is the economic heartland of the realm's core region, the seat of American politics, business, and culture as well as the trading "hinge" between the continent and the rest of the world. Three other primary conurbations also emerged—Lower Great Lakes (Chicago–Detroit–Pittsburgh), California (San Diego–Los Angeles–San Francisco), and Peninsular Florida (Jacksonville–Orlando–[Tampa]–Miami)—and are expected to continue growing along with a number of secondary megalopolitan con-

centrations (Fig. 3–12). And north of the border, the same forces created a nationally predominant conurbation linking Quebec City–Montreal–Toronto–Windsor (called *Main Street*), capping a parallel episode of rapid industrial urbanization in Canada between 1885 and 1940.

Microscale Urbanization

The internal structure of the American metropolis closely reflected the same mixture of forces that shaped the continental urban system. As an industrial city, however, it represented a departure from its European parentage. Whereas Europe's major cities were historically centers of political and military power—onto which industrialization was grafted

almost as an afterthought—most U.S. cities came into being as economic machines to produce the goods and services required to sustain an ongoing Industrial Revolution. Thus, right from the start, the performances of America's cities were judged mainly in terms of their profit-making abilities, and the less successful ones were callously discarded. The chief social function of the city was to receive and process foreign (as well as domestic rural) immigrants for assimilation into the mainstream American nativist culture—the so-called melting pot—which increasingly concentrated in the rapidly growing suburbs after 1920.

At the microscale or intraurban scale, too, transportation technology was a decisive force in shaping geographic patterns. Rails—

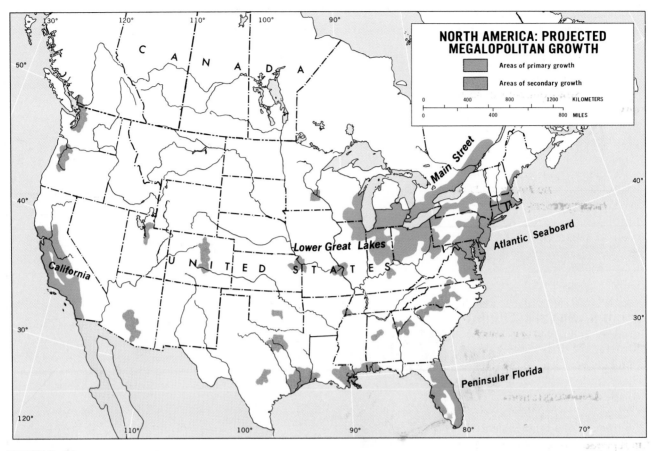

FIGURE 3–12

in this instance, lighter street rail lines—once again shaped spatial structure, as horse-drawn trolleys were succeeded by electric streetcars in the late nineteenth century. The coming of the automobile after World War I changed all that, and America began to turn from building compact cities to the widely dispersed metropolises of the post–World War II highway era. By 1970 the new intraurban expressway network had equalized location costs throughout the metropolis, and the stage was set for suburbia to transform itself swiftly from a residential preserve into a complete *outer city* with amenities and new prestige that proved highly attractive to the business world. As the newly urbanized suburbs started to capture major economic activities and thereby

gain a surprising degree of functional independence, many large cities saw their status diminish to that of coequal, their once thriving CBDs now reduced to serving the less affluent population that dominated the central city's residual rings, sectors, and nucleations (Figs. 3–1 and 3–2).

Intrametropolitan growth can also be organized into **eras of intraurban structural evolution** based on transportation, as described by John Adams (Fig. 3–13). The initial stage (I), the *Walking-Horsecar Era* (prior to 1888), was basically a pedestrian city in which most people had to get around on foot, although after 1850 the not-much-faster horse-drawn trolley began to operate. Urban structure was dominated by compactness (everything had to be within a 30-

minute walk) and very little land-use specialization could occur. The invention of the electric traction motor in 1888, a device that could easily be attached to horsecars, launched the *Electric Streetcar Era* (1888–1920). Speeds of up to 20 miles (32 km) per hour enabled the 30-minute travel radius and the urbanized area to expand considerably along outlying trolley corridors (Stage II in Fig. 3–13), spawning streetcar suburbs and helping to differentiate space within the older core city. The CBD, industrial, transport, and residential land uses emerged in their modern form (see Fig. 3–1). Stage III was the *Recreational Automobile Era* (1920–1945), when the initial impact of cars and highways constantly improved the accessibility of the outer metropolitan ring, thereby launching a wave of mass suburbanization that further extended the urban frontier (and the half-hour time-distance radial). Meanwhile, the still-dominant central city experienced its economic peak and the complete partitioning of residential space into neighborhoods sharply defined by income level, ethnicity, and race. The final stage (IV), the *Freeway Era* (1945 to the present), saw the full impact of automobiles as high-speed expressways pushed the metropolitan-development and time-distance limits in certain sectors to more than 30 miles (50 km) from downtown, thereby spawning round after round of massive new suburbanization. Structurally, the growing distinction between central city and booming suburbia mirrored a steady deconcentration of the metropolis, and after 1970 the emergence of an increasingly independent outer suburban city.

The social geography of the evolving industrial metropolis over the past 150 years has been marked by the development of a residential mosaic that exhibited the congregating of ever-more-specialized groups. Before the arrival of the electric streetcar, which finally introduced "mass" transit affordable by every social class, the heterogeneous city population had been unable to sort itself into ethnically homogeneous neighborhoods. Immigrants pouring into the industrializing cities were forced to live within walking distance of their workplaces in crowded tenements and row houses, whose tiny apartments were literally filled in the order that their tenants arrived in town. Once the inner-city neighborhood pattern was able to form in the 1890s, the familiar residential rings, sectors, and nucleations quickly materialized (see Fig. 3–1).

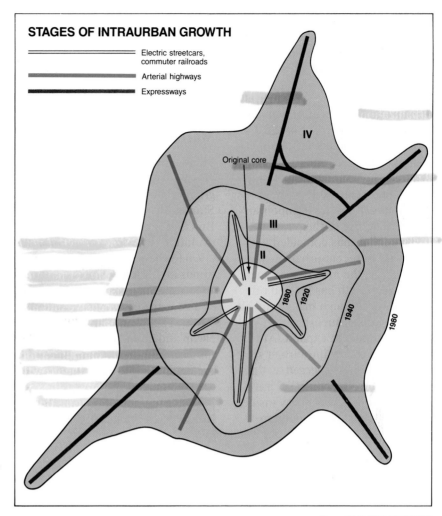

STAGES OF INTRAURBAN GROWTH

Electric streetcars, commuter railroads
Arterial highways
Expressways

IV

Original core

III

II

I

1880 1920 1940 1980

FIGURE 3–13

Source: Adapted from Adams, (1970.)

When the United States all but closed its doors to foreign immigration in the 1920s, industrial managers soon discovered the large black population of the rural Deep South—increasingly unemployed there as cotton-related agriculture declined—and began to recruit them by the thousands to work in the factories of the Manufacturing Belt cities. This new migration had an immediate impact on the social geography of the industrial city because whites refused to share their living space with the racially different newcomers. The result was the involuntary segregation of these newest migrants, who were steered to geographically separate all-black areas that by 1945 became large expanding *ghettoes*, speeding the departure of many white central-city communities in the postwar era and helping to create a racially divided society. The suburban component of the residential mosaic is generally home to the more affluent residents of the metropolis. The social makeup of its congregations is largely determined by income, with the residential turf of affluent suburbia marked by a plethora of minor class- and status-related variables, perfectly suited to a society in which frequent upward social and spatial mobility go hand in hand.

Urbanization in Postindustrial America Since 1970

By 1970 the long arc of industrial urbanization had been traversed. In 1800, only 5 percent of the agriculture-dominated U.S. population lived in cities; today, less than 2 percent live on the farms while over 90 percent reside within a two-hour drive of a metropolitan center. The rise of the postindustrial economy has coincided with a number of recent urban spatial changes. At the macroscale, metropolitan growth has leveled off nationally, although many Sunbelt cities as well as suburbs everywhere continue to expand. At the microscale, a new intraurban spatial structure has emerged as three decades of deconcentration have now turned the central city inside out, producing a metropolis so widely dispersed that large portions of it have subdivided into self-contained functional areas. We now examine current trends at each of these levels of generalization.

Macroscale Urbanization

Many current national- and regional-level changes have been grouped by Phillip Phillips and Stanley Brunn as the fifth stage of Borchert's model of urban-system evolution: the *Slow-Growth Epoch* (1970 to the present). Some key trends at work are slower natural population increases, the opening of a new age of expensive and uncertain energy supplies (always cheap and abundant before 1970), the substitution of communication for transportation, and the growth of the Sunbelt economy at the expense of the Manufacturing Belt. However, in the late 1980s, not all of these forces were so clearly apparent. Energy costs and supplies had, at least temporarily, ceased to be major worries, and the birth rate was higher than 15 years ago; moreover, the substitutability of communication for transport has so far had a greater impact on the suburbanization of business at the intraurban level. Only the regional shift toward the Sunbelt persists, propelled as much by the decline of certain heavy industries in the Manufacturing Belt as by the continued boom of the southern tier. Several Sunbelt cities were surprisingly hard hit by the Great Recession of the early 1980s, and the economies of Texas and neighboring states remain depressed by the persistence of lower oil prices.

This transformation signals the arrival of interregional equalization in location costs, especially for the white-collar, office-based

FIGURE 3—14 The heart of the Silicon Valley research-development-manufacturing complex, the most important innovation center of its kind in North America.

industries. Noneconomic factors, therefore, now play a larger role in locational decision-making, with geographical prestige, local amenities, and proximity to recreational activities heading the list. Perhaps no other place better symbolizes the postindustrial age in the Sunbelt than *Silicon Valley* (Fig. 3–14), headquarters for the microprocessor industry (which produces components for computers and electronic appliances), located near Stanford University in the San Francisco Bay Area's high-amenity Santa Clara Valley. Its success has also spawned a number of similar complexes in the outer cities of San Diego, Los Angeles, Dallas, Miami–Ft. Lauderdale, Raleigh–Durham (N.C.), Denver—and even outside Boston, where the electronics industry was born in the 1940s.

Overall, macrospatial changes in the post-1970 epoch have still not emerged with any great clarity. The national urban system still basically functions as it did a quarter-century ago, although certain Manufacturing Belt cities will likely yield high positions in the hierarchy in the foreseeable future to such up-and-coming metropolises as San Antonio, Phoenix, and San Diego.

Microscale Urbanization

The most dramatic changes in postindustrial urban geography to date have occurred at the scale of the metropolis. The completion of the intraurban expressway system, in effect, destroyed the region-wide centrality advantage of the central city's CBD, making most places on the freeway network just as accessible to the rest of the metropolis as only downtown had been before the 1970s. Industrial and commercial employers quickly realized that most of the advantages of being located in the CBD were now eliminated. Some companies in the office-based services

sector chose to remain and even enlarge their presence downtown, but many others, together with myriad firms in every other sector, chose to respond to the new economic-geographic reality by voting with their feet—or rubber tires—and headed for outlying sites. As far back as 1973, the suburbs surpassed the central cities in total employment; by the opening of the 1990s, even Sunbelt metropolises were experiencing the suburbanization of a critical mass of jobs (greater than 50 percent of the urban-area total).

As the outer city grew rapidly, the volume and level of interaction with the central city began to lessen and suburban self-sufficiency increased. This shift toward functional independence was heightened as new suburban nuclei sprang up to serve the new local economies, the major ones locating near key freeway intersections. These multipurpose activity nodes developed around big regional shopping centers, whose prestigious images attracted scores of industrial parks, office campuses and high-rises, hotels, restaurants, entertainment facilities, and even major league sports stadiums and arenas, that together formed burgeoning new **suburban downtowns** that are an automobile-age version of the CBD (Fig. 3–15). As suburban downtowns flourished, they attracted tens of thousands of local residents to organize their lives around these workplaces, shopping centers, leisure activities, and all the other elements of a complete urban environment—thereby further loosening ties not only to the central city, but also to other sectors of the suburban ring.

FIGURE 3–15 A portion of Las Colinas, a massive planned suburban downtown situated astride the freeway connecting the Dallas CBD (visible on the horizon in the center beyond Texas Stadium) and the Dallas-Fort Worth Airport. This complex has been called the first city of the twenty-first century, and continues to expand rapidly in the early 1990s as Exxon moves its corporate headquarters here from Manhattan.

These spatial elements of the contemporary metropolis are assembled in the model displayed in Fig. 3–16, which should be regarded as an updating and extension of the classical models of intraurban structure (Fig. 3–1). The rise of the outer city has now produced a *multicentered* metropolis consisting of the traditional CBD plus a set of increasingly co-equal suburban downtowns, with each activity center serving a discrete and self-sufficient surrounding area. James Vance has called these new tributary areas **urban realms**, recognizing in his studies that each such realm maintains a separate, distinct economic, social, and political significance and strength. The urban realms of Los Angeles are mapped in Fig. 3–17; a similar regionalization scheme could easily be drawn for other large U.S. metropolises.

The position of the U.S. central city within the new multinodal metropolis of realms is eroding. No longer the dominant metro-politanwide center for urban goods and services, the CBD is being reduced to serving the less affluent residents of the innermost realm and those working downtown. As inner-city manufacturing employment declined precipitously, many large cities adapted successfully by shifting toward the growing service industries. Accompanying this switch is downtown commercial revitalization, which has been widespread since 1970; but in many cities, for each shining new skyscraper that goes up several old commercial buildings are abandoned. Moreover, in a number of Sunbelt cities such as Houston and Los Angeles, a whole forest of new office towers contains suburban commuters whose only contact with the city below is the short drive between the freeway exit and their building's parking garage.

Residential revitalization in and near the CBD has also occurred in many central cities since 1970. However, the number of people involved is not as significant as first believed; in fact, most reinvestment is undertaken by those already residing in the central city, so a "return-to-the-city" movement by suburbanites is not taking place. Such downtown-area neighborhood redevelopment involves *gentrification*—the upgrading of residential areas by new higher-income settlers. However, in order to succeed, this development usually requires the displacement of established lower-income residents—an emotional issue that has sparked many conflicts. Beyond the CBD zone, the vast inner city remains the problem-ridden domain of low- and moderate-income people, with most forced to reside in ghettoes. Financially ailing big-city governments are unable to fund adequate schools, crime-prevention programs, public housing, and sufficient social services, and so the downward spiral, including abandonment in the old industrial cities, continues unabated into the 1990s (Fig. 3–18).

FIGURE 3–16

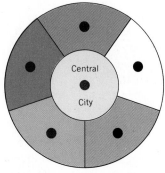

IDEAL FORM OF MULTICENTERED URBAN REALMS MODEL

FIGURE 3–17

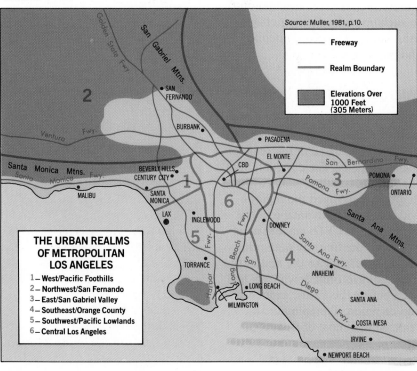

THE URBAN REALMS OF METROPOLITAN LOS ANGELES
1 — West/Pacific Foothills
2 — Northwest/San Fernando
3 — East/San Gabriel Valley
4 — Southeast/Orange County
5 — Southwest/Pacific Lowlands
6 — Central Los Angeles

FIGURE 3–18 New York's devastated inner city, showing numerous signs of abandonment. This is part of Harlem on Manhattan's upper east side, just a few blocks from the northeastern corner of Central Park.

CULTURAL GEOGRAPHY

North America constitutes one of the world's newer geographic realms, yet the contributions of a wide spectrum of immigrant groups over the past two centuries have shaped—and continue to shape—a rich and varied cultural mosaic. Linguistic bifurcation between Anglophones (English-speakers) and Francophones (French-speakers) has inhibited the formation of a single overriding culture in Canada, although other unifying elements continue to bind that country together. In the United States, however, newcomers were far more willing to set their original cultural baggage aside in favor of assimilation into the emerging culture of their adopted homeland, which in itself was a hybrid nurtured by constant infusions of new influences. For most upwardly mobile immigrants, this plunge into the much-touted melting pot prom-

ised a ticket for acceptance into mainstream American society. But for millions of other would-be mainstreamers, the melting pot has proven to be a lumpy stew. Whether by choice or not, they continue to stick together in ethnic communities (especially within the inner cities of the Manufacturing Belt), with a large proportion of those neighborhoods inhabited by racial minorities who do not fully participate in the national culture.

American Cultural Bases

As the nativist culture of the United States matured, it came to develop a set of powerful values and beliefs: love of newness; a desire to be near nature; freedom to move; individualism; societal acceptance; aggressive pursuit of goals; and a firm sense of destiny. Brian Berry has discerned these cultural traits in the behavior of people throughout the evolution of

urban America. A "rural ideal" has prevailed throughout U.S. history and is still expressed in a strong bias against residing in cities. When industrialization made urban living unavoidable, those able to afford it soon moved to the emerging suburbs (*newness*) where a form of country life (*close to nature*) was possible in a semi-urban setting. The fragmented metropolitan residential mosaic, composed of a myriad of slightly different neighborhoods, encouraged frequent *unencumbered mobility* as middle-class life revolved around the *individual* nuclear family's *aggressive pursuit* of its aspirations for *acceptance into the next higher stratum of society*. These accomplishments confirmed to most Americans that their goals could be attained through hard work and perseverance, and that they possessed the ability to realize their *destiny* by achieving the "American Dream" of home-ownership, affluence, and total satisfaction.

Language

Although linguistic variations play a far more important role in Canada, one-eighth of the U.S. population spoke a primary language other than English in 1990. Differences in English usage are also evident at the subnational level in the United States, where regional variations (*dialects*) are still widespread, despite the recent trend toward a truly national society. The Deep South and New England immediately come to mind as areas that possess distinctive accents. An even closer connection between language and landscape is established through *toponymy*, the naming of places. U.S. place-name geography provides important clues to the past movements of cultural influences and national groups: for instance, the preponderance of such Welsh place names as Cynwyd, Bryn Mawr, and Uwchlan to the west of Philadelphia is the final vestige of an erstwhile colony of Celtic-speaking settlers from Wales.

Religion

North America's Christian-dominated kaleidoscope of religious faiths contains important spatial variations. Many major Protestant denominations are clustered in particular regions, with Baptists localized in the southeastern quadrant of the United States, including Texas; Lutherans in the Upper Midwest and northern Great Plains; and Mormons in Utah and southern Idaho. Roman Catholics are most visibly concen-

trated in two locations: the Manufacturing Belt metropolises as well as nearby New England, which received huge infusions of Catholic Europeans over the past century; and the entire southwestern borderland zone that is home to a burgeoning Hispanic-American population. Judaism is the most highly agglomerated major religious group on the continent, its largest congregations clustered in the cities and suburbs of Megalopolis, Southern California, South Florida, the Midwest, and Canada's Main Street conurbation.

Ethnicity

Ethnicity (which means "nationality") was a decisive influence in the shaping of American culture,

which, in turn, also reshaped the cultural traditions of newcomers who were assimilated into mainstream society. The current U.S. ethnic tapestry, as ever, is characterized by a complex mosaic (see *box*). The spatial distribution of ethnic minorities is mapped in Fig. 3–19, which also includes the Native American (Indian) population. Because the latter largely occupy tribal lands on reservations ceded by the federal government, they are undergoing very little distributional change. The Hispanic population, on the other hand, is growing rapidly through in-place natural increase and in-migration from Middle America (much of it illegal). Spatially, Hispanics are both increasing in density along the southwestern border and are fanning out toward large met-

FIGURE 3–19

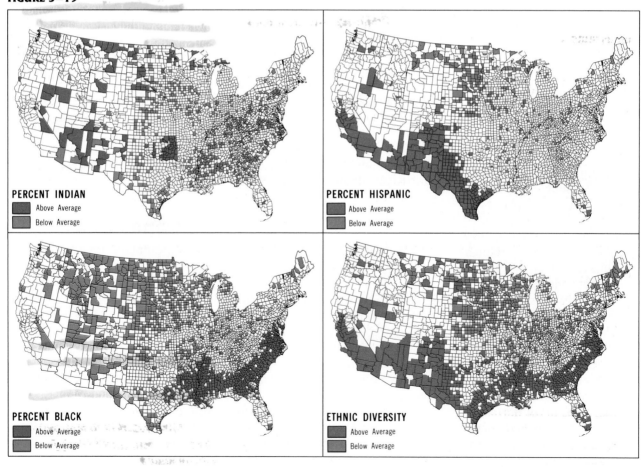

PERCENT INDIAN
- Above Average
- Below Average

PERCENT HISPANIC
- Above Average
- Below Average

PERCENT BLACK
- Above Average
- Below Average

ETHNIC DIVERSITY
- Above Average
- Below Average

THE ETHNIC TAPESTRY OF THE UNITED STATES

The diversity and complexity of America's ancestral makeup was last reported by the 1980 census, which showed that over 83 percent of the U.S. population identified with one of 134 different national backgrounds. There were more Americans of full or partial English descent than the total population of England, more than one-half as many Americans of German stock as currently reside in the two Germanys, and the Irish-Americans outnumber the population of the Republic of Ireland by a ratio of 12 to 1. In addition, so many Americans listed themselves as Afro-Americans that only four Subsaharan African states recorded a larger national population total. Other significant ethnic ancestries include French, Italian, Scottish, Polish, and Mexican. Those of Native American background accounted for 3 percent of the U.S. total. The degree of inter-ethnic mixing is also considerable; 52 percent of those born in the United States of American-born parents reported multiple ancestry.

Geographically, the dominant European ancestries—French, Irish, and those originating in the United Kingdom—were dispersed throughout the United States. Most of the others showed an affinity for a particular region: Italians, Portuguese, and Russians clustered in the Northeast, whereas Scandinavians and Czechs were localized in the north-central states. California, the most populous state, best exhibited the nation's diverse ethnicity (see Fig. 3–19), with more Americans of English, German, Irish, French, Scottish, Dutch, Swedish, and Danish origin concentrated there than in any other state.

Although the Afro-American population remains fairly stable in relative size (about 12 percent of the U.S. total), two other minority groups are substantially increasing their presence in the national ethnic tapestry; Asians and Hispanics. During the 1970s the number of Asians and Pacific Islanders more than doubled—and nearly doubled again in the 1980s—remaining largely concentrated in the cities of the Far West. The biggest single Asian group today is the Filipinos, followed by the Chinese, Vietnamese, Koreans, and Japanese. The Hispanic population, which will surpass the black population to become the nation's largest minority by the end of this decade, grew by more than 60 percent between 1970 and 1980, and by at least another 40 percent since 1980. During the 1970s, the fast-growing Mexican component accounted for 62 percent of the U.S. Hispanic population, Puerto Ricans 13 percent, Central and South Americans 12 percent, and Cubans 5 percent. Pending confirmation by the 1990 census, it is likely that these proportions have not changed significantly since 1980.

more affluent areas of the South, although settling in communities that remain highly segregated. The newest element in the ethnic tapestry, as indicated in the *box*, is the arrival of Asian-Americans, who heavily cluster in major West Coast metropolitan areas.

The Mosaic Culture

American cultural geography continues to evolve. What is now taking place is a new fragmentation into a *mosaic culture*, an increasingly heterogeneous complex of separate, uniform "tiles" that cater to more specialized groups than ever before. No longer based solely on such broad divisions as income, race, and ethnicity, today's residential communities are also forming along the dimensions of age, occupational status, and especially lifestyle. Their success and rapid proliferation reflects an obvious satisfaction by a majority of Americans. But such balkanization—which is fueled by people choosing to interact only with others exactly like themselves—threatens the survival of some important democratic values that have prevailed throughout the evolution of U.S. society.

Environmental Perception and Mental Maps

The cultural geographer Jan Broek has written:

Each society in each era perceives and interprets its physical surroundings and its relations to other [places] *through the prism of its own way of life. . . . It is always and only in relation to culture that the parts of the earth receive specific meaning.*

ropolitan areas to the north and east. Blacks are not as mobile as Hispanics, undoubtedly constrained by racially segregated housing markets. Nonetheless, upwardly mobile blacks are increasing their presence both in suburbia and, via return migration, in the

Recent research on cultural appraisals of the environment has focused on the way a person's perceptions shape his or her spatial decision-making and behavior. This work has revealed that the impressions and images people develop about various places shape their attitude toward them, and that this information is structured within personal **mental maps** that all of us carry around in our minds. Because American society is both affluent and predisposed toward frequent mobility, interpreting those mental maps can be a useful tool in the current understanding and forecasting of migration flows. There are two separate but closely interrelated dimensions of mental mapping.

The first dimension involves *designative* mental mapping, in which spatial information is received and objectively placed on the "map" in a person's mind. Places often mentioned in the news are widely recognized, but sometimes an obscure place comes to our attention and remains fixed in mind (Three Mile Island and Mt. St. Helens are two recent examples). People also learn a great deal more about nearby places that they encounter frequently. At the national level, Wilbur Zelinsky has carefully mapped the "perceptual regions" of North America (Fig. 3–20), which represent a summarization of individual perceptions as to where such traditional regions as "South," "Midwest," and "West" are situated. However, these popularly perceived, or *vernacular*, regions, which were derived by studying local telephone directories and interviewing people throughout the nation, do not readily coincide with the formal regionalization scheme that is developed later in this chapter (see Fig. 3–27).

The second dimension involves *appraisive* mental maps, in which spatial information is subjectively processed according to an individual's personal biases. A good example is the souvenir postcard that can be purchased throughout Texas, which displays a humorous map entitled "A Texan's View of the United States"; it shows Texas occupying about three-quarters of the country, while every other state is belittled (Florida, for example, is called "Swampland"). Although everybody possesses place biases, the process leading to their formation is not well understood. Why should Cleveland and Philadelphia suffer persistently negative images, while Boston and Denver do not? And how did Dallas, the scene of a presidential assassination in 1963,

FIGURE 3–20

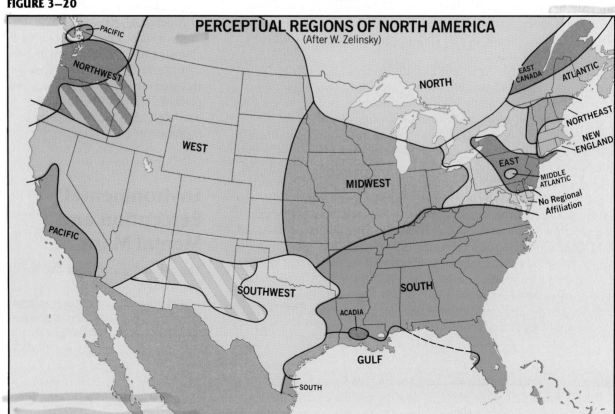

PERCEPTUAL REGIONS OF NORTH AMERICA
(After W. Zelinsky)

manage to reverse its image from bad to glamorous only a few years later? These questions cannot yet be answered with much certainty, though it is likely that an explanation rests with some combination of such locational variables as employment opportunities, weather patterns, political climate, social setting, and recreational possibilities.

A well-developed application of appraisive mental mapping has been the preparation of maps that reveal the collective future residential preferences of college students. In 1984 students at the University of Miami in Coral Gables, Florida (which enrolls a sizable out-of-state student population) were asked to rank the 48 conterminous states in response to the question "Where would you like to live?" Figure 3–21 shows their choices. Follow-up class-

room discussions of this map elicited the following: (1) Florida, North Carolina, Texas, and the West Coast states were picked for their employment opportunities, mild climate, and recreational offerings; (2) Colorado, Wyoming, Vermont, and New Hampshire were chosen for their wide open spaces and mountain scenery; (3) Virginia and New York represented the job opportunities of the Washington and New York City metropolitan areas; (4) the five Deep South states were rejected for their social and political climate; (5) Ohio, Michigan, and New Jersey were seen as undesirable, overindustrialized places with chronic economic and employment problems; and (6) Minnesota, North Dakota, Utah, and Nevada were regarded as "dull" places, with harsh weather for at least part of the year.

THE CHANGING GEOGRAPHY OF ECONOMIC ACTIVITY

The economic geography of North America in the late twentieth century is the product of all the foregoing, as bountiful environmental, human, and technological resources have been cumulatively blended together to create one of the world's most advanced economies. Perhaps the greatest triumph was the overcoming of the "tyranny" of distance—a geographic quality that the Soviets can only view with the greatest of envy—as people and activities were organized into a truly continental spatial economy that took maximum advantage of

FIGURE 3–21

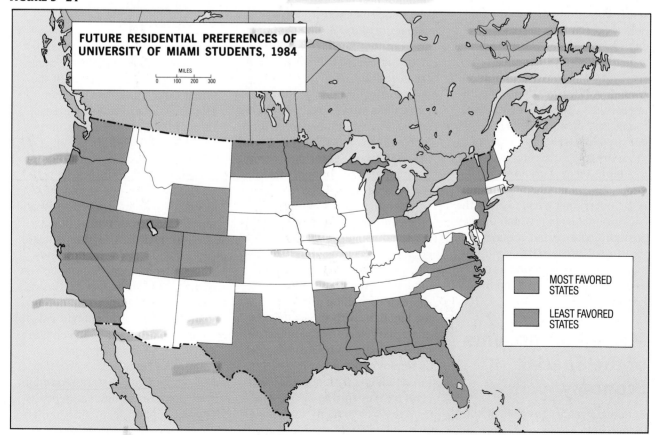

FUTURE RESIDENTIAL PREFERENCES OF UNIVERSITY OF MIAMI STUDENTS, 1984

MILES
0   100   200   300

MOST FAVORED STATES

LEAST FAVORED STATES

agricultural, industrial, and urban development opportunities. Yet, despite these past achievements, American economic geography today is once again in upheaval as it completes the swift transition from industrial to postindustrial society.

As we have seen, a new set of locational forces has been unleashed wherein the distribution of key economic activities is now being shaped by noneconomic variables. Thus, business-location decisions are increasingly dominated by the same concerns that would govern a person's choice of residence if he or she had vast financial resources to draw on—finding a site and building that maximizes geographical prestige, amenities, commuting convenience, and access to recreational activities. Judging by the recent spatial behavior of U.S. companies, these prized locations are most frequently found in the outer suburban city nationwide, and in selected central cities of the Sunbelt. As economic activities become more spatially footloose, the major casualty is the industrial city, especially those within the Manufacturing Belt. With the disappearance of the concentrative forces that created the high-density industrial city, many of these urban centers are struggling to adapt to a new era in which economic-spatial deconcentration is the overriding trend. This is the background against which the new postindustrial spatial economy of the 1990s is materializing, with the metropolitan connection as crucial as ever to its structure and functioning.

Major Components of the Spatial Economy

Economic geography is mainly (although not exclusively) concerned with the locational analysis of productive activities. Four major sets may be identified:

Primary Activity
The extractive sector of the economy, in which workers and the environment come into direct contact, especially *mining* and *agriculture*.

Secondary Activity
The *manufacturing* sector, in which raw materials are transformed into finished industrial products.

Tertiary Activity
The *services* sector, including a wide range of activities from retailing to finance to education to routine office-based jobs.

Quaternary Activity
The fast-growing sector involving the collection, processing, and manipulation of *information*; a subset, sometimes referred to as **quinary activity**, is managerial or control-function activity associated with *decision-making* in large organizations.

Historically, each of these activities has successively dominated the American labor force for a time over the past 200 years, with the quaternary sector now coming to dominate the economy for the foreseeable future. Agriculture was dominant until late in the nineteenth century, giving way to manufacturing by 1900. The steady growth of services after 1920 finally surpassed manufacturing industry in the 1950s, but now shares a dwindling portion of the limelight with the fast-rising quaternary sector. The approximate breakdown by major sector of employment in the U.S. labor force today is agriculture—2 percent; manufacturing—15 percent; services—18 percent; and quaternary—65 percent (with about 10

percent in the quinary sector). We now treat these major productive components of the spatial economy in the following coverage of resource use, agriculture, manufacturing, and the postindustrial revolution.

Resource Use

The North American continent was blessed with abundant deposits of industrial and energy resources. Fortunately, these were usually concentrated in large enough quantities to make long-term extraction an economically feasible proposition, and most of the richest raw material sites are still the scene of major drilling or mining operations. Moreover, the continental and offshore mineral storehouse may still contain outstanding resources for future exploitation.

Industrial Mineral Resources

North America's rich mineral deposits are localized in three zones: the Canadian Shield north of the Great Lakes, the Appalachian Highlands, and scattered areas throughout the mountain ranges of the West. The Shield's most noteworthy minerals are iron ore (Minnesota's Mesabi Range just west of Lake Superior and the eastern Shield in Quebec and Labrador), nickel (around Sudbury, Ontario, near Lake Huron's north shore), and gold, uranium, and copper (from upper Saskatchewan north to the Arctic coast). Besides vast deposits of soft (bituminous) coal, the Appalachian region also contains hard (anthracite) coal in northeastern Pennsylvania and iron ore in central Alabama. The U.S.–Canadian western mountain zone contains significant deposits of coal, copper, lead, zinc, molybdenum, uranium, silver, and gold.

Fossil Fuel Energy Resources

The most strategically important resources of North America are its coal, petroleum (oil), and natural gas supplies—the *fossil fuels*, so named because they were formed by the geologic compression and transformation of plant and tiny animal organisms that lived hundreds of millions of years ago. These energy supplies are mapped in Fig. 3–22 and reveal abundant deposits and distribution networks.

The realm's *coal* reserves are among the greatest anywhere on earth, the U.S. portion alone containing at least a 400-year supply. Three coal regions are evident: (1) Appalachia, which produces about half of all U.S. coal, is still the largest region but is declining be-

FIGURE 3–22

NORTH AMERICA: MAJOR DEPOSITS OF FOSSIL FUELS

cause its high-sulfur coal must be expensively treated to meet the standards set by federal clean-air statutes; (2) the Western coal region, centered on the Great Plains within 500 miles north and south of the U.S.–Canada boundary, is expanding because of its vast low-sulfur supplies in thick, near-surface seams that can be strip-mined; and (3) the Midcontinent coalfields, a large arc of high-sulfur, strippable deposits centered on southern Illinois and western Kentucky, which is also declining as the West moves ahead.

North America's major *oil*-production areas are located along and offshore from the Texas–Louisiana Gulf Coast; in the Mid-continent district, extending through western Texas–Oklahoma–eastern Kansas; and along the front of the Canadian Rockies in central and northern Alberta. Lesser oilfields are also found in Southern California, west-central Appalachia, the northern Great Plains, and the southern Rockies. Both the United States and Canada have recently turned their attention to the Arctic, where supplies exist along Alaska's North Slope and adjacent territory in northwestern Canada. In 1977 the Trans-Alaska Pipeline was opened, a triumph over the harsh Arctic environment that enabled the pumping of North Slope petroleum south 800 miles (1300 km) to the warm-water Pacific port of Valdez for transshipment by tanker to the lower 48 states. The distribution of *natural gas* deposits resembles the geography of oilfields because petroleum and gas are usually found in similar geologic formations (the floors of ancient shallow seas). Accordingly, major gasfields are located in the Gulf, Midconti-nent, and Appalachian districts. However, when subsequent geo-logic pressures are exerted on underground oil, the liquid is con-

verted into natural gas; this has happened frequently in mountain-ous zones, so that western gas de-posits in and around the Rockies tend to stand apart from oilfields.

Agriculture

Despite the twentieth-century em-phasis on urbanization and the development of the nonprimary sectors of the spatial economy, ag-riculture remains an important ele-ment in North America's human geography. Because it is the most extensive (space-consuming) eco-nomic activity, vast expanses of the U.S. and Canadian landscape are clothed by fields of grain. Moreover, great herds of livestock are sustained by pastures and fod-der crops because this wealthy realm can afford the luxury of feeding animals from its farmlands and has come to demand vast quantities of red meat in its diet. The increasing application of high-

technology mechanization to farm-ing (Fig. 3–23) has steadily in-creased both the volume and value of total agricultural production, and has been accompanied by a sharp reduction in the number of those actively engaged in agricul-ture (to less than 2 percent of the U.S. work force today). Small family farms have borne the brunt of this change: those unable or un-willing to modernize and enlarge their operations find it impossible to survive.

The regionalization of U.S. ag-ricultural production, which has been well established for over a century, is shown in Fig. 3–24. Its overall spatial organization devel-oped largely within the framework of the Von Thünen model (review pp. 74–76); as in Europe (Fig. 1–11), the early nineteenth century original-scale model of town and hinterland—in a classical dem-onstration of time-space con-vergence—expanded outward,

FIGURE 3–23 Ultrasophisticated technology has now arrived on the American farm. Here, in an Indiana cornfield, a trencher is guided by a laser to precisely control the direction and angle of subterranean tiling and drainage pipes, which are being installed to better drain the flat land surface.

with constantly improving transportation technology, from a locally "isolated state" to encompass the entire continent by 1900. As the macrogeographical structure formed, the greatly enlarged original Thünian production zones (Fig. 1–10) were modified: (1) the first ring now differentiated into an inner fruit/vegetable zone and a surrounding belt of dairying; (2) the forestry ring was displaced to the outermost limits of the regional system because railroads could now transport wood quite cheaply; (3) the field crops ring subdivided into an inner mixed crop-and-livestock ring to produce meat (the *Corn Belt*, as it came to be called) and an outer zone that specialized in the mass production of wheat grains; and (4) the ranching area remained in its outermost position, a grazing zone that supplied young animals to be fattened

in the meat-producing Corn Belt as well as supporting an indigenous sheep-raising industry. The "supercity" anchoring this macro-Thünian regional system was the northeastern Megalopolis, well on its way toward coalescence and already the dominant food market and transport focus of the entire country.

Although the circular rings of the model are not apparent in Fig. 3–24, many spatial regularities can be observed (remember that Von Thünen also applied his model to reality and thereby introduced several distortions of the theoretically ideal pattern). Most important is the *sequence* of farming regions as distance from the national market increased, especially westward from Megalopolis toward central California, which was the main directional thrust of the historic inland penetration of the United

States. The Atlantic Fruit and Vegetable Belt, Dairy Belt, Corn Belt, Wheat Belts, and Grazing Region are indeed consistent with the model's logical structure, each zone successively farther inland astride the main transcontinental routeway. Deviations from the scheme may be attributed to irregularities in the environment and to unique conditions: (1) central Appalachia and the dry mountains and plateaus of the West cannot support anything other than isolated-valley (General) farming or irrigated cropping; (2) the nearly year-round growing seasons of California and the Gulf Coast–Florida region permit those distant areas (with the help of efficient refrigerated transport) to produce fruits and vegetables in competition with the innermost zone; (3) the bifurcation into two wheat regions results from hilly and sandy

FIGURE 3–24

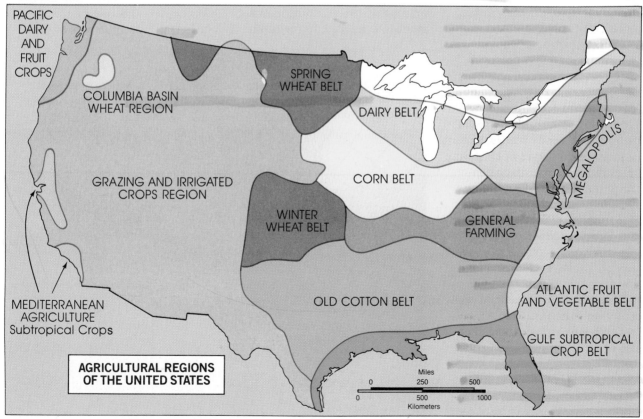

AGRICULTURAL REGIONS OF THE UNITED STATES

PACIFIC DAIRY AND FRUIT CROPS

COLUMBIA BASIN WHEAT REGION

SPRING WHEAT BELT

DAIRY BELT

MEGALOPOLIS

GRAZING AND IRRIGATED CROPS REGION

CORN BELT

WINTER WHEAT BELT

GENERAL FARMING

MEDITERRANEAN AGRICULTURE Subtropical Crops

OLD COTTON BELT

ATLANTIC FRUIT AND VEGETABLE BELT

GULF SUBTROPICAL CROP BELT

Miles
0   250   500
0   500   1000
Kilometers

country in western Nebraska, with the milder-climate winter wheat zone concentrated to the south, and spring wheat deflected north to the much colder northern Great Plains and Canadian prairie provinces beyond (see Fig. 3–30); (4) the small farming regions of the Pacific Northwest serve local populations, sell specialty crops nationwide, and raise wheat for export to Pacific Basin customers; and (5) the Old Cotton Belt is a remnant of the bygone South, today specializing in the production of beef, poultry, soybeans, and timber.

Manufacturing

The geography of North America's industrial production has long been dominated by the Manufacturing Belt (Fig. 3–11). As was noted earlier, the internal structure of the Belt is built around a dozen urban-industrial districts, linked together by a dense transportation network, which also interconnects the manufacturing centers with such major resource deposits as the Appalachian coalfields and the Lake Superior-area iron mines. This highly agglomerated spatial pattern developed because of the economic advantages of clustering industries in cities, which contained the largest concentrations of labor and investment capital, constituted major markets for finished goods, and were the most accessible nodes on the national transport network for assembling raw materials and distributing manufactures. As these industrial centers expanded rapidly during the late nineteenth century, they also achieved economies of scale, savings accruing from large-scale production in which the cost of manufacturing a single item was further reduced as individual com-

panies mechanized assembly lines, specialized their work forces, and purchased raw materials in huge quantities.

This efficient production pattern served the nation well throughout the remainder of the industrial age. Because of the effects of *historical inertia*—the need to continue using hugely expensive manufacturing facilities for their full multiple-decade lifetimes to cover initial long-term investments—the Manufacturing Belt is not about to disappear for a long time to come. However, as aging factories are retired, a process well under way, the distribution of American industry will change. As transportation costs even out among U.S. regions, as energy costs increasingly favor the south-central oil- and gas-producing states, as high-technology manufacturing advances reduce the need for lesser-skilled labor, and as locational decision-making intensifies its attachment to noneconomic factors, industrial management is signaling its willingness to relocate to regions it

perceives as more desirable in the South and West.

The Decline of Traditional Industry

The decline of traditional heavy industry has been a major trend in the Manufacturing Belt since 1980. One recent study estimates that smokestack-industry employment will plunge from 20 percent of the U.S. work force in 1980 to 8 percent by the mid-1990s. Within the Belt today, problems of obsolescence, labor resources, and foreign competition are widespread. The region's aging physical plant is seldom worth reinvesting in, and major factory closures are often reported in the evening newscasts; on the blighted landscape of the inner industrial city, in particular, the abandoned manufacturing complex is a depressing, all-too-familiar sight (Fig. 3–25). Clearly, as employment opportunities shrink, it will become necessary to retrain large numbers of blue-

FIGURE 3–25 Portions of the Manufacturing Belt have appropriately been called the "Rustbelt" as factory closures have multiplied: an abandoned plant on Chicago's South Side.

collar workers for alternative careers in the services sector—a looming crisis that government has all but totally ignored. Competing high-quality industrial imports from abroad increased their share of the U.S. market from 9 percent in 1970 to about 25 percent today. Once based exclusively on cheap labor, America's chief foreign competitors in Europe and non-Communist Asia now possess superior technologies and productive efficiencies in a growing list of industries. The Japanese, for example, use five times as many robots as the Americans in running some of the world's most sophisticated factories.

The reduced global position of the United States is particularly well demonstrated in the case of steel, one of the most basic heavy industries and the very foundation on which the Manufacturing Belt was built. The world leader in production as recently as the late 1960s, American steelmaking then fared so poorly that by 1986 sales had declined by over 40 percent and employment by nearly 60 percent. Half of that decline was due to lowered worldwide demand, and much of the rest was attributable to imports, whose lower prices found many U.S. customers. But the American steel industry also fell behind its competitors because it was too slow to modernize its old-fashioned manufacturing methods.

The Revitalization Effort

Belatedly, U.S. manufacturers, led by the humbled steelmakers (Fig. 3–26), are learning from their pace-setting foreign rivals. A major new effort has been under way since the late 1980s to create high-technology "factories of the future," with much of this activity concentrated in the Midwest portion of the Manufacturing Belt. High-skilled jobs by the tens of thousands have been created in research and development and new equipment manufacturing, emphasizing robotics and other state-of-the-art automation technologies. The goal is nothing less than the modernization of the region's old manufacturing base and the regaining of its once formidable position in the global marketplace. But there is a downside to this progress: the more successful high-tech manufacturing becomes, the fewer the factory jobs to be had. Thus, it is *not* the American manufacturing sector that is declining—industrial production in 1990 was over 65 percent higher than in 1970. Rather, it is

FIGURE 3–26   The trend toward modernization of the steel industry is not limited to the Manufacturing Belt: this is the North Star Steel Company's new minimill at Beaumont on the Gulf Coast of Texas.

the rapid shrinkage of blue-collar *employment* opportunities that has caused most of the pain, a human problem certain to intensify as the dynamics driving the **postindustrial revolution** continue to mandate substantial change in the structure and composition of the North American labor force.

The Postindustrial Revolution

The signs of postindustrialism are visible throughout the North American realm today, and they are popularly grouped under such names as "the information society" and "the electronic era." The term *postindustrial* by itself, of course, tells us mainly what the theme of the American economy is no longer; yet the term is also used by many scholars to refer to a specific set of societal traits that comprise an historic break with the recent past (see *box*). Many of the urban-geographic expressions of the currently transforming society have already been highlighted; the discussion here focuses on some broader economic spatial patterns.

High-technology activities are the leading growth industries of the postindustrial economy, and they are highly prized by local area-development agencies. The model for such development is Silicon Valley (Fig. 3–14), which has spawned competitors in many other metropolises that possess the right blend of locational qualities. Developers often cite the conditions that are likeliest to attract a critical mass of high-tech companies to a given locality:

1. A nearby major university that offers an excellent graduate engineering program.
2. Close proximity to a cosmopolitan urban center.
3. A large local pool of skilled and semiskilled labor.
4. Three hundred days of sunshine a year.
5. Recreational water within an hour's drive.
6. Affordable nearby housing.
7. Even closer prestigious luxury housing for top executives.
8. Start-up capital worth at least $1 billion to lure new high-tech firms.
9. Lower-than-normal risk for establishment of profitable high-tech businesses.
10. Cooperative spirit among landowners, lenders, government, and business.

Despite their glamour, there are growing concerns about the stability of certain high-tech industries,

THE LINEAMENTS OF POSTINDUSTRIAL SOCIETY

In *The Coming of Postindustrial Society*, sociologist Daniel Bell sketched the distinguishing features of the emerging economy and its impact on American life. The new society and economy is marked by a fundamental change in the character of technology use—from fabricating to processing—in which telecommunications and computers are vital for the exchange of knowledge. Information becomes the basic product, the key to wealth and power, and is generated and manipulated by an intellectual rather than a machine technology. Yet, postindustrial does not displace industrial society, just as manufacturing did not obliterate agriculture. Instead, in Bell's words, "the new developments overlie the previous layers, erasing some features and thickening the texture of society as a whole."

Several hallmarks of postindustrialism can be identified. Knowledge is central to the functioning of the economy, which is led by science-based and high-tech industry. The technical/professional or "knowledge" class increasingly dominates the work force: quaternary and quinary activity already employed 40 percent of the U.S. labor force in 1975, a proportion that surpassed 65 percent in the early 1990s. The nature of work in the information society focuses on person-to-person interaction rather than person-product or

which may be unusually sensitive to economic fluctuations and foreign competition. Silicon Valley in the late 1980s was a good case in point: overcapacity and fierce competition from Japanese and South Korean semiconductor manufacturers resulted in numerous business failures and the loss of thousands of jobs.

The geographic impact of the postindustrial revolution has been uneven over the past two decades. States offering high-amenity environments have generally fared best, especially Florida, Texas, California, North Carolina, Virginia, and those of the Pacific Northwest and northern New England. Not surprisingly, the Manufacturing Belt states as a group performed weakly, although the Midwest has enjoyed a resurgence since 1985. Most interesting has been the overall performance of

the highly touted Sunbelt. The notion that the U.S. southern tier of states is a uniformly booming area is largely a myth. Even at the state level, significant internal variations are obvious; within each Sunbelt state, stark contrasts are common as the burgeoning economy of one county all too frequently leaves a dormant neighbor completely untouched. It is worth pointing out as well that even prosperous corners of the Sunbelt are not immune to changing economic currents—as Texas, Oklahoma, and Louisiana discovered with the onset of the lingering oil-industry slump that began in the early 1980s. But however spotty the overall economic development pattern, it is still very likely that many southern-tier metropolises will continue to capture an expanding share of major U.S. business activity.

person-environment contacts. With the eclipsing of manufacturing, more and more women participate in today's labor force. A meritocracy now prevails in which individual advancement is based on education and acquired job skills.

Among the many spatial implications of this socioeconomic transformation, Bell observed that postindustrial occupations would gravitate toward five major types of "situses" or locations: economic enterprises, government, universities, social-service complexes, and the military. Central North Carolina's famous *Research Triangle Park*, located near the center of the "triangle" circumscribed by the nearby cities of Raleigh, Durham, and Chapel Hill, is the quintessential example of a prestigious Sunbelt high-technology manufacturing and research complex. Even a small sampling of the list of tenants fully supports Bell's hypothesis: (1) among its business enterprises are such corporate giants as IBM, TRW, and the headquarters of Burroughs Wellcome Pharmaceuticals; (2) government is represented by the U.S. Environmental Protection Agency, the Southern Growth Policies Board, and North Carolina's Board of Science and Technology; (3) three local universities are prime movers in the Park's research operations—Duke, North Carolina, and North Carolina State—which have also attracted the National Humanities Center and the Sigma Xi Scientific Research Society; (4) social services are the business of the National Center for Health Statistics Laboratory and the International Fertility Research Program; and (5) the military presence is embodied by the U.S. Army Research Office.

REGIONS OF THE NORTH AMERICAN REALM

The ongoing transformation of North America's human geography is also reflected in its internal regional organization. As we pointed out earlier (see Fig. 3–20), traditional popularly-perceived regions are increasingly irrelevant to understanding the new spatial realities of postindustrial America. Whereas many established regions still exhibit strong ties, those bonds may be assuming some new dimensions (New England and the U.S. South are good examples). Elsewhere, long-dormant areas are now vigorously asserting new regional identities (such as the U.S. Southwest). And in other cases,

particularly the Anglo-American Core, old regions are struggling to find new answers to problems that threaten to intensify as the postindustrial revolution favors other more dynamic areas.

Quite clearly, as this century draws to a close, North America's regions are in a period of deep-seated change as new forces uproot and redistribute people and activities. Although a number of old locational rules no longer apply, the varied character of the continent's physical, cultural, and economic landscapes does assure that meaningful regional differences will persist. The current areal arrangement of the United States and Canada is presented within a framework of eight major regions (Fig. 3–27). Each will be briefly reviewed, and then the chapter will conclude with an

overview of a provocative new regionalization scheme based on the hypothesis that the realm now consists of a mosaic of nine diversifying "nations."

1. The Anglo-American Core

This continental core region (Fig. 3–11)—synonymous with the American Manufacturing Belt—has been discussed earlier. Serving as the historic workshop for the closely linked spatial economies of the United States and Canada, this region was the unquestioned leader and centerpiece during the century between the Civil War and the close of the industrial age (1865–1970). With the coming of the postindustrial revolution, however, that linchpin regional role is declining today as the Anglo-American Core increasingly shares a growing number of major functions with fast-rising areas to the west and south.

Unfortunately for the Manufacturing Belt, its continuing decline in blue-collar employment is not being matched by the creation of sufficient replacement jobs in other sectors. Nor is there much progress toward the retraining of thousands of displaced factory workers, who will probably never find another industrial job. Thus, parts of the region are becoming home to an ever-larger group of the permanently unemployed, problems most apparent in the inner cities and innermost suburbs of Chicago (8.4 million), Detroit (4.7 million), Philadelphia (6.0 million), Pittsburgh (2.3 million), Cleveland (2.8 million), and Baltimore (2.5 million).

Manufacturing remains a highly important activity within the transformed American economy, but the productivity and obsolescence problems that elevate production

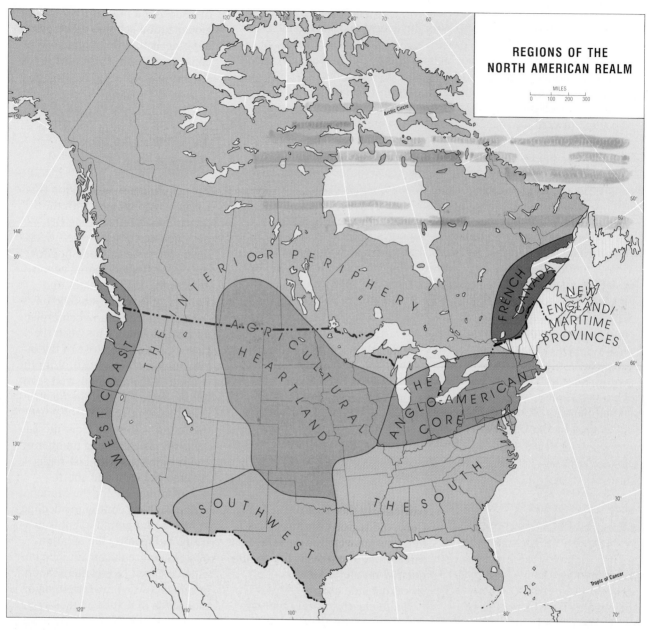

FIGURE 3–27

costs in the core region present obstacles in an era when its traditional industries are much more able to respond to locational forces that operate in favor of newer regions. To remain competitive, a key concern for the Manufacturing Belt is the attraction of new investment capital. As we have noted, parts of the Midwest are succeeding in pursuing the high-tech upgrading of an aged industrial base. How quickly that ef-

fort spreads will be an important indicator as the Belt struggles to forge a new role for itself in a wholly new age.

In certain parts of the core region, where manufacturing has traditionally played a secondary role, postindustrial development has produced a number of new growth centers. The major metropolitan complexes of the northeastern Megalopolis—already the scene of much quaternary and quinary eco-

nomic activity—are adjusting well. The Boston area (4.1 million), well endowed with research facilities and local investment capital, has become a focus of innovative high-tech businesses, particularly along Route 128's freeway corridor that rings the central city. New York City (7.4 million) remains the national leader in finance and advertising, and it houses the broadcast media; yet, even here, the city increasingly shares its once-exclusive decision-making leadership with its own outer suburban city (11.1 million), which has surpassed Manhattan in total number of corporate headquarters facilities.

Perhaps the core-area metropolis that has gained the most from the emergence of postindustrialism is Washington (4.0 million). As the information and control-function sectors have blossomed, and as the U.S. federal government extends its connections ever deeper into America's business operations, the District of Columbia (605,000) together with its surrounding outer city of hyperaffluent Maryland and Virginia suburbs (3.4 million) has amassed an enormous complex of office, research, trade-organization, lobbying, and consultant firms. Suburbanization of facilities has been heightened by a lack of space in the District's center, avoidance of sprawling black ghettoes near downtown, and particularly the lure of the Capital Beltway—a 66-mile (105-km) freeway that encircles Washington and connects its most prestigious suburbs. Indeed, this thriving curvilinear outer city underscores that nonresidential suburbia constitutes the healthiest category of economic subareas throughout the Anglo-American Core.

2. New England/ Maritime Canada

New England, one of the continent's historic culture hearths, has retained a powerful regional identity for almost 400 years. Although the urbanized southern half of New England has been the northeastern anchor of the Anglo-American Core since the mid-nineteenth century—where it must be regionally classified—its six states (Maine, New Hampshire, Ver-

mont, Rhode Island, Massachusetts, and Connecticut) still share many common characteristics. Besides this overlap with the Manufacturing Belt (and Megalopolis) in its south, the New England region also extends northeastward across the Canadian border to encompass the three maritime provinces of New Brunswick, Nova Scotia, and Prince Edward Island as well as the large outer island of Newfoundland.

A long association based on economic and cultural similarities has tied northern New England to Maritime Canada. Both are rural in character, possess rather difficult environments in which land resources are limited, and were historically bypassed in favor of more fertile inland areas. Thus, economic growth here has always lagged behind the rest of the realm. Development has centered on primary activities, mainly fishing the rich offshore banks of the

nearby North Atlantic, forestry in the uplands, and farming in the few fertile valleys available. Recreation and tourism have boosted the regional economy in recent times, with scenic coasts and mountains attracting millions from the neighboring core region; the growth of skiing has also helped, extending the tourist season through the harsh winter months.

A rich sense of history and tradition permeates the close cultural affinity of New England and the Maritimes. Both possess homogeneous English-based cultures that have long withstood persistent Francophone incursions from adjacent Quebec, and both have given rise to a staunchly self-sufficient, pragmatic, and conservative population. Village settlement is overwhelmingly preferred (Fig. 3–28) in sharp contrast to the dispersed rural population of the continental interior that clings to its individual farmsteads.

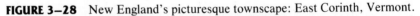

FIGURE 3–28   New England's picturesque townscape: East Corinth, Vermont.

3. French Canada

Francophone Canada comprises the effectively settled (southern) portion of the province of Quebec, straddling the lower St. Lawrence Valley from where the river crosses the Ontario-Quebec boundary just upstream from Montreal to its mouth in the Gulf of St. Lawrence; also included are sizable concentrations of French speakers who reside just across the provincial border in New Brunswick and the U.S. border in northernmost Maine (Fig. 3–27). Significantly, this is the only North American region that is defined by culture alone, but the reasons are compelling: more than 80 percent of French Canada's population speaks French as a mother tongue.

Most of the Anglophones (many of whom use French as a second language) are clustered in the Montreal metropolitan area (2.9 million). The old-world charm of Quebec's cities (Fig. 3–29) is matched by an equally unique rural settlement landscape introduced by the French—narrow rectangular farms known as *long lots* are laid out in sequence perpendicular to the St. Lawrence and other rivers, allowing each farm access to the waterway.

The economy of French Canada is no longer rural (although dairying remains a leading agricultural pursuit), exhibiting urbanization rates similar to those of the rest of the country. Industrialization is widespread, supported by cheap hydroelectric energy gen-

erated at huge dams in northern Quebec. Tertiary and postindustrial commercial activities are centered in Montreal; tourism and recreation are also important.

As mentioned earlier in this chapter, heightened nationalism in Quebec during the 1970s first raised the specter of that province's secession from the Canadian confederation. Although secession was rejected in a 1980 referendum, the provincial government enacted new laws that strengthened the French language and culture within Quebec. With English domination ended, the French Canadians in the 1980s channeled their energies into developing Quebec's economy. A by-product of that successful effort was the rapid urbanization of

FIGURE 3–29 An aerial view of Quebec City on the banks above the St. Lawrence River, with the magnificent Château Frontenac Hotel surrounded by the old city.

young families and a loosening of their ties to Roman Catholicism, producing a sharp decline in the provincial birth rate—which the Quebec government tries to offset by encouraging (with cash payments) large families and Francophone immigration.

By 1990, however, secession again became an even more serious threat in the wake of the failure of the so-called Meech Lake accord. This 1987 agreement, which could not be ratified by all 10 provinces within the mandated three-year period, would have brought Quebec into Canada's new constitution by officially recognizing its right to "preserve and promote its distinct society."

4. The Agricultural Heartland

In the heart of the continent, agriculture becomes the predominant feature of the landscape. While the innermost fruit-and-vegetable and dairying belts in the macro-Thünian regional system (Fig. 3–24) are in competition with other economic activities in the Anglo-American Core, by the time one reaches the Mississippi Valley, meat and grain production prevail for much of the 1000 miles (1600 km) west to the base of the Rocky Mountains. Because the eastern half of the Agricultural Heartland lies in Humid America and closer to the national food market on the northeastern seaboard, mixed crop-and-livestock farming wins out over less competitive wheat raising, which is relegated to the fertile but semiarid environment of the central and western Great Plains on the dry side of the 100th meridian. The latter area also contains Canada's agricultural heartland north of the 49th-parallel border; however, although these fertile prairie provinces are some-

what less subject to serious dry spells than the U.S. high plains to the south, they are situated at a more northerly latitude that makes for a shorter growing season.

The distribution of North American corn and wheat production is shown in Fig. 3–30, which rather neatly defines the boundaries of the Heartland region. The Corn Belt to the east is focused on Illinois and Iowa, with extensions into neighboring states (see Fig. 3–24). The area of north-central Illinois, which displays a major cluster of corn farming, is a classic example of transition along a regional boundary: the Agricultural Heartland/Anglo-American Core dividing line connecting Milwaukee (1.6 million) and St. Louis (2.6 million) passes right through this zone (Fig. 3–27), which contains a number of sizable urban-industrial concentrations interspersed by some of the most productive corn-producing counties in the na-

FIGURE 3–30

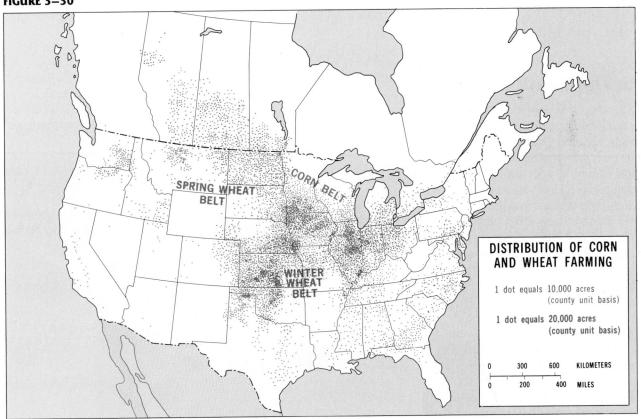

tion. On its western margins, the Corn Belt quickly yields to the dryland Wheat Belts. The recent growth of foreign wheat-sale opportunities has prompted the intensification of production in the Winter Wheat Belt through the use of center-pivot irrigation (Fig. 3–5), a means of overcoming the moisture deficit as long as groundwater supplies hold up.

Throughout the Heartland region, nearly everything is oriented to agriculture. Its leading metropolises—Kansas City (1.7 million), Minneapolis (2.5 million), Winnipeg (590,000), Omaha (640,000), and even Denver (2.0 million)—are major processing and marketing points for beef packing, flour milling, and pork production. People in this region are generally of northern European ancestry, and conservative. Yet, rapid technological advances require that farmers keep abreast of increasingly scientific agricultural techniques and business methods if they are to survive in a hypercompetitive atmosphere.

the urban South. Cities such as Atlanta (3.1 million), Houston (3.8 million), Miami (2.0 million), Tampa (2.2 million), and New Orleans (1.4 million) turned into booming metropolises practically overnight, and conurbations swiftly formed in such places as southern and central Florida, the Carolina Piedmont, and along the Gulf coast from Houston to Mobile, Alabama. This "bulldozer revolution" was matched in the more favored rural areas by an agricultural renaissance that stressed such high-value commodities as beef, soybeans, poultry, and lumber. And on the social front, institutionalized racial segregation was dismantled; although certain problems in minority relations persist, inequalities in the South today are no worse than in other parts of the country; in fact, the opening of the 1990s saw black mayors in Atlanta, Birmingham (940,000), and New Orleans, and women in Houston and Tampa.

Yet, for all the growth that has taken place since the 1960s, the

South remains a region beset by many economic problems because the geography of its development has been quite uneven. Although certain urban and farming areas have benefited, many others containing large populations have not. Even places adjacent to boom areas have often been left unaffected, and the juxtaposition of progress and backwardness is frequently encountered in the Southern landscape (Fig. 3–31). The geographic selectivity of nonagricultural growth is noteworthy in another regard: with the exception of centrally located Atlanta, the most important development has occurred near the region's periphery: in the Washington suburbs of northern Virginia, the North Carolina Piedmont, the Houston area, and Florida's central and southern-coast corridors. And as the checkerboard pattern of development intensifies, even places within boom areas may be falling behind—especially large Southern central cities, whose prospects appear to be swiftly converging with

5. The South

The American South is the most rapidly changing of the realm's eight regions. Choosing to pursue its own sectional interests practically from the advent of nationhood, this region's economic and cultural isolation from the rest of the United States deepened during the long and bitter aftermath of the ruinous Civil War. For over a century the South languished in economic stagnation; but the 1970s witnessed a reassessment in the nation's perception of the region that launched a still-ongoing wave of growth and change unparalleled in Southern history.

Propelled by the forces that created the Sunbelt phenomenon, people and activities streamed into

FIGURE 3–31   Jarring contrasts characterize the landscape of today's South where the old and the new often stand side by side—even here in Atlanta's CBD.

their northern counterparts as burgeoning outer suburban cities around Atlanta, Memphis (1.0 million), Houston, Miami, Tampa, Orlando (1.1 million), and New Orleans capture a larger share of total metropolitan employment.

In retrospect, Southern development over the past two decades has produced much beneficial change and has helped to remove the region's longstanding inferiority complex. But one must also keep in mind that the South had a long way to go, and much of it still lags behind the times. One of its foremost students, the historian C. Vann Woodward, offers this revealing insight about the "New South":

The old Southern distinction of being a people of poverty among a people of plenty lingers on. There is little prospect of closing the gap overnight.

6. The Southwest

As recently as the 1960s, North America geography textbooks did not identify a distinct southwestern region, classifying the Mexican borderland zone into separate southward extensions of the Great Plains, Rocky Mountains, and Intermontane Plateaus. Today, however, this area must be recognized as a major regional entity, albeit the youngest of this realm. The emerging Southwest is also unique in the United States because it is a bicultural regional complex peopled by in-migrating Anglo-Americans atop the crest of the Sunbelt wave as well as by the quickly expanding Mexican-American population anchored to the sizable long-time resident Hispanic group, which traces its origins to the Spanish colonial territory that once extended from the Texas Gulf coast to San Francisco. In fact, if one counts the

large Native American (Indian) population, then the Southwest is actually a *tricultural* region.

Recent rapid development in Texas, New Mexico, and Arizona is essentially built on a three-pronged foundation: (1) availability of huge amounts of generated *electricity* to power air conditioners through the long, brutally hot summers; (2) sufficient *water* to supply everything from swimming pools to irrigated crops so that large numbers may reside in this dry environment; and (3) the *automobile*, so that affluent newcomers may spread themselves out at much-desired low densities. The first and third of the foundation prongs have been rather easily attained because the eastern flank of the Southwest is abundantly endowed with oil and natural gas (Fig. 3–22). The future of water supplies, however, is far more problematical; for instance, Phoenix (2.2 million) and Tucson (660,000) could not keep growing at their current pace without the massive new Central Arizona Project canal, the state's last major source (the Colorado River) of drinking water.

So far, the generation of new wealth in this growing region makes it an undeniable success. However, since much of this achievement is linked to the fortunes of the oil business, the Southwest may well be entering the future beset by unpredictable economic turns if the experiences of the energy industry since 1970 are an indicator. Much of Texas was staggered by the falling fortunes of oil in the 1980s, only rebounding slowly as the decade ended. However, the postindustrial revolution is also prominently represented in this region—with a specialized activity complex of electronic and space-technology facilities located in the eastern Texas triangle formed by connecting Houston, San Antonio (1.4 mil-

lion), and Dallas–Fort Worth (4.0 million); and the state capital of Austin (825,000) near its center is becoming a leading high-tech research complex in a partnership between private manufacturers and the University of Texas.

7. The Interior Periphery

Despite its apparent contradiction, the Interior Periphery is an appropriate name for North America's largest region by far. It covers most of Canada, all of Alaska, the northern salients of Minnesota–Wisconsin–Michigan, New York's Adirondacks, and the inland U.S. West between (and including) the Sierra Nevada–Cascades and the Rocky Mountains. There can be no doubt about the interior position of this vast slice of the North American realm; its peripheral nature stems from its isolation and rugged environment, which have attracted only the sparsest of populations relative to the other seven regions. In the U.S. portion, even after counting substantial recent growth, population density is only 12 per square mile (5 per square km) in contrast to 69/26 overall. Yet, its disadvantages notwithstanding, the Interior Periphery contains great riches because it is one of the earth's major storehouses of mineral and energy resources.

Accordingly, the region's history has been a frustrating one of boom-and-bust cycles, determined by technological developments, economic fluctuations, and corporate and government decisions made in the Anglo-American Core. The events of the past several years underscore this bittersweet development process. Following the oil shortage of the early 1970s, there was a virtual invasion in the search for additional petroleum and natural gas, with efforts concentrating on the Alaskan North Slope and southwestern Wyoming's Overthrust Belt; the latter area and neighboring Colorado were also overrun by energy companies seeking to perfect methods for obtaining gasoline from huge local deposits of oil shale. With the opening of the Trans-Alaska Pipeline in 1977, optimism reached an all-time high, and western Wyoming and Colorado were inundated by tens of thousands of migrants who taxed local community facilities to the breaking point. But, as has happened so many times before, the boom soon fizzled out with a bewildering suddenness. Everything from oilfield homebuilding to the wave of skyscraper construction in nearby Denver was predicated on a continuing rise in the price of oil. Therefore, when that price leveled off in the early 1980s, the repercussions were swift and painful. Oil companies soon halted most of their exploration activity, and the newcomers found themselves faced with unemployment.

Elsewhere in the region, economic recession impacted negatively on the mining industry: parts of Minnesota's Iron Range were closed down, and once-productive copper mines in Montana, Utah, and Arizona faced the biggest slowdown in their history. Nonetheless, the future still beckons brightly. Rocky Mountain coal- and uranium-extraction operations were relatively untouched by these economic misfortunes, and the mining of silver, lead, zinc, and nickel endured at scattered sites throughout western North America and on the Canadian Shield to the east. And surely, with the steady decline of fossil-fuel supplies worldwide, oil and gas drillers will one day return in force to the central Rockies.

The recent sharp swings of the extractive economy also mask a steadier influx of population and nonprimary activities to other parts of the Interior Periphery. In fact, from 1970 to 1990, the U.S. segment grew by more than 60 percent, thereby advancing its relative position from 4 to 7 percent of the national population. The guiding force, once again, was the search for high-amenity locations, with the clean, stress-free, wide-open spaces of certain parts of the Rockies and Intermontane Plateaus a particular attraction. Accordingly, Nevada, Alaska, Arizona, Utah, Wyoming, Colorado, and Idaho all ranked among the 10 fastest-growing states between 1970 and 1985.

8. The West Coast

The Pacific Coast of the conterminous United States and southwestern Canada has been a powerful lure to migrants since the Oregon Trail was pioneered 150 years ago. Unlike the remainder of the North American west, the narrow strip of land between the Sierra-Cascade mountain wall and the sea receives adequate moisture; it also possesses a far more hospitable environment with generally delightful weather south of San Francisco, highly productive farmlands in California's Central Valley, and such scenic glories as the Big Sur coast, Washington's Olympic Peninsula, and the spectacular waters surrounding San Francisco (6.3 million), Seattle–Tacoma (2.6 million), and Vancouver (1.9 million). Most major development here took place during the post–World War II era, accommodating enormous population and economic growth, and the West Coast is just now beginning to face the less pleasant consequences of regional maturity.

Forty-plus years of unrelenting growth have especially taken their toll in California, because the massive development of America's most populated state has been overwhelmingly concentrated in the teeming conurbation extending south from San Francisco through San Jose (1.5 million), the San Joaquin Valley, the Los Angeles Basin, and the southwestern coast into San Diego (2.7 million) at the Mexican border (Fig. 3–32). Environmental hazards threaten this entire corridor, including inland droughts, coastal flooding, mudslides, brush fires, and earthquakes—with the ominous San Andreas Fault practically the axis of megalopolitan coalescence. To all this, humans have added their own abuses of the fragile natural habitat, from overuse of water supplies (requiring vast aqueduct systems to import water from hundreds of miles away) to the incredible air pollution of Los Angeles (see Fig. 3–7) caused by the emissions of over 9 million automobiles, which are vital to movement within this particularly dispersed metropolis of 14.7 million. Lost somewhere in the shuffle is yesterday's glamorous image of Southern California—communicated so effectively in the motion pictures of the postwar period—that is based on relaxed outdoor living in luxurious horticultural suburbs amid one of the most agreeable climates to be found anywhere on earth. Undeniably, economic development has brought California new prosperity, and the state continues its leadership as a national innovator. But some of this affluence has also been subject to sudden shifts because the Southern California economy is tied to the cyclical aerospace industry. Clearly, the years of innocence and unbridled optimism are over, and the future will probably include a struggle to maintain the state's still-considerable advantages in the face of growing competition from would-be new Californias elsewhere on the continent.

FIGURE 3–32   Panoramic view of the curving San Diego waterfront, highlighting the locational amenities of California's second-largest city.

The northern portion of the West Coast region is the Pacific Northwest, focused on Oregon's Willamette Valley, the Cowlitz–Puget Sound Lowland of western Washington, and the British Columbia coast of southwesternmost Canada. Originally built on timber and fishing—primary activities that still thrive here—the impetus for industrialization came from the massive Columbia River dam projects of the 1930s and 1950s that created cheap hydroelectricity. This, in turn, attracted aluminum and aircraft manufacturers, and the huge Boeing aerospace complex around Seattle makes that metropolis one of the world's biggest company towns. Unique environmental amenities—zealously safeguarded here—have lured hundreds of growth companies, and the Pacific Northwest should have little trouble adjusting to the postindustrial economy. Perhaps one of its greatest advantages, which is shared with urban California to the south, is as a gateway to the emerging Pacific Basin. As North America is increasingly enmeshed in the global-scale economy, that part of the world will be a critically important marketplace, and Portland (1.5 million) and Seattle, as well as Vancouver in

Canada, are bound to benefit because their airports are the closest to East Asia of any in the conterminous United States.

THE EMERGING "NINE NATIONS" OF NORTH AMERICA

At the outset of our regional survey, we pointed out that the American-Canadian regional system was in the throes of change, a dynamism highlighted in the profile of each region. An important overview of the direction and significance of all this change is contained in *Washington Post* editor Joel Garreau's 1981 book, *The Nine Nations of North America*. His **nine-nations hypothesis** is that the realm is reorganizing into nine separate "nations":

Each has its capital and its distinctive web of power and influence. A few are allies, but many are adversaries. . . . Some are close to being raw frontiers; others have four centuries of history. Each has a peculiar economy; each commands a certain emotional allegiance from its citizens. These nations look different, feel different, and sound different from each other, and few of their boundaries match the political lines drawn on current maps. . . . Each nation has its own list of desires. Each nation knows how it plans to get what it needs from whoever's got it. Most important, each nation has a distinct prism through which it views the world.

Garreau's regionalization scheme is mapped in Fig. 3–33, and it covers not only the United States and Canada but the Caribbean Basin as well. Briefly reviewing the essence of each "nation," *The Foundry* is dominated by a declining industrial infrastructure of aging gritty cities and faces a bleak future as its human and economic resources relocate to areas possessing lower production costs. *New England*, which has survived innumerable economic crises, is turning local amenities, outstanding educational resources, and a much-admired regional tradition into a haven for new growth activi

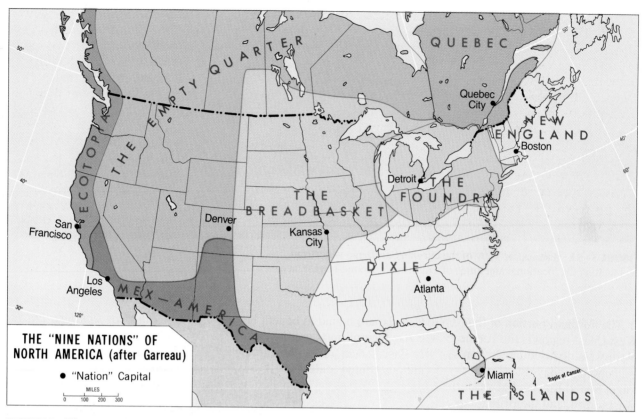

THE "NINE NATIONS" OF
NORTH AMERICA (after Garreau)

● "Nation" Capital

MILES
0   100   200   300

FIGURE 3—33

ties. *Quebec* is asserting its cultural identity as never before and gaining new leverage within Canada. *Dixie* is now permeated with change, but the "Old South" is not being replaced by a clear regional personality and sense of direction. *The Islands* is the newest nation, built on growing economic ties between Miami and nearby Middle America, based, at least in part, on illicit drug flows. *The Breadbasket*, the "nation that works best," is characterized by an abiding agricultural tradition and a conservative population at peace with itself. *The Empty Quarter* is endowed with an abundance of energy resources that will lead to much, albeit scattered, development. *Ecotopia* is the preserve of environmentalists who guard the status quo but, ironically, it is also pioneering new technologies of the postindustrial future. And *Mex-America* is swiftly being shaped

into a distinctive region by its burgeoning Hispanic populations, who are helping to generate new wealth in the border zone from South Texas clear across to central California.

The geographic pattern of this regionalization scheme invites comparison with ours (Fig. 3–27). Certainly there are many elements in common, and it is quite possible that our eight regions are evolving into something resembling Garreau's spatial classification of the continent. In fact, the only "nation" in our scheme that is "missing" is The Islands, which could well be in a formative stage (as residents of South Florida, we acknowledge that this area is becoming more and more different from the rest of the South and is already the financial capital of the Caribbean).

Proceeding across the map, the Anglo-American Core is somewhat

more rectangular than the square-shaped Foundry, incorporating the important southern Illinois coalfields and St. Louis manufacturing complex and taking in all of Megalopolis in the northeast. Our New England/Maritimes region is, therefore, somewhat smaller, as is French Canada, which Garreau treats as the political entity of Quebec. The South and Dixie concur closely, differing only slightly around the edges and on the matter of the incipient Islands "nation." Agricultural Heartland and Breadbasket also agree about regional cores, with variations again observed at the fringes: actually, Garreau may well have a stronger case vis-à-vis the Breadbasket/Empty Quarter boundary in Wyoming and Montana—coal stripmining is quickly becoming the leading activity just to the west of the dividing line, which offsets places where "hydrocarbons become more important than carbohydrates." The Interior Periphery and Empty Quarter are similarly in accord, varying only in the minor matters of interior Quebec, the upper Great Lakes area, and the southern and western margins. The West Coast region does not separate Ecotopia from Mex-America, but it does recognize internal California differences, which could indeed lead to a full regional split in the foreseeable future. Finally, the ripening Southwest was viewed as an increasingly robust region contain-

ing a decidedly Hispanic flavor, but we have reservations about projecting it west to the Pacific because the Southern California international-banking and high-tech boom—particularly in burgeoning suburban Orange County (2.4 million) between Los Angeles and San Diego—is more an expression of West Coast economic geography than the assertiveness of an up-and-coming ethnic group (which may soon be more *Asian* in composition than Latin).

As we already know from previous chapters, no two regionaliza-

tion schemes will ever be exactly the same. The differences here are relatively minor ones, based on superficial form rather than supporting logic. Only one thing is certain: North America's dynamic regions will continue to change, in some cases quite rapidly. The *nine-nations* hypothesis is a valuable and thought-provoking contribution to the fascinating business of monitoring the geographical restructuring of this remarkable realm and could well prove to be correct in its central arguments. Time will tell.

PRONUNCIATION GUIDE

Broek (*brook*)
Bryn Mawr (*brinn*-mar)
Chaparral (shap-uh-*ral*)
Cynwyd (*kin*-widd)
Francophone (*frank*-uh-phone)
Garreau (*garrow*)
Loess (*lerss*)
Megalopolis (mega-*lop*-uh-liss)
Montreal (mun-tree-*awl*)

Nova Scotia (no-va-*skoh*-shuh)
Ogallala (oh-guh-*lah*-luh)
Ottawa (*ot*-tah-wuh)
Puget (*pyoo*-jet)
Puerto Rico (pwair-toh-*ree*-koh)
Quaternary (kwa-*ter*-nuh-ree)
Quebec (kwih-*beck*)
Quinary (*kwy*-nuh-ree)
San Joaquin (san-wah-*keen*)

Saskatchewan (suss-*katch*-uh-wunn)
Tertiary (*ter*-shuh-ree)
Toponymy (toh-*ponn*-uh-mee)
Uwchlan (*yoo*-klan)
Valdez (val-*deeze*)
Von Thünen (fon-*too*-nun)
Willamette (wuh-*lam*-ut)

REFERENCES AND FURTHER READINGS

(BULLETS [●] DENOTE BASIC INTRODUCTORY WORKS ON THE REALM OR SYSTEMATIC ESSAY TOPIC)

Adams, John S. *Housing America in the 1980s* (New York: Russell Sage Foundation, 1988).

Adams, John S. "Residential Structure of Midwestern Cities," *Annals of the Association of American Geographers*, 60 (1970): 37–62. Model diagram adapted from p. 56.

● Adams, John S., ed. *Contemporary Metropolitan America* (Cambridge, Mass.: Ballinger, 4 vols., 1976).

Allen, James P. & Turner, Eugene J. *We the People: An Atlas of America's Ethnic Diversity* (New York: Macmillan, 1987).

Atlas of North America: Space Age Portrait of a Continent (Washington: National Geographic Society, 1985).

● Atwood, Wallace W. *The Physiographic Provinces of North America* (New York: Ginn, 1940).

Bell, Daniel. *The Coming of Postindustrial Society* (New York: Basic Books, 1973). Quotation taken from p. xvi of the 1976 [paperback] Foreword.

Berry, Brian J. L. "The Decline of the Aging Metropolis: Cultural Bases and Social Process," in Sternlieb, George & Hughes, James W., eds., *Post-Industrial America: Metropolitan Decline and Inter-Regional Job Shifts* (New Brunswick, N.J.: Center for Urban Policy Research, Rutgers University, 1975), pp. 175–185.

● Birdsall, Stephen S. & Florin, John W. *Regional Landscapes of the United*

States and Canada (New York: John Wiley & Sons, 3 rev. ed., 1985).

Borchert, John R. "American Metropolitan Evolution," *Geographical Review*, 57 (1967): 301–332.

Broek, Jan O. M. *Geography: Its Scope and Spirit* (Columbus: Charles E. Merrill, 1965). Quotation taken from p. 72.

● Brown, Ralph H. *Historical Geography of the United States* (New York: Harcourt, Brace & World, 1948).

● Christian, Charles M. & Harper, Robert A., eds. *Modern Metropolitan Systems* (Columbus: Charles E. Merrill, 1982).

● Clark, David. *Post-Industrial America: A Geographical Perspective* (London & New York: Methuen, 1985).

"Energy: A Special Report in the Public Interest," *National Geographic*, February 1981, pp. 1–114.

Frey, William H. & Speare, Alden, Jr. *Regional and Metropolitan Growth and Decline in the United States* (New York: Russell Sage Foundation, 1988).

Garreau, Joel. *The Nine Nations of North America* (Boston: Houghton Mifflin, 1981). Quotation taken from pp. 1–2.

Gastil, Raymond D. *Cultural Regions of the United States* (Seattle: University of Washington Press, 1975).

Gottmann, Jean. *Megalopolis: The Urbanized Northeastern Seaboard of the United States* (New York: Twentieth Century Fund, 1961).

Gould, Peter R. & White, Rodney. *Mental Maps* (Winchester, Mass.: Allen & Unwin, 2 rev. ed., 1986).

Harris, Chauncy D. & Ullman, Edward L. "The Nature of Cities," *Annals of the American Academy of Political and Social Science*, 242 (1945): 7–17.

• Hartshorn, Truman A. *Interpreting the City: An Urban Geography* (New York: John Wiley & Sons, 1980).

• Hunt, Charles B. *Natural Regions of the United States and Canada* (San Francisco: W. H. Freeman, 2 rev. ed., 1974).

Knox, Paul L. et al. *The United States: A Contemporary Human Geography* (London & New York: Longman, 1988).

Lewis, Peirce F. "Learning from Looking: Geographic and Other Writing About the American Cultural Landscape," *American Quarterly*, 35 (1983): 242–261.

Louv, Richard. *America II* (Los Angeles: Jeremy Tarcher, 1983).

Malcolm, Andrew H. *The Canadians* (New York: Times Books, 1985).

Mayer, Harold M. "Geography in City and Regional Planning," in John W. Frazier, ed., *Applied Geography: Selected Perspectives* (Englewood Cliffs, N.J.: Prentice-Hall, 1982). Quotation taken from p. 27.

• McCann, Lawrence D., ed. *Heartland and Hinterland: A Geography of Canada* (Scarborough, Ont.: Prentice-Hall Canada, 1982).

Meinig, Donald W. *The Shaping of America: A Geographical Perspective on 500 Years of History, Volume 1, Atlantic America, 1492-1800* (New Haven: Yale University Press, 1986).

• Mitchell, Robert D. & Groves, Paul A., eds. *North America: The Historical Geography of a Changing Continent* (Totowa, N.J.: Rowman & Littlefield, 1987).

Muller, Peter O. *Contemporary Suburban America* (Englewood Cliffs, N.J.: Prentice-Hall, 1981).

• *Oxford Regional Economic Atlas of the United States and Canada* (New York: Oxford University Press, 2 rev. ed., 1975).

• Paterson, John H. *North America: A Geography of Canada and the United States* (New York: Oxford University Press, 8 rev. ed., 1989).

Phillips, Phillip D. & Brunn, Stanley D. "Slow Growth: A New Epoch of American Metropolitan Evolution," *Geographical Review*, 68 (1978): 274–292.

• Putnam, Donald F. & Putnam, Robert G. *Canada: A Regional Analysis* (Toronto: Dent, 2 rev. ed., 1979).

• Rooney, John F., Jr., et al., eds. *This Remarkable Continent: An Atlas of United States and Canadian Society and Cultures* (College Station, Tex.: Texas A&M University Press, 1982).

Ross, Thomas E. & Moore, Tyrel G., eds. *A Cultural Geography of North American Indians* (Boulder, Colo.: Westview Press, 1987).

Vance, James E., Jr. *The Continuing City: Urban Morphology in Western Civilization* (Baltimore: Johns Hopkins University Press, 1990).

Vance, James E., Jr. *This Scene of Man: The Role and Structure of the City in the Geography of Western Civilization* (New York: Harper's College Press, 1977). Urban realms model discussed on pp. 411–416.

Wheeler, James O. & Muller, Peter O. *Economic Geography* (New York: John Wiley & Sons, 2 rev. ed., 1986). Map adapted from p. 330.

• White, C. Langdon et al. *Regional Geography of Anglo-America* (Englewood Cliffs, N.J.: Prentice-Hall, 6 rev. ed., 1985).

Woodward, C. Vann. "The South Tomorrow," *Time*, The South Today—Special Issue, September 27, 1976. Quotation taken from p. 99.

Yeates, Maurice H. *Main Street: Windsor to Quebec City* (Toronto: Macmillan of Canada, 1975).

• Zelinsky, Wilbur. *The Cultural Geography of the United States* (Englewood Cliffs, N.J.: Prentice-Hall, 1973).

Zelinsky, Wilbur. "North America's Vernacular Regions," *Annals of the Association of American Geographers*, 70 (1980): 1–16. Map adapted from p. 14.

PRODIGIOUS JAPAN:
Triumph of Technology

IDEAS AND CONCEPTS

modernization

relative location (2)

exchange economy

industrial location (2)

areal functional organization

physiologic density

postindustrial revolution (2)

In the non-Western, non-European world, there is no country quite like Japan. None of the metaphors of underdevelopment applies here: Japan is an industrial giant, a technological pacesetter, a fully urbanized society, a political power, a thriving affluent nation. Probably no city in the world today is without Japanese cars in its streets; few photography stores lack Japanese cameras and film; many laboratories use Japanese optical equipment. From microwave ovens to VCRs, from oceangoing ships to stereo systems, Japanese manufactures flood the world's markets. The Japanese seem to combine the precision skills of the Swiss with the massive industrial power of prewar Germany, the forward-looking designs of the Swedes with the innovations of the Americans. How have the Japanese done it? Does Japan have the kind of resources and raw materials that helped boost Britain into its early, revolutionary industrial leadership? Are there locational advantages in Japan's position off the eastern coast of the Eurasian landmass, just as Britain benefited from its situation off the western coast? Could Japan's rise to power and prominence have been predicted, say, a century ago?

Japan consists of four large and mostly mountainous islands (plus innumerable small islands and islets) whose total area amounts to a mere 146,000 square miles (377,000 square kilometers), a territory smaller than California. Japan's population of 124 million is mostly crowded onto the limited coastal plains, valleys, and lower slopes of these islands, principally on the largest (Honshu) and southernmost (Kyushu and Shikoku) (Fig. J-1). The northernmost island, Hokkaido, still possesses frontier characteristics. And had it not been for Russian imperialism, Japan might also control Sakhalin Island (just north of Hokkaido), the fifth largest island of the Japanese archipelago (island chain).

Is Japan a discrete geographic realm or should it be viewed as an insular extension of the Chinese world? Japan was peopled from East Asia mainly via the Korean Peninsula, and over many centuries it received numerous cultural infusions from the Asian mainland. By the sixteenth century, a highly individualistic culture had already evolved. But since the 1860s Japan has taken new directions—developmental, technological, cultural, and political—so unlike those prevailing in mainland Asia that there is ample justification for its designation as a separate realm. By the turn of the twentieth century, Japan was an industrial force and a colonial power. Through a unique blend of tradition and modernization, borrowing and innovation, discipline and ambition, and with enormous energy and organizational ability, the Japanese transformed their country, overcame defeat and disaster, and created a society unlike any other in Asia.

The Japanese realm is a reality despite its comparatively small territorial size and relatively modest population numbers (although Japan does contain more than six times as many inhabitants as Australia and New Zealand combined). Not the numbers, but the remarkable homogeneity of that population constitutes one of the

TEN MAJOR GEOGRAPHIC QUALITIES OF JAPAN

1. Japan is an archipelago of mountainous islands off the eastern coast of Eurasia.
2. Japan buys raw materials and sells finished products the world over. It lies remote from most of the locations with which it has close economic contact.
3. Japan is the prime example in the world of modernization in a non-Western society.
4. Japan was the non-Western world's major modern colonial power. During the colonial era, it strongly influenced the development of the western Pacific.
5. Although recent interaction has been limited, Japan lies poised for a major role in the development of East Asia, including eastern Siberia as well as China.
6. Japan's modern industrialization depends on the acquisition of raw materials (including energy resources) from distant locales. Japan's own mineral-resource base is limited.
7. Modern and traditional society exist intertwined in Japan.
8. Japan has one of the highest physiologic population densities in the world.
9. Japan's agriculture, given conditions of climate, soil, and slope, is the most efficient and productive in Asia.
10. Japan's areal functional organization, based on regional specialization, is maturely developed.

main reasons why Japan is a separate realm. Centuries of insulation and isolation coupled with government policies designed to perpetuate that condition have given Japan the most homogeneous population of its size in the world. One telling example of the government's concern in this matter involved the outflow of hundreds of thousands of boat people escaping from communist-dominated Vietnam in the 1970s and early 1980s. Although other countries accepted nearly 1 million of these refugees, Japan permitted fewer than 200 to enter, leaving no doubt about the reason. "Japan is a country of Japanese," said a government representative, "we are not a nation of minorities and subgroups."

Closely linked with this ethnic homogeneity is a cultural uniformity that confirms and emphasizes Japan's geographical identity. From Hokkaido to Kyushu, from one end of Honshu to the other, Japan exhibits a sameness that is pervasive and even baffling. Japan has its huge modern cities and its small traditional villages—in that sense, there is contrast. But so do most other societies. In Japan, uniformity is assured by history and custom, by inherited pattern and controlled behavior. From the universal language to the Shinto national belief system, from the strength of family ties to the role of education and the schools (everyone in Japan wears the approved school uni-

form), from diets to diversions, the Japanese exhibit a cultural unity unmatched in modern times.

A further confirmation of Japan's status as a separate realm lies in its unmatched technological development and, hence, its economic geography. Using raw materials from virtually all corners of the planet, the Japanese have created a manufacturing complex whose production far outstrips that of giant China, the two Koreas, and Taiwan combined. On this basis alone (more will be said about it later), Japan is no mere corner of an East Asian geographic realm, but merits its own discrete status.

In the mid-nineteenth century, Japan hardly seemed destined to become Asia's leading power. For 300 years, the country had been closed to outside influences, and Japanese society was stagnant and tradition-bound. Very early during the colonial period, European merchants and missionaries had been tolerated and even welcomed; but as the European presence on the Pacific's coastlines grew greater, the Japanese began to shut their doors. By the end of the sixteenth century, Japan's overlords, fearful of Europe's imperialism, decided to expel all foreign traders and missionaries. Christianity (especially Catholicism) had gained a considerable foothold in Japan, but it was now viewed as a prelude to colonial conquest; therefore, in the first part of the seventeenth century, it was practically stamped out in a massive bloody crusade. Determined not to share the fate of the Philippines, which had by then fallen to Spain, the Japanese permitted only the most minimal contact with Europeans. Thus, a few Dutch traders, confined to a little island near the city of Nagasaki, were for many decades the sole representatives of the European realm. So Japan retreated into a

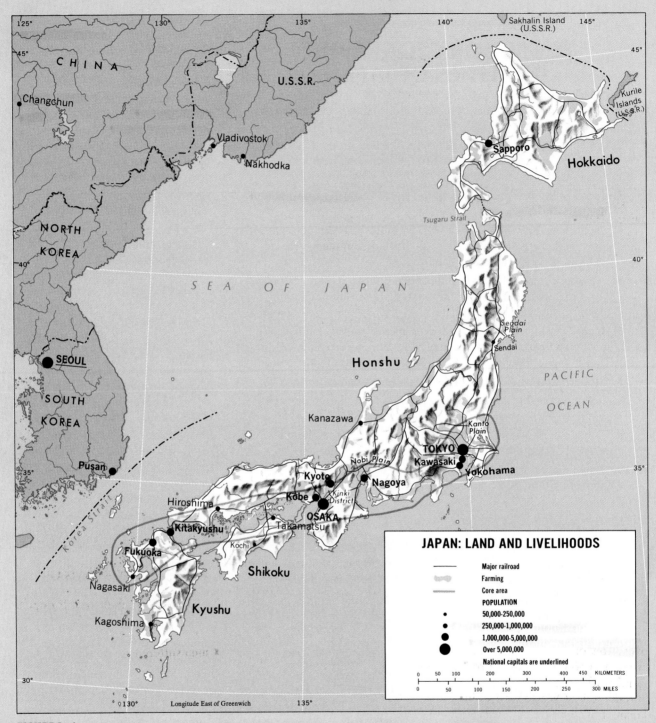

FIGURE J–1

long period of isolation, which lasted past the middle of the nineteenth century.

Japan could maintain its aloofness while other areas were falling victim to the colonial tide because of a timely strengthening of its central authority, because of its insular character, and because of its remote position along East Asia's difficult northern coast. There were other factors as well—Japan's isolation was far less splendid than that of China, whose silk and tea and skillfully made wares always attracted hopeful traders. But in the eighteenth and early nineteenth centuries, the modernizing influences brought by Europe to the rest of the world passed by Japan as they did China.

When Japan finally came face to face with the steel "black ships" and the firepower of the United States as well as Britain and France, it was no match for its enemies, old or new.

In the 1850s, the United States first showed the Japanese to what extent the balance of power had shifted. American naval units—beginning with Commodore Perry's flotilla in 1853—sailed into Japanese harbors and extracted trade agreements through a show of strength. Soon the British, French, and Dutch were also on the scene seeking similar treaties. When there was resistance in local areas to these new associations, the Europeans and Americans quickly demonstrated their superiority by shelling parts of the Japanese coast. By the late 1860s, no doubt remained that Japan's protracted isolation had come to an end.

IMPERIAL JAPAN

In the century following 1868, Japan emerged from its near-colonial status to become one of the world's major powers, finally overtaking a number of its old European adversaries in the process. The year 1868 is important in Japanese history, for it marks the overthrow of the old rulers that brought to power a group of reformers whose objective was the introduction of Western ways in Japan. The supreme authority officially rested with the emperor, whose role during the previous militaristic period had been pushed into the background. Thus, the 1868 rebellion came to be known as the *Meiji Restoration*—the return of enlightened rule, centered around Emperor Meiji. But despite the divine character ascribed to the royal imperial family, Japan

was, in fact, ruled by the revolutionary leadership, a small number of powerful men whose chief objective was to modernize the country as rapidly as possible in order to make Japan a competitor, not a colonial prize.

The Japanese success is written on the map; even before the turn of the twentieth century, a Japanese empire was in the making, and the country was ready to defeat encroaching Russia. Japan claimed the Ryukyu Islands in 1879, took Formosa (Taiwan) from China in 1895, occupied Chosen (Korea) in 1910, and established a sphere of influence in Northeast China (Manchukuo [Manchuria]). Various archipelagoes in the Pacific Ocean were acquired by annexation, conquest, or mandate (Fig. J–2). In the early 1930s, Manchukuo (Manchuria) was finally conquered, and the Tokyo government began calling for a Greater East Asia Co-Prosperity Sphere, which would combine—under Japanese leadership—all of China, Southeast Asia, and numerous Pacific island territories. By the late 1930s, deep penetrations had been made into China, but Japan got its big chance in the early 1940s during World War II. Unable to defend their distant Asian possessions while engaged by Germany at home, the European colonial powers offered little resistance to Japan's takeover in Indochina, Burma, Malaya, and Indonesia; neither was the United States able to protect the Philippines. Japan's surprise 1941 attack on Pearl Harbor, a severe blow to the U.S. military installations there, was a major success—and is still admired in Japanese literature and folklore.

The Japanese war machine was evidence of just how far Japan's projected modernization had gone: airplanes, tanks, warships, guns, and ammunition were all produced by Japanese industries, which

were now a match for their Western counterparts. Japan, as we know, was ultimately defeated by superior American power (Hiroshima and Nagasaki were all but annihilated by nuclear bombs in 1945), but this defeat coupled with the loss of its prewar as well as its wartime empire still failed to destroy the country's progress. Postwar Japan rebounded with such vigor that its overall economic development now ranks among the most advanced in the world.

With the Meiji Restoration over a century ago, Japan began the phenomenal growth that brought it an empire, involved it in a disastrous war, and eventually saw the country achieve an industrial capacity unmatched not only in the non-Western world, but in much of the Western world as well. As if to symbolize their rejection of the old isolationist position of the country, the new Meiji rulers moved the capital from Kyoto to Tokyo (meaning "Eastern Capital"), the new name for the already populous city of Edo on the Kanto Plain. Kyoto had been an interior capital; Tokyo lay on the Pacific Coast alongside an estuary that was soon to become one of the busiest harbors on earth. The new Japan looked to the sea as an avenue to power and empire, and it looked to industrial Europe for the lessons that would help it achieve those ends. Whereas China suffered as conservatives and modernizers argued the merits of an open-door policy to Westernization, Japan's leaders committed themselves—with spectacular results.

MODERNIZATION

What has happened in Japan during the past century is a process that is far-reaching, transforming

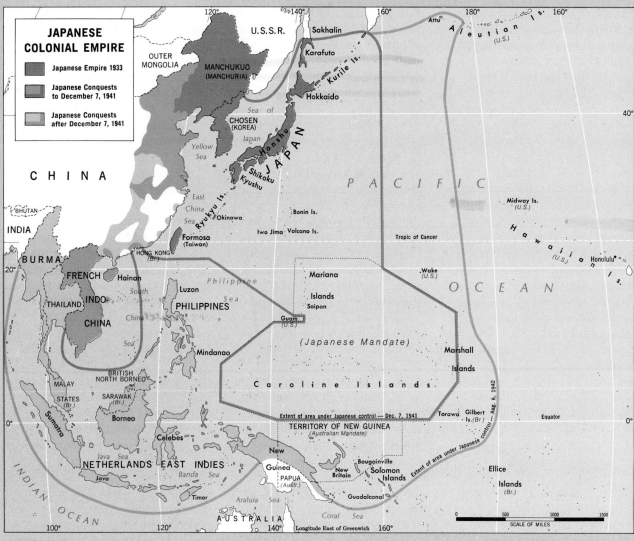

FIGURE J-2

society from "traditional" to "modern," but the process is complex, difficult to define, and subject to much debate among social scientists. We in the Western world tend to view **modernization** as identical to Westernization: urbanization, the spread of transport and communications facilities, the establishment of a market (money) economy, the breakdown of local (tribal) communities, the proliferation of formal schooling, and the acceptance and adoption of foreign innovations. In the non-Western world, however, "modernization" is often seen as an outgrowth of colonial imperialism, the perpetuation of a system of wealth accumulation introduced by alien invaders who were driven by greed. The local elites who have replaced the colonizers in the newly independent states, in that view, only carry on disruption of traditional societies, not their modernization. The capital cities of Africa and Asia, with their crowded avenues and impressive skyscrapers, continue to grow—but they do not signal real *development*. Traditional societies can be modernized without being Westernized.

Japan's modernization, in this context, is unique in many ways. Having long resisted foreign intru-

sion, the Japanese did not achieve the transformation of their society by importing a Trojan horse; it was done by Japanese planners, based on the existing Japanese infrastructure, and it fulfilled Japanese objectives. Certainly, Japan imported foreign technologies and adopted innovations from outside, but the Japan that was built, a unique combination of modern and traditional elements, was basically an indigenous achievement. Like the European powers they often emulated, the Japanese created a large colonial empire to supply the homeland with raw materials. The ripple effect of the Japanese period still continues in the economies of such former dependencies as Taiwan and Korea—and there the fundamental similarities between imperial Japan and imperialist Europe can still be discerned. But modernization in Japan itself defies the generalizations that apply to the process in the underdeveloped world.

LIMITED ASSETS

Japan's success might lead to the presumption that the country's raw material base was sufficiently rich and diverse to support the same kind of industrialization that had characterized England. Britain's industrial rise, after all, was accomplished largely because the country possessed a combination of resources that could sustain an industrial revolution. But Japan is not nearly so well off. Although Japan still has coal deposits in Kyushu and Hokkaido, there is little of the good coking coal that is required for steel production. The most accessible coal seams were soon exhausted; before long, the cost of coal mining was rising steadily. There was a time when Japan was a net coal exporter, when coal was shipped to the coastal cities of China, but today Japan must import over 85 percent of this commodity. Nonetheless, Japan did have enough coal, fortuitously located near its coasts, to provide a vital stimulus for initial industrialization during the late nineteenth century. Both in northern Kyushu and Hokkaido, the coal deposits lie so near the coast that transport is not a problem, and, by the nature of Japan's terrain, the major cities and industrial areas also lie on or near the coast. Thus, coal could be provided quite cheaply and efficiently to every developing manufacturing center,

and the fringes of the Seto Inland Sea were among the first areas to benefit from this spatial relationship.

As for the other major industrial raw material—iron ore—Japan has only a tiny domestic supply, limited to small scattered deposits of variable but generally low quality. In this respect, there is simply nothing to compare to what Britain had at its immediate disposal; and neither is Japan rich in the various ferroalloy minerals that are vital to the steelmaking process. The Japanese extract what they can, including ores that are so expensive to mine and refine that—in view of import alternatives—it is hardly worth the effort, and they carefully collect and reprocess all available and imported scrap iron. Nevertheless, iron ore still ranks high on each year's import list.

Japan also possesses negligible energy resources, and has been forced to import over 99 percent of its oil in recent years. With an overall 83 percent of its energy derived from imported fuels in 1987, Japan is now vigorously pursuing policies to reduce that dependence on foreign oil, particularly for the generation of electricity. The government has decided to rely most heavily on nuclear power: this source today accounts for about 30 percent of Japan's electrical output, but is expected to surpass the 50-percent mark within the next two decades. Although the country's topography favors local hydroelectric-power generation, Japan annually consumes so much electricity that existing dams cannot keep pace with demand; with (a shrinking) one-sixth of Japan's energy supplied by hydropower facilities, this source is not expected to play a major role in the future.

This, then, answers one of the questions we posed in the first paragraph of this vignette. Japan really has nothing to compare to

what Britain had during its industrial transformation. In fact, after what has just been discussed, it would seem that we are dealing with just another of those many countries whose underdevelopment can be attributed largely to a paucity of resources and whose economic health is, therefore, deemed to depend on agriculture. Certainly, it would have been difficult in 1868 to predict that the enthusiastic leaders of the Meiji Restoration would really have much success with their plan for the Westernization and strengthening of their country. What, then, gave Japan its opportunity? Obviously, foreign trade and technology transfer played a great part in it. Was Japan's location the key to its fortune?

DOMESTIC FOUNDATIONS

It is tempting to turn immediately to Japan's external connections and to call on those factors of location to account for Japan's great industrial growth. After all, the Japanese built an empire, much of it right across the Korea Strait, and this empire contained many of the resources Japan lacked at home. But before we rush into such a deduction, let us consider what Japan's domestic economy was like in the mid-nineteenth century, because, as we will see, what Japan achieved was primarily based on its internal human— and to a large extent natural— resources.

In the 1860s, manufacturing— light manufacturing of the handicraft type—was already widespread in Japan. In cottage industries and community workshops, the Japanese produced silk and cotton-textile manufactures, porcelain, wood products, and metal goods. At that time, the modest metal ore deposits of the country were sufficient to supply these local industries. Power came from human arms and legs, and from wheels driven by water; the chief source of fuel was charcoal. Thus, there was an industrial tradition and an experienced labor force that possessed appropriate manufacturing skills.

The planners who took over the country's guidance after 1868 realized that this was an inadequate base for industrial modernization, but it might nevertheless generate some of the capital needed for this process. The community and home workshops were integrated into larger units, thermal and hydroelectric power began to be made available to replace more primitive sources of energy, and for the first time Japanese goods—still of the light manufacture variety—began to compete with Western products on the world's markets. Meanwhile, the Japanese continued to resist any infusion of Western capital, which might have accelerated the industrialization process but would have cost Japan its economic autonomy. Instead, farmers were more heavily taxed, and the money thus earned by the state was poured into the industrialization effort.

Now Japan's layout and topography contributed to the modernization process. Coal could be mined and, in the relatively small quantities then needed, transported almost wherever it was required. Throughout the country, a hydroelectric site was usually nearby, and electric power became available anywhere—in the cities as well as the populated countryside. Japan was still managing on what it possessed at home, and the factories and workshops multiplied. Soon the beginnings of a chemical industry emerged, and government subsidies led to the establishment of heavier industries. The period after 1890 was a time of great progress, stimulated by the wars against China (1894–1895) and Russia (1904–1905). Nothing served to emphasize the need for industrial diversification as much as war, and Japan's victories vindicated the ruthlessness with which industrial objectives were sometimes pursued.

Japan's war effort contributed to its improved world economic position as the era of imperialism ended. When the Meiji Restoration took place, Britain lay at the center of a global empire and the Europeanization of the world was in full swing; the United States was still a developing country. The wars of the twentieth century saw the Japanese defeat British and French forces in eastern Asia; the United States and its allies crushed Germany and Italy. Although Japan also came out of World War II a defeated and devastated country, the global situation had changed dramatically. The United States, Japan's trans-Pacific neighbor, had become the world's most powerful and wealthiest state, whereas distant Britain was fading, with its empire on the verge of disintegration. Japan's relative location—its location relative to the economic and political foci of the world—had changed, and therein lay much of Japan's new postwar opportunity.

JAPAN'S SPATIAL ORGANIZATION

Japan is not a very large country, but it was already quite populous when its modern economic revolution began: about 30 million people were crowded into its confined living space. In the 120-plus years since then, despite a habitable area limited to a meager 18 percent of its national territory, Japan's population has more than quadrupled and its productive complex has grown spectacularly. With 1 million people, Tokyo (then still named Edo) may already have been the world's largest city in the 1600s. Today, the capital is about nine times as large and is part of a much bigger urbanized area, which includes Yokohama as well. This is the world's largest urban agglomeration, now containing a population of 27.2 million, which exceeds the size of the second largest conurbation (Mexico City) by about 6 million and third-place São Paulo (Brazil) by over 8 million; fourth-place Seoul (South Korea) trails Tokyo by more than 10 million, and fifth-place metropolitan New York trails by nearly 13 million.

Given such enormous growth, space in Japan has become increasingly scarce, and urban and regional planners have shared with farmers the challenge of how to make every square mile count. Before the onrush of urbanization, it was often said that the most densely populated rural areas in Japan were more crowded than any other such areas in the world. Now the cities, towns, and villages must also vie for the country's limited livable terrain. When Japan's industries were still of the light handicraft variety, the myriad workshop-type establishments were located in the cities as well as in villages; the small-scale smelting of iron could also be carried on in many places (charcoal was widely available), as could the manufacturing of textiles.

Then, the age of the factory arrived, and the modern rules of industrial location went into effect. We have discussed these rules in earlier chapters, notably the advantages of processing bulky raw materials at the spot where they are mined to save movement costs and the transport of more easily moved lighter raw materials to areas containing heavier less-transferable resources. In Britain during the Industrial Revolution, this led to the rapid growth of certain towns into industrial cities, the decline of others, and the founding of still other manufacturing centers. In Japan, on the other hand, urban-industrial growth was straitjacketed. Apart from northern Kyushu's coal, no great reserves of raw materials were found, and it soon was clear that these would have to be imported from elsewhere. Internally, the distribution of coal was mainly by water; externally, iron ore and other required commodities, of course, came by sea. Hence, the coastal cities, where the labor forces were also concentrated, became the major centers of manufacturing activity.

The modern industrial growth of Japan was shaped by the economic development that was already taking place. Although the country did not have the resources to merit any major internal reorientation, some cities did possess advantages over others in terms of their existing facilities and their situation vis-à-vis the source areas of raw materials. Accordingly, there now began the kind of differentiation that has marked regional organization everywhere in the world, governed by the principle that Allen Philbrick called **areal functional organization**. Five interrelated ideas are involved.

First, human activity has spatial focus, in that it is concentrated in some locale—a farm or factory or store. Second, such "focal" activ-

ity is carried on in certain particular places. Obviously, no two establishments can occupy exactly the same spot on the earth's surface, so every one of them has a finite or absolute location; but what is more relevant is that every establishment has a location *relative* to other establishments and activities. Since no human activity is carried on in complete isolation, the third idea is that interconnections develop among the various establishments. Farmers send crops to markets and buy equipment at service centers. Mining companies buy gasoline from oil companies and lumber from sawmills, and they send ores to refineries. Thus, a system of interconnections emerges and grows more complex as human capacities

FIGURE J–3

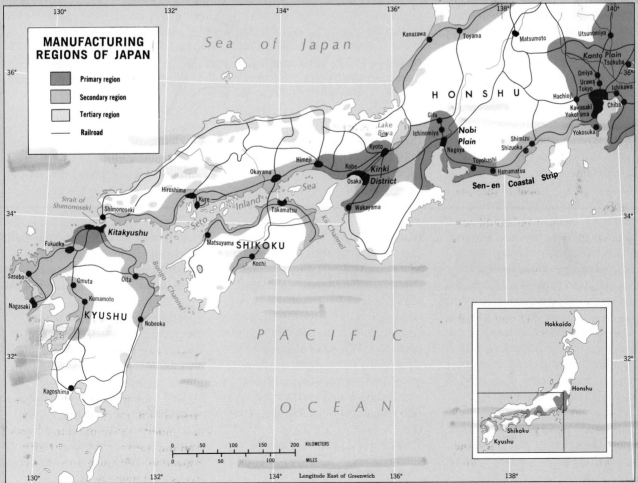

and demands expand, and they are expressed spatially as units of areal organization. Philbrick's fourth idea is that the evolution of these units of areal organization—or regions—is the product of human "creative imagination" as people apply their total cultural experience as well as technological know-how when they decide how to organize and rearrange their living space. Finally, it is possible to recognize *levels* of development in areal organization, a ranking or hierarchy based on type, extent, and intensity of exchange. To quote Philbrick: "The progression of [areal] units from individual establishments to world regions and the world as a whole are formed into a hierarchy of regions of human organization."

In the broadest sense, regions of human organization can be categorized under subsistence, transitional, and exchange types, with Japan's areal organization reflecting the last of these. But within each type of areal organization, especially within the complex exchange type of unit, individual places can also be ordered or ranked on the basis of the number and kinds of activities and interconnections they generate. A map of Japan showing its resources, urban settlements, and surface communications can tell us a great deal about the kind of economy the country has. It looks just like maps of other parts of the world where an exchange type of areal organization has developed—a hierarchy of urban centers ranging from the largest cities to the tiniest hamlets, a dense network of railways and roads connecting these places, and productive agricultural areas near and between the urban centers. In Japan's case, Fig. J–3 shows us something else: it reflects the country's external orientation, its dependence on foreign trade. All primary and secondary regions lie on the coast; of all

FIGURE J–4   The serenity of Shinto (the Japanese ethnic religion closely related to Buddhism) pervades this temple complex, which is part of the preserved landscape of Kyoto (left rear), the premodern capital.

the cities in the million-size class, only Kyoto (1.6 million) lies in the interior. If we were to deduce that Kyoto does not match Tokyo-Yokohama, Osaka-Kobe (11.8 million), or Nagoya (4.8 million), in terms of industrial development, that would be correct—the old capital remains a center of small-scale light manufacturing. Actually, Kyoto's ancient character has been deliberately preserved, and large-scale industries have been discouraged. With its old temples and shrines, its magnificent gardens, and its many workshop and cottage industries, Kyoto remains a link with Japan's pre-modern past (Fig. J–4).

As Fig. J–3 shows, Japan's dominant region of urbanization and industry (along with very productive agriculture) is the *Kanto Plain*, which is focused on the Tokyo-Yokohama metropolitan area (27.2 million). This gigantic cluster of cities and suburbs, interspersed with intensively cultivated farmlands, forms the eastern anchor of the country's elongated and fragmented core area (Fig. J–1). Besides its flatness, the Kanto Plain possesses other advantages: its fine natural harbor at Yokohama, its relatively mild and moist climate, and its central location with respect to the country as a whole. It has also benefited enormously from Tokyo's designation as the modern capital, which coincided with Japan's embarkation on its planned course of economic development. Many industries and businesses chose Tokyo as their headquarters in view of the advantages of proximity to the government's decision-makers (Fig. J–5).

The Tokyo-Yokohama conurbation has become Japan's leading manufacturing complex, producing about one-fifth of the country's annual output. But the raw materials for all this industry come from far away. For example, the Tokyo area is among the chief steel producers in Japan, using iron ores from the Philippines, Malaysia, Australia, India, and even Africa; most of the coal is imported from Australia and North America, and the petroleum from Southwest Asia and Indonesia. The Kanto Plain cannot produce nearly enough food for its massive population. Imports must come from Canada, the United States, and Australia as well as from other areas in Japan. Thus, Tokyo depends completely on its external trading ties for food, raw materials, and markets for its wide variety of products, which run the gamut from children's toys to high-precision optical equipment to the world's largest oceangoing ships.

The second-ranking economic region in Japan's core area is the Osaka-Kobe-Kyoto triangle—also known as the *Kinki District*—located at the eastern end of the Seto Inland Sea. The position of the Osaka-Kobe conurbation, with respect to the old Manchurian empire created by Japan, was advantageous. Situated at the head of the Seto Inland Sea, Osaka was the major Japanese base for the China trade and the exploitation of Manchuria (Manchukuo), but it suffered when the empire was destroyed and its trade connections with China were not regained after World War II. Kobe (like Yokohama, Japan's chief shipbuilding center) has remained one of the country's busiest ports, han-

dling inland-sea traffic as well as extensive overseas linkages (Fig. J–6). Among these trade connections for many decades was the importing of cotton for Osaka's textile factories. With its large skilled labor force and its long history of productive settlement, Osaka until recently was Japan's

first-ranked textile producer; today, textiles (mainly synthetics) form only part of this metropolitan area's huge and varied industrial base. Kyoto, as we noted, remains much as it was before Japan's great leap forward, a city of small workshop industries. Similar to the Kanto Plain, the Kinki District

FIGURE J–5   The heart of Tokyo, high-density focus of the world's largest urban agglomeration.

FIGURE J–6 The bustling port of Kobe on the Seto Inland Sea, which serves an endless stream of domestic and international maritime traffic. Huge cranes load and unload ships in a few hours, an activity that goes on around the clock—symbolizing Japan's productive energy.

is also an important farming area. Rice, of course, is the most intensively grown crop in the warm moist lowlands, but agriculture here is less widespread than on the Kanto Plain. Thus, we observe another huge concentration of people that requires a large infusion of foodstuffs every year.

The Kanto Plain and Kinki District, as Fig. J–3 indicates, are the two leading primary regions within Japan's core. But between them lies the *Nobi Plain*, focused on the industrial metropolis of Nagoya, the city that not long ago ousted Osaka from first place among Japan's textile producers. The map indicates some of the Nagoya area's advantages and liabilities. The Nobi Plain is larger than the lowlands of the Kinki District; thus its agricultural productivity is greater, although not as great as that of the Kanto Plain. But Nagoya has neither Tokyo's centrality nor Osaka's position on the Seto Inland Sea: its connections to Tokyo are via the tenuous Sen-en Coastal Strip. Its westward con-

nections are somewhat better, and there are signs that the Nagoya area is coalescing with the Osaka-Kobe conurbation. Still another disadvantage of Nagoya lies in the quality of its port, which is not nearly as good as Tokyo's Yokohama or Osaka's Kobe, and has been plagued by silting problems.

Westward from the three regions just discussed—which together comprise what is often called the *Tokaido* megalopolis—extends the Seto Inland Sea, along whose shores the remainder of Japan's core area is continuing to develop. The most impressive growth has occurred around the western entry to the Inland Sea (the Strait of Shimonoseki), where *Kitakyushu*—a conurbation of five cities on northern Kyushu—constitutes the fourth Japanese manufacturing complex and primary economic region. Honshu and Kyushu are connected by road and railway tunnels, but the northern Kyushu area does not have an urban-industrial equivalent on the Honshu side of the strait. The Kita-

kyushu conurbation (1.8 million) includes Yawata, site of northwest Kyushu's (rapidly declining) coal mines, and it was on the basis of this coal that the first steel plant in Japan was built there; for many years it was Japan's largest. The advantages of transportation here at the western end of the Seto Inland Sea are obvious: no place in Japan is better situated to do business with Korea and China. As relations with mainland Asia expand, this area will reap many of the benefits. Elsewhere on the inland-sea coast, the Hiroshima-Kure urban area (1.5 million) has a manufacturing base that includes heavy industry. And on the coast of the Korea Strait, Fukuoka (1.9 million) and Nagasaki (550,000) are the principal centers, the former an industrial town, the latter a center of large shipyards.

Only one Japanese manufacturing complex in the area covered by Fig. J–3 lies outside the belt extending from Tokyo in the east to Kitakyushu in the west—the secondary region centered on Toyama (380,000) on the Sea of Japan. The advantage here is cheap power from nearby hydroelectric stations, and the cluster of industries reflects it: paper manufacturing, chemical factories, and textile plants have located here. Of course, our map also gives an inadequate picture of the variety and range of industries that exist throughout Japan, many of them oriented to local (and not insignificant) markets. Thousands of manufacturing plants operate in cities and towns other than those shown on the map, even on the cold northern island of Hokkaido, now connected to Honshu by the Seikan rail tunnel, the world's longest, beneath the Tsugaru Strait.

The map of Japan's areal organization shows the country's core area to be dominated by four primary regions, each of them primary because they duplicate to some degree the contents of the others. Each contains iron and steel plants, each is served by one major port, and each lies in or near a large, productive farming area. What the map does not show is that each also has its own external connections for the overseas acquisition of raw materials and the sale of finished products. These linkages may even be stronger than those among the four internal regions of the core area; only in the case of Kyushu and its coal have domestic raw materials played much of a role in shaping the nature and location of heavy manufacturing, and these resources are all but depleted today. In the structuring of the country's areal functional organization, therefore, more than just the contents of Japan itself is involved. In this respect, Japan is not unique: all countries that have exchange-type organizations must, to some degree, adjust their spatial forms and functions to the external interconnections required for progress.

But it would be difficult to find a country in which this is truer than in Japan.

FOOD PRODUCTION

The rapid modernization of the Japanese economy over the past half-century is reflected in its changing occupational structure. A study by Chauncy Harris has shown that in 1920, prior to the urban-industrial transformation, 55 percent of Japan's workers were employed in the primary sector, 21 percent in manufacturing, and 24 percent in the tertiary sector; his findings for 1980 revealed a complete reversal, with only 11 percent remaining as primary workers, 34 percent engaged in industry, and 55 percent employed in the postindustrial tertiary-quaternary-quinary sectors. The most remarkable changes have occurred in agricultural employment: whereas 51 percent of the labor force were farmers in 1920, the proportion that makes possible today's far more productive domestic agriculture is less than 8 percent—with over 85 percent of them part-timers who derive much of their income from nonagricultural sources. Let us take a closer look.

Japan's economic modernization so occupies the center stage that it is easy to forget that considerable achievements have been made in the venerable field of agriculture. Japan's planners, no less interested today in closing the food gap than in expanding industries, have created extensive networks of experiment stations to promote mechanization, optimal seed selection and fertilizer use, and information services to distribute to farmers as rapidly as possible knowledge that is useful for enhancing crop yields. Although this program has been very successful, Japan faces the unalterable reality of its stubborn topography: there simply is insufficient land to farm.

Less than one-fifth of the country's total area is in cultivation; although there may still be some land that can be brought into production, it is so mountainous or cold that its contribution to the total harvest would be minimal anyway. Japanese agriculture may resemble Asian agriculture in general, but nowhere else do so many people depend on so little land (Fig. J–7). Japan's overall population density is 854 per square mile (330 per square kilometer) to begin with; but when the *arable* land alone is considered, it turns out that 6035 people depend on the average square mile of farmland (2330 per square kilometer). This measure of persons per unit of cultivable land, as we noted on p. 61, is the **physiologic density**, and Japan's is one of the highest on earth. Even Bangladesh (2735 per square mile [1055 per square kilometer]) and Egypt (2300 [890]), notoriously crowded in confined agricultural areas, exhibit far lower physiologic densities than Japan.

Japanese agriculture stands apart from farming elsewhere in rice-growing Asia: it is by far the most efficient, given existing conditions of slope, soil, and climate. Rice yields in Japan are among the highest in the world, and researchers constantly find new ways to improve this productivity; not long ago, for instance, they developed new, better hybrid varieties of rice that mature so quickly that it is now possible to raise two crops every year, even in cooler climates. Japan no longer enjoys its one-time luxury of self-sufficiency in food, but it is remarkable that the country is somehow able to

FIGURE J–7 To augment Japan's meager land resources, farmers have long been required to cultivate steep slopes. This aerial view near the Seto Inland Sea shows the intricate terracing of hillsides that is necessary to create level surfaces and safeguard against ruinous soil erosion.

successful that Japan today is not only self-sufficient in rice production, but the government usually promotes the planting of less rice to avoid large surpluses. As a result, more agricultural space can be devoted to Japan's second and third crops, usually wheat and barley. Wheat is used as a winter crop in rotation with rice in the warmer parts of the country, and barley is grown in the north as a summer crop. Sweet and white potatoes, as the major root crops, occupy less than 10 percent of the farmland. Even less space (much of it on hilly slopes) is given over to cash crops such as tea, grown for local consumption, and mulberry, the plant whose leaves are used to feed the silkworms—the basis of Japan's once-rich silk industry. There are also vegetable gardens that cling to the outskirts of large cities, a land use increasingly threatened by urban encroachment. But even with every usable space in agricultural production, Japan must still rely on foreign food sources, yearly importing substantial quantities of wheat, corn, sugar, and soybeans.

With their national diet so rich in such starchy foods as rice, wheat, barley, and potatoes, the Japanese need protein to balance it. Fortunately, this can be secured in sufficient quantities, not through foreign purchase, but by harvesting it from rich fishing grounds near the Japanese islands. With customary thoroughness, the Japanese have developed a fishing industry that is larger than that of the United States or any of the long-time fishing nations of northwestern Europe, and it now supplies the domestic market with a second staple after rice. Although mention of the Japanese fishing industry brings to mind a fleet of ships scouring the oceans and seas far from Japan, the fact is that most of this huge catch (about one-seventh of the world's annual

produce about 70 percent of its annual needs. This is achieved through a number of parallel methods: devoting more than 90 percent of all farmland to food crops, mechanizing every possible operation, irrigating more than half of Japan's agriculture, painstakingly terracing steep slopes, practicing

multiple cropping (now greatly enhanced by the newest hybrid rice), intercropping, intensive fertilization, and transplanting—that is, the use of seedbeds to raise the young plants of the next crop while its predecessor matures in the field.

These methods have been so

total) comes from waters within a few dozen miles of Japan itself. Where the warm Kuroshio and Tsushima currents meet colder water off Japan's coasts, a rich fishing ground exists that yields sardines, herring, tuna, and mackerel in the warmer waters and cod, halibut, and salmon in the seas to the north. Along Japan's coasts there are about 4000 fishing villages, and tens of thousands of small boats ply the waters offshore to bring home catches that are distributed both locally and in city markets (Fig. J–8). The Japanese also practice *aquaculture*—the "farming" of freshwater fish in artificial ponds and flooded paddy (rice) fields, seaweeds in home aquariums, and oysters, prawns, and shrimp in shallow bays; they are even experimenting with the cultivation of algae for their food potential.

FIGURE J–8 One of the hundreds of prosperous fishing villages that line the shores of Shikoku.

JAPAN IN THE POSTINDUSTRIAL ERA

Japan has now become one of the world's most advanced nations. Its population of 124.4 million is as heavily urbanized as that of Western Europe and North America— 77 percent, and still climbing—a complete reversal since 1920 when only 18 percent of the Japanese lived in cities and towns. Although Western-style prosperity was a bit late in coming, the past two decades have witnessed a significant narrowing of the affluence gap between Japan and the United States; since 1970, the proportion of Japanese families owning automobiles has risen from 17 percent to nearly 80 percent, those owning air conditioners from 6 to about 50 percent, and those owning color televisions from 25 to more than 99 percent.

As Japan's advancing economy shifts ever more steadily away from primary and secondary activities, there is little doubt that it is fast approaching a full-scale North-American-style **postindustrial revolution**. The rise of an ultramodern, information-based economy is evident throughout the country: ubiquitous computerization, rapid automation of the work place, the establishment of new communications networks (including the world's largest fiber-optic cable system), and a major research program to develop ever more sophisticated computer technology. The new thrust into research and development is best symbolized by the construction of a new community in Tokyo's outer suburbs called Tsukuba Science City (Fig. J–9). This 70,000-acre (28,000-hectare) complex, reminiscent of North Carolina's Research Triangle Park (*box* p. 215), is built around two new major university campuses and more than 50 scientific institutes and government research agencies that have relocated from central Tokyo, 40 miles (65 kilometers) to the southwest. More than 12,000 scientists, engineers, and technicians work here, performing experiments in every field from high-energy physics to earthquake simulation. Tsukuba is well on its way to becoming a self-sufficient urban center of 250,000 residents, and the government is proceeding with its ambitious "Technopolis Plan" to build another 18 new high-technology cities across Japan to serve as the engines of economic growth in the next century.

If past performance is any guide to the future, Japan can be expected to share (and even aggressively promote) the new products of its technological breakthroughs with the rest of the world. Yet, despite its vigorous marketing and exporting efforts, Japan will probably not substantially increase its trading activities. In fact, the overall foreign interaction of the Japanese is often exaggerated by outside observers because the country's economy is not all that heavily dependent on sales in the international marketplace. For example, Japan in 1986 exported just under 8 percent of its gross national product (GNP), with only the United States exhibiting a lower proportion (5.6 percent) among the most developed nations. Included among the far more active exporters were West Germany (27 percent), Canada (22 percent), and the United Kingdom (19 percent). What is most remarkable about Japan's foreign sales activity is its success, which is largely based on a superior reputation for product quality and fair pricing—the result of survival in a hypercompetitive domestic market that is one of the toughest on earth.

The selective new trading ties that the Japanese will encourage in the future are the ones that will most enhance their economic and political stability. With its heavy reliance on imported oil, Japan can be expected to maintain cordial re-

lations with the suppliers of the Persian Gulf region. Future trade with the United States (by far Japan's leading trading partner) will intensify as well, but more heavily on the importing side. Not only do the Americans supply such critical raw materials as cotton and scrap metal, but as Japanese consumerism continues to grow in an increasingly wealthy domestic market, only the U.S. has the range and surplus quantity of products to supplement local production if these widening demands are to be satisfied without long delays.

Closer ties with the communist world may also be in the offing. The immense market of nearby China does offer an enormous opportunity, but the normalization of trade relations since 1972 has progressed at a leisurely pace. The

Soviet Union also approached Japan in the 1980s to develop joint projects in eastern Siberia. Although some energy and transportation ventures were initiated, a major obstacle to wider cooperation remains the politico-geographical situation wherein Japan continues to oppose Soviet control over two former Japanese islands located between Hokkaido and the Kurile archipelago to the northeast.

Perhaps the biggest adjustment required of the Japanese in their dawning postindustrial age will be in the way they view themselves. Despite its enormous postwar advances, Japan retains an understandable inferiority complex. After all, this was until quite recently a tradition-bound nation of farmers (many living at the edge of

FIGURE J–9 A panoramic view of Tsukuba, located on the outer edge of metropolitan Tokyo. This showcase high-technology city, devoted to scientific research and development, was the site of Japan's last world's fair (Expo-85).

FIGURE J–10 Mimicry overrides innovation as a cultural force: Japan's Disneyland just outside Tokyo, which opened in 1983.

their unkind habitat produced a culture that has coped with the problem of numbers by emphasizing rigid adaptation, consensus, and miniaturization (even today there is little sense that rising national wealth should bring greater personal entitlement). Therefore, as contacts with the West were constantly strengthened after 1850, the Japanese have overwhelmingly preferred to copy new ideas rather than undertake their innovation. The opening of their own Disneyland in an eastern Tokyo suburb in 1983—which not only recreates the ''Magic Kingdom'' theme park, but also replicates a little piece of the United States itself by banning Japanese signs and refreshments—exquisitely demonstrates that mimicry remains a cultural force to be reckoned with (Fig. J–10).

But postindustrial Japan in the 1990s confronts a wholly new reality. This transforming nation has reached its pinnacle of success and affluence by copying and improving the inventions of others, and it has now achieved the unintended position of world leader in many endeavors. Further success will depend on the seizing of this mantle of leadership in the near future; thus far, most Japanese have demonstrated a decided reluctance to do so.

poverty) that became a prosperous urban society practically overnight. Thus, the Japanese are bewildered that outsiders should see them as an economic threat to other nations, because their self-perception is one of vulnerability based on their poor-quality land, meager natural resource base, and a continuing dependency on foreign countries for vital food and fuel supplies. In fact, the vast majority of the Japanese do not live like the affluent people they are. Housing is dominated by tiny uncomfortable homes, and the government as recently as 1985 reported that 51 percent of the population resided in quarters smaller than the national average standard (even newly-built dwellings averaged 36 percent less space than their U.S. counterparts); moreover, such basics as central heating and sewage systems are still considered luxuries.

Japan's cultural traditions, which also reflect these shortcomings, reinforce this view. The crowding engendered for centuries by

PRONUNCIATION GUIDE

Chosen (*choe-sen*)
Edo (*edd*-oh)
Fukuoka (foo-kuh-*woh*-kuh)
Hiroshima (hirra-*shee*-muh/huh-*roh*-shuh-muh)
Hokkaido (hah-*kye*-doh)
Honshu (*honn*-shoo)
Kanto (*kan*-toh)
Kinki (kin-*kee*)

Kitakyushu (kee-*tah-kyoo*-shoo)
Kobe (*koh*-bay)
Korea (*career*)
Kure (*kooh*-ray)
Kurile (*cure*-reel)
Kuroshio (koor-oh-*shee*-oh)
Kyoto (kee-*yoh*-toh)
Kyushu (kee-*yoo*-shoo)
Manchukuo (mahn-*joh*-kwoh)

Manchuria (man-*choory*-uh)
Meiji (may-*ee*-jee)
Nagasaki (nah-guh-*sockee*)
Nagoya (nuh-*goya*)
Nobi (*noh*-bee)
Osaka (oh-*sah*-kuh)
Ryukyu (ree-*yoo*-kyoo)
Sakhalin (*sock*-uh-leen)
Seikan (say-*kahn*)

Sen-en (sen-*nenn*)
Seto (*set*-oh)
Shikoku (shick-*koh*-koo)
Shimonoseki (shim-uh-noh-*seckee*)

Taiwan (tye-*wahn*)
Tokaido (toh-*kye*-doh)
Tokyo (*toh*-kee-oh)
Toyama (toh-*yah*-muh)
Tsugaru (tsoo-*gah*-roo)

Tsukuba (tsoo-*koob*-uh)
Tsushima (tsoo-*shee*-muh)
Yawata (yuh-*wah*-tuh)
Yokohama (yo-kuh-*hah*-muh)

REFERENCES AND FURTHER READINGS

(BULLETS [●] DENOTE BASIC INTRO-DUCTORY WORKS)

Allinson, Gary D. "Japanese Urban Society and Its Cultural Context," in Agnew, John A. et al., eds., *The City in Cultural Context* (Winchester, Mass.: Allen & Unwin, 1984), pp. 163–185.

Association of Japanese Geographers. *Geography of Japan* (Tokyo: Teikoku-Shoin, 1980).

Beasley, W.G. *The Rise of Modern Japan* (New York: St. Martin's Press, 1990).

● Burks, Ardath W. *Japan: A Postindustrial Power* (Boulder, Colo.: Westview Press, 2 rev. ed., 1984).

Christopher, Robert C. *The Japanese Mind: The Goliath Explained* (New York: Linden Press/Simon & Schuster, 1983).

Cybriwsky, Roman A. "Shibuya Center, Tokyo," *Geographical Review*, 78 (January 1988): 48–61.

Dempster, Prue. *Japan Advances: A Geographical Study* (New York: Barnes & Noble, 1967).

Emmott, Bill. *The Sun Also Sets: The Limits of Japan's Economic Power* (New York: Times Books, 1989).

Eyre, John D. *Nagoya: The Changing Geography of a Japanese Regional Metropolis* (Chapel Hill: University of North Carolina, Studies in Geography, No. 17, 1982).

● Hall, Peter. "Tokyo," in *The World Cities* (New York: St. Martin's Press, 3 rev. ed., 1984), pp. 179–197.

Hall, Robert B. *Japan: Industrial Power of Asia* (New York: Van Nostrand Reinhold, 2 rev. ed.,1976).

Hane, Mikiso. *Modern Japan: A Historical Survey* (Boulder, Colo.: Westview Press, 1986).

● Harris, Chauncy D. "The Urban and Industrial Transformation of Japan," *Geographical Review*, 72 (January 1982): 50–89.

Hendry, Joy. *Understanding Japanese Society* (London & New York: Routledge, 1988).

Ishinomori, Shotaro. *Japan, Inc.: Introduction to Japanese Economics* (Berkeley: University of California Press, 1988).

Ito, T. & Nagashima, C. "Tokaido—Megalopolis of Japan," *GeoJournal*, 4 (1980): 231–246.

● "Japan: A Nation in Search of Itself," *Time*, Special Issue, August 1, 1983.

● *Japan: A Regional Geography of an Island Nation* (Tokyo: Teikoku-Shoin, 1985).

● Kornhauser, David H. *Japan: Geographical Background to Urban-Industrial Development* (London & New York: Longman, 2 rev. ed., 1982).

Kornhauser, David H. "A Selected List of Writings on Japan Pertinent to Geography in Western Languages With Emphasis on the Work of Japan Specialists," University of Hiroshima, 1979.

● MacDonald, Donald. *A Geography of Modern Japan* (Ashford, U.K.: Paul Norbury, 1985).

Murata, Kiyogi & Ota, Isamu, eds. *An Industrial Geography of Japan* (New York: St. Martin's Press, 1980).

Pezeu-Massabuau, Jacques. *The Japanese Islands: A Physical and Social Geography* (Rutland, Vt.: Charles E. Tuttle, trans. Paul C. Blum, 1978).

Philbrick, Allen K. "Principles of Areal Functional Organization in Regional Human Geography," *Economic Geography*, 33 (1957): 299–336.

Pollack, Andrew. "Japan's Science Showplace [Tsukuba]: Expo '85 Set at New City," *New York Times*, September 11, 1984, pp. 29, 31.

Popham, Peter. *Tokyo: The City at the End of the World* (Tokyo: Kodansha International Ltd., 1985).

● Reischauer, Edwin O. *The Japanese Today: Change and Continuity* (Cambridge, Mass.: Belknap/Harvard University Press, 1988).

Seidensticker, Edward. *Tokyo Rising: The City Since the Great Earthquake* (New York: Alfred A. Knopf, 1990).

Smith, Michael. "Japan," in Smith, Michael et al., *Asia's New Industrial World* (London & New York: Methuen, 1985), pp. 4–36.

Tatsuno, Sheridan. *The Technopolis Strategy: Japan, High Technology, and the Control of the Twenty-First Century* (New York: Prentice-Hall Press, 1986).

● Trewartha, Glenn T. *Japan: A Geography* (Madison: University of Wisconsin Press, 2 rev. ed., 1965).

PART TWO

UNDERDEVELOPED REALMS

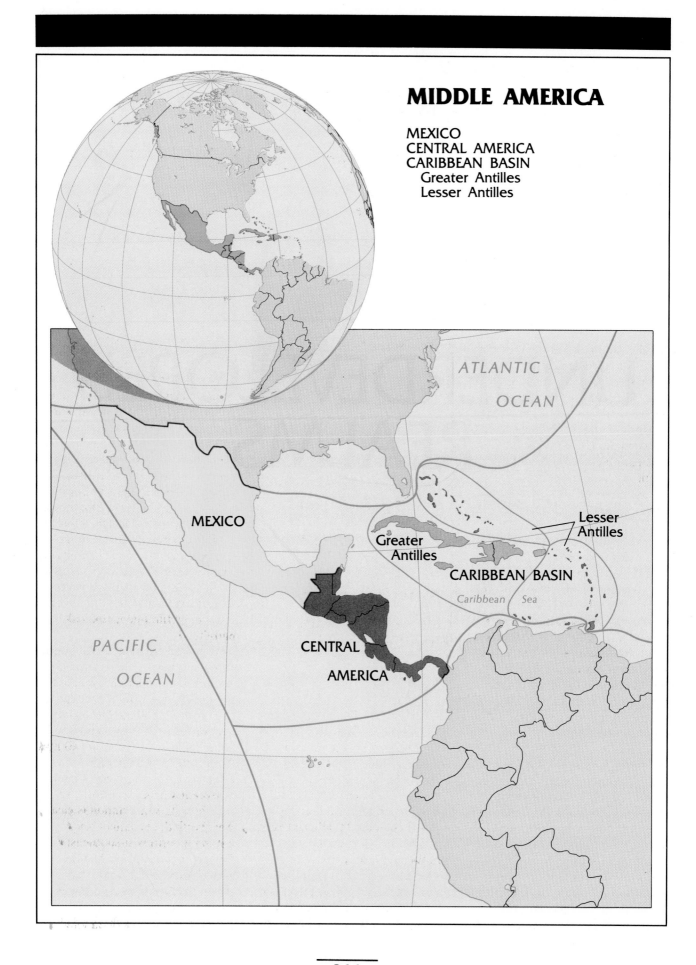

MIDDLE AMERICA

MEXICO
CENTRAL AMERICA
CARIBBEAN BASIN
    Greater Antilles
    Lesser Antilles

ATLANTIC
OCEAN

MEXICO

Lesser
Antilles

Greater
Antilles

CARIBBEAN BASIN

Caribbean Sea

PACIFIC
OCEAN

CENTRAL
AMERICA

MIDDLE AMERICA:
COLLISION OF CULTURES

Middle America is a realm of vivid contrasts, turbulent history, political turmoil, and an uncertain future. The realm's diversity is reflected by the map in Fig. 4–1. It consists of all the lands and islands between the United States to the north and South America to the south. This includes the substantial landmass of Mexico, the narrowing strip of land extending from Guatemala to Panama, and the many large and small islands of the Caribbean Sea to the east. Middle America is a realm of soaring volcanoes and forested plains, of mountainous islands and flat coral cays. Moist tropical winds sweep in from the sea, watering windward coasts while leaving leeward areas dry. Soils vary from fertile volcanic to desert-barren. Spectacular scenery abounds, and tourism is one of the realm's leading industries.

Culturally, Middle America's contrasts are stark. Teeming, poverty-afflicted cities grow ever larger. People in search of a better existence cross into the United States by the hundreds of thousands, transforming the cultural geography of states from California to Texas as well as South Florida (see Fig. 3–19). Northern South America, too, carries the imprints of Caribbean cultures and

IDEAS AND CONCEPTS
land bridge
culture hearth (2)
historical geography
cultural landscape (2)
environmental determinism
mainland-rimland framework
plural society (2)
transculturation
altitudinal zones
tropical deforestation
insurgent state

economies (see Fig. 5–8); those imprints range from Afro-Caribbean to Hispanic-American and include Indian, European, and even Asian elements.

Is Middle America a discrete geographic realm? Some geographers combine Middle and South America into a realm they call "Latin" America, citing the dominant Iberian (Spanish and Portuguese) heritage and the prevalence of the Roman Catholic religion. But these criteria apply far more

strongly to South America. In Middle America, large populations exhibit African and Asian as well as European ancestries. Nowhere in South America has Indian culture contributed to modern civilization as strongly as it has in Mexico. The Caribbean area is a patchwork of independent states, territories in political transition, and dependencies. The Dominican Republic speaks Spanish, adjacent Haiti uses French; Dutch is spoken in Curaçao and on neighboring islands, while English is spoken in Jamaica. Middle America, therefore, gives vivid definition to concepts of cultural-geographical pluralism.

Compared to continental South America, the Middle American realm is divided and fragmented. Even its funnel-shaped mainland, a 3800-mile (6000-kilometer) connection between North and South America, narrows to a slim 40-mile (65-kilometer) ribbon of land in Panama. Here, this strip of land, or *isthmus*, bends eastward, so that Panama's orientation is east-west (with the Panama Canal cutting it northwest-southeast). On the map, mainland Middle America looks like a bridge between the Americas, and this is exactly what physical geographers call such an isthmian link:

a **land bridge**. North and South America were not always, in their geological history, connected this way. Land bridges form, exist for a time, and then disappear, either through geologic processes or because sea levels rise and submerge them.

If you examine a globe, you can see other present and former land bridges: the Sinai Peninsula between Asia and Africa, the (now-broken) Bering land bridge between northeasternmost Asia and Alaska, and the former connection between the island chain off Southeast Asia and Australia. Such land bridges have played crucial roles in the dispersal of animals and humans across the globe. Many scholars believe that the Bering land bridge was the gateway for the first humans to enter the Americas. Migration routes into South America probably lay along the Middle American land bridge, and if dense forests made that passage difficult, the coast afforded a frame of reference for the ancient migrants. Even though mainland Middle America forms a land bridge, its fragmentation inhibits movement to this day. Mountain ranges, volcanoes, swampy coastlands, and dense rainforests make contact and interaction difficult, especially where they all combine in much of Central America (see *box*, p. 254).

The islands of the Caribbean Sea stretch in a vast arc from Cuba to Trinidad, with numerous outliers outside (such as Barbados) and inside (for example, the Cayman Islands) the main chain. The large islands (Cuba, Hispaniola, Puerto Rico, Jamaica) are called the *Greater Antilles;* the smaller islands are referred to as the *Lesser Antilles.* Again, the map gives a hint of the physiography: the island arc consists of the crests and tops of mountain chains that rise from the floor of the Caribbean. Some of these crests are rel-

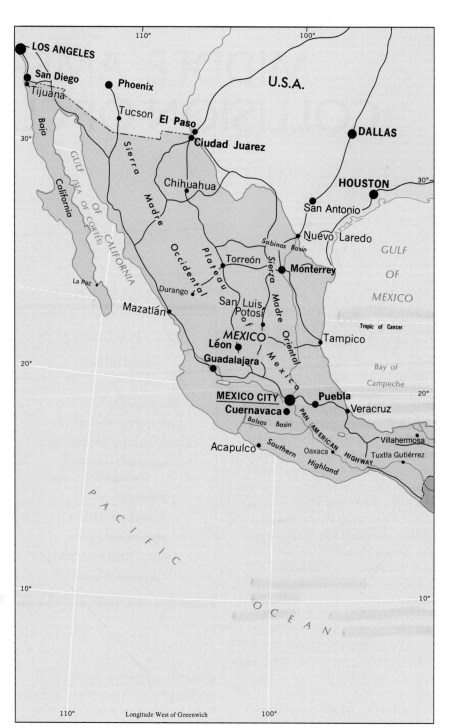

FIGURE 4–1

atively stable, but elsewhere they consist of active volcanoes. Almost everywhere, earthquakes are a hazard; in the islands as well as on the mainland, the crust is unstable where the Caribbean, Cocos, and North American Plates come together (Fig. I–6). Add to this the seasonal risk of hurricanes spawned and nurtured by the realm's abundant tropical waters, and the Middle American environment ranks among the world's most difficult and dangerous.

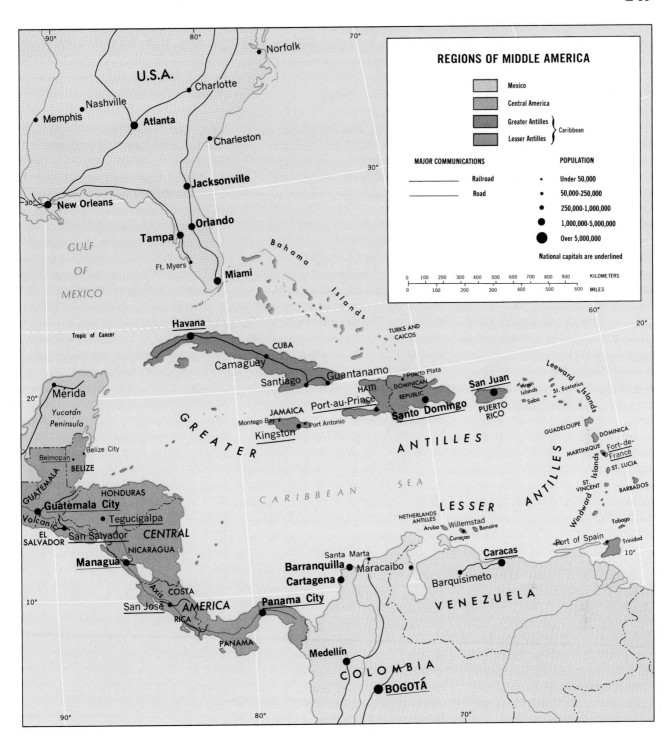

LEGACY OF MESOAMERICA

Mainland Middle America was the scene of the emergence of a major ancient Indian civilization. Here lay one of the world's true **culture hearths** (see Fig. 6–3, p. 343), an area where population could increase and make significant advancements. Agricultural specialization, urbanization, and transport networks developed, and writing, science, art, architecture, religion, and other spheres of achievement saw major progress. Anthropologists refer to the Middle American culture hearth as *Mesoamerica,* and what is espe-

S Y S T E M A T I C   E S S A Y

HISTORICAL GEOGRAPHY

Even though our survey of world realms is only one-third complete, it has become evident that the geographical present is a product of the past. This is just as true of underdeveloped realms as it was in the developed world regions already covered. The study of how a region evolves and continues to change is the domain of **historical geography**. Because it treats the all-encompassing dimension of time, the concepts and methods of historical geography—like those of regional geography—are applicable to every branch of the discipline. Historical geography, however, is clearly distinct from the field of history: it is dominated by the work of geographers whose main task is the interpretation of spatial change on the earth's surface.

The rich historical dimension of regional geography is indispensable to an understanding of contemporary landscapes and societies. For example, the shifting distribution of economic activities inside U.S. metropolitan areas, as we saw in Chapter 3, is a response to changing geographic forces as the ending industrial age gives way to its postindustrial successor. Those dynamics remind us again of the central role of *spatial processes*, the causal forces that act and unfold over time to shape the spatial distributions we observe today. In turn, these geographical patterns may be regarded as individual frames in a relentlessly advancing film. Thus, spatial organization at any given moment of time represents a point in a long process and is clearly related to what went before (as well as to what will develop in the future).

Emphasizing this central concern, historical geographers today approach their diverse subdiscipline from a number of different (although not mutually exclusive) perspectives.

The first of these is the study of *the geographic past*, wherein earlier spatial patterns are reconstructed and compared for time periods that were crucial in the formation of the present regional structure. The raw materials for assembling these snapshots of the past come from many sources: historical maps, books, censuses, archives, and interviews of long-time residents of the study area. This approach can also be used to build evolutionary models of spatial organization: the four-stage model of U.S. intraurban growth (Fig. 3–13) is a good example, with the structural pattern of each era shaped by its prevailing transportation technology.

The second perspective focuses on *landscape evolution*, the historical analysis and interpretation of existing **cultural landscapes**. This search for "the past in the present" takes a number of paths. The study of individual relics, such as Hadrian's Wall (Fig. I–18), is one of them. Another is comparative regional analysis of landscape artifacts such as house types, which facilitates the mapping of cultural change.

A third approach to historical geography can be called *perception of the past*, the application of modern behavioral perspectives to recreate and study the attitudes of previous generations toward their environments. Migration decision-making is of particular interest where it involved the acquisition

of accurate spatial information on the "mental maps" of potential migrants. For instance, before the U.S. Civil War, the Great Plains area was widely perceived as part of a "Great American Desert," a myth that helped delay its settlement until the late nineteenth century. Abstractions are valuable, too; a recent example is Donald Meinig's analysis of symbolic historical landscapes, from which he derived three idealized American community types that still exert a powerful hold on residential preferences throughout the United States—the New England village, the small-town Main Street, and the California horticultural suburb. More formal modeling approaches exist as well, such as the *time-geography* perspective developed by the Swedish geographer Torsten Hägerstrand, in which allocations of time for performing tasks in space are measured and regionally analyzed.

One of the most active areas of historical-geographical research—which has effectively integrated various approaches to the subject—is the study of *settlement geography*, the facilities people build while occupying a region. Because these facilities usually survive beyond the era of their original functions, they often provide some of the clearest expressions of the past in the contemporary landscape. The colonial landscapes of the New World have received considerable attention, especially the layout of Spanish colonial towns in Middle America.

The layout of this urban-settlement type is shown in Fig. 4–2, and its historical-geographical significance is described by Charles Sargent:

The morphology, or form, of any city reflects its past, and the clearest reflections are found in the street pattern, the size of city blocks, the dimensions of urban lots, and the surviving colonial architecture. In the typical Spanish American city, this colonial past is today on view in what is now the city core, commonly laid out in a gridiron of square or rectangular blocks subdivided into long, narrow lots along equally narrow streets.

This particular spatial pattern hardly came about by accident—it was decreed by royal regulations down to the smallest details of street, building, and central-plaza construction—and was repeated throughout Hispanic America (including former Spanish-controlled areas of the southwestern U.S.

such as New Mexico). Since most Spaniards were urban dwellers, it was hardly surprising that they chose to build towns in their colonial territories. But why this particular town form? The answer rests with that other major quality of an urban settlement—*function.*

The Spanish colonial town possessed several functions that were best suited to the compact, gridiron-street-pattern layout. It is often said that the threefold aims of the rulers of New Spain were ''God, glory, and gold.'' Because the main function of the settlement was administrative (a point developed in this chapter), town sites were chosen to maximize accessibility to regional trade routes and sources of tribute from local Indians. This control function also extended to the internal structure of the town: everything was tightly

focused on the central *plaza*, or market square, under the watchful eye of government authorities in adjacent buildings (*G* symbols in Fig. 4–2). Another leading town function was expressed in the central role of the church, which always faced the plaza as well (Fig. 4–2). The Roman Catholic Church sought to convert as many Indians to its faith as possible, and the easiest way to do this was to resettle the dispersed aboriginal population forcibly in Spanish towns, where the collection of tribute, the recruitment of mine workers, and the farming of land surrounding the town were also facilitated. Because of the gridiron street plan, any insurrections by the resettled Indians could be contained by having a small military force seal off the affected blocks and root out the troublemakers. The Greeks and Romans learned this lesson when they established far-flung empires, and the grid-plan tradition was passed down to their Mediterranean European successors. This idea even has modern applications: the battle for Saigon (a non-gridiron city) in the 1960s and 1970s favored guerrillas who could move at will through twisting streets and alleyways, whereas the inner-city riots of 1967 in gridded Newark and Detroit were quickly squelched by the U.S. Army, which systematically surrounded and pacified block after block.

Historical geography will continue to be a prominent theme in upcoming chapters. Two additional important concepts that involve the meshing of time and space will also be discussed: *spatial diffusion* in Chapter 6 (pp. 348–351) and *sequent occupance* in Chapter 7 (p. 427).

FIGURE 4–2

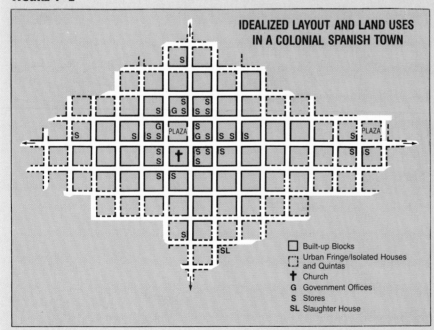

IDEALIZED LAYOUT AND LAND USES IN A COLONIAL SPANISH TOWN

□ Built-up Blocks
┌─┐ Urban Fringe/Isolated Houses
└─┘ and Quintas
† Church
G Government Offices
S Stores
SL Slaughter House

cially remarkable about this development is that it happened in very different geographic environments. In the low-lying tropical plains of Honduras, Guatemala, Belize, and Mexico's Yucatán Peninsula, the Maya civilization arose; on the high plateau of present-day Mexico, the Aztecs founded their well-organized civilization. In the process, the Maya and the Aztecs overcame some serious environmental obstacles. Mayan Yucatán may not have been as hot and humid as it is today, but the integration of so large an area was a huge accomplishment; the Aztecs also solved problems of distance and managed to unify people over a wide area, despite the topographic barriers of the interior upland (Fig. 4–3).

In the 1910s and 1920s, a school of thought developed in American geography that favored the view that human cultural and economic progress could only occur under certain environmental conditions. Pointing to the present concentration of wealth and power in the middle latitudes of the Northern Hemisphere, these geographers postulated that the tropics were simply not conducive to human productivity. Mostly, they based their argument on the role of climate. The monotonous heat and humidity of tropical climates were supposed to retard cultural progress. Mid-latitude climates, on the other hand, with their variable weather, were alleged to stimulate human achievement. This school of **environmental determinism** (or simply, *environmentalism*) held that the natural environment to a large extent dictated the course of civilization. The leading proponent and popularizer of this hypothesis was Ellsworth Huntington (1876–1947), who believed that human progress rested on three bases: climate, heredity, and culture. Not surprisingly, his controversial conclusions came under attack as ill founded and supportive of "master race" philosophies. Although he may have generalized carelessly, Huntington did pose crucial questions that are still unanswered today, and now sociobiologists, approaching them from another angle, face similar obstacles. Human societies and natural environments interact, but how? When does a combination of particular environmental circumstances, inherited capacities, and cultural transmissions stimulate a new cultural explosion?

The Lowland Maya

Certainly, few environmentalists could easily account for the rise of Maya civilization in the lowland tropics of Middle America. Some

FIGURE 4–3

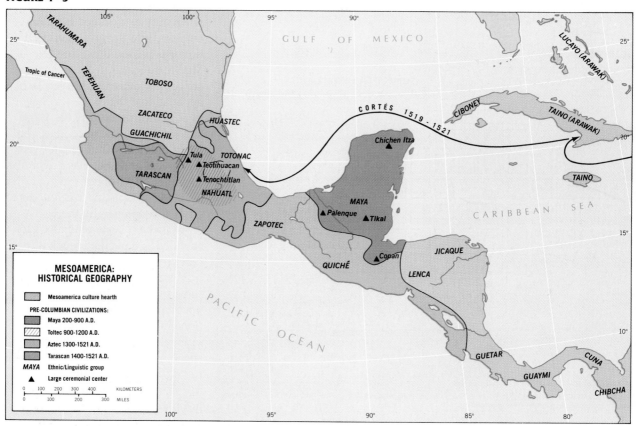

MESOAMERICA: HISTORICAL GEOGRAPHY

Mesoamerica culture hearth

PRE-COLUMBIAN CIVILIZATIONS:

Maya 200-900 A.D.

Toltec 900-1200 A.D.

Aztec 1300-1521 A.D.

Tarascan 1400-1521 A.D.

MAYA   Ethnic/Linguistic group

▲   Large ceremonial center

TEN MAJOR GEOGRAPHIC QUALITIES OF MIDDLE AMERICA

1. Middle America is a fragmented realm that consists of all the mainland countries from Mexico to Panama and all the islands of the Caribbean Sea to the east.
2. Middle America's mainland constitutes a crucial barrier between Atlantic and Pacific waters. In physiographic terms, this is an intercontinental land bridge.
3. Middle America's tropical location and climates are in important places ameliorated by altitude and its resulting vertical zonation of natural environments.
4. Middle America is a realm of intense cultural and political fragmentation. The political geography defies unification efforts, and instability is now at an all-time high.
5. Middle America's cultural geography is complex. African influences dominate the Caribbean, whereas Indian traditions survive on the mainland.
6. Middle America's historical geography is replete with involvement by its powerful neighbor, the United States.
7. Underdevelopment is endemic in Middle America. The realm contains the Americas' least-developed territories.
8. In terms of area, population, and economic strength, Mexico dominates the realm.
9. Out-migration of population for economic and political reasons continues.
10. The realm (notably Mexico) contains major actual and potential reserves of mineral fuels.

FIGURE 4—4 These ruins at Tikal, Guatemala testify to the scale of the Mayan achievement.

scholars have suggested that cultural stimuli from ancient Egypt actually reached Middle American shores and that the pyramid-like stone structures of Mayan cities represent imitations and variations of Egyptian achievements. More likely, Maya civilization arose spontaneously and independently. It experienced successive periods of glory and decline, reaching its zenith in present-day Guatemala from the fourth to the tenth centuries A.D.

The Maya civilization unified an area larger than any of the modern Middle American states except Mexico. Its population was probably somewhere between 2 and 3 million; the Maya language—the local *lingua franca*—and some of its related languages remain in use in the area to this day. The Maya state was a theocracy (ruled by religious leaders) with a complex religious hierarchy, and the great cities that today lie in ruins served in the first instance as ceremonial centers. Their structures testify to the architectural capabilities of the Maya and include huge pyramids and magnificent palaces (Fig. 4–4). We also know that Maya culture had skilled artists, writers, mathematicians, and astronomers. No doubt the Maya civilization also had poets and philosophers, but it had a practical side too. In agriculture and trade these people achieved a great deal. They grew cotton, had a rudimentary textile industry, and even exported finished cotton cloth by sea-going canoes to other parts of Middle America in return for, among other things, the cacao that they prized so highly.

The Highland Aztecs

About the same time that the Maya civilization emerged in the tropical lowland forests, the Mexican highland also witnessed the rise of a great Indian culture, similarly focused on ceremonial centers and marked as well by major developments in agriculture, architecture, religion, and the arts. These achievements were comparable to those of the Maya, but they were to be overshadowed by what followed. An important successor of the early highland civilization were the Toltecs, who moved into this area from the north, conquered and absorbed the local Indian peoples, and formed a powerful state centered on one of the first true cities of Middle America: Tula. The Toltecs' period of hegemony was relatively brief, lasting for less than three centuries after their rise to power around A.D. 900, but they conquered parts of the Maya domain, absorbed many Mayan innovations and customs, and introduced them on the plateau. When the Toltec state was, in turn, penetrated by new elements from the north, it was already in decay, but its technology was readily adopted and developed by the conquering Aztecs (or Mexicas).

The Aztec state, the pinnacle of organization and power in Middle America, is thought to have originated in the early fourteenth century when a community of Nahuatl- (or Mexicano-) speaking Indians founded a settlement on an island in one of the many lakes that lay in the Valley of Mexico. This ceremonial center, named Tenochtitlán, was soon to become the greatest city in the Americas and the capital of a large and powerful state. Through a series of alliances with neighboring peoples, the early Aztecs gained control over the entire Valley of Mexico, the pivotal geographic feature of Middle America that is still the heart of the modern state of Mexico. This 30-by-40-mile (50-by-65-kilometer) region is, in fact, a mountain-encircled basin positioned about 8000 feet (nearly 2500 meters) above sea level. Elevation

MIDDLE AMERICA AND CENTRAL AMERICA

Middle America, as we define it, includes all the mainland and island countries and territories that lie between the United States of America and the continent of South America. Sometimes the term *Central America* is used to identify the same realm, but Central America is actually a region within Middle America. Central America comprises the republics that occupy the strip of mainland between Mexico and Panama: Guatemala, Belize, Honduras, El Salvador, Nicaragua, and Costa Rica. Panama itself is regarded here as belonging to Central America as well; however, it should be noted that many Central Americans do not consider Panama to be part of their realm, because that country was, for the most of its history, a part of South America's Colombia.

and interior location both affect its climate; for a tropical area, it is quite dry and very cool. The region's lakes formed a valuable means of internal communication for the Aztec state. The Indians of Middle America never developed wheeled vehicles, so they relied heavily on porterage and, where possible, the canoe for the water transportation of goods and people. The Aztecs connected several of the Mexican lakes by canals and maintained a busy canoe traffic on their waterways, bringing agricultural produce to the cities and tribute paid by their many subjects to the headquarters of the ruling nobility.

Throughout the fourteenth century, the Aztec state strengthened its position by developing a strong military force, and by the early fifteenth century the conquest of neighboring peoples had begun. The Aztec drive to expand the empire was directed primarily eastward and southward. To the sparsely settled north, the rugged and unproductive land quickly became drier; to the west lay the powerful competing state of the Tarascans, with whom the Aztecs sought no quarrel, given the opportunities to the east and south.

Here, then, they penetrated and conquered almost at will. The Aztec objective was neither the acquisition of territory nor the spreading of their language and religion, but the subjugation of peoples and towns in order to extract taxes and tribute. They also carried off thousands of people for purposes of human sacrifice in the ceremonial centers of the Valley of Mexico, which required a constant state of enmity with weaker peoples in order to take their human prizes.

As Aztec influence spread throughout Middle America, the goods streaming back to the Valley of Mexico included gold, cacao beans, cotton and cotton cloth, the feathers of tropical birds, and the skins of wild animals. The state grew ever richer, its population mushroomed, and its cities expanded. At its peak Tenochtitlán had over 100,000 inhabitants or more—perhaps as many as a quarter of a million. These cities were not just ceremonial centers, but true cities with a variety of economic and political functions and large populations including specialized skilled labor forces.

The Aztecs produced a wide range of impressive accomplishments, although they were better borrowers and refiners than they were innovators. They practiced irrigation by diverting water from streams to farmlands, and they built elaborate walls to terrace slopes where soil erosion threatened. Indeed, when it comes to measuring the legacy of the Mesoamerican Indians to their successors—and to humankind— then the greatest contributions surely came from the agricultural sphere. Corn (maize), various kinds of beans, the sweet potato, a variety of manioc, the tomato, squash, the cacao tree, and tobacco are just a few of the crops that grew in Mesoamerica when the Europeans first made contact.

COLLISION OF CULTURES

We in the Western world all too often are under the impression that history began when the Europeans arrived in some area of the world and that the Europeans brought such superior power to the other continents that whatever existed there previously had little significance. Middle America bears out this perception: the great, feared Aztec state fell before a relatively small band of Spanish invaders in an incredibly short time (1519–1521). But let us not lose sight of a few facts. At first, the Spaniards were considered to be ''White Gods,'' whose arrival was predicted by Aztec prophecy. Having entered Aztec territory, the earliest Spanish visitors could see that great wealth had been amassed in Aztec cities. And Hernán Cortés, for all his 508 soldiers, did not single-handedly overthrow the Aztec authority. What Cortés brought on was a revolt, a rebellion by peoples who had fallen under Aztec domination and who

had seen their relatives carried off for human sacrifice to Aztec gods. Led by Cortés with his horses and artillery, these Indian peoples rose against their Aztec oppressors and followed the band of Spaniards toward Tenochtitlán, where thousands of them died in combat against the Aztec warriors. They fed and guarded the Spanish soldiers, maintained connections for them with the coast, carried supplies from the shores of the Gulf of Mexico to the point Cortés had reached, and secured and held captured territory while the Europeans moved on. Cortés started a civil war; he got all the credit for the results. But it is reasonable to say that Tenochtitlán would not have fallen so easily to the Spaniards without the sacrifice of many thousands of Indian lives.

Actually, in the Americas as well as in Africa, the Spanish, Portuguese, British, and other European visitors considered many of the peoples they confronted to be equals—equals to be invaded, attacked, and, if possible, defeated—but equals nonetheless. The cities and farms of Middle America, the urban centers of West Africa, the great Inca roads of highland South America all reminded the Europeans that technologically they were only a few steps ahead of their new contacts. Thus, the gap that developed between the European powers and the indigenous peoples of many other parts of the world only emerged clearly when the Industrial Revolution came to Europe— centuries after Vasco da Gama, Columbus, and Cortés.

Effects of the Conquest

In Middle America, the confrontation between Hispanic and Indian cultures spelled disaster for the Indians in every conceivable way: a

drastic decline in population, rapid deforestation, pressure on vegetation from grazing animals, substitution of Spanish wheat for Indian corn on cropland, and construction of new Spanish towns. The quick defeat of the Aztec state was followed by a catastrophic decline in population. Of the 15 or 25 million Indians in Middle America when the Spaniards arrived (estimates vary), only a century later just 2.5 million survived. The Spanish were ruthless colonizers, but not much more so than other European powers that subjugated other cultures. True, the Spanish first enslaved the Indians and were determined to destroy the strength of Indian society. But biology accomplished what ruthlessness could not have achieved in so short a time. Nowhere in the Americas did the Indians have immunity to the diseases the Spaniards brought: smallpox, typhoid fever, measles, influenza, and mumps. Neither did they have any protection against the tropical diseases that the white people introduced through their African slaves, such as malaria and yellow fever, which took enormous tolls in the hot, humid lowlands of Middle America.

Middle America's cultural landscape—its great cities, its terraced fields, its dispersed Indian villages—was drastically modified. The Indian cities' functions ceased as the Spanish brought in new traditions and innovations in urbanization, agriculture, religion, and other pursuits. Having destroyed Tenochtitlán, the Spaniards did recognize the attributes of its site and situation and chose to rebuild it as their mainland headquarters. Whereas the Indians had used stone almost exclusively as their building material, the Spaniards employed great quantities of wood and used charcoal for heating, cooking, and smelting metal. The onslaught on the forests was immediate, and expanding rings of deforestation quickly formed around the Spanish towns. Soon, the Indians also adopted Spanish methods of wooden construction and charcoal use, further accelerating forest depletion and the erosional scarring of the land.

The Indians had been planters but had no domestic livestock that made demands on the original vegetative cover. Only the turkey, the dog, and the bee (for honey and wax) had been domesticated in Mesoamerica. The Spaniards, however, brought with them cattle and sheep, whose numbers multiplied rapidly and made increasing demands not only on the existing grasslands but on the cultivated crops as well. Again, the Indians adopted these Spanish practices, putting further pressure on the land. Cattle and sheep became av-

enues to wealth, and the owners of the herds benefited. But the livestock now competed with the people for the available food (requiring the opening up of vast areas of marginal land in higher and drier locations), thereby contributing to a major disruption of the region's food-production balance. Hunger, therefore, quickly became a significant problem in Middle America during the sixteenth century, heightening the susceptibility of the Indian population to many diseases.

The Spaniards also introduced their own crops (notably wheat) and farming equipment, of which the plow was the most important. Thus, large fields of wheat began to make their appearance beside the small plots of corn that the Indians cultivated. The encroaching

wheat fields soon further reduced the Indians' lands. Moreover, because this grain was raised by and for the Spanish, what the Indians lost in farmland was not made up in additional available food. Neither were their irrigation systems spared. The Spaniards needed water for their fields and hydropower for their mills, and they had the technological know-how to take over and modify regional drainage and irrigation systems. This they did, leaving the Indian fields insufficiently watered, thereby diminishing even further the Indians' chances for an adequate food supply.

The most far-reaching cultural landscape changes that the Spanish introduced had to do with their traditions as town dwellers. To facilitate control, they relocated

FIGURE 4–5   A Spanish-built urban settlement: the crucible in which the acculturation of the Indians under their new masters took place. This is the town of Taxco, nestled in the dry mountains southwest of Mexico City.

the Indians from their land into nucleated villages and towns that the Spanish established and laid out. In these settlements, the kind of government and administration to which the Spanish were accustomed could be exercised; the focus of each town was the Catholic church (see p. 251). The location of each of these towns was chosen to lie near what was thought to be good agricultural land, so that the Indians could go out each day and work in the fields. Unfortunately, the selection was not always good, and land surrounding a number of villages was not suitable for Indian farming practices. Here food shortages and even famine resulted—but only rarely was a settlement abandoned in favor of a better situation.

Domination

In the towns and villages, the Indians came face to face with Spanish culture (Fig. 4–5). Here they learned the white invaders' religion and paid their taxes and tribute to a new master—or found themselves in prison or in a labor gang according to European regulations. Packed tightly in a concentrated settlement, they were rendered even more vulnerable to the diseases that regularly ravaged the population. Despite all this, the nucleated Indian village survived. Its administration was taken over by the Spaniards (and later by their post-colonial successors), and today it is still a key landscape feature of the Indian areas of Mexico and Guatemala. But anyone who wants to see remnants of the dispersed Indian dwellings and hamlets must travel into the remotest parts of Middle America, where Indian languages still prevail (Fig. 4–6).

Once the indigenous population was conquered and resettled, the Spanish were able to pursue an-

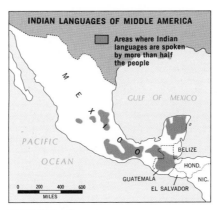

FIGURE 4–6

other primary goal in their New World territory: the exploitation of its wealth for their own benefit. Lucrative trade, commercial agriculture, livestock ranching, and especially mining were the avenues to affluence. The mining of gold held the greatest initial promise, and the Spaniards simply took over the Indian miner work force already in operation. But the small-scale *placering*, or "washing" of gold from streams carrying gold dust and nuggets, yielded a diminishing supply of the precious metal, and with the abolition of Indian slavery Spanish prospectors began searching for other valuable minerals. They were quickly successful, finding enormously profitable silver and copper deposits, particularly in a wide zone north of the Valley of Mexico (see Fig. 4–13). The development of these resources set into motion a host of changes in this part of Middle America. Mining towns drew laborers by the thousands; the mines required equipment, timber, mules; the people in the towns needed food. Since most of these settlements were located in dry country, irrigated fields were laid out wherever possible. Mule trains connected farm-supply areas to the mining towns, and these towns to the coastal ports. Thus was born a new urban system, one that not only integrated and organized the Spanish domain in Middle

America, but also extended effective economic control over some far-flung parts of New Spain. Mining, truly, was the mainstay of colonial Middle America.

MAINLAND AND RIMLAND

In Middle America outside Mexico, only Panama, with its twin attractions of inter-oceanic transit and gold deposits, became an early focus of Spanish activity. The Spaniards founded the city of Panama in 1519. Apart from their use of the corridor of the modern Panama Canal as an Atlantic–Pacific link, their main interest lay on the Pacific side of the isthmus. From here, Spanish influence began to extend northwestward into Central America. Indian slaves were taken in large numbers from the densely peopled Pacific lowlands of Nicaragua and shipped to South America via Panama. The highlands, too, fell under Hispanic control; before the middle of the sixteenth century, Spanish exploration parties based in Panama met those moving southeastward from Mesoamerica.

However, the leading center of Spanish activity remained in what is today central and southern Mexico, and the major arena of international competition in Middle America lay not on the Pacific side but on the islands and coasts of the Caribbean Sea. (Only the British ever gained a foothold on the mainland, controlling a narrow, low-lying coastal strip that extended from Yucatán to what is now Costa Rica.) As the colonial-era map (Fig. 4–7) shows, in the Caribbean the Spaniards faced the British, French, and Dutch, all interested in the lucrative sugar

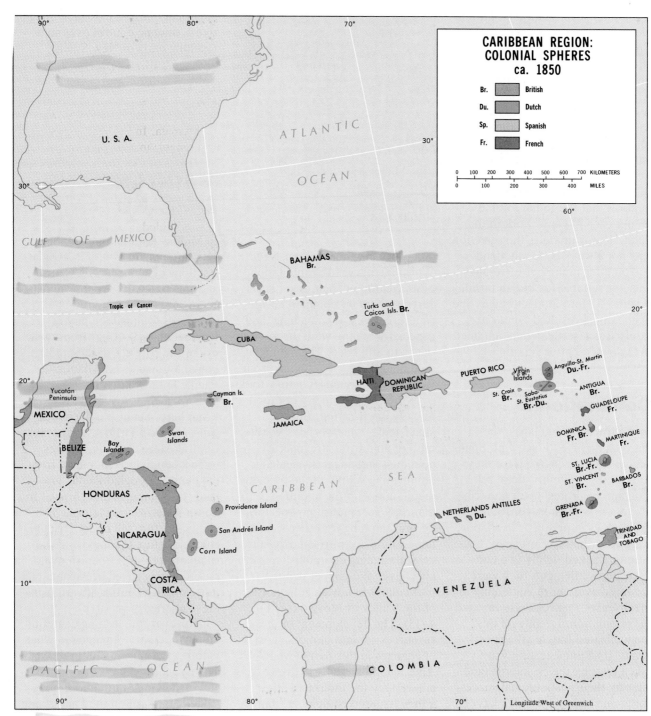

CARIBBEAN REGION: COLONIAL SPHERES ca. 1850

Br.	British
Du.	Dutch
Sp.	Spanish
Fr.	French

FIGURE 4–7

trade, all searching for instant wealth, and all seeking to expand their empires. Later, after centuries of European colonial rivalry in the Caribbean Basin, the United States entered the picture and made its influence felt in the coastal areas of the mainland—not

through colonial conquest, but by the introduction of widespread large-scale plantation agriculture.

The effects of these plantations were as far-reaching as the impact of colonialism on the Caribbean islands. The economic geography of the Caribbean coastal zone was

transformed, as hitherto unused alluvial soils in the many river lowlands were planted with thousands of acres of banana trees. Since the diseases the Europeans had brought to the New World had been most rampant in these hot and humid areas, the Indian popu-

THE ISLAND INDIANS

Mesoamerican Indian cultures have in some measure survived the European invasion. Indian communities remain, and Indian languages continue to be spoken by about 3 million people in southern Mexico and Yucatán and another million in Guatemala (Fig. 4–6). But on the islands of the Caribbean, the Indian communities were smaller and more vulnerable. At about the time of Columbus's arrival, there probably were 3–4 million Indians living in the Caribbean, a majority of them in Cuba, Jamaica, Puerto Rico, and Hispaniola (the island containing modern-day Haiti and the Dominican Republic). These larger islands (the *Greater Antilles*) were peopled by the Arawaks, whose farming communities raised root crops, tobacco, and cotton. In the eastern Caribbean, the smaller islands of the *Lesser Antilles* (Guadeloupe, Martinique, Dominica, and dozens of others) had more recently been peopled by the adventurous Caribs, who traversed the waters of the Caribbean in huge canoes that carried several dozen persons. When the European sailing ships began to arrive, the Caribs were in the process of challenging the Arawaks for their land, just as the Arawaks centuries earlier had ousted the Ciboney.

The Europeans quickly laid out their sugar plantations and forced the Indians into service. But Arawaks and Caribs alike failed to adapt to the extreme rigors of the labor to which they were subjected, and they perished by the thousands. Some fled to smaller islands where the European arrival was somewhat delayed. But after only half a century, just a few hundred survived in the dense interior forests of some of the islands. There they were soon joined by runaway African slaves (brought to the Caribbean to replace the dwindling Indian labor force), and, with their mixture, the last pure Caribbean Indian strain disappeared forever.

lation that survived was too small to provide a sufficient labor force; the Indians of the Caribbean islands had faced an even harsher fate (see *box*). Consequently, tens of thousands of black laborers were brought to the mainland coast from Jamaica and other islands (more came when the Panama Canal was dug between 1904 and 1914), completely altering the demographic mix. Physically, in many ways, the coastal belt already resembled the islands more than the Middle American plateau, and now the economic and cultural geographies of the islands were extended to it as well.

These contrasts between the Middle American highlands on the one hand, and the coastal areas and Caribbean islands on the other, were conceptualized by John Augelli into a **Mainland-Rimland framework** (Fig. 4–8). Augelli recognized (1) a Euro-Indian *Mainland*, consisting of mainland Middle America from Mexico to Panama, with the exception of the Caribbean coast from mid-Yucátan southeastward, and (2) a Euro-African *Rimland*, that included this coastal zone and the islands of the Caribbean. The terms *Euro-Indian* and *Euro-African* underscore the cultural heritage of each region: on the Mainland, European (Spanish) and Indian influences are paramount; in the Rimland, the heritage is European and African. As the map shows, the Mainland is subdivided into several areas on the basis of the strength of the Indian legacy. In southern Mexico and Guatemala, Indian influences are prominent; in northern Mexico and parts of Costa Rica, Indian influences are limited; between these areas lie sectors with moderate Indian influence. The Rimland, too, is subdivided. The most obvious division is between the mainland-coastal plantation zone and the islands. But the islands themselves can be classified according to their cultural heritage. Thus, there is a group of islands with Spanish influence (Cuba, Puerto Rico, and the Dominican Republic on old Hispaniola) and another group with other European influences, including the former British West Indies, the various French islands, and the Netherlands Antilles.

These contrasts of human habitat are supplemented by regional differences in outlook and orientation. The Rimland was an area of sugar and banana plantations, of high accessibility, of seaward exposure, and of maximum cultural contact and mixture. The Mainland, on the other hand, being farther removed from these contacts, was an area of greater isolation. The Rimland was the region of the great plantation, and its commercial economy was therefore susceptible to fluctuating world markets and tied to overseas investment capital. The Mainland was the region of the hacienda, more self-sufficient, and considerably less dependent on external markets.

The Hacienda

In fact, this contrast between plantation and hacienda land tenure in itself constitutes strong evidence

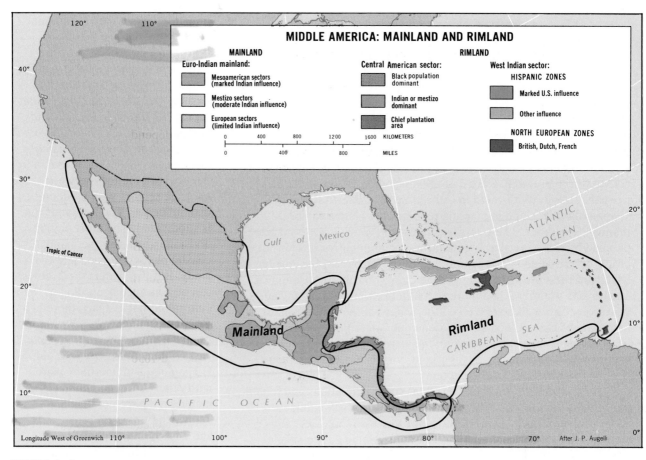

FIGURE 4–8

for a Rimland-Mainland division. The *hacienda* was a Spanish institution, but the modern plantation, Augelli argued, was the concept of Europeans of more northerly origin. In the *hacienda*, Spanish landowners possessed a domain whose productivity they might never push to its limits: the very possession of such a vast estate brought with it the social prestige and comfortable lifestyle they sought. The workers lived on the land—which may once have been *their* land—and had plots where they could grow their own subsistence crops. Traditions survived: in the Indian villages incorporated into the early haciendas, in the methods of farming, and in the means of transporting produce to markets. All this is written as though it is mostly in the past, but

the legacy of the hacienda, with its inefficient use of land and labor, is still visible throughout mainland Middle America.

The Plantation

The *plantation*, on the other hand, was conceived as something totally different. In their book, *Middle America: Its Lands and Peoples*, Robert West and John Augelli listed five characteristics of Middle American plantations that always apply. These clearly illustrate the differences between hacienda and plantation: (1) plantations are located in the humid tropical coastal lowlands of the realm; (2) plantations produce for export almost exclusively—usually a single crop; (3) capital and

skills are often imported, so that foreign ownership and an outflow of profits occur; (4) labor is seasonal—needed in large numbers during the harvest period but often idle at other times—and such labor has been imported because of the scarcity of Indian workers; and (5) with its "factory-in-the-field" operation, the plantation is more efficient in its use of land and labor than the hacienda. The objective was not self-sufficiency but profit, and wealth rather than social prestige was a dominant motive in the plantation's establishment and operation.

During the past century, both systems of land tenure have changed a great deal. The massive U.S. investment in the Caribbean coastal belt of Guatemala, Honduras, Nicaragua, Costa Rica, and

Panama transformed that area and brought a whole new concept of plantation agriculture to the region. On the Mainland, the hacienda has been under increasing pressure from national governments that view it as an economic, political, and social liability. Indeed, some haciendas have been parceled out to small landholders, whereas others have been pressed into greater specialization and productivity. Still other land has been placed in *ejidos*, where it is communally owned by groups of families. Both the hacienda and the plantation have for centuries contributed to the different social and economic directions that gave the Mainland and Rimland their respective regional personalities.

seen, the Caribbean is hardly an area of exclusive Spanish influence: whereas Cuba has an Iberian heritage, its southern neighbor, Jamaica (population 2.6 million, mostly black), has a legacy of British involvement, and Haiti's strongest imprints have been African and French. The crowded island of Hispaniola (total population 14.1 million) is shared between Haiti and the Dominican Republic, where Spanish influence survives (it also predominates in nearby Puerto Rico).

In the Lesser Antilles, too, there is great cultural diversity. There are the (once Danish) American Virgin Islands; French Guadeloupe and Martinique; a group of British-influenced islands, including Barbados, St. Lucia, St. Vincent, and Grenada; and

Dutch St. Maarten (shared with the French), Saba, St. Eustatius, and the A-B-C islands—Aruba, Bonaire, Curaçao—off the northwestern Venezuelan coast. Standing apart from the Antillean arc of islands is Trinidad, another former British dependency that, with its smaller neighbor Tobago, became a sovereign state in 1962.

Spain was the first colonial power in Middle America and was also the first to be forced to yield power in the face of independence movements. But independence came to the mainland colonies much earlier than in the Caribbean because the islands were easier to hold. Mexico proclaimed sovereignty in 1810; the Central American republics emerged during the 1820s and 1830s. In the Greater Antilles, however, Spain held

POLITICAL DIFFERENTIATION

Mainland Middle America today is fragmented into eight different countries, all but one of which (Belize, the former British Honduras) have Hispanic origins. Largest of them all—the giant of Middle America—is Mexico, whose 762,000 square miles (1,972,500 square km) constitute over 70 percent of the entire land area of Middle America (the islands included) and whose 91 million people outnumber those of all the other countries and islands of Middle America combined.

The cultural variety in Caribbean Middle America is much greater. Here Cuba dominates: its area is almost as large as that of all the other islands put together, and its population of 10.7 million is well ahead of the next-ranking country (the Dominican Republic, with 7.4 million). But, as we have

FIGURE 4–9   The Dutch colonial impress, vividly etched in the townscape of Willemstad, Curaçao.

Cuba and Puerto Rico until the Spanish-American War of 1898. By that time, these two islands had come to resemble the mainland republics in the composition of their population and in their cultural imprint. Spain's colonial arch-rival, Britain, also had a share of the Greater Antilles in Jamaica, and it gained several footholds in the Lesser Antilles to the southeast. The sugar boom and the strategic character of the Caribbean Sea had attracted the European competitors to the West Indies. By the opening of the twentieth century, the United States was making its presence felt as well—although the plantation crop was now the banana and the American strategic interest even more immediate than that of the European powers.

In parts of Caribbean Middle America, the period of colonial control is only just ending. Jamaica attained full independence from Britain in 1962, as did Trinidad and Tobago. An attempt by the British to organize a Caribbean-wide West Indies Federation failed, but other long-time British dependencies, including Barbados, St. Lucia, Dominica, and Grenada, were steered toward a precarious independence nonetheless (with disastrous results in Grenada and a U.S. invasion in 1983). France, on the other hand, has made no moves to end the status of Martinique and Guadeloupe as Overseas *Départements* of the French national state; and the Dutch A-B-C islands remain within the Netherlands empire (although Aruba now holds "separate" semiautonomous status until the achievement of full independence in 1996). Thus, the European imprint in the cultural landscape is strong and abiding, with Willemstad—Dutch Curaçao's capital—a particularly striking example (Fig. 4–9).

CARIBBEAN REGIONAL PATTERNS

Caribbean America today is a land crowded with so many people that, as a region (encompassing the Greater and Lesser Antilles), it is the most densely populated part of the Americas. It is also a place of grinding poverty and, in all too many localities, unrelenting misery with little chance for escape. In some respects, U.S.–affiliated Puerto Rico (see *box*) and communist Cuba constitute exceptions to any such generalizations made about Caribbean America. But, on most of the other islands, life for the average person is difficult, often hopeless, and tragically short.

All this is in jarring contrast to the early period of riches based on the sugar trade. But that initial wealth was gained while an entire ethnic group (the Indians) was being wiped off the Caribbean map and while another (the Africans) was being imported in bondage; the sugar revenues, of course, always went to the planters, not the laborers. Subsequently, the regional economy faced rising competition from other tropical sugar-producing areas; it soon lost its monopoly of the European market, and difficult times prevailed. Meanwhile, just as they did in other parts of the world, the Europeans helped stimulate the rapid growth of the island population. Death rates were lowered, but birth rates remained high and explosive population increases resulted.

With the decline of the sugar trade, millions of people were pushed into a life of subsistence, malnutrition, and hunger. Many sought work elsewhere: tens of thousands of Jamaican laborers went to the plantations of the Rim-land coast; large numbers of British West Indians went to England in search of a better life; and Puerto Ricans streamed to the United States mainland. But this outflow has failed to stem the tide of population growth: today, there are over 34 million people on the Caribbean islands, a total expected to increase by 50 percent over the next three decades.

Opportunities at Home

The Caribbean islanders just have not had many alternatives in their search for betterment. Their island habitat is fragmented by both water and mountains; the total amount of flat cultivable land is only a small fraction of the whole archipelago. Although there has been some economic diversification, agriculture remains the region's mainstay. Sugar is still the leading product, and it continues to head the export lists of the Dominican Republic and Cuba; in Haiti, coffee has become the chief export. Trinidad, just off the northeastern Venezuelan coast, is fortunate in possessing sizable oilfields, but the global oil depression of the 1980s produced hard times for a country where petroleum accounts for almost 90 percent of that country's exports. In the Lesser Antilles, sugar has retained a somewhat less prominent position, having been supplanted by such crops as bananas, sea-island cotton, limes, and nutmegs. Even here, however, sugar continues to dominate, particularly in Barbados, St. Kitts, St. Lucia, and Antigua.

All the crops grown in the Caribbean—Haiti's coffee, Jamaica's bananas, the Dominican Republic's cacao, the Lesser Antilles' fruits as well as the pivotal sugar industry—constantly face

PUERTO RICO: CLOUDED FUTURE?

The largest, most populous U.S. domain in Middle America is Puerto Rico, easternmost and smallest of the Greater Antilles chain. The 3400-square-mile (8800-square-kilometer) island, now home to 3.4 million people, fell to the United States during the 1898 Spanish-American War; at the same time the United States also acquired the Philippines and Guam in the Pacific.

Thus Puerto Rico's struggle for independence from Spain ended, but American administration was at first difficult. Not until 1948 were Puerto Ricans allowed to elect their own governor. In 1952, following a referendum, the island became the autonomous Commonwealth of Puerto Rico; San Juan and Washington share governmental responsibilities in a complicated arrangement. Puerto Ricans have American citizenship but pay no federal taxes on local incomes. The political situation fails to satisfy everyone: there still is a small pro-independence movement, a sizable number of voters who favor statehood, and, according to the most recent plebiscite in 1967, a majority who want to continue the Commonwealth status but with certain modifications (another vote on these alternatives is likely in 1991).

Puerto Rico stands in sharp contrast to Hispaniola and Jamaica. Long dependent on a single-crop economy (sugar), Puerto Rico during the 1950s and 1960s industrialized rapidly as a result of tax breaks for corporations, comparatively cheap labor, governmental incentives of various kinds, political stability, and special access to the U.S. market. Today, chemicals—not bananas or sugar—rank as the leading export. Puerto Rico does not have substantial mineral resources and it lies far from the U.S. core area (San Juan is about 1600 miles [2600 kilometers] from New York), but these disadvantages have largely been overcome by a development program that began in the 1940s. Nonetheless, the Puerto Rican economy responds to and reflects the U.S.-mainland economic picture. Times of recession and high unemployment generate an exodus of islanders to the mainland, where they can qualify for welfare support even if they cannot find work. During the 1960s, the Puerto Rican population of New York City alone grew by some 450,000, accounting for the comparatively low population growth rate of the island. Association with the United States has undoubtedly speeded Puerto Rico's development. The question now is whether the island's social order and political system will be able to withstand the pressures that lie ahead.

severe competition from other parts of the world and have still not become established at a scale that could begin to have an effect on improving standards of living. Those minerals that do exist in this region—Jamaica's bauxite (aluminum ore), Cuba's iron and chromium, Trinidad's oil—do not support any significant industrialization within the Caribbean Basin itself. As in other parts of the developing world, these resources are exported for use elsewhere.

Problems of Widespread Poverty

Therefore, the vast majority of the people in this area continue to eke out a precarious living from small plots of ground, mired in poverty and threatened by disease. Food is not always adequate on Caribbean islands, and some countries—most notably Haiti—are chronically food-deficient. What foodstuffs there are tend to be sold in small markets whose produce is dominated by a few local staples that hardly offer a balanced diet. Farm tools are still primitive, and cultivation methods have undergone little change over the generations. Land inheritance customs have divided and redivided peasant families' plots until they have become so small that the owner must share-crop some other land or seek work on a plantation in order to supplement the meager harvest. Bad years of drought, winter cold waves, or hurricanes can spell disaster for the peasant family. Moreover, soil erosion constantly threatens; much of the Jamaican countryside is scarred by gulleys and ravines, and Haiti's land has become so ravaged by spectacular erosion (frequently down to bedrock level) that the entire country may soon become an ecological wasteland (Fig. 4–10). Where soils are not eroded, their nutrients are depleted, and only the barest yields are extracted. The good land lies under cash crops for the export trade, not in food crops for local consumption. Those expanses of sugarcane and banana trees symbolize the disadvantaged position of the Caribbean countries, dependent on uncertain markets for their revenues and trapped in an international economic order they cannot change.

With such problems, it would be unlikely for Caribbean America to have many large cities; after all,

FIGURE 4–10 Along the border between the two countries, the ravaged landscape of Haiti (left) stands in stark contrast to the thick forests of the neighboring Dominican Republic (right). Hungry Haitians have been stripping the land of its trees for centuries to clear space for raising crops. Trees, however, bind the soil; with the country 98 percent deforested, tropical rains are now steadily washing away the soils, leaving behind a completely denuded land surface.

FIGURE 4–11 Nowhere can the most extreme poverty of the Third World be seen better than in the miserable slums of Haiti's capital, Port-au-Prince.

these countries have little basis for major industry, little capital, little local purchasing power. Indeed, the figures reflect this situation, with only slightly more than half the population classified as urban. Cuba and Puerto Rico, however, have more than two-thirds of their populations in urban areas; the region's largest and third largest cities, Cuba's Havana (2.1 million) and Puerto Rico's San Juan (1.8 million), owe a great deal of their development to earlier U.S. influences. The second largest city is the Dominican Republic's Santo Domingo (2.0 million); next on the list are the capitals of Haiti and Jamaica—Port-au-Prince and Kingston, respectively—each containing about 800,000 inhabitants. Cities on these less fortunate islands, however, often exhibit even more miserable living conditions than the poorest rural areas. The slums of Port-au-Prince are among the Third World's worst (Fig. 4–11); no wonder such abysmal conditions drive the most desperate Haitians away as "boat people" in search of a better life elsewhere.

Tourism: The Irritant Industry

The cities of Caribbean America constitute a potential source of income as tourist attractions and ports of call for cruise ships. The tourist industry (based in Miami and Ft. Lauderdale) experienced a spectacular growth boom in the 1980s, and places such as Ocho Rios (Jamaica) and Puerto Plata (Dominican Republic) have now been added to cruise itineraries that already included San Juan, Port-au-Prince, and Bahamian Nassau. The Caribbean has long been known for its magnificent beaches and beautiful island landscapes, but visitors also are attracted by the night life and gambling of San Juan, the cuisine and shopping of Martinique's Fort-de-France, and the picturesque architecture of Curaçao's Willemstad (Fig. 4–9).

Certainly, Caribbean tourism is a prospective money-earner, and it already ranks at or near the top on many of the islands of the Lesser Antilles. But tourism has serious drawbacks. The invasion of overtly poor communities by wealthier visitors at times leads to hostility, even actual antagonism among the hosts. For some island residents, tourists have a "demonstration effect," which leads locals to behave in ways that may please or interest the visitors but are disapproved by the larger community. Free-spending, sometimes raucous tourists contribute to a rising sense of local anger and resentment. Moreover, tourism has the effect of debasing local culture, which is adapted to suit the visitors' tastes. Anyone who has witnessed hotel-staged "culture" shows has seen this process. And many workers

say that employment in the tourist industry is dehumanizing; expatriate hotel and restaurant managers demand displays of friendliness and servitude that locals find difficult to sustain. Nevertheless, the Caribbean's tourist trade has stimulated several island industries to produce for the visitors. A typical handicraft "industry" that has mushroomed with the tourist business is Haiti's visual arts: when cruise ships are in port, dockside areas are bedecked with countless paintings and carvings for sale by local artists (Fig. 4–12).

Tourism does generate income in the Caribbean where alternatives are few, but the flood of North American tourists cannot be said to have a beneficial effect on the great majority of Caribbean residents. In the popular tourist areas, the intervention of island governments and multinational corporations has removed opportunities from local entrepreneurs in favor of large operators and major resorts; tourists are channeled on prearranged trips in isolation from the local society. There are

FIGURE 4–12 Abundant local artwork is for sale at tourist stops throughout the Caribbean.

some cultural advantages to this because many tourists do not enhance international understanding when they invade Caribbean communities; but such practices undoubtedly deprive local small establishments and street vendors of potential income.

Tourism, then, is a mixed blessing for the underdeveloped Caribbean Basin. Given the region's limited options, it provides revenues and jobs where otherwise there would be none. But there is a negative cumulative effect that intensifies contrasts and disparities: gleaming hotels tower over substandard housing, luxury liners glide past poverty-stricken villages, opulent meals are served where, down the street, children suffer from malnutrition. The tourist industry contributes positively to island economies but strains the fabric of the local communities involved; it is also an outstanding example of the tensions that frequently arise when direct contact occurs between residents of the First and Third Worlds—a subject explored in the introductory chapter's Opening Essay (pp. 41–42).

African Heritage

The Caribbean region is also a legacy of Africa, and there are places where cultural landscapes strongly resemble those of West and Equatorial Africa. In the construction of village dwellings, the operation of rural markets, the role of women in rural life, the preparation of certain kinds of food, methods of cultivation, the nature of the family, artistic expression, and in an abundance of other traditions, the African heritage can be read throughout the Caribbean-American scene.

Nonetheless, in general terms, it is still possible to argue that the European or white person is in the best position in this island chain, politically and economically; the *mulatto* (mixed white-black) ranks next; and the black person ranks lowest. In Haiti, for instance, where 95 percent of the population is "pure" black and only 5 percent mulatto, this mulatto minority holds a disproportionate share of power. On nearby Jamaica, the 15 percent "mixed" sector of the population plays a role of prominence in island politics far out of proportion to its numbers. In the Dominican Republic, the pyramid of power puts the white sector (16 percent) at the top, the mixed group (73 percent) next, and the black population (11 percent) at the bottom; this country clings tenaciously to its Spanish-European legacy in the face of a century and a half of hostility from neighboring Afro-Caribbean Haiti. In Puerto Rico, likewise, Spanish values persist despite American cultural involvement and a non-Hispanic sector accounting for about one-tenth of the island's 3.4 million people. In Cuba, too, the 11 percent of the population that is black has found itself less favored than the white sector (37 percent), the mulatto sector (36 percent), and the *mestizo* (mixed white-Indian) population (15 percent).

The composition of the population of the islands is further complicated by the presence of Asians from both China and India. During the nineteenth century, the emancipation of slaves and the ensuing local labor shortages brought some far-reaching solutions. Some 100,000 Chinese emigrated to Cuba as indentured laborers (today they still comprise 1 percent of the island's population); and Jamaica, Guadeloupe, Martinique, and especially Trinidad saw nearly 250,000 East Indians arrive for similar purposes. To the Afro-modified forms of English and French heard in the Caribbean, therefore, can be added several Asian languages; Hindi is particu-larly strong in Trinidad, whose overall population is now about 40 percent South Asian. The ethnic and cultural variety of the **plural societies** of Caribbean America is indeed endless.

THE MAINLAND MOSAIC

Mainland Middle America consists of two regions, Mexico and Central America, the former constituted by a single country and the latter by seven. Mexico is Middle America's giant in virtually every way, a region by virtue of its physical size, population, cultural qualities, resource base, and relative location. Mexico's geographic position adjacent to the United States has been a blessing as well as a curse—a blessing because it has facilitated economic interaction, a curse because it has led to neighborly friction over illegal emigration and illicit drug traffic (Mexico is the leading source of heroin and marijuana coming into the United States). Economic conditions in Mexico have stimulated a growing cross-border emigration into the United States, a process that involves millions of illegal immigrants. The huge market for illicit drugs in the United States has generated a burgeoning network of opportunity in Mexico (as well as in South America, as we will see in Chapter 5), and coordination of anti-drug operations has proved difficult.

The seven republics of Central America cannot match Mexico in terms of total population or territory, but they constitute a geographic region nonetheless—a region afflicted by conflict (with big-power involvement), economic stagnation, major refugee flows, military coups, and environmental crises. We focus first on Mexico, and then turn to Central America.

Mexico: Land of Troubled Revolution

Mexico is a land of distinction. It is the colossus of Middle America, with a 1991 population of 91 million—exceeding the combined total of all the other countries and islands of the realm by 28 million—and a territory more than twice as large (Fig. 4–13). Indeed, in all of Spanish-influenced "Latin" America, no country has even half as large a population as Mexico (Argentina and Colombia are next). Moreover, Mexico has grown so rapidly over the past two decades that its 1970 population will have *doubled* before the end of 1993.

The United Mexican States—the country's official name—consists of 31 states and the Federal District of Mexico City, the capital. Urbanization, as was noted in the introductory chapter, has rapidly expanded in the developing world, and no less than 68 percent of the Mexican people now reside in towns and cities (only 50 percent were urbanites as recently as 1960). Almost one-fourth of the national total (21 million people) are crammed into the conurbation centered on Mexico City alone, which is growing at the astonishing rate of about 750,000 per year (see *box*). The country's second city, Guadalajara, located approximately 300 miles (470 kilometers) west of the capital, has about 3.4 million inhabitants.

The physiography of Mexico is reminiscent of the western United States, although environments are more tropical. Figure 4–1 shows several prominent features: the elongated Baja (Lower) California Peninsula in the northwest, an extension of California's Coastal Ranges, and separated from the mainland by the Gulf of California

FIGURE 4—13

(which the Mexicans call the Sea of Cortés); the Yucatán Peninsula in the southeast, jutting out into the Gulf of Mexico; and the Isthmus of Tehuantepec, southeast of Veracruz, where Mexico's landmass becomes narrowest. Here in the southeast, Mexico most resembles Central America physiographically; a mountain backbone forms the isthmus and extends northwest toward Mexico City. Shortly before reaching the capital, this mountain range divides into two

chains, the Sierra Madre Occidental (in the west) and Sierra Madre Oriental (in the east). These diverging ranges frame the funnel-shaped Mexican heartland, the center of which consists of a great central plateau (the Valley of Mexico lies near its southeastern end). This rugged tableland is some 1500 miles (2400 kilometers) in length and up to 500 miles (800 kilometers) wide. The plateau is highest in the south, near Mexico City, where it is about 8000 feet (2450

meters) in elevation; from there it gently declines northwest toward the Rio Grande. As Fig. I–10 reveals, Mexico's climates are marked by dryness, particularly in the broad, mountain-flanked north. It has been estimated that only 12 percent of the country receives adequate rainfall throughout the year. Most of the better-watered areas lie in the south and east— the areas where Mesoamerica's great indigenous civilizations arose.

The Indian imprint on Mexican culture remains extraordinarily strong. Today, 60 percent of all Mexicans are mestizos, and 29 percent are Indians; only about 9 percent are Europeans. There is a Mexican saying that Mexicans who do not have Indian blood in their veins nevertheless have the Indian spirit in their minds. Certainly, the Mexican Indian has been Europeanized, but the Indianization of modern Mexican society is so powerful that it would be inappropriate here to speak of one-way, European-dominated acculturation. Clearly, what took place in Mexico is **transculturation**, the two-way exchange of culture traits between societies in close contact. About 1 million Mexicans still speak only an Indian language, and as many as 5 million still use Indian languages in everyday conversation, even though they also speak Mexican Spanish (Fig. 4–6). Mexican Spanish itself has been strongly influenced by Indian languages, but this is only one aspect of Mexican culture that has received an Indian impress. Distinctive Mexican modes of dress, foods and cuisine, sculpture and painting, architectural styles, and folkways also vividly reflect the Indian contribution.

Mexico is a large country, but its population is substantially concentrated in a zone that centers on Mexico City and extends across the southern "waist" from Veracruz in the east through Guadalajara in the west. This zone contains over half the Mexican people, as is shown on the world population distribution map (Fig. I–14). One-third of Mexico's population lives in rural areas associated with the country's better farmlands. Although some notable development has recently occurred in agriculture, there still are enormous difficulties to be overcome. After achieving independence from Spain in the early

TEEMING MEXICO CITY

FIGURE 4–14   A portion of greater Mexico City.

The Mexico City conurbation has borne the brunt of the recent migratory surge toward urban areas. With a total population of 21 million (second in the world in 1991), it is already home to nearly one out of every four Mexicans, and it continues to grow at an alarming rate (Fig. 4–14). Each day, about 1000 people move to Mexico City; when added to the 1000 or so babies born there daily, that produces a truly staggering addition of approximately 750,000 people every year. However, birth rates in urban Mexico are higher than the national level; with half its population 18 years of age or younger, some demographers have forecast an astounding total of between *40 and 50 million* residents for greater Mexico City by the year 2010. In any case, by 2005 Mexico City will have become the world's largest single population agglomeration, surpassing Tokyo–Yokohama's 30-odd million.

Even in a well-endowed natural environment, such an enormous cluster of humanity would severely strain local resources. But Mexico City is hardly located in a favorable habitat; in fact, it lies squarely within one of the most hazardous surroundings of any city on earth—and human abuses of the immediate environment are constantly aggravating the potential for disaster. The conurbation may be located in the heart of the scenic Valley of Mexico, whose situational virtues led to the building of Tenochtitlán by the Aztecs, but two serious geologic problems loom: the vulnerability of the basin to major volcanic and seismic activity (such as the devastating earthquake in 1985), and the overall instability resulting from the weak, dry-lakebed surface that underlies much of the metropolis (aggravated by land subsidence as groundwater supplies are pumped out in vast quantities). Water availability presents another problem in this semiarid climate: dwindling local supplies must be augmented by the expansion of long-distance transportation of drinking water from across the mountains, an enormously expensive undertaking that must be accompanied by the growth of a parallel network to pipe sewage *out* of the waste-choked basin. Mexico City's worsening air pollution, however, poses the greatest health hazard and is conceded to be the world's most serious, exacerbated by the thin air that contains 30 percent less oxygen than at sea level (the city's elevation is 7350 feet/2240 meters). The conurbation's 3 million-plus cars and 7500 diesel buses produce about 75 percent of the smog, with the remainder caused by the daily spewing of 15,000 tons of chemical pollutants into the atmosphere by the area's 37,000 factories; on any given day, the pollution of Mexico City's air exceeds *100 times* the acceptable level.

For the affluent and tourists, Mexico City is undoubtedly one of the hemisphere's most spectacular primate cities, with its grand boulevards, magnificent palaces and museums, vibrant cultural activities and night life, and luxury shops. But most of its residents dwell in a world apart from the glitter of the Paseo de la Reforma, an increasing majority of them forced to live in the miserable poverty and squalor of the conurbation's 500 slums plus the innumerable squatter shacktowns that form the burgeoning metropolitan fringe (the notorious *ciudades perdidas*, or "lost cities"). This is the domain of the newcomers, the peasant families who have abandoned the hard life of the difficult countryside, lured to the urban giant in search of a better life. With Mexico's underemployment rate hovering above 30 percent in recent years, decent jobs and upward mobility quickly become elusive goals for most of the new arrivals—and these families are forced to scratch out an existence on less than U.S. $5.00 per day. Yet, despite the overwhelming odds, a surprising number of migrants eventually do enjoy some success by becoming part of the so-called *informal sector*. This is a primitive form of capitalism that is now common in many Third World countries, and takes place beyond the control—and especially the taxation—of the government. Participants are unlicensed sellers of homemade goods (such as arts and crafts, clothing, food specialties) and services (auto repair, odd jobs, and the like), and their willingness to engage in this hard work has transformed many a slum into a beehive of activity that can propel resourceful residents toward a middle-class existence. For its part, although officially discouraging the growth of squatter settlements, the government has recently made life on the Mexico City outskirts more comfortable by improving schools, roads, and other municipal services; moreover, it still permits squatters who settle on public lands to get free title to those properties after a period of five years.

The ongoing growth of Mexico City (and its problems of overcrowding and environmental degradation) is not a unique phenomenon. In 1985, according to a U.N. study, there were 35 metropolitan areas worldwide whose populations exceeded 5 million; by 2025, that total is predicted to rise to 93. Of these 58 additions to the 5-million-plus category, 57 are located in developing countries. Moreover, 12 of these urban clusters will contain more than 20 million people by 2025: Mexico City; Tokyo–Yokohama (still the developed world's only entry on this list); São Paulo, Brazil; Cairo, Egypt; Lagos, Nigeria; Karachi, Pakistan; India's Bombay, Delhi, and Calcutta; Dhaka, Bangladesh; Shanghai, China; and Jakarta, Indonesia. These growth leaders, of course, are simply the tip of the iceberg: urbanization rates are skyrocketing throughout the Third World. This is a topic to be explored in Chapter 5, with emphasis on the contemporary urban geography of Middle and South America.

1800s, Mexico failed for nearly a century to come to grips with the problems of land distribution, which were a legacy of the colonial period. By the opening of the twentieth century, the situation had worsened to the point where 8245 haciendas covered nearly 40 percent of Mexico's entire area; moreover, about 96 percent of all rural families owned no land whatsoever, with these landless people working as *peones* (landless, constantly indebted serfs) on the haciendas. There was deprivation and hunger, and the few remaining Indian lands and small holdings owned by mestizos or whites could not produce enough food to satisfy the country's needs. Meanwhile, thousands of acres of arable land lay idle on the haciendas, which blanketed just about all the good farmland in Mexico.

Revolution

Not surprisingly, a revolution began in 1910 and set into motion a sequence of events that is still unfolding today. One of its major objectives was the redistribution of Mexico's land, and a program of expropriation and parceling out of the haciendas was made law by the Constitution of 1917. Since then, about half the cultivated land of Mexico has been redistributed, mostly to peasant communities consisting of 20 families or more. Such lands are called *ejidos*; the government holds title to the land, and use rights are parceled out to villages and then individuals for cultivation. Most of the *ejido* lands carved out of haciendas lie in central and southern Mexico, where Indian traditions of landownership and cultivation survived and where the adjustments were most successfully made.

With such a far-reaching program, it is understandable that agricultural productivity temporarily declined. The miracle is that land reform has been carried off without a major death toll and that the power of the wealthy landowning aristocracy could be broken without ruin to the state. Mexico alone among the region's countries with large Indian populations has made major strides toward solving the land question, although there is still widespread malnutrition and poverty in the countryside. But the revolution that began in 1910 did more than that. It resurrected the Indian contribution to Mexican life, and blended Spanish and Indian heritages in the country's social and cultural spheres. It brought to Mexico the distinctiveness that it alone possesses in "Latin" America.

The revolution could change the distribution of land, but it could not change the land itself or the methods by which it was farmed. Corn (maize), beans, and squash continue to form the subsistence food of most Mexicans, with corn the chief staple and still occupying over half the cultivated land. It used to be more than that, and even this proportion suggests that some crop diversification is taking place. But corn is grown all too often where the conditions are not right for it, so that yields are low; wheat might do better, but the people's preference, not soil suitability, determines the crop. And if the people's preference has not changed a great deal, neither have farming methods over much of the country, in a vertical as well as horizontal geographic setting (see box).

Economic Geography

Commercial agriculture in Mexico has diversified and made major strides in recent decades with respect to both the home market and export. The greatest productivity still emanates from the hands of private cultivators, although much of the land involved has been subdivided into *ejidos*. The central

---

ALTITUDINAL ZONATION

Mainland Middle America and the western margin of South America are areas of high relief and strong local contrasts. People live in clusters in hot tropical lowlands, in temperate intermontane valleys, and even on high plateaus just below the snow line in the Andes. In each of these various zones, distinct local climates, soils, crops, domestic animals, and modes of life prevail. Such **altitudinal zones** are known by specific names as if they were regions with distinguishing properties—as, in reality, they are.

The lowest vertical zone, from sea level to 2500 feet (about 750 meters), is known as the *tierra caliente*, the "hot land" of the coastal plains and low-lying interior basins where tropical agriculture (including banana plantations) predominates. Above this lowest zone lie the tropical highlands containing "Latin" America's largest population clusters, the *tierra templada* of temperate land reaching up to about 6000 feet (1850 meters). Temperatures here are cooler; prominent among the commercial crops is coffee, and corn (maize) and wheat are the staple grains. Still higher, from about 6000 feet to nearly 12,000 feet (3600 meters) is the *tierra fría*, the cold country of the higher Andes where hardy crops such as potatoes and barley are the people's mainstays. Only small parts of the Middle American highlands reach into the *fría* zone, but in South America this environment is much more extensive in the Andes. Above the tree line, which marks the upper limit of the *tierra fría*, lies the *puna* (also known as the *páramos*); this fourth altitudinal zone, extending from about 12,000 to 15,000 feet (3600 to 4500 meters), is so cold and barren that it can only support the grazing of sheep and other hardy livestock. The highest zone of all is the *tierra helada* or "frozen land," a zone of permanent snow and ice that reaches to the peaks of the loftiest Andean mountains.

These elevation ranges are for highlands lying in the equatorial latitudes. Understandably, as one moves poleward of the tropics beyond 15 degrees of latitude, the sequence of five vertical zones is ratcheted downward, with the breaks occurring at progressively lower altitudes.

---

plateau is geared mainly to the domestic production of food crops, but in the north large irrigation projects have been built on the streams flowing down from the interior highlands. Cotton is cultivated for the domestic market as well as for export (Mexico leads Middle and South America in this commodity); for the home market, wheat and winter vegetables are grown. Cattle raising is another major pursuit, recently expanding onto the Gulf Coast lowlands from

its long-time base in the north. The unlikely boom of large-scale agriculture in Mexico's arid north is due in large measure to the adoption of U.S. irrigation and mechanized-farming technology.

Mexico's metal mining industries are less important today than they once were. The country still exports a major share of the world's silver, and other important commodities include copper, zinc, and lead. Mining activity is scattered throughout northern and central Mexico (Fig. 4–13), but many of the mines that were important in the colonial period—and near which urban centers of some size developed—have been exhausted.

More recently, Mexico has enjoyed the advantages—and suffered the problems—of a large and productive petroleum industry. Centered on the southern Gulf Coast's Bay of Campeche around the city of Villahermosa, these oilfields brought Mexico huge revenues when the world oil price was high and serious economic difficulties when the price fell. Discoveries of oil and natural gas reserves in this area (Fig. 4–13)—the largest found in the world since 1970—have made Mexico self-sufficient in these fossil fuels, an energy situation reinforced by major additional production and reserves located in oilfields lying along the Gulf Coast between Veracruz and Tampico. Additional discoveries in the past few years, in the area south of Villahermosa and northeastward into the Yucatán, now help to rank Mexico as the world's sixth largest petroleum producer.

As the world oil price rose to unprecedented heights during the late 1970s, Mexico's economic geography began to be transformed. The government had great plans for the economy, based on expected future income from oil and natural gas (much of it to be sold to the United States). Mexico borrowed and spent as never before, running up a huge debt to foreign countries and banks. But then the price of oil plunged—and with it Mexico's income. The government could not pay the interest on the loans it had taken, and a long economic crisis ensued as the foreign debt became ever larger. This experience, of course, is not unique; many countries in the underdeveloped world, counting on huge future incomes, strapped themselves with enormous debts. But

in Mexico's case, serious efforts were made by the government over the past decade to reduce this burden. Unfortunately, progress has been minimal, and the Mexican foreign debt still stood in the vicinity of a staggering U.S. $100 billion at the outset of the 1990s.

The importance of manufacturing in the Mexican economy is also rising. Mexico possesses a wide range of raw materials, many of which are located in the north (Fig. 4–13). An iron and steel industry, dating back to 1903, was built in Monterrey (2.9 million) in the northeast, using iron ore located near Durango and coking coal from the Sabinas Basin just north of Monterrey. A second complex was developed in the 1950s at Monclova near the source of coking coal. Most of Mexico's industrial production takes place in cities, which means that there is a particularly heavy concentration (just under 50 percent) of manufacturing (and its pollution) in and around Mexico City.

The newest development in Mexico's manufacturing geography, and a highly significant one, is the rapid recent growth of *maquiladora* plants in the northern border zone. The maquiladoras are foreign-owned factories (mainly by large U.S. companies) that assemble imported, duty-free components and/or raw materials into finished industrial products. At least 80 percent of these goods are then re-exported to the United States, whose import tariffs are limited to the value added to the products during their Mexican fabrication stage. All parties benefit from this industrial system: the Mexicans gain tens of thousands of well-paying jobs, and the foreign owners take advantage of Mexico's wage rates that are well below those north of the border. Although this development program was initiated in the 1960s, the number of maquiladoras grew to only a modest 588 (with 122,000

employees) as recently as 1982. But then the numbers suddenly took off: by 1989 more than 1600 assembly plants were employing over 400,000, and the maquiladoras were accounting for a robust 17 percent of Mexico's industrial labor force and fully 5 percent of its gross domestic product.

Among the goods being assembled were electronic equipment, electrical appliances, auto parts, clothing, plastics, and furniture. As the 1990s began, these were being joined by white-collar activities, especially routine data-processing operations, with the massive relocation of such back-office jobs from Southern California considered likely in the years ahead. Two "twin" borderland metropolises have enjoyed the greatest growth surge: Ciudad Juarez/El Paso and Tijuana/San Diego. Tijuana (1.3 million) has experienced a particularly swift transformation from a honky-tonk border enclave into one of the Third World's most prosperous cities: it is now the world's busiest border-crossing point, admitting millions of free-spending U.S. tourists each year. Its 500-plus maquiladoras employ 70,000 workers who already assemble more television sets than anywhere else on earth (Japanese manufacturers have been arriving in droves). So many Mexicans and Americans want to live there that by 2000 Tijuana is expected to be the continent's second largest west-coast city after Los Angeles. The Mexican government is now trying to capitalize on this enormously successful program by encouraging maquiladoras to move deeper into the country. However, with a few exceptions in the near-border northern cities of Monterrey and Chihuahua, this has so far proven to be a difficult challenge.

Mexico has made impressive gains in industrialization; it is addressing itself with new determination to agrarian reform; and it seeks to integrate all sectors of the population into a truly Mexican nation. After a century of struggle and oppression, this country has taken long strides toward representative government. But its progress is threatened by the explosive growth of its population, for no amount of reform ultimately can keep pace with a growth rate (now 2.4 percent annually) that will, if unchecked, produce a Mexican population of more than 107 million (versus 91 million in 1991) by the turn of the century. Increasingly, this growth is being steered toward the country's burgeoning urban areas, especially

Mexico City and Tijuana. In the countryside, there is already local strife over land allocations, and tens of thousands of Mexicans move illegally across the U.S. border each year in search of a means of survival. Small areas are made unsafe by outlaw bands that prey on highway travelers, and they may foreshadow serious regional insurgencies. No political or economic system in the underdeveloped world could long withstand the impact of population growth faced in Mexico, and the nation's latest accomplishments remain under the cloud of this threat.

The Central American Republics

Crowded onto the narrow segment of the Middle American land bridge between Mexico and the South American continent are seven countries collectively known as the Central American republics (Fig. 4–15). Territorially, they are all quite small: only one, Nicaragua, is larger than the Caribbean island of Cuba. Populations range from Guatemala's 9.5 million down to Panama's 2.5 million in the six Hispanic republics, whereas the sole former British

FIGURE 4–15

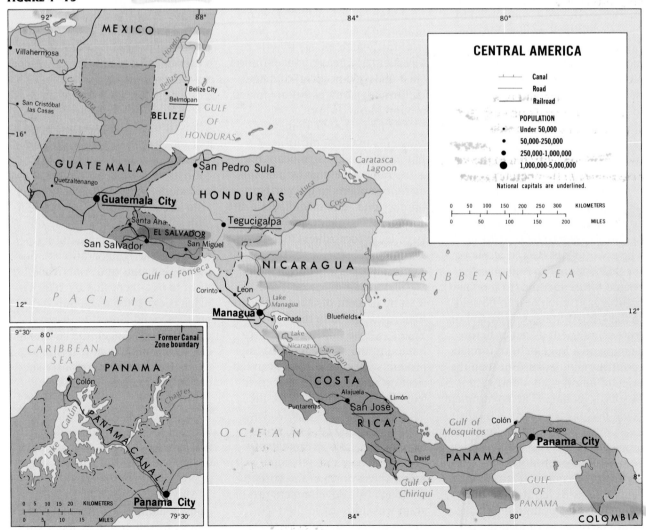

territory, Belize (which until 1981 was British Honduras), has only about 187,000 inhabitants. As elsewhere in Middle America, the ethnic composition of the population is varied, with Indian and white minorities and a mestizo majority. The exceptions are in Guatemala, where 45 percent of the population remains relatively "pure" Indian (Maya and Quiché—see Fig. 4–6) with the remainder of strongly Indian mestizo ancestry, and in Belize, where just under 50 percent of the population is black or mulatto, a situation reminiscent of the social geography of the Caribbean. Demographic complexity is at its least in Costa Rica, where there is a large white majority of Spanish and other relatively recent European immigrants; in a population of 3.1 million, the black component constitutes only 3 percent of the total and Indians 1 percent.

The narrowing land bridge on which these republics are situated consists of a highland belt flanked by coastal lowlands on both the Caribbean and Pacific sides, and from the earliest times the people have been concentrated in the upland *templada* (temperate) zone. Here tropical temperatures are moderated by elevation (see *box*, p. 270), and rainfall is adequate for the cultivation of a variety of crops. As noted earlier, the Middle American highlands are studded with volcanoes, and local areas of fertile volcanic soils are scattered throughout the region (Fig. 4–16). The old Indian agglomerations were located in these more fertile parts of the highlands, and this human distribution persisted during the Spanish period. Today, the capitals of Guatemala (Guatemala City, 1.3 million), Belize (Belmopan, 4500), Honduras (Tegucigalpa, 750,000), El Salvador (San Salvador, 1.1 million), Nicaragua (Managua, 900,000), and Costa Rica (San José, 1.1 million), all lie in the interior, most of them at

FIGURE 4–16   An active Nicaraguan volcano, with a town at its base that is laid out in typical Spanish fashion (cf. Fig. 4–2). Although lethal emissions from the mountain could reach inhabitants in a matter of moments, farm fields creep up the hazardous slopes in order to take advantage of the productive volcanic soil.

3500 feet (1060 meters) or higher in elevation. In all of mainland Middle America, Panama City (725,000) is the only coastal capital (Fig. 4–15). The size of these cities, in countries whose populations average about 4 million, is a reflection of their primacy, identi-

cal to the dominance of Mexico City over the rest of Mexico. On average, the next-ranking town is only one-fifth as large as the capital city.

The distribution of population within Central America, besides its concentration in the region's uplands, also exhibits greater densities toward the Pacific than toward the Caribbean coastlands (Fig. I–14). El Salvador, Belize, and to some degree Panama, are exceptions to the rule that people in mainland Middle America are concentrated in the *templada* zone. Most of El Salvador is *tierra caliente*, and the majority of its 5.4 million people are crowded, hundreds to the square mile, in the intermontane plains lying less than 2500 feet (750 meters) above sea level. In Nicaragua, too, the Pacific-side areas are the most densely populated; the early Indian centers lay near Lake Managua, Lake Nicaragua, and in the adjacent highlands. The frequent activity of volcanoes in this Pacific zone is accompanied by the emission of volcanic ash, which settles over the countryside and quickly weathers into fertile soils (Fig. 4–16). By contrast, the Caribbean coastal lowlands—hot, wet, and awash in leached soils—support comparatively few people. In the most populous republic, Guatemala, the heartland also has long been in the southern highlands. Although the large majority of Costa Rica's population is concentrated in the central uplands around San José, the Pacific lowlands have been the scene of major in-migration since banana plantations were first established there. Even in Panama, there is a strong Pacific orientation: more than half of all Panamanians (and this means over 70 percent of the rural population) live in the southwestern lowlands and on adjoining mountain slopes; another 25 percent live and work in the Panama Canal corridor; and of the remainder, a majority, many of them descendants of black immigrants from the Caribbean, live on the Atlantic or Rimland side of the isthmus.

With the single exception of Costa Rica, Middle America's mainland republics face the same problems as Mexico, only more

TROPICAL DEFORESTATION

As the world vegetation map (Fig. I–12) indicates, two-thirds of Central America was covered by tropical rainforests. We deliberately use the word *was* because destruction of this precious woodland resource is proceeding so swiftly that a typical landscape scene today shows not a rainforest but a scraggly, tree-cleared wasteland ripe for environmental disaster (Fig. 4–17).

Tropical deforestation in mainland Middle America, as was pointed out, began with the Spanish colonial era in the sixteenth century. But the pace has accelerated incredibly in recent decades. Since 1950, 75 percent of Central America's forests have been decimated; during the 1980s, the region's annual deforestation rate was 2.5 percent—or approximately 1600 square miles/4140 square kilometers (about the size of the state of Delaware) of woodland loss *each year*. According to biologist Norman Myers, Central America exhibits the world's fastest depletion rate. El Salvador has already lost its entire rainforest area, and the six other republics will reach that stage around the turn of the century. Costa Rica's present situation is typical: the 80 percent of the country that was covered by forests in 1940 shrank to 17 percent in 1988 and will be reduced to zero by 2002 if the current tree-removal rate is sustained.

The causes of tropical deforestation are several, and they are related to the persistent economic and demographic problems of the Third World. In Central America, the leading cause has been the need to clear rural lands for cattle pasture as many countries (notably Costa Rica) attempt to become meat producers and exporters. Although some gains have been recorded, the price of environmental degradation has been enormous. Because tropical soils are so nutrient-poor, newly deforested areas can only function as pastures for a few years at most. These fields then are summarily abandoned for other freshly cutover lands and quickly turn into the devastated landscape seen in Fig. 4–17. Without the protecting trees, local erosion and flooding immediately become problems, affecting (and reducing the output capacity of) still-productive nearby areas.

A second cause of deforestation is the rapid logging of tropical woodlands as the lumber industry increasingly turns away from the exhausted forests of the mid-latitudes to harvest the rich tree resources of the equatorial zones, responding to accelerating global demands for new housing, paper, and furniture.

The third major contributing factor is directly related to the population explosion in developing countries: as more and more peasants are required to extract a subsistence from inferior lands, they have no choice but to begin cutting the forest for both firewood and additional crop-raising space, and their intrusion prevents most of the trees from regenerating.

Although deforestation is a depressing event, it does not seem to tropical pastoralists, farmers, and timber producers to be life-threatening, and perhaps it even seems to offer some short-term economic advantages. Why, then, should there be such an outcry from the scientific community? And why should the World Resources

FIGURE 4–17 This scene in Costa Rica shows how badly the land can be scarred in the wake of deforestation. Without roots to bind the soil, tropical rains swiftly erode the unprotected topsoil.

Institute call this "the world's most pressing land-use problem"? The answer is that unless immediate large-scale action is taken, by 2000 the tropical rainforest will be reduced to two large patches—the western Amazon Basin of South America and the middle Zaïre (Congo) Basin of Central Africa—and these will disappear completely by 2035. Thus, the tropical forest must be a very important part of our natural world—and indeed it is. Biologically, the rainforest is by far the richest, most diversified arena of life on our planet: even though it covers only a shrinking 6 percent of the earth's land area, it contains well over half of all plant and animal species. Its loss would not only mean the extinction of millions of species, but also the end of birth because the evolutionary process that produces new species would be terminated. Since the rainforests already yield countless valuable medicinal, food, and industrial products, how many potential disease-combatting drugs or new crop varieties to feed undernourished millions would be irretrievably lost?

Perhaps the most ominous (and as yet unknown) consequence of all would be the impact of rainforest disappearance on the planet's climates. Environmental scientists are quite concerned about this coming crisis, and some forecast a global warming trend as the burning of the remaining forests adds vast quantities of carbon dioxide to the air, forming a thickening layer of the gas that would prevent excess heat from escaping the atmosphere. This trapping of warm air—the so-called *greenhouse effect*, because of the obvious analogy—is further heightened by the loss of the carbon dioxide–absorbing trees, and could lead to higher temperatures at every latitude, polar-icecap melting, and significantly raised sea levels that would imperil the world's crowded coastal zones.

so; they also share many of the difficulties confronting the Caribbean islands. Although efforts are being made to improve the landownership situation, the colonial legacy of hacienda and peon hangs heavily over the region. Generations of people have seen little change in their way of life and have had even less opportunity to bring about any real improvement. Dwellings are still built of mud and straw, sanitary facilities have not yet reached them, schools are badly overcrowded, and hospital facilities are inadequate. Each year brings a renewed struggle to extract a subsistence livelihood, with little hope that the next year will bring anything better. And in recent years, a new threat has overshadowed their fragile existence—the rapid destruction of the tropical forest, which portends future environmental disaster (see *box*). One of the most jarring contrasts in Middle America is that between the attractive capital city and its immediate surroundings, and the desolation of outlying rural areas. Another is that between the splendor, style, and refined culture of the families who own the coffee farms and other enterprises and the destitute, rag-wearing peasants.

Insurgent States

Such inequities, repressive government, and big-power involvement can lead to insurgency, and this is what has happened in Central America. Events follow a similar pattern, whatever the country. First there are reports of armed rebellions in remote places where the government has little presence. Soon these localized uprisings begin to attain permanence, and the rebels manage to control parts of the country on a more or less continuous basis. Full-scale civil war follows, and the government may fall. Political scientists identify these three stages as: (1) *conten-*

THE INSURGENT STATE: MODEL AND REALITY

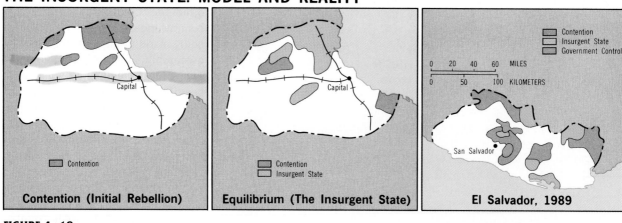

Contention (Initial Rebellion)

Contention

Equilibrium (The Insurgent State)

Contention
Insurgent State

El Salvador, 1989

Contention
Insurgent State
Government Control

0   20   40   60   MILES
0   50   100   KILOMETERS

San Salvador

FIGURE 4–18

tion, the period of initial rebellion; (2) *equilibrium*, when critical areas of the country are in rebel hands; and (3) *counteroffensive*, the stage of ground and air war that will decide the future.

These stages were translated into spatial terms by Robert McColl, who suggested that the equilibrium stage actually marks the emergence of an **insurgent state**, a state within a state (Fig. 4–18). The rebel-controlled insurgent state may not be contiguous, that is, it may be composed of several separate cores; but it undeniably has a headquarters, boundaries, governmental structure and systems of administration, hospitals, schools, and even direct communications with (possibly sympathetic) neighboring countries. In the 1960s and 1970s, Southeast Asia's Vietnam followed this pattern. The Communists established a growing insurgent state that coexisted for several years with the part of the country still controlled by the government in Saigon (now Ho Chi Minh City) and its U.S. allies. Closer to the United States, this also was the sequence of events in Cuba more than 30 years ago, when then-rebel Fidel Castro managed to establish a stable and expanding insurgent state on the island. And in Nicaragua during the 1970s, a rebel group known as the Sandinistas challenged a dic-

tatorial government, first in remote locales, then in a substantial insurgent state, and finally in a full-scale conflict that led to Sandinista rule in Managua.

Today Central America is in turmoil, and the political map (Fig. 4–15) conceals many politico-geographical developments. Nicaragua's Sandinista government, with Soviet and Cuban support, faced a United States–backed force for most of the 1980s that sought to establish its own insurgent state. With the defeat of the Sandinistas in the election of 1990, that rebel military campaign by the so-called *Contras* (which never achieved more than contention) has ended—but instabilities remain. In El Salvador, leftist rebels have come closer to creating an insurgent state, notably in the northeastern part of the country. In Guatemala, reform after decades of repressive government may be too late: a guerrilla movement is active and approaching contention in the western highlands of the country.

The roots of these insurgencies are old and deep. Central America is not a large region, but because of its physiography it includes many isolated, comparatively inaccessible locales. Conflicts between Indian population clusters and mestizo groups are endemic to the region, and contrasts between the privileged and the poor are harsh.

Dictatorial rule by local elites followed authoritarian rule by Spanish colonizers. The collision course has been long, and the ongoing upheaval is not the first outbreak. What *is* new is the role of outside forces: what once was clearly an area of U.S. dominance, an unchallenged sphere of influence, is now the scene of superpower competition. Soviet support for the Sandinista government and alleged Communist assistance to El Salvador's rebels were met by U.S. aid to the Contras and strong support for El Salvador's beleaguered government and military. In this manner local rebellions can swiftly expand into major ideological confrontations, with flows of arms out of all proportion to the basic capacities of the combatants.

One unprecedented side effect of these conflicts has been an enormous flow of refugees. These are not people who, like Haiti's boat people or Mexico's cross-border emigrants, are trying to escape poverty; nor are they comparable to the floods of emigrants who left Cuba for Florida after the Cuban Revolution and again during the 1980 Mariel boatlift. Central America's refugees by the tens of thousands have escaped combat and civil war, from the ashes of their villages, from rebel threats and death-squad terrorism. From city as well as countryside they have come, often leaving behind broken families and shattered lives. Many of these refugees have found their way to the United States, but thousands of others cannot leave the region and are forced to exist in the misery of squalid and overcrowded refugee camps. The insurgent state may not appear on the official map of Central America, but its presence is etched in the dislocation of peoples and economies.

Guatemala: On the Doorstep

Guatemala, land of the ancient Maya, began its modern existence as an entity of the Mexican Empire following independence from Spain in 1821. Bordered by Mexico to the west and north, and by Belize, Honduras, and El Salvador to the east, Guatemala has a lengthy Pacific coastline but a mere window on the Caribbean Sea (Fig. 4–15). Only one of Guatemala's neighbors, El Salvador, has felt the full onslaught of modern insurgency. Geography has shielded the country: Mexico, as yet unaffected; Belize, a British dependency until a decade ago; and Honduras, among the more stable countries in the region.

But Guatemala contains the ingredients that led to insurgency elsewhere in Central America. Outside the core area effective national control dwindles, notably in the northwest and north. During the past half-century, turbulence has dominated the political scene. In 1944 Guatemala seemed headed for reform after a brutal dictatorship, but 10 years later a military force organized in Honduras captured the capital and took over the government. Ever since, there has been a rebellion against the entrenched rulers in Guatemala City, a rebellion that has reached into the capital itself. Successive military governments (and the pro-military civilian regime elected in

1985) have tried various means to defeat the rebels, who presently are strongest in the western highlands. Death-squad assassinations of suspected collaborators, kidnapping, and military action have been combined with attempts at economic and social reform in what Guatemalans call the "rifles-and-beans" policy. But in the countryside, the military is feared and the rebels are often viewed as agents of change—where any change is seen as being for the better.

In the meantime, the economy of Guatemala suffers under the burden of huge military expenditures and from the effects of continuing strife. Coffee from the highlands and cotton from the drier areas are the main sources of income. Tourism was once a leading industry, but the political situation has destroyed it; foreign investment has all but dried up. Certainly, Guatemala has potential: it produces a wide range of farm crops and has major stands of exploitable timber and substantial nickel deposits. There is evidence that oil lies beneath the northeastern coastal plain. But Guatemala is polarized, and the poles are drifting farther apart. Until the future is clearer, the opportunities lie dormant.

Belize: In the Rimland

Strictly speaking, Belize is not a Central American republic in the same tradition as the other six. This country, a wedge of land between northern Guatemala, Mexico's Yucatán Peninsula, and the Caribbean, was a British dependency (British Honduras) until 1981; thus English, not Spanish, is the *lingua franca* here. Slightly larger than Massachusetts with a population of just 187,000, many of African descent, Belize is more reminiscent of a Caribbean island than of a mainland Middle American state.

But in other ways, Central American patterns do prevail. The largest city (and former capital), Belize City (60,000), is the only substantial urban center. Sugar is the principal export, and other commercial crops also generate external income; lobsters and fish are sold on foreign markets as well. Belize has not escaped political problems. A treaty between Britain and Guatemala that defined the territory's boundaries, signed in 1859, has been disputed by Guatemala's military rulers. From time to time, the Guatemalans have asserted that the 1859 treaty is void and that Belize ultimately belongs to Guatemala. Even when Belize attained independence a decade ago, that dispute was not finally resolved. Belizeans sometimes express the fear that their country might one day become the Falkland Islands (Malvinas) of Central America (see p. 315), and they worry that Guatemala's insurgencies may spill over into their country. Central America's turbulence leaves no country untouched.

Honduras: On the Front Line

Honduras occupies a critical place in the political geography of Central America, flanked as it is by Sandinist Nicaragua, strife-torn El Salvador, and tense Guatemala. As the map shows, Honduras, in direct contrast to Guatemala, has a lengthy Caribbean coastline and a small window on the Pacific. But that window on the Pacific's Gulf of Fonseca separates Nicaragua from El Salvador and puts Honduran territory and waters between alleged donor and recipients of communist weapons.

Honduras has a democratically elected government, but the military wields considerable power. With 5.3 million inhabitants, about 90 percent mestizo, Honduras is the region's poorest, least developed country. Agriculture, livestock, forestry, and some mining (lead and zinc) form the mainstays of the economy. The familiar Central American products—bananas, coffee, lumber, sugar—earn most of the external income. During the 1960s and 1970s, there was a promising development of light industry in the San Pedro Sula area near the northwest coast. But Honduras fell victim to regional strife, investors were afraid to risk their funds, and tourism declined. In recent years, refugees streamed into Honduras and American arms flowed through the country to the Nicaraguan border—and prospects have remained bleak. This is especially tragic because social and ethnic divisions, which so strongly mark Honduras's neighbors, are not serious here, and the gap between rich and poor, although evident, is not as wide. But Honduras does not, by itself, have the strength to keep outside enemies from violating its sovereignty.

El Salvador: Beyond Contention

El Salvador is Central America's smallest country territorially, smaller even than Belize—but, with a population almost 30 times as large (5.4 million), the most densely peopled. Again, as with Belize, it is one of only two mainland republics that do not have coastlines on the Atlantic as well as Pacific sides. El Salvador adjoins the Pacific in a chain of volcanic mountains. The country's heartland lies behind those mountains in the central interior, where the capital, San Salvador, is located. North of this core area lies another zone of mountains, through which the boundary with Honduras was delimited in the 1820s and 1830s. This mountainous interior also has always contained areas beyond effective governmental control.

Unlike neighboring Guatemala,

El Salvador's population is quite homogeneous (90 percent mestizo and just 10 percent Indian). But ethnic homogeneity did not translate into social or economic equality, or even opportunity. Other Central American countries were called "banana republics"; El Salvador was a coffee republic, and the coffee was produced on the huge landholdings of a few landowners and on the backs of a subjugated peasant labor force. During the first half of the twentieth century, an elite group known as the "fourteen families" ran the country almost as a feudal holding. The military supported this system and repeatedly suppressed violent and desperate peasant uprisings.

Were it not for great-power involvement, the events of the past several years might be viewed as simply another contest between ruling junta and peasant. But the nationwide outbreak of rebellion that began in the late 1970s was magnified by alleged Nicaraguan support (with Soviet weapons) for the rebels and U.S. support for the government, which made belated moves toward democracy. In El Salvador, an insurgent state has existed for over a decade, its representatives even meeting in scheduled discussions with members of the San Salvador administration. The stalemate, however, has continued, at disastrous cost to the country and its economy. As right-wing death squads and left-wing guerrillas carried out their campaigns, it became obvious that a century of oppression cannot be remedied by a few years of reform. As recently as 1980, 70 percent of all farm workers were laborers or sharecroppers, and most of the land remained in the hands of a few landlords. A sweeping program of land reallocation was started, despite opposition by both conservative landowners and leftist rebels. Now the owners, fearing further redistribution, re-

fuse to invest in their farmlands, and agriculture is in alarming decline all over El Salvador. As the country fragmented, the entire economy began to unravel and those who could, fled—on foot into Honduras, and by air and sea to countries farther away. In the early 1990s the map still shows a Republic of El Salvador, but in reality this republic has shrunk within its own boundaries, a remnant now of the coffee factory it once was (see right-hand map in Fig. 4–18).

Nicaragua: Heart of the Region

When studying Nicaragua, it is well to look again at the map (Fig. 4–15), which underscores the country's pivotal central position. Flanked by Honduras, poorest of the region's countries, to the north, and by Costa Rica, its richest, to the south, Nicaragua occupies the heart of Central America. The Pacific coast follows a southeasterly direction, but the Caribbean coast is oriented north–south, so that Nicaragua forms a triangle of land with its capital of Managua located on the mountainous Pacific side. Indeed, the core of Nicaragua always has been in this western zone. The Atlantic side, where the mountains and valleys give way to a coastal plain of (disappearing) rainforest, pine savanna, and swampland, has for centuries been home to Indian peoples such as the Miskito who have been relatively remote from the focus of national life.

Nicaragua was the typical Central American republic, ruled by a dictatorial government and exploited by a wealthy landowning minority, its export agriculture dominated by huge foreign-owned plantations (run by such large U.S. banana producers as United Brands and Standard Fruit). It was a situation ripe for contention, and the Sandinista rebels achieved that by the mid-1970s. An insurgent state was established, and in 1979 the Somoza government in Managua suffered the same fate as that of Batista in Cuba 20 years earlier. But Sandinista rule produced its own excesses, and opposition, although weak, persisted. The Miskito and other Indian peoples resisted Sandinista repression, and in the borderlands—chiefly along the boundary with Honduras—the Contras fought the Sandinista forces. Once again, big-power support magnified the conflict. Undoubtedly, the U.S. support for the Contras gave an ideological advantage to the Sandinistas, who successfully rallied nationalistic fervor at home against the "giant of the north" and its agents. The Nicaraguan economy was a major casualty of this conflict, and the country today (despite the election of a more democratic regime) is desperately trying to produce sufficient food and other basic necessities—a struggle made all the more difficult by a population-increase rate that now stands at a catastrophic 3.5 percent annually (Middle America's highest).

Costa Rica: Durable Democracy

As if to confirm what was said about Middle America's endless variety and diversity, Costa Rica differs in very significant ways from its neighbors—and from the norms of Central America. Bordered by Sandinist Nicaragua to the north and by volatile Panama to the east, Costa Rica is a nation with an old democratic tradition and, in this cauldron, *no* standing army! The country is, in fact, the oldest democracy in Middle and South America, enjoying a freely elected government (except for two brief periods) since 1889. Although the initial Hispanic imprint was similar to that found elsewhere on the Mainland, early inde-

pendence, the good fortune to lie remote from regional strife (which fostered an enduring posture of neutrality), and a leisurely pace of settlement allowed Costa Rica the luxury of concentrating on its economic development. Perhaps most important, internal political stability has prevailed over much of the past 175-odd years; the last brush with conflict, in the late 1940s, left the nation resolved to avoid further violence, and the armed forces were abolished in 1948 (along with a military establishment, so often the source of trouble throughout Central America).

Costa Rica, like its neighbors, is also divided into environmental zones that parallel the coasts. The most densely settled is the central highland zone, lying in the cooler *tierra templada*. Volcanic mountains prevail in much of this zone, but the heartland is the *Valle Central* (Central Valley), a fertile 40-by-50-mile/65-by-80-kilometer basin that contains the leading population cluster focused on the capital city and the country's main coffee-growing area. The capital, San José, is atypical of Middle America, a clean and slumless metropolis that is the most cosmopolitan urban center between Mexico City and South America. To the east of the highlands, are the hot and rainy Caribbean lowlands, a sparsely populated segment of Rimland where many plantations have now been abandoned and replaced by subsistence farmers. Between 1930 and 1960, the (U.S.–based) United Fruit Company shifted most of the country's banana plantations from the crop-disease-ridden Caribbean littoral to Costa Rica's third zone—the plains and gentle slopes of the Pacific coastlands. This gave the Pacific zone a major boost in its economic growth, and today it is the scene of diversifying and expanding commercial agriculture (often requiring irrigation) as well

as successful colonization schemes in previously undeveloped valleys and basins.

The long-term development of Costa Rica's economy has given it the region's highest standard of living, literacy rate, and life expectancy. Agriculture continues to dominate, with coffee, bananas, beef, and sugar the leading exports. These commodities, however, have been vulnerable to sharp price fluctuations on the world market since 1980. Low prices and heavy borrowing abroad (to cover the costs of expensive domestic social programs and fuel imports) have now combined to saddle Costa Rica with an enormous foreign debt, giving the country the dubious distinction of leading Central America in that category. Thus, lowered national aspirations are now the order of the day, and the hope is that economic problems will not spill over into the political sphere. The latter remains quite stable, despite the proximity to the region's trouble spots and the presence of thousands of refugees from El Salvador and Nicaragua. Costa Rica itself harbors no rebel activity at all, and the overwhelming majority of its peace-loving people prefer the country to maintain its neutrality as "the Switzerland of Central America."

Panama: Strategic Canal, Volatile Corridor

The Republic of Panama owes its birth to an idea: the construction of an artificial waterway to connect the Atlantic and Pacific oceans and thereby avoid the lengthy circumnavigation of South America. In the 1880s, when Panama was still an extension of neighboring Colombia, a French company directed by Ferdinand de Lesseps (1805–1894)—builder of the Suez Canal in the 1860s—tried and failed to build such a canal here; thousands of workers died of yellow fever, malaria, and other tropical diseases, and the company went bankrupt. By the turn of the century, U.S. interest in a Panama canal (which would shorten the sailing distance between the east and west coasts by 8000 nautical miles) rose sharply, and the United States in 1903 proposed a treaty that would permit a renewed effort at construction across Colombia's Panamanian isthmus. When the Colombian Senate refused to go along, Panamanians rebelled and the United States supported this uprising by preventing Colombian forces from intervening. The Panamanians, at the behest of the United States, declared their independence from Colombia, and the new republic immediately granted the United States rights to the Canal Zone averaging about 10 miles (16 kilometers) in width and just over 50 miles (80 kilometers) in length.

Soon canal construction commenced, an epic struggle documented by historian David McCullough in his classic book, *The Path Between the Seas*. This time, the project succeeded as American engineering and technology—and medical advances—triumphed over a formidable set of obstacles. The Panama Canal (see inset map, Fig. 4–15) was opened in 1914, a symbol of U.S. power and influence in the Caribbean and Middle America. The Canal Zone was held by the United States under a treaty that granted it "all the rights, powers, and authority" in the area "as if it were the sovereign of the territory."

Such language might suggest that the United States held rights over the Canal Zone in perpetuity, but the treaty nowhere stated specifically that Panama permanently yielded its own sovereignty in that transit corridor. In the 1970s, as the canal was transferring 20,000 ships per year (Fig. 4–19) and generating hundreds of

FIGURE 4–19 Opened in 1914, the Panama Canal is still a vital waterway for world shipping. These are the locks west of the Gaillard Cut (at the rear), the man-made gorge in which the canal passes through the isthmus's continental divide.

millions of dollars in tolls, Panama sought to terminate U.S. control in the Canal Zone. Delicate negotiations began; in 1977, an agreement was reached on a staged withdrawal by the United States from the territory, first from the Canal Zone and then, by 2000, from the Panama Canal itself. This agreement took the form of two treaties, and following the signing by Presidents Carter and Torrijos, they were ratified in spite of stubborn opposition in the U.S. Senate.

Today, Panama shares some but not all the usual geographic features of the Central American republics. Its population of 2.5 million is about 70 percent mestizo, but also contains substantial black, white, and Indian sectors. Spanish is the official (and majority) language, but English is in widespread use. Ribbon-like and oriented east–west, Panama's topography is mountainous and hilly, with some mountains reaching higher than 10,000 feet (3000 meters). Eastern Panama, especially Darien Province adjoining Colombia, is densely forested, and here is the only remaining gap in the otherwise complete intercontinental Pan American Highway. Most of the rural population lives in the uplands west of the canal (much of the urban population is concentrated in the vicinity of the waterway, anchored by the cities at each end of the canal, Panama City and Colón). There, Panama produces bananas, rice, sugarcane, and coffee—and from the sea, shrimp and fishmeal.

Panama today finds itself rebuilding from the excesses of the Noriega regime, which in the late 1980s brought on international economic sanctions and the U.S. military invasion that ousted the dictator. Although damage to the country's economy and political stability was considerable, the canal remains Panama's focus, its lifeline, its future. It has now been augmented by an oil pipeline that crosses the isthmus, facilitating the interoceanic flow of petroleum, and by the Colón Free Zone (at the Caribbean end of the canal), one of the world's largest free trade areas. Relative location was Panama's seminal advantage; it has been its ally ever since its birth. Separated by Costa Rica from the turmoil of the other Mainland republics and by the impassable Darien Gap from turbulent Colombia, Panama still has a future potential that few Middle American countries can match.

PRONUNCIATION GUIDE

Antigua (an-*tee*-gwuh)

Antilles (an-*till*-eeze)

Arawak (*arra*-wak)

Archipelago (ark-uh-*pell*-uh-go)

Aruba (uh-*roo*-buh)

Augelli (aw-*jelly*)

Baja (*bah*-hah)

Balsas (*bahl*-suss)

Barbados (bar-*bay*-dohss)

Bauxite (*bawks*-ite)

Belize (buh-*leeze*)

Belmopan (bell-moe-*pan*)

Bering (*berring*)

Bonaire (bun-*air*)

Cacao (kuh-*kay*-oh)

Campeche (kahm-*pay*-chee)

Caribbean (kuh-*rib*-ee-un/karra-*bee*-un)

Cay (*kee*)

Chihuahua (chuh-*wah*-wah)

Ciboney (see-*boh*-nay)

Ciudad Juarez (see-you-*dahd* wah-rez)

Ciudades perdidas (see-you-*dah*-dayss pair-*dee*-duss)

Cocos (*koh*-kuss)

Colón (kuh-*loan*)

Cortés, Hernán (kor-*tayss*, air-*nahn*)

Costa Rica (koss-tuh-*ree*-kuh)

Curaçao (koor-uh-*sau*)

Darien (dar-*yen*)

de Lesseps, Ferdinand (duh-leh-*sepps*, faird-*nahn*)

Ejido[s] (eh-*hee*-doh[ss])

Fonseca (fahn-*say*-kuh)

Fort-de-France (for-duh-*frahss*)

Gaillard (gil-*yard*)

Grenada (gruh-*nay*-duh)

Guadalajara (gwah-duh-luh-*hahr*-uh)

Guadeloupe (*gwah*-duh-loop)

Guatemala (gwut-uh-*mah*-lah)

Hacienda (ah-see-*en*-duh)

Hägerstrand, Torsten (*hay*-gherr-strand, *tor*-stun)

Haiti (*hate*-ee)

Hispaniola (iss-pahn-*yoh*-luh)

Honduras (hon-*dure*-russ)

Isthmus/isthmian (*iss*-muss/*iss*-mee-un)

Jamaica (juh-*make*-uh)

Littoral (*lit*-oh-rull)

Maize (*mayz*)

Managua (mah-*nah*-gwuh)

Maquiladora (mah-kee-yuh-*dorr*-uh)

Martinique (mahr-tih-*neek*)

Maya[n] (*my*-uh[un])

Mesoamerica (*mezzoe*-america)

Mestizo (meh-*stee*-zoh)

Mexica (meh-*shee*-kuh)

Monterrey (mahnt-uh-*ray*)

Mulatto (moo-*lah*-toe)

Nahuatl (nah-*wattle*)

Nassau (*nass*-saw)

Nicaragua (nick-uh-*rah*-gwuh)

Ocho Rios (oh-choe-*ree*-ohss)

Páramos (*pah*-ruh-mohss)

Paseo de la Reforma (puh-*say*-oh day-luh ray-*for*-muh)

Peones (pay-*oh*-nayss)

Placer[ing] (*plass*-uh[ring])

Port-au-Prince (por-toh-*pranss*)

Puerto Plata (pwair-toh-*plah*-tuh)

Puerto Rico (pwair-toh-*ree*-koe)

Puna (*poona*)

Quiché (kee-*chay*)

Saba (*say*-buh/*sah*-buh)

Sabinas (sah-*bee*-nuss)

Sandinista (sahn-dee-*nee*-stuh)

San José (sahn-hoe-*zay*)

San Juan (sahn-*hwahn*)

San Pedro Sula (sahn-pay-droe-*soo*-yuh)

Sierra Madre (see-*erra mah*-dray)

Occidental (oak-see-den-*tahl*)
Oriental (aw-ree-en-*tahl*)
St. Eustatius (saint yoo-*stay*-shuss)
St. Lucia (saint *loo*-shuh)
St. Maarten (sint-*mahrt*-un)
Tampico (tam-*peek*-oh)
Tarascans (tuh-ruh-*skahnz*)
Tegucigalpa (tuh-goose-ih-*gahl*-puh)

Tehuantepec (tuh-*whahn*-tuh-pek)
Tenochtitlán (tay-noh-chit-*lahn*)
Tierra caliente (tee-*erra* kahl-*yen*-tay)
Tierra fría (tee-*erra free*-uh)
Tierra helada (tee-*erra* ay-*lah*-dah)
Tierra templada (tee-*erra* tem-*plah*-dah)

Tijuana (tee-*whahn*-uh)
Tobago (tuh-*bay*-goh)
Torrijos (tor-*ree*-hohss)
Tula (*too*-lah)
Valle Central (*vah*-yay sen-*trahl*)
Villahermosa (vee-yuh-air-*moh*-suh)
Willemstad (*vill*-um-staht)
Yucatán (yoo-kuh-*tahn*)
Zaïre (zah-*ear*)

REFERENCES AND FURTHER READINGS

(BULLETS [•] DENOTE BASIC INTRO-DUCTORY WORKS ON THE REALM OR SYSTEMATIC ESSAY TOPIC.)

• Augelli, John P. "The Rimland-Mainland Concept of Culture Areas in Middle America," *Annals of the Association of American Geographers*, 52 (1962): 119-129.

Blakemore, Harold & Smith, Clifford T., eds. *Latin America: Geographical Perspectives* (London & New York: Methuen, 2 rev. ed., 1983).

Blakemore, Harold et al., eds. *The Cambridge Encyclopedia of Latin America and the Caribbean* (London: Cambridge University Press, 1985).

• Blouet, Brian W. & Blouet, Olwyn M., eds. *Latin America: An Introductory Survey* (New York: John Wiley & Sons, 1982).

Boehm, Richard G. & Visser, Sent, eds. *Latin America: Case Studies* (Dubuque, Iowa: Kendall/Hunt, 1984).

Brown, Ralph H. *Mirror for Americans: Likeness of the Eastern Seaboard 1810* (New York: American Geographical Society, 1943).

Butlin, Robin A. *Historical Geography: Retrospect, Context and Prospect* (London & New York: Routledge, 1989).

• Clark, Andrew H. "Historical Geography," in James, Preston E. & Jones, Clarence F., eds., *American Geography: Inventory and Prospect* (Sy-racuse, N.Y.: Syracuse University Press, 1954), pp. 70–105.

• Crowley, William & Griffin, Ernst C. "Political Upheaval in Central America," *Focus*, September–October 1983.

Davidson, William V. & Parsons, James, eds. *Historical Geography of Latin America* (Baton Rouge: Louisiana State University Press, 1980).

Dunkerley, James. *Power in the Isthmus: A Political History of Modern Central America* (New York: Verso/Routledge, 1989).

Gourou, Pierre. *The Tropical World: Its Social and Economic Conditions and Its Future Status* (London & New York: Longman, 5 rev. ed., trans. Stanley H. Beaver, 1980).

• Griffin, Ernst C. & Ford, Larry R. "Cities of Latin America," in Brunn, Stanley D. & Williams, Jack F., eds., *Cities of the World: World Regional Urban Development* (New York: Harper & Row, 1983), pp. 198–240.

Hall, Carolyn. *Costa Rica: A Geographical Interpretation in Historical Perspective* (Boulder, Colo.: Westview Press, 1985).

Helms, Mary W. *Middle America: A Culture History of Heartland and Frontiers* (Englewood Cliffs, N.J.: Prentice-Hall, 1975).

Huntington, Ellsworth. *Mainsprings of Civilization* (New York: John Wiley & Sons, 1945).

• James, Preston E. & Minkel, Clarence W. *Latin America* (New York: John Wiley & Sons, 5 rev. ed., 1986), Chaps. 2–16.

Kandell, Jonathan. *La Capital: The Biography of Mexico City* (New York: Random House, 1988).

Knight, Franklin W. & Palmer, Colin A., eds. *The Modern Caribbean* (Chapel Hill, N.C.: University of North Carolina Press, 1989).

Levy, Daniel C. & Szekely, Gabriel. *Mexico: Paradoxes of Stability and Change* (Boulder, Colo.: Westview Press, 2 rev. ed., 1987).

Lewis, Peirce F. "Axioms for Reading the Landscape: Some Guides to the American Scene," in Meinig, Donald W., ed., *The Interpretation of Ordinary Landscapes: Geographical Essays* (New York: Oxford University Press, 1979), pp. 11–32.

• Lowenthal, David. *West Indian Societies* (New York: Oxford University Press, 1972).

"[Maquiladora] Manufacturing in Mexico: On Uncle Sam's Coat-Tails," *The Economist*, September 16, 1989, p. 82.

MacPherson, John. *Caribbean Lands* (London & New York: Longman, 4 rev. ed., 1980).

McColl, Robert W. "The Insurgent State: Territorial Bases of Revolution," *Annals of the Association of American Geographers*, 59 (1969): 613–631.

McCullough, David G. *The Path Between the Seas: The Creation of the Panama Canal, 1870–1914* (New York: Simon & Schuster, 1977).

Meinig, Donald W. "Symbolic Landscapes: Some Idealizations of American Communities,'' in Meinig, Donald W., ed., *The Interpretation of Ordinary Landscapes: Geographical Essays* (New York: Oxford University Press, 1979), pp. 164–192.

"Mexico City: The Population Curse,'' *Time*, August 6, 1984, pp. 26–35.

Myers, Norman. *The Primary Source: Tropical Forests and Our Future* (New York: W.W. Norton, 1984).

Nietschmann, Bernard Q. *Caribbean Edge: The Coming of Modern Times to Isolated People and Wildlife* (Indianapolis: Bobbs-Merrill, 1979).

Pearce, Douglas. *Tourism Today: A Geographical Analysis* (London & New York: Longman, 1986).

Richardson, Bonham C. *Caribbean Migrants: Environment and Human Survival on St. Kitts and Nevis* (Knoxville: University of Tennessee Press, 1983).

Riding, Alan. *Distant Neighbors: A Portrait of the Mexicans* (New York: Alfred A. Knopf, 1985).

Rumney, Thomas. *Mexico and Central America: A Selected Bibliography on the Geography of the Region* (Monticello, Ill.: Vance Bibliographies, No. P-1947, 1986).

Sargent, Charles S., Jr. "The Latin American City,'' in Blouet, Brian W. & Blouet, Olwyn M., eds., *Latin America: An Introductory Survey* (New York: John Wiley & Sons, 1982), pp. 201–249. Quotation taken from p. 221; diagram adapted from p. 223.

Sauer, Carl O. "Foreword to Historical Geography,'' *Annals of the Association of American Geographers*, 31 (1941): 1–24.

Scott, Ian. *Urban and Spatial Development in Mexico* (Baltimore: Johns Hopkins University Press, 1982).

Sealey, Neil. *Tourism in the Caribbean* (London: Hodder & Stoughton, 1982).

Stanislawski, Dan. "Early Spanish Town Planning in the New World,'' *Geographical Review*, 37 (1947): 94–105.

Turner, Bill L., II. *Once Beneath the Forest* (Boulder, Colo.: Westview Press, 1983).

Watts, David. *The West Indies: Patterns of Development, Culture and Environmental Change Since 1492* (New York: Cambridge University Press, 1987).

Weaver, Muriel P. *The Aztecs, Maya, and Their Predecessors: Archaeology of Middle America* (New York: Academic Press, 2 rev. ed., 1981).

• West, Robert C., Augelli, John P. et al. *Middle America: Its Lands and Peoples* (Englewood Cliffs, N.J.: Prentice-Hall, 3 rev. ed., 1989).

• Wilkie, Richard W. *Latin American Population and Urbanization Analysis: Maps and Statistics, 1950–1982* (Westwood, Calif.: UCLA Latin American Center, 1984).

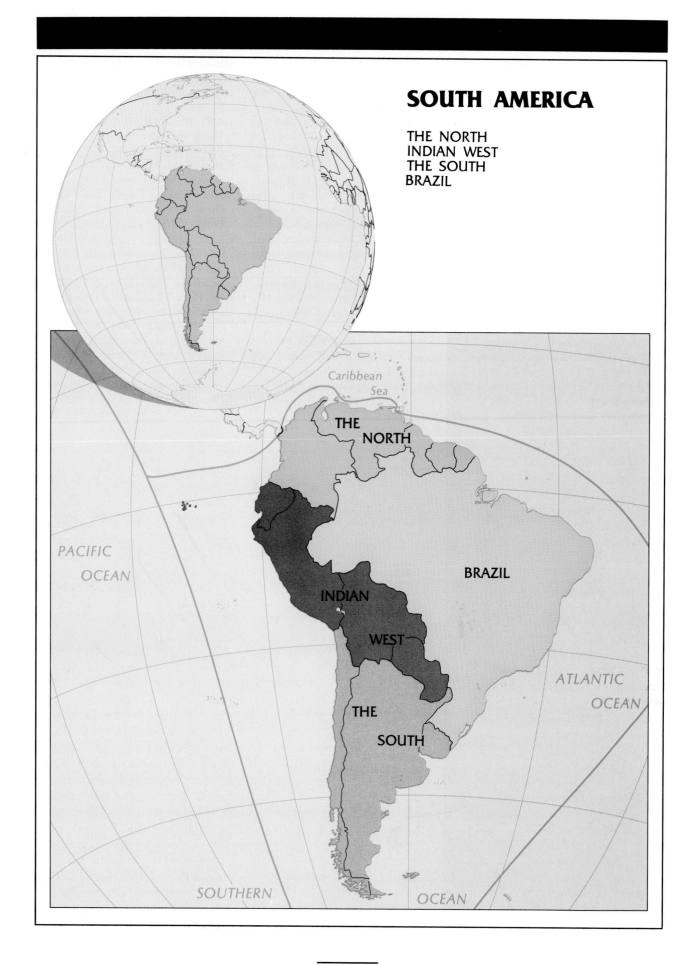

SOUTH AMERICA

THE NORTH
INDIAN WEST
THE SOUTH
BRAZIL

Caribbean Sea

THE NORTH

PACIFIC OCEAN

INDIAN

WEST

BRAZIL

THE SOUTH

ATLANTIC OCEAN

SOUTHERN OCEAN

CHAPTER 5

SOUTH AMERICA: TRADITION AND TRANSITION

South America, of all the continents, has the most familiar shape, that giant triangle connected by mainland Middle America's tenuous land bridge to its sister continent in the north. What we less often realize about South America is that it lies not only south, but also mostly east of its northern counterpart as well. Lima, the capital of Peru—one of the continent's westernmost cities—lies farther east than Miami, Florida. Thus, South America juts out much more prominently into the Atlantic Ocean than does North America, and South American coasts are much closer to Africa and even to southern Europe than the coasts of Middle and North America. Lying so far eastward, South America's western flank faces a much wider Pacific Ocean than does North America. From its west coast to Australia is nearly twice as far as from San Francisco to Japan, and South America has virtually no interaction with the Pacific world of Australasia—not just because of vast distances, but also because both lie in the insular and less populous Southern Hemisphere, whereas western North America faces Japan and the crowded East Asian mainland.

As if to reaffirm South America's northward and eastward orientation, the western margins of

---

IDEAS AND CONCEPTS

economic geography

agricultural systems

land alienation

isolation

south american culture spheres

third world urbanization

the "latin" american city

rural-to-urban migration

---

the continent are rimmed by one of the world's longest and highest mountain ranges, the Andes, a gigantic wall that extends from Tierra del Fuego near the southern tip of the triangle to Venezuela in the north (Fig. 5–1). Every map of world physical geography clearly reflects the existence of this mountain chain—in the alignment of isohyets (lines connecting places of equal precipitation totals) (Fig. I–9, p. 14), in the elongated zone of highland climate (Fig. I–10, p. 18), and in the regional distributions of vegetation and soils (Figs. I–12 and I–13, pp. 22 and 24, respectively). Moreover, as Fig. I–14 (p. 28) reveals, South America's biggest population clusters

are located along the eastern and northern coasts, overshadowing those of the Andean west.

THE HUMAN SEQUENCE

Although modern South America's largest populations are situated in the east and north, there was a time during the height of the Inca Empire when the Andes Mountains contained the most densely peopled and best organized state on the continent. Although the origins of Inca civilization are still shrouded in mystery, it has become generally accepted that the Incas were descendants of ancient peoples who came to South America via the Middle American land bridge (possibly following earlier migrations from Asia to North America via the Bering land bridge). But even this is not totally beyond doubt; some scholars maintain that the first settlers in this part of the Western Hemisphere may have reached the Chilean and Peruvian coasts directly by sea from distant Pacific islands. In any case, for thousands of years before the Europeans arrived in

TEN MAJOR GEOGRAPHIC QUALITIES OF SOUTH AMERICA

1. South America's physiography is dominated by the Andes Mountains in the west and the Amazon Basin in the central north. Much of the remainder is plateau country.
2. Half the realm's area and just over half its population are concentrated in one country—Brazil.
3. South America's population remains concentrated in peripheral zones. Most of the interior is sparsely peopled.
4. The leading population trend is cityward migration. No realm has urbanized more rapidly since 1980.
5. Regional economic contrasts and disparities, both in the realm as a whole and within individual countries, are strong. In general, the south is the most developed, the northeast the least.
6. Interconnections among the states of the realm remain comparatively weak. External ties are frequently stronger.
7. Strong cultural pluralism exists in the majority of the realm's countries, and this pluralism is often expressed regionally.
8. With the exception of three small countries in the north, the realm's modern cultural sources lie in a single subregion of Europe, the Iberian Peninsula. Spanish is the *lingua franca*, except in Portuguese-speaking Brazil.
9. Lingering politico-geographical problems beset the realm. Boundary disputes and territorial conflicts persist.
10. The Catholic church dominates life throughout the realm and constitutes one of its unifying elements.

the sixteenth century, Indian communities and societies had been developing in South America.

About 1000 years ago, a number of regional cultures thrived in Andean valleys and basins, and at places along the Pacific Coast. The llama had been domesticated as a beast of burden, a source of meat, and a producer of wool. Religions flourished and stimulated architecture as well as the construction of temples and shrines. Sculpture, painting, and other art forms were practiced. Over these cultures the Incas extended their authority from their headquarters in the Cuzco Basin of the Peruvian Andes, beginning late in the twelfth century, to forge the greatest em-

pire in the Americas prior to the coming of the Europeans.

Nothing to compare with the cultural achievements of this central Andean zone existed anywhere else in South America. Beyond the Andean civilizations, anthropologists recognize three other major groupings of Indian peoples: (1) those of the Caribbean fringe, (2) those of the tropical forest of the Amazon Basin and other lowlands, and (3) those called "marginal," whose habitat lay in the Brazilian Highlands, the headwaters of the Amazon River, and most of southern South America (Fig. 5–3). It has been estimated that these Caribbean, forest, and marginal peoples together con-

stituted only about one-quarter of the continent's total indigenous (native) population.

The Incan Empire

When the Inca civilization is compared to that of ancient Mesopotamia, Egypt, the old Asian civilizations, and the Aztecs' Mexica Empire, it quickly becomes clear that this was an unusual achievement. Everywhere else, rivers and waterways provided avenues for interaction and the circulation of goods and ideas. Here, however, an empire was forged out of a series of elongated basins (called *altiplanos*) in the high Andes, created when mountain valleys between parallel and converging ranges filled with erosional materials from surrounding uplands. These *altiplanos* are often separated from one another by some of the world's most rugged terrain, with high snowcapped mountains alternating with precipitous canyons. Individual *altiplanos* accommodated regional cultures; the Incas themselves were first established in the intermontane basin of Cuzco (Fig. 5–3). From that hearth, they conquered and extended their authority over the peoples of coastal Peru and other *altiplanos*. Their first thrust, apparently late in the fourteenth century, was southward.

More impressive than the Incas' military victories was their subsequent capacity to integrate the peoples and regions of the Andean realm into a stable and smoothly functioning state. The odds would seem to have been against them, because as they progressed as far south as central Chile, their domain became ever more elongated, making effective control much more difficult. The Incas, however, were expert road and bridge builders, colonizers, and administrators; in an incredibly short time, they consolidated these new ter-

FIGURE 5–1

ritories. And just before the Spanish arrival (1531), they even conquered territories to the north, including Ecuador and part of southern Colombia.

The early sixteenth century was a critical period in the Inca Empire because the conquest of Ecuador and areas to its north for the first time placed stress on the existing administrative framework. Until that time, the dominant center of the state had been Cuzco; but now it was decided that the empire should be divided into two units— a southern one ruled from Cuzco and a northern sector focused on Quito. This decision was related to the problem of control over the rebellious north and the possibilities for expansion deeper into Colombia. Thus, the empire was now beset by a number of difficulties, notably the uncertain northern

S Y S T E M A T I C   E S S A Y

ECONOMIC GEOGRAPHY

In the preceding chapters and vignettes, we have discussed several concepts, principles, and examples of **economic geography**. In the introductory chapter, this subdiscipline was defined (p. 38) as being concerned with the diverse ways in which people earn a living, and how the goods and services they produce are expressed and organized spatially. In Chapter 3 (p. 208), we identified four sets of productive activities as the major components of the *spatial economy*: (1) primary activities (agriculture, mining, and other extractive industries); (2) secondary activities (manufacturing); (3) tertiary activities (services); and (4) quaternary activities (information and decision-making). Since most of the world's secondary, tertiary, and quaternary activities are located within the developed realms, we focus here on *agriculture*, the dominant livelihood of the hundreds of millions of workers who inhabit the remaining underdeveloped realms.

The global distribution of **agricultural systems** is displayed in Fig. 5–2. The spatial organization of agriculture in the advanced commercial economies of Europe and the United States has already been explained in the context of the Von Thünen model, expanded, thanks to modern transportation technology, to the continental scale. In fact, the macro-Thünian framework can even be extended to the world as a whole. The "global city" would be the European and North American edges of the North Atlantic Basin, and many of the colonially generated farming systems would fit a sequence of concentric and increas-

WORLD AGRICULTURE

1 Dairying

2 Fruit, Truck and Specialized Crops

3 Mixed Livestock and Crop Farming

4 Grain Farming

5 Subsistence Crop and Livestock Farming

6 Mediterranean Agriculture

7 Diversified Tropical Agriculture -chiefly plantation

8 Intensive Subsistence Farming -chiefly rice

9 Intensive Subsistence Farming -other crops

10 Rudimental Sedentary Cultivation

11 Shifting Cultivation

12 Livestock Ranching

13 Nomadic and Semi-Nomadic Herding

Nonagricultural Areas

FIGURE 5–2

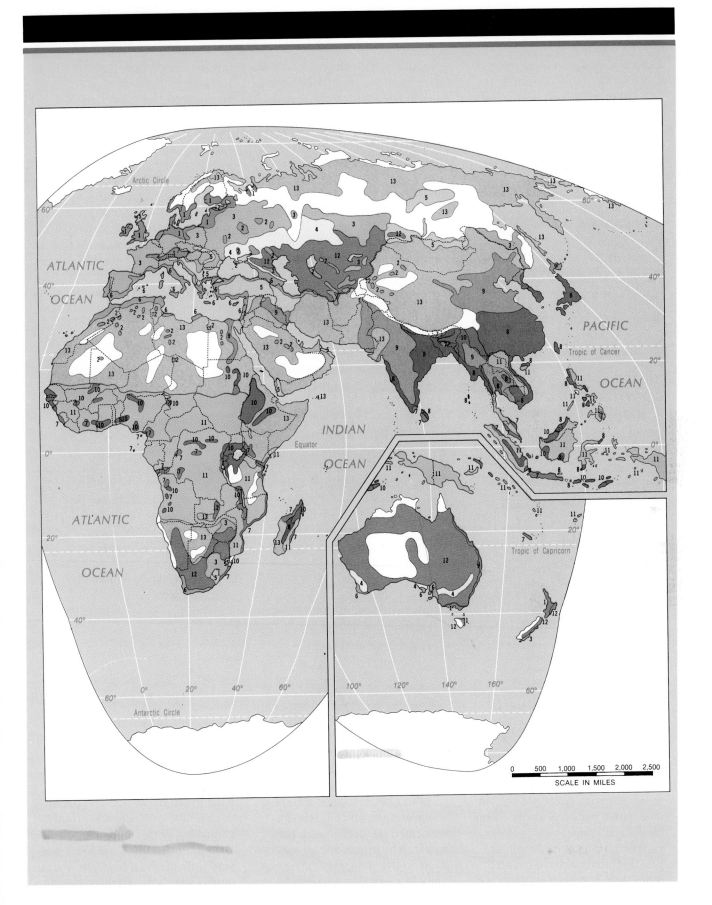

SCALE IN MILES

ingly distant agricultural zones (such as Middle American fruit and sugar, Argentine beef, Australian wheat, and New Zealand wool). Although this worldwide Von Thünen structuring has eroded with the demise of colonialism, many of that era's features are still apparent (for example, plantation farming), reminding us that most nonsubsistence Third World agriculture remains firmly oriented to the markets of the Western powers that first reorganized primary production in countries of the underdeveloped realms. The spatial pattern of those commercial farming systems can be observed in Fig. 5–2 in the distributions of Categories 1 through 4 and 6, 7, and 12.

The six remaining categories are all associated with the various types of noncommercial, subsistence agriculture. Shifting cultivation and nomadic/seminomadic herding (Categories 11 and 13) are the least intensive of these activities, relegated to marginal environments that—together with vast (unnumbered/white) nonagricultural areas—occupy more than two-thirds of the planet's land surface. Categories 5, 8, 9, and 10 entail more intensive forms of

subsistence agriculture. As we saw in the introductory chapter, the intensive rice farming of Category 8 supports the earth's largest and densest nonurban population clusters on the alluvial soils of Asia's great river valleys and coastal plains; the intensive subsistence cultivation of other crops (Category 9), especially such grains as wheat, can also support huge population concentrations (northern China and northwestern India are two leading examples). Categories 5 and 10 are occasionally associated with sizable populations as well, but these farming systems are far less productive and often represent a transition between intensive and extensive agriculture (the latter embodied in Categories 11 and 13).

Looking at the South American portion of Fig. 5–2, commercial and subsistence farming exist side by side here to a greater degree than in any other realm (where one of the two always geographically dominates the other). This, of course, does not represent a planned "balance" between the two; rather, it reflects the continent's deep internal cultural and economic divisions, on which this chapter elaborates. The commercial agricultural side of South

America is expressed in: (1) a huge cattle-ranching zone near coasts (Category 12) that stretches southwest from northeastern Brazil to Patagonia; (2) Argentina's wheat-raising Pampa (Category 4), which is comparable to the U.S. Great Plains; (3) a Corn Belt–type crop and livestock zone (Category 3) in northeastern Argentina, Uruguay, southern Brazil, and south-central Chile; (4) a number of seaboard tropical plantation strips (Category 7) located in Brazil, the Guianas, Venezuela, Colombia, and Peru; and (5) a Mediterranean agricultural zone (Category 6) in Middle Chile. In stark contrast to these commercial systems, subsistence farming covers the rest of the realm's arable land: (1) primitive shifting cultivation (Category 11) occurs in the rainforested Amazon Basin and its hilly perimeter; (2) rudimental sedentary cultivation (Category 10) dominates the Andean plateau country from Colombia in the north to the Bolivian Altiplano in the south; and (3) a ribbon of mixed subsistence farming (Category 5) courses through most of eastern Brazil between the coastal plantation and interior grazing zones.

frontier and rising tensions between Cuzco and Quito. And just as the Aztec Empire had been ripe for internal revolt when Cortés and his party entered Mexico, so the Spanish arrival in western South America happened to coincide with a period of stress within the Inca Empire.

When it was at its zenith, the Inca Empire may have counted more than 20 million subjects. Of

course, the Incas themselves were always in a minority in this huge state, and their position became one of a ruling elite in a rigidly class-structured society. The Incas, representatives of the emperor in Cuzco, formed a caste of administrative officials who implemented the decisions of their monarch by organizing all aspects of life in the conquered territories. They saw to it that harvests were

divided between the church, the community, and individual families; they maintained the public granaries; they made investments to construct and maintain roads, terrace hillsides, and expand irrigation works. The life of the empire's subjects was strictly controlled by this bureaucracy of Inca administrators, and there was very little personal freedom. Farm quotas were set, and there was no

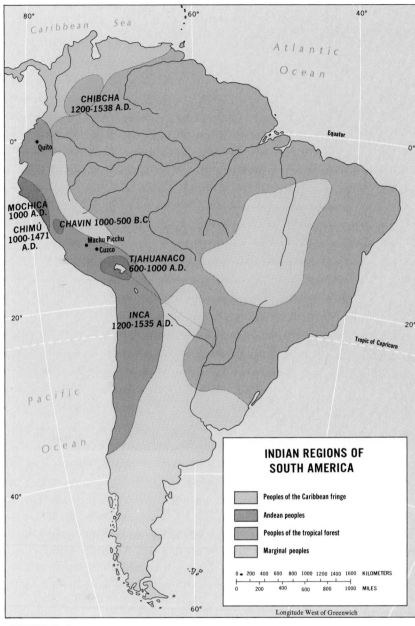

FIGURE 5–3

Perhaps the swiftness of its development contributed to its fatal weakness. At any rate, besides spectacular ruins, like those at Peru's Machu Picchu (Fig. 5–4), the empire left behind numerous social values that have remained a part of Indian life in the Andes to this day and continue to contribute to fundamental divisions between the Iberian and Indian populations in this part of South America. For example, the Inca state language, Quechua, was so firmly rooted that it is still spoken by more than 5 million Indians living in the highlands of Peru, Ecuador, and Bolivia.

The Iberian Invaders

In South America, as in Middle America, the location of Indian peoples determined to a considerable extent the direction of the thrusts of European invasion. The Incas, like Mexico's Maya and Aztec peoples, had accumulated gold and silver in their headquarters, possessed productive farmlands, and constituted a ready labor force. Not long after the 1521 defeat of the Mexica Empire of the Aztecs, the Spanish conquerors crossed the Panamanian isthmus and sailed southward along the continent's northwestern coast. On his first journey Francisco Pizarro heard of the existence of the Inca Empire. After a landfall in 1527 at Tumbes, on the northern coastal extremity of Peru, he returned to Spain to organize the penetration of the Incan domain. Pizarro returned to Tumbes with 183 men and two dozen horses in 1531, a time when the Incas were preoccupied with problems of royal succession and strife in the northern provinces. The events that followed are well known, and less than three years later the party rode victorious into Cuzco. Initially, the Spaniards kept intact

true market economy; the crops, like the soil on which they were grown, belonged to the state. Marriages were officially arranged, and families could live only where the Inca supervisors would permit. Indeed, the family (as a productive entity within the community), not the individual, was considered to be the basic unit of administration. Inca rule was thoroughly effective, and obedience was the only course

for its subjects. So highly centralized was the state and so complete the subservience of its tightly controlled population that a takeover at the top was enough to gain power over the entire empire—as the Spaniards quickly proved in the 1530s.

The Inca Empire, which had risen to greatness so rapidly, disintegrated abruptly with the impact of the Spanish invaders.

FIGURE 5—4 Machu Picchu's famous Inca ruins in their spectacular mountain setting, located not far from Cuzco in the high Peruvian Andes.

the Incan imperial structure by permitting the crowning of an emperor who was, in fact, under their control; but soon the land- and gold-hungry invaders were fighting among themselves, and the breakdown of the old order began.

The new order that eventually emerged in western and southern South America placed the Indian peoples in serfdom to the Spaniards. Great haciendas were formed by **land alienation** (taking over former Indian lands), taxes were instituted, and a forced-labor system was introduced to maximize the profits of exploitation. As in Middle America, most of the Spanish invaders had little status in Spain's feudal society, but they brought with them the values that

prevailed in Iberia: land meant power and prestige, gold and silver meant wealth. Lima, the west-coast headquarters of the Spanish conquerors, was founded by Pizarro in 1535, about 375 miles (600 kilometers) northwest of the Andean center of Cuzco. Before long it was one of the richest cities in the world, reflecting the enormity of the wealth that the ravaged Inca Empire was yielding. Lima soon became the capital of the viceroyalty of Peru, as the authorities in Spain integrated the new possession into their colonial empire (Fig. 5–5). Subsequently, when Colombia and Venezuela came under Spanish control, and later on when Spanish settlement began to expand in the coastlands

of the Rio de la Plata estuary in present-day Argentina and Uruguay, two additional viceroyalties were added to the map: New Granada in the north and La Plata in the south.

Meanwhile, another vanguard of the Iberian invasion was penetrating the east-central part of the continent, the coastlands of present-day Brazil. This area had become a Portuguese sphere of influence, because Spain and Portugal had agreed in the Treaty of Tordesillas (1494) to recognize a north-south line (drawn by Pope Alexander VI) 370 leagues west of the Cape Verde Islands as the boundary between their New World spheres of influence. This border ran approximately along the meridian of 50°W longitude, thereby cutting off a sizable triangle of eastern South America for Portugal's exploitation (Fig. 5–5).

A brief look at the political map of South America (Fig. 5–6), however, shows that the 1494 treaty did not succeed in limiting Portuguese colonial territory to the east of the agreed-on 50th meridian. True, the boundaries between Brazil and its northern and southern coastal neighbors (French Guiana and Uruguay) both reach the ocean near 50°W, but then the Brazilian boundaries bend far inland to include almost the entire Amazon Basin as well as a good part of the Paraná-Paraguay Basin. Thus, Brazil came to be only slightly smaller than all the other South American countries combined. In population, too, Brazil now accounts for about half the total of the entire continent. The successful, enormous westward thrust was the work of many Brazilian elements—missionaries in search of converts, explorers in search of quick wealth—but no group did more to achieve this penetration than the so-called *Paulistas*, the settlers of São Paulo. From early in its colonial history São Paulo had been a thriving settlement, with highly profitable plantations and an ever-growing need for labor. The *Paulistas* organized huge expeditions into the interior, seeking Indian slaves, gold, and precious stones; at the same time, they were intent on reducing the influence of Jesuit missionaries over the far-flung Indian population.

The Africans

As Fig. 5–5 shows, the Spaniards initially got very much the better of the territorial partitioning of South America—not just quantitatively, but qualitatively as well.

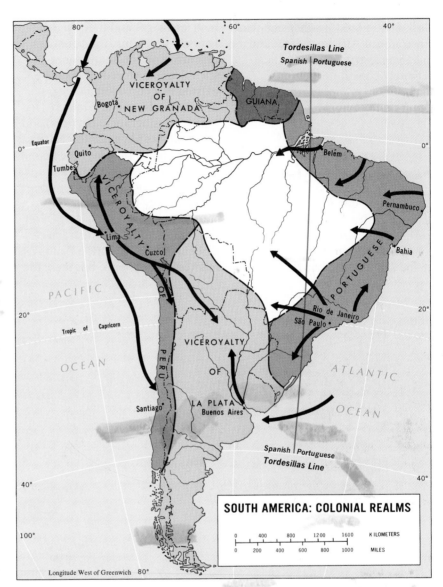

FIGURE 5–5

There were no rich Indian states to be conquered and looted east of the Andes, and no productive agricultural land was under cultivation. The comparatively few eastern Indians constituted no usable labor force; it has been estimated that the entire area of present-day Brazil was inhabited by no more than 1 million aboriginal people. When the Portuguese finally began to develop their New World territory, they turned to the same lucrative activity that their Spanish rivals had pursued in the Carib-

bean—the plantation cultivation of sugar for the European market. And they, too, found their labor force in the same source region, as millions of Africans were brought in slavery to the tropical Brazilian coast north of Rio de Janeiro. Not surprisingly, Brazil now has South America's largest black population, which is still heavily concentrated in the country's poverty-stricken northeastern states (Fig. 5–7). Today, with the overall population of Brazil at 154 million, about one-eighth of the people are black and at least another one-third are of mixed African, white, and Indian ancestry. Africans, then, definitely constitute the third major immigration of foreign peoples into South America (see Fig. A–2).

Persistent Isolation

Despite their common cultural heritage (at least insofar as their European-mestizo population is concerned), their adjacent location on the same continent, their com-

mon language, and their shared national problems, the countries that arose out of South America's Spanish viceroyalties have existed in a considerable degree of **isolation** from one another. Distance, physiographic barriers, and other factors have reinforced this separation. To this day, the major population agglomerations of South America adhere to the coast, mainly the eastern and northern coasts (Fig. I–14). Of all the continents, only Australia has a population distribution that is as markedly peripheral, but there are

only some 17 million people in Australia against 302 million in South America.

Compared with other world realms, South America may be described as underpopulated, not just in terms of its modest total for a continental area of its size but also because of the resources available or awaiting development. This continent never drew as large an immigrant European population as did North America. The Iberian Peninsula could not provide the numbers of people that western and northwestern Europe did, and Spanish colonial policy had a restrictive effect on the European inflow. The New World viceroyalties existed primarily for the purpose of extracting riches and filling Spanish coffers; in Iberia, there was little interest in developing the American lands for their own sake. Only after those who had made Spanish and Portuguese America their permanent home and who had a stake there rebelled against Iberian authority did things begin to change, and then very slowly. South America was saddled with the values, economic outlook, and social attitudes of seventeenth-and eighteenth-century Iberia—not the best tradition from which to begin the task of forging modern nation-states.

FIGURE 5–6

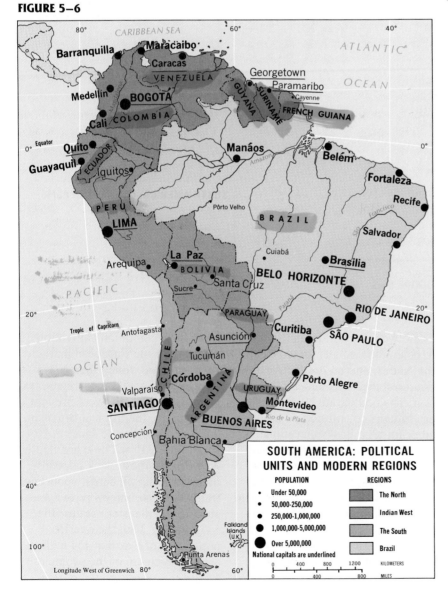

Independence

Some isolating factors had their effect even during the wars for independence. Spanish military strength was always concentrated at Lima, and those territories that lay farthest from this center of power—Argentina and Chile—were the first to establish their independence from Spain in 1816 and 1818, respectively. In the north, Simón Bolívar led the burgeoning independence movement, and in 1824 two decisive military defeats there spelled the end of

FIGURE 5–7 Brazilians of African descent constitute a major component of the population along the country's tropical northeastern coast. This is an outdoor market in central Recife, the large city located just to the south of Brazil's easternmost point.

Spanish power in South America. Thus, in little more than a decade, the Hispanic countries had fought themselves free. But this joint struggle did *not* produce unity: no fewer than nine countries emerged out of the three former viceroyalties. One, Bolivia, hitherto known as Upper Peru, was named after Bolívar when it was declared independent in 1825. Bolívar's confederacy, the Republic of Gran Colombia, which had achieved independence in 1819, broke up in 1829–1830 into Venezuela, Ecuador, and New Granada (the last was renamed Colombia in 1861). Uruguay was temporarily attached to Brazil, but it too attained separate political identity by 1828. Paraguay, once a part of Argentina, also appeared on the map as a sovereign state before 1830 (Fig. 5–6).

It is not difficult to understand why this fragmentation took place. With the Andes intervening between Argentina and Chile, and the Atacama Desert between Chile and Peru, overland distances seem even greater than they really are, and these obstacles to contact

have proven quite effective. Thus, the countries of South America began to grow apart, a separation process sometimes heightened by uneasy frontiers. Friction and even wars have been frequent, and a number of boundary disputes remain unresolved to this day. For example, Bolivia at one time had a direct outlet to the sea through northern Chile but lost this access corridor in a series of conflicts involving Chile, Peru, and indirectly, Argentina. Chile and Argentina themselves were locked in a long-standing dispute over their Andean boundary. Peru and Ecuador both laid claim to the upper Amazon Basin and the key Peruvian town of Iquitos.

Brazil attained independence from Portugal at about the same time the Spanish possessions in South America were struggling to end overseas domination, but the sequence of events was quite different. In Brazil, too, there had been revolts against Portuguese control. But the early 1800s, instead of witnessing a decline in Portuguese authority, actually brought the Portuguese govern-

ment (headed by Prince Regent Dom João) from Lisbon to Rio de Janeiro. Incredibly, Brazil in 1808 was suddenly elevated from colonial status to the seat of empire; it owed its new position to Napoleon's threat to overrun Portugal, which was then allied with the British. Although many expected that a new era of Brazilian progress and development would follow, things soon fell apart. By 1821, with the Napoleonic threat removed, Portugal decided to demote Brazil to its former colonial status. But Don Pedro, Dom João's son and the new regent, defied his father and led a successful struggle for Brazilian independence. With overwhelming popular support, he was crowned emperor of Brazil in 1822, and loyalist Portuguese forces still in the country were ignominiously deported to Lisbon.

The post-independence relationships between Brazil and its Spanish-influenced neighbors have been similar to the limited interactions among the individual Hispanic republics themselves. Distance, physical barriers, and cultural contrasts served to inhibit positive contact and interaction. Thus, Brazil's orientation toward Europe, like that of the other republics, remained stronger than its involvement with the countries on its own continent.

CULTURE AREAS

When we speak of the "orientation" or "interaction" of South American countries, it is important to keep in mind just who does the orienting and interacting, for there is a tendency to generalize the complexities of these countries away. The fragmentation of colonial South America into ten individual republics and the nature of their subsequent relationships was the work of a small minority of the people in each country. The black people in Brazil at the time of independence had little or no voice in the course of events; the Indians in Peru, numerically a vast majority, could only watch as their European conquerors struggled with each other for supremacy. It would not even be true to say that the European minorities *in toto* governed and made policy: the wealthy, landholding, upper-class elite determined the posture of the state. These were—and in some cases still are—the people who made the quarrels with their neighbors, who turned their backs on wider American unity, and who kept strong the ties with Madrid and Paris (Paris had long been a cultural focus for Middle and South America's well-to-do, and their children were usually sent to French rather than Spanish schools).

So complex and heterogeneous are the societies and cultures of Middle and South America that practically every generalization has to be qualified. Take the one so frequently used—the term *Latin* America. Apart from the obvious exceptions that can be read from the map, such as Jamaica, Guyana, and Suriname, which are clearly not "Latin" countries, it may be improper even to identify some of the Spanish-influenced republics as "Latin" in their cultural milieu. Certainly the white, wealthy upper classes are of Latin European stock, and they have the most influence at home and are most visible abroad; they are the politicians and the businesspeople, the writers and the artists. Their cultural environment is made up of the Spanish (and Portuguese) language, the Catholic church, and the picturesque Mediterranean architecture of Middle and South America's cities and towns. These

FIGURE 5–8

things provide them with a common bond and strong ties to Iberian Europe. However, in the mountains and villages of Ecuador, Peru, and Bolivia, there live millions of people to whom the Spanish language is still alien, to whom the white people's religion is another unpopular element of acculturation, and to whom decorous Spanish styles of architecture are meaningless when a decent roof and a solid floor are still unattainable luxuries.

South America, then, is a continent of pluralistic societies, where Indians of different cultures, Euro-

peans from Iberia and elsewhere, blacks from western tropical Africa, and Asians from India, Japan, and Indonesia have produced a cultural and economic kaleidoscope of almost endless variety. Certainly to call this human spatial mosaic "Latin" America is not very useful. Is there a more meaningful approach to a regional generalization that would better represent and differentiate the continent's cultural and economic spheres? John Augelli, who also developed the Rimland-Mainland concept for Middle America, made such an attempt. His map (Fig.

5–8) shows five **South American culture spheres**—internal cultural regions—within the South American realm. This scheme is quite useful if one keeps in mind that the realm is undergoing economic change in certain areas today and that these culture spheres are generalized and subject to further modification as well.

Tropical-Plantation Region

The first culture sphere, the *tropical-plantation* region, in many ways resembles the Middle American Rimland. It consists of several separated areas, of which the largest lies along the northeastern Brazilian coast, with four others along the Atlantic and Caribbean coastlands of northern South America. Location, soils, and tropical climates favored plantation crops, especially sugar. The small indigenous population led to the introduction of millions of African slave laborers, whose descendants today continue to dominate the racial makeup and strongly influence the cultural expression of these areas. The plantation economy later failed, soils became exhausted, slavery was abolished, and the people were largely reduced to poverty and subsistence—socioeconomic conditions that now dominate much of the region mapped as tropical-plantation.

European-Commercial Region

The second region on Augelli's map, identified as *European-commercial*, is perhaps the most truly "Latin" part of South America. Argentina and Uruguay, each

with a population that is more than 85 percent "pure" European and with a strong Hispanic cultural imprint, constitute the bulk of the European-commercial region. Two other areas also lie within it: most of Brazil's core area and the central core of Chile. Southern Brazil shares the temperate grasslands of the Pampa and Uruguay (see Fig. I–12), and this area is important as a zone of livestock raising as well as corn production. Middle Chile is an old Spanish settlement zone, home to the one-quarter of the Chilean population who claim pure Spanish ancestry; here, in an area of Mediterranean climate (Fig. I–10), cattle and sheep pastoralism plus mixed farming are practiced. In general, then, the European-commercial region is economically more advanced than the rest of the continent. A commercial economy rather than a subsistence way of life prevails, living standards are better, literacy rates are higher, transportation networks are superior, and, as Augelli has pointed out, the overall development of this region surpasses that of several parts of Europe itself.

Indo-Subsistence Region

The third region is identified as *Indo-subsistence*, and it forms an elongated zone along the length of the central Andes from southern Colombia to northern Chile/northwestern Argentina, closely approximating the area occupied by the old Inca Empire. The feudal socioeconomic structure that was established here by the Spanish conquerors still survives. The Indian population forms a large landless peonage, living by subsistence or by working on haciendas far removed from the Spanish culture that forms the primary force in the national life of their country. This region includes some of South

America's poorest areas, and what commercial activity there is tends to be in the hands of whites or mestizos. The Indian heirs to the Inca Empire live, often precariously, at high elevations (as much as 12,500 feet/3800 meters) in the Andes. Poor soils, uncertain water supplies, high winds, and bitter cold make farming a constantly difficult proposition; what little exchange exists takes place at remote upland Indian markets.

Mestizo-Transitional Region

The fourth region, *mestizo-transitional*, surrounds the Indo-subsistence region, covering coastal and interior Peru and Ecuador, much of Colombia and Venezuela, most of Paraguay, and large parts of Argentina, Chile, and Brazil (including the valleys of the Amazon and certain of its tributary rivers). This is the zone of mixture between European and Indian—or African in Brazil, Venezuela, and Colombia. The map thus reminds us that countries like Bolivia, Peru, and Ecuador are dominantly Indian and mestizo. In Ecuador, for example, these two groups make up about 80 percent of the total population, and a mere 10 percent can be classified as white (the remaining 10 percent is black and mulatto). The term *transitional* has an economic connotation, too, because, as Augelli puts it, this region "tends to be less commercial than the European sphere but less subsistent in orientation than dominantly Indian areas."

Undifferentiated Region

The fifth region on the map is marked as *undifferentiated* because its characteristics are hard to classify. Some of the Indian peoples in the interior of the Amazon Basin had remained almost completely isolated from the momentous changes in South America since the days of Columbus. Although isolation and lack of change are still two notable aspects of this subregion, the ongoing development of Amazonia is reversing that situation. The most remote Amazon backlands plus the Chilean and Argentinean southwest are also sparsely populated and exhibit only very limited economic development; poor transportation and difficult location continue to contribute to the unchanging nature of these areas.

The framework of the five culture spheres just described is necessarily a generalization of a rather complex geographic reality. Nevertheless, even in its simplicity, it underscores the diversity of South America's peoples, cultures, and economies.

URBANIZATION

As in other parts of the developing world, people in South America today are leaving the land and moving to the cities. This **urbanization** process intensified sharply after 1950, and Table I–1 (p. 34) shows that it persists so strongly that "Latin" America's percentages are much more typical of the developed world than those of the other Third World realms. The South American component of that aggregate figure has consistently been at least 10 percentage points higher than that of Middle America, and a measure of the pace of urbanization is provided by the following indexes. In 1925, about one-third of South America's peoples lived in cities and towns, and

as recently as 1950 the percentage was just over 40. In 1975, by contrast, the continent-wide figure had surpassed 60 percent, and by the early 1990s, three out of four South Americans lived in urban areas. Of course, these percentages mask the actual numbers, which are even more dramatic. Between 1925 and 1950, the continent's towns and cities grew by about 40 million residents as the urbanized percentage rose from 33 to 40. Then between 1950 and 1975 more than 125 million people crowded into the teeming metropolitan areas—more than *three times* the total for the previous quarter-century—and an additional 85 million since 1975 swelled the continental total to nearly 220 million by 1991.

South America's population of 302 million (20 percent larger than that of the United States) has a high growth rate, but nowhere are the numbers increasing faster than in the towns and cities. We usually assume that the populations of rural areas grow more rapidly than urban areas because farm families traditionally have more children than city dwellers. But overall, the urban population of South America has grown annually by nearly 5 percent since 1950, while the rural areas increased yearly by only about 1.5 percent. These figures reveal the dimensions of the **rural-to-urban migration**, from the countryside to the cities—still another migration process that affects modernizing societies throughout the underdeveloped realms.

The generalized spatial pattern of South America's urban transformation is seen in Fig. 5–9, which shows a *cartogram* of the continent's 1980 population (of the type presented for the world in Fig. I–15 on pp. 30–31). But this cartogram goes further than the one in Fig. I–15: it not only shows the 13 South American countries in

"population-space" relative to each other, but also displays the proportionate sizes of individual large cities within their total national populations.

Regionally, the realm's highest urban percentages are found in southern South America. In 1991 Argentina, Chile, and Uruguay had about 85 percent of their people residing in cities and towns (statistically on a par with Europe's most urbanized countries). Ranking next among the most heavily urbanized populations is Brazil at 75 percent, now growing so rapidly that it records increases of more than a 1 percent each year. That may not sound very significant, but once again percentages mask the real numbers; in a population of 154 million, 1 percent indicates that an additional 1.5 million annually are somehow squeezed into Brazil's already overcrowded urban centers. The next highest group of countries, averaging 72 percent urban, border the Caribbean in the north: Venezuela leads with 82 percent, Colombia follows with 68 percent, and even the Guianas' Suriname records 66 percent. The Indo-subsistence-dominated Andean countries, not surprisingly, comprise the realm's least urbanized zone. Peru, because of its strong Spanish imprint, is the region's leader by far, with fully 70 percent of its population agglomerated in towns and cities. Ecuador, Bolivia, and transitional Paraguay, on the other hand, lag well behind the rest of the continent, with all three exhibiting urban proportions in the 43-to-55-percent range. Figure 5–9 also tells us a great deal about the relative positions of major metropolises in their countries. Three of them—Brazil's São Paulo (18.7 million) and Rio de Janeiro (11.7 million), and Argentina's Buenos Aires (11.6 million)—today rank among the world's ten largest urban concentrations.

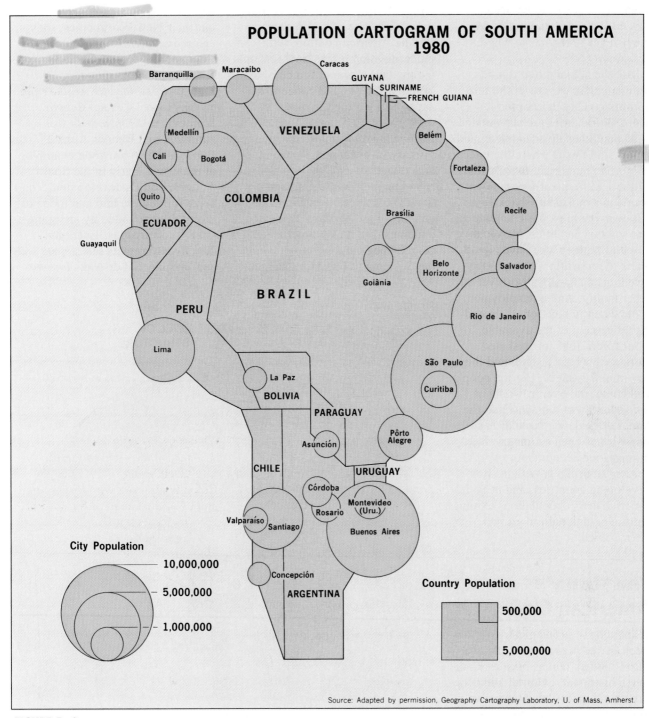

POPULATION CARTOGRAM OF SOUTH AMERICA 1980

City Population

10,000,000
5,000,000
1,000,000

Country Population

500,000

5,000,000

Source: Adapted by permission, Geography Cartography Laboratory, U. of Mass, Amherst.

FIGURE 5–9

In South America—as in Middle America, Africa, and Asia—people are attracted to the cities and driven from the poverty of the rural areas. Both *pull* and *push* factors are at work. Rural land reform has been slow in coming, and every year many farmers simply give up and leave, seeing little or no possibility for economic improvement. The cities lure because they are perceived to provide opportunity—the chance to earn a regular wage. Visions of education for their children, better medical care, upward social mobility, and the excitement of life in a big city draw hordes to places such as São Paulo and Lima. Road and rail connections continue to improve, so that access is easier and exploratory visits can even be made. City-based radio stations beckon the listener to the locale where the action is.

But the actual move can be traumatic. As we saw in the box on Mexico City in Chapter 4, Third World cities are surrounded and sometimes invaded by squalid slums, and this is where the uncertain urban immigrant usually finds a first (and frequently permanent) abode in a makeshift shack without even the most basic amenities and sanitary facilities. Many move in with relatives who have already made the transition, but whose dwelling can hardly absorb yet another family. And unemployment is persistently high, often exceeding 20 percent of the available labor force. Jobs for unskilled workers are hard to find, and they pay minimal wages. But the people still come, the overcrowding in the shacktowns worsens, and the threat of epidemic-scale disease rises. It has been estimated that in-migration accounts for over 50 percent of urban growth in some developing countries, and in South America urban populations exhibit unusually high natural growth rates as well.

The "Latin" American City

Although the urban experience has been varied in the South and Middle American realms because of diverse historical, cultural, and economic influences, there are many common threads that have prompted geographers to search for useful generalizations. One is the model of the intraurban spatial structure of **the "Latin" American city** proposed by Ernst Griffin and Larry Ford (Fig. 5–10) that may have even wider application to cities throughout the geographic realms of the Third World.

The basic spatial framework of

city structure, which blends traditional elements of South and Middle American culture with modernization forces now reshaping the urban scene, is a composite of radial sectors and concentric zones. Anchoring the model is the thriving CBD, which, like its European counterpart, remains the primary business, employment, and entertainment focus of the surrounding metropolitan agglomeration. The landscape of the CBD contains many modern high-rise buildings, but it also mirrors its colonial beginnings. As we saw in the Systematic Essay in Chapter 4,

when the Spanish colonizers laid out their New World cities, they created a central square, or *plaza*, dominated by a church and flanked by imposing government buildings. Lima's *Plaza de Armas* (Fig. 5–11), Bogotá's *Plaza Bolívar*, Montevideo's *Plaza de la Constitución*, and Buenos Aires's *Plaza de Mayo* are good examples of the genre. Early in the South American city's development, the plaza formed the hub and focus of the city, surrounded by shopping streets and arcades. Eventually the city outgrew its old center, and new commercial districts formed

FIGURE 5–10

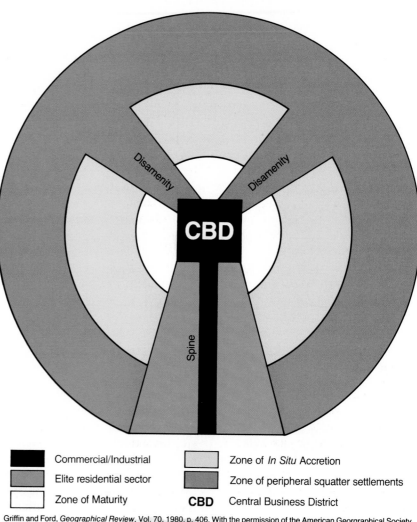

A GENERALIZED MODEL OF LATIN AMERICAN CITY STRUCTURE

Griffin and Ford, *Geographical Review*, Vol. 70, 1980, p. 406. With the permission of the American Geographical Society.

FIGURE 5–11 Lima's *Plaza de Armas*, heart of the Peruvian capital's old city. This open square is a classic central plaza—focus of both the church and the government—the hub of the planned Spanish city in its New World colonies (see Fig. 4–2).

Upper- and Middle-Class Areas

Emanating outward from the urban core along the city's most prestigious axis is the commercial *spine*, which is surrounded by the *elite residential sector* (shown in green). This widening corridor is essentially an extension of the CBD, featuring offices, shopping, high-quality housing for the upper and upper-middle classes, restaurants, theaters, and such amenities as parks, zoos, and golf courses that give way to wealthy suburbs, which carry the elite sector beyond the city limits.

The three remaining concentric zones are home to the less fortunate residents of the city (who comprise the great majority of the urban population), with socioeconomic levels and housing quality decreasing markedly as distance from the city center increases. The *zone of maturity* in the inner city contains the best housing outside of the spine sector, attracting the middle classes who invest sufficiently to keep their solidly built but aging dwellings from deteriorating. The adjacent *zone of in situ accretion* is one of much more modest housing interspersed with unkempt areas, representing a transition from inner-ring affluence to outer-ring poverty. The residential density of this zone is usually quite high, reflecting the uneven assimilation of its occupants into the social and economic fabric of the city.

Poverty Areas

The outermost *zone of peripheral squatter settlements* is home to the impoverished and unskilled hordes that have recently migrated to the city from rural areas. Although housing in this ring mainly consists of teeming, high-density shantytowns, residents here are surprisingly optimistic about finding work and eventually bettering their living conditions—a realistic aspiration documented by researchers, who confirm a process of gradual upgrading as squatter communities mature.

A final structural element of many "Latin" American cities is the *disamenity sector* that contains relatively unchanging slums, known as *favelas* or *barrios*. The worst of these poverty-stricken areas often include sizable numbers of people who are so poor that they are forced literally to live in the streets. Thus, South America's cities present almost inconceivable contrasts between poverty and affluence, squalor and comfort; this harsh juxtaposition is frequently observed in the cityscape, as Fig. 5–12 underscores in the case of Venezuela's capital city, Caracas.

As with all models, this particular construct can be criticized for certain shortcomings. Among criticisms directed at this generalization are the need for a sharper sectoring pattern, a stronger time element involving a multiple-stage approach, and the observation that the form perhaps too closely resembles that of the U.S. city. Others view this work as not only a successful abstraction of the "Latin" American city but of the Third World city in general. Later on in Chapter 10 (p. 566), we will explore this topic further when we consider a similar model developed for the Southeast Asian city, which incorporates the lasting spatial imprints of colonialism on contemporary urban structure.

THE REPUBLICS: REGIONAL GEOGRAPHY

We turn now to the countries of South America and the principal

FIGURE 5–12 Extreme social contrasts, such as this scene near the edge of Caracas, often mark South America's teeming cityscapes where the elite residential sector adjoins lower-income zones.

characteristics of their regional geography. In general terms, it is possible to group the realm's countries into regional units (Fig. 5–6), because several have qualities in common. Accordingly, the northernmost Caribbean countries, Venezuela and Colombia, form a regional unit that might also include neighboring Guyana, Suriname, and French Guiana. On the basis of their Indian cultural heritage, Andean physiography, and modern populations, the western republics of Ecuador, Peru, and Bolivia constitute a regional entity, to which Paraguay may be added on grounds to be discussed later. In the south, Argentina, Uruguay, and Chile have a common regional identity as the realm's mid-latitude, most strongly European states. And Brazil by itself constitutes a geographic region in South America because it

contains about half the continent's land and people; in fact, we will treat that huge country in a separate vignette following the end of this chapter.

The North: Caribbean South America

As another look at Fig. 5–8 confirms, the countries of the northern shore have something in common besides their coastal location: each has a tropical plantation area, signifying early European plantation development, the in-migration of black laborers, and the absorption of this African element into the population matrix. Not only did myriad black workers arrive; many thousands of Asians (from India) also came to South Amer-

ica's northern shores as contract laborers. The pattern is familiar: in the absence of large local labor sources, the colonists turned to slavery and indentured workers to serve their lucrative plantations. Between Spanish Venezuela and Colombia on one hand, and the non-Hispanic "three Guianas" on the other, there is this difference: in the former, the population center of gravity soon moved into the interior and the plantation phase was followed by a totally different economy. In the Guianas, on the other hand, coastal settlement and the plantation economy still predominate (see *box*).

Venezuela and Colombia have what the Guianas lack (Fig. 5–13): their territories and populations are much larger, their natural environments more varied, their economic opportunities greater. Each has a share of the Andes Mountains (Colombia's being larger), and each produces oil from an adjacent joint reserve that ranks among the world's major deposits (here Venezuela is the leading beneficiary).

Venezuela

Much of what is important in Venezuela is concentrated in the northern and western parts of the country, where the Venezuelan Highlands form the eastern spur of the north end of the Andes system. Most of Venezuela's 20 million people are concentrated in these uplands, which include the capital of Caracas (3.2 million), its early rival Valencia (850,000), the commercial and industrial center of Barquisimeto (700,000), and San Cristóbal (300,000) near the Colombian border.

The Venezuelan Highlands are flanked by the Maracaibo Lowlands and Lake Maracaibo to the northwest, and by a region of savanna country called the *llanos* in the Orinoco Basin to the south and east. The Maracaibo Lowland, once a disease-infested, sparsely peopled coastland, is today one of the world's leading oil-producing areas; much of the oil is drawn from reserves that lie beneath the shallow waters of the lake itself. Actually, Lake Maracaibo is a misnomer, for the "lake" is open to the ocean and is, in fact, a gulf with a very narrow entry. Venezuela's second city, Maracaibo (1.2 million), is the focus of the oil industry that transformed the Venezuelan economy in the 1970s. But since the early 1980s, the country has suffered the consequences of the global oil depression. Like Mexico, Venezuela borrowed heavily against its future oil revenues and now faces the economic problems of meeting a huge foreign debt.

The llanos on the southern side of the Venezuelan Highlands and the Guiana Highlands in the country's southeast are two of those areas that contribute to South America's image as "underpopulated" and "awaiting development." Whereas the llanos are beginning to share in Venezuela's oil production (large reserves have been discovered here), the superior commercial agricultural potential of these savannas and of the *templada* areas (see *box,* p. 270) of the Guiana Highlands has hardly begun to be realized. Economic integration of this interior zone with the rest of Venezuela has been encouraged by the dis-

FIGURE 5–13

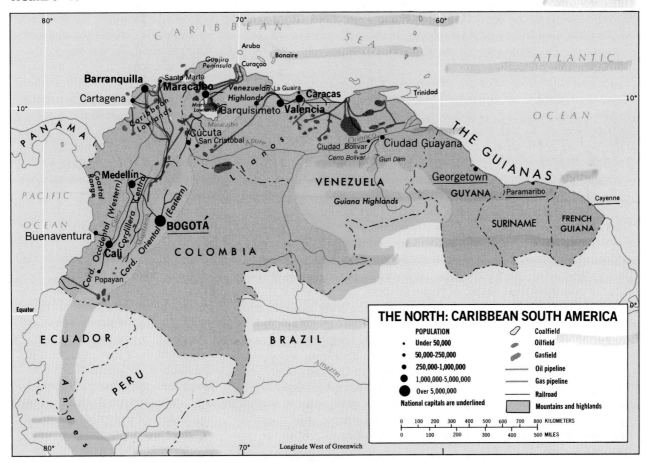

THE GUIANAS

On the north coast of South America, three small countries lie east of Venezuela and next to northernmost Brazil (Fig. 5–13). In "Latin" South America, these territories are anomalies of a sort: their colonial heritage is British, Dutch, and French. Formerly, they were known as British, Dutch, and French Guiana, and called the three *Guianas*. But two of them are independent now: Guyana, Venezuela's neighbor, and Suriname, the Dutch-influenced country in the middle. The easternmost territory, French Guiana, continues under colonial rule.

None of the three countries has a population over 1 million. Guyana is the largest with 800,000, Suriname has 425,000 inhabitants, and French Guiana a mere 100,000. Culturally and spatially, patterns here are Caribbean: Asian (Indian) and black peoples are in the majority, whites a small minority. In Guyana, Asians make up just over half the population, blacks and mixed Africans 43 percent, and the others, including Europeans, are tiny minorities. In Suriname, the ethnic picture is even more complicated, for the colonists brought not only Indians from South Asia (now 37 percent of the population) and blacks (31 percent), but also Indonesians (15 percent) to the territory in servitude; the remaining 10 percent of the Surinamese population consists of a separate category of black communities, peopled by descendants of African slaves (known as Bush Negroes) who escaped from the coastal plantations and fled into the forests of the interior. French Guiana is the most European of the three countries; three-quarters of its small population speak French, and the ethnic majority is of mixed African, Asian, and European ancestry—the *creoles*.

French Guiana, an overseas *département* of France, is the least developed of the three Guianas. There is a small fishing (shrimping) industry, and some lumber is exported; but food must be brought in from abroad. By contrast, Suriname has progressed more rapidly, despite political instabilities during the 1980s. The plantation economy has given way to production on smaller farms where enough rice, the country's staple crop, is grown to make Suriname self-sufficient. Bananas and citrus fruits are exported, and a variety of fruits and vegetables for the home market are raised on farms in the coastal zone where once sugar plantations prevailed. But Suriname's major income earner is bauxite, mined in a zone across the middle of the country. British-influenced Guyana became independent in 1966 amid internal conflict that basically pitted people of African origins against people of Asian descent. Such problems have continued to afflict this country in which the great majority of the people live in small villages near the coast. Bauxite and plantation products earn much of the country's annual revenues.

Unsettled boundary problems affect all three countries. Venezuela has laid claim to all of mineral-rich western Guyana—the Essequibo region, purportedly "stolen" by the British in 1899—a lingering territorial dispute that shows no sign of resolution. Guyana, in turn, claims a corner of southwestern Suriname. And part of the border between Suriname and French Guiana also is under contention.

covery of rich iron ores on the northern flanks of the Guiana Highlands, chiefly near Cerro Bolívar. Local railroads connect with the Orinoco River, and from there the ores are shipped directly to the steel plants of the U.S. Northeast. Ciudad Guayana, less than four decades old, has nearly 700,000 inhabitants; neighboring Ciudad Bolívar today is home to over 250,000 residents. The nearby Guri Dam has been put into service and now supplies much of Venezuela with electricity; two modern paved roads link this eastern frontier with Caracas.

Colombia

Colombia also has a vast area of llanos, covering about 60 percent of the country. Here, too, as in Venezuela, this is comparatively empty land far from the national core and much less productive than it could be. Eastern Colombia consists of the upper basins of major tributaries of the Orinoco and Amazon rivers, and it lies partly under savanna and partly under rainforest vegetation. Although the Colombian government is promoting settlement east of the Andes, it will be a long time before any part of eastern Colombia matches the Andean part of the country or even the Caribbean Lowlands of the north. These two regions contain the vast majority of Colombians, and here lie the major cities and productive areas.

Western Colombia is dominated by mountains, but there is a regularity to this terrain. For the most part, there are four parallel ranges, generally aligned north–south and separated by wide valleys. The westernmost of these ranges is a coastal belt, less continuous and lower than the other three, which constitute the true Andean mountain chains of Colombia. In this country the Andes separate into three ranges: the Eastern (Oriental), Central, and Western (Occidental) Cordilleras (Fig. 5-13). The valleys between these Cordilleras open northward onto the Caribbean Lowland, where the two important Colombian ports of Barranquilla (1.2 million) and Cartagena (700,000) are located.

Colombia's population of 33 million consists of more than a dozen separate clusters, some of them in the Caribbean Lowland, others in the inter-Cordilleran valleys, and still others in the intermontane basins within the Cordilleras themselves. This insular distribution existed before the Spanish colonizers arrived, with the Chibcha civilization (Fig. 5–3) concentrated in the intermontane basins of the Eastern Cordillera. The capital city, Bogotá (corrupted from the Chibchan word *Bacatá*), was founded in one of the major basins in this Andean range

at an elevation of 8500 feet (2700 meters). For centuries, the Magdalena Valley between the Cordillera Oriental and the Cordillera Central was a crucial link in the key cross-continental route that began in Argentina and ended at the port of Cartagena, and Bogotá benefited by its position along this artery. Today, the Magdalena Valley is still Colombia's leading transport corridor, but Bogotá's connections with much of the rest of Colombia still remain quite tenuous. Nonetheless, the capital centers a major metropolis that is home to 5.9 million.

Colombia's physiographic variety is matched by its demographic diversity. In the Andean south, it has a major cluster of Indian inhabitants, and here the country begins to resemble its southern neighbors. In the northern coast-

lands, traces remain of the plantation period and the African population it brought. Bogotá is a great "Latin" cultural headquarters, whose influence extends beyond the country's borders. In the Cauca Valley between the Cordillera Central and Cordillera Occidental, the city of Cali (1.7 million) is the commercial focus for a hacienda district where sugar and tobacco are grown. Farther north, the Cauca River flows through a region comprising the provinces of Antioquia and Caldas, whose urban focus is the textile manufacturing city of Medellín (1.9 million) but whose greater importance to the Colombian economy entails the production of coffee. With its extensive *templada* areas along the Andean slopes, Colombia in the late 1980s was the world's largest producer of coffee (Fig. 5–14).

FIGURE 5–14 Unlike the *tierra-fría*-dominated Andean settlement areas to the south, northern and central Colombia's lower-lying and less rugged mountain terrain offers ample agricultural opportunities in a broad *templada* zone. Coffee, seen here under cultivation, is the most lucrative commercial crop, and Colombia regularly ranks among the world's leading producers.

In Antioquia-Caldas, it is grown on small farms by a remarkably unmixed European population cluster; elsewhere, it is produced on the haciendas so common in this geographic realm.

Following coffee, oil and coal are Colombia's other two leading exports. Major oil deposits have been worked for decades in the north, adjacent to the (still-disputed) Venezuelan border where the Maracaibo oilfields extend into Colombia. But in 1984, significant discoveries were made in the northeast in the province of Arauca, and development began immediately. A 500-mile pipeline was quickly completed across the northern Andes to the Caribbean coast, and Colombia was turned into a net oil exporter overnight. Unfortunately, local guerrilla groups have repeatedly bombed the pipeline, interrupting operations and raising uncertainties about the future.

Colombia has had better luck exploiting its other recently discovered fossil fuel—coal—a resource that is virtually absent from the South American continent. These rich coal deposits are fortuitously located on the Guajira Peninsula, which juts out into the Caribbean between Lake Maracaibo and the coastal strip that contains Colombia's leading ports. In fact, production here has expanded so rapidly that, in the early 1990s, the Cerrejón North operation has become one of the world's largest bituminous coal mines. Colombia, of course, is also very well known as a major source of illicit drugs, particularly cocaine, which undoubtedly ranks among its leading exports as well (see *box*).

Venezuela and Colombia both exhibit a pronounced clustering of often-isolated populations, share a relatively empty interior, and depend on a small number of products for the bulk of their export

THE GEOGRAPHY OF COCAINE

It is impossible to discuss northwestern South America today without reference to one of its most widespread activities: the production of illegal narcotics. Of the enormous flow of illicit drugs into the United States each year, one-third of the heroin and over 80 percent of the marijuana originate in Middle America, mainly in Mexico; but the most widely used of these illegal substances undoubtedly is cocaine, all of which comes from South America, mainly Bolivia, Peru, and Colombia (the sources of about 90 percent of the total world supply). Within these three countries, cocaine annually brings in billions of (U.S.) dollars and "employs" thousands of workers, constituting an industry that functions as a powerful economic force. Moreover, the notorious Medellín Cartel and other wealthy drug lords, who operate the industry, have accumulated considerable power through bribery of politicians, threats of violence, and even alliances with guerrilla movements in outlying areas beyond governmental control. The cocaine industry itself is structured within a tightly organized network of territories that encompass the various stages of this narcotic's production.

The first stage of cocaine production is the extraction of coca paste from the coca plant, a raw-material-oriented activity that is located near the areas where the plant is grown. The coca plant was domesticated in the Andes by the Incas centuries ago for use as a stimulant (and in certain rituals), and it is still cultivated and used widely today by the Incas' descendants. Millions of upland peasants chew coca leaves for stimulation and to ward off altitude sickness; they also brew the leaves into coca tea (the area's leading beverage), which they use as a general-purpose medicinal.

The leading zone of coca-plant cultivation is along the eastern slopes of the Andes and adjacent tropical lowlands in Bolivia and Peru. A lesser variety of the plant is now grown in the same physiographic setting in Ecuador and southern Colombia—and increasingly in the nearby upper Amazon Basin of Brazil, which is emerging as a major source area. Today, three main areas dominate in the growing of coca leaves for narcotic production (Fig. 5–15): Bolivia's Chaparé district (in the marginal Amazon lowlands northeast of the city of Cochabamba), the Yungas highlands (north of the Bolivian capital, La Paz), and Peru's upper Huallaga Valley around the town of Tingo Maria (about 200 miles north of Lima). These areas thrive because they combine many favorable conditions: local environments are conducive to high leaf yields that produce four crops per year (instead of the usual two), and plants here have developed near-total immunity to both disease and insect ravages. The success of these areas has enabled their Indian populations to turn coca plants into a cash crop, and operations often reach the scale of plantations, thereby inducing subsistence farmers from other areas to migrate into these now-specialized coca-raising regions. In fact, profitable coca cultivation has led thousands of peasants to abandon the far less lucrative production of food crops, further reducing the capability of nutrition-poor Bolivia to feed itself. Another undesirable side-effect is environmental damage in Peru and Bolivia—notably deforestation and soil erosion—as large areas of wilderness are swiftly transformed into coca-plant "farms." The coca leaves harvested in these eastern Andean regions make their way to nearby centers where coca paste is extracted and prepared; these centers are located at the convergence of rivers and trails, including the Beni rainforests of northernmost Bolivia and particularly the central Bolivian city of Santa Cruz (Fig. 5–15).

The second stage of production involves refining the coca paste (about 40 percent pure cocaine) into *cocaine hydrochloride* (more than 90 percent pure), a lethal concentrate that is diluted with substances such as sugar or flour before being sold on the streets to consumers. Cocaine refining requires sophisticated chemicals, carefully controlled processes, and a labor force skilled in their supervision, and here Colombia predominates with most of this activity taking place within its borders—sometimes in cities such as Medellín, but increasingly in the ultramodern forest complexes in the southeastern lowlands where alliances between drug lords and guerrillas help protect against the intrusion of the Bogotá government. Colombia also has other geographic advantages: proximity to the remote upper Amazon rainforests across the Brazilian border (where the cocaine industry is rapidly expanding), and smuggling access to the United States, which is a short hop by air or sea across the Caribbean.

The final stage of production entails the distribution of cocaine to the U.S. marketplace, which depends upon an efficient (but clandestine) transportation network. One favorite avenue for smugglers is to use commercial air and sea links between South America and the United States, concealing the product in countless imaginative ways. More likely, however, is the use of private planes that operate directly out of remote airstrips near the refineries in places beyond effective government control. United States drug interdiction records show that about 75 percent of all cocaine coming into the country travels by noncommercial plane or boat, with Colombia most often the origin, but with Brazil steadily increasing its role as a distribution center for outbound "coke."

The transport of cocaine to the United States is often a two-step process, making use of such intermediate Middle American transshipment points as the Bahamas, Jamaica, Cuba, and lately, Panama on the mainland. Although South Florida is frequently believed to be a leading port of illegal entry, other coastal points in the U.S. Southeast as well as cities along the Mexican border in the Southwest have recorded significant increases in cocaine smuggling over the past several years. Judging by recent drug seizures in Southern California, the overland route through mainland Middle America has experienced a major upsurge in trafficking since 1985.

revenues. The majority of the people in these countries practice subsistence agriculture and labor under the social and economic inequalities common to most of Iberian America.

The Indian West: Andean South America

The second regional grouping of South American states encompasses Peru, Ecuador, Bolivia, and Paraguay (Fig. 5–15), four contiguous countries that include South America's only landlocked republics (the latter two). The map of culture spheres (Fig. 5–8) shows the Indo-subsistence region extending along the Andes Mountains, indicating that these countries have large Indian components in their populations. Just over half the people of Peru (population: 22.4 million) are of Indian stock, and in Ecuador and Bolivia the figure is also around 50 percent; in Paraguay it is over 90 percent. However, all these percentages are only approximate because it is often impossible to distinguish between Indian and "mixed" people of strong Indian character. But there are other similarities between these countries: their incomes are low; they are comparatively unproductive; and, unhappily, they exemplify the grinding poverty of the landless peonage—a problem that looms large in the future of Ibero-America. Moreover, as noted earlier, these are South America's least urbanized countries; only Lima (6.8 million), the capital of Peru, ranks with Bogotá and Rio de Janeiro as a major-scale metropolis.

Peru

In terms of territory as well as population, Peru is the largest of

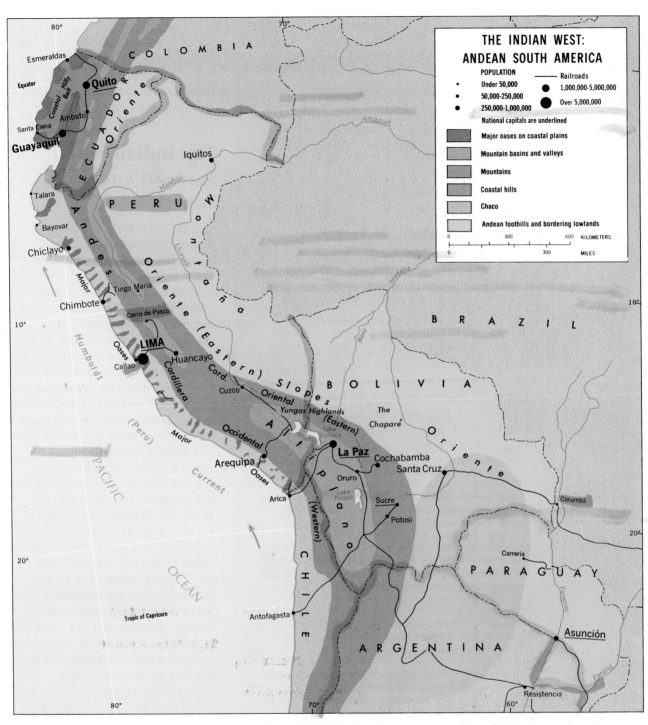

FIGURE 5–15

the four republics. Its half-million square miles (1.3 million square kilometers) divide both physiographically and culturally into three subregions: (1) the desert coast, the European-mestizo region; (2) the Andean highlands, or Sierra, the Indian region; and (3) the eastern slopes and adjoining *montaña*, the sparsely populated Indian-mestizo interior (Fig. 5–15). It is symptomatic of the cultural division still prevailing in Peru that Lima is not located centrally in a populous basin of the Andes but in the coastal zone. In 1986, Peru's president did propose shifting the capital—to the lush eastern Andean slopes—but no legislative action followed. At Lima, the Spanish avoided the greatest of the Indian empires, choosing a site some 8 miles (13 kilometers) inland from a suitable anchorage which became the modern outport of Callao. From an economic point of view, the Spanish choice of a coastal headquarters proved to be sound, for the coastal region has become commercially the most productive part of the country. A thriving fishing industry based on the cool productive waters of the Peru (Humboldt) Current offshore contributes significantly to the export trade. Irrigated agriculture in some 40 oases distributed all along the arid coast produces cotton, sugar, rice, vegetables, fruits, and wheat; the cotton and sugar are important export products, and the other crops are grown mostly for the domestic market.

The Andean (Sierra) region occupies about one-third of the country, and contains the majority of Peru's Indian peoples, most of them Quechua-speaking. However, despite the size of its territory and population (nearly half of all Peruvians reside here), the political influence of this region is slight, as is its economic contribution (except for the mines). In the high valleys and intermontane basins, the Indian people are clustered either in isolated villages, around which they practice a precarious subsistence agriculture, or in the more favorably located and fertile areas where they are tenants, peons on white- or mestizo-owned haciendas. Most of the these people never receive an adequate daily caloric intake or balanced diet of any sort; the wheat produced around Huancayo, for instance, is sent to Lima's European market and would be too expensive for the Indians themselves to buy. Potatoes (which can be grown at altitudes up to 14,000 feet/4250 meters), barley, and corn are among the subsistence crops

here in this *fría* zone, and in the *puna* of the higher basins the Indians graze their llamas, alpacas, cattle, and sheep (these vertical zones are discussed in the box on p. 270). The major mineral products from the Sierra are copper, silver, lead, and several other metallic minerals, with the largest mining complex centered on Cerro de Pasco.

The Peruvian Andes are also the setting for one of South America's more ominous insurgencies, the Communist *Sendero Luminoso* (Shining Path). This strengthening guerrilla movement, which gains support by exploiting the desperate poverty of the Indians, today increasingly spills out of the mountains to threaten Lima, other coastal cities, and much of the rest of central Peru.

Of Peru's three subregions, the *Oriente* or East—the inland slopes of the Andes and the Amazon-drained, rainforest-covered *montaña*—is the most isolated. A look at the map of permanent (as opposed to seasonal) routes shows how disconnected Peru's regions still are. However marvelous an engineering feat, the railroad that connects Lima and the coast to Cerro de Pasco and Huancayo in the Andes does not even begin to link the country's east and west.

The focus of the eastern region, in fact, is Iquitos (300,000), a city that looks east rather than west and can be reached by oceangoing vessels sailing 2300 miles (3700 kilometers) up the Amazon River across Brazil. Iquitos grew rapidly during the Amazon wild-rubber boom early in this century and then declined; but now it is growing again and reflects Peruvian plans to begin development of the east. Petroleum was discovered west of Iquitos in the 1970s, and since 1977 oil has flowed through a pipeline built across the Andes to the Pacific port of Bayovar. Mean-

while, traders plying the navigable Peruvian rivers above Iquitos continue to collect such products as chicle, nuts, rubber, herbs, and special cabinet woods.

Ecuador

Ecuador, smallest of the four republics, on the map appears to be just a corner of Peru. But that would be a misrepresentation because Ecuador possesses a full range of regional contrasts: it has a coastal belt, an Andean zone that may be narrow (under 150 miles/ 250 kilometers) but by no means of lower elevation than elsewhere; and an *Oriente*—an eastern region that is just as empty and just as undeveloped as that of Peru. As in Peru, the majority of the people of Ecuador are concentrated in the Andean intermontane basins and valleys, and the most productive region is the coastal belt. But here the similarities end. Ecuador's coastal zone consists of a belt of hills interrupted by lowland areas, of which the most important lies in the south between the hills and the Andes, drained by the Guayas River and its tributaries. The largest city and commercial center of the country (but not the capital), Guayaquil (1.9 million), forms the focus for this subregion. Ecuador's lowland west, moreover, is not desert country; it consists of fertile tropical plains not bedeviled by excessive rainfall (see Fig. I–9).

Ecuador's west-coast lowland is also far less Europeanized than Peru's Pacific-facing plain, because the white component of the total population of 11 million is a mere 8 percent. A sizable portion of the latter is engaged in administration and hacienda ownership in the interior, where most of the 42 percent of the Ecuadorians who are Indian also reside. Of the remaining national population, 10

tween Santa Cruz and Corumbá (Brazil) in the southeastern lowlands, where commercial agriculture (notably soybean production) is on the rise.

Bolivia has had a turbulent history. Apart from endless internal struggles for power, the country first lost its window on the Pacific coast in a disastrous conflict with Chile, then lost its northern territory of Acre to Brazil in a dispute involving the rubber boom in the Amazon Basin, and finally lost 55,000 square miles (140,000 square kilometers) of southeastern Gran Chaco territory to Paraguay. The most critical, by far, was the loss of its outlet to the sea (see *box*, p. 311).

Although Bolivia has rail connections to the Chilean ports of Arica and Antofagasta, it is permanently disadvantaged by its landlocked situation. Since the Cordillera Occidental and the Altiplano form the country's inhospitable western margins, one might suppose that Bolivia would look eastward and that its *Oriente* might be somewhat better developed than that of Peru or Ecuador, but such is not the case (the booming cocaine industry notwithstanding). The densest settlement clusters are in the valleys and basins of the Cordillera Oriental, where the mestizo sector is also stronger than elsewhere in the country. Cochabamba (400,000), Bolivia's third largest city, lies in a basin that forms the country's largest concentration of population; Sucre, the legal capital (La Paz is the de facto capital), is located in another. Here, of course, lie major agricultural districts between the barren Altiplano to the west and the just-opened tropical savannas to the east.

Paraguay

Paraguay is the only non-Andean country in this region, but it is no

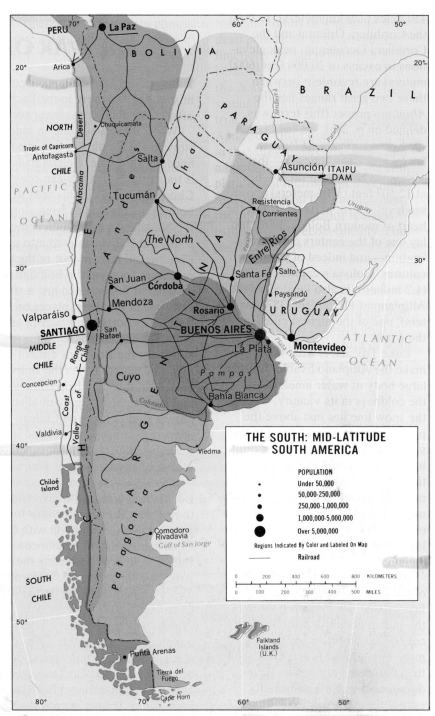

FIGURE 5–17

less Indian. Of its 4.4 million people, more than 95 percent are mestizo, but with so pervasive an Indian influence that any white ancestry is almost totally submerged. Although Spanish is Paraguay's official language, Indian Guaraní

is more commonly spoken (in fact, the country is probably the world's most completely bilingual). By any measure, Paraguay is the poorest of the four countries of Indian South America, although it does have opportunities for pastoral and agricultural industries that have thus far gone unrealized. One of the reasons for this must be isolation—the country's land-locked position. Paraguay's exports, in their modest quantities, must be shipped through Argentina's Buenos Aires, a long haul via the Paraguay-Paraná River from the capital of Asunción (900,000). Meat (dried and canned), timber (sold to Uruguay and Argentina), oilseeds, quebracho extract (for tanning leather), cotton, and some tobacco reach foreign markets. Grazing in the northern Chaco region is the most important commercial activity, but here cattle generally do not compare favorably to those of Argentina. Hoped-for development in the fertile Paraguayan east around Itaipu Dam on the Paraná border with Brazil (see p. 326) has not materialized.

The South: Mid-Latitude South America

Argentina

South America's three southern countries—Argentina, Chile, and Uruguay—are grouped into one region. By far the largest in terms of both population and territory is Argentina, whose 1.1 million square miles (2.8 million square kilometers) and 32.8 million people rank second only to Brazil in this geographic realm. Argentina exhibits a great deal of physical-environmental variety within its boundaries, and the vast majority of the population is concentrated in the physiographic subregion known as the Pampa. Figure I–14 indicates the degree of clustering of Argentina's inhabitants on the

land and in the cities of the Pampa (the word means "plain"). It also shows the relative emptiness of the other six subregions (mapped in Fig. 5–17)—the scrub-forest Chaco in the northwest, the mountainous Andes in the west (along whose crestline lies the boundary with Chile), the arid plateaus of Patagonia south of the Rio Colorado, and the undulating transitional terrain of intermediate Cuyo, Entre Rios (also known as "Mesopotamia"), and the North.

The Argentine Pampa is the product of the past 125 years. During the second half of the nineteenth century, when the great grasslands of the world were being opened up (including those of the United States, Russia, and Australia), the economy of the long-dormant Pampa began to emerge. The food needs of industrializing Europe grew by leaps and bounds,

FIGURE 5–18 The street scene in central Buenos Aires, here on the Avenida Florída, bears a striking resemblance to the landscape of the CBD of any large European city.

and the advances of the Industrial Revolution—railroads, more efficient ocean transport, refrigerated ships, and agricultural machinery—helped make large-scale commercial meat and grain production in the Pampa not only feasible but highly profitable. Large haciendas were laid out and farmed by tenant workers who would prepare the virgin soil and plant it with wheat and alfalfa, harvesting the wheat and leaving the alfalfa as pasture for livestock. Railroads radiated ever farther outward from the capital of Buenos Aires, and brought the entire Pampa into production. Today, Argentina has South America's densest railroad network and

once-dormant Buenos Aires is now the world's tenth-largest metropolitan complex (11.7 million). Yet the Pampa itself has hardly begun to fulfill its productive potential, which could easily double with more efficient and intensive agricultural practices.

Over the decades, several specialized agricultural areas appeared on the Pampa. As we would expect, a zone of vegetable and fruit production became established near the huge Buenos Aires conurbation located beside the estuary of the Rio de la Plata. To the southeast is the predominantly pastoral district, where beef cattle and sheep are raised. In the drier west, northwest, and southwest,

wheat becomes the important commercial grain crop, but half the land remains devoted to grazing. Among the exports, meat usually leads by value, followed by cereals, vegetable oils and oilseeds, hides and skins, and wool.

Argentina's wealth and vigor are reflected in its fast-growing cities, which epitomize the European-commercial cultural character of the realm's southernmost countries (Fig. 5–18). No less than 85 percent of the Argentine population may be classified as urbanized, an exceptionally high figure for a country in a Third World realm. More than one-third of all Argentineans now live in the Greater Buenos Aires conurbation

FIGURE 5–19 The navigational perils of South Chile's mountainous coast. This is the important Strait of Magellan that separates Tierra del Fuego from the mainland—and also links the Atlantic with the Pacific, thereby allowing ships to avoid the far more treacherous passage around Cape Horn (see Fig. 5–17).

alone, which also contains most of the industries, many of them managed by Italians, Spaniards, and other immigrants. This dominance over national life, however, has disillusioned many Argentines, who now support the movement to shift the capital from Buenos Aires 600 miles (965 km) south to Viedma on the Patagonian frontier (Fig. 5–17). The necessary legislation has been passed, but economic difficulties in the late 1980s forced the government to delay this costly relocation. The Córdoba area (1.5 million) is also a focus of industrial growth. Much of the manufacturing in the major cities is associated with the processing of Pampa products and the production of consumer goods for the domestic market. One out of every five wage earners in the country is engaged in manufacturing—another indication of Argentina's somewhat advanced economic standing.

Argentina's population shows a high degree of clustering and a decidedly peripheral distribution. The Pampa subregion covers only a little more than 20 percent of Argentina's territory, but with three-fourths of the people concentrated here the rest of the country cannot be densely populated. Outside the Pampa, pastoralism is an almost universal pursuit, but the quality of the cattle is much lower than in the Pampa; in semiarid Patagonia, sheep are raised. Some of the more distant areas are of actual and potential significance to the country: in Patagonia an oilfield is in production near coastal Comodoro Rivadavia; in the far northwest Chaco, Argentina may share with Paraguay significant oil reserves that are yet to be exploited. Yerba maté, a local tea, is produced in the North and Entre Rios subregions, and the quebracho-tree extract that both Paraguay

FALKLANDS OR MALVINAS?

Off the southern tip of Argentina, about 300 miles (500 kilometers) from the mainland, lies a small archipelago that in 1982 suddenly became the scene of a bitter war between the Argentines and the islands' owners, the British. This conflict over the Islas Malvinas—as the Falkland Islands are called in Argentina—was a reminder of the continuing potential for strife in the aftermath of the colonial era. On the world politico-geographical map, there are numerous locales capable of causing similar crises: Gibraltar, Guantánamo Bay (Cuba), Hong Kong, and many less familiar places. And in this age of superpowers, the consequences could be far more serious.

Judging from the map (Fig. 5–17), the Falklands hardly seem likely to occasion a full-scale war. The archipelago consists of two main islands (East and West Falkland) and about 100 smaller ones, with a total area of just over 4600 square miles (12,000 square kilometers), slightly smaller than Connecticut. The countryside reflects the high latitude: behind the rugged coastline lies a treeless, rolling landscape that looks like the tundra of the Northern Hemisphere. The weather is cold and damp, with much wind and cloud. When the war broke out in 1982, there were only about 1800 permanent residents on the islands, nearly all of them in some way dependent on the main industry, sheep farming.

The Falklands' tiny population (still about 1800) is mainly of British ancestry. England first took control of the Falklands in the 1760s, but was soon ousted by Spain. When Spanish power in southern South America weakened and the new republic of Argentina emerged, the Spanish abandoned the area, leaving it to the Argentines. In 1820, Argentina claimed the Malvinas and followed this up by establishing a base on East Falkland. But in 1833, a British naval force arrived and expelled the Argentines; British sea-power was at its zenith. In Argentina, that action has never been forgotten, and British rule has never been acknowledged.

Negotiations over the future of the islands had been continuing without progress when on April 2, 1982, an Argentine force invaded the Falklands, overpowered the small British garrison, and hoisted Argentina's flag over Stanley, the capital. In the United Kingdom 7500 miles (12,000 kilometers) away, this act had an impact similar to the Japanese attack on Hawaii in the United States of 1941. The British government launched a counterattack by sea and air, and in a war that cost about 200 British and more than 700 Argentinean lives, defeated the invaders by mid-June.

The British victory, however, did not do anything to solve the fundamental politico-geographical problem. Argentina continues to insist that the British takeover of the Malvinas in 1833 remains an unresolved act of war and annexation. As the legitimate successors of the Spanish rulers over what is today Argentina, the Argentines say that the Malvinas, part of that legacy, are now legally theirs. The islands' nearness to Argentina and their potential importance in maritime boundary-making (see Chapter 10) also figure in Argentina's claim. The British respond that if all forcibly acquired territories in the world were given back to their ''original'' owners, there would be global chaos. They also point to the English character of the Falklands' population, six generations of close ties with Britain, and the expressed desires of the settlers to remain under the British flag. And the United Kingdom also has claims and objectives in nearby Antarctica.

Further negotiations are obviously needed, but the chances for a peaceful resolution of this lingering dispute are not good. The British, after their victory, have considerably strengthened the defensive capabilities of the islands—at enormous cost. A commission was charged with the task of finding ways to bolster the local economy. But much less was done to seek a negotiated settlement to a conflict in which both sides have strong arguments. And lately, the stalemate has not been helped by Britain's claiming a 150-mile-wide (240-kilometer-wide) ''exclusive economic zone'' around the islands in order to protect the Falklands' fisheries. This proclamation went into effect in 1987, creating overlapping zones with Argentinean coastal waters and thereby heightening the risk of incidents between the Argentine navy and foreign fishing vessels as a result of differing interpretations of who controls various marine areas.

and Argentina export comes from the Paraguay-Paraná Valley.

In addition, the streams that flow eastward from the Andes provide opportunities for irrigation. Tucumán (650,000), the focus of Argentina's major sugar-producing district, developed in response to a unique set of physical circumstances and a rapidly growing market in the Pampa cities, which the railroad made less remote. At Mendoza (775,000) and San Juan to the north and San Rafael to the south, vineyards and fruit orchards reflect the economy of the Cuyo subregion. But despite these sizable near-Andean outposts, effective Argentina remains the area within a radius of 350 miles (560 kilometers) of Buenos Aires.

The 1980s was a particularly stressful decade for Argentina, involving the ''disappearances'' of thousands of non-rightist political activists, a humiliating military defeat by the British in the disastrous war for the offshore Falkland Islands (see box), the unraveling of

the inept and brutal military junta that had ruled since the mid-1970s, and ever-deepening economic crisis that followed the 1984 return to what passes as ''democracy'' in this part of the world.

In sum, Argentina is a land of opportunity and challenge—opportunity because its territory is large, its environments varied, and its resources rich and plentiful; challenge because those advantages have been diminished by political instabilities, economic policy problems, and especially a paralyzing inability to get anything done. Practically every enterprise in Argentina has been bedeviled by these adversities, and the country will have to overcome its disorganization if it is even to begin to realize anything approaching its full potential.

Uruguay

Uruguay, unlike Argentina or Chile, is compact, small, and fairly densely populated. This buffer

state of old became a fairly prosperous agricultural country, in effect a smaller-scale Pampa (though possessing less favorable soils and topography); Figs. I–5 and I–10 show the similarity of physical conditions on the two sides of the Plata estuary. Montevideo, the coastal capital with 1.5 million residents, contains 50 percent of the country's population; from here, railroads and roads radiate out into the productive agricultural interior. In the immediate vicinity of Montevideo lies Uruguay's major farming area, which produces vegetables and fruits for the city as well as wheat and fodder crops. Just about all of the rest of the country is used for grazing sheep and cattle, with wool, hides, and meat dominating the export trade. Of course, Uruguay is a small country, and its 68,000 square miles (176,000 square kilometers), less territory even than Guyana, do not leave much room for population clustering. But it is, nevertheless, a special quality of the land area of Uruguay that this republic is rather evenly peopled right up to the boundaries with Brazil and Argentina. Of all the countries in South America, Uruguay is the most truly European, notably lacking the racial minorities that mark even Chile and Argentina, but with a sizable non-Spanish European component in its population.

Chile

For 2500 miles (4000 kilometers) between the crestline of the Andes and the coastline of the Pacific lies the narrow strip of land that is the Republic of Chile. On the average just 90 miles/150 kilometers wide (and only rarely over 150 miles/250 kilometers in width), Chile is the textbook case of what political geographers call an *elongated state*, one whose shape tends to contribute to external political, internal administrative, and general economic problems. In the case of Chile, the Andes Mountains do form a barrier to encroachment from the east and the sea constitutes an avenue of north-south communication; history has shown the country to be quite capable of coping with its northern rivals, Bolivia and Peru.

As Figs. I–10 and 5–17 indicate, Chile is a three-subregion country. About 90 percent of Chile's 13.4 million people are concentrated in what is called Middle Chile, where Santiago (5.4 million), the capital and largest city (one of the world's smoggiest), and Valparaíso (350,000), the chief port, are located. North of Middle Chile lies the Atacama Desert, wider, drier, and colder than the coastal desert of Peru. South of Middle Chile, the coast is broken by a plethora of fjords and islands, the topography is mountainous (Fig. 5–19), and the climate—wet and cool near the Pacific—soon turns drier and colder against the Andean interior. South of the latitude of the island of Chiloé there are no permanent overland routes, and there is hardly any settlement.

These three subregions are also clearly apparent on the map of culture spheres (Fig. 5–8), which displays a mestizo north, a European-commercial zone in Middle Chile, and an undifferentiated south. In addition, a small Indo-subsistence zone in northern Chile's Andes is shared with Argentina and Bolivia. The Indian component in the two-thirds of the Chilean population that is mestizo largely originated from the million

or so Indians who lived in Middle Chile.

Despite the absence of a large landless Indian class, Chile has tenant farmers (known as the *rotos*) on its haciendas, people who are no better off than their peon counterparts elsewhere in South America. However, there were always a few means of escape from this system; a popular one was to leave the land and head for the cities to seek work, a migration flow that continues in the 1990s.

Some intraregional differences exist between northern and southern Middle Chile, the country's core area. Northern Middle Chile, the land of the hacienda and of Mediterranean climate with its dry summer season, is an area of (usually irrigated) crops that include wheat, corn, vegetables, grapes, and other Mediterranean products; livestock-raising and fodder crops also take up much of the productive land. But agricultural methods are not very efficient, and the area could undoubtedly yield a far greater volume of food crops than it does. Southern Middle Chile, into which immigrants from both the north and from Europe (especially Germany) have pushed, is a better-watered area where raising cattle predominates but where wheat and other food crops are also widely cultivated.

Few of Chile's agricultural products, however, reach external markets. Some specialized items such as wine, grapes, and raisins are exported, but Chile remains a net importer of food. Thus, despite the fact that nine out of ten Chileans live in what has been defined

here as Middle Chile, the Atacama region in the north provides most of the country's foreign revenues. The Atacama contains the world's largest exploitable deposits of nitrates, and at first these provided the country's economic mainstay. But this mining industry soon declined after the discovery of methods of synthetic nitrate production early in the twentieth century. Subsequently, copper became the chief export (Chile possesses the world's largest reserves); it is found in several places, but the main concentration lies on the eastern margin of the Atacama Desert near the town of Chuquicamata, not far from the port of Antofagasta.

Copper has been mined for well over a century, and U.S. investment in this Chilean industry (which began in 1915) has been heavy. The Chilean government, anxious for a greater share of the profits, expropriated some mining properties in the early 1970s. The resulting disagreement over compensation created a political storm that illuminated some of the problems of direct foreign investment in an economy.

Soon, however, more pressing political difficulties overrode the economic: the leftist Chilean government fell and was replaced by a military regime. The priorities of this rightist government differed from those of its predecessor, and new issues arose. This combination of foreign involvement, political instability, and changing national directions is a dominant quality of the geography of South America today.

PRONUNCIATION GUIDE

Acre (*ah*-kray)

Altiplano (ahl-tee-*plah*-noh)

Antioquia (ahn-tee-*oh*-kee-ah)

Antofagasta (untoh-fah-*gahss*-tah)

Arauca (ah-*rau*-kah)

Arica (ah-*ree*-kah)

Asunción (ah-soohn-see-*oan*)

Atacama (ah-tah-*kah*-mah)

Augelli (aw-*jelly*)

Barquisimeto (bar-key-suh-*may*-toh)

Barranquilla (bah-rahn-*kee*-yah)

Barrio[*s*] (*bar*-ree-oh[ss])

Bayovar (bye-*yoh*-vahr)

Beni (*bay*-nee)

Bogotá (boh-goh-*tah*)

Bolívar, Simón (boh-*lee*-vahr, see-*moan*)

Buenos Aires (*bway*-nohss *eye*-race)

Cacao (kuh-*kah*-oh)

Caldas (*kahl*-dahss)

Cali (*kah*-lee)

Callao (kah-*yah*-oh)

Caracas (kah-*rah*-kuss)

Cartagena (kar-tah-*hay*-nah)

Caste (*cast*)

Cauca (*kow*-kah)

Central (sen-*trahl*)

Cerrejón (serray-*hone*)

Cerro Bolívar (serro-boh-*lee*-vahr)

Cerro de Pasco (serro-day-*pah*-skoh)

Chaco (*chah*-koh)

Chaparé (chah-pah-*ray*)

Chibcha (*chib*-chuh)

Chile (*chilli*/*chee*-lay)

Chiloé (chee-luh-*way*)

Chuquicamata (choo-kee-kah-*mah*-tah)

Ciudad Guayana (see-you-*dahd* gwuh-*yahna*)

Cochabamba (koh-chah-*bum*-bah)

Comodoro Rivadavia (comma-*dore*-oh ree-vah-*dah*-vee-ah)

Cordillera (kor-dee-*yerra*)

Córdoba (*kord*-oh-bah)

Corumbá (kor-room-*bah*)

Creole[s] (*kree*-ole[z])

Cuyo (*koo*-yoh)

Cuzco (*koo*-skoh)

Département (day-part-*maw*)

Ecuador (*eck*-wah-dor)

Entre Rios (en-truh-*ree*-ohss)

Esmeraldas (ezz-may-*rahl*-dahss)

Essequibo (essa-*kwee*-boh)

Falkland[s] (*fawk*-lund[z])

Favela[*s*] (fah-*vay*-lah[ss])

Fría (*free*-uh)

Guajira (gwah-*hear*-ah)

Guantánamo (gwahn-*tah*-nah-moh)

Guaraní (gwah-rah-*nee*)

Guayaquil (gwye-ah-*keel*)

Guayas (*gwye*-ahss)

Guiana[s] (ghee-*ah*-nah[z])

Guri (*goor*-ree)

Guyana (guy-*ahna*)

Hacienda (ah-see-*en*-duh)

Huallaga (wah-*yah*-gah)

Huancayo (wahn-*kye*-oh)

Iberia (eye-*beery*-uh)

Iquitos (ih-*kee*-tohss)

Isohyet (*eye*-so-hyatt)

Itaipu (ee-*tie*-pooh)

João, Dom (*zhwow*, dom)

Junta (*hoon*-tah)

La Paz (lah-*pahz*)

Lima (*lee*-mah)

Llama (*lah*-muh)

Llano[s] (*yah*-noh[ss])

Machu Picchu (*mah*-choo *peek*-choo)

Magdalena (mahg-dah-*lay*-nah)

Malvinas, Islas (mahl-*vee*-nahss, *eece*-lahss)

Maracaibo (mah-rah-*kye*-boe)

Medellín (meh-deh-*yeen*)

Mestizo (meh-*stee*-zoh)

Mexica (meh-*shee*-kuh)

Montaña (mon-*tahn*-yah)

Montevideo (moan-tay-vee-*day*-oh)

Occidental (oak-see-den-*tahl*)

Oriental (orry-en-*tahl*)

Oriente (orry-*en*-tay)

Orinoco (orry-*noh*-koh)

Pampa (*pahm*-pah)

Paraguay (*pahra*-gwye)

Paraná (pah-rah-*nah*)

Patagonia (patta-*goh*-nee-ah)

Paulista[*s*] (pow-*leash*-tah[ss])

Pizarro, Francisco (pea-*sahro*, frahn-*seece*-koh)

Plata, Rio de la (*plah*-tah, ree-oh-day-luh)

Plaza de Armas (*plah*-sah day-ar-mahss)

Plaza de la Constitución (*plah*-sah day-luh con-stee-too-see-*yoan*)

Plaza de Mayo (*plah*-sah day-*mye*-oh)

Potosí (poh-toh-*see*)

Puna (*poona*)

Quebracho (kay-*brotch*-oh)

Quechua (*kaytch*-wah)

Quito (*kee*-toh)

Rio de Janeiro (*ree*-oh day zhah-*nair*-roh)

San Cristóbal (sahn-kree-*stoh*-bahl)

San Juan (sahn-*hwahn*)

San Rafael (sahn-rah-fye-*ell*)

Santa Cruz (sahnta-*krooz*)

Santiago (sahn-tee-*ah*-goh)

São Paulo (sau-*pau*-loh)

Sendero Luminoso (sen-*dare*-oh loo-mee-*no*-so)

Sucre (*soo*-kray)

Suriname (soor-uh-*nahm*-uh)

Templada (tem-*plah*-dah)

Tierra del Fuego (tee-*erra* dale *fway*-goh)

Titicaca (tiddy-*kah*-kuh)
Tordesillas (tor-day-*see*-yahss)
Tucumán (too-koo-*mahn*)
Tumbes (*toom*-base)

Uruguay (*oo*-rah-gwye)
Valparaíso (vahl-pah-rah-*ee*-so)
Venezuela (ve-ne-*sway*-lah)
Viedma (vee-*yed*-mah)

Von Thünen (fon-*too*-nun)
Yerba maté (yair-bah-*mah*-tay)
Yungas (*yoong*-gahss)

REFERENCES AND FURTHER READINGS

(BULLETS [•] DENOTE BASIC INTRO-
DUCTORY WORKS ON THE REALM
OR SYSTEMATIC ESSAY TOPIC)

Augelli, John P. "The Controversial Image of Latin America: A Geographer's View," *Journal of Geography*, 62 (1963): 103–112. Quotation taken from p. 111.

Berry, Brian J. L. et al. *Economic Geography: Resource Use, Locational Choices, and Regional Specialization in the Global Economy* (Englewood Cliffs, N.J.: Prentice-Hall, 1987).

• Blakemore, Harold & Smith, Clifford T., eds. *Latin America: Geographical Perspectives* (London & New York: Methuen, 2 rev. ed., 1983).

• Blakemore, Harold et al., eds. *The Cambridge Encyclopedia of Latin America and the Caribbean* (London: Cambridge University Press, 1985).

• Blouet, Brian W. & Blouet, Olwyn M., eds. *Latin America: An Introductory Survey* (New York: John Wiley & Sons, 1982).

Boehm, Richard G. & Visser, Sent, eds. *Latin America: Case Studies* (Dubuque, Iowa: Kendall/Hunt, 1984).

• Bromley, Rosemary D. F. & Bromley, Ray. *South American Development: A Geographical Introduction* (New York: Cambridge University Press, 1982).

Caviedes, Cesar N. *The Southern Cone: Realities of the Authoritarian State* (Totowa, N.J.: Rowman & Allanheld, 1984).

Child, Jack. *Geopolitics and Conflict in South America: Quarrels Among Neighbors* (Westport, Conn.: Praeger, 1985).

"Cocaine Wars: South America's Bloody Business," *Time*, February 25, 1985, pp. 26–35.

Denevan, William M., ed. *Hispanic Lands and Peoples: Selected Writings of James J. Parsons* (Boulder, Colo.: Westview Press, 1988).

Drakakis-Smith, David. *The Third World City* (Boston: Routledge & Kegan Paul, 1987).

Gilbert, Alan et al., eds. *Urbanization in Contemporary Latin America: Crucial Approaches to the Analysis of Urban Issues* (Chichester, U.K.: John Wiley & Sons, 1982).

Griffin, Ernst C. & Ford, Larry R. "Cities of Latin America," in Brunn, Stanley D. & Williams, Jack F., eds., *Cities of the World: World Regional Urban Development* (New York: Harper & Row, 1983), pp. 198–240.

• Griffin, Ernst C. & Ford, Larry R. "A Model of Latin American City Structure," *Geographical Review*, 70 (1980): 397–422. Model diagram adapted from p. 406.

Grigg, David B. *The Agricultural Systems of the World: An Evolutionary Approach* (London: Cambridge University Press, 1974).

Gwynne, Robert N. *Industrialization and Urbanization in Latin America* (Baltimore: Johns Hopkins University Press, 1986).

Harris, David R., ed. *Human Ecology in Savanna Environments* (New York & London: Academic Press, 1980).

Hudson, Tim. "South American High: A Geography of Cocaine," *Focus*, January 1985, pp. 22–29.

• James, Preston E. & Minkel, Clarence W. *Latin America* (New York: John Wiley & Sons, 5 rev. ed., 1986), Chaps. 17–28, 37.

Knox, Paul L. & Agnew, John A. *The Geography of the World-Economy* (London & New York: Routledge, 1989).

• Lowder, Stella. *The Geography of Third World Cities* (Totowa, N.J.: Barnes & Noble, 1986).

Maos, Jacob O. *The Spatial Organization of New Land Settlement in Latin America* (Boulder, Colo.: Westview Press, 1984).

Miller, E. Willard & Miller, Ruby M. *South America: A Bibliography on the Third World* (Monticello, Ill.: Vance Bibliographies, No. P-1064, 1982).

Miller, Tom. *The Panama Hat Trail: A Journey from South America* (New York: William Morrow, 1986).

Mörner, Magnus. *The Andean Past: Land, Societies and Conflict* (New York: Columbia University Press, 1985).

Morris, Arthur S. *Latin America: Economic Development and Regional Differentiation* (Totowa, N.J.: Barnes & Noble, 1981).

• Morris, Arthur S. *South America* (Totowa, N.J.: Barnes & Noble, 3 rev. ed., 1987).

• Odell, Peter R. & Preston, David A. *Economies and Societies in Latin America* (Chichester, U.K.: John Wiley & Sons, 2 rev. ed., 1978).

• Preston, David, ed. *Latin American Development: Geographical Perspectives* (London & New York: Longman, 1987).

"The Cocaine Economies: Latin America's Killing Fields," *The Economist*, October 8, 1988, pp. 21–24.

Theroux, Paul. *The Old Patagonian Express: By Train Through the Americas* (Boston: Houghton Mifflin, 1979).

Wallace, Iain. *The Global Economic System* (Winchester, Mass.: Unwin Hyman, 1989).

Weil, Connie H., ed. *Medical Geographic Research in Latin America* (Elmsford, N.Y.: Pergamon, 1982).

• Wheeler, James O. & Muller, Peter O. *Economic Geography* (New York: John Wiley & Sons, 2 rev. ed., 1986).

• Wilkie, Richard W. *Latin American Population and Urbanization Analysis: Maps and Statistics, 1950–1982* (Westwood: UCLA Latin American Center, 1984).

EMERGING BRAZIL:
Potentials and Problems

Brazil, South America's long-dormant giant, is wide awake. The Brazilian economy, now the world's eighth-largest, has the capacity to grow at rates that far exceed those of underdeveloped countries. The population, recently one of the world's fastest-growing, has now slowed its increase, but the final decade of this century will still see the addition of about 30 million more Brazilians. Symbolized by its modern forward capital of Brasília (Fig. B–l), the vast Amazon-Basin-dominated interior has been opened up by new roads that annually lure at least a quarter of a million new settlers. The nation's cities are mushrooming and its industries are thriving. Despite having been staggered by severe economic problems in the 1980s, Brazil today shows unmistakable signs of taking off toward a future in the First World.

With its 3.3 million square miles (8.5 million square kilometers) of equatorial and subtropical South

FIGURE B–1   Part of the federal government complex in central Brasília. Three decades after its completion, a certain rawness still pervades the landscape, and along the lake at the rear the ubiquitous squatter shacktowns have crept up to this showcase city's edge.

more than doubled the annual output of Florida, thanks to a climate free of winter freezes, ultramodern processing plants, and a fleet of specially equipped tankers to ship the concentrate to markets in the United States—and more than 40 other countries from Japan to the Soviet Union.

Matching this prodigious agricultural output is the state's industrial strength. The city of São Paulo (Fig. B–5) is now the country's leading manufacturing center. The state does not have a mineral base to compare to that of Minas Gerais, but it has nevertheless become the leading industrial region not only of Brazil, but of all "Latin" America. The revenues derived from the coffee plantations provided the necessary investment capital, hydroelectric power from the slopes of the coastal escarpment produced the needed energy, and immigration from Portugal, Italy, Japan, and elsewhere contributed the labor force. São Paulo State lay juxtaposed between raw-material-producing Minas Gerais and the states of the South (Fig. B–2). Communications were improved, notably with the outport of Santos, but also with the interior hinterland. As the capacity of the domestic market grew, the advantages of central location and agglomeration secured São Paulo's primacy. Metropolitan São Paulo today is truly the pulse of Brazil; its burgeoning population of 18.7 million also makes it the realm's biggest urban complex—and the third largest in the world after Tokyo and Mexico City.

4. The South

Three states make up the southernmost Brazilian region: Paraná, Santa Catarina, and Rio Grande do Sul. The contribution of recent European immigrants to the agricultural development of southern Brazil, as in neighboring Uruguay and Argentina, has been considerable. Many came not to the coffee country of São Paulo State, but to the available lands farther south. Here they occupied fairly discrete areas. Portuguese rice farmers clustered in the valleys of the ma-

FIGURE B–5 Panoramic view of São Paulo, the dynamic pulse of late-twentieth-century Brazil. South America's largest metropolis (and the world's third biggest) continues to grow rapidly and will surpass the 20-million mark before 1994.

jor rivers of Rio Grande do Sul, and that state now produces one quarter of Brazil's annual rice crop. The Germans, on the other hand, occupied the somewhat higher areas to the north and in Santa Catarina, where they were able to carry on the type of mixed farming with which they were familiar: corn, rye, potatoes, and hogs as well as dairying. The Italians selected the highest slopes and established thriving vineyards. The markets for this produce, of course, are the large urban areas to the north. Paraná, on the other hand, exports its coffee harvest to overseas markets.

Unlike the Northeast and Southeast, the South has never been a boom area. It does, however, have a stable, progressive, and modern agricultural economy; farming methods here are the most advanced in Brazil. The diversity of its European heritage is still reflected in the regional towns, where German and other European languages are preserved. More than 23 million people live in the three southern states, and the region's importance is increasing. Coal from Santa Catarina and Rio Grande do Sul, shipped north to the steel plants of Minas Gerais, was a crucial element in Brazil's industrial emergence. Local industry is growing as well, especially in Pôrto Alegre (3.1 million) and Tubarão, where South America's single largest steelmaking facility opened in 1983.

The Brazilian South today is also the scene of one of the world's greatest construction projects, the just-completed Itaipu Dam (Fig. B–6). Located astride the Paraná River on the Brazil–Paraguay border in the southwest corner of Paraná State (Fig. B–2), Itaipu's dimensions are truly awesome: at 600 feet (180 meters) in height and 5 miles (8 kilometers) in length, it is six times bigger than Egypt's Aswan High Dam, and its

FIGURE B–6   Mighty Itaipu Dam, which became fully operational at the end of the 1980s.

peak electricity output equals more than 50 percent of Brazil's total electrical consumption in 1982 (the equivalent of a half-million barrels of oil a day). Begun in 1978 in response to the oil crisis of the 1970s, the first of the dam's 18 massive turbines was opened in 1984; the rest came on line at regular intervals until Itaipu's full capacity was achieved in 1990. Much of this energy is being transmitted to the São Paulo metropolis, 500 miles (800 kilometers) to the east, and will fuel new rounds of industrialization. It was hoped that the Itaipu area itself would also become a growth center and enjoy considerable benefits as well; in anticipation, hundreds of thousands migrated there, but opportunities have fallen well below expectations, particularly in Para-

guay where the Itaipu work force has been disbanded.

5. The Interior

Interior Brazil is often referred to as the Central-West, or *Centro-Oeste*. This is the region Brazil's developers hope to make a part of the country's productive heartland, and in the 1950s the new capital of Brasília was deliberately situated on its margins (Figs. B–1, B–2). By locating the new capital city in the untapped wilderness 400 miles (650 kilometers) inland from its predecessor, Rio de Janeiro, the country's leaders dramatically signaled the opening of Brazil's age of development and a new thrust toward the west. As we saw earlier, in Japan's shifting of its capital from Kyoto to Tokyo when it entered its modernization stage, governments have used the relocation of their own headquarters to emphasize the beginning of new eras and new geographic linkages. Moreover, a number of countries have shifted their capitals in recent years (among them Pakistan and Nigeria), and we noted that in South America both Argentina and Peru have lately considered such moves.

Brasília is also noteworthy in another regard: it represents what political geographers call a **forward capital**. There are times when a state will relocate its capital to a sensitive area, perhaps near a zone under dispute with an unfriendly neighbor, partly to confirm its determination to sustain its position in that contested zone. A recent example is the decision by Pakistan to move its capital from coastal Karachi to northern Islamabad, near the disputed territory of Kashmir. These are called "forward" capitals because of their position in an area that would be first to be engulfed by conflict in case of strife with a neighbor. Brasília, of course, does not lie in or near a contested zone, but Brazil's interior has been an internal frontier, one to be conquered by a developing nation; in that drive, the new capital occupies a decidedly forward position.

Despite the growth of Brasília, the economic integration of the Interior into the rest of the country will take considerable time. This is a vast upland area, largely a plateau at an elevation over 3000 feet (900 meters) covered with savanna vegetation (see Fig. I–12). Like Minas Gerais, the Interior was the scene of gold rushes and some discoveries, but unlike its eastern neighbor, it did not present agricultural alternatives when those mineral searches petered out. Today, the three large states

of the Interior (Goiás, Mato Grosso, and Mato Grosso do Sul) have a combined population of about 11 million, so that the average density is under 15 per square mile (6 per square kilometer). Nevertheless, this represents a noteworthy increase, for in 1960 the Centro-Oeste contained just 2.9 million people. Thus, this Brazilian region's population has nearly quadrupled in 30-plus years—a pace well ahead of the national growth rate.

Despite these gains, the Centro-Oeste continues to face the problems common to tropical savanna regions everywhere—soils are not especially fertile and the vegetation is susceptible to damage by overgrazing. In the absence of major mineral finds, pastoralism remains the chief economic activity. But there is always the fear that what happened to the Northeast could happen here, and that Brazil's determination to open the Interior might reproduce the serious environmental problems of Amazonia (discussed in the next section). But at the moment that day seems far off: one can fly for hours above this huge area and see only an occasional small settlement and a few widely spaced roads (communications are best in the south). What the Interior needs is investment—in the clearing and opening of the alluvial soils in the region's river valleys, in experimental farms, in the provision of electric power, in further mineral exploration. The first step was the relocation of the national capital— an enormously expensive venture, but a mere beginning in view of what the Interior really requires.

6. The Amazonian North

The largest and most rapidly developing region—whose popula-

tion has more than doubled to 12 million since 1980—is also the most remote from the core of Brazilian settlement: the five states and single federal territory of the Amazon Basin (Fig. B–2). This was the scene of the great rubber boom at the turn of the century, when the wild rubber trees in the *selvas* (tropical rainforests) produced huge profits and the central Amazon city of Manáos (1991 population: 1.3 million) enjoyed a brief period of wealth and splendor. But the rubber boom ended in 1910 when plantations elsewhere (notably in Southeast Asia) began to produce rubber more cheaply, efficiently, and accessibly. For most of the following seven decades, Amazonia was a stagnant hinterland, but all that changed dramatically during the 1980s. This region is now bursting with development and has become the scene of the world's largest migration into virgin territory as over 250,000 new settlers arrive every year. Most of this activity is occurring south of the Amazon River, in the tablelands between the major waterways and along the Basin's wide rim. Two new development schemes in the early 1990s are especially worth noting, because they are quintessential expressions of what is going on here.

The first is the Grande Carajás Program in southeastern Pará State, a huge multifaceted scheme that is centered on the world's largest-known deposits of iron ore in the Serra dos Carajás hills (Fig. B–2). In addition to a vast mining complex, other new construction here includes the Tucuruí Dam on the nearby Tocantins River and a 535-mile/850-kilometer railroad to the Atlantic port of São Luis. This ambitious development project also emphasizes further mineral exploitation (including bauxite, manganese, and copper), cattle raising, farming, and forestry, and if expansion plans are implemented,

Grande Carajás will one day cover one-sixth of all Amazonia.

Understandably, tens of thousands of settlers have descended on this area in the past few years. Those seeking business opportunities have been in the vanguard, but they have been followed by masses of lower-income laborers and farmers in search of jobs and landownership. Surveying the initial stage of this colossal enterprise in his recent book, *Passage Through El Dorado*, Jonathan Kandell has compared it to the westward surge of nineteenth-century pioneers in the United States and used the words "energy, hope, greed, and savagery" to describe the turmoil of the competition to succeed in this remote and often hostile environment.

The second leading development scheme, known as the Polonoroeste Plan, is located about 500 miles (800 kilometers) to the southwest in the 1500-mile-long (2400-kilometer-long) Highway BR-364 corridor that parallels the Bolivian border and connects the western Brazilian towns of Cuiabá, Pôrto Velho, and Rio Branco (Fig. B–2). Although the government had planned for the penetration of the North to proceed via the east-west Trans-Amazon Highway, the migrants of the 1980s and 1990s have preferred to follow BR-364 and settle within the Basin's southwestern rim zone, mostly in Rondônia State. Agriculture has been the dominant activity here, attracting affluent growers and ranchers from the South, plantation workers from São Paulo and Paraná states displaced by agrarian mechanization, and subsistence farmers from all over the country. The common denominator has been the quest for land, and bitter conflicts continue to break out between peasants and landholders as the Brazilian government cautiously pursues the persistent and volatile issue of land reform.

The usual pattern of Amazonian settlement is something like this. As main and branch highways are cut through the wilderness, settlers (enticed by cheap land) follow and move out laterally to clear spaces for farming. Crops, usually maize (corn) or upland rice are planted, but within three years the heavy equatorial-zone rains leach out soil nutrients and accelerate surface erosion. As soil fertility declines, pasture grasses are then planted, and the plot of land is soon sold to cattle ranchers (most are associated with big agribusiness corporations based in Rio or São Paulo). The peasant farmers then move on to newly opened areas, clear more land for planting, and the cycle repeats itself. As long as open spaces remain, this is a profitable pursuit for all parties, but it assures the widespread establishment of a low-grade land use (Fig. B–7) that will ultimately concentrate most of the earnings in the hands of large landowners.

Perhaps worst of all is the environmental impact of this extensive grazing system that involves so little attachment between farmers and their lands: it requires clearing enormous stands of tropical woodland. In the late 1980s, between 20,000 and 30,000 square miles (52,000 and 78,000 square kilometers) of rainforest were disappearing *annually* in Amazonia—an area larger than the state of West Virginia (Fig. B–8). Moreover, the deforestation crisis here in the Brazilian North is exacerbated by the scattered development, so that the remaining selvas are dotted with a myriad of expanding clearings that are constantly coalescing. And because it also accounts for over half the total worldwide tropical deforestation, Amazon woodland destruction has much wider implications; as the *box* on pp. 274–275 describes, this is an environmental crisis of global proportions, with future consequences that could negatively impact every form of terrestrial life.

FIGURE B–7 Newly cleared land in the wide Amazon-rim zone usually reverts to cattle ranching following the predictable failure of crops in the heavily-leached tropical soil. Here, a wealthy landowner, who obviously resides at some distance, inspects a herd from the seat of a private plane.

FIGURE B—8 The awesome power of an Amazon rainforest fire, the fastest and cheapest way to clear the tropical woodlands for farming. This photograph also underscores the atmospheric effects of deforestation, discussed in the *box* on p. 275.

POPULATION PATTERNS

Brazil's population of 153.5 million is as diverse as that of the United States, and Brazilian society has been a melting pot, perhaps to an even greater degree than the United States itself. In a pattern that is familiar in the Americas, the original Indian inhabitants of the country were decimated following the European invasion; estimates of the number of Indians who survive today in small communities deep in the Amazonian interior vary, but their total is not likely to exceed 200,000—well under 10 percent of the number thought to have been there when whites first arrived. Africans came in great numbers, too, and today there are more than 10 million blacks in Brazil. Significantly, however, there was also much mixing, and nearly 70 million Brazilians (about 45 percent of the national population) have combined European, African, and minor Indian ancestries. The remaining 75 million or so, now no longer in the majority, are of European descent.

Until Brazil became independent in 1822, the Portuguese were virtually the only Europeans to settle in this country. But after independence other European settlers were encouraged to come, and many Italians, Germans, and Eastern Europeans arrived to work on the coffee plantations, farm in the south, or try their luck in business. Immigration reached a peak during the 1890s, a decade in which nearly 1.5 million newcomers reached Brazilian shores. The complexion of the population was further diversified by the later arrival of Lebanese and Syrians, many of whom opened small shops. Even the Japanese have

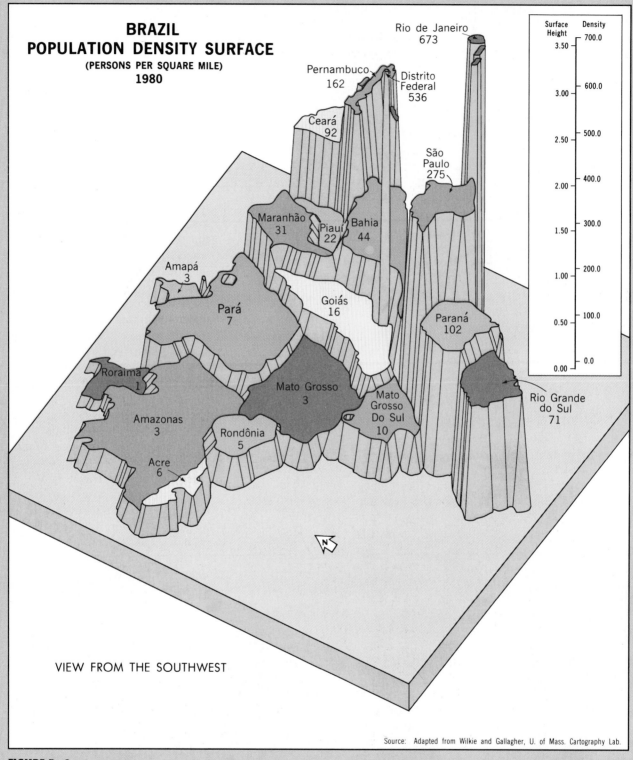

BRAZIL
POPULATION DENSITY SURFACE
(PERSONS PER SQUARE MILE)
1980

Rio de Janeiro 673

Pernambuco 162

Distrito Federal 536

Ceará 92

São Paulo 275

Maranhão 31

Piauí 22

Bahia 44

Amapá 3

Pará 7

Goiás 16

Paraná 102

Roraima 1

Mato Grosso 3

Mato Grosso Do Sul 10

Rio Grande do Sul 71

Amazonas 3

Rondônia 5

Acre 6

Surface Height	Density
3.50	700.0
3.00	600.0
2.50	500.0
2.00	400.0
1.50	300.0
1.00	200.0
0.50	100.0
0.00	0.0

VIEW FROM THE SOUTHWEST

Source: Adapted from Wilkie and Gallagher, U. of Mass. Cartography Lab.

FIGURE B–9

had an impact on Brazilian social geography, and today they constitute a bustling community of over a million (the largest anywhere outside Japan) that grew from the nucleus of 781 pioneers who arrived in 1908. At first they performed menial jobs on the coffee *fazendas*, but now the fully integrated ethnic Japanese of São Paulo State have risen to the highest ranks in Brazil's easy-going immigrant society as highly successful farmers, urban professionals, business leaders—and traders with Japan.

Brazilian society, to a greater degree than is true elsewhere in the Americas, has made progress in dealing with its racial divisions. To be sure, blacks are still the least advantaged among the country's population groups, except for the Indians. But ethnic mixing in Brazil is so pervasive that hardly any group is unaffected, and official statistics about "blacks" and "Europeans" are meaningless (census data show blacks comprise only about 7 percent of the population). What the Brazilians do have is a true national culture, expressed in an adherence to the Catholic faith (it is the world's largest Roman Catholic country), in the universal use of a modified form of Portuguese as the common language, and in a set of lifestyles in which vivid colors, distinctive music, and a growing national consciousness and pride are fundamental ingredients.

This is not to suggest an absence of regional variety in Brazilian socio-geographical patterns. The black component of the population in the Northeast, for instance, remains far stronger than it is in the Southeast. After their initial concentration in the Northeast, black workers were taken southward to Bahia and Minas Gerais as the economic heartland moved in that direction. Today the black population remains strongest in these areas and in Rio de Janeiro; it is weak in the states of the South, more recently settled and more completely taken over by Europeans. Nor has Brazil escaped problems of racial prejudice, and black leaders and others in the country have in recent years publicly voiced their objections against discrimination. Yet, although Brazil may not be the multiracial society it is sometimes

portrayed to be, it also does not have the history of overt and legally sanctioned racism that has characterized certain other plural societies.

Another population problem that has beset Brazil—one that could still resurface and jeopardize the overall development potential—was its persistently high birth rate. As recently as the late 1970s, the country grew at the very rapid pace of 2.8 percent annually. But the 1980s saw a turnaround and then a steady drop, and by the opening of the 1990s this rate was down to (a still high) 2.0. Whereas a reversal is possible at any time, demographers expect the decline to continue and then level off through the foreseeable future. One of the most noteworthy aspects of this post-1980 trend has been a dramatic plunge in the fertility rate, from 4.4 children per woman in 1980 to about 3.0 today. Surprisingly, the slowdown in population increase has taken place in the absence of an active birth control policy by the Brazilian government (and also in the face of disapproval by the Roman Catholic church). Although the phenomenon needs closer study, the initial assessment is that Brazil's self-induced decline is the result of three factors: a rapid spread in contraceptive usage, the negative influence of recent economic stagnation on family formation, and newly widespread access to television (which seems to have reshaped the attitudes and aspirations of the Brazilian people).

A final glimpse at the population geography of Brazil underscores the country's unevenness—a concluding theme in the section on economic development that follows. Figure B–9 is a map of the population density of the Brazilian states and territories in 1980 (a distribution that is still quite current), with the density of each of the country's 27 subnational political

units shown as a flat platform "pushed up" from a base plane. Its height is determined by the magnitude of its density, measured on the scale shown next to the population "surface." The "peaks" and "troughs" of this three-dimensional surface etch as sharply as possible the density contrasts of a nation in which 90 percent of the people are still crowded into 10 percent of the available space, yielding a high average density on that tenth of land of 410 people per square mile (157 per square kilometer).

The map in Fig. B–9 is also a good example of some recent technological advances in cartography. The computer-drawn perspective we see here is that of an astronaut looking down on Brazil from a point in space approximately above Santiago, Chile. The total surface appears as a series of steps that are joined to form a number of analogous mountains, ridges, depressions, and plains that represent variations in population density across Brazil. The SYMVU computer program used to generate the map permits us to view this surface from any direction and angle, and make any number of manipulations; software packages of this kind are constantly increasing in sophistication and are also available for personal computers.

DEVELOPMENT PROBLEMS

Brazilian development has followed a roller coaster pattern for the past 25 years. Between 1968 and 1973, the country embarked on a period of spectacular growth, its economy expanding at an annual rate of 9 percent, its exports doubling in value every two years, and its new investments making real headway in improving health standards, agriculture, and living conditions. But the global energy crisis of 1974 quickly wrote an end to these years of the Brazilian "economic miracle" as the nation's huge new imported-fuel bills thwarted its forward drive.

After marking time in the economic doldrums that endured through the rest of the 1970s, Brazil's military rulers in 1980 vowed to resume the developmental push by borrowing heavily abroad, thereby breaking a long-standing guideline to keep away from foreign financial involvements to preserve full control over the country's economy. All too soon, however, Brazilians found themselves seriously mired in the same economic morass as Mexico, Argentina, and a number of other South American countries, having accumulated a staggering debt in foreign loans that surpassed U.S. $90 billion by 1984.

What had gone wrong? Basically, the answer lay with the worldwide economic recession of the early 1980s: falling commodity prices on the international market suddenly and sharply reduced Brazil's income and caused a sizable decline in exports. The government, deeply committed to several development programs and unwilling to trim them back, redoubled its borrowing abroad (at ever-higher interest rates) to make up the difference. As runaway deficits mounted up at home from continued massive governmental spending, inflation soared from 80 to nearly 150 percent annually between 1980 and 1983.

Increasingly, Brazil found itself unable to keep up payments on its rapidly climbing overseas debt. In order to stave off bankruptcy, the International Monetary Fund agreed to supply emergency funds so that Brazil could begin to repay its loans, but only if a severe domestic austerity program was initiated at once. Stiff economic and

energy conservation measures were swiftly imposed, and the government stuck to them despite sporadic outbreaks of social unrest among the poor of the Northeast and Rio de Janeiro State. Almost immediately things began to turn around; by 1986, a spectacular economic growth rate of 10 percent was recorded, led by the rising output of steel, aircraft, coffee, and other farm products.

By 1985, civilian rule had been restored, but ineffective leadership led to upward-spiralling inflation rates. In 1987, the foreign debt reached U.S. $110 billion, a level it hovered at into the early 1990s. The newly elected regime that took office in 1990 inspired greater confidence, and hopes were high that the Brazilian "economic miracle" would soon be back on track.

Agriculture presents Brazil with great opportunities, as we have seen. Problems continue as well, but these have not prevented the country from becoming the second-biggest farm producer in the Western world. Besides the ample revenues derived from coffee sales overseas, Brazil has also been selling ever-greater quantities of sugar, soybeans, and orange-juice concentrate on the international market. The range of crops Brazilian farmers can grow is considerable; but much land that could be cultivated is not, agricultural methods still need improvement, and land reform in the all-important coffee *fazendas* has come too slowly. Millions of peasants in Brazil still practice shifting cultivation when environmental conditions do not really demand it, so that returns from the soil too often are minimal. As we noted in the overview of the Amazon region, widespread cattle ranching is too low-grade an activity on lands that could be farmed much more productively. It has been estimated that the productivity per farm worker in the United States is at least 50 times higher than in

Brazil; mechanization has barely begun in some Brazilian regions. However, despite the rush to the cities, the labor force engaged in agriculture grew during the 1980s, reflecting Brazil's still underdeveloped condition.

In this context, we should keep in mind just how far Brazil has yet to go from its present **takeoff stage of development**. The total area of Brazil is nearly twice that of Europe, but the total market value of its goods and services produced in 1986 was only 54 percent higher than that of the Netherlands alone (with a population of only 15 million versus Brazil's 153.5 million); calculated on a per capita basis, however, the Brazilian figure is a mere 17 percent of the Dutch level of output. Brazil's awakening is sometimes compared to the first stage of Japan's modernization, but such analogies may still be premature.

Growth Poles

When the military regime took control of Brazil in 1964, it embarked on a development program that is still referred to as the *Brazilian Model*. Essentially, this program involved the government in shared participation with private enterprise so that public interests, private concerns, and foreign investors would not operate at mutual disadvantage—and to the disadvantage of Brazil as a whole. Agriculture and industry were supported and promoted, but under certain priorities and guidelines.

Among the problems always facing national development planners is the remote region where opportunities lie, but where the investments needed to exploit them are very high. If an area is already developing and attracting immigrants, should money be spent to improve already existing facilities, or should a new, more distant location be stimulated? In such decisions, the **growth-pole concept** becomes relevant. The term is almost self-explanatory: a growth pole is a location where a set of industries, given a start, will expand and spawn ripples of development in the surrounding area. Certain conditions must exist, of course. There would be little point in selecting Cuiabá as a growth pole in the middle of the Centro-Oeste region where there is no real prospect for development in the immediate hinterland and where there are relatively few people and only minor agricultural or industrial activities to stimulate. But growth-pole theory did help to shape the decision to build Brasília in the 1950s, for in underpopulated Goiás State a new market of 700,000 people constituted a stimulus for many activities (over 6 million people reside there today).

Multinational Economic Influences

As we noted above, the Brazilian Model involves governmental intervention in the economy, including various forms of control over foreign investors. This was a significant dimension of Brazilian development policy, for large corporations played a major role in Brazil's economic surge between the late 1960s and 1980. In fact, the power of multinational corporations in underdeveloped countries has recently become a matter for concern among the leaders of those countries, because these global corporations, backed by enormous financial resources, can

influence the economics as well as the politics of entire states (see Opening Essay on pp. 41–42).

Brazil's leaders had long welcomed foreign investment, but as economic progress accelerated in the 1960s, they perceived the risks involved in foreign control over Brazilian firms. Multinational corporations can introduce and spread technological advances, they can provide capital and increase employment—but in the process, they gain control over the industrial and agricultural export sectors of the economy, and through their efficiency they can throttle local competition and damage business oriented toward local markets. Aware of such impacts, the government of Brazil for a time prohibited foreign commercial banks from entering the Brazilian economy, and imposed regulations requiring multinational corporations active in Brazil to keep more of their huge profits inside the country. But the severe economic recession of the early 1980s forced the government to abandon its strictures against foreign borrowing, which subsequently accelerated so rapidly that by 1985 Brazil had become the world's biggest debtor.

Internal Unevenness

Our survey of South America's largest country underscored its highly variable character and the sharp divisions between modern and traditional Brazil. This is truly a nation of profound contrasts, with one of the widest gaps separating the rich and the poor to

be found anywhere in the Third World. The wealthiest 2 percent of its citizens control 70 percent of the land and annually earn as much as the poorest 65 percent of the population. Even though it now ranks among the world's leading food exporters, two-thirds of all Brazilians suffer from chronic malnutrition. And, as we have observed, even though considerable regional development is taking place, the benefits are overwhelmingly channeled toward already affluent individuals and powerful corporations, thereby further widening that enormous gulf between the haves and the have-nots.

Significantly, 36 percent of the population is under 15 years of age, with the lion's share of this demographic cohort living in or near poverty. Therefore, unless Brazil's resources and access to its opportunities are redistributed in a more equitable manner, the future will continue to hold potentially dangerous social problems. Civilian governments since 1985 have already demonstrated their reluctance to move boldly against these challenges, caving in to the political pressures of the powerful landowners to move very slowly on the crucial issue of land reform, and refraining from launching major new social programs. Thus, dynamic Brazil may now have the eighth-largest national economy— and could even surpass Canada and Italy on this list in the near future—but its move toward membership in the developed world is clouded by the reality that well over half of its people have been shut out from participating in this historic crusade.

PRONUNCIATION GUIDE

Amazonas   (ahma-*zone*-ahss)

Bahia   (bah-*ee*-yah)

Belo Horizonte   (*bay*-loh haw-ruh-*zonn*-tee)

Brasília   (bruh-*zeal*-yuh)

Centro-Oeste   (sentro oh-*ess*-tee)

Cuiabá   (koo-yuh-*bah*)

Fazenda   (fah-*zenn*-duh)

Goiás   (goy-*ahss*)

Grande Carajás   (*grun*-dee kuh-ruh-*zhuss*)

Islamabad   (iss-*lahm*-uh-bahd)

Itaipu   (ee-*tye*-pooh)

Karachi   (kuh-*rah*-chee)

Kyoto   (kee-*yoh*-toh)

Lafaiete   (lah-fuh-*yay*-tuh)

Manáos   (muh-*nauss*)

Mato Grosso   (mutt-uh-*groh*-soh)

Mato Grosso do Sul   (mutt-uh-*groh*-soh duh-*sool*)

Minas Gerais   (*mee*-nuss zhuh-rice)

Pará   (puh-*rah*)

Paraguay   (*pahra*-gwye)

Paraíba   (pah-rah-*ee*-buh)

Paraná   (pah-rah-*nah*)

Pernambuco   (pair-num-*boo*-koh)

Polonoroeste   (polloh-nuh-roh-*ess*-tee)

Pôrto Alegre   (*por*-too uh-*leg*-ruh)

Pôrto Velho   (*por*-too *vell*-yoo)

Recife   (ruh-*see*-fuh)

Rio Branco   (*ree*-oh *brung*-koh)

Rio Grande do Norte   (*ree*-oh *grun*-dee duh-*nortah*)

Rio Grande do Sul   (*ree*-oh *grun*-dee duh-*sool*)

Rio de Janeiro   (*ree*-oh day zhah-*nair*-roh)

Rondônia   (roh-*doan*-yuh)

Salvador   (*sull*-vuh-dor)

Santa Catarina   (sun-tuh-kuh-tuh-*ree*-nuh)

Santos   (*sunt*-uss)

São Luis   (sau-loo-*eece*)

São Paulo   (sau-*pau*-loh)

Serra dos Carajás   (serra doo kuh-ruh-*zhuss*)

Sertão   (sair-*towng*)

Sisal   (*sye*-sull)

Tocantins   (toke-un-*teens*)

Tubarão   (too-buh-*raung*)

Tucuruí   (too-koo-roo-*ee*)

Uruguay   (*oo*-rah-gwye)

Volta Redonda   (vahl-tuh-rih-*don*-duh)

REFERENCES AND FURTHER READINGS

(BULLETS [•] DENOTE BASIC INTRO-DUCTORY WORKS)

Brooke, James. "Decline in Births in Brazil Lessens Population Fears," *New York Times*, August 8, 1989, pp. 1–8.

Burns, E. Bradford. *A History of Brazil* (New York: Columbia University Press, 2 rev. ed., 1980).

Carter, James R. *Computer Mapping: Progress in the 80s* (Washington: Association of American Geographers, Resource Publications in Geography, 1984).

Cole, John P. *Latin America: An Economic and Social Geography* (Totowa, N.J.: Rowman & Littlefield, 2 rev. ed., 1975).

de Castro, Josué. *Death in the Northeast* (New York: Vintage Books, 1969).

• Dickenson, John P. *Brazil* (London & New York: Longman, 1983).

Dickinson, Robert E., ed. *The Geophysiology of Amazonia: Vegetation and Climate Interactions* (New York: John Wiley & Sons, 1987).

Fearnside, Philip M. *Human Carrying Capacity of the Brazilian Rainforest* (New York: Columbia University Press, 1986).

Hall, Anthony L. *Developing Amazonia: Deforestation and Social Conflict in Brazil's Carajás Programme* (Manchester, U.K.: Manchester University Press, 1989).

• Haller, Archibald O. "A Socioeconomic Regionalization of Brazil," *Geographical Review*, 72 (1982): 450–464.

Hemming, John, ed. *Change in the Amazon Basin: Volume I, Man's Impact on Forests and Rivers/Volume II, The Frontier After a Decade of Colonization* (Manchester, U.K.: Manchester University Press, 1985).

• Henshall, Janet D. & Momsen, Richard P. *A Geography of Brazilian Development* (Boulder, Colo.: Westview Press, 1974).

• James, Preston E. & Minkel, Clarence W. *Latin America* (New York: John Wiley & Sons, 5 rev. ed., 1986), Chaps. 29–36.

Kandell, Jonathan. *Passage Through El Dorado* (New York: William Morrow, 1984).

Katzman, Martin T. *Cities and Frontiers in Brazil: Regional Dimensions of Economic Development* (Cambridge, Mass.: Harvard University Press, 1977).

Lisansky, Judith. *Migrants to Amazonia: Spontaneous Colonization in the Brazilian Frontier* (Boulder, Colo.: Westview Press, 1989).

Margolis, Mac. "Amazon Ablaze," *World Monitor Magazine*, February 1989, pp. 20–29.

Merrick, Thomas W. & Graham, Douglas H. *Population and Economic Development in Brazil: 1800 to the Present* (Baltimore: Johns Hopkins University Press, 1979).

• Moran, Emilio F., ed. *The Dilemma of Amazonian Development* (Boulder, Colo.: Westview Press, 1983).

Roett, Riordan. *Brazil: Politics in a Patrimonial Society* (Westport, Conn.: Praeger, 3 rev. ed., 1984).

Schumann, Debra A. & Partridge, William L., eds. *The Human Ecology of Tropical Settlement* (Boulder, Colo.: Westview Press, 1987).

Simons, Marlise. "Brazil's Blacks

Feel Prejudice 100 Years After Slavery's End," *New York Times*, May 14, 1988, pp. 1, 6.

Smith, Nigel J.H. *Rainforest Corridors: The Transamazonian Colonization Scheme* (Berkeley: University of California Press, 1982).

Stepan, Alfred, ed. *Democratizing Brazil: Problems of Transition and Consolidation* (New York: Oxford University Press, 1989).

Taylor, Michael & Thrift, Nigel, eds.

The Geography of Multinationals: Studies in the Spatial Development and Economic Consequences of Multinational Corporations (New York: St. Martin's Press, 1982).

"Torching the Amazon: Can the Rain Forest Be Saved?," *Time*, September 18, 1989 (Cover Story), pp. 76–85.

Uys, Errol L. *Brazil* (New York: Simon & Schuster, 1986).

• Weil, Connie H. "Amazon Update:

Developments Since 1970," *Focus*, March–April 1983.

• Wilkie, Richard W. *Latin American Population and Urbanization Analysis: Maps and Statistics, 1950–1982* (Westwood: UCLA Latin American Center, 1984).

Wood, Charles H. & Magno de Carvalho, José A. *The Demography of Inequalities in Brazil* (New York: Cambridge University Press, 1988).

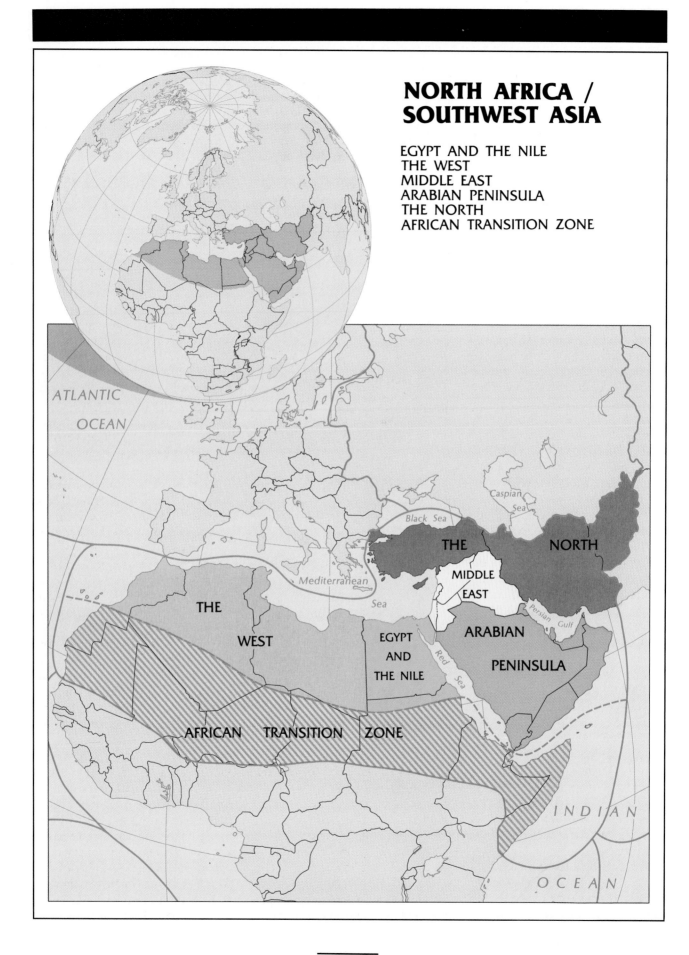

NORTH AFRICA / SOUTHWEST ASIA

EGYPT AND THE NILE
THE WEST
MIDDLE EAST
ARABIAN PENINSULA
THE NORTH
AFRICAN TRANSITION ZONE

ATLANTIC
OCEAN

Caspian Sea

Black Sea

THE

NORTH

MIDDLE

EAST

Mediterranean

Sea

Persian Gulf

THE

WEST

ARABIAN

EGYPT
AND
THE NILE

Red Sea

PENINSULA

AFRICAN   TRANSITION   ZONE

INDIAN

OCEAN

CHAPTER 6

NORTH AFRICA/ SOUTHWEST ASIA: FUNDAMENTALISM VERSUS MODERNIZATION

From Morocco on the shores of the Atlantic Ocean to Afghanistan deep inside Asia, and from Turkey between the Black and Mediterranean Seas to Somalia in the Horn of Africa, lies a vast realm of enormous historical and cultural complexity. It lies at the crossroads where Europe, Asia, and Africa meet, and is part of all three; throughout history, its influences have radiated to these continents and to practically every other part of the world as well. This is one of humankind's primary source areas. On the Mesopotamian Plain between the Tigris and Euphrates rivers (in modern-day Iraq) and on the banks of the Egyptian Nile arose civilizations that must have been among the very earliest. In the soils plants were domesticated that are now grown from the Americas to Australia. Along its paths walked prophets whose religious teachings are still followed by hundreds of millions of people. And in the final decade of the twentieth century, the heart of this realm is beset by some of the most bitter and dangerous conflicts on earth— ones that could still provoke a direct confrontation between the superpowers.

It is tempting to characterize this geographic realm in a few words, to stress one or more of its dominant features. It is, for in-

---

IDEAS AND CONCEPTS

cultural geography

culture hearth (2)

spatial diffusion

energy resources

remote sensing

irrigation

nomadism

ecological trilogy

von thünen isolated state (2)

---

stance, often called the "Dry World," containing as it does the vast Sahara and Arabian deserts. But most of the realm's people live where there is water—in the Nile Delta, along the Mediterranean coastal strip (or *tell*) of northwesternmost Africa, along the Asian eastern and northeastern shores of the Mediterranean Sea, in the Tigris–Euphrates Basin, in far-flung desert oases, and along the Iranian mountain slopes south of the Caspian Sea. True, we know this world region as one where water is almost always at a premium, where peasants often struggle to make soil and moisture yield a small harvest, where nomadic peoples and their animals circulate

across dust-blown flatlands, where oases are islands of sedentary farming and trade in a sea of aridity. But it is also is the land of the Nile, the lifeline of Egypt and the mainstay of Sudan's impressive Gezira irrigation scheme, and the crop-covered tell of the northern coast of Morocco, Algeria, and Tunisia.

North Africa/Southwest Asia is also often referred to as the "Arab World." However, once again, this implies a uniformity that does not actually exist. In the first place, the name *Arab* is applied loosely to the peoples of this area who speak Arabic and related languages, but ethnologists normally restrict it to certain occupants of the Arabian Peninsula—the Arab "source." In any case, the Turks are not Arabs, and neither (for the most part) are the Iranians or the Israelis. Moreover, although it is true that Arabic is spoken over a wide region that extends from Mauritania in the west across all of North Africa to the Arabian Peninsula, Syria, and Iraq in the east, there are many areas within the realm where it is not used by most of the people. In Turkey, for example, Turkish is the major language, and it has Ural-Altaic rather than Semitic or Hamitic roots. In Iran, the Iranian language belongs to the Indo-European lin-

S Y S T E M A T I C   E S S A Y

CULTURAL GEOGRAPHY

The essence of **cultural geography** is captured by Joseph Spencer in his 1978 survey of this subdiscipline:

[It] *focuses on the ways in which human achievement produces many different living systems among the varied peoples of the whole earth. Interest lies in the spatial distribution and the functioning patterns of all culture systems that become operative on earth.*

In the introduction to their 1962 edited volume, *Readings in Cultural Geography*, Philip Wagner and Marvin Mikesell describe the field as the application of the idea of culture to geographic problems. They define its core as consisting of five interrelated themes: (1) culture, (2) culture area, (3) cultural landscape, (4) culture history, and (5) cultural ecology. This quintet provides the framework for the following overview.

The concept of *culture* was discussed in the Introduction (pp. 6–9), and subsequent chapters have shown it to be a key to the systematic understanding of differences and similarities among human societies. In terms of structure, each culture may be viewed as an interconnected whole that can be dissected into a number of interacting components. The basic building block is the *culture trait*, a single element of normal practice in a culture; the collective individual traits of each culture form a larger discrete combination referred to as a *culture complex*; these, in turn, are aggregated into a national-scale *culture system*, a set of cul-

ture complexes united by common characteristics and strong historical-cultural bonds.

Language is a leading cultural component, the medium wherein people communicate by spoken and written words, facial expressions, and numerous other interpersonal gestures. As the accepted belief system, religion is another major component, one that can demonstrate the integration of cultural traits and (in certain societies) even shape daily life. In the North Africa/Southwest Asia realm, religion can influence clothing styles (women in public must be cloaked and veiled in many Islamic countries), shopping days (many businesses are closed on the Muslim and Jewish sabbaths), dietary practices (fasting on religious holidays; food avoidances), and even governmental administration (Iran's fundamentalist Shi'ite mullahs have a major say in the exercise of political power). Among the many other components of culture—most possessing clear spatial dimensions—are settlement patterns, architecture, legal institutions, family structures, adornment, and the arts.

The study of *culture areas* entails the identification and mapping of phenomena to recognize and de limit territories occupied by human communities that share a particular culture. By aggregating individual traits into culture complexes and systems, hierarchical geographic classifications can be developed. In the Introduction (p. 4), the three-tier hierarchy of culture realm/region/subregion was outlined; in fact, a culture system on the map constitutes a cul-

ture region, and an assemblage of culture systems creates a culture realm. Cultural geography's concern with distributions of cultural phenomena is especially sensitive to changes over time, as innovations originate in source areas and subsequently disperse across great distances. It studies transmitting agents and channels, as well as cultural barriers. For example, the spatial expansion of Islam was far more rapid than the spreading of the Arabic language, because recipient cultures were relatively amenable to religious change but resisted attempts to modify their linguistic traditions.

Cultural landscape interpretation focuses on the physical imprint of material culture on the land surface that forms the geographic content of a culture region. In addition to the artifacts of material culture, many geographers now also study the landscapes of popular culture. This represents a broadening of the field to include vernacular culture—culture in the context of its sustainers, the people who maintain and nurture it. *Popular culture* is the ever-changing "mass" culture of urbanized and industrialized society that is less tradition-bound, more open, individualistic, and class-structured than strongly traditional *folk culture*, the durable way of life found in the comparatively isolated rural areas. North Africa/Southwest Asia, like most underdeveloped realms, is dominated by the latter (the 1979 overthrow of the shah's regime in Iran was an object lesson in the limitations to bucking this status quo). Among the topics

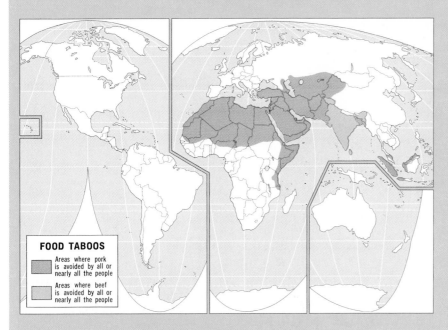

FOOD TABOOS

Areas where pork is avoided by all or nearly all the people

Areas where beef is avoided by all or nearly all the people

FIGURE 6–1

recently researched by vernacular-culture geographers are music styles, sports, food and drink, and personal-preference patterns.

Culture history is explicitly concerned with the cultural-geographical past. In this chapter, we formally treat the time-space interface through an examination of *spatial diffusion* in Model Box 3. Here we briefly consider the topic of food taboos—a practice that affects tens of millions in the realms of the three upcoming chapters, and an outstanding example of current cultural behavior that evolved over the long course of human development in the Old World.

The interrelated systems of food production, distribution, and consumption form a fundamental component of every culture. The way land is allocated to individuals or families (or bought and sold), the manner in which it is farmed, the

functions of livestock, and the consumption of food from crops and animals are all aspects of culture. Moreover, food consumption itself is often associated with religious beliefs and practices. Adherents of Islam and Judaism avoid pork; Hindus do not consume beef or other kinds of meat (Fig. 6–1). Various other forms of partial or total abstinence occur among human cultures, including periodic fasts. Such proscriptions, like the religions that generated them, tend to be old and persistent and change only very slowly.

Cultural ecology involves the multiple relationships between human cultures and their physical environments. However, as we saw in Chapter 4, this tradition of geography in the past attracted proponents of environmental determinism and similar viewpoints, and the idea of a mutual and balanced human-habitat in-

teraction did not gain currency until the 1930s. Environmentalism in American geography was countered in the 1920s from two directions. From Western Europe came the short-lived doctrine of *possibilism*, which argued against climatic and physiographic determinism by claiming that people, through their cultures, are free to choose from a number of environmental "possibilities." The second, and far more persuasive, attack came from within, through the scholarship of Carl Sauer, this century's most distinguished cultural geographer. His position, the one all but unanimously adopted today, regards humans and the natural environment as co-equal partners in an interacting unity, with humankind the agency of the systematic modification of the physical landscape.

guistic family. In Ethiopia, Amharic is spoken by the ruling plateau people; although it is more closely related to Arabic than is Iranian, it nonetheless remains a distinct language as well. The same is true of Hebrew, which is spoken in Israel. Other "Arab World" languages that have separate ethnological identities are spoken by the Tuareg people of the Sahara, the Berbers of northwestern Africa, and the peoples of the transition zone between North Africa and Subsaharan Africa to the south.

Another name given to this realm is the "World of Islam." The prophet Muhammad (Mohammed) was born in Arabia in A.D. 571, and in the centuries that followed his death in 632, Islam spread into Africa, Asia, and Europe. This was the age of Arab conquest and expansion; their armies penetrated southern Europe, caravans crossed the deserts, and ships plied the coasts of Asia and Africa. Along these routes they carried the Muslim (Islamic) faith, converting the ruling classes of the states of the West African savanna, threatening the Christian stronghold in the highlands of Ethiopia, penetrating the deserts of inner Asia, and pushing into India and even the island extremities of Southeast Asia. Islam was the religion of the marketplace, the bazaar, the caravan. Where necessary, it was imposed by the sword, and its protagonists aimed directly at the political leadership of the communities they entered. Today, the Islamic religion with its more than 900 million followers extends well beyond the limits of the realm discussed here (Fig. 6–2): it is the major religion in northern Nigeria, in Pakistan, and in Indonesia; it is strong in the Soviet Union's Muslim South, and it survives even in Eastern Europe (notably in Albania and Yugoslavia). On the other hand, the "World of Islam"

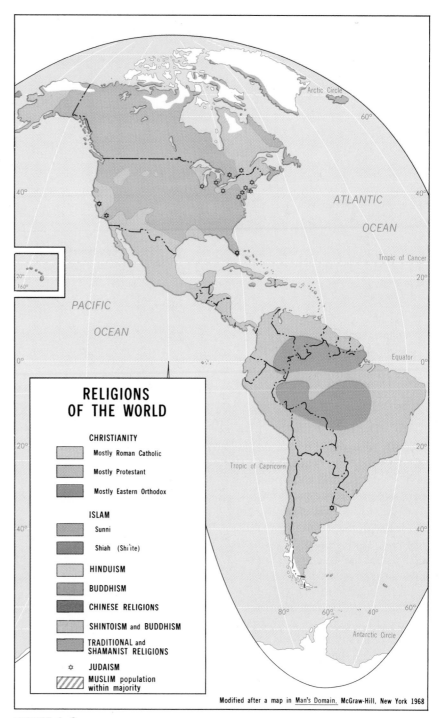

FIGURE 6–2

is not entirely Muslim either. In Israel, Judaism is the prevailing faith; in Lebanon, a sizable segment of the population adheres to old forms of Christianity; in Ethiopia, the Amharic-speaking ruling class, having managed centuries ago to stave off the Muslim onslaught, also practices an ancient Christian religion, and Coptic Christian churches still exist in Egypt. Thus, the connotation "World of Islam" for North Africa/Southwest Asia is far from satisfactory—the religion prevails far beyond these areas, and within

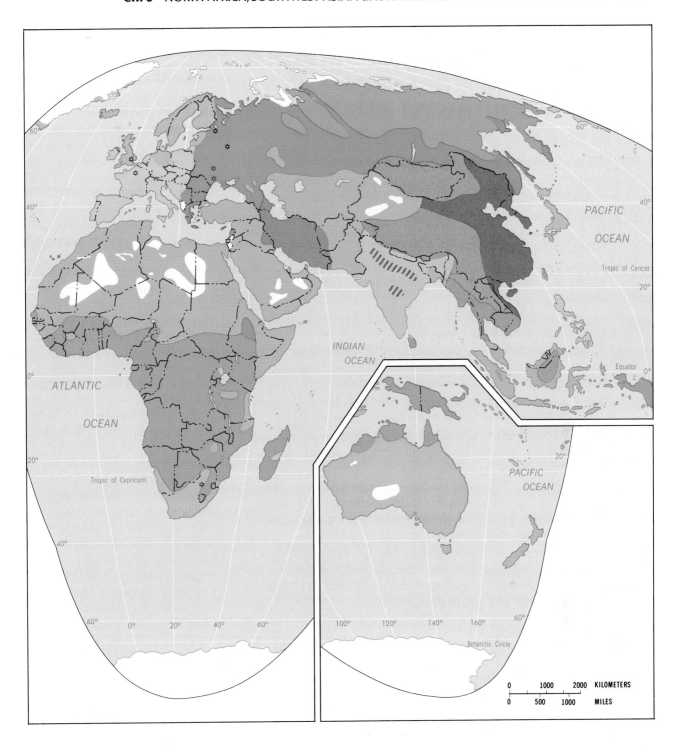

the realm there are important countries in which Islam is not the majority's faith.

Finally, this realm is frequently called the ''Middle East.'' That must sound quite odd to someone, say in India, who might think of a Middle West rather than a Middle East! The name, of course, reflects the biases of its source: the ''Western'' world, which saw a ''Near East'' in Turkey, a ''Middle'' East in Egypt, Arabia, and Iran, and a ''Far'' East in China and Japan. Still, the term has taken hold, and it can be seen and heard in every-day usage by journalists as well as members of the United Nations. In view of the complexity of this realm, its transitional margins, and its far-flung areal components, at least the name *Middle East* need be faulted only for being imprecise—it does not make a single-factor re-

TEN MAJOR GEOGRAPHIC QUALITIES OF NORTH AFRICA/ SOUTHWEST ASIA

1. This realm contains several of the world's great ancient culture hearths and some of its most durable civilizations.
2. North Africa/Southwest Asia is the source region of several world religions, including Judaism, Christianity, and Islam.
3. This realm is predominantly but not exclusively Islamic (Muslim). That faith pervades cultures from Morocco in the west to Afghanistan in the east.
4. North Africa/Southwest Asia is the "Arab World," but significant population groups in this realm are not of Arab ancestry.
5. The population of North Africa/Southwest Asia is widely dispersed in discontinuous clusters.
6. Natural environments in this realm are dominated by drought and unreliable precipitation. Population concentrations occur where the water supply is adequate to marginal.
7. The realm contains a pivotal area in the "Middle East," where Arabian, North African, and Asian regions intersect.
8. North Africa/Southwest Asia is a realm of intense discord and bitter conflict, reflected by frequent territorial disputes and boundary frictions.
9. Both the United States and the Soviet Union have vital interests and key client-states in this realm, thereby magnifying internal stresses as potential global problems in the relationship between the two superpowers.
10. Enormous reserves of petroleum lie beneath certain portions of the realm, bringing wealth to these favored places. But overall, oil revenues have raised the living standards of only a small minority of the total population.

gion of North Africa/Southwest Asia, as do the terms *Dry World*, *Arab World*, and *World of Islam*.

A GREATNESS PAST

In our discussions of North and Middle America, we made refer-ence to the concept of **culture hearth**—a source area or innova-tion center from which cultural traditions are transmitted. In such regions of comparative success, growth clusters of population de-veloped, and within them natural increase was supplemented by the immigration of people attracted from afar. New ways were found to exploit local resources, and power was established over re-sources located farther away. Farming techniques improved, and so did crop yields. Settlements could expand and begin to acquire urban characteristics. The circula-tion of goods and ideas intensified. Traditions emerged in various spheres of life, and these tradi-tions, along with inventions and innovations, radiated outward into the areas beyond. Among the ideas that took hold and developed were political ideas, especially the the-ory and practice of the political or-ganization necessary to cope with society's growing complexity.

Such developments occurred on several continents (Fig. 6–3). North Africa/Southwest Asia was a pre-eminent source area, where culture hearths lay in the Tigris-Euphrates Plain (Mesopotamia) of what is now Iraq and the Nile Valley of Egypt (as well as in the neighboring Indus Valley of present-day Pakistan). The main hearth of Chinese culture was cen-tered on the confluence of the Wei and Huang He rivers. South Asia's Ganges Delta was another key cul-tural spawning ground. As we noted in earlier chapters, in Mid-dle America the Yucatán Peninsula and later the Mexican Plateau were significant culture hearths; in South America, the central Andes Mountains witnessed the rise of a major civilization. And, as Fig. 6–3 shows, a culture hearth also emerged in West Africa's savanna-lands.

Mesopotamia

The so-called Middle East, as it now exists as a geographic realm, thus included several of the ear-liest (if not the first) hearths of hu-man culture. Mesopotamia, the "land between the two rivers," anchored one end of the Fertile Crescent, which stretched to the eastern Mediterranean coast. This was one of the places where peo-ple first learned to domesticate plants and gather harvests in an organized way. Mesopotamia be-

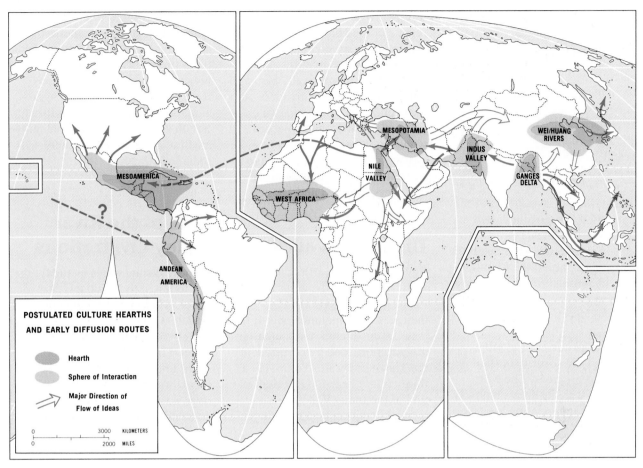

FIGURE 6–3

came a crossroads for a whole network of routes of trade and movement across Southwest Asia, and its accessibility enhanced its adoption of numerous new ideas and inventions. Many innovations originated in Mesopotamia itself, to be spread to other regions of development. Besides the organized and planned cultivation of grain crops such as wheat and barley, the Mesopotamians knew how to make implements out of bronze, had learned to use draft animals to pull vehicles (including plows designed to prepare the fields), and employed the wheel, a revolutionary invention, to build carts and wagons.

The ancient Mesopotamians also built some of the world's earliest cities. This development was made possible by their accomplishments in many spheres, espe-

cially agriculture, whose surpluses could be stored and distributed to the city dwellers. Essential to such a system of allocation were decision-makers and organizers, an elite group who controlled the lives of others. Such an urban-based elite could afford the luxury of leisure and could devote time to religion and philosophy. Out of such pursuits came the concept of writing and record-keeping, which made possible the codification of tradition; it was a crucial element in the development of systematic administration in urbanizing Mesopotamia and in the evolution of its religious-political ideology. The rulers in the cities were both priests and kings, and the harvest that the peasants brought to be stored in the urban granaries was a tribute as well as a tax.

Mesopotamia's cities emerged about 6000 years ago, and some may have had more than 10,000 residents. Today, these urban places are extinct, but excavations tell the story of their significant and sometimes glorious past. Mesopotamia's cities had temples and shrines, priests and kings; there were also wealthy merchants, expert craftspeople, and respected teachers and philosophers. But Mesopotamia was no unified political state: each city centered a hinterland over which its power and influence prevailed. Despite the strife in political relationships that this system generated, Mesopotamia still made progress in politics; eventually, regional unification was achieved during the fourth millennium B.C. in at least two early states named Sumer and Elam.

Ancient Egypt

Another area that witnessed very early cultural and political development was Egypt. Egypt's evolution may have started even earlier than Mesopotamia's; it certainly has possessed urban centers for at least 5000 years. The focus of the ancient Egyptian culture hearth lay above (south of) the Nile Delta and below (north of) the Nile's first cataract. This segment of the river valley lies surrounded by rather inhospitable desert country (see Fig. I–1, p. 2). Thus, with the region open only to the Mediterranean, in contrast to Mesopotamia which was something of a marchland, the Nile Valley was a natural fortress.

Here, the ancient Egyptians converted the security of their isolation into progress. The Nile waterway was the area's highway of trade and association, its true lifeline. It also sustained agriculture by irrigation, and the cyclical regime of the Nile's water level was far more predictable than that of the Tigris or Euphrates. By the time ancient Egypt finally began to fall victim to outside invaders, after about 1700 B.C., a full-scale urban civilization had emerged, whose permanence and continuity are reflected in the massive stone monuments that its artist-engineers designed and created.

The Indus Valley

Egypt's culture hearth lay on the western fringe of the great Middle Eastern source area. As Fig. 6–3 indicates, Mesopotamia was at the heart of this region. On the eastern margin, separated by desert from the Tigris-Euphrates Basin, lay a third culture hearth—the civilization of the Indus Valley in what is today Pakistan (see map on p. 458). By modern criteria, this eastern hearth lies outside the realm under discussion here, but in ancient times it had effective ties with lands to its west. Probably, Mesopotamian innovations reached the cities of the Indus region, power centers of a civilization that at its height reached into much of present-day northern India.

Gifts of the River Valley Civilizations

In those distant times as today, the key to life in the Middle East was water; where it is available in sufficient quantity—as in the Nile Delta—the "Dry World" turns green (Fig. 6–4). The Mesopotamians and Egyptians achieved breakthroughs in the control of river water for irrigation and in the domestication of several staple crops. Today, people throughout

FIGURE 6–4 Water availability is a crucial ingredient in the human geography of this realm. These are the fertile fields of the Nile Delta—also shown from an orbiting satellite in Fig. 6–17.

the world continue to benefit from these contributions, which include cereal crops (such as wheat, rye, and barley), vegetables and fruits (peas, beans, grapes, apples, and peaches), and domesticated animals (horses, pigs, and sheep). And the range of indirect contributions is quite beyond measure. The ancient Mesopotamians advanced not only irrigation and agriculture, but also calendrics, mathematics, astronomy, governmental administration, engineering, metallurgy, and a host of other technologies. As time went on, many of their innovations were adopted and then modified by other cultures in the Old World, and eventually even in the New World. Europe was the greatest beneficiary of the legacies of Mesopotamia and Pharaonic Egypt, whose achievements constituted the very foundations of ''Western'' civilization.

DECLINE AND REBIRTH

Today, the great cities of this realm's culture hearths are archaeological curiosities. In some instances, new cities have been built on the sites of the old, but the great ancient cultural traditions of North Africa/Southwest Asia went into a deep decline after many centuries of continuity. It cannot escape our attention that a large number of the ruins of these ancient urban centers are located in what is now desert. Presuming that they were not built in the middle of these drylands, it is tempting to conclude that climatic change, associated with shifting environmental zones in the wake of the last Pleistocene glacial retreat, destroyed the old civilizations. Indeed, some geographers

have suggested that the momentous innovations in agricultural planning and irrigation technology may have been made in response to changing environmental conditions as the riverine communities tried to survive. The scenario is not difficult to imagine: as outlying areas began to fall dry and farmlands were destroyed, people congregated in the already crowded river valleys—and every effort was made to increase the productivity of lands that could still be watered. Eventually overpopulation, destruction of the watershed, and perhaps reduced rainfall in the rivers' headwater areas combined to deal the final blow. Towns were abandoned to the encroaching desert; irrigation canals filled with drifting sand; remaining farmlands dried up. Those who could, migrated to areas reputed to still be productive. Others stayed, their numbers dwindling, increasingly reduced to subsistence.

As old societies disintegrated, new power emerged elsewhere. First the Persians, then the Greeks, and later the Romans imposed their imperial designs on the tenuous lands and disconnected peoples of the Middle East. Roman technicians converted North Africa's farmlands into irrigated plantations whose products went by the boatload to Roman Mediterranean shores. Thousands of people were carried in bondage to the cities of the new conquerors. Egypt was quickly colonized; but the Arab settlements on the Arabian Peninsula lay more remote and somewhat more secure in their isolation.

The Rise of Islam

In one of these relatively isolated places on the Arabian Peninsula, an event occurred that was to change the course of history and

affect the destinies of people all over the world. About A.D. 613, in Mecca, a town 45 miles (72 kilometers) from the Red Sea coast in the Jabal Mountains, according to his followers, a man named Muhammad began to receive the truth from Allah (God) in a series of revelations. Muhammad (A.D. 571–632) already was in his early forties by then, and he had barely 20 years to live. But during those two decades began the transformation of the Arab world. Convinced after some initial self-doubt that he was indeed chosen to be a prophet, Muhammad committed his life to the fulfillment of the divine commands he believed he had received. The Arab world was in social and cultural disarray; in the feudal chaos, Muhammad soon attracted enemies who feared his new personal power. He fled Mecca for the safer haven of Medina (but Mecca's place as a holy center was soon assured), and from there he continued his work.

The precepts of Islam, in many ways, constituted a revision and embellishment of Judaic and Christian beliefs and traditions. There is but one god, who occasionally communicates with prophets; Islam acknowledges that Moses and Jesus were such prophets. What is earthly and worldly is profane; only Allah is pure. Allah's will is absolute; Allah is omnipotent and omniscient. All humans live in a world created for their use, but only to await a final judgment day.

Islam brought to the Arab world not only a unifying religious faith, but also a whole new set of values, a new way of life, a new individual and collective dignity. Islam dictated observance of the five *pillars of Islam*—repeated expressions of the basic creed, daily prayer, a month of daytime fasting, almsgiving, and at least one pilgrimage to Mecca—the faith prescribed and proscribed in other spheres of life as well. Alcohol, smoking, and gambling were forbidden. Polygamy was tolerated, although the virtues of monogamy were acknowledged. Mosques appeared in Arab settlements, not only for the (Friday) sabbath prayer, but also as social gathering places to bring communities closer together. Mecca became the spiritual center for a divided, widely dispersed people for whom a collective focus was something new.

The Arab Empire

The stimulus that Muhammad provided, spiritual as well as political, was so great that the Arab world was mobilized almost overnight.

The prophet died in 632, but his faith and fame continued to spread like wildfire. Arab armies formed, they invaded and conquered, and Islam was carried throughout North Africa. By the early ninth century, the Muslim world included emirates or kingdoms extending from Egypt to Morocco, a caliphate occupying most of Spain and Portugal, and a unified region encompassing Arabia, the Middle East, Iran, and much of Pakistan (Fig. 6–5). Muslim influences had attacked France and Italy, and penetrated what is now the Soviet Muslim South. Ultimately, the Arab empire extended from Morocco to tropical Asia and from Turkey to the southern margins of the Sahara (Fig. 6–6). The original capital was at Medina in Arabia,

but in response to these strategic successes it was moved, first to Damascus and then to Baghdad. In the fields of architecture, mathematics, and science, the Arabs far overshadowed their European contemporaries, and they established institutions of higher learning in many cities, including Baghdad, Cairo, and Toledo in Spain. The faith had spawned a culture, and it is still at the heart of that culture today.

The spread of the Islamic faith throughout North Africa/Southwest Asia (and beyond) occurred in waves radiating from Medina and the nearby holy city of Mecca. Islam went by camel caravan and by victorious army; it was carried by pilgrim and sailor, scholar and sultan. Its dissemination

FIGURE 6–5

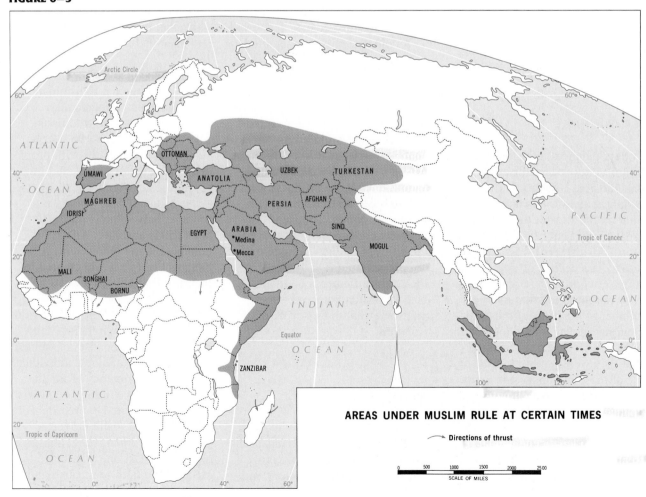

AREAS UNDER MUSLIM RULE AT CERTAIN TIMES

Directions of thrust

SCALE OF MILES

through so vast and discontinuous a realm is an example of the process of **spatial diffusion**, which is discussed in Model Box 3.

The worldwide diffusion of Islam continues today. In the United States, it is emerging in the religious movement commonly called the Black Muslims but officially known as the American Muslim Mission. There are Islamic communities in South Africa, in the Philippines, and other far-flung places. This is not the first time that the Middle East has generated an idea that affected much of the world. Agricultural methods, metallurgical techniques, architectural styles, and countless other innovations that developed in this realm have, throughout the course of recorded history, been adopted by societies across the globe.

FIGURE 6–6   Islam's impact has spread far beyond its Arabian hearth, a dispersal underscored by this thriving mosque on the Southeast Asian island of Borneo, well over 5000 miles/8000 kilometers from Mecca.

BOUNDARIES AND BARRIERS

If Islam constituted such a strong unifying force and if Middle Eastern armies could penetrate Europe, Asia, and Africa south of the Sahara, why is North Africa/Southwest Asia today a realm of boundaries and barriers, tension and conflict, hostility and disunity?

Religious Divisions

In part, the answer lies in the rise of another religion, one that emerged even earlier than Islam—Christianity. Until the challenge of Islam, the Christian faith (which began as a movement within another Middle Eastern religion, Judaism) had won acceptance in the Roman Empire, could afford the luxury of an east–west ideological split, and diffused to the tribes of Europe north of Rome's boundaries. Then, the Arabs rode on Muhammad's teachings into Iberia and threatened the Roman core of Italy itself. Europeans mobilized to meet the challenge; as soon as Arab power showed signs of weakening (as early as the eleventh century), the first of many crusades by Chris-

FIGURE

vation i
ner tha
This "t
to inter

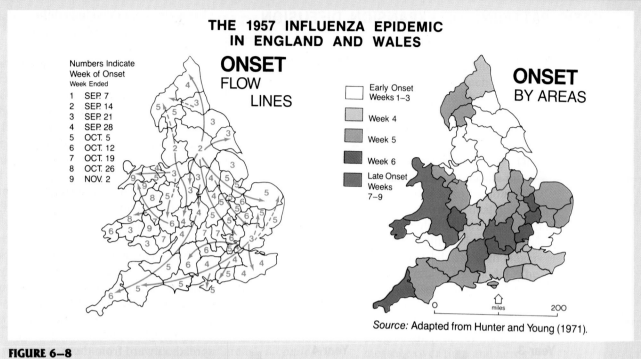

THE 1957 INFLUENZA EPIDEMIC IN ENGLAND AND WALES

ONSET FLOW LINES

Numbers Indicate
Week of Onset
Week Ended
1 SEP. 7
2 SEP. 14
3 SEP. 21
4 SEP. 28
5 OCT. 5
6 OCT. 12
7 OCT. 19
8 OCT. 26
9 NOV. 2

ONSET BY AREAS

Early Onset Weeks 1–3
Week 4
Week 5
Week 6
Late Onset Weeks 7–9

0 — miles — 200

Source: Adapted from Hunter and Young (1971).

FIGURE 6–8

FIGURE 6–9

DIFFUSION OF THE PLANNED REGIONAL SHOPPING CENTER IN THE UNITED STATES, 1949-1968

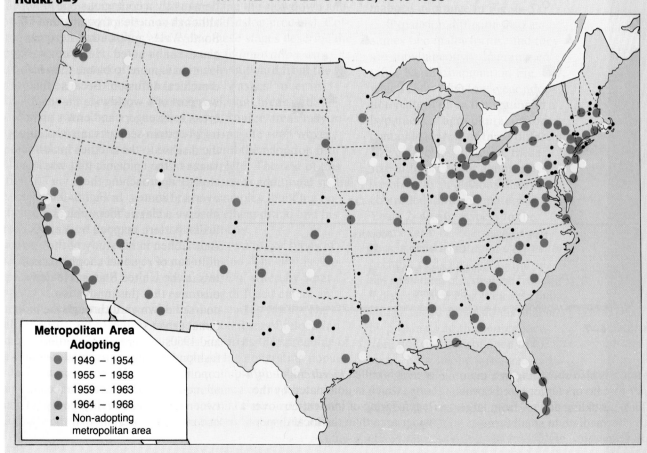

Metropolitan Area Adopting
1949 – 1954
1955 – 1958
1959 – 1963
1964 – 1968
Non-adopting metropolitan area

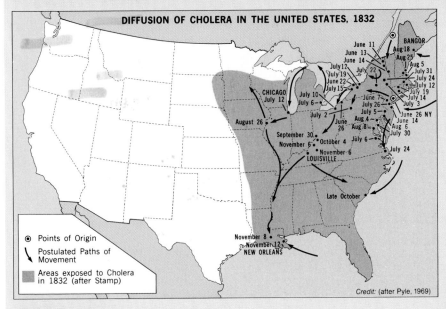

DIFFUSION OF CHOLERA IN THE UNITED STATES, 1832

June 11
June 13
June 14
July 12
July 19
June 22
July 15
July 22
Aug 18
Aug 25
Aug 5
July 31
July 24
July 12
July 19
July 14
July 3
June 26 NY
June 14
BANGOR

CHICAGO
July 12
July 10
July 6
June 7
July 26
July 5
Aug 4
Aug 8
July 6
Aug 6
July 30

August 26
July 2
June 26
September 30
November 6
October 4
November 6
LOUISVILLE

July 24

Late October

November 8
November 12
NEW ORLEANS

◉ Points of Origin

↘ Postulated Paths of
    Movement

▓ Areas exposed to Cholera
    in 1832 (after Stamp)

Credit: (after Pyle, 1969)

FIGURE 6–10A

FIGURE 6–10B

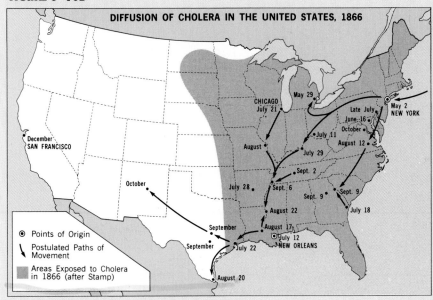

DIFFUSION OF CHOLERA IN THE UNITED STATES, 1866

December
SAN FRANCISCO

October

May 29
CHICAGO
July 21
Late July
June 16
October
July 11
August 12
August
July 29
Sept. 2
Sept. 6
Sept. 9
Sept. 9
July 28
July 18
August 22
September
August 17
July 12
NEW ORLEANS
September
July 22
August 20
May 2
NEW YORK

◉ Points of Origin

↘ Postulated Paths of
    Movement

▓ Areas Exposed to Cholera
    in 1866 (after Stamp)

distance along such main water routes as the Erie Canal and the Great Lakes (the railroad had not yet arrived). The 1866 map shows a radically different picture: from a similar source area, cholera now advances far more swiftly along a maturing long-distance rail network that connects the nation's largest cities. Therefore, distant Detroit receives the epidemic before nearby Baltimore, and the disease even arrives in the lower Ohio Valley before it reaches Virginia. These maps thus provide "before" and "after" snapshots vis-à-vis the emergence of the U.S. national urban system, which became firmly established around the middle of the nineteenth century.

These carrying processes of diffusion notwithstanding, the spreading of innovations is also affected by barriers of various kinds. In extreme cases, these may be *absorbing* barriers in which the impulse is completely halted. More commonly, spreading innovations encounter *permeable* barriers, which slow down or "filter" the diffusion wave. A physical barrier is a good example: a mountain range or unbridged river will act in this way because communication across such an obstacle is more difficult. Cultural barriers function similarly, often impeding the adoption of innovations from other cultures. Where internal linguistic or religious differences exist, diffusion inside a country may also be shaped by cultural barriers. Among other kinds of diffusion barriers are psychological barriers. In every population, no matter how culturally homogeneous, one finds avant-garde groups that readily adopt new things as well as laggards who only accept changes after everyone else has.

biggest city (New York) did not participate in the earliest years, whereas such smaller centers as Sacramento, Buffalo, and Elyria, Ohio (a suburb of Cleveland) did. The historical significance of the urban hierarchy is displayed in

two maps (Fig. 6–10), based on Gerald Pyle's research of nineteenth-century cholera epidemics. In the 1832 map, the disease diffuses contagiously from New York State into the interior, its rate and direction of spread controlled by

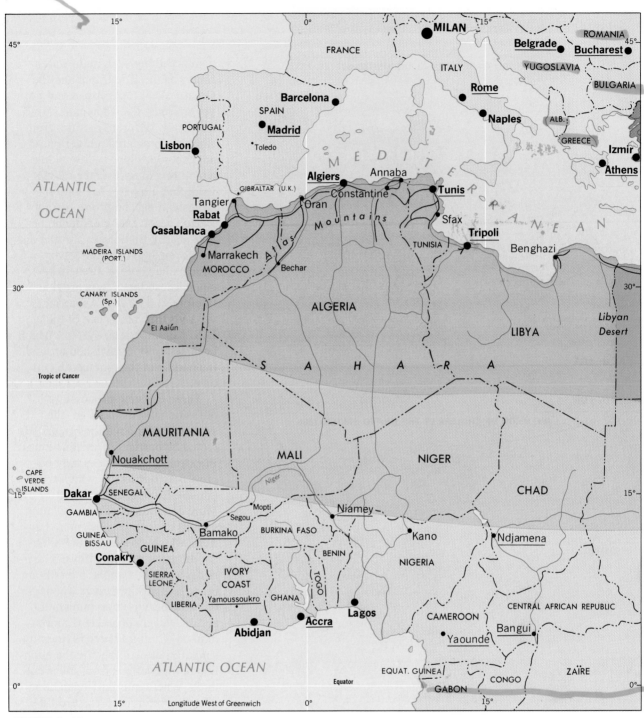

FIGURE 6–12

tians to the eastern Mediterranean took place. The aftermath of that contest left parts of the Middle East converted to Christianity; today, a large segment of the population of Lebanon still adheres to an early Christian faith. Eight centuries later, in 1948, Judaism re-

gained territorial expression as the state of Israel. The city of Jerusalem is a holy place for Jews, Muslims, and Christians alike: fragmented and embattled, it exquisitely personifies the Middle East's divisions (Fig. 6–11).

But the realm is vast, and ad-

herents to faiths other than Islam form only small minorities. Most of the boundaries on the map of North Africa/Southwest Asia today (Fig. 6–12) were established after the last of the great Islamic empires collapsed. The Muslims' last hurrah in Europe came when

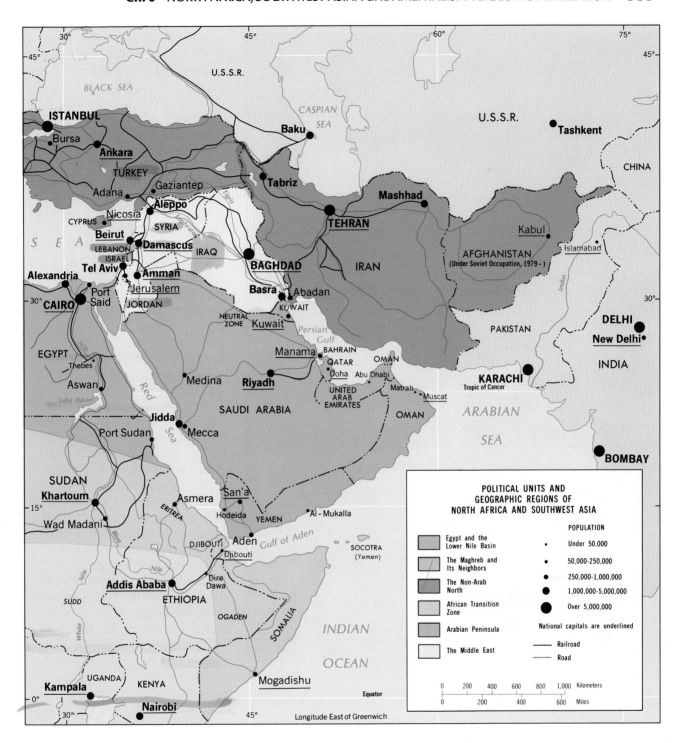

the Ottoman Empire, centered on the Turkish city of Constantinople (now Istanbul), extended its power over the Balkans and beyond. Greece, Bulgaria, Albania, Yugoslavia, Hungary, Romania, and the Black Sea area fell to the sultans, whose empire also extended far

to the west along Mediterranean shores and eastward into Persia. As the twentieth century opened, the Ottoman Empire was already decaying, and when the Turks chose the (losing) German-Austrian side in World War I, its fate was sealed.

Political Boundaries

From the fragments of the Ottoman Empire, the modern political divisions of the Middle East emerged. Turkey survived as a discrete entity, but other areas were assigned to the victorious

FIGURE 6–11 Two of Jerusalem's most sacred shrines: the Wailing Wall of the Jews adjoining the Muslim Dome-of-the-Rock complex. This embattled and fragmented holy city of Judaism, Islam, and Christianity typifies the realm's bitter divisions.

scores, this realm's population of 368 million is clustered, fragmented, and strung out in narrow coastal belts, confined valleys, and crowded oases. It is useful to compare Fig. 6–12 and Fig. I–14, for it is clear that thousands of miles of political boundaries in North Africa/Southwest Asia lie in virtually uninhabited terrain. Geographers classify boundaries on various grounds, one of which is morphological. Does the boundary coincide with a cultural break or transition in the landscape? Does it coincide with a physical feature such as a river or the crest of a mountain range? Is the boundary visibly straight? As our maps show, North Africa contains a framework of straight-line boundaries, long segments of which lie in the nearly empty Sahara Desert. These are *geometric* boundaries (Egypt is largely enclosed by such borders), drawn to coincide with lines of latitude or longitude (or at an angle, from point to point), mostly through territory over which, so it seemed, no one would ever quarrel. However, that was before some of the realm's oil reserves were discovered. Other boundaries such as the Sudan-Ethiopia border along the foot of the Ethiopian Highlands and the Israel-Jordan boundary along the Jordan River, coincide with natural features and are *physiographic* boundaries. Political boundaries and their evolution will be discussed further in the Systematic Essay that opens Chapter 10.

European powers. Syria and Lebanon went to France, adding to the colonies the French had already acquired in Algeria, Tunisia, and Morocco. Palestine wound up in the hands of the British, who also administered Iraq, Sudan, and Egypt. The Italians established North African spheres of influence in Libya and along the southern Red Sea coast in Eritrea; they also controlled part of Somalia in Africa's Horn. A combination of European imperialism and the final breakdown of Muslim power thus produced a mosaic of political boundaries to suit the colonial powers, not the interests of the Arabs.

As Fig. I–14 (p. 29) under-

PERSIAN GULF OIL BONANZA

The political fragmentation of this geographic realm into two dozen countries has major ramifications

in relation to the distribution of its leading export commodity—oil. Figure 6–13 displays current estimates of the world's fossil fuel reserves. Many of this realm's countries possess substantial oil and natural gas reserves, notably the states and emirates bordering the Persian Gulf as well as Libya and Algeria in North Africa. As new discoveries are made, estimates change, but as a whole, the states of North Africa/Southwest Asia still probably contain over 50 percent of the world's petroleum reserves. Oil has become the realm's most valuable export commodity, providing Saudi Arabia and even such small countries as Kuwait with huge revenues and giving the Middle East substantial world influence. As the bar graph on Fig. 6–13 indicates, the great consumers of energy are not the countries of the Middle East, but the nations of the developed world. Production in the United States and several European countries continues to lag behind consumption, and Western dependence on Middle Eastern **energy resources** was underscored by the supply disruptions that caused two serious petroleum shortages during the 1970s.

If the oil-exporting states of the North Africa/Southwest Asia realm were united ideologically, the advantages accruing from their petroleum wealth would be far greater. But the area's traditional disunity once again worked to its disadvantage. Of the 13 member states of the Organization of Petroleum Exporting Countries (OPEC), 8 lie in this realm (Fig. 6–13), but their efforts to agree on petroleum policy and to develop an effective cartel have thus far been only minimally successful. Moreover, during the 1980s, global recession helped create an oil glut that resulted in sharply reduced prices. Nonetheless, the organization survived; by the early 1990s,

prices were rising again, internal bickering had abated, and a new generation of more moderate leaders was searching for greater unity while reasserting OPEC's economic position as the controller of 80 percent of the world's oil reserves.

What has been the impact of the realm's "black gold" on those countries and societies fortunate enough to possess substantial oil reserves? In small states such as Kuwait (2.2 million), a true transformation has taken place: in the 1970s, this country possessed the world's highest per capita income, and it still ranks in the top ten. Currently, the topmost position is claimed by the United Arab Emirates (U.A.E.) at the other end of the Persian Gulf. Formerly known as British-administered "Trucial Oman" and "The Seven Sheikdoms," this tradition-bound country is today a union of seven emirates on the Persian Gulf coast of the Arabian Peninsula—Abu Dhabi, Dubai, Ajman, Sharjah, Umm al-Qaiwain, Ras al-Khaimah, and Fujairah. By far the largest in area of these emirates is Abu Dhabi, with 26,000 square miles (68,000 square kilometers) and a population of approximately 800,000. The U.A.E.'s total area is only 32,000 square miles (84,000 square kilometers); total population is about 1.8 million. Oil income, though no longer at the level it was in the 1970s, is still enormous: in 1987 the U.A.E. averaged U.S. $19,120 per person (while the U.S. itself ranked fourth in the world at $16,400).

In countries with larger populations, the picture is quite different. In Iran, with a population of 58 million, the deposed shah's government (which was overthrown in 1979) had poured vast sums of its enormous petroleum-derived income into a major military complex. Sizable investments also allegedly disappeared into the per-

sonal coffers of the monarch and into the accounts of those in favor. The funds that remained were still quite sufficient to support a number of major development programs, including an export-oriented group of industries, a series of reforms in agriculture, and an attack on illiteracy. Modernization during this period made its mark on the capital, Tehran, and on such other cities as the Persian Gulf oil-refining center of Abadan. But in the Iranian countryside, there was much less evidence of progress. In the last year of the shah's rule, only about 3000 villages out of a national total of 60,000 had running water available. Because Iran has been in near chaos since 1979, it is difficult to assess the revolutionary changes that have occurred under the Ayatollah Khomeini and his successors. But it is unlikely that the lot of the average Iranian has improved; tellingly, the 1986 estimate of average per capita income was a paltry U.S. $1667.

Neither has Iraq, with its 20 million inhabitants, found its petroleum income to be a shortcut to general national welfare. To modernize tradition-bound agriculture is an enormously expensive proposition, and Iraq has lacked what Iran's authoritarian government did achieve before 1979—political continuity and stability. Internal political struggles, corruption, and factionalism together with urban-oriented investment policies, which persuaded hundreds of thousands of villagers to leave the countryside and come to Baghdad, deprived Iraq of much of the development its oil revenues should have brought. Moreover, a very costly war between Iraq and Iran dragged on for most of the 1980s, further blocking progress in both countries.

Figure 6–14 shows a system of oil and gas pipelines and export locations that very much resembles the exploitative railroads in a mineral-rich colony. Such a pattern spells disadvantage for the exporter, whether colony or independent country. In North Africa, Libya once again reflects the advantages of scale: its mere 4.3 million people have felt the impact of their oil bonanza far more strongly than neighboring Algeria's 27 million, and the Libyan government has used its oil revenues to gain a measure of political influence in the realm far beyond its modest dimensions.

Oil-derived wealth has transformed parts of the Arab world, and it has brought the twentieth century into sometimes rude contact with local traditions. High-rise buildings and luxury automobiles vie for space with mud-walled huts and donkey carts; in the desert, refineries shimmer in sun-drenched expanses, which until recently were disturbed only by a passing camel caravan. It is also evident that the impact of the realm's oil wealth is felt most strongly in the less populous oil-rich countries than in the larger states. Furthermore, petroleum revenues have had the effect of intensifying urban dominance and regional disparities, even where, as in Saudi Arabia, a conservative government implements national-development programs funded by oil revenues and designed to limit the abrasive effects of modernization.

Supranationalism in the Middle East

Their many divisions and quarrels notwithstanding, the countries of North Africa/Southwest Asia have forged some effective international organizations. One of these is the League of Arab States, founded in 1945 "to strengthen relationships between members and to promote

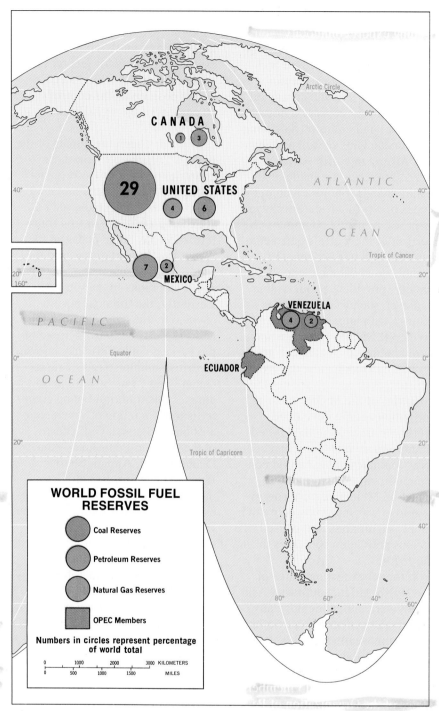

FIGURE 6–13

Arab aspirations." The Arab League today counts 21 members: Algeria, Bahrain, Djibouti, Egypt, Iraq, Jordan, Kuwait, Lebanon, Libya, Mauritania, Morocco, Oman, Palestine Liberation Or-

ganization (PLO), Qatar, Saudi Arabia, Somalia, Sudan, Syria, Tunisia, United Arab Emirates, and Yemen. Egypt, although a charter member, was suspended from 1979 to 1990 because it

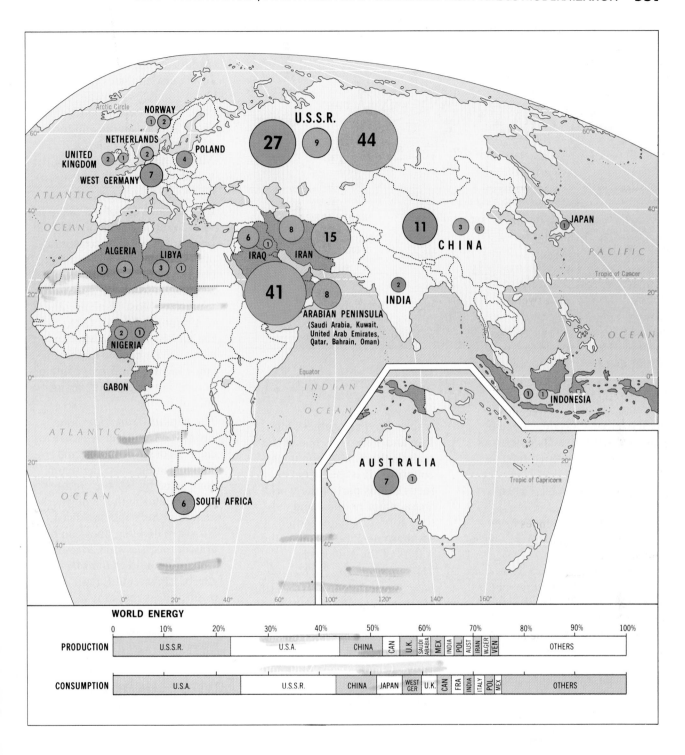

signed a peace treaty with Israel; the League's headquarters is scheduled to be returned to Cairo by 1991, following its temporary relocation to Tunis during Egypt's suspension.

Member states have some-times been in conflict. Since 1976, Morocco and Algeria have dis-puted (and warred) over the future of the former Spanish Sahara (fol-lowing partition between Morocco and Mauritania, all of Spanish Sa-hara became a Moroccan area); and Syria and Iraq have argued over the sharing of the waters of the Euphrates River. The League has also been helpful in the media-tion of several international prob-lems, has advanced such Arab causes as the Palestinian case for a

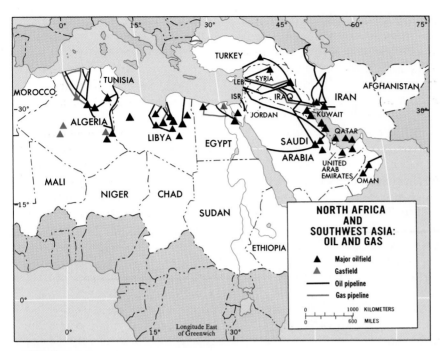

FIGURE 6–14

4. *The Arabian Peninsula*. Dominated by the large territory of Saudi Arabia, the Arabian Peninsula also includes the United Arab Emirates, Kuwait, Bahrain, Qatar, Oman, and Yemen. Here lies the source and focus of Islam, the holy city of Mecca; here too lie many of the world's greatest oil deposits.

5. *The Non-Arab North*. Across the northern tier of the realm lie four countries, each of which in one way or another constitutes an exception to the prevailing norms. Turkey, Cyprus, Iran, and Afghanistan are not Arab countries. Most Iranians adhere to a minority Islamic faith, the Shi'ah; these *Shi'ite* Muslims do not share certain important values with their *Sunni* contemporaries, who form the majority sect of Islam.

6. *The African Transition Zone*. From Mauritania in West Africa clear across to Ethiopia and Somalia in the east, the realm dominated by Islamic culture interdigitates with the northern margin of Subsaharan Africa. No sharp dividing lines can be drawn; this is a wide zone of transition south of the Maghreb, Libya, and Egypt. As a result, this is the least well defined of the six regions of the North Africa/Southwest Asia realm.

REGIONS AND STATES

This vast sprawling realm is not easily divided into geographic regions. Not only are population clusters widely scattered, but cultural transitions—internal as well as external—make it difficult to discern a regional framework. If we constructed a series of maps based on environmental aridity, religion, language, and ethnicity, there would be few areas of agreement. Certainly the regionalization that follows is debatable, but it does highlight the six major geographic components of the realm (mapped in Fig. 6–12):

1. *Egypt and the Nile Basin*. This region, in many ways, constitutes the heart of the realm as a whole. Egypt (together with Iran and Turkey) is one of the realm's three most populous countries. It is the historic focus of this part of the world and a major political and cultural force. It shares with its southern neighbor, Sudan, the waters of the Nile River.

2. *The Maghreb and Its Neighbors*. Western North Africa (the Maghreb) and the areas that border it also form a region, consisting of Algeria, Tunisia, and Morocco at the center and Libya, Chad, Niger, Mali, and Mauritania along the broad periphery. The last four of these countries also lie astride or adjacent to the broad transition zone where the Arab-Islamic realm of northern Africa merges into Subsaharan Africa.

3. *The Middle East*. This region includes Israel, Jordan, Lebanon, Syria, and Iraq. In effect, it is the crescent-like zone of countries that extends from the eastern Mediterranean coast to the head of the Persian Gulf.

Egypt: Anchor of the Lower Nile Basin

Egypt occupies Africa's northeastern corner, and mostly geometric boundaries separate its 387,000 square miles (1 million square kilometers) from Libya to the west and Sudan to the south. Alone among states on the African continent, Egypt extends into Asia through its foothold on the Sinai Peninsula—a presence that has in recent times proved rather tenu-

homeland, and has developed a communications satellite network among its member countries.

ous. Egypt lost its entire Sinai territory during the Six-Day War with Israel in 1967; it then regained a toehold across the Suez Canal in the ensuing conflict of 1973; and finally, following the implementation of the 1978 (U.S.-supervised) Camp David peace agreement with Israel, Egypt's Sinai lands were returned in 1982.

The Greek scholar Herodotus described Egypt as the gift of the Nile, but Egypt also was a product of natural protection (at least before the age of air warfare). The middle and lower Nile lie enclosed by inhospitable country, open to the Mediterranean Sea but otherwise all but inaccessible to overland contact. To the south, the upstream Nile is interrupted by a series of cataracts that begin near the present boundary of Egypt and Sudan. To the northeast, the Sinai Peninsula never afforded an easy crossing, and the southwestern arm of the ancient Fertile Crescent lay some distance beyond. To the west, there is the endless Sahara. The ancient Egyptians thus had a natural fortress, and in their comparative isolation they converted security into progress. Internally, the Nile was then what it remains to this day—the country's lifeline and highway of trade. Externally, the Egypt of the pharaohs had to trade, for there was little wood and metal, but most of this trade was left to the Phoenicians and Greeks. Egypt's cultural landscape still carries the record of antiquity's accomplishments in the great pyramids and stone sculptures that bear witness to the rise of a culture hearth 5000 years ago.

The Nile

Perhaps 95 percent of Egypt's 58 million people live within a dozen miles of the Nile or one of its deltaic distributaries. The majestic river is the product of headwaters that rise far to the south in two dif-

ferent parts of Africa: Ethiopia, where the Blue Nile originates, and the great lakes region of equatorial East Africa, source of the White Nile (the two streams join at Khartoum in Sudan). Before dams were constructed across the river, the Nile's dual origins assured a fairly regular natural flow of water, making the annual floods predictable both in terms of timing and intensity. The Nile normally is at its lowest level in April and May, but rises during the summer to its flood stage at Cairo in October, when it may be more than 20 feet (6 meters) above its low stage. After a rapid fall during November and December, the river recedes gradually until its springtime minimum.

In ancient times, this regime made possible the invention of **irrigation**, for the silt left behind by the Nile floods was manifestly cultivable and fertile. Eventually, a system of cultivation known as *basin irrigation* developed, whereby fields along the low banks of the Nile were partitioned off by earth ridges into a large number of artificial basins. The silt-rich river waters would pour into these basins during flood time; then the exits would be closed so that the water would stand still, depositing its fertile load of alluvium. Later, after six to eight weeks, the exit sluices were opened and the water drained away to leave a rejuvenated soil ready for sowing. This method, although revolutionary in ancient times, had disadvantages. The susceptible lands must lie near or below the flood level, only one crop could be grown annually, and if the floods were less intense (as they were in some years) some basins remained unirrigated because floodwaters could not reach them. Still, traditional basin irrigation prevailed all along the Egyptian Nile for thousands of years; not until a century ago did more modern methods develop. And even as

late as 1970, as much as one-tenth of all irrigation in Egypt was basin irrigation, most of it practiced in Upper Egypt (the southern region nearest the Sudanese border).

The construction of dams, begun during the nineteenth century, made possible the *perennial irrigation* of Egypt's farmlands. By building a series of artificial barriers across the river (with locks for navigation), engineers were able to control the floods, raise the Nile's water level, and free the farmers from their dependence on the sharp seasonal fluctuations of the river's natural regime. Not only did the country's cultivable area expand substantially, but it also became possible to grow more than one crop per year. By the 1980s, all farmland in Egypt finally came under perennial irrigation, a transformation completed within a single century.

The greatest of all Nile control projects, the Aswan High Dam (Fig. 6–15), was begun in 1958 and opened in 1971. The dam is located some 600 miles (1000 kilometers) upstream from Cairo in a comparatively narrow, granite-sided section of the valley. The dam wall, 365 feet (110 meters) high, creates Lake Nasser, one of the largest artificial lakes in the world. This reservoir inundates over 300 miles (500 kilometers) of the Nile's valley not only in Upper Egypt, but in Sudan as well, and the cooperation of this southern neighbor was required because some 50,000 Sudanese had to be resettled. The impact of the dam on Egypt's cultivable area was momentous. Before construction, the Nile's waters could irrigate some 6.25 million acres (2.53 million hectares) of farmland; to this, the Aswan High Dam added 2.9 million acres (1.2 million hectares). Moreover, the dam now also supplies Egypt with about 40 percent of its electricity requirements.

FIGURE 6–15 The flood-control and hydropower facilities of the Aswan High Dam, looking south toward Lake Nasser which extends hundreds of miles upstream from here. The Nile Valley has been dammed at Aswan since 1902, with the present barrier functioning since the early 1970s.

Nonetheless, by regulating annual floods, the Aswan Dam has also had negative impacts. The deposition of fertile silt below Aswan has been sharply reduced, forcing the usage of costly fertilizers; river waters are now insufficient to dissolve the salts left by flooding; and in the delta, the lowered volume of Nile sediments is no longer able to replenish the Mediterranean shoreline against coastal erosion. It has also become evident in recent years that Lake Nasser has itself accumulated sediments far more rapidly than expected and that droughts to the south cause its water level to drop substantially; both developments portend a shorter life for the lake than planners had predicted—and raise the specter of chronic water and energy shortages in the twenty-first century.

Egypt has often been described as one elongated oasis, and any map of its human geography confirms the appropriateness of that perspective in a country where almost all of the people are concentrated on the livable 4 percent of the land area (Fig. 6–16). In Upper and Middle Egypt, the strip of abundant intense green, 3–15 miles (5–25 kilometers) wide, lies in stark contrast to the barren, harsh, dry desert immediately adjacent (see photo on p. 2), a reminder of what Egypt would be without its Nile lifeline. But north of Cairo, the great river fans out across its wide delta—100 miles (160 kilometers) in length and 155 miles (250 kilometers) in width—along the Mediterranean coast between Alexandria and Port Said. In ancient times, the river's waters reached the sea via several chan-

nels and the delta was floodprone, inhospitable country; thus, Egypt in antiquity was Middle and Upper Egypt, the Egypt of the Nile Valley.

Today, however, the delta waters are diverted through two controlled channels, the Rosetta and Damietta distributaries. The delta contains twice as much cultivable land as Middle and Upper Egypt combined; it also contains some of Africa's most fertile soils and is able to produce over one-third of Egypt's food supply. But the increasing use of Nile waters upstream and the diminished flow of the river, with its vital supply of silt, now pose a threat to all the progress that has been made in the delta. And because the delta is geologically subsiding, there is growing danger that brackish water will invade the channels from

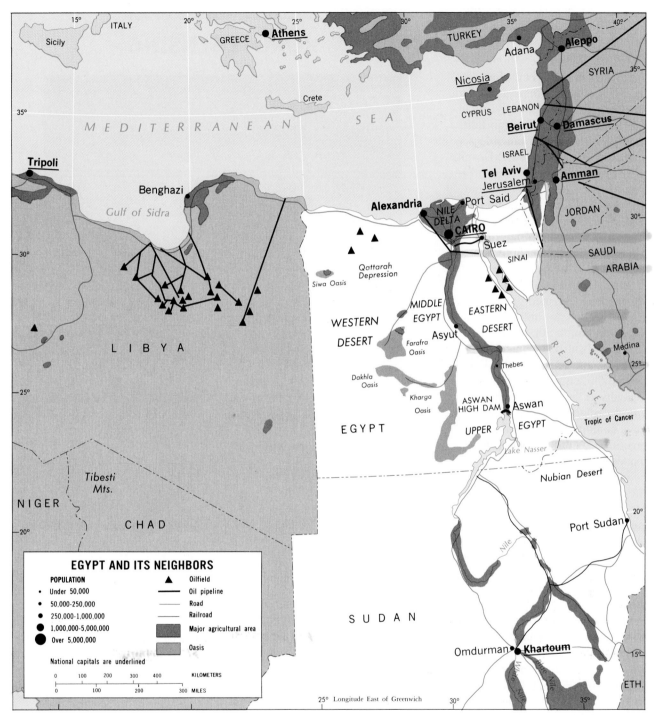

FIGURE 6–16

the sea, seep into the soil, and once again reduce much of the area to an unproductive state.

A Pivotal Location

Egypt has changed markedly in modern times and is today a more populous and urbanized country (47 percent) than ever before. But Egypt's subsistence farmers, the *fellaheen*, still struggle to make their living off the land, as did the peasants in the Egypt of 5000 years ago. In a society that at one time was the source of countless innovations, the tools of too many of these farmers often are as old as any still in use—the hand hoe, the wooden buffalo-drawn plow, and the sickle. Water continues to be drawn from wells by the wheel and bucket, and the hand-turned Archimedean screw (to move water over riverbank ridges) can still be seen in service. And the peasant still lives in a small mud dwelling, which would look very familiar to the farmer of many centuries ago; neither would that distant ancestor be surprised at the poverty, recurrent disease, high infant mortality rates, and the lack of tangible change in the countryside. Despite the Nile dams and irrigation projects, Egypt's available farmland per capita has steadily declined during the past two centuries as uncontrolled population growth keeps nullifying gains in crop productivity.

Yet, of all the states in the realm loosely termed the Middle East, Egypt is not only one of the largest in population but by many measures is also the single most influential country. Indeed, Egypt is spatially, culturally, and ideologically at the heart of the Arab world. Among the factors that have combined to place it in this position, location is clearly a primary element in Egypt's eminence. The country's position at the southeastern corner of the Mediterranean Sea presented an early impetus as Phoenicians and Cretans linked the Nile Valley with the east-coast Levant and other parts of the Mediterranean Basin. Egypt lay protected but immediately opposite the Arabian Peninsula; in the centuries of Arab power and empire-building, it sustained the full thrust of the new wave. The Arab victors founded Cairo in A.D. 969 and made it Egypt's capital (replacing Alexandria), selecting a fortuitous site and situation at the place where the Nile Valley opens onto the delta.

Egypt also lies astride the land bridge between Africa and Southwest Asia—and between the Mediterranean and the Red Seas. In modern times, the latter became a major asset to the country as the Suez Canal, completed in 1869,

REMOTE SENSING

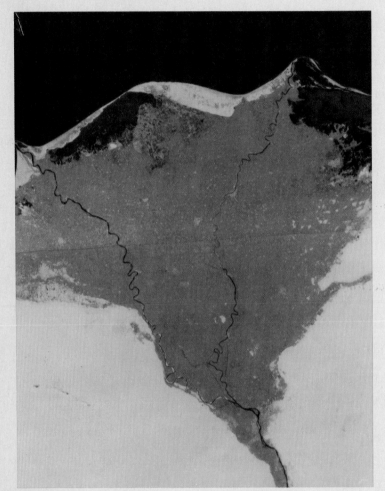

FIGURE 6–17 A satellite image of the entire Nile Delta (north is at the top), with each color indicating a specific land use.

Geographers have for centuries been improving their methods of observing and interpreting the earth's surface. The achievement of aerial photography within the past century has enabled analysts to

became the vital bottleneck link in the shortest route between Europe and South and East Asia. Built by foreign interests and with foreign capital, the Suez Canal also brought a stronger foreign presence to Egypt. Until 1922, Egypt was a British protectorate; even when Egyptian demands for sovereignty led to the later establishment of a kingdom, British influence remained paramount and the Suez Canal a foreign operation.

The 1952 uprising against King Farouk and the proclamation of an Egyptian republic in the following year presaged the showdown over the Suez Canal that was now all

gather large amounts of data and study the arrangements of objects on the ground. During the past two decades, new technologies have greatly and rapidly expanded the use of this tool for monitoring the environment and human spatial organization. The term now used to refer to these techniques is **remote sensing**, which Arthur and Alan Strahler (in their book *Modern Physical Geography*) define as the gathering of geographic information from great distances and over broad areas, usually through instruments mounted on high-flying aircraft or space vehicles, such as Landsat satellites.

Highly sophisticated instruments called remote sensors, which are far more sensitive and versatile than conventional cameras, collect the necessary data. They are designed to detect both nonvisible and visible light across the entire electromagnetic spectrum as emitted from the surface, and many of today's remote sensing installations use multispectral scanners (including radar) that are capable of automatically producing color imagery for detailed interpretation by analysts. Since each surface feature emits its own *spectral signature* (unique waveband pattern), it is now possible to identify everything from a blacktop parking lot to a cluster of desert shrubbery. Hundreds of spatial phenomena around the globe are today being monitored by remote sensing, among them soil and vegetation change, crop health (diseased fields have different temperatures from normal ones and can easily be detected), locust infestations, potential oil and gas deposits, air pollution, land-use change, population shifts, and heat losses by buildings.

A pertinent example of remotely sensed imagery is Fig. 6–17, which shows the Nile Delta from the Landsat satellite 560 miles (900 kilometers) above the earth. The well-watered delta (red-orange) is easily distinguished from the surrounding desert (light tan). Within the delta region, population centers are clearly visible in solid gray (the Cairo metropolis is at the southern apex of the delta) as is the variation of crop types shown by the different shadings of red-orange; the Nile's two channels—the Damietta to the east (right) and the Rosetta to the west (left)—are readily apparent as two dark blue threads. Since comprehensive geographic investigations also include information on the landscape of the study area—so-called *ground truth*—this satellite imagery should be compared to the view of the delta's surface provided in Fig. 6–4.

maintained that respect; but Egypt's 1978 peace accord with the Israelis brought harsh Arab criticism as well as a suspension of its membership in the Arab League, which endured into the 1990s.

Subregions

As Fig. 6–16 indicates, Egypt contains six subregions: (1) the Nile Delta, or Lower Egypt; (2) Middle Egypt, consisting of the Nile Valley from Cairo to Thebes; (3) Upper Egypt, the Nile Valley from Thebes to the border with Sudan, including Lake Nasser; (4) the Western Desert, which contains several large oases; (5) the Eastern Desert and Red Sea coast; and (6) the Sinai Peninsula. The overwhelming majority of the Egyptian people live in Lower and Middle Egypt, which are highlighted here.

The Nile Delta (see *box* to the left) covers an area of just under 10,000 square miles (25,000 square kilometers)—roughly equivalent to the size of Vermont. For thousands of years, only the southernmost land nearest the Nile Valley was farmed, because the main part of the delta was flood-prone, sandy, and excessively salty; seven Nile distributaries found their way to the Mediterranean Sea. Today, however, the two Nile channels are controlled as modern engineering converted the region into one of perennial irrigation, fertile farmland, and multiple cropping of rice (the staple) and the cultivation of cotton (the chief cash crop). The completion of the Aswan High Dam now makes possible the reclamation and eventual cultivation of an additional 1 million acres (400,000 hectares) of delta land; today, close to half of the Egyptian population resides in this region.

The urban focus of Lower Egypt is Alexandria, Egypt's lead-

but inevitable. In 1956, President Gamal Abdel Nasser announced the nationalization of the canal and the expropriation of the company that controlled it. Three months later, Israel, Britain, and France undertook a miscalculated and strategically disastrous invasion of the Sinai Peninsula. Pressured by the United States and the United Nations and pushed into retreat, the invaders soon abandoned the canal area, and Egyptian stature in the Arab world (and in the Third World generally) was immeasurably enhanced. Subsequently, the initiation of two wars with arch-enemy Israel (in 1967 and 1973)

ing seaport, home to 2.9 million residents. Industrial growth accelerated between the two World Wars (Alexandria is now Egypt's leading industrial center), and during the royal period the city served as second capital. Today, Alexandria has a canal connection to the Nile as well as efficient road and rail connections to Egypt's heartland. The city has also become a resort and, of course, is a magnet in the contemporary Third World migration toward burgeoning urban areas.

Middle Egypt begins at the "throat" of the Nile Delta—specifically at Cairo, the teeming capital and interface between valley and delta. This primate city's location is quintessentially nodal: its delta connections are good and a railroad extends along the entire length of Middle Egypt and beyond to Aswan. A string of towns lies along this route, and centrally positioned Asyut is Middle Egypt's subregional focus.

Metropolitan Cairo has 10.1 million residents, and the city's rapid growth rate of recent decades continues unabated. Cairo is now the world's 13th largest metropolis, and it shares with other giant cities of the underdeveloped world the staggering problems of crowding, poverty, sanitation, health, and unemployment of its huge numbers. Cairo is a city of stunning contrasts. Along the Nile waterfront, elegant skyscraper hotels rise above surroundings that are frequently carefully manicured and look somewhat Parisian. But beyond this facade lies the maze of depressing ghettoes, narrow alleys, overcrowded slums, and a low skyline dominated by dozens of minarets, the towers of mosques pointing skyward. Hundreds of architectural achievements, many of them mosques, tombs, and shrines, are scattered throughout Cairo, but seemingly everywhere one encounters the mud huts and hovels of the very poor; in fact, well over a million squatters now live in the sprawling cemetery known as the City of the Dead (Fig. 6–18). Still, Cairo is a truly cosmopolitan urban center, the cultural capital of the Middle East and Arab world, with a great university, magnificent museums, a symphony orchestra, national theater, and opera. Although Cairo has always been primarily a center of government and administration, it is also a river port and an industrial complex (textiles, food processing, and iron and steel production). Countless thousands of small handicraft industries are scattered throughout the neighborhoods of the city, and the grand

FIGURE 6–18 The desert sun bakes the City of the Dead, a vast complex of tombs on the eastern edge of the old city of Cairo, which today is forced to serve as home for hundreds of thousands of impoverished squatters.

bazaar (traditional market square) throbs daily with the trade in small items.

From the waterfront skyscrapers of Cairo to the monuments of Thebes, ancient capital of the pharaohs, the Nile River gives life to the long oasis that is Middle Egypt. In places, the walls of the valley approach the river so closely that the farmlands of the floodplain are interrupted, but along most of the river's course in Middle Egypt the strip of cultivation is between 5 and 10 miles (8–16 kilometers) wide. Everywhere, however, the sharply outlined contrast between luxuriant green and barren desert is intense and startling (see Fig. I–1). Clover (*berseem*, a fodder crop), cotton, corn, wheat, rice, millet, sugarcane, and lentils are among the crops that thrive on fields now under perennial irrigation.

Despite the expansion of its irrigated farmlands, Egypt nonetheless must import more than half of its food. This reliance on foreign sources will undoubtedly increase through the 1990s, because the country's population is growing too quickly (at a rate of 2.8 percent yearly). Large families have always been an Egyptian tradition, and resistance from the Islamic clergy is further slowing the diffusion of birth-control practices. Unfortunately, development gains are being wiped out in the losing demographic battle: land reform, increased crop yields, and expanded cultivation are simply not enough to overcome Egypt's greatest obstacle to economic advancement.

The Maghreb and Libya

West of Egypt lies Libya and beyond it the *Maghreb*, the western region of the realm's North African component (Fig. 6–19). The three countries of northwestern-

most Africa are collectively called the Maghreb, based on the Arab term meaning *western isle* and suggesting the physiographic basis of the region, with the great Atlas Mountains rising like an immense island from the waters of the Mediterranean Sea and the sandy flatlands of the Saharan interior. Taken together, the four states of Morocco, Algeria, Tunisia, and Libya have a population not much larger than Egypt's. In 1991, Morocco's population is 27.1 million; Algeria, the largest Maghreb state territorially, has 26.6 million inhabitants; Tunisia, the smallest country in North Africa, has 8.3 million people; and Libya's population is 4.3 million.

Whereas Egypt is the gift of the Nile, the Atlas Mountains form the nucleus of the settled Maghreb. These high ranges wrest from the rising air the orographic rainfall that sustains life in the intervening valleys, which contain good soils and sometimes rich farmlands. From the vicinity of Algiers eastward along the coast into Tunisia, annual rainfall averages more than 30 inches (75 centimeters), a total more than three times as high as that recorded for Alexandria in Egypt's delta. Even 150 miles (240 kilometers) inland, the slopes of the Atlas still receive over 10 inches (25 centimeters) of rainfall. The effect of the topography can be read on the world map of precipitation (Fig. I–9): where the highlands of the Atlas terminate, desert conditions immediately begin.

The Atlas Mountains are structurally an extension of the Alpine system that forms the orogenic backbone of Europe, of which Switzerland's Alps and Italy's Appennines are also parts. In northwestern Africa, these mountains trend southwest-northeast and commence in Morocco as the High Atlas, with elevations close to 13,000 feet (4000 meters). East-

ward, two major ranges appear that dominate the landscapes of Algeria proper: the Tell Atlas to the north facing the Mediterranean and the Saharan Atlas to the south overlooking the great desert. Between these two mountain chains, each consisting of several parallel ranges and foothills, lies a series of intermontane basins (analogous to South America's Andean altiplanos but at lower elevations), markedly drier than the northward-facing slopes of the Tell Atlas. In these valleys, the rain shadow effect of the Tell Atlas is reflected not only in the steppe-like natural vegetation, but also in land-use patterns: pastoralism replaces cultivation and stands of short grass and bushes blanket the countryside.

The countries of the Maghreb are sometimes referred to as the Barbary states, in recognition of the region's oldest inhabitants, the Berbers. The Berbers' livelihoods (nomadic pastoralism, hunting, some farming) changed as foreign invaders, first the Phoenicians and then the Romans entered their territory. The latter built towns and roads, laid out farm fields and irrigation canals, and introduced new methods of cultivation. Then came the Arabs, conquerors of a different sort. They demanded the Berbers' allegiance and their conversion to Islam, radically changed the political system, and organized an Arab-Berber alliance (the *Moors*) that pushed across the Straits of Gibraltar into Spain and colonized a large part of southwestern Europe. After the Moors' power declined, the Ottoman Turks established a sphere of influence along North Africa's coasts. But the most pervasive foreign intervention came during the nineteenth and twentieth centuries when the European colonial powers—chiefly France, but also Spain—established control and (in the now-familiar sequence of

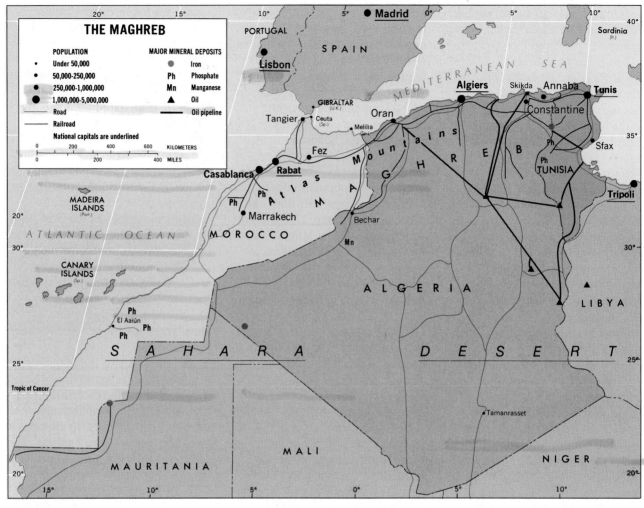

FIGURE 6–19

events) delimited the region's political boundaries.

During the colonial era, well over a million Europeans came to settle in North Africa—most of them French, and a large majority to Algeria—and these immigrants soon dominated commercial life. They stimulated the renewed growth of the region's towns, and Casablanca (3.0 million today), Algiers (2.1 million), and Tunis (1.4 million) rose to become the urban foci of the colonized territories. Although the Europeans dominated trade and commerce and integrated the North African countries with France and the European Mediterranean world, they did not confine themselves to the cities and towns. They recog-

nized the agricultural possibilities of the favored parts of the *tell* and established thriving farms. Agriculture here, not surprisingly, is of the Mediterranean variety: Algeria soon became known for its vineyards and wines, citrus groves, and dates; Tunisia has long been one of the world's leading exporters of olive oil; Moroccan oranges went to many European markets. And staples such as wheat and barley were also among the exports.

Despite the proximity of the Maghreb to Mediterranean Europe and the tight integration of the region's territories within the French political framework, nationalism emerged as a powerful force here. Morocco and Tunisia secured in-

dependence mainly through negotiation, but in Algeria a costly revolutionary war was fought between 1954 and 1962. It was not difficult for the nationalists to recruit followers for their campaign; the justification for it was etched in the very landscape of the country, in the splendid shining residences of the landlords and the miserable huts of the peasants. But the revolution's success brought new troubles. Hundreds of thousands of French people left Algeria and the country's agricultural economy fell apart, making an orderly transition impossible. Productive farms went to ruin, exports declined badly, and needed income was lost.

But there are some bright spots

in the Maghreb. Algeria has one resource to compensate for its losses in agriculture—oil, the same oil that led the French to resist Algerian independence through the 1950s. Petroleum has now become dominant among Algeria's export commodities; it is extracted from the Sahara Desert and piped to Algerian as well as Tunisian ports for transshipment to Europe (Fig. 6–14). Libya's production from its Saharan oil-fields has risen spectacularly to place that country among the world's leading exporters.

Furthermore, unlike most of the remainder of the Middle East, the countries of the Maghreb are endowed with a diversity of mineral resources. Chief among these is phosphate of lime, used to manufacture fertilizer. Morocco is a leading exporter of this commodity, which also occurs in Algeria and Tunisia. Both Morocco and Algeria also possess Atlas-related iron ores that are exported to the United Kingdom from their favorably located sources. Manganese, lead, and zinc are mined as well in all three countries, but sufficient coal is not locally available. Despite the new sources of energy from oil and natural gas, this impressive range of minerals has stimulated very little domestic industrial development.

The oil boom, which decidedly slackened during the 1980s, has yet to improve life substantially for the majority of the people, who remain poor, unemployed, undernourished, and far removed from the centers of change and progress. Tens of thousands of Algerians have followed the European emigrants and gone to France's cities in search of jobs. A large number of those who stayed at home remain trapped in the *bidonvilles*, the sprawling, poverty-stricken shantytowns that surround every Maghreb city.

The three countries of the Ma-

ghreb, together with Libya, reflect the ideological division that so strongly marks the entire realm. Morocco, farthest west and overlooking the Atlantic Ocean rather than Mediterranean Sea, has remained a kingdom, a relatively conservative force in Arab-world politics. Algeria, heart of the Maghreb and directly opposite France across the Mediterranean, is a one-party "Democratic and Popular Republic," nonaligned, socialist, and vigorous in its support of Arab causes. Tunisia, smallest and in many ways the poorest among North Africa's states, became a multi-party democracy under the powerful leadership of its longtime, pre-1987 president and national hero, Habib Bourguiba. And Libya—the "Socialist People's Libyan Arab Jamahiriya"— emerged as the region's radical state, often at odds with its neighbors.

Libya, almost rectangular in shape, is a country of four corners. Along the Mediterranean coast in the west lies Tripolitania, centered on the capital, Tripoli (1.6 million); in the east lies Cyrenaica, with Benghazi (700,000) as its urban focus. Between these two coastal sectors lies the Gulf of Sidra, a deep indentation of the Mediterranean Sea. Libya's revolutionary leader, Muammar Gadhafi, has proclaimed most of this gulf to be Libyan territorial sea, but he has not managed to sustain that claim. Libya's two southern corners are the desert Fezzan, a mountainous land on the southwestern border with Algeria and Niger, and the Kufra Oasis, a sparsely populated zone centered on the oasis of Al Jawf in the southeast. During the 1980s, Libya was a source of recurring controversy in its military challenges to its southern neighbor (Chad), its loud confrontations with the United States, and its open support of terrorist violence.

How has this mostly desert country, with a population of under 5 million people, become a regional power and an international issue? The answer has two parts: the discovery of oil in 1959 and the overthrow of the kingdom by military coup ten years later. The new leader, Colonel Gadhafi, not only transformed the government but also set the country on a course in support of revolutionary Arab causes elsewhere in the realm and even beyond. He was able to do this with money derived from the sale of crude oil drawn from huge reserves south of Benghazi (Fig. 6–16). A vast program of construction (roads, schools, clinics, ports) was launched in order to link and unify the country's dispersed population; moreover, work was begun on a massively expensive (ca. U.S. $25 billion) 2500-mile/4000-kilometer pipeline network that, by the year 2000, will tap aquifers deep beneath the Sahara to bring adequate water to the arid populated coastal strip. Much money was also spent on military equipment and training, and by 1980 Libya had become a threat to its neighbors. But Gadhafi's greatest impact in the world at large resulted from his stated foreign policy: Arab unity, the elimination of Israel, the advancement of Islam, the eradication of Western and other foreign influences from the realm, support for Palestinian causes, and assistance to revolutionary movements. In practice, from Western viewpoints, this has involved support for international terrorism and subversion against moderate governments. By the opening of the 1990s, however, Libya was increasingly isolated, reeling from humiliating defeat by the Chadians as well as the 1986 U.S. bombing raids on Tripoli and Benghazi; and the persistence of lowered oil prices on the world market served to dampen Libyan aspirations further.

The Middle East

The Middle East, as a region within the North Africa/Southwest Asia realm, consists chiefly of the pivotal area positioned between Turkey and Iran to the north and east, Saudi Arabia and Egypt to the south and southwest, and the Mediterranean Sea to the west (Fig. 6–12). Five countries lie in this region: Iraq, largest in population (just under 20 million) as well as territory; Syria, next in both categories; and Jordan, Lebanon, and Israel.

Israel

Israel lies at the very heart of the Arab world (Fig. 6–20). Its neighbors are Lebanon and Syria to the north and northeast, Jordan to the east, and Egypt to the southwest—all in some measure still resentful of the creation of the Jewish state in their midst. Since 1948, when Israel was created as a homeland for the Jewish people on recommendation of a United Nations commission, the Arab-Israeli conflict has overshadowed all else in the Middle East.

Indirectly, Israel was the product of the collapse of the Ottoman Empire. Britain gained control over the Mandate of Palestine, and British policy supported the aspirations of European Jews for a homeland in the Middle East, embodied in the concept of Zionism. In 1946, the British granted independence to the territory lying east of the Jordan River, and "Transjordan" (now the state of Jordan) came into being. Shortly afterward, the territory west of the Jordan River was partitioned by the United Nations, and the Jewish people got slightly more than half of it—including, of course, some land that had been occupied by Arabs. The original UN plan proposed to allot 55 percent of all Palestine to the Jewish sector, al-

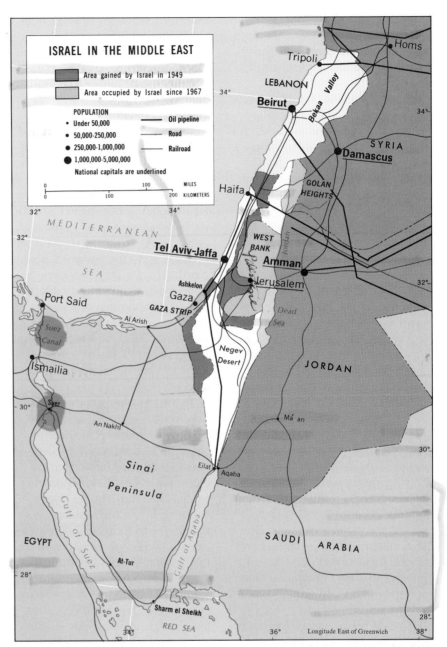

FIGURE 6–20

though only 8 percent of the land was actually owned by Jews (who comprised about one-third of the total population of Palestine), but this partition plan was never implemented as intended. As soon as the Jewish people declared the independent state of Israel on May 14, 1948, the new country was attacked by its Arab neighbors who rejected the scheme. In the ensuing battle, Israel not only held its

own but gained some crucial territory in its central and northern areas as well as in the Negev Desert to the south (mapped in lavender in Fig. 6–20). At the end of this first war in 1949, the Jewish population controlled 80 percent of what had been Palestine west of the Jordan River. Of course, this success was not won by overnight organization: at the time of Israeli independence, there already were

750,000 Jews in Palestine. Indeed, the world Zionist movement had been assisting Jews in their return to Palestine since the late nineteenth century.

For over four decades now, a state of latent—and at times actual—war has existed between Israel and the Arab world. In 1967, a week-long military conflict resulted in Israeli victory and occupation of Syrian, Jordanian, and Egyptian territory, including the Sinai Peninsula to the edge of the Suez Canal—a facility that the Egyptians had not allowed Israeli vessels to use. In 1973, another brief war led to Israel's withdrawal from the Suez Canal to truce lines in the Sinai, but it also extended Israel's control over Syria's Golan Heights, creating a further source of friction. Since then, peace has been sustained largely because of Egypt's willingness to maintain cordial relations with Israel; however, the level of hostility between Israel and other Arab states has remained intense.

So many issues divide the Jews and the Arabs that a permanent solution or even a peace imposed by the superpowers seems unattainable. One prominent issue has been the displacement of Palestinians and their subsequent quest for a homeland: when Israel was created, over 600,000 Palestinian Arabs were forced to leave the area to seek refuge in Jordan, Egypt, Syria, and elsewhere (see box). Another problem has been the city of Jerusalem, a holy place for Jews, Muslims, and Christians. In the original UN blueprint for Palestine, the city was to have become internationalized; however, Jerusalem was then divided between the Jewish state and Jordan, a source of friction with neither side satisfied. But the 1967 war saw the holy city fall into Israeli hands as the entire West Bank portion of Jordan was captured (Fig. 6–20). Israel has since begun

to settle these lands west of the Jordan River, scattering its new settlements throughout the area so as to make future territorial concession difficult. Moreover, the rate of immigration has accelerated in recent years: in 1977 only 5000 Jews lived in the West Bank territory, but by 1990 over 80,000 resided there. Thus, the Israeli percentage of the total population grew from less than 1 percent to over 10 percent today.

A major irritant, too, has been Israel's rapid rise to strength and prosperity amid the underdevelopment so common to the Middle East. In terms of physical geography, there is nothing particularly productive about most of Israel's land; neither is its territory that large (8000 square miles/20,000 square kilometers—smaller than Massachusetts). But Israel has been transformed by the energies of its settlers and, importantly, by heavy investments and contributions made by Jews and Jewish organizations elsewhere in the world, especially the United States. It is often said that in Israel the desert has been converted into farmland, and it is increasingly so. The irrigated acreage has been enlarged to many times its 1948 extent, and water is now carried even into the Negev Desert itself where new food-production technologies are achieving some astonishing results in agricultural output. Particularly important are advances in computer-controlled *fertigation*, whereby brackish desert water mixed with fertilizers is piped directly to the roots of each plant (Fig. 6–22). Fruit, vegetable, and grain yields have been spectacular under the harshest extremes of temperature and aridity, and Israeli technicians are now training agronomists in over 60 countries in the use of automated desert agricultural technology.

Without an appreciable resource base, industrialization also

presents quite a challenge to Israel. Evaporation of waters in the zone bordering the Dead Sea has left deposits of potash, magnesium, and salt, and there is rock phosphate in the Negev. But there is very little fuel available within Israel; no coal deposits are known, although some oil has been found in the Negev. To circumvent the Suez Canal, an oil pipeline leads from the port of Eilat on the Gulf of Aqaba to the refinery at Haifa; a second, larger pipeline has been constructed to link Eilat and Ashkelon. Imported oil formerly came from Iran, but since the Iranian revolution in 1979, Israel has been forced to turn elsewhere for its petroleum. One source is Egypt. When Israel ceded a section of the occupied Sinai Peninsula where oilfields had been found and developed, Egypt agreed to sell oil to Israel in return. The small steel plant at Tel Aviv uses imported coal and iron ore, but was built for strategic reasons; it is not an economic proposition. Thus, the only manufacturing industry for which Israel has any domestic raw materials is the chemical industry, and it has seen substantial growth. But for the rest, Israel must depend on the considerable skills of its labor force; the specialized field of diamond-cutting, in which the Israelis have a global reputation, is an outstanding example. Many technicians and highly skilled craftspeople have been among the hundreds of thousands of Jewish immigrants who came to Israel since its creation, and the best course was to make maximum use of this personnel.

In effect, then, Israel is a Western-type, developed country within the Middle East. Its population of 4.8 million is highly urbanized, with just over 90 percent of the people living in towns and cities. Israel's core includes the two major metropolises, Tel Aviv–Jaffa (1.8 million) and Haifa (600,000), and the coastal strip between them. In total, this core area incorporates over three-quarters of the country's population, although the proportion is less if the territories conquered in the 1967 war are added (including East Jerusalem's Arab sector, with about 200,000 residents).

Despite the persistence of antagonisms, time may be slowly eroding some barriers between Israel and the Arabs. There are behind-the-scenes contacts between the Jewish state and some of its neighbors, tourist exchanges with Egypt, and Israel's Arab residents (about one-sixth of the population) are now being received cordially by the Saudis on pilgrimages to Mecca. In Jerusalem, now Israel's largest central city (550,000), tensions and physical barriers among Jews, Muslims, and Christians are beginning to dissipate.

As some external problems ease, however, divisions within Israel loom ever more threateningly. Today, the country's 4.1 million Jews are divided by a social gap based on ethnicity, income level, and cultural heritage. On one hand, there are the majority Sephardic Jews of Asian–North African origin who largely occupy the lower echelons of Israeli society; on the other, there are the minority Ashkenazim Jews of European and American background who dominate the professions and higher-income groups. Constant frictions affect the relationships between these groups, reflecting a clash between the Islamic and Western cultural values of their source areas, and residential and educational segregation mark the social geography of the two populations. Thus, the younger generations of the Sephardim and Ashkenazim are not encouraged to integrate and commingle, perpetuating a schism that could one day threaten national unity. Moreover, Orthodox Jews, who

THE PALESTINIAN DILEMMA

Ever since the creation of the state of Israel in 1948 in what had been the British Mandate of Palestine, hundreds of thousands of Arabs who called Palestine their homeland have lived as refugees in neighboring countries. A large number have been assimilated into the societies of Israel's neighbors, but many still live in refugee camps. The Palestinians call themselves a nation without a state (much as the Jews were before Israel was founded), and they demand that their grievances be heard. Until Palestinian hopes for a national territory in the Middle East are considered by the powers that have influence in the region—and until Israel's right to exist is acknowledged by all parties—a full settlement between Israel and the Arab countries may remain unattainable. Estimates (1989) of Palestinian populations in the realm are:

Israel and Occupied Territories		2,250,000
Israel	800,000	
West Bank	900,000	
Gaza Strip	550,000	
Jordan		1,700,000
Kuwait		400,000
Lebanon		350,000
Saudi Arabia		250,000
Syria		225,000
Iraq		70,000
Egypt		60,000
Libya		25,000
Other Arab states		425,000
TOTAL		5,755,000

For over a quarter-century, the militant Palestine Liberation Organization (PLO) has claimed to represent Palestinian views. But the PLO is actually an umbrella organization whose main group, Fatah, has often been involved in bitter policy disputes with at

advocate the dominance of religious law over civil Israeli law, are often at odds with nonreligious Jews who prefer to live in a secular society—a division that has sometimes resulted in violence.

Neighbors and Adversaries

Not only does Israel have the misfortune to be located in the heart of the Arab world rather than in

one of the realm's peripheral or transition zones, but it also has an inordinate number of neighbors for so small a country with so much coastline—five (counting Saudi Arabia). Of these five, three lie along Israel's northern and eastern boundary: Lebanon, Syria, and Jordan.

Lebanon

Lebanon, Israel's northern coastal neighbor on the Mediterranean Sea, is one of the exceptions to the

FIGURE 6–21 The Israeli army and rock-throwing Palestinian youths confront each other in the historic West Bank town of Bethlehem, as the late 1980s *intifada* uprising enters the 1990s.

least a dozen smaller factions. Thus, Palestinian forces have not only fought Israeli troops and various Lebanese factions, but also among themselves; moreover, Palestinian terrorists continue to attack targets in Israel and around the world.

Various territorial solutions have been proposed. The suggestion heard most often is that a Palestinian state be created from the West Bank—the Israeli-occupied component of Jordan west of the Jordan River—plus the Gaza Strip. But it is here that the Palestinian uprising (known as the *intifada*) was launched against the Israeli occupiers in late 1987, adding yet another complication (Fig. 6–21). Thus, the territorial objectives of the Palestinians still face some formidable obstacles.

outdated. In the 1930s, Muslims and Christians in Lebanon were at approximate parity; but over the ensuing half-century, the Muslims increased their numbers at a much faster rate than the more urbanized and generally wealthier Christians (many of whom have emigrated from Lebanon in recent years). The Muslims' displeasure with an outdated political arrangement (developed during the French occupation) was expressed during several outbreaks of rebellion prior to full-scale civil war. By then Lebanon had also become a base for over 300,000 Palestinian refugees; these people, many of them living in squalid camps, were never satisfied with Lebanon's moderate posture toward Israel. When the first fighting between Muslims and Christians broke out in the northern coastal city of Tripoli, the Palestinians joined the conflict on the Muslim side. In the process, Lebanon was wrecked.

Beirut, the capital, once a city of great architectural beauty and often described as the Paris of the Middle East, was heavily damaged as Christian, Sunni Muslim, Shi'ite Muslim, Druze, and Palestinian militias—as well as the Lebanese and Syrian armies—fought for control. By the beginning of the 1990s, Beirut faced total destruction, and only the poorest 150,000 war-ravaged residents remained of the 1.5 million who had lived there as recently as the late 1970s (Fig. 6–23). As the Muslims' strength intensified, the Christians concentrated in an area along the coast between Beirut and Tripoli. Although attempts have been made since 1980 by various outside peacemakers to help the combatants work out a lasting political solution, these efforts have been rebuffed and the cycles of ferocious violence continue.

Syria

Syria's role as a force to be reckoned with in any solution of the

rule that the Middle East is the world of Islam: one-fourth of the population of 3.4 million adheres to the Christian rather than Muslim faith. Only a little over half of Israel's territorial size, Lebanon has a long history of trade and commerce, beginning with the Phoenicians of old who were based here. Lebanon must import much of its staple food, wheat; the coastal belt below the mountains, although intensively cultivated,

normally cannot produce enough grain to feed the entire population.

But normality has not prevailed in Lebanon during recent times. The country fell apart in 1975 when a civil war broke out between Muslims and Christians. It was a conflict with many causes. Lebanon for several decades had functioned with a political system that divided power between these two leading communities, but the basis for that system had become

FIGURE 6–22 The barren Negev Desert is the country's least-favored land resource, but the Israelis are making it bloom by pioneering ultrasophisticated fertigation techniques. Clearly, this is a food-production breakthrough of the utmost importance for water-deficient farming areas the world over.

Lebanese crisis has enhanced that country's status since its loss of the Golan Heights to Israel. For a long time, Syria was politically unstable, a poor second to Egypt in their abortive attempt at union, the short-lived United Arab Republic (1958–1961). But the government of Hafez al-Assad, after taking power in 1970, began to reap the benefits of continuity. In 1976, Syrian armed forces intervened and brought a measure of peace to embattled Lebanon; by backing

various client factions, Syria has now so entrenched itself there that it has become a key to any peace settlement.

Syria, like Lebanon and Israel, has a Mediterranean coastline where unirrigated agriculture is possible. Behind this densely populated coastal belt, Syria has a much larger interior than its neighbors, but the areas of productive capacity are quite dispersed. Damascus, the capital (2.0 million), was built on an oasis and is considered to be the world's oldest continuously inhabited city; although in the dry rain shadow of the coastal mountains, it is surrounded by a district of irrigated agriculture. It lies in the southwestern corner of the country (Fig. 6–12), in close proximity to Israel. In the far north, near the Turkish border, lies another old caravan-route center, Aleppo (1.3 million), at the northern end of Syria's important cotton-growing area; here the Orontes River, whose headwaters flow through northern Lebanon's strategic Bekaa Valley, is the chief source of irrigation water. Syria's wheat belt, stretching east along the northern border, also focuses on Aleppo. In the eastern part of the country, the Euphrates Valley and the far northeast (the Jazirah) are being developed for large-scale mechanized wheat and cotton farming with the aid of pump irrigation systems. Although future water supplies are uncertain (upstream, Turkey has built dams on the Euphrates that affect the river's flow in both Syria and Iraq), production of these crops is rising, and more than half the Syrian harvest now comes from this area. Recent discoveries of oil also add greatly to the importance of this long-neglected part of the country.

Southward and southeastward, Syria turns into desert, and the familiar sheep, camel, and goat herders move endlessly across the parched countryside (see *box*). There may be as many as a half

FIGURE 6–23 Death of a once-great city: the awesome destruction of Beirut since 1980 has forced 90 percent of its population to flee and most of the remaining survivors to huddle in bomb shelters.

PEOPLE ON THE MOVE

FIGURE 6–24   Camp has been broken, the camel caravan formed, and the purposeful journey of these nomads resumes—here in the Ahaggar Mountains of southernmost Algeria in the very heart of the Sahara.

Countless thousands of people in North Africa/Southwest Asia are on the move; movement is a permanent part of their lives. They travel with their camels, goats, and other livestock along routes that are almost as old as human history in this realm. Most of the time, they do so in a regular seasonal pattern, visiting the same pastures year after year, stopping at the same oases, pitching their elaborate tents near the same stream. This is a form of cyclical migration—**nomadism**.

Nomadic movement, then, is not simply an aimless wandering across boundless dry plains. Nomadic peoples know their domain intimately; they know when the rains have regenerated the pastures they will make their temporary locale, and they know when it is necessary to move on. Some nomadic peoples remain in the same location for several months every year, and their portable settlements take on characteristics of permanence—until, on a given morning, an incredible burst of activity accompanies the breaking of camp. Amid ear-piercing bleating of camels, the clanking of livestock's bells, and the shouting of orders, the whole community is loaded onto the backs of the animals and the journey resumes (Fig. 6–24). The leaders of the groups know where they are going; they have been making this circuit all their lives.

Among the members of the community there is a division of labor, and some are skilled at crafts and make leather and metal objects for sale or trade when contact is made with the next permanent settlement. A part of the herd of livestock traveling with the caravan may belong to townspeople, who pay for their care. Nomads are not aimless wanderers, nor are they free from the tentacles of the cities. Although nonsedentary peoples have steadily declined in number during recent decades, nomadic pastoralism remains a significant way of cultural and economic life in the Syrian Desert, in the Arabian Peninsula, and elsewhere in the sparsely populated areas of the realm.

million of them, a noticeable proportion of Syria's 13 million people. In contrast to Israel and, to a lesser degree, to Lebanon as well, Syria is very much a country of farmers and peasants; only 51 percent of the people live in cities and towns of any size. But again, Syria produces adequate harvests of wheat and barley, and normally does not need staple imports. It also exports these two grains, but its biggest source of external revenue after oil remains cotton. In addition, Syria is a country where opportunities for the expansion of agriculture still exist; their realization would improve the cohesion of the state and bring its separate subregions into a tighter spatial framework.

Jordan

None of this can be said for Jordan (4.3 million), the desert kingdom that lies east of Israel and south of Syria. It, too, was a product of the Ottoman collapse, but it suffered heavily when Israel was created, more than any other Arab state. In the first place, Jordan's trade used to flow through Haifa, now an Israeli port, so that Jordan has to depend on destabilized Lebanon's harbors or the tedious route via Aqaba in the far south. Second, Jordan's final independence in 1946 was achieved with a total population of only about 400,000, including nomads, peasants, villagers, and a few urban dwellers. Then, with the partition of Palestine and the creation of Israel, Jordan received more than half a million Arab refugees; it soon also found itself responsible for another half million Palestinians who, although living on the western side of the Jordan River, were incorporated into the state. Thus, refugees outnumbered residents by more than two to one, and internal political problems were added to external ones—not to mention the economic difficulties of begin-

UNRESOLVED DISPUTES IN THE 1990s

The Turkish Ottoman Empire long dominated Arab peoples in the Middle East. When the French and British sought to end Turkish rule following World War I, they enlisted the help of the Arabs. Combined European and Arab forces defeated the Turks, but for the Arabs the outcome was mixed. The former Ottoman domain was first colonized by the Europeans and later divided into a patchwork of countries in which ruling royal (Arab) families took control. Kuwait, for example, had been administered from Basra in present-day Iraq during colonial times. But when the British withdrew in the early 1960s, they defined a boundary that created an independent sheikdom that separated Kuwait from Iraq; this newly-created mini-state very nearly landlocked Iraq from its natural outlet to the sea at the head of the Persian Gulf (Fig. 6–25). Another example is Jordan, which was carved out of the desert (following World War II) by a set of straight-line boundaries and connected to the Gulf of Aqaba by the narrowest of corridors; this country is separated from the Mediterranean Sea by Israel, and must depend on the Suez Canal for much of its maritime traffic (Fig. 6–20).

Several of the old Arab kingdoms have been overthrown and were replaced by revolutionary regimes. Today, historic kingdoms, sheikdoms, and emirates adjoin modern, sometimes militaristic republics. Add to this the uneven distribution of oil reserves (many of which were unknown when the boundaries were drawn), Western dependence on these resources, and Arab-Israeli tension, and it is obvious that the potential for conflict is strong. The current political geography of the realm leads to the following conclusion: in addition to the ever-present possibility of renewed Arab-Israeli strife, the locales most vulnerable to territorial disputes and even annexation by force include Lebanon (by Syria), Kuwait (Iraq), Oman (Yemen), and Yemen (Saudi Arabia).

ning national life as a very poor country.

Nonetheless, Jordan has survived with U.S., British, and other aid, but its problems have hardly lessened. Many Jordanian residents still have only minimal commitment to the country, do not consider themselves its citizens, and give little support to the hard-pressed monarchy. Dissatisfied groups constantly threaten to drag the country into another war with Israel; the 1967 war was disastrous for Jordan, which lost the West Bank (its claim was formally surrendered to the PLO in 1988) as well as its sector of Jerusalem (the kingdom's second largest city). Where hope for progress might lie—for example, in the development of the Jordan River Valley—political conflicts intrude. The capital city, Amman (1.5 million), reflects the limitations and poverty of the country. Without oil, without much farmland, without unity or strength, and overwhelmed with refugees, Jordan presents one of the bleaker pictures in the Middle East.

Iraq

Iraq, by comparison, is well endowed, containing the lower valleys of both the historic Euphrates and Tigris rivers. Because its agricultural potential is far greater than what is now used, Iraq is a rarity in the Middle East: it can be described as *under*populated in that it could feed a far larger number of people than it presently does. This is a legacy of the decline that Mesopotamia went through during the Middle Ages, but steps have been taken to improve conditions and raise standards of living. Iraq was also a major beneficiary of the oil reserves of the Middle East, and it needs the income from petroleum to make the necessary investments in industry and agriculture; oil accounts for over 90 percent of the country's export revenues. Unfortunately, almost all of Iraq's oil income was diverted to the fighting of a bitter war with neighboring Iran from 1980 through 1988. That conflict made exporting oil via the Persian Gulf hazardous and piping crude to Mediterranean ports an impossibility, because intervening Syria (Iran's ally) shut off the pipeline; however, a new pipeline, built across Saudi Arabia, opened in the mid-1980s (Fig. 6–14). As the 1990s opened, Iraq remained a very heavily armed state located in the portion of the realm most vulnerable to further territorial disputes (see *box*).

How can a country that exports food crops and has a sizable income from oil have maintained such a low standard of living? There are many answers to this question. For over five decades since its independence (in 1932, after a decade as a British Mandate), Iraq has suffered from administra-

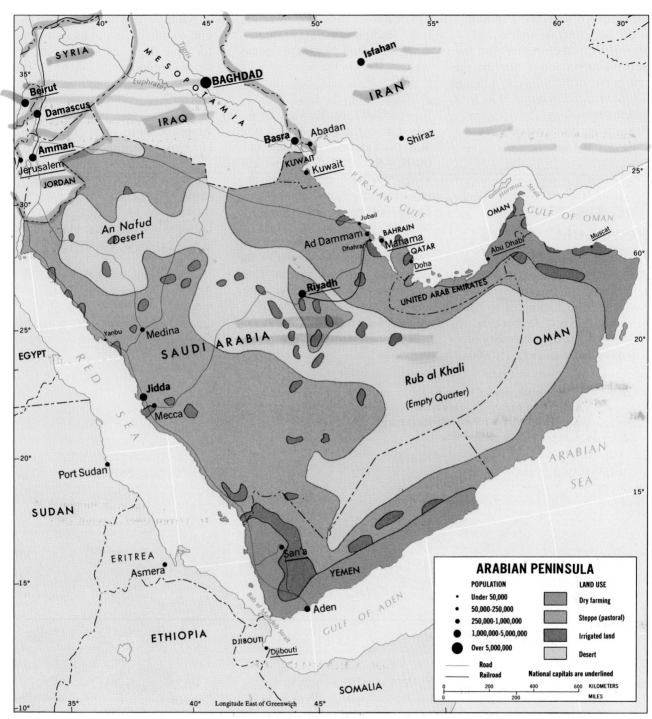

FIGURE 6–25

tive inefficiency, corruption in government, misuse of the national income, inequities in land tenure—a set of problems that practically defines the notion of underdevelopment. Apart from western Iraq, where nomads herd-

ing camels, sheep, and goats traverse the desert of Jordan-Syria-Iraq, most of the people live in small villages strung along the riverine lowland from the banks of the Shatt-al-Arab (the joint lowest course of the Tigris-Euphrates sys-

tem) to the land of the Kurds near the Turkish border. The Kurds, who number over 3.5 million of Iraq's 19.5 million people, have at times opposed the government headquartered at Baghdad (4.1 million), and there still is a serious minority problem (and a sporadic but persistent insurgency) of strong regional character. But the general impression of rural Iraq reminds one of rural Egypt, although Egypt is ahead in terms of its irrigation technology; the peasants face similar problems of poverty, malnutrition, and disease.

The Arabian Peninsula

South of Jordan and Iraq lies the Arabian Peninsula, environmentally dominated by a desert habitat and politically dominated by the Kingdom of Saudi Arabia (Fig. 6–25). With its huge territory of 830,000 square miles (2,150,000 square kilometers), Saudi Arabia is the realm's third biggest state; only Sudan and Algeria are somewhat larger. On the peninsula, Saudi Arabia's neighbors (moving clockwise from the head of the Persian Gulf) are: Kuwait, Bahrain, Qatar, the United Arab Emirates, the Sultanate of Oman, and the new Republic of Yemen (created in 1990 through the unification of former North Yemen and South Yemen). Together, these countries on the eastern fringes of the peninsula contain about 17 million inhabitants; the largest by far is Yemen, with 10.1 million. The interior boundaries of these states, however, are still inadequately defined, here in one of the world's last remaining frontier-dominated zones.

Saudi Arabia itself has only 15.8 million inhabitants in its vast territory, but the kingdom's importance is reflected by Fig. 6–13: the Arabian Peninsula contains the earth's largest concentration of known petroleum reserves. Saudi Arabia occupies most of this area, and by some estimates may possess as much as one-quarter of all the world's remaining oil. These reserves lie in the eastern part of the country, particularly along the Persian Gulf coast and in the Rub al Khali (Empty Quarter) to the south.

The national state that is Saudi Arabia was only consolidated as recently as the 1920s through the organizational abilities of King Ibn Saud. At the time, it was a mere shadow of its former greatness as

FIGURE 6–26 An aerial view of booming Jubail, one of the world's most modern, state-of-the-art planned cities. This is the focus of Saudi Arabia's budding petrochemical industry, with a major manufacturing facility under construction along the Persian Gulf shore (center rear).

the source of Islam and the heart of the Arab world. Apart from some permanent settlements along the coasts and in scattered oases, there was little to stabilize the country; most of it is desert, with annual rainfall almost everywhere under 4 inches (10 centimeters). The land surface rises generally from east to west, so that the Red Sea is fringed by mountains that reach nearly 10,000 feet (3000 meters). Here the rainfall is slightly higher, and there are some farms (coffee is a cash crop). These mountains also contain known deposits of gold, silver, and other metals, and the Saudis hope to diversify their exports by adding minerals from the west to the oil from the east.

Figure 6–25 reveals that most economic activities in Saudi Arabia are concentrated in a wide belt across the "waist" of the peninsula, from the boom town of Dhahran on the Persian Gulf through the national capital of Riyadh (2.0 million) in the interior to the Mecca-Medina area near the Red Sea. A fully effective internal transportation and communications network, one of the world's most modern, has recently been completed. But in the more remote zones of the interior, Bedouin nomads still ply their ancient caravan routes across the vast deserts. For several decades, Saudi Arabia's aristocratic royal families were virtually the sole beneficiaries of their country's incredible wealth, and there was hardly any impact on the lives of villagers and nomads.

That is now changing, however. Agriculture in particular is receiving major government investments, because the Saudis want to prevent the food weapon from being used against them (as they themselves have occasionally wielded the oil weapon). As a result, widespread well drilling has significantly enhanced water supplies to support crops, and surpluses were quickly produced. These achievements, however, have been enormously expensive, for they necessitate the pumping of water from aquifers far below the desert surface as well as the construction of extensive center-pivot irrigation systems (see Fig. 3–5). Despite these efforts, there is now a growing realization that the Saudi Arabian goal of attaining self-sufficiency in food production is elusive. To begin with, the underground water supplies are proving to be a one-time-only, nonrenewable resource. But even more importantly, the Saudi population is today increasing at a much faster pace (3.4 percent yearly) than the domestic food supply can be expanded.

In recent years, the country's rulers have also begun to institute reforms in housing, medical care, and education, and they have spent hundreds of billions of dollars since 1970 on a multifaceted national development program. Although sharply lowered oil prices in the 1980s slowed certain social programs, the overall results are impressive and living standards have improved significantly.

On top of all this, the Saudis have made important gains in their drive to industrialize. The leading priority is a huge petrochemical manufacturing complex that is being developed for future diversification of the economy. The centerpiece is the meticulously planned new city of Jubail (Fig. 6–26), located on the Persian Gulf 55 miles (90 kilometers) north of Dhahran. Begun in 1977, this huge construction project is turning a sleepy fishing port of 8000 into an industrial metropolis of 300,000 (the projected population by the year 2000). Based on an agglomeration of oil refineries, petrochemical and metal-fabrication plants, Jubail is already home to well over 100,000 residents. Together with Yanbu, a smaller (target population: 150,000) new city on the Red Sea coast 185 miles (300 kilometers) north of Jidda, Jubail personifies the country's plans to develop a diversified industrial base to sustain high living standards after the petroleum revenues end sometime in the next century. Still, much of Saudi Arabia remains quite tradition-bound, and in most of the villages—even in Medina and Mecca—the modernizing impacts of the oil bonanza seem far away in place as well as time.

The Non-Arab North

Across the northern tier of the North Africa/Southwest Asia realm lie three geographically distinct countries: Turkey, Iran, and landlocked Afghanistan (Fig. 6–27). Turkey, although a Muslim country, has been strongly oriented toward Europe in recent decades (it is a member of NATO) and has not concerned itself with many of the issues that excite Arab countries. Iran, prior to its 1979 revolution, had also long kept its ideological distance. When the Suez Canal was closed to Israeli shipping by the Arab countries, Iranian oil flowed, on orders of the shah, through Israel's Gulf of Aqaba pipelines. And Afghanistan, although a Muslim country as well, lies remote and peripheral, devastated by the aftermath of the nine-year Soviet occupation of its mountainous Asian interior.

Turkey

Turkey at the beginning of the twentieth century was at the center of a decaying and corrupt Ottoman Empire whose sphere of influence had extended across much of the Middle East and southeastern Europe. At that time, conditions

FIGURE 6–28 Istanbul, situated astride the European-Southwest Asian interface, exhibits landscape features of both realms.

(Fig. I–10, the area marked *BSk*). Here the people live in small villages, grow subsistence cereals, and raise livestock. But the best farmlands lie in the coastal zones, small as they are. There is little coastal lowland along the Mediterranean—the areas around Antalya, Adana, and Iskenderun are the only places where the mountains retreat somewhat from the sea; offshore, Cyprus is another important agricultural concentration, despite the long-standing conflict between Turks and Greeks in that island country (see *box*). The northern Black Sea coast is also quite narrow, but it is comparatively moist; it gets winter as well as summer rainfall (though with a cool-season maximum) totaling 30–105 inches (75–260 centimeters) yearly—uncommonly high for this water-deficient realm. Thus, as Fig. I–14 reflects, these coastal areas, especially the Aegean-Marmara lowlands, are the most densely populated parts of

Turkey and the most productive ones as well.

In population size, Turkey is one of the three largest countries in the realm, with 58 million inhabitants (both Egypt and Iran contain the same total). As Fig. 6–27 suggests, Turkey has a superior surface communications network; it is also the realm's most industrialized country, despite the dominance of agriculture in its export economy. The production of cotton has stimulated a textile industry, and a small steel industry has been established based on a coalfield near Zonguldak, near the Black Sea, and an iron ore deposit several hundred miles away in east-central Turkey. In the southeast, Turkey has found some oil, and it may share in the zone of oilfields of which the famed Kirkuk reserve in nearby Iraq is a part. With its complex geology, Turkey has proven to have a variety of mineral deposits, which provide ample opportunity for future

development. Spatially, this development seems destined to reinforce the country's westward orientation. The Mediterranean coastal zone is already the economic focus of the country; of what remains, the western half is again the more advanced.

Iran

Iran constitutes another exception to the claim that this geographic realm is the Arab world. Iranians are not Arabs; nor are they adherents to the majority (Sunni) Muslim faith. Most Iranians are members of the Shi'ite group, who represent about one-seventh of all Islam, mainly in Iran and neighboring Iraq (see Fig. 6–2). Persia (as Iran was formerly called) was a kingdom as far back as 2500 years ago, but its royal succession has faltered repeatedly during the twentieth century. The most recent period of monarchy ended in 1979, when an Islamic fundamentalist revolution that had long been intensifying achieved the overthrow of the shah. Ironically, the year 1971 had witnessed the 2500th anniversary of Persia's first monarchy amid celebrations and scenes of royal splendor—and without indications of the upheaval that lay just ahead.

Iran is a country of mountains and deserts. The heart of the country is an upland, the Plateau of Iran, that lies surrounded by even higher mountains including the Zagros Mountains in the west, the Elburz Mountains along the southern shores of the Caspian Sea, and the mountains of the Khurasan region to the northeast. The Iranian Plateau, therefore, is actually a highland basin marked by salt flats and wide expanses of stone and sand. On the hillsides, where the topography wrests a modicum of moisture from the air, lie some fertile soils. Elsewhere, only oases break the arid monotony—oases

THE PROBLEM OF CYPRUS

In the northeastern corner of the Mediterranean Sea, much farther from Greece than from Turkey or Syria, lies Cyprus (population: 730,000). Since ancient times, this island has been predominantly Greek in its population, but in 1571 it was conquered by the Turks, under whose control it remained until 1878. Following the decay of the Ottoman state, the British soon took control of the island. By the time Britain was prepared to offer independence to much of its empire in the years following World War II, it had a problem in Cyprus because the Greek majority among the 600,000 population preferred *enosis*—union with Greece. It is not difficult to understand that the 20 percent of the Cypriot population that was Turkish wanted no such union.

By 1955, the dispute had reached the stage of violence because differences between Greeks and Turks are deep, bitter, and intense. It had become impossible to find a solution to a problem in which the residents of this island country think of themselves as Greeks or Turks first rather than Cypriots. Nonetheless, in 1960 Cyprus was granted independence under a complicated constitution designed to permit majority rule but with a guarantee of minority rights.

The fragile order finally broke down in 1974. As a civil war engulfed the island, Turkish armed forces intervened (29,000 soldiers were still present in 1990), and a major redistribution of population occurred. The northern 40 percent of Cyprus became the stronghold of more than 100,000 Turkish Cypriots; only a few thousand Greeks remained there. The rest of the island—south of the UN-patrolled "Green Line" that was demarcated across the country to divide the two ethnic communities—became the domain of the Greek majority, now including nearly 200,000 in refugee camps. As in Lebanon, partition looms as a very serious prospect; the fundamental differences between Greek and Turkish Cypriots are still unresolved, hatreds have deepened, and a form of de facto apartheid now exists on the island.

In 1983, again bringing the stalemated conflict to the attention of the world, the Turkish community (today comprising 18 percent of the population) seceded by declaring itself the new independent Turkish Republic of Northern Cyprus. The situation is especially difficult because of the many factions involved. The Turkish Cypriots, many of whom have expressed a desire for reintegration, are discrete from the Turkish nation and express a strong attachment to their island homeland. The Greek Cypriot majority is ideologically divided, and the crisis of 1974 began not as a Greek-Turkish conflict, but resulted from a coup by Greek Cypriot soldiers against the Greek-dominated government of the island. Internal factionalism and external involvement have made Cyprus a pawn of foreign powers—and a casualty of history.

that for countless centuries have been stopping places on the area's caravan routes. With so little usable land (such as the moist narrow ribbon along the Caspian Sea coast and the war-torn zone east of the Shatt-al-Arab around Abadan in the southwest) and with most of the people depending directly on agricultural or pastoral subsistence, it is noteworthy that Iran's population is as large as 58 million (and growing at an annual rate of 3.6 percent).

In ancient times, Persepolis in southern Iran was the focus of the powerful Persian kingdom. Then, as now, people clustered in and around the oases or depended on *qanats* (underground tunnels leading from the mountains) for their water supply, as Persepolis did. The focus of modern Iran, Tehran (9.8 million), lies far to the north on the southern slopes of the Elburz Mountains; it also remains partially dependent on the *qanat* system for its water, and rose from a caravan station to become the capital of a modernizing state.

Iran indeed modernized during the regime of Shah Muhammad Reza Pahlavi (1941–1979), although its myriad villages and nomadic communities are reminders that change does not come quickly or easily to a country as large and tradition-bound as Iran. Thus, the modernization process operated to intensify local and regional contrasts, contributing to the success of the revolution that was to come, an upheaval that had its roots in the Islamic traditions of the great majority of the people. Like Atatürk and his successors in Turkey, the shah sought to bring major reforms to Iran (a task greatly facilitated by the country's enormous oil income). Apart from the social changes brought by this "White Revolution," industrialization was introduced, agriculture was modernized, and a kind of domestic peace corps was established

FIGURE 6—29 The 1989 funeral of Ayatollah Khomeini, with a massive Tehran crowd of mourners in attendance. For a decade, this magnetic Shi'ite religious leader masterminded the fundamentalist transformation of Iran, a legacy that continues to prevail in the early 1990s.

to improve medical services and literacy in rural areas. But the shah's reign was sustained by the intervention of foreign interests, and his policies ran counter to the Islamic traditions of a multitude of his people. Muslim leaders opposed him, among them the exiled Ayatollah Khomeini. In time, this fanatical religious leader became the symbol of the Islamic revolution that exploded around his return in 1979 and continued to run its course beyond his death ten years later (Fig. 6–29).

Among the circumstances that brought Khomeini to power were the searing material inequities that the modernization effort had engendered. The contrasts between the advantaged elite in Tehran and

the poverty of urban slum dwellers, the haves in cities like Isfahan and Shiraz and the have-nots nearly everywhere else who remained mired in a web of debt and dependency—all this produced wide opposition to the shah's policies. Ruthless political repression, reports of torture, and a lack of avenues for the expression of alternative ideas also contributed to the breakdown of the order the shah's authoritarian rule had wrought.

The wealth generated by petroleum could not transform Iran in ways that might have staved off the revolution. Modernization remained but a veneer: in the villages away from Tehran's polluted air, the holy men continued to

dominate the lives of ordinary Iranians. As elsewhere in the Muslim world, urbanites, villagers, and nomads remained enmeshed in a web of production and profiteering, serfdom, and indebtedness that has always characterized traditional society here. This **ecological trilogy**, as Paul Ward English called it, is not unique to Iran: it is characteristic of much of this entire realm. It ties the people who live in cities and towns to villagers and nomads; the urbanites dominate because they have the money and own the land and the livestock. Thus, as English wrote:

Society is divided into three mutually dependent types of communities—the city, the village, and the tribe—each with a distinctive life-mode, each operating in a different setting, each contributing to the support of the other two sectors and thereby to the maintenance of total society. . . . The [wealthy and powerful] *city dwellers are principally engaged in collecting and processing raw materials from the hinterland—wool for carpets and shawls, vegetables and grain to feed the urban population, and nuts, dried fruit, hides, and spices for export. In return, the urbanites supply peasants and nomads with basic economic necessities such as sugar, tea, cloth, and metal goods as well as cultural imperatives such as religious leadership, entertainment, and a variety of services. This concept of* **urban dominance** *is basic to the idea of an interdependent ecological trilogy.*

Again, Iran's twentieth-century oil-based affluence did little to change these relationships. The peasants tilling the fields near their villages still worked on land that someone else owned. Neither were the nomadic herdspeople moving their flocks along centuries-old routes in search of grazing grounds freed from the tentacles of the city; most of the animals belonged to someone living in a distant town—far away, but in control of the herders' existence nonetheless. In the shah's modernizing of Iran, the new and the old were never brought into a harmony that would have averted the turbulence of the fundamentalist revolution and its aftermath.

Afghanistan

The easternmost country in the realm's northern tier—and the only landlocked state in the realm—is Afghanistan (Fig. 6–27). Its territory is compact in shape, except for a narrow but lengthy land extension in the northeast (the Wakhan Corridor) that has the

effect of adding a significant neighbor, China, to its other bordering states: the U.S.S.R., Iran, and Pakistan.

For centuries, Afghanistan has been a battleground for outsiders, including Turks and Indians, Tatars and Persians; but the country became a recognized political unit because two other competitors—the Russians and the British—agreed to guarantee its integrity. For a buffer state (see *box*, p. 526), Afghanistan's high mountains, remote valleys, and inaccessible frontiers were ideal. The nineteenth-century boundary created a country of 250,000 square miles (650,000 square kilometers) that today incorporates nearly 16 million people, about half of them Pathans (also called Pushtuns or Pakhtuns) and the remainder Uzbeks, Tajiks (Tadzhiks), Turkomans, Hazaras, and smaller ethnic groups. Thus, Afghanistan was hardly a nation-state, although virtually its entire population adheres to the Sunni form of Islam.

Even though Afghanistan does not possess oil reserves (some natural gas has been found in the north) and remains a country of herders, farmers, and nomads, it does have one coveted asset it cannot sell—its relative location. Southern Afghanistan lies only 250 miles (400 kilometers) from the Indian Ocean's Arabian Sea, not far from the mouth of the all-important Persian Gulf. Along the country's eastern flank lies a province of Pakistan that has displayed separatist tendencies (see pp. 486–487), a situation exacerbated by the presence of more than 3 million recent Afghan refugees. Afghanistan's western border with Iran does not reflect a clear break in the cultural landscape; Dari, a Persian language, has official status inside Afghanistan. And along Afghanistan's northern border lies the Soviet Union, which from 1979 to 1989 became the lat-

est of the country's foreign occupiers.

Although foreigners may attempt to erect puppet governments in the capital, Kabul (1.9 million), the fact remains that Afghanistan is one of the realm's least developed countries. Urbanization is still below 20 percent, nonmilitary communications are minimal, agriculture and pastoral subsistence remain the dominant livelihoods, and there is little national integration. The recent Soviet intervention supported an unpopular Marxist regime whose modernizing intentions were viewed with suspicion by millions of traditional and conservative Muslim Afghans. Following the Soviet pullout in 1989, that regime held shaky control, but like the Red Army before it, was unable to subdue the entrenched *mujahedin* insurgency.

Geographically, Afghanistan not only lies in a historic buffer zone, but is also a transitional region. It differs culturally from Iran and Pakistan; its economic geography is unlike either. On various grounds, Afghanistan may be viewed as part of each of the three world realms here juxtaposed (see Fig. I–19). A similar transitional situation exists on a wider scale all across the realm's southern periphery, and we now turn to that final regional component of North Africa/Southwest Asia.

The African Transition Zone

All across Africa south of the Sahara, from Mauritania and Senegal in the west to Ethiopia and Somalia in the east, the Islamic domain yields to the cultures of the Subsaharan Africa realm. However, it is a transition zone quite different from the northern tier. Turkey may be Europe-oriented, but it is a Muslim country; so is individualistic, non-Arabic Iran. But in Africa, the transition exhibits local variations in several countries. Northern Nigeria is Muslim, the south is not. Northern Sudan, including the core area between the White and the Blue Nile focused on the capital of Khartoum (1.9 million), is Muslim, but the south is converted to Christianity or pursues traditional animist religions. Ethiopia's heartland, organized around its capital of Addis Ababa (1.7 million), is the land of the Amharic people, who profess the Ethiopian Orthodox (Coptic Christian) faith. Its northern coastal province of Eritrea and its eastern region centered on Harar, however, are Muslim.

Life in the Transition Zone has never been easy. The heart of this region is the notorious *Sahel* (Arabic for "border"), where a multi-year drought in the 1970s—exacerbated by widespread overgrazing practices—starved perhaps 300,000 people and 5 million head of cattle. The Sahel itself is a 200-mile-wide (330-kilometer-wide) zone that stretches completely across Africa along the southern margin of the Sahara. This steppe environment is clearly discernible in Fig. I–10 as an east-west band of *BSh* climate; annual rainfall is highly variable here—ranging from 4 inches (10 centimeters) to 24 inches (60 centimeters)—making livestock-raising an always risky proposition that courts disaster if too many animals upset the delicate ecological balance (Fig. 6–31).

Sedentary agriculture is therefore rare (see Fig. 5–2), limited to those few Transition Zone locales where highlands such as in Ethiopia (see *box*) or *exotic rivers* (which flow through arid areas unable to generate their own streams) such as the Sudan Nile offer sufficient water supplies. But even here other environmental

VON THÜNEN IN THE UNDERDEVELOPED WORLD

The Von Thünen model—treated on pp. 74–76 and diagrammed in Fig. 1–10—can also be applied to many farming areas in the developing realms. Unlike the case of the United States (pp. 210–212), where contemporary Thünian effects are observed at the continental scale, the far more localized movement systems of the Third World provide a transportation setting much like that of Von Thünen's time in early-nineteenth-century Europe. Therefore, the agricultural lands surrounding large cities would be the likeliest places to find evidence of Thünian spatial organization.

Ronald Horvath discovered just such a pattern around Addis Ababa. The original **Von Thünen Isolated State** model showed a forest zone in the second land-use ring, from which the pre-industrial city drew wood for both construction and fuel. In his study, Horvath found striking evidence of such an inner wood-producing ring in the form of the continuous belt of eucalyptus forest that surrounded the Ethiopian capital (Fig. 6–30). The spatial configuration of this forest nicely conforms to Von Thünen's own empirical application of his model. Instead of a circular ring, a girdling wedge-shaped zone develops that reflects the greater accessibility to the urban marketplace along radial road corridors. Horvath also captured some interesting dynamics: the rapid outward expansion of the eucalyptus zone between 1957 and 1964 represents the improvement of surface transport in the Addis Ababa area, permitting the importing of wood from a greater distance (and the freeing up of land nearer the city to raise more vegetables—a leading activity of the innermost ring in the ideal Thünian scheme). Additional research has documented the existence of Thünian production patterns in the hinterlands of a number of other Third World cities, and this variation of the model may well apply to the vicinity of any major pre-industrial urban area.

hazards threaten: the worst locust infestation in a quarter-century broke out in 1988, presenting the possibility of sizable crop losses lasting well into the 1990s.

The African Transition Zone has not been a peaceful arena in recent years. Ethiopia's northern provinces of Eritrea and Tigre—home to 12 percent of the country's 52 million inhabitants—continue to fight a long war for independence against the Marxist Addis Ababa government. In neighboring Sudan (26 million),

a civil war has intensified as anti-Islamic forces in the south battle the Muslim rulers of the north to achieve greater regional autonomy and a bigger role in national development programs.

Yet another trouble spot is Sudan's western neighbor, sparsely populated Chad (5.1 million), where a civil war also pitted Muslim northerners against non-Islamic southerners. When expansionist Libya sent ground troops and air power to help the rebels push south in 1983, the Chad gov-

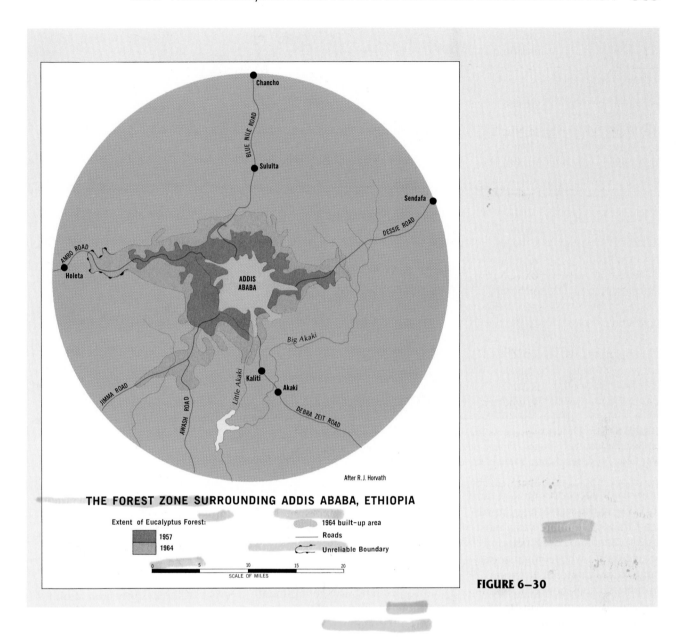

After R. J. Horvath

THE FOREST ZONE SURROUNDING ADDIS ABABA, ETHIOPIA

Extent of Eucalyptus Forest:

1957

1964

1964 built-up area

Roads

Unreliable Boundary

0   5   10   15   20
SCALE OF MILES

FIGURE 6–30

ernment persuaded France to return to its former colony with troops and planes, and the battle lines were subsequently shoved back to the north, producing a major Libyan defeat. All this regional conflict within many of the countries of the African Transition Zone came about in the first place because the European colonial invasion occurred after the diffusion of Islam had penetrated deep into West and East Africa, and the Europeans subjugated Muslims and non-Muslims alike.

The resulting problems are far from being resolved, and the African Horn is a good case in point. Here, Somalia (8.7 million), where Islam is the state religion, lies adjacent to Ethiopia's Muslim east. Periodically, Somalia aspires to a Greater Somali Union, and that has led to strong **irredentism** in the neighboring Ogaden region of eastern Ethiopia. This, in turn, reinforces long-standing local political instability because the Somalia-Ethiopia boundary has been in dispute since its superimposition

during the colonial heyday of the late nineteenth century. In the 1970s, Ethiopia tried to strengthen its control over the rebellious Somali east, where conditions were reaching the contention stage (see p. 276). As this campaign progressed, hundreds of thousands of Somalis were driven out of the Ogaden and have been forced to live in refugee camps in western Somalia.

All this turmoil, however, was overshadowed in the 1980s when a devastating drought struck the re-

FIGURE 6–31 Desert sand meets the shrinking pastureland in the heart of the Sahel in southern Niger. This is one of the most dramatic photographs ever taken of desertification caused by overgrazing.

gion, its effects concentrated most severely in the war zone of northern Ethiopia. Famine was almost instantaneous for the 8 million people there, whose plight soon came to international attention through the rock concerts (and especially the song ''We Are the World'') that were organized to raise millions of dollars and distribute emergency relief supplies. Although avoiding starvation was an immediate concern for much of the civilian population of Eritrea, Tigre, and adjoining Welo Province (who began a mass westward migration on foot toward sanctuary in Sudan), intervention by the Marxist Ethiopian government was what they feared above all else. By regulating the inflow of emergency food shipments, however, the government distributed these supplies only at towns that it controlled in rebel territory. This forced people to cross enemy

lines; many were presumed to be rebels and were imprisoned. Others faced an equally harsh fate—compulsory mass relocation.

Claiming that the three desiccated northern provinces were unfit for farming, the government has forced hundreds of thousands of peasants to migrate to southern and western Ethiopia, flying them by cargo plane and often separating families in the process. Their destinations were newly opened lands containing minimal facilities in an alien tropical climate, which presented major new problems of disease and poor soil fertility. The radical Marxist government now pursues an even more ambitious resettlement program in Ethiopia's eastern region. Called ''villagization,'' it has forcibly uprooted several million peasants from their scattered farmlands and placed them under tight control in centralized towns and villages in

an attempt to collectivize agriculture along the lines of the early Soviet model. As the peasantry became aware of this scheme, another exodus was triggered that has further aggravated the refugee crisis of the eastern Transition Zone. In this case, people fled northeast across the border into northern Somalia, adding significantly to the population of the already overcrowded refugee camps of that country. And on top of all this, in the late 1980s a fierce civil war broke out between northern and southern Somalia, which by 1990 had sent over 300,000 new Somali refugees scurrying across the Ethiopian border! Clearly, Muslim and non-Muslim coexistence in this southernmost region of the North Africa/Southwest Asia realm is a deteriorating proposition. That it has already failed in the African Horn and Sudan is beyond doubt.

PRONUNCIATION GUIDE

Abadan   (ahbah-*dahn*)

Abu Dhabi   (ah-boo-*dah*-bee)

Adana   (*ahd*-uh-nuh)

Addis Ababa   (adda-*sab*-uh-buh)

Aden   (*aid*-nn)

Aegean-Marmara [Lowlands]   (uh-*jee*-un *mah*-muh-rah)

Ahaggar   (uh-*hahg*-er)

Ajman   (adge-*mahn*)

Aleppo   (uh-*lep*-poh)

Algeria   (al-*jeery*-uh)

Algiers   (al-*jeerz*)

Al Jawf   (ahl-*jowff*)

Allah   (*ahl*-ah)

Amharic   (am-*hah*-rick)

Amman   (uh-*mahn*)

Ankara   (*ang*-kuh-ruh)

Antalya   (ant-ull-*yah*)

Aqaba   (*ah*-kuh-buh)

Archimedean   (arka-*meedy*-yun)

Ashkelon   (*osh*-kuh-lahn)

Ashkenazim [Jews]   (osh-kuh-*nah*-zim)

Assad, Hafez al-   (uh-*sahd*, huh-*feeze* ull)

Aswan   (as-*swahn*)

Asyut   (ah-see-*oot*)

Atatürk   [see Kemal Atatürk]

Bahrain   (bah-*rain*)

Bedouin   (*beh*-dooh-in)

Beirut   (bay-*root*)

Bekaa   (buh-*kah*)

Benghazi   (ben-*gah*-zee)

Bidonvilles   (*bee*-daw-veal)

Borneo   (*bore*-nee-oh)

Bosporus   (*bahss*-puh-russ)

Bourguiba, Habib   (boor-*ghee*-buh, huh-*beeb*)

Cairo   (*kye*-roh)

Caliphate   (*kalla*-fate)

Constantinople   (kon-stant-uh-*nople*)

Cretans   (*kreet*-tunz)

Cyprus   (*sye*-pruss)

Cyrenaica   (sear-uh-*nay*-icka)

Damietta   (dam-ee-*etta*)

Dari   (*dah*-ree)

Dhahran   (dah-*rahn*)

Djibouti   (juh-*boody*)

Druze   (*drooze*)

Dubai   (doo-*bye*)

Eilat   (*ay*-laht)

Elam   (*ee*-lum)

Elburz   (el-*boorz*)

Elyria   (uh-*leary*-uh)

Emirate   (*emma*-rate)

Enosis   (ee-*noh*-sis)

Eritrea   (erra-*tray*-uh)

Ethiopia   (eeth-ee-*oh*-pea-uh)

Eucalyptus   (yuke-uh-*lip*-tuss)

Euphrates   (yoo-*frate*-eeze)

Farouk   (fuh-*roohk*)

Fellaheen   (fella-*heen*)

Fujairah   (fuh-*jye*-rah)

Gadhafi, Muammar   (guh-*dahfi*, *mooh*-uh-mar)

Ganges   (*gan*-jeeze)

Gaza   (*gah*-zuh)

Gezira   (juh-*zeer*-uh)

Golan   (goh-*lahn*)

Hägerstrand, Torsten   (*hayger*-strand, *tor*-stun)

Haifa   (*hye*-fah)

Hamitic   (ham-*mittick*)

Harar   (*har*-rar)

Hazaras   (huh-*zar*-ahss)

Herodotus   (heh-*rodda*-tuss)

Hierarchical   (hire-*ark*-uh-kull)

Huang He   (*hwahng*-huh)

Ibn Saud   (ib'n-sah-*ood*)

Intifada   (intee-*fah*-duh)

Iran   (ih-*ran*/ih-*rahn*)

Iranian   (ih-*rain*-ee-un)

Iraq   (ih-*rahk*)

Isfahan   (iz-fuh-*hahn*)

Iskenderun   (iz-ken-duh-*roon*)

Islam   (iss-*lahm*)

Israel   (*iz*-rail)

Istanbul   (iss-tum-*bool*)

Jabal   (*jab*-ull)

Jazirah   (juh-*zeer*-uh)

Jerusalem   (juh-*rooh*-suh-lum)

Jubail   (joo-*bile*)

Judaism   (*joody*-ism)

Kabul   (*kah*-bull)

Kemal Atatürk, Mustafa   (keh-*mahl* atta-tyoork, moosta-*fah*)

Khartoum   (kar-*toom*)

Khomeini, Ayatollah   (hoh-*may*-nee, eye-uh-*toh*-luh)

Khurasan   (koor-uh-*sahn*)

Kibbutz   (kih-*boots*)*

Kirkuk   (keer-*cook*)

Kufra   (*kooh*-fruh)

Kurdistan   (*kerd*-duh-stahn)

Kurds   (*kerdz*)

Kuwait   (kuh-*wait*)

Levant   (luh-*vahnt*)

Maghreb   (*mahg*-rahb)

Mali   (*mah*-lee)

Mauritania   (maw-ruh-*taney*-uh)

Medina   (muh-*deena*)

Mesopotamia   (messo-puh-*tay*-mee-uh)

Monogamy   (muh-*nog*-ah-mee)

Mosque   (*mosk*)

Muhammad   (mooh-*hah*-mid)

Mujahedin   (mooh-*zhah*-heh-deen)

Muslim   (*muzz*-lim)

Nasser, Gamal Abdel   (*nass*-er, guh-*mahl* ab-dul)

Niger   (nee-*zhair*)

Nigeria   (nye-*jeery*-uh)

Ogaden   (oh-gah-*den*)

Oman   (oh-*mahn*)

Orontes   (aw-*rahn*-teeze)

Pahlavi, Shah Muhammad Reza   (puh-*lah*-vee, shah mooh-*hah*-mid ray-zuh)

*Double "o" pronounced as in "book."

Pakhtuns (puck-*toonz*)
Palestine (*pal*-uh-stine)
Pashtuns [see Pushtuns]
Pathans (puh-*tahnz*)
Persepolis (per-*sepp*-uh-luss)
Pharaonic (fair-ray-*onnick*)
Phoenicians (fuh-*nee*-shunz)
Polygamy (puh-*lig*-ah-mee)
Port Said (port-sah-*eed*)
Pushtuns (*pah*-shtoonz)
Qanats (*kah*-nahts)
Qatar (*kotter*)
Ras al Khaimah (rahss ahl *kye*-muh)
Riyadh (ree-*ahd*)
Rosetta (roh-*zetta*)
Rub al Khali (rube ahl *kah*-lee)
Sahara (suh-*harra*)
Sahel (suh-*hell*)
Saudi Arabia (*sau*-dee uh-*ray*-bee-uh)

Semitic (seh-*mittick*)
Sephardic [Jews] (suh-*far*-dik)
Sharjah (*shar*-juh)
Shatt-al-Arab (shot-ahl-uh-*rahb*)
Sheikdom (*shake*-dum)
Shiah/Shi'ite [Muslims] (*shee*-uh/*shee*-ite)
Shiraz (shih-*rahz*)
Sinai (*sye*-nye)
Somalia (suh-*mahl*-yah)
Suez (*sooh*-ez)
Sumer (*sooh*-mer)
Sunni [Muslims] (*soo*-nee)
Syria (*seary*-uh)
Tajiks (*tudge*-ix)
Tehran (tay-uh-*rahn*)
Tel Aviv-Jaffa (tella-*veeve joff*-uh)
Thebes (*theebz*)
Tigre (*tee*-gray)
Tigris (*tye*-gruss)

Tripoli (*trippa*-lee)
Tripolitania (trip-olla-*taney*-yuh)
Tuareg (*twah*-reg)
Tunis (*too*-niss)
Tunisia (too-*nee*-zhah)
Umm al-Qaiwain (*oom* ahl-kye-*wine*)
Ural-Altaic (*yoor*-ull al-*tay*-ick)
Uzbeks (*ooze*-bex)
Von Thünen (fon-*too*-nun)
Wakhan (wah-*kahn*)
Wei (*way*)
Welo (*way*-loh)
Yanbu (*yan*-boo)
Yemen (*yemmon*)
Zagros (*zah*-gruss)
Zionism (*zye*-un-ism)
Zonguldak (zawngle-*dahk*)

REFERENCES AND FURTHER READINGS

(BULLETS [●] DENOTE BASIC INTRO-DUCTORY WORKS ON THE REALM OR SYSTEMATIC ESSAY TOPIC)

Ahmed, Akbar S. *Discovering Islam: Making Sense of Muslim History and Society* (Boston: Routledge & Kegan Paul, 1987).

al Fārūqi, Ismail R. & al Fārūqi, Lois L. *The Cultural Atlas of Islam* (New York: Macmillan, 1986).

Beaumont, Peter. *Environmental Management and Development in Drylands* (London & New York: Routledge, 1989).

Beaumont, Peter & McLachlan, Keith, eds. *Agricultural Development in the Middle East* (New York: John Wiley & Sons, 1986).

● Beaumont, Peter et al. *The Middle East: A Geographical Study* (New York: Wiley/Halsted, 2 rev. ed., 1988).

Blake, Gerald H. & Lawless, Richard I., eds. *The Changing Middle Eastern City* (Totowa, N.J.: Barnes & Noble Books, 1980).

Blake, Gerald H. & Schofield, R.N., eds. *Boundaries and State Territory in the Middle East and North Africa* (Cambridge, U.K.: Menas Press, 1987).

● Blake, Gerald H. et al., eds. *The Cambridge Atlas of the Middle East and North Africa* (New York: Cambridge University Press, 1988).

Brawer, Moshe. *Atlas of the Middle East* (New York: Macmillan, 1988).

Brown, Lawrence A. *Innovation Diffusion: A New Perspective* (London & New York: Methuen, 1981).

Campbell, James B. *Introduction to Remote Sensing* (New York: Guilford Press, 1987).

Chapman, Keith. *People, Pattern and Process: An Introduction to Human Geography* (New York: Wiley/Halsted, 1979). Diagram adapted from p. 141.

Cloudsley-Thompson, J.L., ed. *Sahara Desert* (Oxford, U.K.: Pergamon Press, 1984).

Cohen, Yehoshua S. *Diffusion of an Innovation in an Urban System* (Chicago: University of Chicago, Department of Geography, Research

Paper No. 140, 1972). Map adapted from p. 40.

● Cook, Earl F. *Man, Energy, Society* (San Francisco: W.H. Freeman, 1976).

● Cressey, George B. *Crossroads: Land and Life in Southwest Asia* (Philadelphia: J.B. Lippincott, 1960).

Dohrs, Fred E. & Sommers, Lawrence M., eds. *Cultural Geography: Selected Readings* (New York: Thomas Y. Crowell, 1967).

● Drysdale, Alasdair & Blake, Gerald H. *The Middle East and North Africa: A Political Geography* (New York: Oxford University Press, 1985).

English, Paul Ward. ''Urbanites, Peasants and Nomads: The Middle Eastern Ecological Trilogy,'' *Journal of Geography*, 66 (1967): 54–59. Quotation taken from pp. 54–55.

● Fisher, William B. *The Middle East: A Physical, Social and Regional Geography* (London & New York: Methuen, 7 rev. ed., 1978).

Fromkin, David. *A Peace to End All Peace: Creating the Modern Middle East, 1914–1922* (New York: Henry Holt, 1989).

- Gould, Peter R. *Spatial Diffusion* (Washington: Association of American Geographers, Commission on College Geography, Resource Paper No. 4, 1969).

  Heathcote, Ronald L. *The Arid Lands: Their Use and Abuse* (London & New York: Longman, 1983).

- Held, Colbert C. *Middle East Patterns: Places, Peoples, and Politics* (Boulder, Colo.: Westview Press, 1989).

  Horvath, Ronald J. "Von Thünen's Isolated State and the Area Around Addis Ababa, Ethiopia," *Annals of the Association of American Geographers*, 59 (1969): 308–323.

  Hunter, John M. & Young, Johnathan C. "Diffusion of Influenza in England and Wales," *Annals of the Association of American Geographers*, 61 (1971): 637–653. Map adapted from p. 645.

  Lamb, David. *The Arabs: Journeys Beyond the Mirage* (New York: Random House, 1987).

- Longrigg, Stephen H. *The Middle East: A Social Geography* (Chicago: Aldine, 2 rev. ed., 1970).

  Mikesell, Marvin W. "Tradition and Innovation in Cultural Geography," *Annals of the Association of American Geographers*, 68 (1978): 1–16.

  Miller, E. Willard & Miller, Ruby M. *The Middle East (Southwest Asia): A Bibliography on the Third World* (Monticello, Ill.: Vance Bibliographies, No. P-938, 1982).

  Miller, E. Willard & Miller, Ruby M. *Northern and Western Africa: A Bibliography on the Third World* (Monticello, Ill.: Vance Bibliographies, No. P-818, 1981).

- Mostyn, Trevor, ed. *The Cambridge Encyclopedia of the Middle East and North Africa* (New York: Cambridge University Press, 1988).

  Peters, Joan. *From Time Immemorial: The Origins of the Arab-Israeli Conflict Over Palestine* (New York: Harper & Row, 1984).

  Prescott, J.R.V. *Political Frontiers and Boundaries* (Winchester, Mass.: Allen & Unwin, 1987).

  Pyle, Gerald F. "The Diffusion of Cholera in the United States in the Nineteenth Century," *Geographical Analysis*, 1 (1969): 59–75. Maps adapted from pp. 63, 72.

  Rahman, Mushtaqur, ed. *Muslim World: Geography and Development* (Lanham, Md.: University Press of America, 1987).

  Rogers, Everett M. *Diffusion of Innovations* (New York: Free Press, 3 rev. ed., 1983).

  Sauer, Carl O. *Agricultural Origins and Dispersals* (New York: American Geographical Society, 1952).

- Spencer, Joseph E. "The Growth of Cultural Geography," *American Behavioral Scientist*, 22 (1978): 79–92. Quotation taken from p. 79.

  Starr, Joyce R. & Stoll, Daniel C., eds. *The Politics of Scarcity: Water in the Middle East* (Boulder, Colo.: Westview Press, 1988).

  Strahler, Arthur N. & Strahler, Alan H. *Modern Physical Geography* (New York: John Wiley & Sons, 3 rev. ed., 1987). Definition on p. 68.

- Wagner, Philip L. & Mikesell, Marvin W., eds. *Readings in Cultural Geography* (Chicago: University of Chicago Press, 1962).

  Wagstaff, J. Malcolm, ed. *Landscape and Culture: Geographical and Archaeological Perspectives* (New York: Basil Blackwell, 1987).

  Wright, Robin B. *In the Name of God: The Khomeini Decade* (New York: Simon & Schuster, 1989).

  Wright, Robin B. *Sacred Rage: The Crusade of Modern Islam* (New York: Linden Press/Simon & Schuster, 1985).

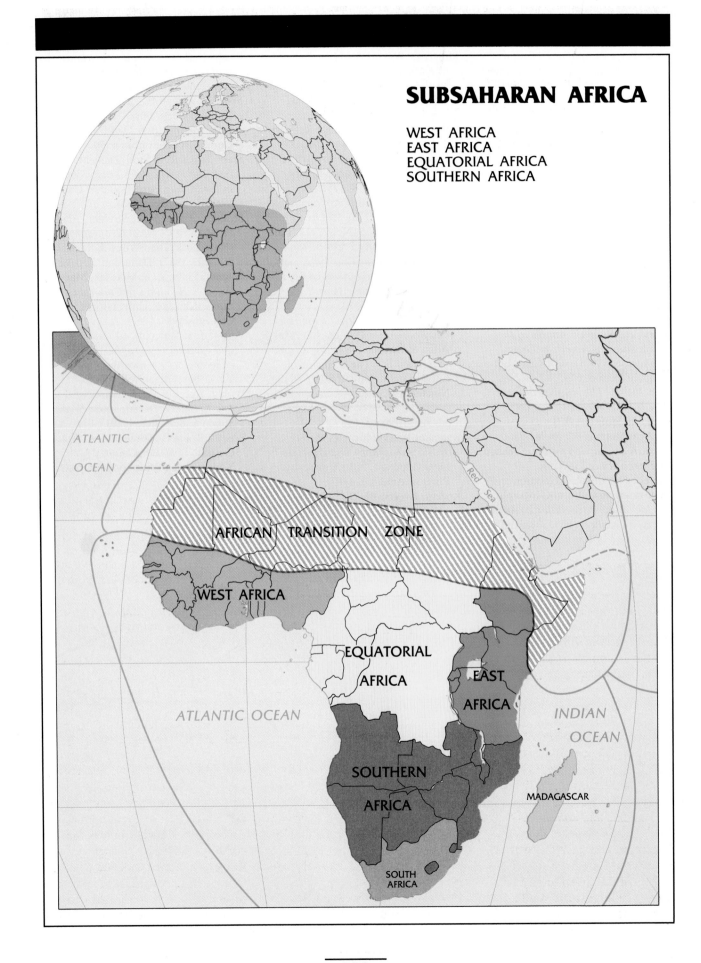

SUBSAHARAN AFRICA

WEST AFRICA
EAST AFRICA
EQUATORIAL AFRICA
SOUTHERN AFRICA

ATLANTIC

OCEAN

Red Sea

AFRICAN   TRANSITION   ZONE

WEST AFRICA

EQUATORIAL

AFRICA

EAST

AFRICA

ATLANTIC OCEAN

INDIAN

OCEAN

SOUTHERN

AFRICA

MADAGASCAR

SOUTH
AFRICA

SUBSAHARAN AFRICA: REALM OF REVERSALS

The African continent lies at the heart of the earth's landmasses. Situated astride the equator, Africa looks westward to the Americas, northward to Europe and Asia, eastward to Australia, and southward to Antarctica. Flanked by the Atlantic and Indian Oceans, Africa stretches from the Mediterranean Sea in the north to the subpolar West Wind Drift current of the Southern Ocean in the south. Africa contains the world's greatest desert and some of its mightiest rivers, snow-capped mountains and lake-filled valleys, parched steppes and dripping rainforests. It is a landmass of escarpments and plateaus, where altitude moderates the tropical heat. Africa may have been the source of humanity, the scene of the emergence of the first communities. Africa is also a realm of great problems. Ravaging diseases afflict its peoples. Environmental deterioration is widespread. In some areas, population growth is the highest in the world, and the food supply cannot keep up. After slave-raiding and colonial exploitation, Africa now suffers from numerous political conflicts that have made millions homeless.

The African geographic realm extends from near the southern edge of the Sahara southward

IDEAS AND CONCEPTS

medical geography

geography of language

relative location (2)

rift valleys

continental drift

endemic diseases

green revolution

regional complementarity (2)

colonial spatial organization

periodic markets

sequent occupance

population spiral

landlocked state

to the Cape of Good Hope. Of course, as we know, the Saharan boundary is not sharp; languages, religions, and ways of life change across a wide transition zone. But the Africa of Algeria and Egypt stands a world apart from the Africa of Nigeria and Zaïre. Our focus in this chapter is on the latter—Subsaharan Africa.

REGIONS AND BOUNDARIES

We have seen that geographic realms and regions are parts of a global system of spatial classification and are defined to help us understand our complex world. Few of the realm boundaries shown in Fig. I–19 are as difficult to establish as is this northern limit of the African realm. And yet, when we examine such maps as those of world religions, African languages, and human racial groups, the line shown on Fig. I–19 appears on every one. Islam, as we observed earlier, penetrated Africa by various routes, but in Subsaharan Africa it loses the dominance it enjoys in North African countries (Fig. 6–2). Note how the realm boundary crosses Sudan, the country immediately south of Egypt: northern Sudan is overwhelmingly Islamic, but the south is almost entirely non-Islamic. On the language map (Fig. 7–1), note that northern Sudan speaks Arabic, whereas the south speaks Sudanic languages. And observe the transition from Caucasian to African racial groups on Fig. 7–2. Add to this the contrasting ways of life between north

S Y S T E M A T I C   E S S A Y

MEDICAL GEOGRAPHY*

Medical geography is that branch of the discipline concerned with people's health. Practitioners of medical geography study the spatial aspects of health and illness. Where are diseases found? How do they spread? Are there specific kinds of environments in which certain illnesses are located? How are disease and environment related? How do the changes that societies undergo affect the health of their populations? Where do people go when they seek health care? How does the location of a health-care facility affect the opportunity to obtain care? These and many other related questions are of interest to medical geographers.

Insights into disease occurrence or disease origin can sometimes be gained simply by mapping the distribution of that disease. In a famous early case, Dr. John Snow was able to relate the deadly effects of cholera in 1854 to a contaminated water source by mapping the locations of deaths in a portion of London (Fig. I–24, p. 54); the spatial pattern clearly showed that many deaths occurred within a few blocks of a pump used as a major source of drinking water. Although we now know much more about this disease and its causes, the medical community in the mid-nineteenth century was still arguing whether such invisible things as "germs" actually existed! It took a map to provide evidence that, in this case at least, something about an apparently healthful water supply was fatal to many who drank from it.

This example illustrates a major concern among medical geogra-

phers with *disease ecology*. Disease ecology studies the manner and consequences of interaction between the environment and the causes of morbidity and mortality. *Morbidity* is the condition of illness, whereas *mortality* is the occurrence of death. By "environment" we mean here the varied aspects of nature and society that bear on people's lives; geographers have a long tradition of dealing with the full array of physical and human environmental factors. Medical geographers study in an integrated manner the complex interaction of a disease *agent* (the pathogen or illness-causing organism itself), a possible disease *vector* (the intermediate transmitter of the pathogen), the physical and social environments, and even the cultural behavior patterns of individuals potentially at risk.

Geographers map the distributions of disease (as did Dr. Snow) in order to identify the spatial context in which the illness occurs and to monitor changes in these patterns. For instance, even when a primary agent is known and can be treated, as in the case of rubella (German measles), outbreaks continue to occur. Medical geographers studied the location of recurrent appearances of rubella and suggested that one of the secondary causes might well be the lack of information by the population at risk about the availability of vaccines. More difficult is the study of *degenerative* diseases (such as heart disease, cancer, and stroke), because of the multiplicity of factors that lie behind these illnesses. The contributions of medical geographers—with their ecological, integrated, and descriptive approaches—are especially

useful for suggesting hypotheses to other medical researchers who deal directly with the causes of these diseases and their effects on individuals.

Analysis of the diffusion of illnesses has also been given a great deal of attention by medical geographers. The notion of *contagious diffusion* (see p. 348) is drawn directly from our experience with diseases. In Africa, for example, the gradual spread of cerebrospinal meningitis, from its early appearance in Ethiopia in 1927 across the Sahel to Senegal in 1941, was associated with the main east-west transport routes traversing the continent. Just as important, however, was the more rapid spread that occurred during the dry seasons, when contact at water sources was at its highest.

In a separate study of the diffusion of cholera in Africa during the 1970s, Robert Stock identified at least four distinct route patterns in the complex sequence of cholera diffusion that affected much of West and East Africa. Each of the four—coastal, riverine, urban-hierarchical, and radial contact diffusion—was identified by a careful mapping of the direction and timing of the disease's spread. Work such as Stock's is especially useful for those attempting to prevent disease diffusion; policies designed to interrupt one type of disease spread may be unsuccessful if applied in areas where other route patterns are more important.

Not all studies of disease ecology by medical geographers suggest natural environmental causes for disease occurrence. When studying cancer of the esophagus in Malawi and Zambia during the 1960s and 1970s, Neil McGlashan

*Contributed by Stephen S. Birdsall

noted a higher rate among people living in the eastern part of the region. After mapping a number of physical and cultural features and considering a variety of explanations, he argued that the distribution of this form of cancer was matched by the distribution of a particular form of locally distilled spirit. Therefore, the argument went, the cancer was related to either the distillation process or the materials used in this particular beverage. The actual cause of the cancer remains unknown, but McGlashan's early association of several mapped patterns provided a focus for subsequent medical research.

There are times, too, when studies focus on secondary patterns of activity as the source of the illness. During the 1970s, when most gasoline produced in the United States contained a lead additive to boost octane, a number of researchers examined the possible relationship between this fuel additive and patterns of lead poisoning among children. One study, for example, found a much higher level of lead in the blood of children who lived within 300 feet (90 meters) of a main road than among those who lived at greater distances. Another study found that lead levels in the blood of youthful residents of a northern U.S. city varied seasonally: levels were significantly higher during the summer months when children spent more time outdoors.

If studies of disease ecology are one major concern for medical geographers, a second involves the delivery of health care. At a very basic level, equitable health-care delivery is made difficult by the uneven spatial distribution of those who seek care and those who provide it. If people live close

to a hospital or clinic, they are more likely to use those facilities when illness strikes—or so one would expect. But how far will an individual travel to obtain care? The answer, studies show, depends on the individual's physical characteristics (age, sex), social context (marital status, proximity to family or neighbors), economic resources (ability to pay for travel or service), attitude toward "illness" (Is a doctor necessary?), barriers to treatment (racial or religious discrimination), and much more. An answer to the question "Where does one go to obtain health care?" is not obvious. A great deal of research has been done to identify the importance of each variable in terms of its effect on the health care that is sought or obtained.

In some countries such as the United States, doctors exercise considerable personal choice in deciding where they will practice medicine. Some researchers have pointed to the heavy concentration of doctors in urban areas—more specifically in the more affluent sections of metropolitan areas—as an example of inequitability in health care; thus, patients least able to afford it must bear high travel costs or do without the care. In other countries, the government may regulate the distribution of health care to some degree, but many parts of the world have too few medical professionals for the population's need. In the early 1980s, there were more than 12,000 people per physician in Nigeria, and about 100,000 people per physician in Ethiopia. These figures compared with 510 people per physician in the United States.

Related to the study of health-care delivery, medical geographers are also involved at the planning stage. One of the major research

efforts by geographers has been to develop methods of determining where activities *should* be located under ideal conditions. Planners seeking to identify the optimum locations for hospitals, clinics, medical offices, and even sites for emergency medical service centers have used such techniques of medical geography. These so-called "location allocation" models have been developed primarily for the developed world; elsewhere, questions about the number of practitioners and the available alternatives to "Western" medicine may be more important than their spatial distribution.

Different types of questions are pressing in low-income countries. How does economic development affect the health of a population? Although the ability to pay for health care may improve on the average, new diseases may also be introduced or spread by development projects. Irrigation schemes, for example, have been found to provide new habitats for disease vectors and spread the incidence of malaria and schistosomiasis. The labor migration encouraged by urban and industrial growth has been shown to spread contagious diseases such as hepatitis when workers return home after contracting the illness, or they may bring it to the urban center when they arrive looking for work.

Clearly, there are no easy or obvious paths to the resolution of health-care questions, however vital that resolution may be. Medical geography makes its contribution to what must be a truly multidisciplinary approach through the identification and analysis of spatial patterns, locational associations, and the full environmental contexts within which morbidity and mortality occur.

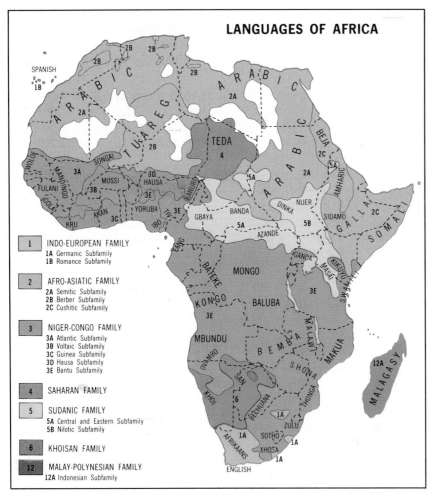

FIGURE 7-1

heart of Ethiopia and then turns eastward to separate Somalia from Kenya.

One of the most interesting aspects of the Subsaharan Africa realm is its **geography of languages**. Here, in an area about one-seventh of the inhabited world, is spoken one-third of all languages by a population comprising less than one-tenth of all humankind. This is one of Africa's most distinct regional properties, part of the richness of its mosaic of cultures—and one of its problems in modern times. In the classification of languages, we use terms employed in biology so that languages thought to have a shared but possibly distant origin are grouped into a *family*; when their relationship is closer, they belong together in a *subfamily*. The Niger-Congo family of languages (Category 3 on the map), therefore, contains five subfamilies, and the Bantu subfamily (3E) is the most extensive of these (Fig. 7-1). Scholars studying African languages do not yet agree on the regional distribution of African language clusters. It is agreed that the Khoisan family (6) represents the oldest surviving African languages, spoken over a far larger area of the continent before the Niger-Congo diffusion. It is also evident that Madagascar's languages belong to a non-African, Malay-Polynesian family (12), revealing the Southeast Asian origins of a large sector of that island's population. Afrikaans (1A) is an Indo-European language, a derivative of Dutch spoken by a majority of the 7 million whites living in South Africa, the realm's southernmost state.

African Languages

The linguistic evidence is the strongest of all, not only in Sudan, but across all of Africa. This certainly has validity west of Sudan where, as Fig. 7-1 shows, the Afro-Asiatic languages (such as widely used Arabic) give way to the Niger-Congo languages spoken throughout most of the heart of Africa. Crossing this linguistic frontier is a vivid geographic experience. Quite suddenly, the *lingua franca* of Arabic ceases to be useful; a desert-margin trade language,

Hausa, replaces it. But where Hausa fades out, the intricate linguistic mosaic of Africa takes over—dozens of languages in a single country, hundreds in one region, as many as a thousand in the realm as a whole. East of Sudan, the linguistic basis for the realm's limit is less evident. Amharic and Galla, spoken in Ethiopia, are quite closely related to Arabic and are part of the realm where Afro-Asiatic languages prevail. But the peoples of the Ethiopian Highlands are not Muslims; Islam dominates only north and east of this historic Coptic Christian area (Fig. 6-2). Based on the religious criterion and on traditions and ways of life, the realm's boundary bends southward to include the

Racial Patterns

The concept of race remains under constant debate among anthropologists and other scholars. Attempts

TEN MAJOR GEOGRAPHIC QUALITIES OF AFRICA

1. The physical geography of Africa is dominated by the continent's plateau character, variable rainfall, soils of low fertility, and persistent environmental problems in farming.

2. The majority of Africa's peoples remain dependent on farming for their livelihood. Urbanization is accelerating, but most countries' populations remain below 40 percent urban.

3. The people of Africa continue to face a high incidence of disease, including malaria, sleeping sickness, and river blindness.

4. Most of Africa's political boundaries were drawn during the colonial period without regard for the human and physical geography of the areas they divided. This has caused numerous problems.

5. Considerable economic development has occurred in many scattered areas of Africa, but much of the realm's population continues to have little access to the goods and services of the world economy.

6. The realm is rich in raw materials vital to industrialized countries.

7. Patterns of raw-material exploitation and export routes set up during the colonial period still prevail in most of Subsaharan Africa. Interregional connections are poor.

8. Africa has increasingly been drawn into the competition and conflict between the world's major powers. The continent contains about one-third of the world's refugee population.

9. Africa's population growth rate is by far the highest of any continent's in spite of a difficult agricultural environment, numerous hazards and diseases, and periodic food shortages. Some of the best land is used to produce such cash crops as coffee, tea, cocoa, and cotton for sale overseas.

10. Even though post-independence dislocations, civil wars, and massive losses of life have plagued some parts of Africa, other areas have shown relative stability, cohesion, and economic growth.

to define race on purely genetic bases have not proven satisfactory. Race, like culture, is a complex and multifaceted concept. In addition, race is frequently misused in everyday thinking as a substitute for preconceptions about social potential. Categories of individual physical characteristics are erroneously assumed to represent visible indicators of behavior. A great deal of evidence to the contrary notwithstanding, this socialized approach to the concept of race also continues to create a great deal of disagreement. In spite of all these difficulties, it is useful to treat human races as broad categories of genetically defined features and to compare

Fig. 7–2 with Fig. I–19. The comparison reveals the coincidence between the African racial group and the realm as defined by other criteria.

Separated by the Sahara from lands to the north, and by vast oceans from lands to the west and east, Africa is an almost insular continent. External influences reached Africa more by sea than over land; even Islam came to East Africa by Persian and Arab boats. Europe's colonial powers came first to West Africa's coasts, then rounded the southern Cape and advanced along the eastern shores in search of adequate harbors. No Tatar hordes invaded from afar; no Mongol armies swept across the countryside. Until Islam penetrated from the north, and until European colonizers pushed inland from their coastal stations, black Africa lay isolated. Only its northernmost peoples interacted regularly with North African societies, and even these contacts were limited by the capacities of desert caravans.

ENVIRONMENTAL BASES

Africa has an unusual location. No other landmass is so squarely positioned astride the equator, reaching almost as far to the north as to the south; this location has much to do with Africa's vegetation, soils, agricultural potential, and population distribution. Africa also is a very large landmass, containing about one-fifth of all the earth's land surface. It is 4800 miles (7700 kilometers) from the north coast of Tunisia to the southern coast of South Africa, and 4500 miles (7200 kilometers) from coastal Senegal in the extreme western *Bulge* of Africa to the tip of the *Horn* in easternmost Somalia. This size, too, has its environmental and human implications. Much of Africa is far from marine sources of moisture. Many of Africa's peoples live remote from routes to the outside world.

Physiography

Not only is Africa's **relative location** unusual—its physical geography is unique. Take, for example, the distribution of the world's linear mountain ranges. Every major landmass has at least one mountainous backbone: South America's Andes, North America's Rocky Mountains, Europe's Alps, and Asia's Himalayas. Yet Africa, covering one-fifth of the land surface of the earth, has nothing comparable. The Atlas Mountains of the far north occupy a mere corner of the landmass, and the Cape Ranges of the far south are not of continental dimensions. And where Africa does have high mountains, as in Ethiopia and South Africa, these are really deeply eroded plateaus—or, as in East Africa, high snowcapped volcanoes (Fig. 7–3). Missing in Africa are those elongated, parallel ranges of the Andes or Alps.

Lakes and Rift Valleys

This discovery should stimulate us to look more closely at the physiographic map of Africa (Fig. 7–4). What else is unusual about Africa's landscapes? As the map shows, Africa has a set of great lakes, concentrated in the east-central portion of the continent.

With the single exception of Lake Victoria, these lakes are re-

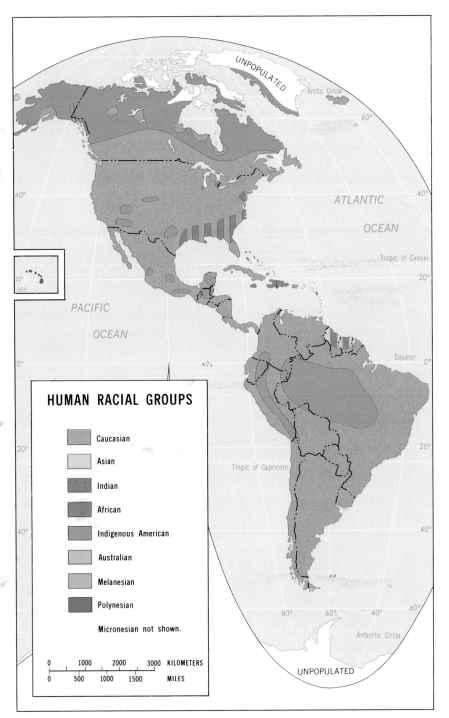

HUMAN RACIAL GROUPS

- Caucasian
- Asian
- Indian
- African
- Indigenous American
- Australian
- Melanesian
- Polynesian

Micronesian not shown.

0   1000   2000   3000   KILOMETERS
0   500   1000   1500   MILES

FIGURE 7–2

markably elongated, from Lake Malawi in the south to Lake Turkana in the north. What causes this elongation and the persistent north-south alignment that can be observed in these lakes? The lakes occupy portions of deep trenches cutting through the East African plateau, trenches that can be seen to extend well beyond the lakes themselves. Northeast of Lake Turkana, such a trench cuts the

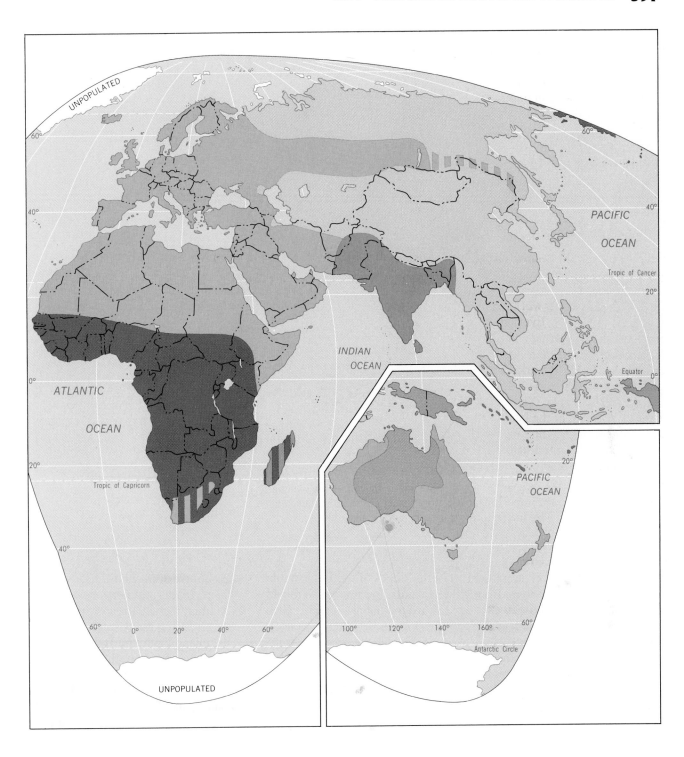

Ethiopian Highlands into two sections, and the entire Red Sea looks
much like a northward continuation of it. On both sides of Lake
Victoria, smaller lakes lie in similar trenches, of which the western

one runs into Lake Tanganyika
and the eastern one extends completely across Kenya, Tanzania,
and Malawi (Fig. 7–4). The technical term for these trenches is **rift
valleys**. As the name implies, they

are formed when huge parallel
cracks or faults appear in the
earth's crust, and the strips of
crust between them sink or are
pushed down to form great linear
valleys (Fig. 7–5). Altogether,

FIGURE 7–3 The twin volcanic peaks of Kilimanjaro rise above the plateau surface of East Africa near the Tanzania-Kenya border. Kibo, the taller cone (elevation 19,340 feet/5895 meters), carries permanent ice and snow.

these rift valleys stretch more than 6000 miles (9600 kilometers) from the north end of the Red Sea to Swaziland in Southern Africa. In general, the rifts from Lake Turkana southward are between 20 and 60 miles (30 and 90 kilometers) wide, and the walls, sometimes sheer and sometimes step-like, are well defined.

River Courses

Next, our attention is drawn to Africa's unusual river systems (Fig. 7–4). Africa has several great rivers, with the Nile and Zaïre (or Congo) ranking among the most noteworthy in the world. The *Niger* rises in the far west of Africa,

on the slopes of the Futa Jallon Highlands, but first flows inland toward the Sahara Desert. Then, after forming an interior delta, it suddenly elbows southeastward, leaves the desert area, plunges over falls as it cuts through the plateau area of Nigeria, and creates another large delta at its mouth. The *Zaïre* (*Congo*) River begins as the Lualaba River on the Zaïre-Zambia boundary; for some distance, it actually flows northeast before turning north, then west and southwest, finally cutting through the Crystal Mountains to reach the ocean. Note that the upper courses of these first two rivers appear quite unrelated to the continent's coasts where they eventually exit. In the case of the

Zambezi River, whose headwaters lie in Angola and northwestern Zambia, the situation is the same; the river first flows south, toward the inland delta known as the Okovango Swamp, then it turns northeast and southeast, eventually to reach its delta south of Lake Malawi. Finally, there is the famed erratic course of the *Nile* River, which braids into numerous channels in the Sudd area of southern Sudan, and in its middle course actually reverses direction and flows southward before resuming its flow toward the Mediterranean delta in Egypt. With so many peculiarities among Africa's river courses, could it be that all have been affected by the same event at some time in the

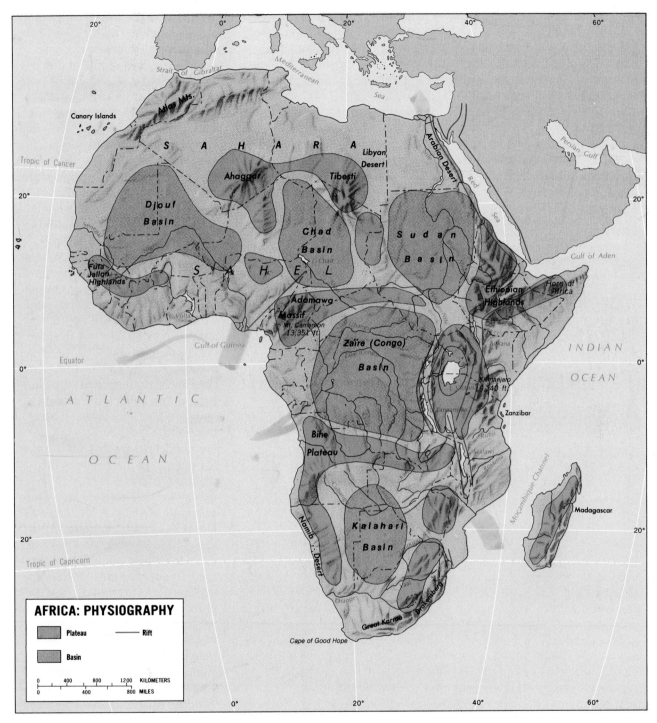

AFRICA: PHYSIOGRAPHY

Plateau   Rift

Basin

| 0 | 400 | 800 | 1200 | KILOMETERS |
| 0 | | 400 | | 800 | MILES |

FIGURE 7–4

continent's history? Perhaps—but first let us look further at the map.

Plateaus and Escarpments

All continents have low-lying areas; witness the interior lowlands of North America or the coastal and river lowlands of Eurasia and Australia. But as the map shows, coastal lowlands are few and of limited extent in Africa. In fact, it is reasonable to call Africa a *plateau continent*: except for coastal Moçambique and Somalia, and along the northern and western coasts, almost the entire continent lies above 1000 feet (300 meters) in elevation and fully half of it is over 2500 feet (800 meters) high. Even the Zaïre (Congo) Basin, Equatorial Africa's huge tropical lowland, lies well over 1000 feet above sea level in contrast to the much lower-lying Amazon Basin across the Atlantic.

Although Africa is mostly plateau, this does not mean that the surface is completely flat and unbroken. In the first place, the rivers have been eroding the surface for millions of years and have made some pretty good dents in it. For example, Victoria Falls on the Zambezi (Fig. 7–6) is one mile (1600 meters) wide and over 300 feet (90 meters) high. Volcanoes and other types of mountains, some of them erosional leftovers, stand well above the landscape in many areas (Fig. 7–3); the Sahara Desert is no exception, where the Ahaggar and Tibesti Mountains both reach about 10,000 feet (3000 meters) in elevation. In several places, the plateau has sagged down under the weight of accumulating sediments. In the Zaïre Basin, for example, rivers have transported sand and sediment downstream for tens of millions of years and, for some reason, dropped their erosional loads into what was once a giant lake the size of an interior sea. Today the lake is gone, but the thick sediments that press this portion of the African surface into a giant basin are proof that it was there. And this was not the only inland sea. To the south, the Kalahari Basin was filling with sediments that now comprise that desert's sand; and far to the north, in the Sahara, three similar basins lie centered on Sudan, Chad, and what today is Mali (the Djouf Basin).

The margins of Africa's plateau are of significance, too. Much of the continent, because it is a plateau, is surrounded by an escarpment. In Southern Africa where this feature is especially pronounced, the *Great Escarpment* (as it is called there) marks the plateau's edge along many hundreds of miles; here the land drops precipitously from more than 5000 feet (1500 meters) in elevation to a narrow, hilly coastal belt. From Zaïre to Swaziland, and intermittently on or near most of the African coastline, a scarp bounds the interior upland. Such escarpments are found in other parts of the world, too: Brazil at the eastern margins of the Brazilian Highlands, and India at the western edge of its Deccan Plateau. But Africa, even for its size, has a disproportionately large share of this topographic phenomenon.

Continental Drift and Africa

Africa's remarkable and unusual physiography was one of the

FIGURE 7–5 Rift valleys fracture the East African landscape: the edge of the rift in northern Kenya's semiarid highlands.

FIGURE 7–6 The erosive power of fast-moving water has produced a deep, lengthy gorge below the Zambezi River's spectacular Victoria Falls.

pieces of evidence that the geographer Alfred Wegener used, early in this century, to construct his hypothesis of **continental drift**. According to this idea (first introduced on p. 9), all the landmasses on earth were assembled into one giant continent named *Pangaea*; the southern continents constituted *Gondwana*, the southern part of this supercontinent (Fig. 7–7). After a long (geologic) period of unity, this huge landmass began to break up more than 100 million years ago. Africa, which lay at the heart of Gondwana, attained the approximate shape we see today when North and South America, Antarctica, Australia, and India drifted radially away. Later, geophysicists gave the name *plate tectonics* to this process, and it is now understood that the breakup of Pangaea was

only the most recent in a series of continental collisions and separations that spans much of the earth's history (see Fig. I–6).

Africa's situation at the heart of the supercontinent explains much of what we see in its landscapes today. The Great Escarpment is a relic of the giant faults (fractures) that formed when the neighboring landmasses split off. The rift valleys are only the most recent evidence of the pulling forces that affect the African plate; the Red Sea is an advanced stage of such rifts, and we may expect East Africa to separate from the rest of Africa much as Arabia did earlier (and Madagascar before that). Africa's rivers once filled lakes within the continent, but they did not reach the distant ocean; today, the lakes are drained by rivers that eroded inland when the oceans

began to wash Africa's new shorelines.

And why does Africa not have those mountain chains that led us to look at the map more closely? The answer may lie in the direction and distance of plate motion. South America moved far westward, its plate colliding with the plate under the waters of the Pacific Ocean; the Andes crumpled up, accordion-like, in this collision. India moved northeastward, wedging into the Asian landmass; similarly, the Himalayas rose upward. But Africa moved comparatively little, being more affected by pulling tensional forces that create rifts rather than the pressures of plate collision that generate the uplifting of mountains. Decidedly, the physiographic map reveals more than just the location of rivers, plains, and lakes.

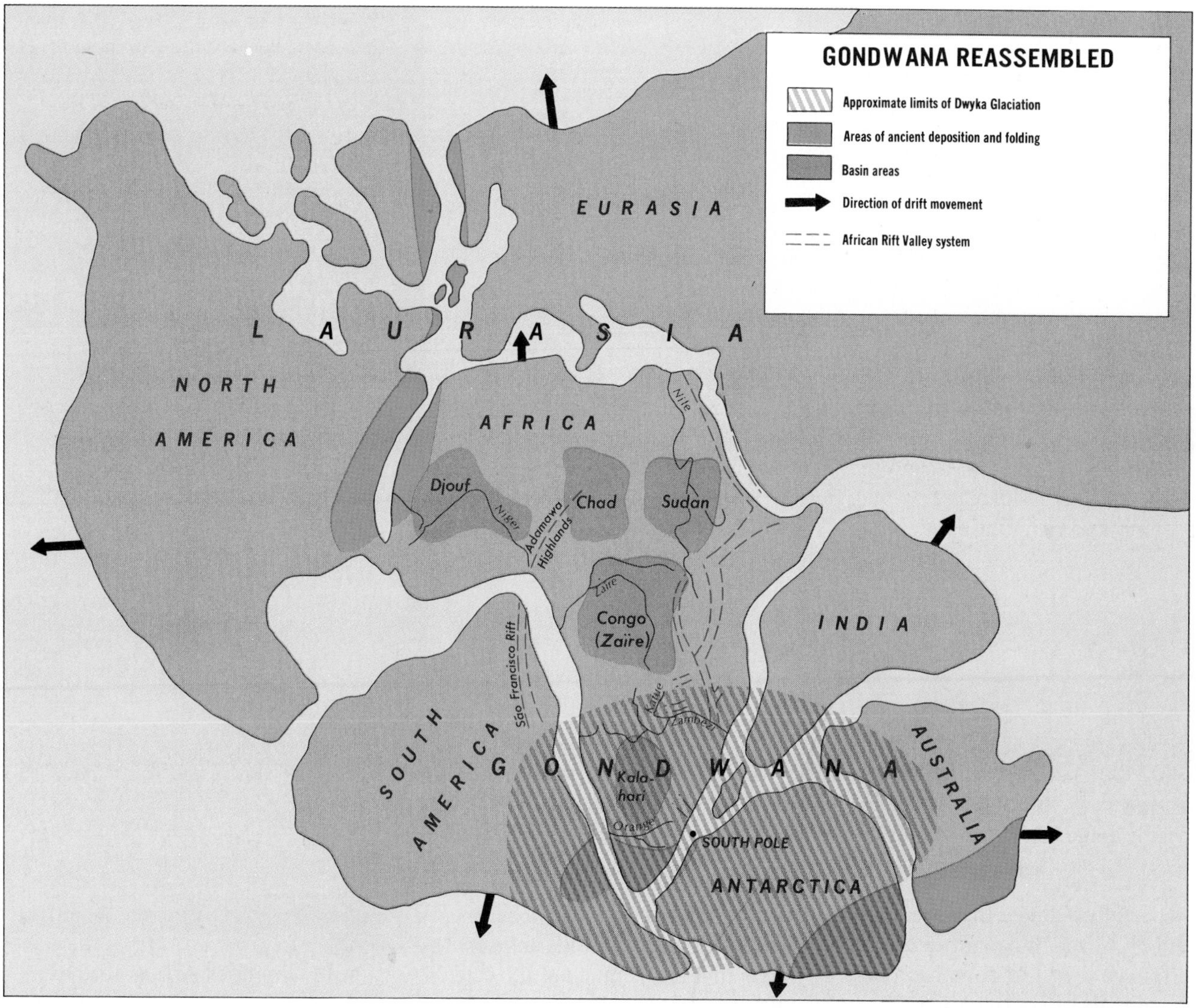

FIGURE 7–7

Climate and Vegetation

Africa's climatic and natural vegetation distributions are almost symmetrical about the equator (Figs. I–10 and I–12). With the equator virtually bisecting the continent, the atmospheric conditions that affect the African surface tend to be similar in both halves. The hot rainy climate of the Zaïre (Congo) Basin merges gradually, both north and south, into climates with distinct winter dry seasons; outside the Basin, therefore, trees are less majestic and then less numerous as one moves away from the equator. Eventually, the humid equatorial region is left entirely behind as annual rainfall becomes less abundant and less reliable. Now the tree-studded grasslands of the savanna dominate the landscape, a wide zone of sometimes park-like vegetation that has long been the principal home of the continent's great herds of wildlife. Moving poleward, the savanna gives way to the drier tropical steppe, the sparse grasslands that are so easily overgrazed by livestock (Fig. 7–8). Beyond the steppe lies the desert: the Sahara in the north, the Kalahari in the south. A close look at Fig. I–10 reveals that even the dry-summer subtropical climate, known as the Mediterranean climate, occurs on both the north and the south coasts of Africa, although the zone at the southern tip is very small.

It is interesting to compare this pattern to that of annual rainfall (Fig. I–9). Note how much larger the region of heavy precipitation is in the equatorial zone of South America, in contrast to that of

Africa. Only the western zone of Equatorial Africa and a strip along the southern West African coast receive substantial rainfall. Much of eastern and southern Africa receive modest rainfall amounts, especially when we remember the considerable evaporation that takes place. Africa's topography—its high eastern plateau, escarpments, and generally elevated interior—coupled with its bulk, creates large areas where insufficient moisture is received. Here the soils are poor, the natural vegetation is fragile, and the environment endangered by burgeoning human demands.

ENVIRONMENT, PEOPLE, AND HEALTH

Africa south of the Sahara, in average terms, is not a densely peopled realm. Less than one-tenth of the world's population—about 525 million people—inhabit this subcontinent of difficult environments, and a majority of them remain farmers. The situation is reflected in Fig. I–14: note that much of the realm is moderately populated and that large clusters are few. The largest human concentrations occur in Nigeria, around Lake Victoria in East Africa, and in several smaller areas of Southern Africa.

What Fig. I–14 does *not* reveal, however, is the overall health of the African population. The tragic truth is that Africa suffers from local and regional famines, that diets are generally not well balanced, and that life expectancies are lower here than they are in any other world geographic realm (Fig. 7–9). Why is so much of Africa so afflicted? The realm's physical geography has much to do with it.

Tropical areas are breeding grounds for organisms that carry disease, such as flies, mosquitoes,

FIGURE 7–8 Cattle feeding in the steppelands of eastern Kenya. Overgrazing is always a danger in such environments—as the nearly exhausted grassy vegetation in this scene underscores.

fleas, and even snails. Many people depend on river and well water that may carry these *vectors* (disease transmitters) or may be contaminated. Inadequate or imbalanced diets weaken the body and make it susceptible to illness; many Africans are thus at risk because food availability is limited. Africa's soils and climates limit the range of foods that can be produced locally, and people generally do not have the money to spend on meats or other expensive supplements. As a result, millions of Africans spend their entire lives in a state of poor health. Children who do not get enough protein in their diets, for instance, develop such food-deficiency disorders as kwashiorkor and marasmus; they may survive, but their resistance to diseases of older age will be reduced.

Diseases strike populations in different ways. A local or regional outbreak, affecting many, is known as an *epidemic*. When a disease spreads worldwide, as some forms of influenza have done in recent years, the phenomenon is called a *pandemic*. AIDS may have begun as an epidemic in a part of Equatorial Africa in the 1970s; but by the 1990s, it had assumed pandemic proportions, killing victims from Kenya to Romania to the United States. A disease can affect a population in still another way. Some illnesses do not come in a violent attack but invade and inhabit the body, establishing a kind of equilibrium with it. These diseases sap energies and shorten lifetimes, but they are not usually the cause of eventual death. Such **endemic diseases** afflict tens of millions of Africans.

Undoubtedly, malaria, endemic to most of Africa and a killer of up to 1 million children per year, is the worst. In recent years malaria has again been on the rise, and in rural areas shows growing resis-

tance to drugs that once kept it under control. The malarial mosquito is the vector of the parasite, and the mosquito prevails in almost all of inhabited Africa. Africans who survive childhood are

likely to suffer from malaria to some degree, with a debilitating effect.

African sleeping sickness is transmitted by the tsetse fly (Fig. 7–10, right map) and now affects

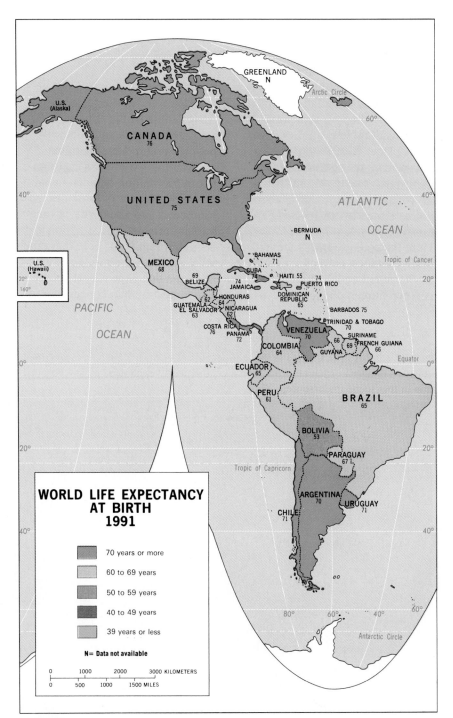

FIGURE 7–9

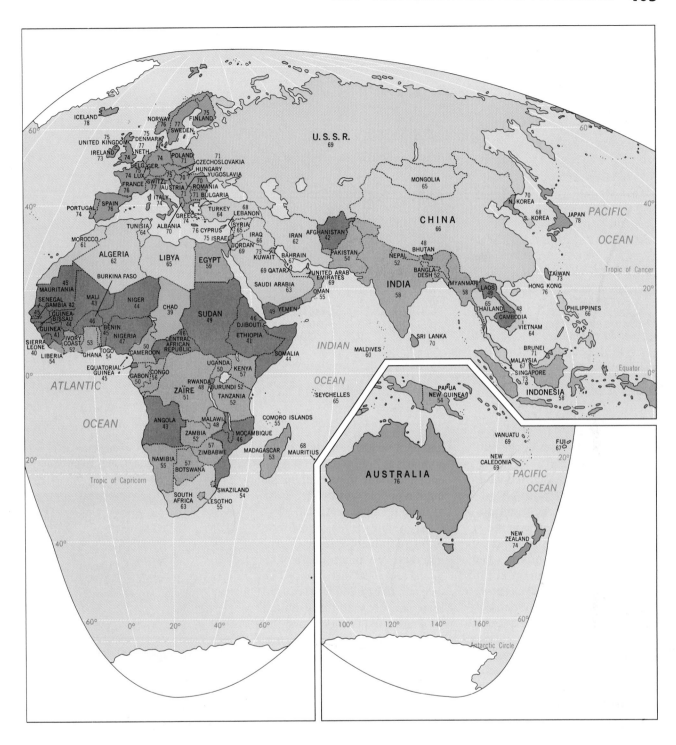

most of tropical Africa. This fly infects not only people, but also their livestock. Its impact on Africa's population has been incalculable. The disease appears to have originated in a West African source area about A.D. 1400; since then, it has inhibited the development of livestock herds where meat would have provided a crucial balance to seriously protein-deficient diets. It channeled the migrations of cattle herders through fly-free corridors in East Africa (Fig. 7–10), destroying herds that moved into infested zones. Most of all, it ravaged the human population, depriving it not

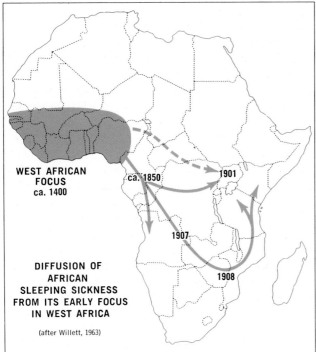

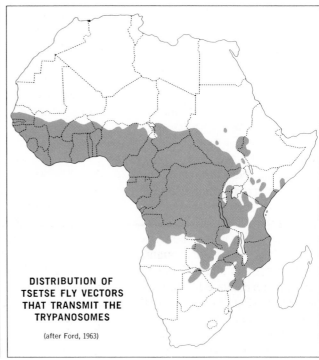

FIGURE 7–10

only of potential livelihoods but also of its health.

Still another serious and widespread disease is yellow fever, also transmitted by mosquitoes. Yellow fever is endemic in the wetter tropical zones of Africa, but it sometimes appears in other areas in epidemic form, as in 1965 in Senegal when 20,000 cases were reported (there were undoubtedly thousands more). This is another disease that strikes children, who, if they survive, acquire a certain level of immunity.

To this depressing list must be added schistosomiasis (also called bilharzia), which is transmitted by snails. The parasites enter via body openings when people swim or wash in slow-moving or standing water infested by the snails. Many development projects in Africa involving irrigation have inadvertently introduced schistosomiasis into populations previously not exposed to these parasites. Internal bleeding, loss of energy,

and pain result, although schistosomiasis is not by itself fatal. It is endemic today to more than 200 million people worldwide, most of them residing in Africa.

These are the major and more or less continentwide diseases, and there are many others of regional and local distribution. The dreaded river blindness, caused by a parasitic worm transmitted by a small fly, is endemic in the savanna belt south of the Sahara from Senegal east to Kenya; in northern Ghana alone, it blinds a large percentage of the adult villagers (Fig. 7–11). And animals and plants as well are attacked by Africa's ravages. Besides sleeping sickness, livestock herds are also afflicted by rinderpest. Crops and pastures are stripped periodically by swarms of locusts (the late 1980s were particularly bad years), which number an average of 60 million insects each, travel thousands of miles, and devour the plant life of entire countrysides.

FIGURE 7–11 A young child acts as the "eyes" of a victim of river blindness.

THE PREDOMINANCE OF AGRICULTURE

The great majority of Africans live today as their predecessors lived—by subsistence farming, herding, or both. Not that the African environment is particularly easy for the cultivator who tries to grow food crops or raise cattle: tropical soils are notoriously unproductive, and much of Africa suffers from excessive drought.

Most African families still depend on subsistence farming for their living (Fig. 7–12). Along the coast (especially the west coast) and the major rivers, some communities depend primarily on fishing. Otherwise, the principal mode of life is farming—with grain crops dominant in the drier areas and root crops in the more humid zones. In the methods of farming, the sharing of the work between men and women, the value and prestige attached to herd animals, and in other cultural aspects, subsistence farming provides the opportunity to gain insight into the Africa of the past. Moreover, the subsistence form of livelihood was changed only very indirectly by the colonial impact; tens of thousands of villages all across Africa were never fully brought into the economic orbit of the European invaders, and life in these settlements went on more or less unchanged.

Africa's pastoralists more often than not mix farming with their livestock-raising pursuits, and very few of them are exclusively herders. In Africa south of the Sahara, there are two belts of herding activity: one extending laterally along the West African savanna-steppe transition zone and connecting with the East African area (where the Masai drive their herds), the other centering on the plateau of South Africa. Especially in East and Southern Africa, cattle are less important as a source of food than they are as a measure of their owners' wealth and prestige in the community. Hence African cattle-owners in these areas have always been more interested in the size of their herds than in the quality of the animals. The staple foods here are the grain crops of corn, millet, and sorghum (together with some concentrations of rice), which overlap with the

FIGURE 7–12 A group of women, some accompanied by their babies, performing the rigorous tasks of subsistence farming in Uganda. Cultural traditions here require men and women to share strenuous manual work in the unending struggle to produce a sufficient supply of food.

herding areas. Probably a majority of Africa's cattle-owners are sedentary farmers, although some pastoralists—such as the Masai—engage in a more or less systematic cycle of movement, following the rains and seeking pastures for their livestock (Fig. 7–13).

In West Africa, those pastoralists who sell their cattle for meat face the problem of considerable distances to the major coastal markets. If the animals are taken by truck across the several hundred miles from the savanna-steppe zone to the coast, the cost is high but the cattle do not lose much weight between farm and market. On the other hand, if the animals are driven to the markets on the hoof, the cost of moving them is low but their weight loss is substantial. And the cattle herder also

faces environmental problems, not the least of which is the dreaded tsetse fly (Fig. 7–10).

Cattle, of course, are not the only livestock in Africa. Chickens are nearly ubiquitous, and there are millions of goats everywhere—in the forest, in the savanna, in the steppe, even in the desert; the goats always seem to survive, and no African village would be complete without them. Where conditions are favorable, goats multiply swiftly—and then they denude the countryside and promote soil erosion. In Swaziland, for example, goats constitute both an asset to their individual owners and a serious liability to the state.

A gradual change in the involvement of Africans in the cash economy has taken place during the past quarter-century. Excluded during

the colonial period from most activities that could produce income for themselves and their families, many Africans have chosen since independence to make changes that promise to bring in money. Farmers have introduced cash crops onto their small plots, at times replacing the subsistence crops altogether. These developments notwithstanding, Africa continues to fall further behind in meeting its food provision needs. Continentwide, agricultural output is declining. Per capita food production in Subsaharan Africa during the late 1980s was more than 10 percent below what it was a decade earlier. Even though total production is higher today, with 160 million more mouths to feed in 1991 than in 1980, all food increases are more than offset by the rapid

FIGURE 7–13   Masai herders and their cattle on the move in central Kenya. The healthy appearance of the animals indicates that the search for nutritious pasture grasses has been a successful one this year.

rate of population growth (see *box* at right on Africa's Green Revolution).

AFRICA'S PAST

Africa is the cradle of humanity. Research in Tanzania, Kenya, and Ethiopia has steadily pushed back the date of the earliest origin of human prototypes by hundreds of thousands, even millions of years. It is, therefore, something of an irony that comparatively little is known about Subsaharan Africa from 5000 to 500 years ago—that is, prior to the onset of European colonialism. This is only partly due to the colonial period itself, during which African history was neglected, many African traditions and artifacts were destroyed, and many misconceptions about African cultures and institutions arose and became entrenched. It is also a result of the absence of a written history over most of Africa south of the Sahara until the sixteenth century—and over a large part of it until much later than that. The best records are those of the savanna belt immediately south of the Sahara Desert, where contact with North African peoples was greatest and where Islam achieved a major penetration.

The absence of a written record does not mean, as some scholars suggested, that Africa does not have a history as such prior to the coming of Islam and Christianity. Nor does it mean that there were no rules of social behavior, no codes of law, no organized economies. Modern historians, encouraged by the intense interest shown by Africans generally, are now trying to reconstruct the African past, not only from the meager written record, but also from folklore, poetry, art objects, buildings, and

A GREEN REVOLUTION FOR AFRICA?

The gap between world population and food production has been narrowed through the **Green Revolution**, the development of more productive, higher-yielding types of grains. Where people depend mainly on rice and wheat for their staples, the Green Revolution pushed back the specter of hunger. But the Green Revolution has had less impact in Africa. In part, this relates to the realm's high rate of population growth (which is substantially higher than that of India or China). Other reasons have to do with Africa's staples. Rice and wheat support only a small part of Africa's population. Corn (maize) supports many more, along with sorghum, millet, and other grains; in moister areas, root crops such as the yam and cassava as well as the plantain (similar to the banana) supply the bulk of calories. These crops were not priorities in Green Revolution research.

Lately there have been a few signs of hope. Even as terrible famines struck several parts of the continent and people went hungry in places as widely scattered as Moçambique and Mali, scientists worked to accomplish two goals: first, to develop strains of corn and other crops that would be more resistant to Africa's virulent crop diseases, and second, to increase the productivity of those strains. But these efforts faced serious problems: an average African acre of corn, for example, yields only about 0.5 tons of corn, whereas the world average is 1.3 tons. Now a virus-resistant corn variety has been developed that also yields better than the old, and it is being introduced in Nigeria and other countries. And hardier types of yams and cassava also are evolving, again raising yields significantly.

Africa needs more than this if the cycle of food deficiency is to be reversed. It has been estimated that since 1970 food production has fallen by 1 percent every year. Inefficient farming methods, inadequate equipment, soil exhaustion, apathy, fatalism, and male dominance also must be overcome if Subsaharan Africa is to feed the 700 million mouths expected to live in the realm by 2000. Military governments (of which Africa has many) have been especially neglectful in their agricultural policies, which require quiet persuasion and effective extension systems rather than glamorous, inspiring campaigns. That there is an alternative, a way to success, has been proven in recent years by Zimbabwe, where the farming sector of the economy thrives and where a surplus of staples is produced. But Zimbabwe has a comparatively small population (10.9 million), inherited a sound agrarian sector from colonial times, and has been spared the worst of the droughts that have afflicted other parts of Africa. Today, there are some bright spots, but the battle for food sufficiency in Africa is far from being won.

other such sources. But much has been lost forever. Almost nothing is known of the farming peoples

who built well-laid terraces on the hillsides of northeastern Nigeria and East Africa or of the communities that laid irrigation canals and constructed stone-lined wells in Kenya; and very little is known about the people who, perhaps a thousand years ago, built the great walls of Zimbabwe. Porcelain and coins from China, beads from India, and other goods from distant sources have been found in Zimbabwe and other points in East and Southern Africa, but the trade routes within Africa itself—let alone the products that circulated on them and the people who handled them—still remain the subject of guesswork.

The Pre-European Prelude

Africa on the eve of the colonial period was in many ways a continent in transition. For several centuries, the habitat in and near one of the continent's most culturally and economically productive areas—West Africa—had been changing. For 2000 years, probably more, Africa had been innovating as well as adopting ideas. In West Africa, cities were developing on an impressive scale; in Central and Southern Africa, peoples were moving, readjusting, sometimes struggling with each other for territorial supremacy. The Romans had penetrated to southern Sudan, North African peoples were trading with West Africans, and Arab *dhows* were plying the waters along the eastern coasts, bringing Asian goods in exchange for gold, copper, and a comparatively small number of slaves.

Consider the environmental situation in West Africa as it relates to the past. As Figs. I–9 to I–13 (pp. 14–25) indicate, the environmental regions in this part of the continent exhibit a decidedly east-west orientation. The isohyets (lines of equal rainfall totals) run parallel to the southern coast (Fig. I–9); the climatic regions, now po-

sitioned somewhat differently from where they were two millennia ago, still trend strongly east-west (Fig. I–10). Soil regions are similarly aligned (Fig. I–13); the vegetation map, although generalized, also reflects this situation (Fig. I–12), with a coastal forest belt yielding to savanna—tall grass with scattered trees in the south, shorter grass in the north—which gives way, in turn, to steppe and desert.

Early Trade

Essentially, then, the situation in West Africa was such that over a north-south span of a few hundred miles there was an enormous contrast in environments, economic opportunities, modes of life, and products. The people of the tropical forest produced and needed goods that were quite different from the products and requirements of the peoples of the dry distant north. As an example, salt is a prized commodity in the forest, where the humidity precludes its formation, but salt is in plentiful supply in the desert and steppe. Hence, the desert peoples could sell salt to the forest peoples, but what could be offered in exchange? Ivory and spices could be sent north, along with dried foods. Thus, there was a degree of **regional complementarity** between the peoples of the forests and the peoples of the drylands. And the savanna peoples—those who were located in between—were beneficiaries of this situation, for they found themselves in a position to channel and handle the trade (that activity is always economically profitable).

The markets in which these goods were exchanged prospered and grew, and there arose a number of true cities in the savanna belt of West Africa. One of these old cities, now an epitome of isolation, was once a thriving center of commerce and learning and one of the leading urban places in the

world—Timbuktu. Others, predecessors as well as successors of Timbuktu, have declined, some of them into oblivion. Still other savanna cities continue to have considerable importance, such as Kano in the northern part of Nigeria.

Early States

States of impressive strength and truly amazing durability arose in the West African culture hearth (Fig. 6–3). The oldest state about which anything at all is known is Ghana. Ancient Ghana was located to the northwest of the coastal country that has taken its name in the post-colonial period. It covered parts of present-day Mali and Mauritania, along with some adjacent territory. It lay astride the upper Niger River and included gold-rich streams flowing off the Futa Jallon Highlands, where the Niger has its origins. For a thousand years, perhaps longer, old Ghana managed to weld various groups of people into a stable state. The country had a large capital city, complete with markets, suburbs for foreign merchants, religious shrines, and, some miles from the city center, a fortified royal retreat. There were systems of tax collection for the citizens and extraction of tribute from subjugated peoples on the periphery of the territory; tolls were levied on goods entering the Ghanaian domain; and an army maintained control. Muslims from the northern drylands invaded Ghana about A.D. 1062 when the state may already have been in decline. Even so, Ghana continued to show its strength: the capital was protected for no less than 14 years. However, the invaders had ruined the farmlands, and the trade links with the north were destroyed. Ghana could not survive, and it finally broke apart into a number of smaller units.

In the centuries that followed, the focus of politico-territorial organization in the West African culture hearth shifted almost continuously eastward—first to ancient Ghana's successor state of Mali, which was centered on Timbuktu and the middle Niger River Valley, and then to the state of Songhai, whose focus was Gao, also a city on the Niger and one that still exists today. One possible explanation for this eastward movement may lie in the increasing influence of Islam; Ghana had been a pagan state, but Mali and its successors were Muslim and sent huge, rich pilgrimages to Mecca along the savanna corridor south of the desert. Indeed, hundreds of thousands of citizens of the modern Republic of Sudan trace their ancestry to the lands now within northern Nigeria, their ancestors having settled there while journeying to or from Mecca.

Unmistakably, the West African savanna region was the scene of momentous cultural, political, and economic developments for many centuries, but it was not alone in its progress in Africa. In what is today southwestern Nigeria, a number of urban farming communities became established, the farmers being concentrated in these walled and fortified places for reasons of protection and defense; surrounding each "city of farmers" were intensely cultivated lands that could sustain thousands of people clustered in the towns. In the arts, too, Nigeria produced some great achievements, and the bronzes of Benin are true masterworks (Fig. 7–14). In the region of the Zaïre River mouth, a large state named Kongo existed for centuries. In East Africa, trade on a large scale with China, India, Indonesia, and the Arab world brought crops, customs, and merchandise from these distant regions. In Ethiopia and Uganda, populous kingdoms emerged. Nevertheless, much of what Africa was in those earlier centuries has yet to be reconstructed. But with all this

FIGURE 7–14 Artistic expression has long been a hallmark of West African cultures: a Benin bronze head from Nigeria, probably above five centuries old.

external contact, it was clearly not isolated.

The Colonial Transformation

The period of European involvement in Subsaharan Africa began in the fifteenth century. This period was to interrupt the path of indigenous African development and irreversibly alter the entire cultural, economic, political, and social makeup of the continent. It started quietly enough, with Portuguese ships groping their way along the west coast and rounding the Cape of Good Hope not long before the opening of the sixteenth century. Their goal was to find a sea route to the spices and riches of the Orient. Soon, other European countries were sending their vessels to African waters, and a string of coastal stations and forts sprang up. In West Africa, the nearest part of the continent to European spheres in Middle and South America, the initial impact

was strongest. At their coastal control points, the Europeans traded with African middlemen for the slaves who were wanted on American plantations, for the gold that had been flowing northward across the desert, and for ivory and spices.

Suddenly, the centers of activity lay not with the cities of the savanna, but in the foreign stations on the Atlantic coast. As the interior declined, the coastal peoples thrived. Small forest states rose to power and gained unprecedented wealth, transferring and selling slaves captured in the interior to the European traders on the coast. Dahomey (now called Benin) and Benin (now part of neighboring Nigeria) were states built on the slave trade; when the practice of slavery eventually came under attack in Europe, abolition was vigorously opposed in both continents by those who had inherited the power and riches it had brought.

Although it is true that slavery was not new to West Africa, the kind of slave trading introduced by the Europeans certainly was. In the savanna states, African families who had slaves usually treated them comparatively well, permitting marriage, affording adequate quarters, and absorbing them into the family. The number of slaves held in this way was small; probably the largest number of persons in slavery in pre-colonial Africa were in the service of kings and chiefs. In East Africa, however, the Arabs had introduced (long before the Europeans) the sort of slave trading that was first brought to West Africa by whites: African middlemen from the coast raided the interior for slaves and marched them in chains to the Arab *dhows* that plied the Indian Ocean. There, packed by the hundreds in specially built vessels, they were carried off to Arabia, Persia, and India. It is sad but true that Europeans, Arabs, and Africans combined to ravage the Black Continent, forcing perhaps as many as 30 million persons away from their homelands in bondage (Fig. 7–15). Families were destroyed, as were whole villages and cultures; and those affected suffered a degree of human misery for which there is no measure.

The European presence on the West African coast brought about a complete reorientation of trade routes, for it initiated the decline of the interior savanna states and strengthened the coastal forest states. Moreover, it ravaged the population of the interior through its insatiable demand for slaves; but it did not lead to any major European thrust toward the interior, nor did it produce colonies overnight. The African middlemen were well organized and strong, and they managed to maintain a standoff with their European competitors, not just for a few decades but for centuries. Although the European interests made their initial appearance in the fifteenth century, West Africa was not carved up among them until nearly 400 years later, in many areas not until after the beginning of the twentieth century.

As fate would have it, European interest was to grow strongest—and ultimately most successful—where African organization was weakest. In the middle of the seventeenth century, the Dutch chose the shores of South Africa's Table Bay, where Cape Town lies today (Fig. 7–16), as the site for permanent settlement. Their initial purpose was not colonization, but rather establishment of a resupply station for the months-long voyage to and from Southeast Asia and their East Indies colonies; the southern tip of Africa was the obvious halfway point. There was no intent to colonize here because Southern Africa was not known as a productive area, and more worthwhile East Africa lay in the spheres of the Portuguese and the Arabs. Probably the Hollanders would have elected to build their station at the foot of Table Bay whatever the indigenous population of the interior, but they happened to choose a location about as far away from the centers of Bantu

FIGURE 7–15

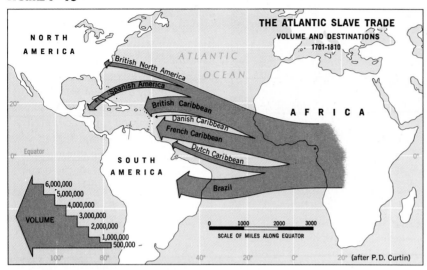

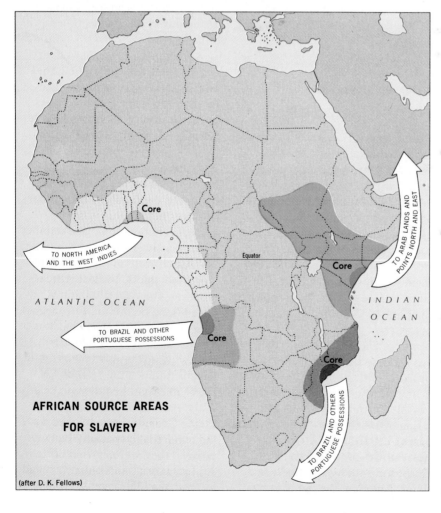

FIGURE 7–16   Table Mountain dominates the site of Cape Town close to Africa's southern tip. It was here that the first Dutch settlers landed in 1652.

settlement as they could have found. Only the San (Bushmen) and their rivals, the Khoikhoi (Hottentots), occupied Cape Town's hinterland. When conflicts developed between Amsterdam and Cape Town and some of the settlement's residents decided to move into the hinterland, they initially faced only the harassment of small groups of these two peoples rather than the massive resistance that the well-organized Bantu states would probably have offered. To be sure, a confrontation eventually did develop between the advancing Europeans and the similarly mobile Bantu Africans, but it began decades after Cape Town was founded and hundreds of miles from it. Unlike some of the West African way-stations, Cape Town was never threatened by African power, and it became a

European gateway into Southern Africa.

Elsewhere in the realm, the European presence remained confined almost entirely to the coastal trading stations, whose economic influence was very strong. No real frontiers of penetration developed; individual travelers, missionaries, explorers, and traders went into the interior, but nowhere else in Africa south of the Sahara was there an invasion of white settlers comparable to Southern Africa's.

Penetration

After more than four centuries of contact, Europe finally laid claim to all of Africa during the second half of the nineteenth century. Parts of the continent had been ''explored,'' but now representatives of European governments

sought to expand or create African spheres of influence for their homelands. Cecil Rhodes for Britain, Karl Peters for Germany, Pierre de Brazza for France, and Henry Stanley for the king of Belgium were some of the leading figures who helped shape the colonial map of the continent. In some areas, such as along the lower Zaïre River and in the vicinity of Lake Victoria, the competition among the European powers was especially intense. Spheres of influence began to crowd each other; sometimes they even overlapped. So in late 1884, a conference was convened in Berlin to sort things out. At this conference, the groundwork was laid for the now-familiar political boundaries of Africa (see *box*).

As the twentieth century opened, Europe's colonial powers were

THE BERLIN CONFERENCE

In November 1884, the imperial chancellor and architect of the German Empire, Otto von Bismarck, convened a conference of 14 powerful states (including the United States) to settle the political partitioning of Africa. Bismarck not only wanted to expand German spheres of influence in Africa, but he also sought to play Germany's colonial rivals off against one another to the Germans' advantage. The major colonial contestants in Africa were the British, who held beachheads along the West, South, and East African coasts; the French, whose main sphere of activity was in the area of the Senegal River and north of the Zaïre (Congo) Basin; the Portuguese, who now desired to extend their coastal stations in Angola and Moçambique deep into the interior; King Leopold II of Belgium, who was amassing a personal domain in the Congo; and Germany itself, active in areas where the designs of other colonial powers might be obstructed as in Togo (between British holdings), Cameroon (a wedge into French spheres), South West Africa (taken from under British noses in a swift strategic move), and East Africa (where the effect was to break the British design for a Cape-to-Cairo axis).

When the conference convened in Berlin, more than 80 percent of Africa was still under traditional African rule. Nonetheless, the colonial powers' representatives drew their boundary lines across the entire map. These lines were drawn through known as well as unknown regions, pieces of territory were haggled over, boundaries were erased and redrawn, and sections of African real estate were exchanged in response to urgings from European governments. In the process, African peoples were divided, unified regions were ripped apart, hostile societies were thrown together, hinterlands were disrupted, and migration routes were closed off. All of this was not felt immediately, of course, but these were some of the effects when the colonial powers began to consolidate their holdings and the boundaries on paper soon became barriers on the African landscape (Fig. 7–17).

The Berlin Conference was Africa's undoing in more ways than one. Not only did the colonial powers superimpose their domains on the African continent: when independence returned to Africa after 1950, the realm had by then acquired a legacy of political fragmentation that could neither be eliminated nor made to operate satisfactorily. The African politico-geographical map, therefore, is a permanent liability that resulted from three months of ignorant, greedy acquisitiveness during the period of Europe's insatiable search for minerals and markets.

a vast realm that reached from Algiers in the north and Dakar in the west to the Zaïre River in Equatorial Africa. King Leopold II of Belgium held personal control over his Belgian Congo (now Zaïre). Germany had colonies scattered in all sections of the continent except the north. The Portuguese controlled two huge territories, Angola and Moçambique, along the flanks of Southern Africa, and a small entity in West Africa known as Portuguese Guinea (now Guinea-Bissau). Italy's possessions in tropical Africa were confined to the Horn, and even Spain got into the act with a small dependency consisting of the island of Fernando Póo (now called Bioko) and the mainland area called Rio Muni (Equatorial Guinea). The only places where the Europeans did not overwhelm African desires to remain independent were Ethiopia, which fought some heroic battles against Italian forces, and Liberia, where Afro-Americans retained control.

The two world wars also had some effect on this colonial map of Africa. In World War I, Germany's defeat resulted in the complete loss of its colonial possessions. The territories in Africa were placed under the administration of other colonial powers by the League of Nations' mandate system. In World War II, fascist Italy launched a briefly successful campaign against Ethiopia, but the ancient empire was restored to independence when the allied forces won the war. Otherwise, the situation in colonial Africa in the late 1940s—after a half-century of colonial control—remained quite similar to the one that arose out of the Berlin Conference.

Policies and Repercussions

Geographers—especially political geographers—are interested in the ways in which philosophies and

busily organizing and exploiting their African dependencies. The British, having defeated the Boers of Dutch heritage in South Africa, came very close to achieving their Cape-to-Cairo axis: only German East Africa interrupted a vast empire that stretched southward from Egypt and Sudan through Uganda, Kenya, Nyasaland (now Malawi), and the Rhodesias (now Zambia and Zimbabwe) to South Africa (Fig. 7–17, 1910 map). The French took charge of

COLONIZATION AND LIBERATION

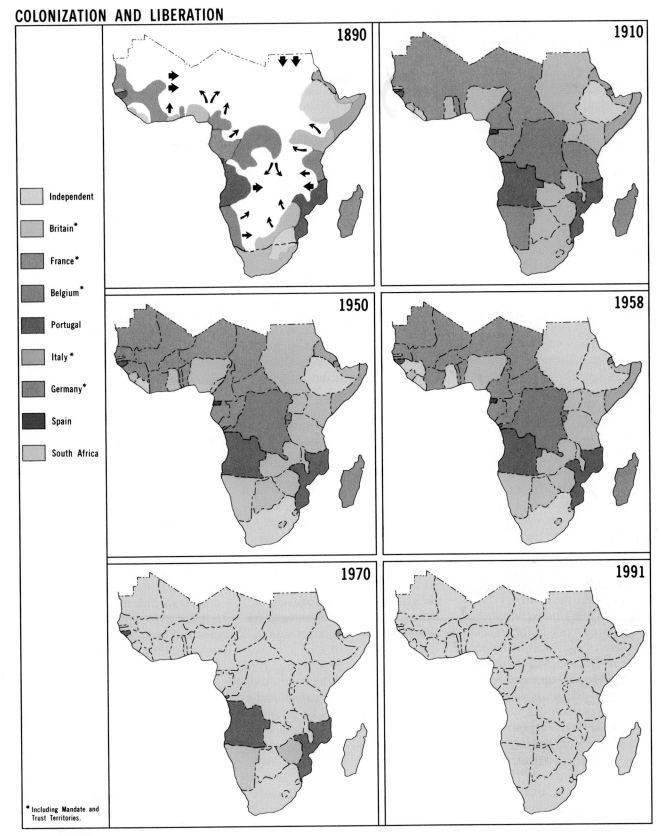

Independent
Britain*
France*
Belgium*
Portugal
Italy*
Germany*
Spain
South Africa

* Including Mandate and Trust Territories.

FIGURE 7–17

policies of the colonial powers were reflected in the spatial organization of the African dependencies. These colonial policies can be expressed in just a few words. For example, Britain's administration in many parts of its vast empire was referred to as *indirect rule*, since indigenous power structures were sometimes left intact and local rulers were made representatives of the crown. Belgian colonial policy was called *paternalism* in that it treated Africans as though they were children who needed to be tutored in Western ways. Although the Belgians made no real efforts to make their African subjects culturally Belgian, the French very much wanted to create an "Overseas France" in their African dependencies. French colonialism has been identified as a process of *assimilation*—the acculturation of Africans to French ways of life—and France made a stronger cultural imprint in the various parts of its huge colonial empire. Portuguese colonial policy had objectives similar to the French, and the African dependencies of Portugal were officially regarded as "Overseas Provinces" of the state. If you sought a one-word definition of Portuguese colonial policy, however, the term *exploitation* would emerge most strongly: few colonies made a greater contribution (in proportion to their known productive capacities) to the economies of their colonial masters than did Moçambique and Angola.

Colonial policies have geographic expressions as well, and the **colonial spatial organization** of the ruling powers has become the infrastructure of contemporary independent Africa. As the map shows (Fig. 7–17), Britain possessed the most far-flung colonial empire in Africa. British colonial policy tended to adjust to individual situations. In colonies, white-settler minorities had substantial autonomy, as in Kenya and Southern Rhodesia (now Zimbabwe); in protectorates, the rights of African peoples were guarded more effectively; in mandate (later trust) territories, the British undertook to uphold League of Nations' (later United Nations') administrative rules; and in one case the British shared administration with another government (Egypt) in Sudan's condominium. Britain's colonial map of Africa was a patchwork of these different systems, and the independent countries that emerged reflect the differences. Nigeria, which had been a colony in its south and an indirectly ruled protectorate in its north, became a federal state containing major internal differences. Kenya, the former colony, became a highly centralized unitary state after the Africans wrested control of the productive core area from the whites.

In contrast to the British, the French placed a cloak of uniformity over their colonial territories in Subsaharan Africa. Contiguous and vast, although not very populous, France's colonial empire extended from Senegal eastward to Chad and southward to the former French Congo (now the Congo Republic). This huge area was divided into two units, French West Africa (focused on Dakar) and French Equatorial Africa (whose headquarters was in Brazzaville).

France itself, as we know, is the classic example of the centralized unitary state, whose capital is the cultural, political, and economic focus of the nation, overshadowing all else. The French brought their concept of centralization to Africa as well. In France, all roads lead to Paris; in Africa, all roads were to lead to France, to French culture, to French institutions. For the purposes of assimilation and acculturation, French West Africa, half the size of the entire United States of America (although with a population of only some 30 million in 1960), was divided into administrative units, each centered on the largest town and all oriented toward the governor's headquarters at Dakar. As the map shows, great lengths of these boundaries were straight lines delimited across the West African landscape; history tells us to what extent they were drawn, not on the basis of African realities but for France's administrative convenience. The present-day state of Burkina Faso, for instance, existed as an entity until 1932 when, because of administrative problems, it was divided among Ivory Coast, Soudan (now Mali), and Dahomey (now Benin). Then in 1947, the territory was suddenly recreated and named Upper Volta. Little did the boundary-makers expect that what they were doing would one day affect the national life of an independent country.

Belgian administration in the Congo (now Zaïre) provides another insight into the results, in terms of spatial organization, of a particular set of policies of colonial government. Unlike the French, the Belgians made no effort to acculturate their African subjects. The policies of paternalism actually consisted of rule in the Congo by three sometimes competing interest groups: the Belgian government, the managements of huge mining corporations, and the Roman Catholic Church. Each of these groups had major regional spheres of activity in this vast country.

As the map shows (see p. 430), Zaïre has a corridor to the ocean along the Zaïre River between Angola to the south and the Congo Republic (former French Congo) to the north. The capital, long known as Leopoldville (now Kinshasa), lies at the eastern end of this corridor; not far from the ocean lies the country's major port, Matadi. It was in this corridor that the administrative and transport core area of Zaïre developed, and this was the place from which the decisions made in Brussels were promulgated by a governor-general. The economic core of the Congo (and now of Zaïre) lay in the country's southeast, in the province then known as Katanga, which was administered from the copper-mining headquarters of Lubumbashi (formerly Elisabethville). This is a portion of the northernmost extension of Southern Africa's great mining belt, the balance of which continues into Zambia as that country's Copperbelt. From hundreds of miles around, African workers streamed toward the mines (Fig. 7–18), a pattern of regional labor migration that has persisted for decades and is still going strong. Between the former Belgian Congo's administrative and economic core areas lay the Congo (Zaïre) Basin, once a profitable source of wild rubber and ivory. The colony's six administrative subdivisions were laid out in such a way that each incorporated part of the area's highland rim and part of the forested basin. As the 1960 date of independence approached, it seemed briefly that each of these Congolese provinces, centered on its own administrative capital, might break away and become an independent African country (as each French African dependency did in West and Equatorial Africa). But in the end the country held together, and Leopoldville became Kinshasa, capital of Zaïre, Subsaharan Africa's largest state in territorial size.

Portugal's rule in Angola and Moçambique was designed to exploit four assets: (1) labor supply for the interior mines, especially in Moçambique; (2) transit functions and port facilities—Moçambique from South Africa and Southern Rhodesia, Angola from the Copperbelt and Katanga; (3) agricultural production, particularly cotton from northern Moçambique and coffee from Angola; and (4) minerals, mainly from diamond-and oil-rich Angola. In this effort, the Portuguese created a system of rigid control that involved forced labor and the compulsory farming of certain crops. In Moçambique particularly, the country was divided into a large number of small districts that were tightly controlled. Movement and communication, even within a single African ethnic area, were kept to a minimum. Accordingly, Portuguese colonial rule was often described as the harshest of all the European systems; for a long time, it seemed unlikely that an independence movement could be mounted. But when independence came to Angola's and Moçambique's neighbors (Zaïre and Tanzania), Portugal's days were numbered. As elsewhere in recently colonial Africa, however, the imprint of colonialism in former Portuguese Africa remains a strong, pervasive element in the regional geography of the continent.

Besides the geographic consequences that flowed from the various forms of colonial policy, the current (and future) map of the realm was affected in many other ways as well. In spite of the differences among individual colonial policies, all were paternalistic, assimilative, and exploitive to some degree. However difficult it was to

FIGURE 7–18

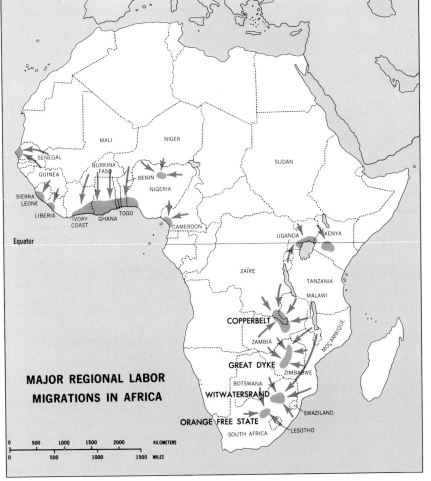

MAJOR REGIONAL LABOR MIGRATIONS IN AFRICA

unify territories within colonial spheres, it has proven virtually impossible to resolve the problems of redefining truly African regional interests without regard to the colonial heritage. With few exceptions, the primary development of capital cities within each colonial territory has been strengthened even further since independence. And much of the transport network at independence reflected the colonial approach to development: railways (if they existed) and major roads (however defined) almost always facilitated the movement of goods between the interior and coastal outlets but offered little basis for interior circulation. The geographic patterns established during the colonial period are more often a hindrance to development than an aid.

AFRICA'S REGIONS TODAY

On the face of it, Africa would seem to be so massive, so compact, and so unbroken that any attempt to justify a contemporary regional breakdown is doomed to failure. No deeply penetrating bays or seas create peninsular fragments as in Europe. No major islands (other than Madagascar) provide the broad regional contrasts we see in Middle America. Nor does Africa really taper southward to the peninsular proportions of South America. And Africa is not cut by an Andean or a Himalayan mountain barrier. Given Africa's colonial fragmentation and cultural mosaic, is regionalization possible? Indeed it is.

FIGURE 7–19

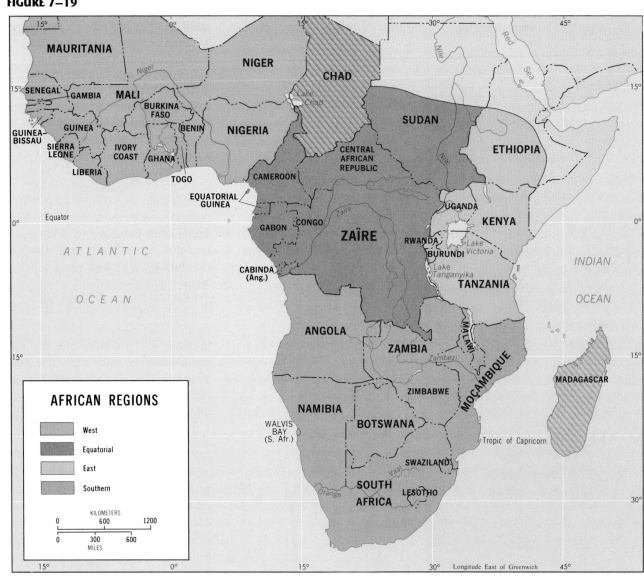

Maps of environmental distributions, population concentrations, ethnic patterns, historic culture hearths, and colonial frameworks yield the four-region structure represented in Fig. 7–19. *West Africa* includes the countries of the western coast and Sahara margin from Senegal and Mauritania in the west to Nigeria and Niger in the east. *Equatorial Africa* is delineated from Nigeria by physiographic as well as cultural breaks: the Adamawa Highlands (the volcanic range containing dangerous gas-emanating lakes) coincide with the border between British-influenced Nigeria and French-acculturated Cameroon. But Chad, located northeast of Cameroon, presents a problem. It possesses both West African and Equatorial African properties; the Sahara-influenced north of Chad is a world apart from the savanna-forested south. So our map shows Chad to be an area of transition between the two regions. Equatorial Africa consists of Zaïre and its neighbors to the north, including the southern part of Sudan. *East Africa* as a geographic region is centered on three countries: Kenya, Tanzania, and Uganda. But two territorially smaller states, Rwanda and Burundi, also lie within East Africa, and the highland portion of Ethiopia is part of this region as well. As the map shows, both Equatorial and East Africa lie astride the equator. But whereas Equatorial Africa is mainly lowland country, with large areas of rainforest and equatorial environments, East Africa is mainly highlands, where cooler and generally drier conditions prevail. *Southern Africa* is the region extending from the southern borders of Zaïre and Tanzania to the continent's southern tip. Madagascar, despite its nearness to Southern Africa, cannot be included in this geographic region without qualification as we will see.

West Africa

West Africa occupies most of Africa's Bulge, extending south from the margins of the Sahara Desert to the coast, and from Lake Chad west to Senegal (Fig. 7–20). Politically, the broadest definition of this region includes all those states that lie to the south of Morocco, Algeria, and Libya, and to the west of Chad (itself sometimes included) and Cameroon. Within West Africa, a rough division is sometimes made between the very large, mostly steppe and desert states that extend across the southern Sahara (Chad could also be included here), and the smaller better-watered coastal states.

Apart from once-Portuguese Guinea-Bissau and long-independent Liberia, West Africa comprises four former British and nine former French dependencies. The British-influenced countries (Nigeria, Ghana, Sierra Leone, and Gambia) lie separated from one another, whereas Francophone West Africa is contiguous. As Fig. 7–20 shows, political boundaries extend from the coast into the interior, so that from Mauritania to Nigeria the West African habitat is parceled out among parallel, coast-oriented states. Across these boundaries, especially across those between former British and former French territories, there is very little trade. For example, in terms of value Nigeria's trade with Britain is about 100 times as great as its trade with nearby Ghana. The countries of West Africa are not interdependent economically, and their incomes are to a large extent derived from the sale of their products on the international market. But the African countries do not control the prices that their goods can command on these world markets, and when prices are low, they face serious problems.

Given these cross-currents of subdivision within West Africa, what are the justifications for the concept of a single West African region? First, there is the remarkable cultural and historical momentum of this part of the realm. The colonial interlude failed to extinguish West African vitality, expressed not only by the old states and empires of the savanna and the cities of the forest, but also by the vigor and entrepreneurship, the achievements in sculpture, music, and dance of peoples from Senegal to Nigeria's southeastern Iboland. Second, West Africa contains a set of parallel east-west ecological belts, clearly reflected in Figs. I–9 to I–13, whose role in the development of the region is pervasive. As the transport-route pattern on the map of West Africa indicates, overland connections within each of these belts, from country to country, are quite poor; no coastal or interior railroad was ever built to connect this tier of countries. On the other hand, spatial interaction is stronger across these belts, and some north-south economic exchange does take place, notably in the coastal consumption of meat from cattle raised in the northern savannas. And third, West Africa received an early and crucial impact from European colonialism, which—with its maritime commerce and the slave trade—transformed the region from one end to the other. This impact was felt all the way to the heart of the Sahara, and it set the stage for the reorientation of the whole area—out of which emerged the present patchwork of states.

The effects of the slave trade notwithstanding, West Africa today is Subsaharan Africa's most populous region (Fig. I–14). In these terms, Nigeria (officially containing 122 million but with perhaps as many as 150 million people) is Africa's largest state, and Ghana (15.5 million) also

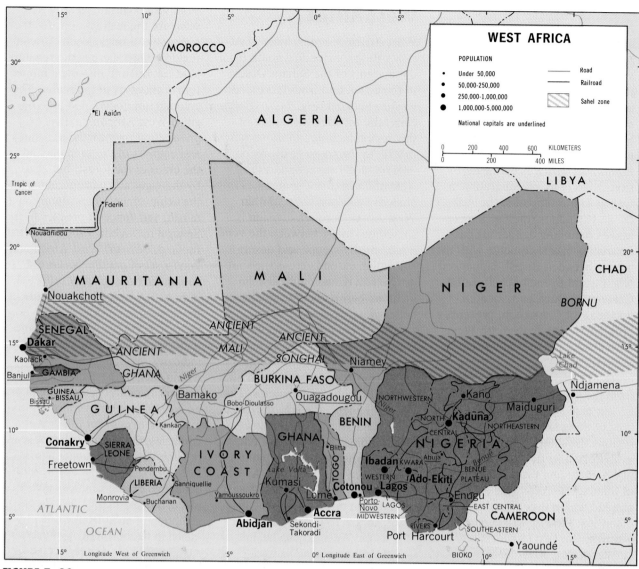

FIGURE 7–20

ranks high. As Fig. I–14 shows, West Africa also claims regional identity in that it constitutes one of Africa's major population clusters. The southern half of the region, understandably, is home to the majority of the people. Mauritania, Mali, and Niger include too much of the unproductive Sahel's steppe and the Sahara Desert to sustain populations comparable to those of Nigeria, Ghana, or Ivory Coast. However, this is not to say that only the coastal areas of West Africa are densely populated; the interior savannalands also contain sizable clusters. The peoples along the coast reflect the modern era in-

troduced by the colonial powers: they prospered in their newfound roles as middlemen in the coastward trade. Later, they were in a position to experience the changes the colonial period brought; in education, religion, urbanization, agriculture, politics, health, and many other endeavors, they adopted new ways. The peoples of the interior, on the other hand, retained their ties with a very different era in African history; distant and often aloof from the main theater of European colonial activity, they experienced significantly less change. But the map reminds us that Africa's boundaries were not drawn to

accommodate such differences. Both Nigeria and Ghana possess population clusters representing the interior as well as the coastal peoples, and in both countries the wide gap between north and south has produced political problems.

Nigeria: West Africa's Cornerstone

When Nigeria achieved full independence in 1960, it was endowed with a federal political structure that consisted of three regions based on the three major population clusters within its borders— two in the south and one in the north. Around the Yoruba core in the southwest lay the Western Region. The Yoruba are a people with a long history of urbanization, but they are also farmers; in the old days, they protected themselves in walled cities around which they practiced intensive agriculture. The colonial period brought coastal trade, increased urbanization, cash crops (the mainstay, cocoa, was introduced from Fernando Póo, now Bioko, in the 1870s), and, eventually, a measure of security against encroachment from the north. Lagos, the country's first federal capital and now a teeming conurbation of 8.0 million people, grew up around port facilities on the region's south coast. A new more centrally located capital is now being completed at Abuja (see Fig. 7–20).

Ibadan (1.7 million), now also one of Subsaharan Africa's largest cities, evolved from a Yoruba settlement founded in the late eighteenth century. At independence, Nigeria's Western Region, more than any other part of Nigeria, had been transformed by the colonial experience.

East of the Niger River and south of the Benue River, the Ibo population formed the core of the Eastern Region. Iboland, although coastal, lay less directly in the path of colonial change, and its history and traditions also differed sharply from those of Western Nigeria. Little urbanization had taken place here, and even today, although over one-third of the Western Region's people live in cities and towns, less than 20 percent of the Eastern population is

urbanized. With more than 20 million people, the rural areas of Eastern Nigeria are densely peopled. Over the years, many Ibo have left their crowded habitat to seek work elsewhere, in the west, in the far north, in Cameroon, and even on the island of Bioko.

The third federal region at independence was at once the largest and the most populous: the Northern Region. It extended across the full width of the country from east to west and from the northern border southward beyond the Niger and Benue rivers. This is Nigeria's Muslim North, centered on the Hausa-Fulani population cluster (Fig. 7–21), where the legacy of a feudal social system, conservative traditionalism, and resistance to change hung heavily over the country.

FIGURE 7–21 Friday prayers attract a multitude of the faithful to a mosque in Kano, a major urban center in Nigeria's Islamic-dominated north.

Nigeria's three original regions, then, lay separated not only by sheer distance, but also by tradition and history, and by the nature of colonial rule. In Nigeria, even physiography and biogeography conspire to divide south from north: across the heart of the country (and across much of West Africa at about the same latitude) stretches the so-called *Middle Belt*—poor, unproductive, disease- and tsetse-ridden country that forms a relatively empty barrier between the northern and southern regions. Not surprisingly, Nigeria's three-region federal system failed. Interregional rivalries and interethnic suspicions led to civil war between 1967 and 1971, when the Eastern Region tried to secede as a separate political entity called Biafra. The original three regions were subdivided, rearranged, and subdivided once again in an attempt to devise a system that would prevent such disasters in the future.

Nigeria today is a cornerstone of the new Africa, a country whose plans for economic growth are based on substantial oil reserves

in the area of the Niger Delta. By the early 1980s, over 90 percent of Nigeria's export revenues were derived from the sale of petroleum and petroleum products, and cities from Port Harcourt (450,000) to Lagos reflected the oil boom of the south. This heavy dependence on petroleum, however, made the country's ambitious development plans a hostage to world oil prices—and lower prices in recent years have thrown the Nigerian economy into disarray. With unemployment rising, the government expelled virtually all non-Nigerian workers and their families (as well as illegal aliens), forcing millions to return to their countries of origin where local economies were no more able to absorb them (Fig. 7–22).

Coastal States

Westward from Nigeria, West Africa's coastline presents a complex and fascinating cultural and political mosaic. Benin, Nigeria's neighbor, takes its name from one of the region's powerful kingdoms

of the past. Many West Africans were carried as slaves from this area to Brazil, and today there is a growing cultural and emotional link between the people of Benin and those of African ancestry in and near Salvador in Bahia State (see map, p. 322) on the opposite side of the Atlantic. Next to Togo, Benin's western neighbor, briefly a German colony and later administered by France, lies Ghana. This country, once known as the Gold Coast, also took its name from an ancient West African state. Ghana was the first West African state to achieve independence (in 1957), but like so many other African countries it has suffered from ineffective government, economic mistakes, corruption, and instability. Optimistic predictions heralding the success of the British "Westminster Model" in Africa were erased by the harsh realities of tribalism and military coups. Not only did Ghana, its economy once prosperous from the sale of cocoa, collapse under Moscow-style planning programs; money was siphoned off to foreign accounts and thousands of the most capable Ghanaians left the country that needed them most. Two of the projects promoted by the first president of Ghana, Kwame Nkrumah, are prominent on the country's map: the great Volta River Dam, and the port of Tema near the capital city of Accra (1.6 million). Neither fulfilled its expectations, and President Nkrumah died in dishonor and exile.

Compared to Ghana's fate, the Ivory Coast fared much better, in large part because the country enjoyed several decades of stability. *Côte d'Ivoire* (the country's official name) was one of France's two most important West African possessions (the other was Senegal), and the president who negotiated independence in 1960 was still at the helm in 1990. During its first 30 years, the Ivory Coast's

FIGURE 7–22 Illegal immigrants congregating at Nigeria's western border with Benin prior to their deportation, one result of widespread Nigerian unemployment triggered by the 1980s drop in world oil prices.

FIGURE 7–23   One of the most astonishing structures ever placed on Africa's human landscape—the recently completed St. Peter's-scale basilica, Notre Dame de la Paix (Our Lady of Peace), in Ivory Coast's capital, Yamoussoukro.

capital, Abidjan (1.8 million), grew into a major urban and industrial center; the cocoa- and coffee-based economy was diversified, and living standards rose. French involvement in the country remained quite strong. But problems also emerged. The Sahelian drought drove cattle herders into northern Ivory Coast, and friction with local farmers whose crops were trampled created a challenge for the government. Worse, President Félix Houphouët-Boigny engineered the movement of the capital to Yamoussoukro (his home town) and embarked on a multimillion dollar project to build there a Roman Catholic basilica to rival that of St. Peter's in Rome (Fig. 7–23); resentment over these costs was among the causes behind unprecedented street disorders as the country's fourth decade of independence opened. As economic growth slowed, the stability of the Ivory Coast was threatened.

France's other major coastal colony, Senegal (capital: Dakar, with 2.0 million people), was the anchor of its West African empire and remains a leading regional state today. As Fig. 7–20 shows, Senegal lay in the path of the Sahel's environmental onslaught, and the country's north suffered especially severely from drought and dislocation. Senegal's economic geography remains fragile: the country depends on the export of peanuts and phosphates, and on its fishing industry. But droughts, reduced market prices for its exports, and a high rate of population growth have inhibited development. It is the story of many an African country.

Between Ivory Coast and Senegal, no fewer than four countries face the Atlantic Ocean: Liberia, the one corner of West Africa that was never colonized and to which many freed slaves from America settled upon their return to Africa; Sierra Leone, a former British dependency that also was the destination of returning slaves; Guinea, one of coastal West Africa's least developed countries from the former French empire; and Guinea-Bissau, a small Portuguese-influenced country. All four suffer severely from the economic, demographic, medical, and political problems that afflict the region (and the realm) as a whole.

Periodic Markets

The great majority of the people of West Africa are not involved in the production of exports for world markets, but subsist on what they can grow and raise—and trade. Their local transactions take place at small markets in villages (Fig. 7–24). These village markets are not open every day, but operate at regular intervals. In this way, several villages in an area get their turn to attract the day's trade and exchange, and each benefits from its participation in the wider network of interactions. People come to these **periodic markets** on foot, by bicycle, on the backs of their animals, or by whatever other means available. Periodic markets are not exclusively a West African phenomenon. They also occur in interior Southeast Asia, in China, and in Middle and South America as well as in other parts of Africa. The intervals between market days vary. In much of West Africa, village markets tend to be held on every fourth day, although some areas have two-day, three-day, or eight-day cycles.

Periodic markets, then, form an interlocking network of exchange places that serves rural areas where there are no roads. As each market in the network gets its turn, it will be near enough to one portion of the area so that the people who live in the vicinity can walk to it, carrying what they wish to sell or trade. In this way, small amounts of produce filter through the market chain to a larger regional market, where shipments are collected for interregional or perhaps even international trade. What is traded, of course, depends on where the market is located. A visit to a market in West Africa's forest zone will produce very different impressions from a similar visit in the savanna zone. In the savanna, sorghum, millet, and shea-butter (an edible oil drawn from the shea-nut) predominate, and you will see some Islamic influences on the local scene. In the southern forest zone, such products as yams, cassava, corn, and palm oil change hands; here, too, one is more likely to see some imported manufactured goods passing through the market chain, especially near the relatively prosper-

FIGURE 7–24   A periodic market turns one of Mali's villages into a beehive of activity.

ous areas of cocoa, coffee, rubber, and palm oil production. But in general, the quantities of trade are very small—a bowl of sorghum, a bundle of firewood—and their value is low; these markets serve people who mostly live at or near the subsistence level.

East Africa

East Africa is highland, plateau Africa, mainly savanna country that turns into steppe toward the drier northeast. Great volcanic mountains rise above a plateau that is cut by the giant rift valleys

(Figs. 7–3 and 7–5). The pivotal physical feature is Lake Victoria, where the three major countries' boundaries come together (Fig. 7–25) and on whose shores lie the primary core area of Uganda and the secondary cores of both Kenya and Tanzania. With limited min-

FIGURE 7–25

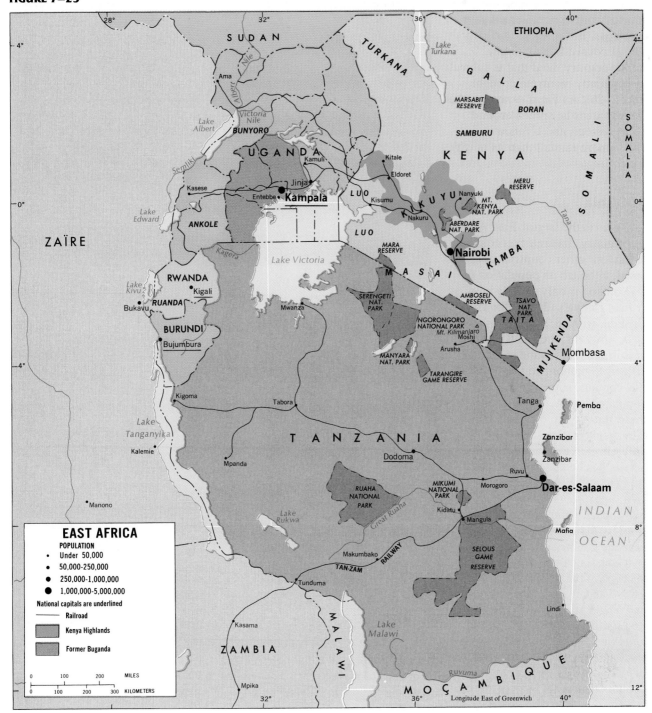

eral resources (the chief ones are diamonds in Tanzania south of Lake Victoria and copper in Uganda to the west of the lake), most people in this region depend on the land—and on the water that allows crops to grow and livestock to graze. In much of East Africa, rainfall amounts are marginal or insufficient. The heart of Tanzania is dry, tsetse and malaria-ridden, and still suffers occasional food shortages. Eastern and northern Kenya consist largely of dry steppe country, land that is subject to frequently recurring drought (Fig. 7–26). As Fig. I–9 shows, the wettest areas lie spread around Lake Victoria and Uganda receives more rainfall than its neighbors.

Tanzania

Tanzania is the largest East African country. Its area exceeds that of the four other countries combined, although its population of 28.3 million is only slightly larger than Kenya's (26.1 million). Uganda, with 18.2 million, ranks next; however, the small countries of Burundi and Rwanda together exceed 13 million, so that East Africa contains some very densely populated areas. Although all these states lie in the same region, they display some strong internal differences. Tanzania, for example, has been described as a country without a primary core area, because its zones of productive capacity and its population (Fig. I–14) lie dispersed, mostly near the country's margins on the east coast, near the shores of Lake Victoria in the northwest, near Lake Tanganyika in the far west, and near Lake Malawi in the interior south. Kenya, on the other hand, has several good-quality agricultural areas and a strongly concentrated core area centered on its capital of Nairobi. Moreover, Tan-

FIGURE 7–26 The same stretch of East African river, two decades apart: drought and vegetation destruction bring desert conditions to the very water's edge.

zania is a country of many ethnic groups, none with a numerical or locational advantage to enable domination of the state.

Tanzania has been viewed as an important example for the rest of Africa. Notwithstanding its limited resources and fragmented, dispersed population, the country achieved political stability in the face of pressures generated by its transition to self-help African socialism. Tanzania has some commercial agriculture (sisal plantations along the north coast, coffee farms on the slopes of Mount Kilimanjaro near the Kenya border, cotton south of Lake Victoria, and tea in the southwest near Lake Malawi), but the great majority of Tanzanians are subsistence farmers. The government of former president Julius Nyerere undertook a major reorganization of agriculture, forming cooperatives, supporting improved farming methods, and creating thousands of new villages, into which about 90 percent of the rural population were relocated. There was opposition, of course, but the new order prevailed. All this occurred while Tanzania was a haven for insurgents fighting the Portuguese in Moçambique, a difficult merger with the island of Zanzibar was accomplished, and the capital was moved from coastal Dar es Salaam (see *box*) to Dodoma in the interior. At the same time, the Chinese were in Tanzania building the Tan-Zam (Tanzania-Zambia) Railway (Fig. 7–25).

The results of Tanzania's social experiment—at least as a model for the Third World to follow—have been disappointing. Much of the country did receive infusions of Western capital and technology to support numerous development schemes, but today the rural landscape is dotted with the abandoned remains of farm equipment, factories, and water-pumping sys-

SEQUENT OCCUPANCE

The African realm affords numerous opportunities to illustrate the geographic concept of **sequent occupance** (first introduced on p. 8). This concept involves the study of an area that has been inhabited—and transformed—by a succession of residents, each of whom has left a lasting cultural imprint. A place and its resources are perceived differently by peoples of different technological and other cultural traditions. These contrasting perceptions are reflected in the contents of their cultural landscapes. The cultural landscape today, therefore, is a composite of these contributions, and the challenge is to reconstruct the contributions made by each successive community. The idea of sequent occupance is applicable in rural as well as urban areas. The ancient San peoples used the hillsides and valleys of Swaziland to hunt and to gather roots, berries, and other edibles. Then the cattle-herding Bantu found the same slopes to be good for grazing, and in the valleys planted corn and other food crops. Next came the Europeans who laid out sugar plantations in the lowlands; but after using the higher slopes for grazing, they planted extensive forests, and lumbering became the major upland industry.

The Tanzanian coastal city of Dar es Salaam (1.8 million) provides a striking urban example of sequent occupance. Its Indian Ocean site was first chosen for settlement by Arabs from Zanzibar to serve as a mainland retreat. Next, it was selected by the German colonizers as a capital for their East African domain, and it was given a German layout and architectural imprint. When the Germans were ousted following their defeat in World War I, a British administration took over in "Dar," and the city entered still another period of transformation; a large Asian population soon left it with a zone of three-and four-story apartment houses that seem transplanted directly from India. Then, in the early 1960s, Dar es Salaam became the capital of newly independent Tanzania, under African control for the first time. Thus, Dar es Salaam in less than one century experienced four quite distinct stages of cultural dominance, and each stage of the sequence remains etched into the cultural landscape. Indeed, a fifth stage has now begun, since national government functions were recently moved from Dar es Salaam to the new interior capital of Dodoma.

Because of Africa's cultural complexity and historical staging, numerous examples of this kind exist. The Dutch, British, Union, and Republican periods can be observed in the architecture and spatial structure of Cape Town and other urban places in South Africa's Cape Province; Mombasa in Kenya carries imprints from early Persian, Portuguese, Arab, British, and several African communities. In the countryside, African traditional farmlands taken over by Europeans and fenced and farmed are now being reorganized by new African proprietors, and contrasting attitudes toward soil and slope are clearly imprinted in the landscape.

tems that the Tanzanians were unable to operate and maintain because they lacked the necessary funding, expertise, and spare parts. Moreover, the "villagization" program has compounded the country's agricultural difficulties in an already fickle natural environment: village sites were picked more for accessibility than availability of fertile soils, and far too many are surrounded by inferior farmlands. Thus, the Tanzanians have experienced a decline in per capita food production and now rely heavily on Western grain and emergency food imports. In the early 1990s, Tanzania's government was experimenting in a new direction, reintroducing open-market principles alongside the remains of the socialist system. But the dis-

location forced upon Tanzania's countryside will take a long time to overcome.

Kenya

Kenya's different path of development is reflected by the tall buildings of Nairobi (1.7 million), the capital and primate city (Fig. 7–27), and the productive farms of the nearby highlands. But Kenya's development is concentrated, not spatially dispersed like Tanzania's, so that the evidence of post-independence development is strong in the core area but very limited in the sparsely peopled interior. As Fig. 7–25 shows, both Tanzania and Kenya have a single-line railroad that traverses the whole coun-

try from the major port in the east (Mombasa [565,000] in the case of Kenya) to the far west, and feeder lines come from north and south to meet this central transport route. In Kenya, that central railroad and its branches in the highlands really represent the essence of the country, but in Tanzania the railroad lies in the "empty heart" of the country, and it is the branch lines that lead to the productive and populated peripheral zones. The Tan-Zam Railway was to provide Tanzania with a link to neighboring landlocked Zambia as well as access to a poorly developed domestic area with relatively good agricultural potential in the southern highlands. Problems of operating costs and maintenance practices, unfortunately, side-

FIGURE 7–27 The modern center of Nairobi is the focus of East Africa's biggest metropolis, which now contains a rapidly growing population that will surpass 2 million by the mid-1990s.

tracked the project, and the Tan-Zam railroad has been a failure.

Kenya does not possess major known mineral deposits. Besides its coffee and tea exports, Kenya earns substantial revenues from a tourist industry that grew rapidly during the 1960s and 1970s, until East African political difficulties that began in the late 1970s reduced the flow of visitors. In the mid-1970s, the total number of foreign visitors to Kenya was approaching half a million, and the industry became the largest single earner of foreign exchange. Most of Kenya's famous wildlife reserves lie in the drier and more remote parts of the country, but population pressure is a growing problem, even in those areas, and poses still another threat to the future of tourism.

Indeed, runaway population growth has become a leading national crisis. Column 6 in Appendix A reveals that Kenya now contains the world's fastest growing population (4.1 percent annually). At today's rate of expansion, the doubling time is only 17 years, so that by 2025 there will be almost 105 million Kenyans—*four* times the population total of 1991! Improved health standards are responsible for much of this population surge: far more infants are now surviving the critical first year, and at the other end of the age scale people are living longer lives. Life expectancy has jumped from 39 thirty years ago to 57 today, and over half of the population is now under age 15. There is little the government can do to combat this **population spiral**. In the mid-1980s, the average Kenyan woman had *8.1 children*; even if the average number of children per woman were to drop from eight to two by the year 2000, the annual population growth rate would still be a very high 3.5 percent. But fewer than 20 per-

cent of Kenya's married women now use birth-control devices, and family-planning programs have been ineffective because they collide with the values and traditions of most of Kenya's diverse cultural groups, many of whom practice polygamy and measure a woman's status in society according to the number of children she has borne. In a dry country, less than 25 percent of which is fertile farmland, the demographic trends of the early 1990s truly portend impending disaster.

Uganda

Uganda contained the most important African political entity in the region when the British entered the scene during the second half of the nineteenth century. This was the Kingdom of Buganda (shown in brown in Fig. 7–25), which faced the north shore of Lake Victoria, had an impressive capital at Kampala, and was stable—as well as ideally suited for indirect rule over a large hinterland. The British established their headquarters at nearby Entebbe on the lake (thus adding to the status of the kingdom), and proceeded to organize their Uganda Protectorate in accordance with the principles of indirect rule. The Baganda (the people of Buganda) became the dominant people in Uganda, and when the British departed, they bequeathed Uganda with a complicated federal system that was designed to perpetuate Baganda supremacy.

Although a **landlocked state** dependent on Kenya for an outlet to the ocean, Uganda at independence (1962) had better economic prospects than many other African countries. It was the largest producer of coffee in the British Commonwealth; it had major cotton exports, and tea, sugar, and other farm products also were exported;

copper was mined in the southwest; an Asian immigrant population of about 75,000 played a leading role in the country's commerce. But political disaster overtook the country. Resentment at Baganda overlordship fueled revolutionary change, and a brutal dictator, Idi Amin, took control in 1971. He ousted the Asians, exterminated his opposition, and destroyed the economy. Eventually, a military campaign supported by neighboring Tanzania drove Amin from power, but by then (1979) Uganda lay in ruins. Recovery has been slow, complicated by the high cost of AIDS which struck Uganda with particular severity. The departure of the British colonizers set in motion a sequence of calamities from which the country may never fully recover.

Equatorial Africa

This region, like East Africa, lies astride the equator, but in Equatorial Africa elevations are lower than in the highland east. As a geographic region (Fig. 7–28), Equatorial Africa consists of Zaïre, Congo, the Central African Republic, Gabon, Cameroon, Equatorial Guinea, and the southern reaches of Chad. (We noted the transitional character of Chad previously; see Fig. 7–19).

Zaïre

The giant of Equatorial Africa is Zaïre, the former Belgian Congo. As Fig. 7–28 shows, Zaïre is one of Africa's biggest countries territorially; its population of 37 million also is among Subsaharan Africa's largest. Zaïre contains a wide range of resources, the region's largest cities, and its greatest potential for development. But the country has been hampered by transport and communication

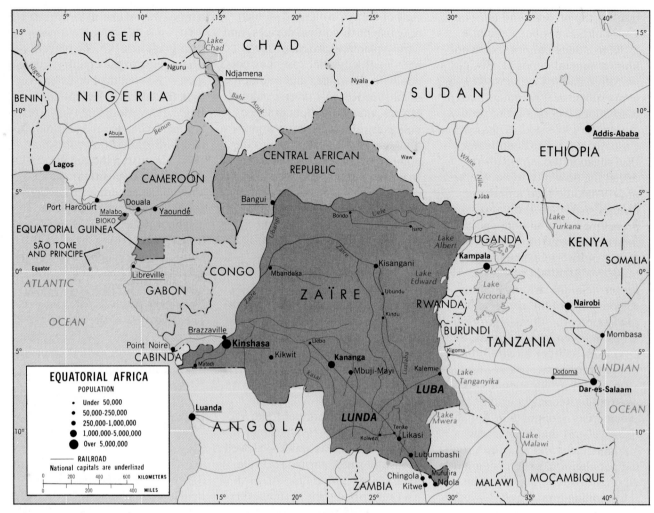

FIGURE 7–28

problems, by a lack of national cohesion, by falling prices for its exports, and by autocratic and corrupt government.

Zaïre's core area lies around the capital, Kinshasa (3.8 million), and along its narrow corridor to the port of Matadi (180,000) near the Atlantic Ocean. But its economic focus has long been in the distant southeast, around the city of Lubumbashi (720,000), where most of the valuable mineral resources (chiefly copper) lie. The great Zaïre River system would seem to create a natural transport network, but a series of rapids interrupts this route and several transshipments are necessary. It has always been easier to send exports through Angola than through

Zaïre itself. Despite major investments in internal transportation links, the situation today is not much better. Internal circulation in Zaïre is plagued by breakdowns and delays, because the huge forested basin that forms the center of the country remains a formidable obstacle even to modern equipment. As the Brazilians know, a large territory can be a liability as well as an asset when it comes to national integration and orderly development.

The Remaining States

Only one of the five remaining countries of the region is landlocked: the Central African Republic, one of Africa's least

developed states. Congo, lying to the west across the river from Zaïre, has few natural resources but does enjoy some locational advantages. Brazzaville (785,000), its capital, was France's Equatorial African headquarters, and Pointe-Noire (440,000) was its major port. This gave Congo a transit function that brought in revenues, but unfortunately the country itself has little to trade. Its western neighbor, Gabon, has brighter prospects: its diversifying economic geography includes oil reserves, manganese, uranium, and iron ores, and a productive (but destructive) logging industry. But the agricultural sector is weak, and large investments in transport systems were necessary (even though

Gabon is only one-ninth the size of Zaïre). Cameroon, north of Gabon, contains tropical forests in the south and open savannas in the north, but the real contrasts are from west to east. Western Cameroon, by Equatorial African standards, has progressed considerably, based on some oil reserves, a diversified agricultural output (Fig. 7–29), exports of forest products, and modest industrial growth. Here lie the capital, Yaoundé (675,000), and the port of Douala (900,000). By comparison, the interior east lags in isolation and remoteness.

By any ranking, Equatorial Africa as a whole remains the least developed of Subsaharan Africa's regions. Subsistence is the main activity, and economic transactions involve mostly the export of raw materials. In only a few places in Zaïre, Cameroon, and Gabon can the beginnings of a new age be detected.

Southern Africa

Southern Africa, as a geographic region, consists of all the countries and territories lying south of Equatorial Africa's Zaïre and East Africa's Tanzania (Fig. 7–30). Thus defined, the region extends from Angola and Moçambique (on the Atlantic and Indian Ocean coasts, respectively) to the turbulent Republic of South Africa, and includes a half-dozen landlocked states. Also marking the northern limit of the region are Zambia and Malawi. Zambia is nearly cut in half by a long land extension from Zaïre, and Malawi penetrates deeply into Moçambique. The colonial boundary framework here, as elsewhere, produced many liabilities.

Southern Africa constitutes a geographic region in physiographic as well as in human terms. Its northern zone marks the southern

FIGURE 7–29 One of Equatorial Africa's few nonsubsistence farming landscapes: the banana-plantation zone around Buea near the Cameroon coast.

limit of the Zaïre Basin in a broad upland that stretches across Angola and into Zambia (the brown corridor extending east from the Bihe Plateau in Fig. 7–4). Lake Malawi is the southernmost of the East African rift valley lakes; Southern Africa has none of East Africa's volcanic and earthquake activity. Much of the region is plateau country, and the Great Escarpment is much in evidence here. There are two pivotal river systems: the Zambezi (which forms the border between Zambia and Zimbabwe) and the Orange-Vaal (South African rivers that combine to demarcate southern Namibia from South Africa).

The social, economic, and political geographies of Southern Africa also confirm the regional definition. Landlocked Zambia, whose economic core area is the Copperbelt northwest of Ndola, always looked southward for its outlets to the sea, its electric power, and its fuels. Malawi's core area lies in the south of the country, and its outlets, too, are southward. Moçambique, with its lengthy Indian Ocean coastline, has served

as an exit not only for Malawi but for Zimbabwe (through the port of Beira) and South Africa (through Maputo) as well. Offshore Madagascar, however, remains a separate entity (see *box,* p. 434).

Southern Africa is the continent's richest region in material terms. A great zone of mineral deposits extends through the heart of the region from Zambia's Copperbelt, through Zimbabwe's Great Dyke and South Africa's Bushveld Basin and Witwatersrand, to the goldfields and diamond mines of the Orange Free State and northern Cape Province in the heart of South Africa. Ever since colonial exploitation of these minerals began, large numbers of migrant laborers have come to work in the mines (Fig. 7–18). The range and volume of minerals mined in this belt are enormous, from the copper of Zambia (Fig. 7–31) and the chrome and asbestos of Zimbabwe to the gold, chromium, diamonds, platinum, coal, and iron ore of South Africa. However, not all of Southern Africa's mineral deposits lie in this central backbone. There is coal in western Zimbabwe at

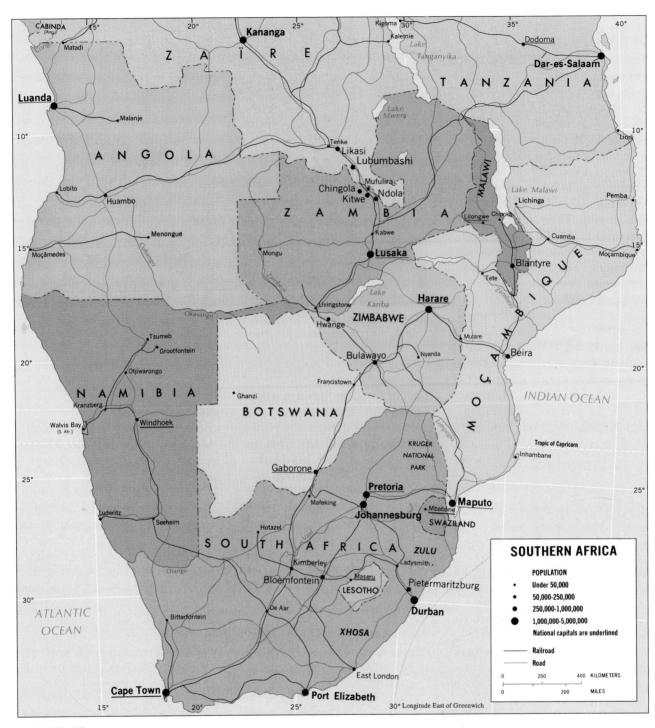

FIGURE 7—30

Hwange (formerly Wankie), and in central Moçambique near Tete. In Angola, petroleum from oilfields along the north coast heads the export list; diamonds are mined in the northeast, manganese and iron on the central plateau. Namibia produces copper, lead, and zinc from a major mining complex anchored to the town of Tsumeb in the north, and diamond deposits are worked along the beaches facing the Atlantic Ocean in the south.

This is a mere summary of Southern Africa's mineral wealth, and it is matched by the variety of crops cultivated in the region. Vineyards drape the valleys of the Cape Ranges in South Africa; apple orchards, citrus groves, banana plantations, and fields of sugarcane, pineapples, and cotton reflect South Africa's diverse natural environments. In Zimbabwe, tobacco has long been the leading commercial crop, but cotton also grows on the plateau and tea thrives along the eastern escarpment slopes. Angola is a major coffee producer. The staple crop for the majority of Southern Africa's farmers is corn (maize), but wheat and other cereals are also grown. Even the pastoral industry is marked by variety: not only are dairying and beef cattle herds large, but South Africa also exports sizable quantities of wool.

Despite this considerable wealth and productivity, the countries of Southern Africa are not well off. Only the most advantaged parts of South Africa can be described as developed; but severely underdeveloped areas exist even within that country.

The Northern Tier

Within the four countries that extend across the northern part of the region—Angola, Zambia, Malawi, and Moçambique—problems abound. Angola (population: 9 million), formerly a Portuguese dependency, had a thriving economy at independence but was engulfed by civil strife that was worsened by outside involvement. Its government initially chose a Marxist course, and when a rebel movement emerged in the south, the South African government supported it financially and militarily. The Angolan leaders called on Moscow for help, and thousands of Cuban troops arrived in the capital of Luanda (1.4 million) to help combat the southern insurgents. In addition, political conflict in neigh-

boring Namibia spilled over into Angola as opponents of the South African administration of Namibia sought refuge there. The impact of all this on the Angolan economy was devastating. Farms were abandoned, railroads mined, port facilities damaged, funds diverted to military use. As the map (Fig. 7–30) shows, a small part of Angola lies separated from the main territory along the coast to the north. This small exclave, Cabinda, contains oil reserves that continued to produce revenues for the Luanda government through the post-independence crisis—or Angola would have been totally bankrupted by the war.

On the opposite coast, Moçambique (16.1 million) was even less fortunate. Also a former Portuguese domain that chose a Marxist

FIGURE 7–31 One of the huge open-pit copper mines of Chingola in the heart of Zambia's ore-rich Copperbelt.

MADAGASCAR

Maps showing the geographic regions of the Subsaharan Africa realm often leave Madagascar unassigned. And for good reason: Madagascar differs strongly from Southern Africa, the region to which it is nearest, just 250 miles away. It also differs from East Africa, from where it has received some of its cultural infusions. Madagascar is the world's fourth-largest island, a huge block of Africa that separated from the main landmass 80 million years ago. About 2000 years ago, the first settlers arrived—not from Africa (although perhaps via Africa), but from Southeast Asia. Malay communities strengthened in the interior highlands of the island, which resembles Africa in a prominent eastern escarpment and a central plateau. Here was formed a powerful kingdom, the empire of the Merina. Its language, Malagasy, of Malay-Polynesian origin, became the indigenous tongue of the entire island (Fig. 7–1).

The Malay and Indonesian immigrants brought Africans to the island as wives and slaves, and from this forced immigration evolved the African component in Madagascar's population of some 12 million. In all, nearly 20 discrete ethnic groups coexist, among which the Merina (3.2 million) and Betsileo (1.5 million)—both plateau peoples—are the most numerous. Like mainland Africa, Madagascar experienced colonial invasion and competition. Portuguese, British, and French colonists appeared after 1500, but the Merina were well organized and long resisted the colonial conquest. Eventually, Madagascar became part of France's colonial empire, and French became the *lingua franca* of the educated elites.

As would be expected, the staple food of Madagascar is rice, not corn. Export crops include coffee, vanilla, and cloves. There is mining of chromite and graphite, and Madagascar has potential oil reserves in coastal and offshore areas. But the period since independence in 1960 has been difficult. Once a rice exporter, the island now requires imports of that staple as population growth increased and agricultural efficiency declined. The flow of exports also shrank as political turmoil engulfed Madagascar. For some years named the Malagasy Republic, Madagascar witnessed authoritarian government, repression, rebellion, and instability. In 1975, following a military coup, a new government was elected and a new constitution written. The country regained its official name: Democratic Republic of Madagascar. During the 1960s and early 1970s Madagascar was an isolated country, almost completely closed to outsiders. But the 1980s have seen a new period of international ties and assistance, research, and reconstruction.

Madagascar's capital, Antananarivo (1.1 million), is the country's primate city, its architecture and atmosphere combining traces of Asia and Africa. The countryside, too, is reminiscent of Southeast Asian villages and terraced farms on an African physical landscape. And in what remains of Madagascar's forests are the remnants of one of the world's most unusual fauna, now threatened with extinction by the rising tide of humanity.

course for development, Moçambique had far fewer resources than Angola. Before independence, its chief sources of income came from its cashew and coconut plantations, and from its relative location. As the map shows, Moçambique's port of Beira (430,000) is ideally situated to handle the external trade of Zimbabwe and southern Malawi, and its capital, Maputo (1.4 million), is the closest port to South Africa's Witwatersrand mining and industrial complex. Goods streamed through these ports, and when hydroelectric facilities on the Zambezi River (the Cabora Bassa Dam) neared completion, prospects seemed fair. But today, Moçambique is often identified as one of the world's most severely underdeveloped countries, subject to famines, social dislocation, political disorder, and economic chaos. Again, a rebel movement, supported for a time by South Africa, contributed to this crisis. The hydroelectric project was damaged; the port facilities at Beira ceased functioning; Maputo's transit function all but ceased. The plantations lay untended, and the tourist industry, once a promising sector, was destroyed. Moçambique by the early 1990s had changed its political course, and an accommodation with the South Africans was reached. But it will take generations for the country to climb from the depth of impoverishment to which it has been condemned.

Between Angola and Moçambique lie three landlocked states that formed part of the British colonial empire: Zambia, Malawi, and Zimbabwe. Zambia (8.8 million) and Malawi (9.3 million) are less developed than Zimbabwe, although Zambia contains the mineral-rich Copperbelt. The decline of world prices for its minerals, plus the problems and costs associated with their long-distance transportation, have hurt the economy of Zambia. Malawi's economic geography is almost completely agricultural, with a variety of crops including tea, cotton, tobacco, and peanuts; this has helped cushion the economy against market swings, and has also involved a larger percentage of the labor force in productive work. Still, more workers labor in the mines of other countries than are gainfully employed within Malawi itself.

Southern States

Six countries constitute Africa's southernmost tier of states and form a distinct subregion within Southern Africa: Zimbabwe, Namibia, Botswana, Swaziland, Lesotho, and the Republic of South Africa. As Fig. 7–30 shows, four of the six are landlocked. Botswana occupies the heart of the Kalahari Desert and surrounding steppe; only 1.3 million people inhabit this Texas-sized country. Lesotho (1.8 million) is encircled by South African territory, and Swaziland (815,000), ancestral home of the Swazi nation, is largely enclosed by the republic as well. All three of these countries depend heavily on the income earned by workers who are employed by South African mines, factories, and farms.

Zimbabwe

Zimbabwe located directly north of South Africa, is an important African country of 10.9 million people. When the British moved this country toward sovereignty, the local white minority declared independence from London in 1965, hoping to retain control. A bitter civil war ensued, which resulted in a settlement and black majority rule in 1980. After independence, friction between the Shona majority and the Ndebele minority clouded the future, but the majority's goal of a one-party state was achieved.

Zimbabwe is a well-defined country. To the north lie the Zambezi River and the giant Kariba Dam. To the east the plateau descends to Moçambique's coastal plain; here, local soils and climates vary and a range of crops is grown. To the west lie Botswana and the Kalahari's arid plains. And to the south, beyond the Limpopo River, lies South Africa, now facing the struggle from which Zimbabwe is emerging.

The core area is defined by the mineral-rich Great Dyke and its environs, extending southwest from the vicinity of the capital, Harare (1.6 million), to the country's second city, Bulawayo (710,000). Copper, asbestos, and chromium are among the major mineral exports, but Zimbabwe is not just an ore-exporting country. Farms produce tobacco, cotton, tea, sugar, and other crops (corn is the local staple), and many whites stayed on their farms after independence, encouraged by the government to do so. Coal is mined in the far west of the country, and hauled by rail to an industrial complex located about midway between Harare and Bulawayo. It is somewhat ironic that the period of UDI ("unilateral declaration of independence," under which the white minority controlled the country), which led to international sanctions against the country, actually stimulated local manufacturing. Cut off from the outside world, small industries were motivated to serve the local market. After independence, some of these industries continued to prosper, reducing newly independent Zimbabwe's dependence on foreign goods.

Zimbabwe still confronts serious social, economic, and political problems, but in virtually every respect it is ahead of its neighbors in the region (except South Africa). The political compromise, the mixed economy, the retention of a productive segment of the white

minority, and an avoidance of involvement in neighbors' conflicts have served Zimbabwe well amid the turbulence of a changing region.

Namibia

Namibia became an independent African state in 1990, after more than 70 years of South African administration and two decades of active black opposition. As Fig. 7–30 shows, Namibia is bounded to the south by the Orange River and to the north and east by mainly geometric (straight-line) boundaries. A finger-like corridor of Namibian territory extends in the northeast between Angola and Botswana to Zambia.

Namibia is named after a desert, and appropriately so. The Kalahari and Namib deserts dominate the physiography; only in the far north do conditions moderate somewhat. This huge country (about as large as Texas and Oklahoma combined) is inhabited by only 1.9 million people, the majority of them concentrated in the north against the Angolan border. The centrally located capital, Windhoek (170,000), lies far from this population concentration. The major port, Walvis Bay, actually lies in an exclave of South African territory. Much of Namibia consists of vast ranches owned by white farmers; the enormous expanses of sheep pastures in the south stand in stark contrast to the village-settled, small-farm-dominated north. Between these areas lies Namibia's mineral belt, southwest of Tsumeb. From mines in this zone, the country exports copper, lead, zinc, and smaller amounts of other minerals. But the most valuable minerals lie elsewhere: along the south coast, where diamonds abound.

Given these assets, Namibia's future would appear bright. But much of the country's productive capacity remains in foreign hands, and many mines would close if wages paid to workers from the north were to be raised substantially. Despite its small population, Namibia needs food imports to supplement what is locally produced. The country's economy, at independence, was tightly linked to South Africa's, a situation that cannot be changed by political action except at the risk of disaster.

For Namibia, the long haul toward true sovereignty has just begun.

The giant of Southern Africa is the Republic of South Africa, the dominant force in the region, now facing the challenge of change. From Namibia to Angola to Moçambique, the republic has influenced and sometimes controlled the sequence of events—even after in-

dependence came to these neighbors. Now, however, South Africa itself confronts the prospect of a new order, and its energies will be spent in channeling that transition rather than on foreign adventures. We discuss this unique African state in the separate vignette that follows.

PRONUNCIATION GUIDE

Abidjan (abbih-*jahn*)

Abuja (uh-*boo*-juh)

Accra (uh-*krah*)

Adamawa (add-uh-*mah*-wuh)

Afrikaans (uff-rih-*kahnz*)

Ahaggar (uh-*hah*-gahr)

Amin, Idi (uh-*meen*, iddie)

Antananarivo (anta-nana-*ree*-voh)

Baganda (bah-*gahn*-duh)

Bantu (ban-*too*)

Beira (*bay*-ruh)

Benin (beh-*neen*)

Benue (*bane*-way)

Betsileo (bet-suh-*lay*-oh)

Biafra (bee-*ah*-fruh)

Bihe (bee-*hay*)

Bilharzia (bill-*harzee*-uh)

Bioko (bee-*oh*-koh)

Boer (*boor*)

Botswana (bah-*tswahn*-uh)

Brazzaville (*brahz*-uh-veel)

Buea (boo-*ay*-uh)

Buganda (boo-*gahn*-duh)

Bulawayo (boo-luh-*way*-oh)

Burkina Faso (ber-keena-*fahsso*)

Burundi (buh-*roon*-dee)

Cabora Bassa (kuh-boar-rah-*bahssa*)

Chingola (ching-*goh*-luh)

Dahomey (duh-*hoh*-mee)

Dakar (duh-*kahr*)

Dar es Salaam (dahr-ess-suh-*lahm*)

de Brazza, Pierre (duh-*brah*-zah, *pyair*)

Deccan (*decken*)

Dhow (*dow*)

Djibouti (juh-*boody*)

Djouf (*joof*)

Dodoma (*dode*-uh-mah)

Douala (doo-*ahla*)

Drakensberg (*drahk*-unz-berg)

Entebbe (en-*tebba*)

Ethiopia (eeth-ee-*oh*-pea-uh)

Fernando Póo (fer-nahn-doh-*poh*)

Fulani (foo-*lah*-nee)

Futa Jallon (food-uh-juh-*loan*)

Gabon (gah-*bong*)*

Gao (*gau*)

Ghana (*gah*-nuh)

Ghanaian (gah-*nay*-un)

Gondwana (gond-*won*-uh)

Guinea (*ghinny*)

Guinea-Bissau (ghinny-bih-*sau*)

Harare (huh-*rah*-ray)

Hausa (*how*-sah)

Houphouët-Boigny, Félix (ooh-*fway* bwah-*nyee*, fay-*leeks*)

Hwange (*wahng*-ghee)

Ibadan (ee-*bahd'n*)

Ibo [land] (*ee*-boh [land])

Kalahari (kallah-*hah*-ree)

Kampala (kahm-*pah*-luh)

Kano (*kah*-noh)

Kariba (kuh-*ree*-buh)

Katanga (kuh-*tahng*-guh)

Kenya (*ken*-yuh)

Khoikhoi (*khoy*-khoy)

Khoisan (khoy-*sahn*)

Kibo (*kee*-boł.)

Kilimanjaro (kil-uh-mun-*jah*-roh)

Kinshasa (kin-*shah*-suh)

Kwashiorkor (kwushy-*oar*-koar)

Lagos (*lay*-gohss)

Lesotho (leh-*soo*-too)

Liberia (lye-*beery*-uh)

Lingua franca (*lean*-gwuh *frunk*-uh)

Lualaba (loo-uh-*lah*-buh)

Luanda (loo-*an*-duh)

Lubumbashi (loo-boom-*bah*-shee)

Madagascar (madda-*gas*-kuh)

Maize (*mayz*)

Malagasy (malla-*gassy*)

Malawi (muh-*lah*-wee)

Malay (muh-*lay*)

Mali (*mah*-lee)

Maputo (muh-*pooh*-toh)

Marasmus (muh-*razz*-muss)

Masai (muh-*sye*)

Matadi (muh-*tah*-dee)

Mauritania (maw-ruh-*tay*-nee-uh)

Merina (meh-*ree*-nuh)

Moçambique (moh-sum-*beak*)

Mombasa (mahm-*bahssa*)

*Final "g" is silent.

Nairobi (nye-*roh*-bee)
Namib (nah-*meeb*)
Namibia (nuh-*mibby*-uh)
Ndebele (en-duh-*beh*-leh)
Ndola (en-*doh*-luh)
Nkrumah, Kwame (en-*kroo*-muh, *kwah*-mee)
Niger [Country] (nee-*zhair*)
Niger [River] (*nye*-jer)
Nigeria (nye-*jeery*-uh)
Nyasaland (nye-*assa*-land)
Nyerere, Julius (nye-ah-*rare*-ree, *joo*-lee-uss)
Olduvai (*ole*-duh-way)
Okovango (oh-kuh-*vahng*-goh)
Pangaea (pan-*jee*-uh)
Pointe-Noire (pwant-nuh-*wahr*)
Polygamy (puh-*lig*-ahmee)
Rhodesia (roh-*dee*-zhuh)
Rinderpest (*rin*-duh-pest)

Rio Muni (ree-oh-*moo*-nee)
Rwanda (roo-*ahn*-duh)
Sahara (suh-*harra*)
Sahel (suh-*hell*)
San (*sahn*)
Schistosomiasis (shisto-soh-*mye*-uh-siss)
Senegal (sen-ih-*gawl*)
Shari (*shah*-ree)
Shona (*shoh*-nuh)
Sisal (*sye*-sull)
Songhai (*sawng*-hye)
Soudan (soo-*dah*)
Sudd (*sood*)*
Swazi [land] (*swah*-zee [land])
Tanganyika (tan-gun-*yeeka*)
Tanzania (tan-zuh-*nee*-uh)

*Double "o" pronounced as in "book."

Tema (*tay*-muh)
Tete (*tate*-uh)
Timbuktu (tim-buck-*too*)
Tsetse (*tsett*-see)
Tsumeb (*soo*-meb)
Turkana (ter-*kanna*)
Uganda (yoo-*gahn*-duh/ yoo-*ganda*)
Vaal (*vahl*)
Wegener (*vay*-ghenner)
Windhoek (*vint*-hook)
Witwatersrand (*witt*-waterz-rand)
Yamoussoukro (yahm-uh-*soo*-kroh)
Yaoundé (yown-*day*)
Yoruba (*yah*-rooba)
Zaïre (zah-*ear*)
Zambezi (zam-*beezy*)
Zimbabwe (zim-*bahb*-way)

REFERENCES AND FURTHER READINGS

(BULLETS [•] DENOTE BASIC INTRODUCTORY WORKS ON THE REALM OR SYSTEMATIC ESSAY TOPIC)

"Africa's Woes," *Time*, Special Section, January 16, 1984, 24–41.

Ajayi, J.F. Ade & Crowder, Michael, eds. *Historical Atlas of Africa* (New York: Cambridge University Press, 1985).

• Bell, Morag. *Contemporary Africa: Development, Culture and the State* (White Plains, N.Y.: Longman, 1986).

• Best, Alan C.G. & de Blij, Harm J. *African Survey* (New York: John Wiley & Sons, 1977).

Boateng, E.A. *A Political Geography of Africa* (New York: Cambridge University Press, 1978).

• Christopher, Anthony J. *Colonial Africa: An Historical Geography* (Totowa, N.J.: Barnes & Noble, 1984).

Clarke, John I. & Kosinski, Leszek A., eds. *Redistribution of Population in Africa* (Exeter, N.H.: Heinemann Educational Books, 1982).

Cohen, Ronald. *Satisfying Africa's Food Needs: Food Production and Commercialization in African Agriculture* (Boulder, Colo.: Lynne Reiner, 1988).

Crush, J. "The Southern African Regional Formation: A Geographical Perspective," *Tijdschrift Voor Economische en Sociale Geografie*, 73 (1982): 200–212.

Curtin, Philip D. *The Atlantic Slave Trade* (Madison: University of Wisconsin Press, 1969).

Curtis, Donald et al. *Preventing Famine: Policies and Prospects for Africa* (London & New York: Routledge, 1988).

Gesler, Wilbert M. *Health Care in Developing Countries* (Washington, D.C.: Association of American Geographers, Resource Publications in Geography, 1984).

Goliber, Thomas J. "Africa's Expanding Population: Old Problems, New Policies," *Population Bulletin*, 44 (November 1989): 1–49.

Gould, Peter R. "Tanzania 1920–63: The Spatial Impress of the Modernization Process," *World Politics*, 22 (1970): 149–170.

Gourou, Pierre. *The Tropical World: Its Social and Economic Conditions and Its Future Status* (London & New York: Longman, 5 rev. ed., trans. Stanley H. Beaver, 1980).

• Griffiths, Ieuan L.L., ed. *An Atlas of African Affairs* (London & New York: Methuen, 2 rev. ed., 1984).

• Grove, Alfred T. *The Changing Geography of Africa* (New York: Oxford University Press, 1989).

Gugler, Josef & Flanagan, William G. *Urbanization and Social Change in West Africa* (New York: Cambridge University Press, 1978).

Hance, William A. *The Geography of Modern Africa* (New York: Columbia University Press, 2 rev. ed., 1975).

Harrison, Paul. *The Greening of Africa* (New York: Viking Penguin, 1987).

Harrison Church, Ronald J. *West Africa: A Study of the Environment and Man's Use of It* (London: Longman, 8 rev. ed., 1980).

Howe, G. Melvyn. *A World Geography of Human Diseases* (New York: Academic Press, 1977).

Howe, G. Melvin, ed. *Global Geocancerology: World Geography of Hu-*

man Cancers (Edinburgh, U.K.: Churchill Livingstone, 1986).

• Knight, C. Gregory & Newman, James L., eds. *Contemporary Africa: Geography and Change* (Englewood Cliffs, N.J.: Prentice-Hall, 1976).

Lamb, David. *The Africans* (New York: Random House, 1983).

Learmonth, Andrew T.A. *Disease Ecology: An Introduction* (New York: Basil Blackwell, 1988).

Lewis, Laurence A. & Berry, Leonard. *African Environments and Resources* (Winchester, Mass.: Unwin Hyman, 1988).

Martin, Esmond B. & de Blij, Harm J., eds. *African Perspectives: An Exchange of Essays on the Economic Geography of Nine African States* (London & New York: Methuen, 1981).

• Meade, Melinda S. et al. *Medical Geography* (New York: Guilford Press, 1987).

Mehretu, Assefa. *Regional Disparity in Sub-Saharan Africa: Structural Readjustment of Uneven Development* (Boulder, Colo.: Westview Press, 1989).

Miller, E. Willard & Miller, Ruby M. *Tropical, Eastern and Southern Africa: A Bibliography on the Third World* (Monticello, Ill.: Vance Bibliographies, No. P-819, 1981).

• Moon, Graham & Jones, Kelvyn. *Health, Disease and Society: An Introduction to Medical Geography* (New York: Routledge & Kegan Paul, 1988).

• Mountjoy, Alan & Hilling, David. *Africa: Geography and Development* (Totowa, N.J.: Barnes & Noble, 1987).

Murdock, George P. *Africa: Its Peoples and Their Culture History* (New York: McGraw-Hill, 1959).

O'Connor, Anthony M. *The Geography of Tropical African Development: A Study of Spatial Patterns of Economic Change Since Independence* (Elmsford, N.Y.: Pergamon, 2 rev. ed., 1978).

• Oliver, Roland & Crowder, Michael, eds. *The Cambridge Encyclopedia of Africa* (Cambridge, U.K.: Cambridge University Press, 1981).

Pritchard, J.M. *Landform and Landscape in Africa* (London: Edward Arnold, 1979).

Rosenblum, Mort & Williamson, Doug. *Squandering Eden: Africa at the Edge* (San Diego, Calif.: Harcourt Brace Jovanovich, 1987).

• Senior, Michael & Okunrotifa, P. *A Regional Geography of Africa* (London & New York: Longman, 1983).

Smith, Robert H.T., guest ed. "Spatial Structure and Process in Tropical West Africa," *Economic Geography*, 48 (1972): 229–355.

Stock, Robert F. *Cholera in Africa: Diffusion of the Disease 1970–1975, with Particular Emphasis on West Africa* (London: International African Institute, 1976).

Stren, Richard & White, Rodney, eds. *African Cities in Crisis: Managing Rapid Urban Growth* (Boulder, Colo.: Westview Press, 1989).

Udo, Reuben K. *The Human Geography of Tropical Africa* (Exeter, N.H.: Heinemann Educational Books, 1982).

Whitaker, Jennifer S. *How Can Africa Survive?* (New York: Harper & Row, 1988).

World Bank. *Sub-Saharan Africa: From Crisis to Sustainable Growth* (Washington, D.C.: World Bank, 1989).

SOUTH AFRICA:
Crossing the Rubicon

IDEAS AND CONCEPTS

plural society

core-periphery relationships

regional inequality

apartheid

separate development

domestic colonialism

regional polarization

The Republic of South Africa occupies the southernmost portion of the African continent, but it lies at the center of world attention. Long in the grip of one of the world's most notorious racial policies (*apartheid* and its derivative, "separate development"), South Africa today is a country in transition. The framework of *apartheid*, the racial separation of South Africa's peoples, is being demolished. The grand design of "separate development," the creation of a mosaic of ethnic states within the republic, is being abandoned. The dimensions of change may not be as great as those affecting the Soviet Union and Eastern Europe, but its significance is no less. And the challenge facing South Africa may be the greatest of all.

South Africa stretches from the warm subtropics in the north to Antarctic-chilled waters in the south. With a land area in excess of 470,000 square miles (1.2 million square kilometers), and with a heterogeneous population totalling 40.5 million, South Africa is the dominant state in Southern Africa. It contains the bulk of the region's minerals, the majority of its good farmlands, its largest cities, best ports, most productive factories, and most developed transport networks (Fig. SA-1). Mineral exports from Zambia and Zimbabwe move through South African ports. Workers from as far away as Malawi and as nearby as Botswana work in South Africa's mines, factories, and fields. South African armed forces have intervened in Angola's civil war even as South African grain went to feed its neighbors.

With such power and influence, South Africa would seem to be the cornerstone of what we have defined as the Subsaharan Africa realm. But this is not the case. South African strength has been built through a system of domina-

FIGURE SA–1 Ultramodern Johannesburg, with a skyline and transportation landscape that matches any in North America, Europe, Australia, or Japan.

TEN MAJOR GEOGRAPHIC QUALITIES OF SOUTH AFRICA

1. South Africa's relative location at the southern tip of the African continent assigns the republic considerable strategic importance.

2. South Africa is Africa's only truly temperate-zone country. Latitude and elevation combine to produce a range of natural environments unmatched on the continent.

3. South Africa contains a wide range of minerals, some of them strategically important. The country is rich in coal, but it has no significant petroleum reserves.

4. The historical geography of South Africa involves the immigration of populations from other parts of Africa, from Europe, and from Asia.

5. South Africa is Africa's sole remaining state in which minority rule is exercised by people of European descent. The country's plural society consists of four major cultural components, each itself divided.

6. South Africa's cultural pluralism is strengthened by its regional expression. A mosaic of "traditional" regions marks the republic.

7. South Africa's white population is larger than the white-settler populations of all other Subsaharan African countries combined, even during the height of the colonial era.

8. South Africa's political geography has been dominated by *apartheid* and its regional expression, "separate development." Although the white-minority government now seeks to retreat from these premises, the system is deeply entrenched.

9. Political and politico-geographical practices by the South African government have caused many countries to cut their economic ties with the republic, isolating South Africa from much of the world.

10. Although South Africa is often identified as Africa's only developed country, the republic consists of juxtaposed, sharply contrasting developed and underdeveloped regions.

tion by the white minority over the black majority. This system has concentrated most wealth, and virtually all power, in the hands of the white minority. Thus, South Africa's policies—political, economic, social—have not been policies of consensus. They have been, as *apartheid* was, policies of the minority. This is now changing, and a new map of South Africa is taking shape.

HISTORICAL GEOGRAPHY

As Fig. 7–17 reminds us, South Africa lay shielded from advancing decolonization by a *buffer zone* (see the 1970 map, bottom left) until the 1970s. Once African independence reached South Africa's borders in Moçambique and Zim-

babwe, and when even Namibia (long ruled by South Africa) became independent, South Africa's geopolitical situation had drastically changed. The republic's comparatively comfortable isolation had come to an end.

South Africa is a **plural society** with the greatest ethnic complexity of all the African countries. Its peoples came not only from other parts of the continent but from Europe and Asia as well. The oldest inhabitants, the San, who speak the "click" language shown as Khoisan on Fig. 7–1, were being overpowered and enslaved by the majority peoples, the Bantu, even before the Europeans arrived to colonize the area. While European settlement developed at the Cape and Cape Town grew into a substantial port town, a major African empire—one of the most powerful ever to emerge in the realm—arose with its core and headquarters in the present-day province of Natal (Fig. SA-2). This empire, the kingdom of the Zulu (see Fig. 7–30), long battled its European enemies but eventually was defeated. The Zulu nation, however, remains a strong force in South Africa, and Zulu leaders know that but for the white invasion they might well be South Africa's rulers today.

To complicate matters still further, the first white settlers in South Africa, the Dutch, who founded Cape Town in 1652, were followed shortly after 1800 by the British. Eventually, the two European communities went to war against each other, and the Boer War (1899–1902) was won by the British. But the Boers, descendants of the Dutch settlers, negotiated a favorable settlement. Out of that settlement came the Union of South Africa (1910). In the decades that followed, the Boers, who came to be known as *Afrikaners*, steadily gained numerical and political strength. In 1948, they

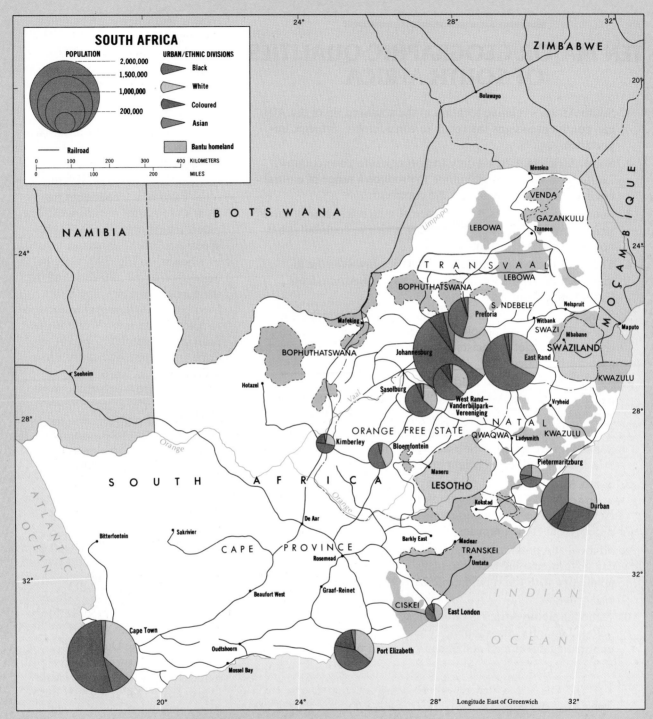

FIGURE SA–2

were able to defeat the British in national (white) elections, and in 1961 they proclaimed the Republic of South Africa (the country's official name today) and left the British Commonwealth.

But Afrikaners, British, Zulu, and other black peoples are not the only players in the social geography of South Africa's drama. At the Cape, from the earliest years of white settlement, there was much intermarriage, and a population cluster of mixed ancestry evolved. Today, these people form the majority population in Cape Town and in a large part of the city's hinterland. They are known in South Africa as the *Coloured* people, and today they number about 4 million—almost 10 percent of the country's population.

The British, following their penetration of South Africa, also changed the demographic mosaic but not by intermarriage. Needing labor for their sugar plantations in Natal, where the defeated Zulu refused to work, the British during the nineteenth century began to bring to South Africa indentured laborers from a distant colony. South Asians (or *Indians*, as they are called in South Africa) came by the thousands and most stayed after their contracts were up. Today there are over 1 million Indians in the republic, heavily concentrated in the city of Durban (which is about one-third Indian) and Natal Province.

The ethnic diversity of South Africa is illustrated by Fig. SA-2. The largest cluster of urbanized population lies in the southern Transvaal, heart of the country's core area. All the cities in this cluster, including the country's largest, Johannesburg (2.1 million), and the administrative capital, Pretoria (1.2 million), have black African majorities and large white minorities, with very small Coloured and Asian sectors. Compare this to Durban (1.4 million) on the east coast in Natal, where African, Asian, and white sectors each form about one-third of the population, with a very small Coloured minority. Next, look at the pie chart for Cape Town (2.3 million), where the Coloured people are in the majority, the white sector is quite large, and the black African minority rather small. Now compare the proportion of black Africans in Cape Town with that of Port Elizabeth (810,000), due east along the coast; in Port Elizabeth, the black Africans form the largest population sector, and the Coloured minority is smaller even than the white sector. South Africa's cities, depending on their location, differ enormously.

South Africa, therefore, is Africa's most pluralistic and heterogeneous society. Of its 40-plus million people, nearly 29 million

are black Africans, representing not only the great Zulu nation but also the Xhosa, Sotho, and other major peoples. About 7 million are white, of European ancestry; a majority of them are Afrikaners, and these are the people who hold political and economic power. Even the Coloured (3.9 million) and Asian (1.2 million) communities are fragmented by ethnic background, language, and religion. Geography and history seem to have conspired to divide, not unite, here at the southern tip of the continent.

PHYSICAL GEOGRAPHY

To gain some understanding of the complexities of South Africa's problems, the physical geography is important. In South African historical geography, landform and landscape have played significant roles. The cradle of the modern state lies where two oceans meet, in the shadow of Table Mountain (see Fig. 7–16). White settlement expanded into the interior, channeled by the valleys that separate the Cape Ranges of which Table Mountain is a part. Here, the seventeenth-century Europeans first contested the land with the San peoples. The southward migration of the Bantu had not yet carried the San's other enemies into this part of what is now Cape Province. Confrontation between white and Bantu was years away.

The Cape Ranges dominate the southern extremity of Africa, but the heart of South Africa is not mountainous—it is a plateau, an elevated flatland with some hilly topography. Much of this plateau is covered by tall natural grass, giving it a prairie-like appearance; in South Africa this is referred to

as the *veld*, a Dutch (also Afrikaans) word meaning ''field.'' The plateau is rimmed by the sometimes sheer, sometimes step-like Great Escarpment (often thousands of feet high), which separates a fairly narrow coastal strip from the extensive veld above. The Great Escarpment, too, played its role in the country's historical geography. The Zulu Empire emerged in the hilly land between scarp and sea, where the Zulu parlayed security into organization and power. The Dutch (Boers), when they embarked on their penetration of the interior, had much trouble ascending the rocky walls to the grasslands above. Even in modern South Africa, a map of surface routes reveals where the escarpment lies. It was and is a barrier, and it separates contrasting environments.

In the interior, the plateau veld slopes and rolls in giant waves of landscape. Much of the surface lies above 5000 feet (1500 meters), and this is the *highveld* with its cold winters and cool, quite untropical summers. In lower areas lies the *middleveld*, as in the basins of major streams; below the escarpment is the *lowveld*, where South Africa takes on more tropical-African qualities. Onto the highveld and middleveld swept the Zulu armies more than 150 years ago, defeating weaker enemies and absorbing lands and peoples. In those days the whites were not yet a factor, and a Zulu era seemed about to unify the region. But the Boers, seeking escape from their British competitors, entered the arena, and the inevitable war ensued. By the late 1830s Zulu power was broken, the Boers controlling the highveld and the British the coast inland to the Great Escarpment.

South Africa's great rivers also figured in the human drama, as Fig. SA-2 confirms. In the high Drakensberg (Dragons Mountains) rises the Orange River, tinted ochre by the local soils. The Orange crosses the country in its westward flow to the Atlantic, forming the boundary between two provinces (Cape and Orange Free State) and, in its lower reaches, the national boundary with Namibia. It is joined by the Vaal (Gray) River, across which lies—not surprisingly—the province of Transvaal. That province, in turn, is bounded by the Limpopo River, the boundary between South Africa and Botswana as well as Zimbabwe. These rivers and their tributaries did more than supply water in an area of chronic water shortage. From the volcanic basalts of the Drakensberg, the Orange carried fragments of rock downstream, including diamonds. Where the Orange and Vaal unite lies one of the world's great diamond accumulations, and nearby arose a city whose name, a century ago, was synonymous with gemstones: Kimberley.

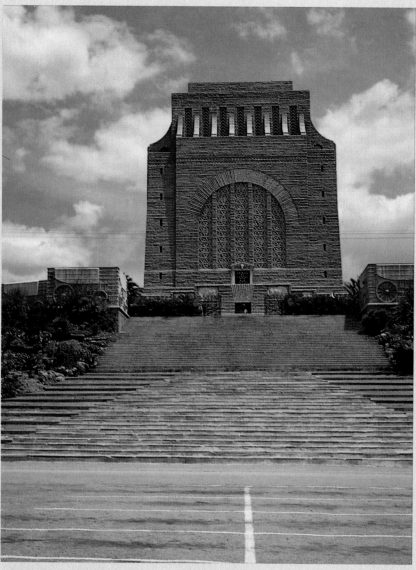

FIGURE SA–3 Pretoria's massive Voortrekker Monument, commemorating the "Great Trek" or migration of the Boers into their highveld stronghold, and an imposing symbol of Afrikaner power, organization, and territoriality.

ECONOMIC GEOGRAPHY

The discovery of diamonds in the late 1860s opened a new age in the country's development. At that time, the British held sway over the Cape (below the Great Escarpment) and Natal (also below the plateau wall). The Boers had established two independent republics on the highveld. In certain areas, such as in present-day Lesotho and Swaziland, African peoples still held on to their lands.

Certainly minerals had been found prior to the diamonds near the Orange-Vaal confluence; in fact, the Bantu mined copper and other ores centuries earlier. But the diamond finds set in motion a new economic geography. Fortune

hunters, capitalists, and black workers came to Kimberley by the thousands. Rail lines were laid from the coast to this growing new market. Equipment was needed; land was surveyed; goods never before seen in South Africa passed through the port of Cape Town. The diamonds also brought new frictions. Until the 1860s, both the colonists and the colonized had been rural people—farmers, pastoralists. Settlements (even Cape Town and Durban) had remained comparatively modest. But now

began an age of mining and industry, of machinery and labor. For the Boers, it was the beginning of the end of their republics, and the end of their isolation.

Just 25 years later, what remained of Boer isolation and aloofness ended finally and totally. In what is today Johannesburg, diggers struck what long was to be the world's greatest goldfield. A new and even larger stream of foreigners and their money came, now to the very heart of the Boers' South African Republic.

Just 30 miles (50 kilometers) from Pretoria, the Boer capital, lay the mushrooming, exploding city of Johannesburg. The Boers tried to keep control; after all, this was their country. But the invaders paid little attention. Conflict was inevitable, and before century's end Boer and Briton were at war. It was a war that soon ended in Boer defeat and the exile of their president, Paul Kruger. It also was a war of which much of the world disapproved. The British Empire was seen as a powerful bully, imposing its will on the underdog Boers. But the Boers knew that the battle was theirs to fight alone, as had been the case earlier against the Zulu. The Boers had no European "homeland" to turn to; they had no allies; they were not colonizers who could abandon the conflict and withdraw from Africa. Their Afrikaner successors today feel much the same way (Fig. SA–3).

Through this era of economic growth and accompanying conflict, South Africa's potential became clear. Its varied climates and soils yielded a wide range of grains, vegetables, and fruits. In addition to diamonds and gold, there were numerous other minerals; from the central Transvaal to the central Orange Free State lay a series of great deposits. Cheap labor was plentiful. When the British and Boers negotiated their Boer War settlement, neither side took the interests of the Africans into account. Both saw the economic potentials.

During the twentieth century, South Africa proved to be even richer than had been foreseen. Additional goldfields were discovered in the Orange Free State. Coal was found to be present in abundance, and so was iron ore; this gave rise to a major iron and steel industry. Other metallic minerals— chromium and platinum among them—yielded large revenues on

world markets. Asbestos, manganese, copper, nickel, antimony, and tin were mined and sold; a thriving metallurgical industry developed in South Africa itself. Capital flowed into the country, white immigration continued, cities and industries mushroomed, and the Union prospered.

In agriculture, too, South Africa's capacities were unmatched elsewhere in Subsaharan Africa. Tropical fruits and sugar became Natal's specialties; other fruits ranging from apples to grapes stood in large orchards at the Cape (Fig. SA–4). Great fields of wheat draped the Cape elsewhere; on the highveld there developed the Maize Triangle, South Africa's Corn Belt. In drier areas of the eastern Cape, thousands of sheep made South Africa one of the world's leading wool producers. Locally produced raw materials from cotton to plastics formed the basis of an expanding industrial complex. South Africa developed Africa's only true networks of internal transportation (as opposed

to export routes). Its ports even handled goods for neighbors and others as far away as Zaïre and Malawi. When World War II cut the country off from its overseas trading partners, there were economic problems, but local industries thrived, exempt from outside competition and benefiting from a large and comparatively wealthy domestic market.

POLITICAL GEOGRAPHY

But economic success carried the seeds of political problems. South Africa had developed **core-periphery relationships** (see pp. 41–42). The great cities, bustling industries, mechanized farms, and huge ranches had begun to resemble a developed country, much like Western Europe or the United States. But beyond the skyscrapers and white-owned corridors lay

FIGURE SA–4   The Mediterranean climate of the southwestern tip of Africa has abetted the rise of a world-class viticulture region. This is the main building of the Groot Constantia estate on the Cape Peninsula south of Cape Town, the region's oldest winery and a gem of Cape-Dutch architecture that now serves as a museum.

quite another South Africa, where conditions were more like those in rural Tanzania or Zambia (Fig. SA–5). From the villages in this periphery came the workers, the cheap labor that kept mines, factories, and commercial farms going—and kept profits high. Many of those workers settled on the edges of the cities where the industries and mines were located, so that South African cities had their downtowns, their white suburbs, and their black "townships" and squatter settlements. (One of these townships outside Johannesburg is well-known *Soweto*, short for *South Western Township*). Thus, there was racial segregation in the Union long before the 1940s—as was the case in Africa's other colonies as well.

But there also was flexibility. In and around Cape Town, the Coloured community had some representation in local (and, through white representatives, even in state) government. In Durban, there had been some progress toward racial accommodation between Asians and other residents. Then in 1948 came a momentous political change, one that was to lead to the politico-geographical transformation of the country. In that year, the Afrikaners defeated the English-speaking people and their party in the (whites-only) election of the national government. Elected as prime minister (like the U.S. president, the most powerful position) was an Afrikaner named Daniel François Malan. His platform had been a one-word solution to the problems of multiracial South Africa—**apartheid**.

Apartheid (an Afrikaans word meaning, literally, apartness or separation) was not just an abstract theory that would be softened by the checks and balances of government. Now that the Afrikaners, nearly 50 years after their Boer War defeat, had control of their country again, they set about transforming it. Just when the rest of the world was examining the fairness of social doctrines, South Africa embarked on a path of opposite direction. "Europeans (Whites) Only" signs appeared where they had not been before. Buses and trains, long informally and habitually segregated, now were segregated as a matter of law. Social services, education, and other institutions were more rigorously segregated than before. Small residential areas near inner cities, where some African, Asian, or Coloured housing remained, were evacuated and the land sold to white investors.

Apartheid, in theory, was not simply the separating of black from white. It was also the separating of Coloured from white, Asian from Zulu. Its chief objective, pursued through a set of strict new laws, was to minimize contact among the many ethnic groups and cultural communities in South Africa. It applied even within the white community. So-called "parallel-medium" schools, where classes were taught in both English and Afrikaans, were divided according to the same principle. White children who spoke Afrikaans would henceforth go to an Afrikaans-language-only school; white children who spoke English went to their own school—elsewhere. Behind apartheid of this kind lay a wider vision, a grand design of a South Africa that would consist of a mosaic of racially-based states, where each ethnic or cultural community would have its own "homeland." Again, it was not just theory. During the 1950s, the government set this plan into motion, and the map of South Africa was transformed as these "homelands" began to appear on it (the bluish areas in Fig. SA–2). Apartheid had now become **separate development**.

Opposition to the "Bantustan" program, as separate development was called, mounted inside as well as outside South Africa. Apartheid itself had already elicited passive

FIGURE SA–5 South Africa's 28-plus million blacks overwhelmingly reside in substandard housing. Conditions usually are at their worst in the poor rural periphery, such as here at Steynsburg in the parched interior of eastern Cape Province.

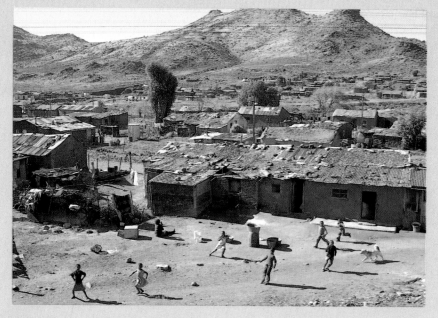

as well as violent opposition, and as early as the 1960s death, imprisonment, and exile were imposed on anti-apartheid activists (one of those imprisoned was Nelson Mandela, a prominent black opponent of the system). Still the program went ahead. The underlying principle was geopolitical: in multiracial South Africa, a Zulu, for instance, would have his or her own national homeland and would be a citizen there. Outside this homeland, in the white part of South Africa (or in another black homeland, or in a future Coloured or Asian homeland), that Zulu would be a foreigner just as any other noncitizen would be. Conversely, a white person would not have any citizen's rights in the Zulu homeland or in any of the other nine black homelands to be created. In 1976, the first of the homelands, the Transkei (Fig. SA–2), homeland of the Xhosa nation, was declared independent. Wedged between the Great Escarpment and the Indian Ocean, and between Natal and the eastern Cape Province, Belgium-sized Transkei was to be the model state-within-a-state. Next came Bophuthatswana (1978), with its infamous capital, Sun City; it was followed by tiny Venda in 1979, and the impoverished Ciskei in 1982.

This politico-geographical reorganization of South Africa was no mere reservation-style warehousing. Enormous investments were made in the creation of national infrastructures—in the capital cities, roadways, agriculture, officialdom, and other trappings of nationhood. But the map reveals an indisputable reality: the total area of the black homelands was to be a mere fraction of all of South Africa. Only about 14 percent of the country was set aside as homelands for the 70 percent of the people who are black Africans. Moreover, much of the land in the homelands is of low productivity; most of the

known minerals lie in the white areas. The conclusion is inescapable: to survive, workers from the homelands would have to apply for temporary work permits in South Africa's mines, in its factories, and on its farms. Many scholars have described the separate development program as an instance of **domestic colonialism**; some likened it to the Soviet empire over its Muslim republics and postwar Eastern Europe.

During the 1980s, a number of circumstances turned the world's attention to South Africa. First, African opposition, long overshadowed by decolonization struggles elsewhere, now united to face one final challenge. An exiled organization, the African National Congress (ANC), could infiltrate the republic from neighboring states. With international support, the ANC was able to raise the consciousness of many people and their governments. The name of Nelson Mandela, not known to many outsiders in the 1960s and 1970s, now became a symbol of black resistance. Second, the schedule of homeland creation ran into trouble inside South Africa. The fifth independent homeland was to be Kwazulu, home of the most powerful African nation. But the Zulu were in a position to refuse their designation—and did so. Through their leader, Chief Mangosuthu Buthelezi, the Zulu rejected homeland status, and the program was thrown into disarray. Third, the South African economy since 1980 has not been nearly as strong as it was when the separate development program was launched. We have seen the problems of oil-rich as well as oil-poor economies in previous chapters, and South Africa lacks this critical energy resource. When rising petroleum costs coincided with declines in the price of gold and other exports, South Africa did not have the money to carry out its plans as

it had before. And fourth, many countries and companies stopped doing business with South Africa. In the United States, "disinvestment" became a rallying point for those opposed to South Africa's social policies; several Western European governments severely cut back their trade connections with South Africa.

The impact of these events in South Africa could be felt in various ways. Within the white power structure, a debate arose over the future of apartheid and separate development. Many changes were made as the "petty" apartheid of the 1950s began to disappear. For those who could afford it, the major hotels in the large cities were "internationalized"—that is, they could accommodate people of all races. The "Whites Only" signs on park benches and at bus stops also began to vanish. More important, the plan whereby citizens of the Transkei (and other homelands) would be foreigners in South Africa was dropped. All residents of South Africa would again have South African citizenship; those with homeland credentials would have dual citizenship.

But the "grand" apartheid design of separate development could not be swept away overnight. The governing party, the Afrikaner-based National Party, began to face opposition from conservative members, who argued that reforms were coming too quickly. A splinter party (the Conservative party) soon formed to oppose any further concessions, whatever the pressures from inside or outside South Africa. This conservative opposition continues to grow today, and it has a spatial dimension. Much of it is concentrated in the small towns and farms of the *platteland*, the rural zones of the Transvaal and the Orange Free State. These are the areas to which Afrikaners moved when they sought to escape British domination; here the Boer War was most vigorously fought, and here the vestiges of Afrikaner orthodoxy are strongest. Thus, **regional polarization** within the white constituency has a significant, and potentially troubling, regional expression.

Crossing the Rubicon

In 1990, momentous developments signaled fundamental change in the political and overall social geography of South Africa. The (still white-controlled) government legalized the outlawed African National Congress. Nelson Mandela, whose name had become synonymous with black opposition to white rule, was released from prison and joined the leadership of the organization despite his advanced age. Moves toward dismantling the structure of apartheid and separate development accelerated. Persons long prevented from public political activity under "banning" laws were permitted to resume their efforts. Senior government officials stated unequivocally that apartheid would be ended and that only the methods and means remained to be determined. In the words of a former president, South Africa had crossed the Rubicon.

The nature of the geography of a post-apartheid South Africa, however, remained far from clear. As white conservative opposition

FIGURE SA–6 Ethnic and political conflict have produced violence among black groups in South Africa. Much of the strife has pitted Zulu members of the Inkatha movement against supporters of the African National Congress (here in Soweto).

to the government's actions grew stronger, strife between black rival groups continued despite Mandela's calls for unity. The old South African adage—that this is a highly pluralistic country whose constituent peoples are themselves divided by culture and ideology—was confirmed again. The Zulu, accused by others of having been too cooperative with white governments in the past, were embroiled in a near-civil war with more radical adversaries that had, by mid-1990, cost as many as 3000 lives (Fig. SA–6). Zulu leaders point out that it is they, not the Xhosa, who successfully resisted being incorporated into the Bantustan scheme. (Mandela is a Xhosa; when he raises his fist and shouts

amandla, the Xhosa word for "power," it sounds like anathema to some Zulu). Thus, the question facing a transforming South Africa is: Who are the negotiating parties? Will the white government speak for the divided white community or will white opponents of its program also have a say? Will the African National Congress speak for the black majority or will Inkatha (the Zulu cultural-political organization) and other groups also participate? And what of the governments of the four established "republics" in the homelands scheme? The challenge before South Africa is daunting, but one thing is clear: the imprints of apartheid are being swept off the map.

PRONUNCIATION GUIDE

Afrikaans   (uff-rih-*kahnz*)
Afrikaner   (uff-rih-*kahn*-nuh)
Amandla   (uh-*mund*-luh)

Apartheid   (*apart*-hate)
Bantu   (ban-*too*)
Bantustan   (ban-too-*stahn*)

Boer   (*boor*)
Bophuthatswana   (boh-pooh-taht-*swahna*)

Botswana (bah-*tswahn*-uh)

Buthelezi, Mangosuthu (boo-teh-*lay*-zee, mungo-*soo*-too)

Ciskei (*siss*-skye)

Drakensberg (*drahk*-unz-berg)

Durban (*der*-bun)

Inkatha (in-*kah*-tah)

Johannesburg (joh-*hannis*-berg)

Khoisan (khoy-*sahn*)

Kruger (*kroo*-guh)

Kwazulu (kwah-*zoo*-loo)

Lesotho (leh-*soo*-too)

Limpopo (lim-*poh*-poh)

Malan, Daniel François (muh-lahn, *dan*-yell frahn-*swah*)

Mandela (man-*della*)

Moçambique (moh-sum-*beak*)

Namibia (nuh-*mibby*-uh)

Natal (nuh-*tahl*)

Ochre (*oh*-ker)

Pretoria (prih-*tor*-ree-uh)

San (*sahn*)

Sotho (*soo*-too)

Soweto (suh-*wetto*)

Steynsburg (*stainz*-berg)

Swaziland (*swah*-zee-land)

Transkei (*trun*-skye)

Transvaal (*trunz*-vahl)

Vaal (*vahl*)

Veld (*velt*)

Voortrekker (*for*-trecker)

Xhosa (*shaw*-suh)

Zimbabwe (zim-*bahb*-way)

Zulu (*zoo*-loo)

REFERENCES AND FURTHER READINGS

(BULLETS [•] DENOTE BASIC INTRO-DUCTORY WORKS)

Berger, Peter L. & Godsell, Bobby, eds. *A Future South Africa: Visions, Strategies, and Realities* (Boulder, Colo.: Westview Press, 1989).

• Best, Alan C.G. "The Republic of South Africa: White Supremacy," *Focus*, March–April 1975, pp. 1–13.

• Best, Alan C.G. & de Blij, Harm J. *African Survey* (New York: John Wiley & Sons, 1977), Chap. 20.

Blumenfeld, Jesmond, ed. *South Africa in Crisis* (Beckenham, U.K.: Croom Helm, 1987).

Board, Christopher et al. "The Structure of the South African Space-Economy: An Integrated Approach," *Regional Studies*, 4 (1970): 357–392.

Carter, Gwendolen M. *Which Way Is South Africa Going?* (Bloomington: Indiana University Press, 1980).

Christopher, Anthony J. "Apartheid Within Apartheid: An Assessment of Official Intra-Black Segregation on the Witwatersrand, South Africa," *The Professional Geographer*, 41 (August 1989): 328–336.

Christopher, Anthony J. *Colonial Africa: An Historical Geography* (Totowa, N.J.: Barnes & Noble, 1984).

• Christopher, Anthony J. *South Africa* (London & New York: Longman, 1982).

Christopher, Anthony J. *South Africa: The Impact of Past Geographies* (Cape Town: Juta & Co., 1984).

• Cole, Monica M. *South Africa* (London: Methuen, 2 rev. ed., 1966).

Crush, J. "The Southern African Regional Formation: A Geographical Perspective," *Tijdschrift Voor Economische en Sociale Geografie*, 73 (1982): 200–212.

• Davenport, T.R.H. *South Africa: A Modern History* (Toronto: University of Toronto Press, 3 rev. ed., 1987).

• Fair, Thomas J.D. *South Africa: Spatial Frameworks for Development* (Cape Town: Juta & Co., 1982).

Leach, Graham. *South Africa: No Easy Path to Peace* (New York: Methuen, 2 rev. ed., 1987).

Leach, Graham. *The Afrikaners* (London: Macmillan, 1989).

• Lelyveld, Joseph. *Move Your Shadow: South Africa, Black and White* (New York: Times Books, 1985).

Mandy, Nigel. *A City Divided: Johannesburg and Soweto* (New York: St. Martin's Press, 1985).

McCarthy, J. & Smit, D. *South African City: Theory in Analysis and Planning* (Cape Town: Juta & Co., 1984).

Meredith, Martin. *In the Name of Apartheid: South Africa in the Postwar Period* (New York: Harper & Row, 1988).

Miller, E. Willard & Miller, Ruby M. *Tropical, Eastern and Southern Africa: A Bibliography on the Third World* (Monticello, Ill.: Vance Bibliographies, No. P-819, 1981).

Murdock, George P. *Africa: Its Peoples and Their Culture History* (New York: McGraw-Hill, 1959).

Pirie, Gordon. "The Decivilizing Rails: Railways and Underdevelopment in Southern Africa," *Tijdschrift Voor Economische en Sociale Geografie*, 73 (1982): 221–228.

Rogerson, C.M., guest ed. "Urbanization in South Africa: Special Issue," *Urban Geography*, 9 (September–October 1989): 549–653.

Saul, John S. *South Africa: Apartheid and After* (Boulder, Colo.: Westview Press, 1990).

• Smith, David M. *Apartheid in South Africa* (Cambridge, U.K.: Cambridge University Press, 3 rev. ed., 1990).

Smith, David M., ed. *Living Under Apartheid: Aspects of Urbanization and Social Change in South Africa* (Winchester, Mass.: Allen & Unwin, 1982).

Thompson, Leonard. *A History of South Africa* (New Haven: Yale University Press, 1990).

Western, John C. *Outcast Cape Town* (Minneapolis: University of Minnesota Press, 1981).

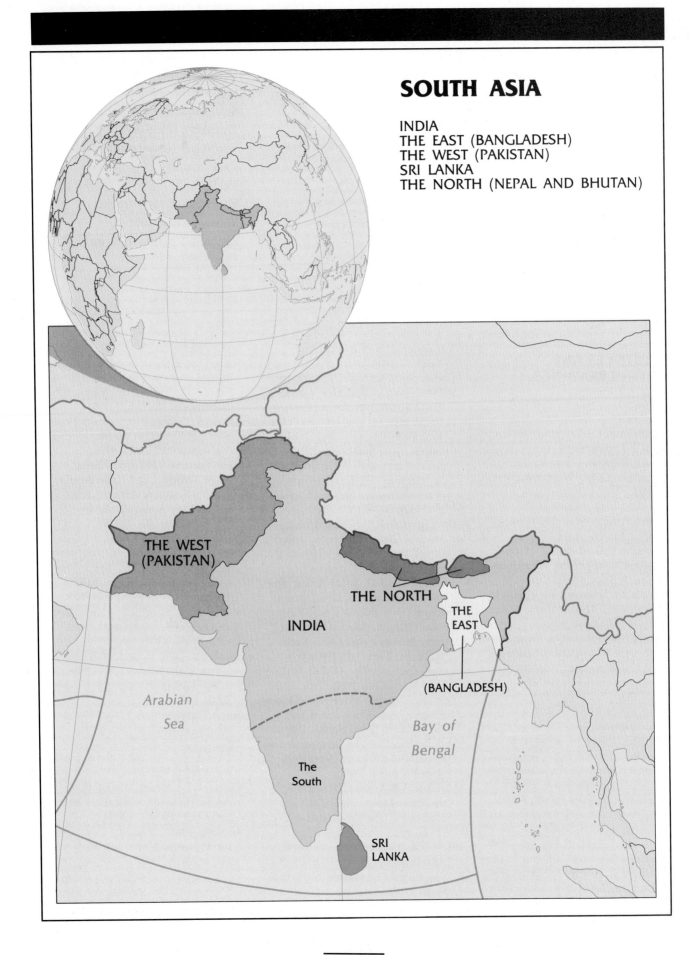

SOUTH ASIA

INDIA
THE EAST (BANGLADESH)
THE WEST (PAKISTAN)
SRI LANKA
THE NORTH (NEPAL AND BHUTAN)

THE WEST
(PAKISTAN)

THE NORTH

THE
EAST

INDIA

(BANGLADESH)

Arabian
Sea

Bay of
Bengal

The
South

SRI
LANKA

CHAPTER 8

SOUTH ASIA: RESURGENT REGIONALISM

Southern Asia consists of three protruding landmasses. In the southwest lies the huge rectangle of Arabia. In the southeast lie the slim peninsulas and elongated islands of Malaysia and Indonesia. Located between these flanking protrusions is the familiar triangle of India—a subcontinent in itself, the centerpiece of South Asia. Bounded by the immense Himalaya ranges to the north, by the mountains of eastern Assam to the east, and by the rugged arid topography of Iran and Afghanistan to the west, the Indian subcontinent is a clearly discernible physiographic region (Fig. 8–1).

The giant of South Asia in politico-geographical terms as well, the state of India lies surrounded by the five smaller countries of the realm. Bangladesh and Pakistan flank India on the east and west, respectively. To the north lie the landlocked states of Nepal and Bhutan as well as the disputed territory of Kashmir. And just southeast of the southern tip of the Indian triangle lies the island state of Sri Lanka (formerly Ceylon). The landlocked countries of the north lie in the historic buffer zone between the former British sphere of influence in South Asia and its Eurasian competitors; India, Pakistan, Bangladesh, and

---

IDEAS AND CONCEPTS

geomorphology

wet monsoon

social stratification

superimposed political boundary

centrifugal forces

centripetal forces

physiological density

green revolution (2)

population growth dynamics

doubling time

demographic transition model (2)

irredentism (3)

Sri Lanka all formed part of what was once known as British India.

This chapter focuses on South Asia's huge human population in a world context (see Figs. I–14 and I–15). India alone, with 871 million people in 1991, has a larger population than Europe and North Africa/Southwest Asia combined. Neighboring Bangladesh (121 million) and Pakistan (117 million) also rank among the world's ten most populous countries. With

well over 1.1 billion inhabitants, this realm constitutes one of the two greatest human concentrations on earth; if present growth rates persist, it will surpass the Chinese realm to become the world's largest population cluster shortly after the twenty-first century opens. South Asia was also the scene of one of the world's oldest known civilizations, and later became the cornerstone of the British colonial empire.

British colonial rule, however, threw a false cloak of unity over a realm that possesses enormous cultural variety. Even before the British departure in 1947, plans had been developed for the separation of a Muslim state, Pakistan, from Hindu-dominated India. Islamic Pakistan was itself divided into a discontinuous state that consisted of two territories, West Pakistan (now Pakistan) and East Pakistan (now Bangladesh); Sri Lanka became an independent state in 1948 (changing its name from Ceylon 24 years later). In 1971, following a bitter war, Bangladesh (with support from Pakistan's rival, India) freed itself from its ties with Pakistan.

Nonetheless, it is something of a miracle that a realm as large and populous as South Asia has generated only four post-colonial states.

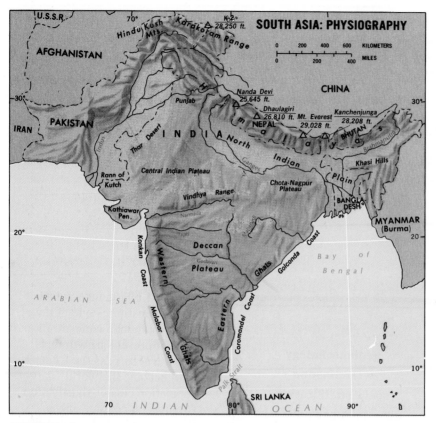

FIGURE 8—1

disorders to express their displeasure at concessions made to minority groups of every stripe.

In Pakistan, too, several internal social issues are reaching the conflict stage. Millions of refugees from neighboring Afghanistan have greatly aggravated long-standing divisions among border-zone residents. The Islamization drive of the entire country threatens to unleash increasingly violent clashes between majority Sunni Muslims and the Shi'ite minority. And ethnic rioting since 1985 in the city of Karachi, by contentious Pakhtuns (migrants from the north) and Mohajirs (migrants from India), has been the worst of the post-independence era. Even poverty-stricken Bangladesh, whose population-growth and food-provision miseries rank among the Third World's most severe, has experienced a widening guerrilla war in its southeastern hill zone.

Within India, regional differences remain strong, but the country has held together for more than 40 years. In Pakistan, separatist movements affect Pakhtunistan in the northwest and Baluchistan in the southwest, but these pressures have largely been accommodated, and the integrity of the state is sustained. In Sri Lanka, the Buddhist Sinhalese majority has clashed from time to time with the minority Hindu Tamils; yet, even though this conflict has markedly intensified in recent years, all-out civil war has been averted.

South Asia's internally diverse states have faced some of their most difficult challenges since the early 1980s. Besides the heightened confrontation between the Sinhalese and Tamils in Sri Lanka, India, Pakistan, and Bangladesh have had to deal with divisive conflicts as well. The most serious so-

cial cleavages have arisen in India, intensified by sheer numbers of people and by the depths of their mutual distrust and dislike. Overshadowing all is the bitter division between majority Hindus and minority Sikhs, stemming from the 1984 assassination of Prime Minister Indira Gandhi by two of her Sikh bodyguards soon after ordering the Indian army to attack extremist Sikh secessionists barricaded inside their faith's holiest shrine. Elsewhere in India, Muslims have battled Hindus in many large cities and in rural Bihar State. Ethnic groups as varied as the Tamils, Gurkhas, and Konkanis have launched uprisings to attain greater recognition. Assamese and tribespeople in the northeastern hills have viciously attacked each other as well as Bangladeshi immigrants. And militant Hindus have sparked frequent

PHYSIOGRAPHIC REGIONS

The Indian subcontinent is a land of immense physiographic variety. Nevertheless, it is possible to recognize three rather clearly defined regions: the northern mountains, the southern peninsular plateau, and in between the belt of river lowlands. In this division, the northern mountains consist of the Baluchistan, Hindu Kush/Karakoram, Himalayan, and Assamese uplands. The southern tableland is constituted mainly by the huge Deccan Plateau. And the intervening tier of river lowlands extends eastward in a vast inverted crescent from Sind (Pakistan's lower Indus Valley) through the Punjab and North Indian

---

TEN MAJOR GEOGRAPHIC QUALITIES OF SOUTH ASIA

1. South Asia is well defined physiographically, extending from the southern slopes of the Himalayas to the island of Sri Lanka.

2. Two river systems, the Ganges-Brahmaputra and the Indus, form crucial lifelines for hundreds of millions of people in this realm. The annual wet monsoon is a critical environmental element.

3. India lies at the heart of the world's second largest population cluster, which by 2010 will be the first.

4. No part of the world faces demographic problems with dimensions and urgency comparable to South Asia.

5. All the states of South Asia suffer from underdevelopment. Food shortages exist, nutritional imbalance prevails, and famines occur.

6. Agriculture in South Asia, in general, is comparatively inefficient and not as productive as it is in other parts of Asia.

7. The great majority of South Asia's peoples live in villages and subsist directly from the land.

8. Strong cultural regionalism marks South Asia. The Hindu religion dominates life in India; Pakistan is an Islamic state; Buddhism thrives in Sri Lanka.

9. The South Asian realm's politico-geographical framework results from the European colonial period, but important modifications took place after the European withdrawal.

10. India constitutes the world's largest and most complex federal state.

(Ganges) Plain and on across the great double delta into Assam's lower Brahmaputra Valley (Fig. 8–1).

But this triad of regions only introduces the varied South Asian land surface. Depending on the degree of detail employed, it would be easy to subdivide each of these three into more specific physiographic subregions. Thus, in the mountain wall that separates India and Pakistan from Asia's interior, we may recognize the desert ranges of Baluchistan and the Afghanistan border, the towering individual ranges of the Hindu Kush, the Karakoram, the Great Himalaya itself, and the jungle-clad mountains of the eastern Assam-Myanmar (Burma) margin. Moreover, these northern mountains do not simply rise out of the river valleys below: there is a continuous belt of transitional foothills between the lofty ranges and the low-lying river basins.

The belt of alluvial lowlands that extends from the Indus Plain to the Brahmaputra Valley also is anything but uniform. This physiographic region is often called the North Indian Plain, and its internal environmental contrasts are clearly reflected in Figs. I–9 to I–13. The Indus, which rises in Tibet and penetrates the Himalayas to the west in its course to the Arabian Sea, receives its major tributaries from the Punjab ("land of five rivers"); this physiographic subregion—situated between the Indus and North Indian plains—extends into Pakistan as well as India. The subregion of the lower Indus is characterized by its minimal precipitation, its desert soils, and its irrigation-based cluster of settlement. This is the heart of Pakistan, and the farmers grow wheat for food and cotton for sale.

Hindustan, the central subregion extending from eastern Punjab near the historic city of Delhi to the Ganges-Brahmaputra Delta at the head of the Bay of Bengal, is wet country. The precipitation here exceeds 100 inches (250 centimeters) in sizable areas, and 40 inches (100 centimeters) almost everywhere (Fig. I–9). Deep alluvial soils cover much of this subregion (Fig. I–13), and the combination of rainfall, river flow, good soils, and a long growing season make this India's most productive area. As Fig. I–14 confirms, this is also India's largest and most concentrated population core: not only do rural densities in places exceed 1000 people per square mile (385 per square kilometer), but also lying between Delhi and delta-side Calcutta is a chain of the country's leading urban centers, connected by the densest portion of the Indian transportation network. In the moister east, rice is the chief food crop; jute is the main commercial product, notably in Bangladesh and the rest of the delta. To the west, near Punjab and around the drier margins of the Hindustan subregion, wheat and such drought-resistant cereals as millet and sorghum are cultivated.

The easternmost extension of the North Indian Plain comprises the lower Brahmaputra Valley in Assam. This valley is much narrower than the Gangetic Plain,

GEOMORPHOLOGY

The geographic study of the configuration of the surface of the earth is called **geomorphology**. The term *physiography* also appears in connection with this work, but it has a wider connotation than geomorphology. The contents of physiographic regions (or *provinces*) include not only terrain but also climate and weather, soils and vegetation. Geomorphology, on the other hand, concentrates on landscapes and landforms alone.

As in other fields of geography, the study of the earth's surface features can be approached at various levels of detail. Some geomorphologists focus on one particular *landform*, trying to learn what forces and processes shape it. Thus, a landform is a single feature; a volcanic mountain, a segment of a river valley, a deposit laid down by a glacier, and a sinkhole are all landforms. Other physical geographers take a broader view and study whole *landscapes* in an effort to discover what led to their formation and present appearance. A landscape, therefore, is an assemblage of landforms. Some knowledge of landform formation may help in the analysis of landscapes, but it is not the entire story.

Obviously, geomorphology is a complicated field. To understand why landscapes appear the way they do, it is necessary to know the forces that bend and break the earth's crust *and* the agents of weathering and erosion that constantly alter the exposed surface. The internal crustal movements are caused by *tectonic* forces, and anyone who has experienced even a mild earthquake will know their power. Whole continents are slowly moved by these forces in response to a greater global pro-

cess, whereby crust is "made" along mid-oceanic ridges and "recycled" where the pieces (*plates*) collide and are pushed under. The continental landmasses are dragged along as these plates move, their contact zones buckling and breaking (forming mountain belts from Alaska to Argentina and across all Eurasia) and their trailing edges warping and bending.

As all this happens, the exposed surface is attacked by *weathering* (rock disintegration by continuous temperature change and by chemical action in the moist atmosphere) and by *erosion* (the removal of weathered rock and its further breakdown by streams, glaciers, wind, and waves). But this is not all. The material loosened by weathering and erosion is carried away and then deposited elsewhere, so that although there is wearing down (*degradation*) in some areas, there is accumulation (*aggradation*) in others. A river system such as the Mississippi and its tributaries thus erodes or degrades the landscapes of its interior and aggrades the landscape in its lower valley and delta. To keep track of all the forces—tectonic, erosional, depositional—involves much scientific detective work.

Tectonic geomorphic processes work at different rates, accentuating surface relief at any given time. Each process is associated with its own assemblage of landforms, and therefore leaves a distinct imprint on the landscape. A good example is volcanism, which can produce conical mountains built by the violent ejection and sudden cooling of molten rock on the surface. Volcanoes, like all landforms, are composed of crustal materials that are sculpted by tectonic and erosional forces. The

type of rock associated with volcanism is known as *igneous*, which is formed by the cooling and solidification of molten material (called *magma* if it hardens underground, *lava* if it solidifies on the surface).

The second major type of rock, and the most widespread, is known as *sedimentary*, because it is formed by the deposition of loose eroded material that is then slowly compacted (lithified) into solid rock by the heat and pressure generated from the increasing weight of newer sediments deposited above. South Asia's Himalayas (Fig. 8–2) are composed of geologically young sedimentary materials that were laid down in a shallow sea (tiny marine fossils are contained in the rocks collected atop the highest peaks); the enormous tectonic forces responsible for this rapid and extreme uplifting were unleashed as the Indian subcontinent, cast adrift as the Gondwana landmass broke apart (Fig. 7–7), smashed into Asia and caused an accordion-like crumpling of the crust, thrusting up the Himalayas in a process that still continues as the Australian Plate grinds against the Eurasian Plate (Fig. I–6). Such transformations of the surface help create the final type of rock, appropriately called *metamorphic*, because it is altered from a pre-existing igneous or sedimentary rock by the reintroduction of great heat and pressure (for example, sedimentary shale is metamorphosed into slate). Metamorphism is often associated with horizontal deformation of the crust, which can make sedimentary rock formations plastic, producing warping or *folding* of parallel rock strata. When these forces are too great, however, fracturing or *faulting* occurs,

FIGURE 8–2   The spectacular terrain of the highest Himalayas along the Nepal-China border. Mount Everest (29,028 feet/8848 meters), the earth's highest, is the black, pyramid-like massif in the center of the photograph.

producing formations like East Africa's rift valleys (Fig. 7–5). Besides vertical faulting, horizontal faulting can also take place (California's San Andreas Fault is a famous example), a dislocation that can trigger sudden and devastating earthquake activity.

Degradational processes wear down the land surface through weathering and erosion, reshaping landscapes and creating new ones where significant quantities of eroded materials are deposited. The progressive actions of erosional agents produce an orderly sequence of landforms. The recently uplifted Himalayas, for instance, will eventually be reduced by erosion to a low rounded mountain chain (much like the U.S. Appalachians), bordered on the south by vast depositional plains and plateaus composed of the eroded Himalayan materials transported by *running water* in streams—the most widespread and effective erosional agent (even in deserts).

The steepness of the slope determines, in part, the work of the river, dragging rock fragments

along, carrying others in suspension, and even dissolving some along the way. Eventually, the stream's velocity slows down as it approaches its mouth, and aggradation becomes the dominant process. Valleys are filled by sediment, and deltas extend from the river mouth into the sea. Where rock strata are soluble, as in limestone areas, the water not only erodes but also dissolves, producing a unique landscape (known as *karst*) pocked by sinkholes and caves. Water in a different form fashions shorelines as *waves* perform degradational functions. Where the coastline coincides with one of those crunching, mountain-creating plate contacts, as along much of the west coast of the United States, the deep-water waves attack with great power and sculpt a spectacular high-relief landscape. Where the coastal landscape is plain-like, as in the southeastern United States, the waves deposit sediment and prove that they also can be aggradational agents, forming wide beaches and offshore sandbars.

Landscapes and landforms associated with *glaciation* usually have complex histories. Where the glaciers formed huge icesheets during the most recent (Pleistocene) glaciation, they scraped the underlying topography into a flat plain, carrying off the rubble, grinding it down, and depositing it in thick layers far to the south, thus burying much of the U.S. Midwest. Where the deepening cold caused mountain glaciers to form, as in the Rocky Mountains and European Alps, the ice descended into valleys that were formerly carved by rivers, deepening and widening them.

The erosive power of *wind* is at its greatest in the drier climates, where it picks up and hurls loose surface particles against obstacles, wearing them away by abrasion. Although thick sand deposits are shaped by the wind into various dune formations in certain deserts, the majority of the world's desert surfaces exhibit rocky landscapes. Wind also creates depositional features, the most important of which is *loess*, a highly fertile windblown silt that can accumulate to great depths (as we will see in Chapter 9).

Geomorphology, a vital inquiry in and of itself, must also be linked to the other elements of physical geography, because all together form a dynamically interacting total environmental system. Climate is the ultimate generator of degradational agents; soils are intimately related to both geomorphic processes and climatic influences, and to vegetation and hydrography in varying degrees as well. Finally, human influences must also be considered because agricultural land use, urban development, water diversion, surface mining, and the like dominate an ever-expanding artificial landscape.

very moist, and suffers from frequent flooding; the river has only limited use for navigation. Up on the higher slopes, tea plantations have been developed, and in the lower elevations rice is grown. Assam is just barely connected with the main body of India through a corridor only a few miles wide, which runs between the southeastern corner of Nepal and the northwestern tip of Bangladesh, and it remains one of South Asia's frontier areas.

Turning now to the peninsular component of the subcontinent, also referred to as plateau India, we find once again that there is more variety than first appearances suggest. There are physiographic bases for dividing the plateau into a northern zone (the Central Indian Plateau would be an appropriate term) and a southern sector, which is already well known as the Deccan Plateau. The dividing line can be drawn on the basis of roughness of terrain (along the Vindhya range) or rock type (much of the Deccan is lava covered), and the Tapti and Godavari rivers—in the west and east, respectively—form a clear lowland corridor between the two regions (Fig. 8–1). The Deccan (meaning "south") has been tilted to the east, so that the major drainage is toward the Bay of Bengal. This plateau is also marked along much of its margin by a mountainous escarpment called the Ghats (the word means "hills"), which descends to the fairly narrow coastal plains below. Parallel to these surrounding coasts are the Eastern and Western Ghats, which meet near the southern tip of the subcontinent.

Peninsular India possesses a coastal lowland zone of varying width. Along the southwestern littoral lies the famous Malabar Coast, and north of it the Konkan Coast. Along the southeastern shore is the Coromandel Coast,

DYNAMICS OF THE WET MONSOON

Land and water surfaces heat and cool at different rates. While the oceans with their various water layers warm and cool rather slowly, solid landmasses heat and cool rapidly in response to the temperature of the air with which they come in contact. In winter, the cold land surface often develops an overlying high-pressure cell from which winds blow toward lower-pressure areas associated with adjacent warmer seas; in summer, the process reverses as the land heats up and generates a low-pressure cell that sucks in moisture-laden air lying above the now higher-pressure (and relatively cooler) ocean surface. The bigger the land area, the more pronounced these effects of "continentality" become—a phenomenon that reaches its grandest scale in the central and eastern portions of the Eurasian landmass.

In southern and eastern Asia, particularly, this mechanism shapes a massive seasonal reversal of onshore and offshore windflows known as *monsoons* (an old Arabic word meaning "season"). During the cool winter months, a high-pressure cell over the land generates an offshore airflow—the *dry monsoon*. But in the hot summer season, a strong onshore flow of rain-bearing winds predominates as low-pressure cells form over the heated continent; it is this *wet monsoon* that is the key to India's agricultural possibilities and seasonal life rhythms, providing the subcontinent with sufficient moisture to support its huge population. Monsoonal windflows, which help balance temperatures between the tropics and mid-latitudes, actually occur over a 10,000-mile-long (16,000-kilometer-long) broad crescent of central Africa and southern Asia, stretching east from the West African coast to India and then northeast as far as Japan. The monsoons function as an enormous heat engine, transferring vast masses of air put into motion by temperature and pressure differences above continents and nearby oceans.

During summer, the solid land surface heats up far more rapidly than the sea (with its many layers of cooler water that constantly mix), and an air parcel in contact with the ground—labeled **X** in diagram **A** of Fig. 8–3—quickly becomes warmer than its offshore counterpart (**Y**). In diagram **B**, parcel **X** begins to rise, cool, and shed its moisture, making room for air parcel **Y** to flow inland to equalize the now-lowered surface air pressure. Parcel **Y** contains even more moisture than **X**, and soon that water content is precipitated as rain as **Y** itself is heated, rises, and cools (remember that as air becomes cooler, it can hold less moisture). These movements are illustrated in diagram **C**, which also shows parcel **Z** moving in to replace **Y** and that original parcel **X** has cooled and dried at high altitude as it moved out over the ocean.

In diagram **D**, this model is applied to India. The subcontinent is influenced by two branches of the wet summer monsoon, both originating in the Indian Ocean to the southwest, with each moisture-laden airstream shaped by regional topography. The Arabian Sea branch ① saturates the Malabar-Konkan coastal strip, surges over the Western Ghats (where orographic uplift removes much of the

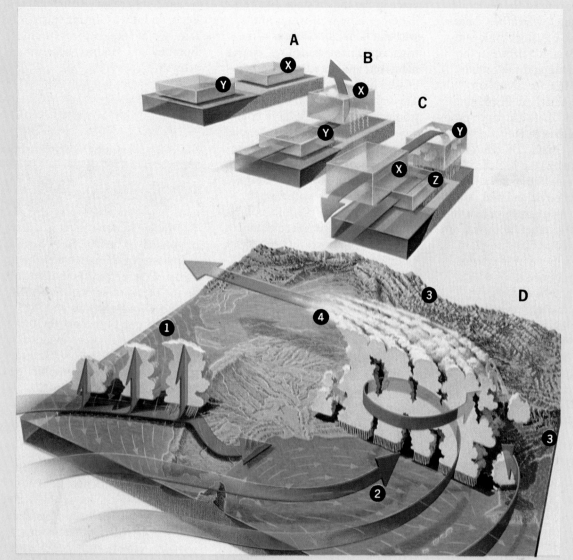

FIGURE 8–3 India's wet monsoon: Model and reality.

remaining moisture), and becomes a relatively dry eastward flow across the interior Deccan Plateau. The Bay of Bengal branch ② comes ashore over the Ganges-Brahmaputra Delta area, but is quickly blocked by the Himalayan wall to the north ③. As diagram **D** illustrates, surface winds now rise in a spiral over Bangladesh and are steered westward by the mountain barrier onto the Gangetic Plain. Massive rainfall occurs throughout this zone, supplying the water necessary to support its intensive agriculture. This wet-monsoon-generated precipitation gradually tapers off across Hindustan ④ as distance from the Bay of Bengal increases.

These critical monsoonal circulation patterns continue for as long as the land surface remains significantly warmer than the surrounding sea. Although this weather system does provide sufficient rainfall in most years, it can vary without warning—producing droughts (such as occurred in 1987) that constitute a serious environmental hazard for the hundreds of millions who must rely on the wet monsoon.

and to its northeast the Golconda Coast leading toward the Ganges Delta. These physiographic sub-regions lie wedged between the interior plateau and the Indian Ocean; the Malabar-Konkan Coast is the more clearly defined, since the Western Ghats are more prominent and higher in elevation than the Eastern Ghats. Thickly forested and steep, the Malabar-Konkan escarpment dominates the Arabian Sea coastland and limits its width to an average of less than 50 miles (80 kilometers); the Coromandel-Golconda coastal plain is wider and its interior margins are less pronounced.

Although not very large in terms of total area, these two coastal subregions have had, and continue to have, great impor-tance in the Indian state. Figure I–9 indicates how well-watered the Malabar-Konkan Coast is—the triggering onshore airflow, or **wet monsoon**, is diagrammed and dis-cussed in the *box* on pp. 456–457. This rainfall supply and the balmy tropical temperatures have com-bined with the coast's fertile soils to create one of India's most pro-ductive areas. On the lowland plain, rice is grown; on the adja-cent slopes spices and tea are cultivated. Of course, this combi-nation of favorable circumstances for intensive agriculture led to the emergence of southern India's major coastal population concen-tration (Fig. I–14). Along these coasts the Europeans made con-tact with India, beginning with the Greeks and Romans. Later they became spheres of British influ-ence, and it was during this period that two of India's greatest cities, Bombay and Madras, began their growth.

THE HUMAN SEQUENCE

The Indian subcontinent is a land of great river basins. Between the mountains of the north and the up-lands of the peninsula in the south lie the broad valleys of the Gan-ges, the Brahmaputra, and the Indus. In one of these—the Indus (see *box*, p. 460)—lies evidence of the realm's oldest civilization, contemporary to, and interacting with, ancient Mesopotamia. Un-fortunately, much of the earliest record of this civilization lies bur-ied beneath the present water table in the Indus Valley, but those archaeological sites that have yielded evidence indicate that here was a quite sophisticated culture with large, well-organized cities. As in Mesopotamia and the Nile Valley, considerable ad-vances were made in the technol-ogy of irrigation, and the civiliza-tion was based on the productivity of the Indus lowlands' irrigated soils (Fig. 8–4).

But the Indus Valley did not es-cape the invasion of the Aryans any more than did Europe or the Mediterranean Basin. After about 2000 B.C., people began to move into the Indus region from west-ern Asia. Culturally, these people were not as advanced as the Indus Valley inhabitants, but they soon destroyed the cities. However, they also adopted many of the in-novations of the Indus civiliza-tion, and they pushed their frontier of settlement eastward beyond the Indus Valley into the Ganges

FIGURE 8–4

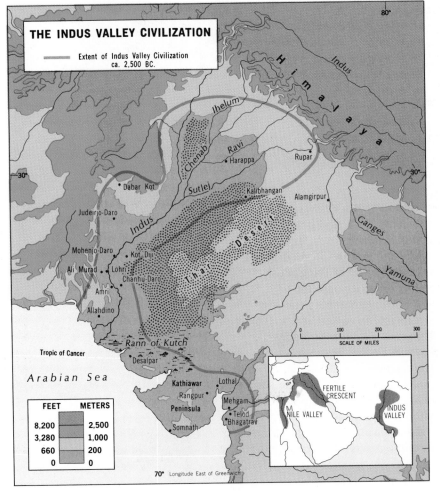

lowland and southward into the peninsula. They conquered and absorbed the tribes they found there, and the language they had brought to India, Sanskrit, began to differentiate into the linguistic complex that is India's today.

In the centuries during and following the Aryan invasion, Indian culture went through a period of growth and development. From a formless collection of isolated tribes and their villages, regional organization began to emerge. Towns developed, arts and crafts blossomed once again, and trade with Southwest Asia was renewed. Most importantly, Hinduism emerged from the religious beliefs and practices brought to India by the Aryans; soon, a whole new way of life was being shaped. **Social stratification** evolved, administered by powerful priests, with the ruling Brahmans at the head of a complex bureaucracy—a *caste system*—in which soldiers, artists, merchants, and all others had their place. Aggressive and expansion-minded kingdoms also arose, always competing with each other for greater power and control.

It was in one of these kingdoms in northeastern India that Prince Siddhartha, better known as Buddha, was born in the sixth century B.C. Buddha voluntarily gave up his princely position to seek salvation and enlightenment through religious meditation. His teachings demanded a rejection of earthly desires and a reverence for all life. Although Buddha had a substantial following during his lifetime, the real impact of Buddhism was to come during the third century B.C. after an interval marked by the end of Persian domination northwest of the Indus Valley and a brief but intense intervention by Alexander the Great (326 B.C.), whose Greek forces pushed all the way to the Hindu Indian heartland of the Ganges Plain.

While northern India was a theater of cultural infusion and innovation, the peninsular south lay removed, protected by distance from much of this change. Here, a very different culture came into being. The darker skins of the southerners today still reflect their direct ties to ancient forebears, such as the Negritos and black Australians who lived here even before the Indus civilization arose far to the northwest. Their languages, too, are distinctive and not related to those of the Indo-Aryan region. Both the peoples and the languages of southern India are known collectively as *Dravidian*. The four major Dravidian languages—Telugu, Tamil, Kanarese (Kannada), and Malayalam—all have long literary histories, and

Telugu and Tamil are the languages of nearly one-fifth of India's 870-odd million inhabitants (Fig. 8–5).

The Mauryan Era

The Mauryan Empire was the first to incorporate most of the subcontinent, emerging with the decline of the Hellenistic influence shortly before 300 B.C. Its heartland lay in the central Ganges Plain; quite rapidly, it extended its power over India as far west as the Punjab and Indus Valley, as far east as Bengal (the Ganges Delta region), and as far south as the modern city of Bangalore. The Mauryan state was led by a series of capable rulers, among whom

FIGURE 8–5

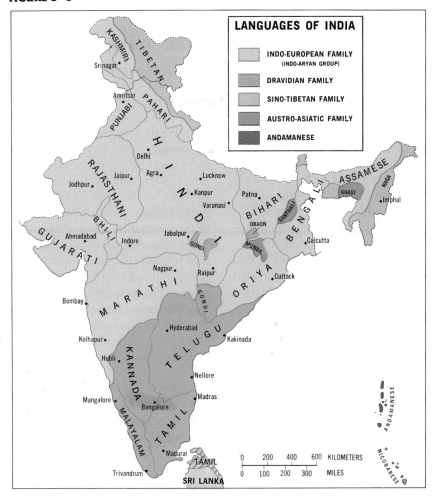

"INDIA"

Our use of the name *India* for the heart of the South Asian realm derives from the Sanskrit word *sindhu*, used to identify the ancient civilization in the Indus Valley. This word became *sinthos* in Greek descriptions of the area, and then *sindus* in Latin. Corrupted to *indus*, which means "river," it was first applied to the region that now forms the heart of Pakistan. Subsequently, it was again modified to *India* to refer generally to the land of river basins and clustered peoples from the Indus in the west to the lower Brahmaputra in the east.

the greatest was undoubtedly Aśoka, who ruled for nearly 40 years during the middle of the third century B.C. and who was a convert to Buddhism. In accordance with Buddha's teachings, Aśoka diverted the state's activities from conquest and expansion to the attainment of internal stability and peace. He also sent missionaries to the outside world to carry Buddha's teachings to distant peoples, thereby contributing to the further spread of Indian culture. As a result, Buddhism became permanently established as the dominant religion in Ceylon (now Sri Lanka), and it achieved temporary footholds even in the eastern Mediterranean lands—although it eventually declined to minor status within India itself.

The Mauryan state, which represented the culmination of Indian cultural achievements to that time, was not soon to be repeated. When this empire collapsed late in the second century A.D., the old forces—regional-cultural disunity, recurrent and disruptive invasions, and failing central authority—again came to prevail. Of course, the cultural disunity of India was a reality even during the era of the Mauryans, who could not submerge it. Although India did not see anything like the Aryan invasions again, there were almost constant infusions of large and

small population groups from the west and northwest. Persians, Afghans, and Turks entered the subcontinent, mostly along the obvious route: across the Indus, through the Punjab, and into the Ganges Plain.

Islamic Imprints

In similar fashion, beginning late in the tenth century, the wave of Islam came rolling like a giant tide across the subcontinent, spreading from Persia and Afghanistan in the northwest. In the Indus region, there was a major influx of Muslims and a nearly total conversion to Islam by the local population; in the Punjab, perhaps two-thirds of the population was converted. Islam crossed the bottleneck in which Delhi is situated, and diffused southeastward into Hindustan (Fig. 8–6). Whereas the Muslim impact here in the Ganges lowland was far less significant, about one-eighth of the local population did become adherents to the new faith. But in the Ganges Delta region to the east, up to three-quarters of the people became Muslims, with Islam spreading largely at the expense of Buddhism. Southward into the peninsula, however, the force of Islam was spent quite rapidly, and the

far south never came under Muslim control.

Islam was an alien faith to India, as it was to southern Europe, and it brought great changes to existing ways of life. It was superimposed by political control; by early in the fourteenth century, a sultanate centered at Delhi controlled all but the extreme southern, eastern, and northern margins of the subcontinent. Nonetheless, there were constant struggles to maintain that control, and out of one of those challenges arose the largest unified Indian state since Aśoka's time. This was the Islam-dominated Mogul Empire, which by about 1690 under the rule of Aurangzeb blanketed almost the entire subcontinent.

But Islam in India was neither the monopoly of the invaders from outside nor the exclusive religion of the rulers and new aristocracy. Islam provided a welcome alternative for Hindus who had the misfortune of being of low caste; it was also an alternative for Buddhists and others who faced absorption into the prevailing Hindu system. In the majority of cases, the Muslims of the Indian subcontinent are racially indistinguishable from their non-Muslim neighbors. Most of the realm's Muslims today are descendants of converts rather than descendants of any invading Muslim ruling elite.

European Intrusion

Into this turbulent complexion of religious, political, and linguistic disunity still another element soon intruded—European powers in search of raw materials, markets, and political influence. The Europeans profited from the Hindu-Muslim contest, and they were able to exploit local rivalries, jealousies, and animosities. British merchants gained control over the trade with Europe in spices, cotton, and

Colonial Transformation

Four centuries of European intervention in South Asia greatly changed the realm's cultural, economic, and political directions. Certainly the British made positive contributions to Indian life, but colonialism also brought serious negative consequences. In this respect, there are important differences between the South Asian case and that of Subsaharan Africa. When the Europeans came to India, they found a considerable amount of industry, especially in metal goods and textiles, and an active trade with both Southwest and Southeast Asia in which Indian merchants played a leading role. The British intercepted this trade, changing the whole pattern of Indian commerce.

India now ceased to be South Asia's manufacturer, and soon the country was exporting raw materials and importing manufactured goods—from Europe, of course. India's handicraft industries declined; after the first stimulus, the export trade in agricultural raw materials also suffered as other parts of the world were colonized and linked in trade to Europe. Thus the majority of India's people, who were farmers then as now, suffered an economic setback as a result of the manipulations of colonialism. Although in total *volume* of trade the colonial period brought considerable increases, the *composition* of the trade India now supported by no means brought a way to a better life for its people.

Neither did the British manage to accomplish what the Mauryans and the Moguls had tried to do: unify the subcontinent and minimize its internal cultural and political divisions. When the crown took over from the East India Company in 1857, about 750,000 square miles (nearly 2 million

FIGURE 8–6 The Taj Mahal tomb, located in Agra, about 100 miles south of Delhi. This architectural jewel is the most famous reminder of the Muslim contribution to the evolution of Hindustan.

silk goods, and in time they ousted their competitors, the French, Dutch, and Portuguese. The British East India Company's ships also took over the intra-Asian sea trade between India and Southeast Asia, long in the hands of Arab, Indonesian, Chinese, and Indian merchants. In effect, the East India Company became India's colonial administration.

As time went on, however, the East India Company faced problems it could not solve. Its commercial activities remained profitable, but it became entangled in a widening effort to maintain political control over an expanding Indian domain. It proved to be an ineffective governing agent at a time when the increasing Westernization of India brought to the fore new and intense frictions.

Christian missionaries were challenging Hindu beliefs, and many Hindus believed that the British were out to destroy the caste system. Changes also came in public education, and the role and status of women began to improve. Aristocracies saw their positions threatened as landowners had their estates expropriated. Finally, in 1857, the three-month Sepoy Rebellion broke out and changed the entire situation. It took a major military effort to put it down, and from that time the East India Company ceased to function as the government of India. Administration was turned over to the British government, the company was abolished, and India formally became a British colony—a status it held for the next 90 years until this *raj* (rule) ended in 1947.

square kilometers) of Indian territory were still outside the British sphere of influence. Slowly, the British extended their control over this huge unconsolidated area, including several pockets of territory already surrounded but never integrated into the previous corporate administration. Moreover, the British government found itself obligated to support a long list of treaties that had been made by the company's administrators with numerous Indian princes, regional governors, and feudal rulers.

These treaties guaranteed various degrees of autonomy for literally hundreds of political entities in India, ranging in size from a few acres to Hyderabad's more than 80,000 square miles (200,000 square kilometers). The British crown saw no alternative but to honor these guarantees, and India was carved up into an administrative framework under which there were more than 600 "sovereign" territories in the subcontinent. These "Native States" had British advisors; the large British provinces such as Punjab, Bengal, and Assam had British governors or commissioners who reported to the viceroy of India who, in turn, reported to Parliament and the monarch in London. In all, this near-chaotic amalgam of modern colonial control and traditional feudalism reflected and in some ways deepened the regional and local disunities of the Indian subcontinent. Although certain parts of India quickly adopted and promoted the positive contributions of the colonial era, other areas rejected and repelled them, thereby adding yet another element of division to an increasingly complicated human spatial mosaic.

Indeed, colonialism did produce assets for India. The country was bequeathed one of the best railroad and highway transport networks of the colonial domain. British engineers laid out irrigation canals through which millions of acres of land were brought into cultivation. Settlements that had been founded by Britain developed into major cities and bustling ports, as did Bombay (12.1 million inhabitants today), Calcutta (11.9 million), and Madras (5.9 million). The latter are still three of India's largest urban centers, and their cityscapes bear the unmistakable imprint of colonialism (Calcutta is shown in Fig. 8–7). Modern industrialization, too, was brought to India by the British on a limited scale. In education, an effort was made to combine English and Indian traditions; the Westernization of India's elite was supported through the education of numerous

FIGURE 8–7 Central Calcutta's cityscape still bears the distinctive imprint of the pre-1947 British colonial *raj* (rule).

Indians in Britain. Modern practices of medicine were also introduced. Moreover, the British administration tried to eliminate features of Indian culture that were deemed undesirable by any standards—such as the burning alive of widows on the funeral pyres of their husbands, female infanticide, child marriage, and the caste system. Obviously, the task was far too great to be achieved successfully in barely three generations of rule, but post-independence India itself has continued these efforts where necessary.

Partition

Even before the British government decided to yield to Indian demands for independence, it was clear that British India would not survive the coming of self-rule as a single political entity. As early as the 1930s, the idea of a separate Pakistan was being promoted by Muslim activists, who circulated pamphlets arguing that British India's Muslims were a distinct nation from the Hindus and that a separate state consisting of Punjab, Kashmir, Sind, Baluchistan, and a portion of Afghanistan should be created. The first formal demand for such a partitioning was made in 1940, and the idea had the almost universal support of the realm's Muslims as subsequent elections proved.

With the colony moving toward independence, a crisis developed: India's majority Congress Party would not consider partition, and the minority Muslims refused to participate in any unitary government. But partition was not simply a matter of cutting off the Muslim areas from the main body of the country. True, Muslims were in the majority of what are now Pakistan and Bangladesh, but other Islamic clusters were scattered throughout the subcontinent (Fig. 8–8). The boundaries between India and Pakistan, therefore, would have to be drawn right through transitional areas, and people by the millions would be dislocated.

In the Punjab, for example, a large number of Sikhs—whose leaders were fiercely anti-Muslim—faced incorporation into Pakistan. Even before independence day, August 15, 1947, Sikh leaders had been talking of revolt, and there were some riots; but no one could have foreseen the horrible killings and mass migrations that followed independence and official partition. Just how many people felt compelled to participate in these migrations will never

FIGURE 8–8

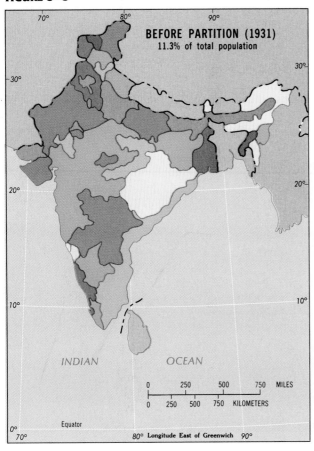

BEFORE PARTITION (1931)
11.3% of total population

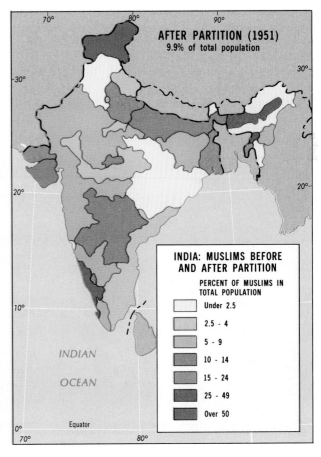

AFTER PARTITION (1951)
9.9% of total population

INDIA: MUSLIMS BEFORE AND AFTER PARTITION

PERCENT OF MUSLIMS IN TOTAL POPULATION

- Under 2.5
- 2.5 - 4
- 5 - 9
- 10 - 14
- 15 - 24
- 25 - 49
- Over 50

be known, but the most common estimate is 15 million, representing a mass of human suffering that is indeed incomprehensible.

Even that huge number of refugees hardly began to purify either India of Muslims or (former East and West) Pakistan of Hindus. After the initial mass exchanges, there were still tens of millions of Muslims in India (right map, Fig. 8–8), and East Pakistan (now Bangladesh) remained about 20 percent Hindu. The remarkable photograph shown in Fig. 8–9 fully captures the frenzied atmosphere of these exchanges; it was taken in late 1947 and shows Hindu refugees jamming the railroad cars of trains arriving at Amritsar from the new Islamic republic of Pakistan.

Facing the difficult alternatives, many people decided to stay where they were and make the best of the situation. For most, this turned out to be a wise choice.

The actual process of partition was accomplished swiftly—and of necessity rather arbitrarily—and was supervised by a joint commission whose chairman was a neutral British representative. Using data from the 1941 census of India, Pakistan's boundaries were defined in such a manner that this Muslim state would incorporate all contiguous civil divisions and territories in which Muslims formed a majority. The commission had to make decisions it knew in advance would be highly unpopular, although no one anticipated the

proportions of the Sikh-initiated violence. What resulted, then, was a **superimposed political boundary** on a cultural landscape in which such a boundary had not previously existed or functioned; this is one of the four genetic boundary types that will be discussed in the essay on political geography that opens Chapter 10.

FEDERAL INDIA

India today is the world's most populous federal state. The Indian federation consists of 25 states and

FIGURE 8–9 Flight was one response to the 1947 partition of what had been British India, resulting in one of the greatest mass population transfers in human history. Here, two trainloads of eastbound Hindu refugees fleeing West Pakistan arrive at the station in Amritsar, the first city inside India.

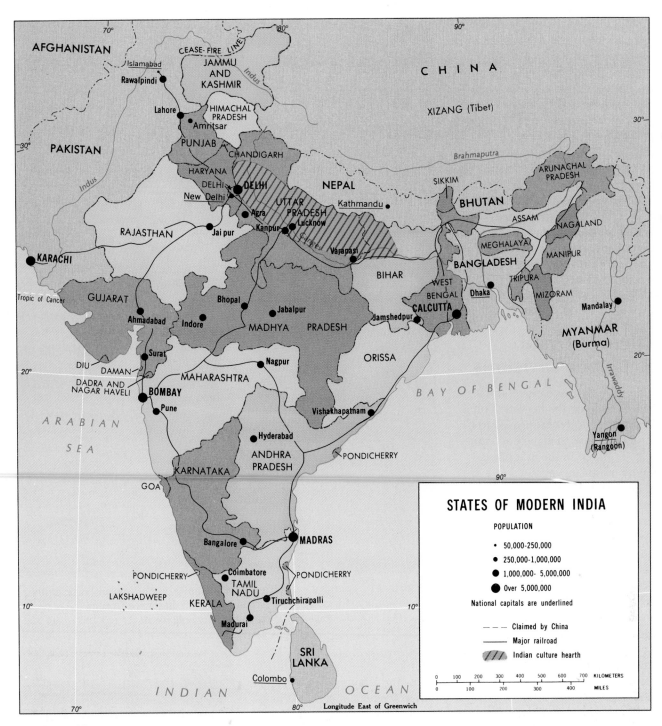

FIGURE 8–10

7 union territories (Fig. 8–10). A comparison between Fig. 8–10 and the map of languages in India (Fig. 8–5) indicates a considerable degree of coincidence between linguistic and state boundaries. The present framework is not the one with which India was born as a sovereign state in 1947, and it is likely to be modified again in the future. Soon after the British withdrew, the several hundred "princely states," whose rights had been protected during the colonial period, were absorbed into the states of the federation (although their rulers' privileged "princely orders" were not terminated until 1972). Next, the Indian government reorganized the country on the basis of its regional languages; Hindi, spoken by more than one-third of all Indians, was designated the official language.

In fact, Hindi was one of 14 languages given national status by the Indian constitution—10 of them spoken in the northern and central part of the country, and 4 in the Dravidian south. English, it was anticipated, would remain a *lingua franca* when Hindi could not serve as a medium of communication at government and administrative levels. This it has proven to be, and English has also become the chief language of the business world of emerging urban India, increasingly regarded as a key to good jobs, financial success, and personal advancement.

Linguistic and Religious Divisions

The politico-geographical framework based on the major regional languages, however, proved to be unsatisfactory to many communities in India. In the first place, many more languages are in use than the 14 that were officially recognized. As early as 1953, the government yielded to demands for the creation of a Telugu-speaking state separate from Tamil-dominated Madras, and the state of Andhra Pradesh was formed. In 1960, the state of Bombay was fragmented into two linguistic states, Gujarat and Maharashtra (Fig. 8–10). In the distant northeastern borderlands, the Naga peoples, numbering less than half a million, put up a struggle against federal authority and local Assamese administration; after Indian armed forces were sent into the area, Nagaland was established as a separate state. In the northwest, the Sikhs demanded the breakup of the original state of Punjab into a Sikh-dominated west (now Punjab State) and a Hindu east (now Haryana State). That action temporarily defused Sikh demands for

FIGURE 8–11 Amritsar's Golden Temple complex, the "Vatican" of Sikhdom. Here a faithful Sikh, in hallmark turban and beard, has just bathed in the holy lagoon that almost completely surrounds the temple itself.

more control over their homeland, but, as was noted earlier, this issue has again come to the forefront since the early 1980s.

The Sikhs (the word means "disciples") are adherents of Sikhism, a religion that was created about 500 years ago to unite warring Hindus and Muslims into a single faith. This alternative religion rejected both the Hindu caste system and the historic expansionism of Islam in favor of monotheism, the accumulation of wealth, and the aggressive defense of the Sikh cause. Sikhism developed in the Punjab region and is still based there; its main focus is the city of Amritsar which contains the Golden Temple (Fig. 8–11), the religion's most sacred shrine (and site of the 1984 Indian army attack that routed the extremist Sikh leadership and led to the assassination of Prime Minister Indira Gandhi).

Given their location in the turbulent zone between Hindustan and the Muslim-dominated Indus Valley, the Sikhs are no strangers to conflict: their very survival has often depended on the bravery of their fierce warriors who were renowned throughout South Asia. When the British arrived in India, the Sikhs were forced to yield to their superior firepower. Soon, many Sikhs came to support the colonial takeover and, in turn, won the respect and trust of the British who employed thousands of them as soldiers and policemen. By 1947, a large segment of the Sikh population had achieved middle-class status. When the Punjab was split between newly created India and Pakistan, many left their rich farmlands, migrated to Indian cities, and entered the professions (from which they still exert a wide influence far beyond the 2 percent of the national population they comprise).

A militant minority, however, resisted partition and helped to unleash the bloody violence mentioned earlier. Following independence, a small group of extremists continued to agitate for a separate Sikh homeland—a new Punjab state to be called "Khalistan." Their efforts were largely ignored for years, but gradually the cause was joined by many Sikhs living outside the Punjab, whose customs and ethnic pride had brought on increasing hostility by the populations into which they were seeking assimilation. With growing support from the Sikh community inside and outside India, a radical movement flourished, extremists began to act with greater boldness, and the situation escalated into the tragic events of 1984. Since then, the intensified polarization between Sikh and Hindu has not lessened, cycles of violence persist, the Punjab remains in a high state of tension, and the Indian govern-

CENTRIFUGAL AND CENTRIPETAL FORCES

Political geographers use the terms *centrifugal* and *centripetal* to identify forces within a state that tend, respectively, to pull that political system apart and to bind it together.

Centrifugal forces are disunifying or divisive. They can cause deteriorating internal relationships. Religious conflict, racial strife, linguistic polarization, and contrasting regional outlooks are among the major centrifugal forces. In recent years, Lebanon, Ethiopia, South Africa—and, of course, India—are among those countries that have experienced such cleavages. Moreover, many newly independent countries find tribalism a leading centrifugal force, sometimes strong enough to threaten the very survival of the whole state system (as the Biafra conflict of the 1960s did in Nigeria).

Centripetal forces tend to bind the state together, to unify and strengthen it. A real or perceived external threat can be a powerful centripetal force, but more important and lasting is a sense of commitment to the governmental system, a recognition that it constitutes the best option. This commitment is sometimes focused on the strong charismatic qualities of one individual—a leader who personifies the state, who captures the population's imagination. (The origin of the word *charisma* lies in the Greek expression meaning "divine gift.") At times, such charismatic qualities can submerge nearly everything else; in India, Mohandas (Mahatma) Gandhi and Jawaharlal Nehru possessed personal qualities that went far beyond political leadership, and their binding effect on the young nation extended far beyond their lifetimes.

The degree of strength and cohesion of the state depends on a surplus of centripetal forces over divisive centrifugal forces. It is difficult to measure such intangible qualities, but some attempts have been made in this direction—for example, by determining attitudes among minorities and by evaluating the strength of regionalism as expressed in political campaigns and voter preferences. When the centrifugal forces gain the upper hand and cannot be checked (even by external imposition) the state may break up—as once-united East and West Pakistan did in 1971.

ment is still trying to find a way to defuse this potentially explosive crisis.

India's Centrifugal Forces

Pressure for greater regional autonomy also continues in several other parts of India (as was pointed out at the beginning of this chapter), and it is remarkable that the federal state has been able to survive all these centrifugal forces (see *box*). India's continued unity is all the more impressive in view of its enormous population and the problems inherent in its demographic condition. The most populous Indian states (Uttar Pradesh, about 140 million; Bihar, over 85 million; Maharashtra, nearly 80 million) have populations larger than most of the world's countries. No country contains greater cultural diversity than India, and variety in India comes on a scale unmatched anywhere else on earth. Even today, after more than four decades of religious partition between Hindus and Muslims in South Asia, India remains 11 percent Muslim; that might seem to be a rather small minority, except that here an 11 percent minority represents almost 100 million people.

A still-pervasive aspect of Indian society is its stratification into *castes*. Under Hinduism, castes are fixed layers in society whose ranks are based on ancestries, family ties, and occupations. The caste system may have evolved from an early social differentiation into priests and warriors, merchants and farmers; it may also have a racial foundation (the Sanskrit term for caste is color). Over the centuries its complexity grew until India came to possess several thousand castes, some with a few hundred members, others containing millions. Thus, in village as well as city, communities were segregated according to caste, ranging from the highest (priests, princes) to the lowest (the untouchables).

A person was born into a caste based on his or her actions in a previous existence. Hence, it would not be appropriate to counter such ordained caste assignment by permitting movement (or even contact) from a lower caste to a higher one. Persons of a particular caste could only perform certain jobs, only wear certain clothes, worship only in prescribed ways at particular places. They or their children could not eat with, play with, or even walk with people of a higher social status. The untouchables occupying the lowest tier were the most debased, wretched members of this rigidly structured social system. Although the British ended the worst excesses of the caste system and modern Indian leaders, including Mohandas Gandhi and Jawaharlal Nehru, have worked to

FIGURE 8–12 Bathing in the holy Ganges River is a venerable Hindu tradition that draws huge crowds to Varanasi each year.

modify it, centuries of class consciousness are not wiped out in a few decades. In traditional India, caste provided stability and continuity; in modernizing India, it constitutes an often painful and difficult legacy.

India's Centripetal Forces

What centripetal forces (*box*, p. 467) have helped keep India unified? Without question, the dominant binding force in India is the cultural strength of Hinduism, its holy writings (read in all languages), its holy rivers, and its many other influences over Indian life (Fig. 8–12). Hinduism is a way of life as much as it is a faith, and its diffusion over virtually the entire country (Muslim, Christian, and Sikh minorities notwithstanding) brings with it a national coherence that constitutes a powerful antidote to regional divisiveness. Moreover, communications in much of densely populated India are better than they are in many other Third World countries, and the constant circulation of people, ideas, and goods helps bind the nation together. Before independence, opposition to British rule was a shared philosophy throughout the country, but India remained divided and separated internally. Independence brought with it a first taste of national planning, national political activity, and national debates over priorities.

India also produced some strong leaders, whose compelling personalities constituted another binding force. Gandhi and Nehru did much to forge the India of today. Nehru's daughter, Indira Gandhi, twice took decisive control (in 1966 and 1980) following episodes of weak national leadership—and *her* son, Rajiv Gandhi,

took over as prime minister from 1984 to 1989.

A final centripetal force is that for the past 45 years federal India has proved capable of accommodating far-reaching change, thereby avoiding the alternative of secession. Although political leaders in some states have on occasion openly endorsed the prospect of secession, India's flexibility on the language issue, its ability to tolerate individuality in its states, and its capacity to modify and remodify the federal map all offer evidence that change *is* possible and that political, economic, educational, and other objectives can, in fact, be achieved within India's complex federal framework.

INDIAN DEVELOPMENT

If India has faced problems in its great effort to achieve political stability and national cohesion, these are more than matched by the difficulties that lie in the way of economic growth and development. The large-scale factories and power-driven machinery of the colonial powers wiped out a good part of India's indigenous industrial base. Indian trade routes were taken over. European innovations in health and medicine sent the rate of population growth soaring, without introducing solutions for the many problems this spawned. Surface communications were improved and food distribution systems became more efficient, but local and regional food shortages occurred (and still do) as droughts frequently caused crop failures. Today, nearly half of India's 870-million-plus people live in abject poverty, and the prospects of re-

ducing that high level of human misery anytime soon are not encouraging.

Agriculture

India's underdevelopment is nowhere more apparent than in its agriculture. Traditional farming methods continue to prevail, and yields per acre and per worker remain low for virtually every crop grown under this low-technology system. The scene in Fig. 8–13, set in a rural area near Madras, is depressingly typical. Water is brought to the surface not by a pump but by oxen pulling a bucket from a well; the animals walk endlessly back and forth along the short path in the center of the photograph, but they are capable of hauling only a comparative trickle of water to the surface (as shown in the ditch at the far left). Moreover, movement of agricultural commodities is hampered by the transportation inefficiencies of the traditional farming system: in 1987, only 36 percent of India's 600,000 villages were accessible by motorable road, and today animal-drawn carts still outnumber motor vehicles nationwide.

As the total population grows, the amount of cultivated land per person declines. Today, this **physiologic density** is approximately 1240 per square mile (480 per square kilometer). However, this is nowhere near as high as the physiologic density in neighboring Bangladesh, where the figure is more than twice as great (2735 and 1055, respectively). But India's farming is so inefficient that this is a deceptive comparison. More than two-thirds of India's huge working population depends directly on the land for its livelihood, but the great majority of Indian farmers are poor, unable to improve their soils, equipment,

FIGURE 8—13 Agriculture in India is bedeviled by problems as inefficient traditional farming methods continue to prevail.

or yields. Those areas in which India has made substantial progress toward the modernization of its agriculture (as in the Punjab's wheat zone) very much remain islands in a sea of agrarian stagnation.

This stagnation has persisted in large measure because India failed, after independence, to implement a much-needed nationwide land reform program. In the 1980s, more than one-quarter of India's entire cultivated area was still owned by less than 5 percent of the country's farming families, and little land redistribution was taking place. Perhaps half of all rural families own either as little as an acre—or no land at all. Independent India inherited inequities from the British colonial period, but the individual states of the federation would have had to coop-

erate in any national land reform program. As always, the large landowners retained much political influence, and so the program never got off the ground.

To make matters worse, much of India's farmland is badly fragmented as a result of local rules of inheritance, thereby inhibiting cooperative farming, mechanization, shared irrigation, and other opportunities for progress. Not surprisingly, land consolidation efforts have had only limited success, except in the states of Punjab, Haryana, and Uttar Pradesh, where modernization has gone farthest. Certainly, official agricultural development policy, at the federal as well as state level, has also contributed to India's agricultural malaise and the uneven distribution of progress. Unclear priorities, poor coordination, in-

adequate information dissemination, and other failures have been reflected in the country's disappointing output.

It is instructive to compare Fig. 8–14, showing the distribution of crop regions and water supply systems in India, with Fig. I–9, which shows mean annual precipitation in India and the world. In the comparatively dry northwest, notably in the Punjab and neighboring areas of the upper Ganges, wheat is the leading cereal crop; here, India has made major gains in annual production through the introduction of high-yielding varieties developed in Mexico. This innovation was part of the so-called **Green Revolution** (see *box*, p. 409) of the 1960s, when strains of wheat and rice were developed that were so much more productive than existing varieties that they were dubbed "miracle" crops. Introducing these new seeds also led to the expansion of cultivated areas, the development of new irrigation systems, and the more intensive use of fertilizer (a mixed blessing, for fertilizers are expensive and the "miracle" crops are more heavily dependent on them).

Toward the moister east, and especially in the wet-monsoon-drenched areas (Fig. 8–3), rice takes over as the dominant staple. About one-fourth of India's total farmland lies under rice cultiva-tion, most of it in the states of As-sam, West Bengal, Bihar, Orissa, and eastern Uttar Pradesh, and along the Malabar coastal strip; these areas receive over 40 inches (100 centimeters) of rainfall annu-ally, and irrigation supplements precipitation where necessary.

FIGURE 8–14

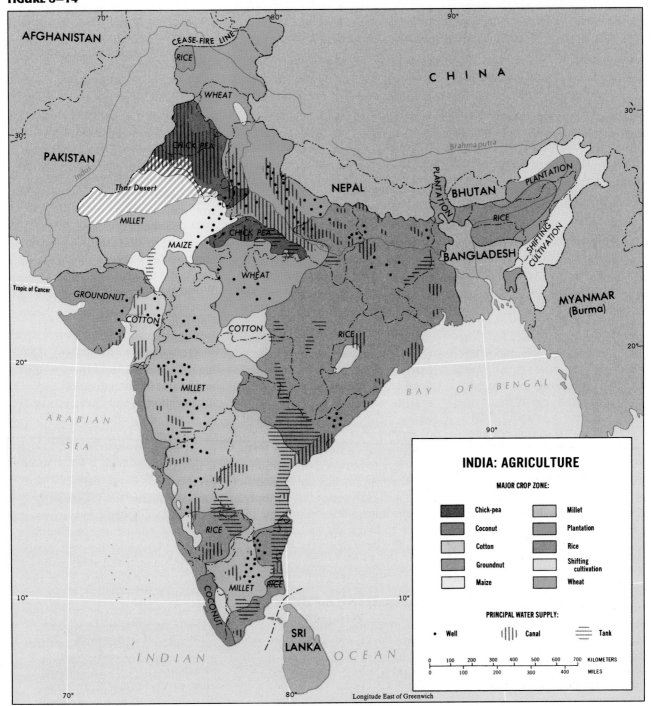

FIGURE 8–15 The same paddyfields before and after the arrival of the monsoon rains in Goa State on the west-central Arabian Sea coast, revealing the stunning contrast in the agricultural landscape of the dry and wet seasons.

The pair of photos in Fig. 8–15 shows diked ricefields in the tiny west-coast state of Goa—and is a vivid reminder of the vast difference in water availability between dry-monsoon May and wet-monsoon August. India does have the largest acreage of rice among all the world's countries, but despite the introduction of "miracle rice" in parts of India, the average yield per acre is still only about 1800 pounds (2000 kilograms per hectare). The world population map (Fig. I–14) reveals a high degree of geographical covariation between India's rice areas and its most densely peopled zones, but overall rice yields per unit area in India continue to rank among the lowest on earth.

As the map of agricultural zones (Fig. 8–14) indicates, India's varied environments support the cultivation of millet, wheat, corn (maize), and other cereals. But subsistence remains a way of life for countless tens of millions of Indian villagers who cannot afford to buy fertilizers, cannot cultivate the new strains of wheat or rice, and cannot escape the cycle of abject poverty. Perhaps as many as 150 million of these Indians do not even have a plot of land and must live as tenants, always uncertain of their fate. This is the enduring reality against which optimistic predictions of better food conditions in the India of the future must be weighed. True, rice and wheat yields have each increased over 50 percent since the arrival of the Green Revolution in the 1960s. But the gap between population and food supply has not closed—the "miracle" crops have simply bought India a little more time in which to find a lasting solution to this persistent dilemma.

Industrialization

Notwithstanding the problems faced by its farmers, agriculture must be the foundation for development in India. Agriculture employs approximately two-thirds of the workers, generates most of the government's tax revenues, contributes many of its chief exports by value (cotton textiles, tea, jute manufactures, and leather goods all rank high), and produces most of the money the country can spend in other sectors of the economy. Add to this the compelling need to grow more and more food crops, and India's heavy investment in agriculture is understandable.

In 1947, India inherited the mere rudiments of an industrial framework. After more than a century of British control over the economy, only 2 percent of India's workers were engaged in industry, and manufacturing and mining combined produced only about 6 percent of the national income. Textile and food-processing industries dominated. Although India's first iron-making plant was opened in 1911 and the first steel mill began operating in 1921, the initial major stimulus for heavy industrialization came after the outbreak of World War II. Manufacturing was concentrated in the largest cities: Calcutta led, Bombay was next, and Madras ranked third.

The geography of manufacturing today still reflects those beginnings, and industrialization and urbanization in India have proceeded slowly, even after independence (Fig. 8–16). Calcutta now forms the center of India's eastern industrial region—the Bihar-Bengal district—where jute manufactures dominate, but engineering, chemical, and cotton industries also exist. On the nearby Chota Nagpur Plateau to the west, coal-mining and iron and steel manufacturing have developed. On the opposite side of the subcontinent, two industrial areas dominate the western manufacturing region: one centered on Bombay and the other focused on Ahmadabad (3.7 million). This region, lying in Maharashtra and Gujarat states, specializes in cotton and chem-

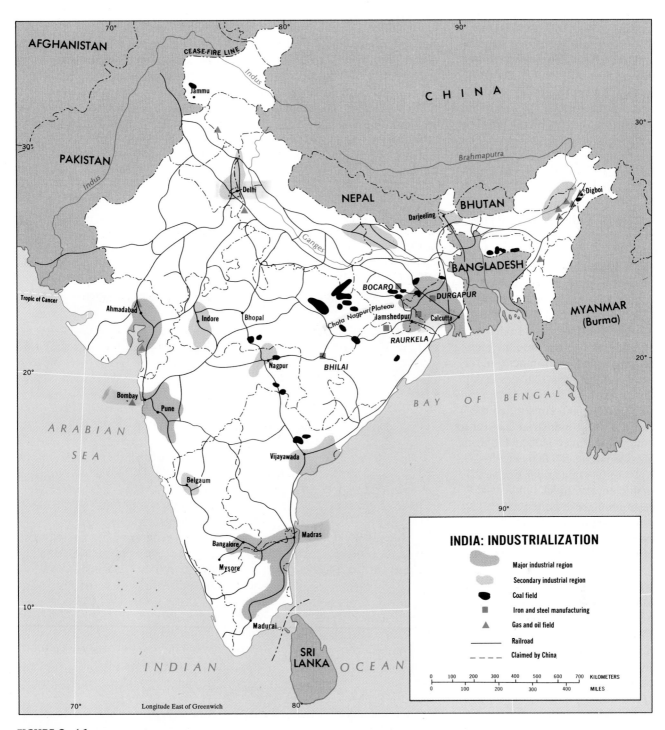

FIGURE 8-16

icals, with some engineering and food processing. Cotton textiles have long been an industrial mainstay in India, and this was one of the few industries to derive some benefit from the nineteenth-century economic order imposed by the British. With the local cot-

ton harvest, the availability of cheap yarn, abundant and inexpensive labor, and the power supply from the Western Ghats' hydroelectric stations, the industry thrived and today outranks Britain itself in the volume of its exports. Finally, the southern industrial re-

gion chiefly consists of a set of linear city-linking corridors focused on Madras, specializing in textile production and light engineering activities.

When India achieved independence, its government immediately set out to develop its own industrial base, both to lessen dependence on imported manufactures and to cease being an exporter of raw materials to the developed world. In the process, emboldened by the early successes of the Green Revolution, Indian planners actually overspent on their industrial programs—but the problem of rapid population growth now bedeviled industry as it did agriculture. Unemployment was to be reduced as industrialization progressed; instead, unemployment rose. Per capita incomes would rise as high-value products rolled off new assembly lines; but income rose very little, and today it is still below U.S. $300 per year. Modern factories would safely and harmoniously blend into the high-density urban environment; unfortunately, the 1984 toxic gas leak at Union Carbide's pesticide plant in Bhopal (740,000) constituted the world's worst industrial disaster to date, killing over 3400 and injuring an astounding 200,000 (Fig. 8–17).

Despite some imbalances and inefficiencies, India's industrial resource base is quite well endowed. Limited high-quality coal deposits are exploited in the Chota Nagpur district; in combination with large lower-grade coalfields elsewhere, the country's total output is high enough to rank it among the world's 10 leading coal producers. In the absence of known major petroleum reserves (some oil comes from Assam, Gujarat, Punjab, and offshore from Bombay), India must spend heavily on energy every year. Major investments have been made in hydroelectric plants, especially multipurpose dams that provide electricity, enhance irrigation, and facilitate flood control. India's iron ores in Bihar State (northwest of Calcutta) and Karnataka State (in the heart of the Deccan) may rank among the largest in the world. Jamshedpur (775,000), located west of Calcutta in the eastern industrial region,

has emerged as India's leading steelmaking and metals-fabrication center. Yet, India still exports iron ore as a raw material to developed countries (mainly Japan)—in the Third World, entrenched patterns are difficult to break.

Indian industrialization was also assisted by a major infrastructural advantage. In contrast to many other former colonial and now underdeveloped countries, India possesses a well-developed network of railroads with over 60,000 miles (100,000 kilometers) of track. The British colonizers built much of this system, but once again they bequeathed India with a liability as well as an asset. The railways were laid out, of course, to facili-

FIGURE 8–17 One of the tens of thousands of the injured being treated in the aftermath of the deadly accidental leak of methyl isocyanate gas, which engulfed part of the city of Bhopal in central India's Madhya Pradesh State in late 1984.

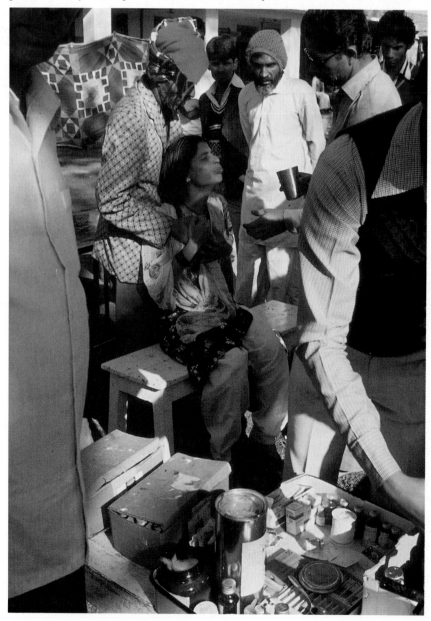

tate exploitation of interior hinterlands and to improve India's governability; this effective transport network permitted the British to move their capital from coastal Calcutta to the deep interior at Delhi. But the railroad system never evolved as a unified whole. Different British colonial companies constructed their own networks, and (reminiscent of Australia) no fewer than four separate railway gauges came into use. The two widest gauges now constitute about 90 percent of the whole system, but many transshipments are still necessary. India's government continues to standardize the rail system when tracks must be replaced, but the expense is considerable and progress remains slow. Yet, India's railway network connects most parts of the country and all its manufacturing complexes (Fig. 8–16). In terms of overall length, the system ranks as the world's fourth largest.

CRISIS OF NUMBERS

The population of South Asia (1.148 billion) constitutes over one-fifth of all humanity and will surpass the Chinese realm shortly after the turn of the century to become the world's biggest population concentration. That reality is alarming enough, especially because so many millions of these people are unable to secure an adequate food supply. But even more frightening are **population growth dynamics**—the *rate* at which the population continues to grow. In the early 1990s, more than 26 million people are being added to the South Asian population *every year*—over 19 million in India alone, and more than 3 million each in Bangladesh and Pakistan. Undoubtedly, here lies the

realm's greatest obstacle to progress: whatever the gains in crop production and industrial output, the demands of an ever-larger population constantly overcome them.

The situation in India reflects an equally disturbing global pattern. Not only is the world's population growing: the rate of expansion is increasing too. This is dramatically illustrated when we take a backward look. At the time of the birth of Christ, world population was probably around 250 million; it subsequently took until about A.D. 1650 to reach 500 million. Then, however, the population grew to 1 billion by 1820, and in 1930 it was 2 billion. In 1975 it surpassed the 4 billion mark, and reached 5.4 billion in 1991; if current growth rates persist, the world's population will total 8 billion shortly after 2010.

We can look at these figures in another way. It took nearly 17 centuries for 250 million to become 500 million, but then that 500 million grew to 1 billion in just 170 years. The billion doubled to 2 billion in 110 years, and 2 billion became 4 billion only 45 years later. The world population's **doubling time** at present growth rates is only 39 years. This steady decrease in the doubling time reflects an accelerating population growth rate, or what mathematicians call an *exponential* increase—growth along an upward curve rather than along a straight line. Geographically, however, it is important to note that the world's population is not growing at the same rate in all realms (see *Appendix A*, column 6). The world rate of natural increase, producing the doubling time of 39 years, is now 1.8 percent. But as Fig. 8–18 shows, the population of many North African and Southwest Asian countries is expanding more rapidly than that—in excess of 3 percent (a level which produces a doubling time of just 23 years). In Subsaha-

ran Africa, too (as we noted in Chapter 7), a number of countries exhibit very high growth rates; Kenya, whose annual natural increase rate is 4.1 percent, now faces a doubling time of only 17 years!

Annual population growth is calculated as the number of excess births over deaths recorded during a given year, usually expressed as a percentage as shown in Fig. 8–18. In Bangladesh, for example, the annual growth rate is given as 2.8. The birth rate in Bangladesh is 43 per thousand, and the death rate 15 per thousand. Thus, for every thousand persons in Bangladesh there were 28 excess births over deaths, or 2.8 percent. In the United States, the population growth rate is 0.7 percent, with 16 births per thousand population and 9 deaths. To compare what such figures mean in terms of doubling time, for Bangladesh's population it is just 25 years, whereas for the United States it is 98.

The Dilemma in India

At the beginning of the twentieth century, India had under 250 million inhabitants. This population fluctuated before 1900 with the vagaries of climate, disease, and war, but historical demographers believe that this average size had remained almost stationary for centuries prior to the European intervention and did not change significantly until the impact of Europe's Industrial Revolution had diffused to the colonies. Then, the gap between birth rates and death rates immediately began to widen. Birth rates remained high, but death rates declined as medical services proliferated, food distribution systems improved, agricultural production expanded, and costly wars were suppressed. The last time death rates (from famines and outbreaks of disease) approached the level of the birth rate was in 1910 when births were near 50 per thousand and deaths in the high 40s.

Since then, the birth rate has declined to 33—it first went below 40 as recently as 1970—but the death rate in India is now only 11. The early-twentieth-century population of 250 million doubled to about 500 million by the mid-1960s (it stood at 342 million when independence was achieved in 1947). Today, only a quarter of a century later, another 375 million people have been added. These staggering gains make one wonder about this planet's ultimate capacity to support human populations (see *box*).

In terms of the country's population growth—as with its history, culture, and economy—there is not just one India but several different and distinct Indias. The growth of population is most rapid (and still truly explosive) north and east of neighboring Bangladesh (Fig. 8–19); in Assam, Nagaland, and Mizoram, the rate of increase exceeds 3.5 percent annually. The rate is over 2.5 percent (as it is in Bangladesh) throughout the rest of northeastern India as well as in a wide area that covers the country's central-north and northwest. In fact, in all of northern India, only the Punjab and the lower Ganges region exhibit rates below 2.5 percent. In the southern peninsula, the growth rate trends nearer the national average of 2.2 percent; it is somewhat higher along the west coast, but the far south and Orissa State in the east record the country's lowest figures.

India obviously must reduce its birth rate, but in a country as vast and heavily rural as this, such a national task confronts virtually insurmountable obstacles. Indian governments have endorsed family planning as part of official development programs, but it has proven

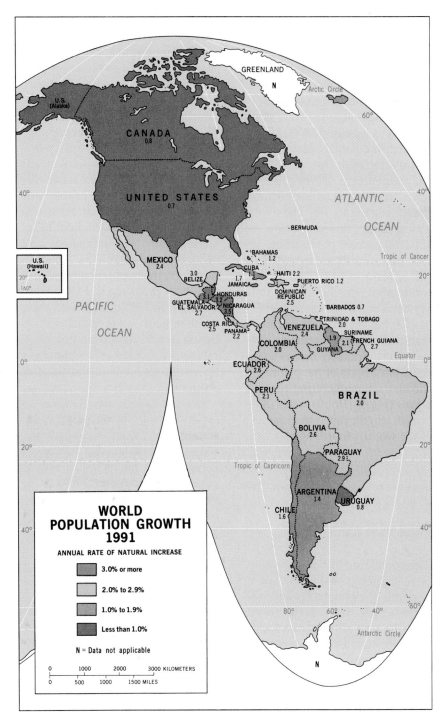

FIGURE 8–18

difficult to disseminate the practice throughout tradition-bound and still largely (64 percent) illiterate Indian society. Moreover, not all the states of the union cooperate equally in the effort. When, in the 1970s, a program of compulsory sterilization of persons with three or more children was instituted in

Maharashtra State, there were riots; after nearly 4 million people had been sterilized, the government was forced to declare that sterilization would no longer be compulsory. By the mid-1980s, however, family planning was reemerging from years of disrepute, and birth control was slowly

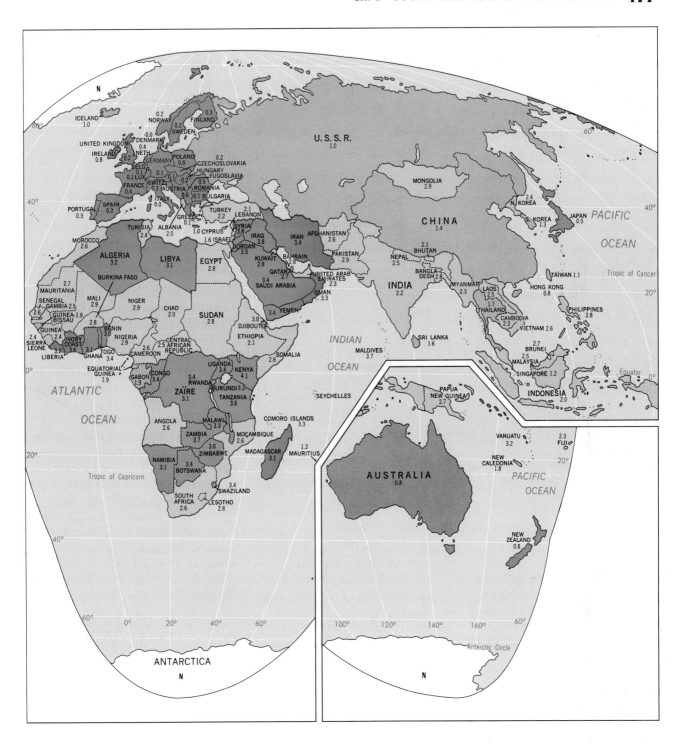

gaining wider acceptance. Today, over 40 percent of India's married couples are believed to be practicing contraception (versus only 10 percent in 1970), a proportion that population planners say must surpass 60 percent by 2000 if India is to get off its demographic treadmill.

Demographic Cycles

Some people predict that India will overcome its population crisis because other countries, too, went through periods of explosive growth only to reach a demographic near-equilibrium. Figure 8–18 indicates that the low-growth

countries are the developed countries (DCs), most of them with an annual population increase of less than 1 percent. In the nineteenth century, many of those countries also experienced explosive population expansion, but when they became industrialized, urbanized, and modernized, their rapid growth

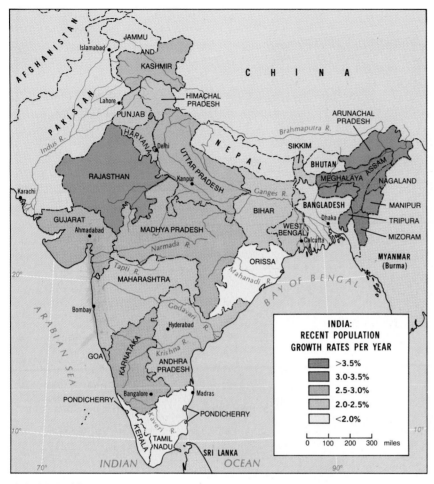

FIGURE 8–19

lion. Moreover, many UDCs do not at present appear to have the raw materials to sustain the kind of development that Europe witnessed. As we noted earlier, all the oil in North Africa/Southwest Asia has not generated genuine development, even though it has produced money with which the products of development can be purchased. Therefore, the Indian dilemma is far from solution.

Urbanization

As we saw in Chapter 5, the swiftly increasing urbanization of underdeveloped countries is one of the world's leading population processes in the 1990s. Although India may be known in part for its teeming cities with their hordes of homeless street dwellers (perhaps 400,000 in Calcutta alone), urbanization has proceeded more slowly here and only reached the 28 percent level in 1991. Once again, however, we must take note of India's massive proportions, and realize that 28 percent of its population involves 244 million people (almost the size of the total United States). The already enormous dimensions of India's urban crisis portended by this figure are now expanding, because there are definite signs that Indian urbanization is accelerating in the 1990s.

The latest findings show an unprecedented upsurge in cityward migration, and urban India today is growing more than twice as fast (about 5 percent yearly) as the country's overall population. Among the reasons for this shift that Indian planners cite are a dramatic loosening of ties between poor peasants and their villages, and the widespread establishment in the cities of villagemen or "caste brothers" who are able to assist their relatives and friends to make similar moves to the burgeoning urban residential colo-

declined. As we saw in the Systematic Essay in Chapter 1 (pp. 61–62), it is possible to discern a population cycle of three major stages that comprises the **Demographic Transition Model** (Fig. 1–3). First, birth as well as death rates are high, and the population fluctuates within a certain low-growth range. Next, as industrial, medical, and technological innovations are introduced, the death rate declines but the birth rate stays high until it begins to decline much later (this is the period of "population explosion"). Finally, birth rates plunge sharply, rejoining the level of the (now-low) death rate, and the overall growth rate is reduced below 1 percent. This is essentially what occurred in Western Europe, and some scholars insist that underdevel-

oped countries (UDCs) will also experience such a sequence.

But the European model may not fit countries of the late-twentieth-century underdeveloped world. To begin with, the population of a country such as England was under 10 million when its rapid-growth stage began. Two doublings would still only have produced 40 million from this base. Furthermore, when Britain went through its explosive period, there was mass emigration to alleviate the population pressure further. Present population totals in countries such as Pakistan and Bangladesh (let alone India) are on a very different scale. In two doubling periods, Pakistan would go from 117 to 468 million, Bangladesh from 121 to 484 million, and India from 871 million to 3.48 bil-

PREDICTING THE NIGHTMARE

In 1650, the world's human population was approximately 500 million; not until 1820 did the world accommodate 1 billion inhabitants. Even at that time, some scientists realized that the world was headed for trouble. The most prominent among them was Thomas Malthus, a British economist who sounded the alarm as early as 1798 in an article entitled *An Essay on the Principle of Population as It Affects the Future Improvement of Society*. In this essay, Malthus reported that population in England was increasing faster than the means of subsistence; he described population growth as geometric and the growth of food production as arithmetic. Inevitably, he argued, population growth would be checked by hunger. Malthus's essay caused a storm of criticism and debate, and between 1803 and 1826 he revised it several times. He never wavered from his basic position, however, and continued to predict that the gap between the population's requirements and the soil's productive capacity would ultimately lead to hunger, famines, and the cessation of population growth.

We know now that Malthus was wrong about several things. The era of colonization and migration vastly altered the whole pattern of world food production and consumption, and global distribution systems made possible the transportation of food surpluses from one region of the world to another. Malthus could not have foreseen this; nor was he correct about the arithmetic growth of agricultural output. Expanded acreages, improved seed varieties, and better farming techniques have produced geometric increases in world food production, but Malthus was correct in his prediction that the gap between need and production would widen. It has— and there are areas in the world in the 1990s where population growth rates are being checked in the way he predicted, as famines have claimed the lives of hundreds of thousands in Africa and Asia.

Today there are still scholars and specialists who essentially adhere to Malthus's position, modified of course by contemporary knowledge and experience. These people are sometimes referred to as prophets of doom by those who take a more optimistic view of the future; they are also called neo-Malthusians, tying their concerns to those of that farsighted Englishman who first warned that there are limits to this earth's capacity to sustain our human numbers.

---

nies of newcomers (increasingly defined by language and custom). Thus, even by Third World standards, Indian cities are places of staggering social contrasts (Fig. 8–20). Not surprisingly, as crowding intensifies, social stresses multiply; sporadic rioting has affected numerous cities in recent years, often attributable to the actions of rootless urban youths unable to find employment.

India's modern urbanization also has its roots in the colonial period, when the British selected Calcutta, Bombay, and Madras as regional trading centers and as coastal focal points for their colony's export and import traffic. All were British military outposts by the late seventeenth century. Madras, where a fort was built in 1640, lay in an area where the British faced few challenges; Bombay (1664) had the advantage of being located closest of all Indian ports to Britain, and Calcutta (1690) was positioned on the margin of India's largest population cluster and had the most productive hinterland. Calcutta lies some 80 miles (130 kilometers) from the coast on the Hooghly River, and a myriad of Ganges Delta channels connect it to its hinterland, a natural transport network that made the city an ideal colonial headquarters. But the British had to contend with Indian rebelliousness in the region; in 1912, they moved the colonial capital from Calcutta to the safer interior city of New Delhi, built adjacent to the old Mogul headquarters of Delhi (today these urban areas have coalesced with a combined population of 8.8 million).

Indian urbanization reveals several regional patterns. In the northern heartland, the west (the wheat-growing zone) is more urbanized than the east (where rice forms the main staple crop). This undoubtedly relates to the differences between wheat farming and the labor-intensive, small-plot cultivation and multiple cropping of rice. In the west, urbanization is around 40 percent (high by Indian standards); in the east only about 20 percent of the population resides in urban centers. Furthermore, India's larger (100,000+) cities are concentrated in three regions: (1) the northern plains from Punjab to the Ganges Delta, (2) the Bombay-Ahmadabad area, and (3) the southern part of the peninsula, which includes Madras and Bangalore (4.8 million). The only interior metropolises with populations over 1.5 million not located within one of these regions are centrally

FIGURE 8—20 The strongest of social contrasts often marks India's severely overcrowded urban scene, here beside the Cooum River in the city of Madras on the southeastern Coromandel Coast.

positioned Nagpur (1.6 million) and the capital of Andhra Pradesh, Hyderabad (3.7 million).

THE EAST: BANGLADESH

The state of Bangladesh was born in 1971 following a brief war of independence against Pakistan, of which it had been a part (as East Pakistan) since 1947. With its economy shattered by months of conflict, Bangladesh began its sovereign existence desperately impoverished, ill fed, and badly overcrowded. Bangladesh is

a comparatively small country (56,000 square miles/144,000 square km), about equal in size to the U.S. state of Wisconsin. Its very large population—now 121 million (population density is an astonishing 2181 per square mile/ 842 per square km)—constituted 55 percent of formerly united Pakistan. Its national territory, nearly surrounded by India (a short stretch of boundary adjoins Myanmar [Burma] in the far southeast), is essentially the deltaic plain of the Ganges-Brahmaputra system, which empties into the Bay of Bengal through numerous channels (Fig. 8–21). Only in the extreme east and southeast does the flat terrain of the floodplains rise into hills and mountains.

The alluvial land of Bangladesh is extremely fertile, and practically every cultivable foot of soil is under crops—rice and wheat for subsistence, jute and tea for cash. The jute industry made a major contribution to the economy of Pakistan prior to the secession of Bangladesh, producing (with other products) well over half of the country's annual export revenues. It was always a bone of contention that the Bangladeshi share of the Pakistani budget was only about 40 percent, so that as East Pakistan Bangladesh served as an exploited colony to West Pakistan in the eyes of many of its people.

The staple food for Bangladesh's enormous population is rice, grown in paddyfields whose fertility is renewed by the silt swept down by the river systems' annual floods. In most places, three harvests of rice per year are possible, and the country normally was able to produce some 80 percent of its food requirements. In the early 1970s, the dislocation brought on by the war of secession threatened widespread starvation as harvests rotted and crops were abandoned. Large-scale emergency imports prevented a major disaster, but Bangladesh has been slow to recover. Despite the arrival of the benefits of the Green Revolution, the country remains the epitome of Malthusian forecasts, constantly facing the specter of mass hunger, malnutrition, and dislocation.

Bangladesh's land is fertile because it is low lying and subject to wet-monsoon-induced river flooding (Fig. 8–22). This condition has negative consequences as well, because to the south the country lies open to the Bay of Bengal where dangerous tropical storms are born and sometimes make landfall. With much of southern Bangladesh less than 12 feet (3 meters) above sea level, the penetration of hurricanes (here called cyclones) can do incalculable damage. In November 1970, a devastating tropical cyclone hit Bangladesh, and the rising waters and high winds exacted a toll of at least 300,000 lives. It was the second greatest natural disaster of the twentieth century (after the 1976 earthquake that killed upwards of 500,000 in Tangshan, China), but it was hardly the last such assault on the land and people of Bangladesh; more than 15,000 were killed during another destructive storm in 1985. Crowding onto low-lying farmlands and delta islands, inadequate escape routes and mechanisms, and insufficient warning procedures continue to prevail.

The people of Bangladesh have in recent years faced disasters of several kinds. The indifference of the Pakistani government during the aftermath of the 1970 cyclone was one of the leading factors in the outbreak of open revolt against established authority. The war of secession brought unspeakable horror to villages and towns; millions left their homes and streamed across the border into neighboring India (which has now begun to erect fences along the boundary to shut off the Bangladeshi immigrant flow that has continued since 1971). Hunger and rampant disease soon followed. It is impossible to be certain of the consequences, but estimates of the total loss of life run as high as 3 million. Since then, destructive seasonal flooding has often besieged the country (1988 was a particularly bad year), but the ever-present threat of mass starvation has lessened a bit.

Although its rice yields per unit area used to be relatively low, Bangladesh has made recent gains with higher-yielding varieties and today approaches self-sufficiency in food production—though at a rather low caloric level (80 percent of the people are still malnourished). New varieties of wheat have also been introduced since 1980; this

FIGURE 8–21

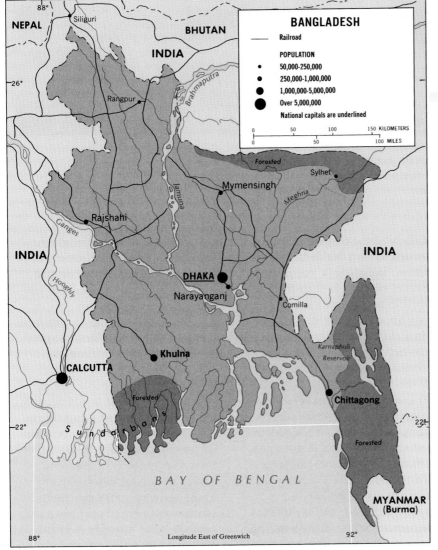

FIGURE 8–22　The intensively-farmed, low-lying alluvial plain and double (Ganges-Brahmaputra) delta of Bangladesh is frequently subjected to flooding, particularly during the peak of the wet-monsoon period. This scene shows the 1988 seasonal flood, one of the century's worst.

crop quickly became successful, acreages have expanded dramatically, and wheat is now increasingly rotated with rice. The longer-term problem, of course, is once again a population that keeps growing too rapidly, thereby limiting the impact of these incremental advances in rice and wheat production to sustaining a nutritional level that is barely adequate.

Thus, Bangladeshi energies are almost totally devoted to fighting the daily battle against chronic malnourishment, and the country's resources of natural gas, coal, timber, and several minerals go unexploited. As in India, effective birth control (involving less than one-third of all married couples) lags

far behind the 60-percent level needed to turn the demographic corner. Bangladesh's appalling level of poverty is such that its late-1980s per capita income was only about U.S. $130. Only Bhutan and Chad (together containing one-eighteenth the Bangladeshi population) ranked lower among all the world's countries—and no nation anywhere suffers more from the effects of underdevelopment. After centuries of British colonialism, Bangladesh reaped a harvest of Pakistani neglect, and the country's economy is still mired in disorder. The jute industry has suffered, and the already-inadequate communications system is in disarray. Governmental

instability, corruption in administration, and a growing insurgency in the southeast have made Bangladesh's first two decades of going it alone even bleaker.

Dhaka, the centrally positioned capital (4.4 million), and the port of Chittagong (1.7 million) are urban islands in a country where only 13 percent of the people live in towns and cities. If one is needed, a further measure of Bangladeshi economic misfortune can be gained from a look at the transport network. There is still no road bridge over the Ganges anywhere in the country, and only one railroad bridge; there is neither a road nor a railroad crossing over the Brahmaputra (Jamuna) anywhere in Bangladesh. Road travel from Dhaka to the eastern town of Comilla involves two ferry transfers across the same river (distributaries of the Meghna); there is hardly any means of surface travel from Dhaka to the western half of the country, except by boat. Thousands of boats ply Bangladesh's ubiquitous waterways, which still form a more effective interconnecting system within the country than do the roads. In struggling Bangladesh, survival is more than ever the leading industry; all else is luxury.

THE WEST: PAKISTAN

The Islamic Republic of Pakistan (Fig. 8–23) is also an underdeveloped country, but it differs in al-

FIGURE 8–23

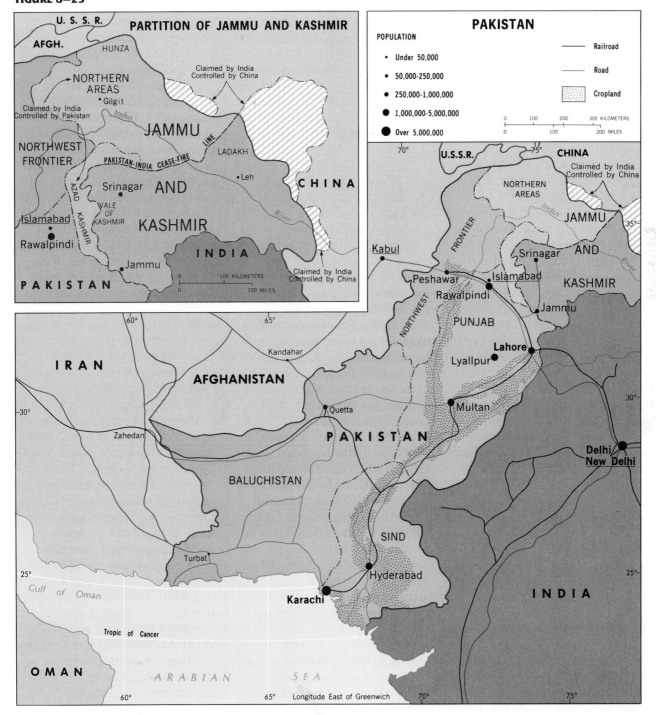

most every respect from its former federal partner. Indeed, so strong are the contrasts between former East Pakistan and the state that now carries the name, that the near-quarter-century survival of two-part Pakistan was in itself something of a miracle. Islam was the unifier (across a thousand miles of hostile Hindu Indian territory), but there were differences even in the paths whereby the Muslim faith reached west and east. In dry-world West Pakistan, it arrived over land, carried eastward by invaders from the west and northwest; in monsoon East Pakistan it came by sea, brought to the double delta by Arab traders.

In fact, it is difficult to find any additional point of similarity between former East and West Pakistan. The two wings of the country lay in different culture regions, with all that this implied (even the official languages—Bengali in the east, Urdu in the west—differed). Pakistan lies on the fringes of Southwest Asia; Bangladesh adjoins Southeast Asia. In Pakistan, the problem has always been drought, and the need is for irrigation; in Bangladesh, the problem is flooding, and the need is for dikes. Pakistanis build their houses of adobe and matting, eat wheat and mutton, and grow cotton for sale; the people of Bangladesh build with grass and thatch, eat rice and fish, and grow jute.

Livelihoods

Contrary to the predictions made at the time Pakistan became a separate state in 1971, it has managed to survive economically and has even made a good deal of progress. This occurred despite the country's poverty in known mineral resources; apart from natural gas in Baluchistan (which may be related to oil reserves) and chromite, Pakistan has only minor iron

deposits, which are being used in a small steel plant at Multan. So the country must make up in agricultural output what it lacks in minerals, and since independence in 1947 it has made considerable strides. Favored by the high prices generated during the Korean War, the economy got an early boost from jute and cotton sales; then the later 1950s brought a period of stagnation, made worse by continuing political conflicts with India. But since the 1960s, a considerable expansion of irrigated agriculture (which now covers two-thirds of Pakistan's arable soils), land reform, and the success of the textile industry have all contributed to noteworthy progress.

Certainly there was room for such progress. Pakistan shares with India the familiar low yields per acre for most crops; per capita annual incomes are quite modest; and manufacturing is only now emerging as a significant contributor to the economy. The low yields for such vital crops as wheat and rice have long been due to the inability of the peasants to buy fertilizers, to the poor quality of seeds, and to inadequate methods of irrigation. In the Sind region of the lower Indus Valley, where large estates farmed by tenant farmers existed before the British began their irrigation programs in the Punjab to the northeast, yields are kept down by outdated irrigation systems and by the paucity of incentives for landless peasants.

Pakistan's virtually new textile industry, based on the country's substantial cotton production, has developed steadily. It now satisfies all of the domestic market, and textiles have risen to become a leading contributor to exports. Other industries have also been stimulated, and Pakistan is increasingly able to produce at home what it formerly had to import from foreign sources. A chemical industry is emerging, and an auto-

mobile assembly plant operates in Karachi, where a small steel-producing plant now functions as well. Thus, Pakistan is trying to reduce its dependency on foreign aid, investing heavily in the agricultural sector, and encouraging local industries as much as possible. But there is a long and difficult road to travel. There are few natural resources, and export products are subject to both sudden price fluctuations on world markets and the increasing pressure of competition.

Pakistan still faces a host of unsolved problems. Illiteracy remains high (74 percent of the population). Family planning is government-approved, but its dissemination and acceptance are really just beginning. Tens of millions of people still live in poverty, subsist on an ill-balanced diet, and have too short a life expectancy (54 years versus 58 in India and 52 in Bangladesh). Of course, there are successful farmers in Punjab, Sind, and elsewhere; but they are still far outnumbered by those for whom life remains a daily struggle for survival.

The Major Cities

Understandably, the newborn, bifurcated state of Pakistan chose Karachi as its first capital in 1947. Clearly, it was desirable to place the capital city in the Muslim stronghold of the country, West Pakistan. But the outstanding center of Islamic culture, Lahore, lay too exposed to the sensitive nearby boundary with India (Fig. 8–23).

Lahore, now home to 4.4 million residents, grew rapidly as a result of the Punjab's partitioning in 1947 (in 1950, its population was less than one-third as large as today's). Founded in the first or second century A.D., Lahore became established as a great Muslim cen-

FIGURE 8–24   Badshahi Mosque in Lahore, one of the world's great Islamic cities. Nearly half a million worshipers fill this shrine's vast enclosed courtyard.

ter during the Mogul period. As a place of royal residence, the city was adorned with numerous magnificent buildings, including a great fort, several palaces, and mosques, which to this day remain monuments of history with their marvelous stonework and tile and marble embellishments (Fig. 8–24). The site of magnificent gardens and an old university, Lahore was also the focus of a large hinterland area in pre-partition times when its connections extended south to Karachi and the sea, north to Peshawar, and east to Delhi and Hindustan. Its Indian hinterland was cut off by the partitioning, but Lahore has retained its importance as the center of one of Pakistan's major population clusters, as a place of diverse in-dustries (textiles, leather goods, and gold and silver lacework), and as the unchallenged historic headquarters of Islam in this region.

Karachi (8.0 million) grew even faster than Lahore. After independence it was favored not only as the first capital of Pakistan, but also as West Pakistan's only major seaport. Its overseas trade rose markedly, new industries were established, and new ties developed between Karachi and former East Pakistan as well as other parts of the world. A flood of Muslim refugees came to the city, presenting serious problems—there simply was not enough housing or food available to cope with the half-million immigrants who had arrived by 1950. Nevertheless, Karachi survived, and in its subsequent growth has reflected the new Pakistan. As a result of the expansion of cultivated acreages in the Pakistani Punjab and in adjacent Sind, Karachi's trade volume has increased greatly, especially in wheat and cotton. Imports of oil and other critical commodities required the construction of additional port facilities.

But Karachi never became the cultural or emotional focus of the nation. It lies symbolically isolated along a desert coast, almost like an island, while the core of Pakistan still lies far inland. In 1959, after just over a decade as the federal capital, Karachi surrendered its headquarters functions to an interior city, Rawalpindi. This was a temporary measure: an entirely new capital city was completed in the early 1970s at Islamabad, a short distance from Rawalpindi. This new planned city, as the maps in Fig. 8–23 show, lies near the boundary of Kashmir. It confirms not only the internal positioning of Pakistan's cultural and economic heartland, but also the state's determination to emphasize its presence in the contested north. It is evidence of a sense of security that Pakistan did not yet possess when it chose Karachi over exposed Lahore. In this context, Islamabad is a classic example of the principle of the *forward capital* (discussed on p. 327).

Irredentism

For many years, Pakistan has been beset by problems in its marginal areas. The far northeastern boundary with India in Kashmir (Fig. 8–23 inset) is a truce line that is again in hot contention (see *box*); in Baluchistan in the southwest and along the boundary with Afghanistan in the northwest, Pakistan faces active **irredentism.** In Baluchistan, a still-remote area where the Iran-Pakistan border has little practical meaning, government efforts to assert its authority were met by opposition from local chiefs. Baluchistan is a region of rugged mountains, severe desiccation, scattered oases, and ancient nomadic routes. It resembles neighboring Iran and is quite unlike the settled farmlands of the Indus Valley. Baluchistan's peoples were not ready to be controlled by the Pakistani government, and in their rebelliousness, they have been supported by the government of Afghanistan.

Afghanistan's majority population, the Pakhtuns (also known as Pathans, Pashtuns, or Pushtuns), constitute about 50 percent of that country's 16 million people and are concentrated in the area centered on the capital (Kabul), posi-

THE PROBLEM OF KASHMIR

Kashmir is a territory of high mountains surrounded by Pakistan, India, China, and, along several miles in the far north, by Afghanistan (Fig. 8–23). Although known simply as *Kashmir*, the area actually consists of several political divisions, including the state properly referred to as Jammu and Kashmir (one of the 562 Indian states at the time of independence) and the administrative areas of Gilgit in the northwest and Ladakh (including Baltistan) in the east. The main conflict between India and Pakistan over the final disposition of this territory has focused on the southwest, where Jammu and Kashmir are located.

When partition took place in 1947, the existing states of British India were asked to decide whether they would go with India or Pakistan. In most of the states, this issue was settled by the local authority, but in Kashmir there was an unusual situation. There were about 5 million inhabitants in the territory at that time, nearly half of them concentrated in the basin known as the Vale of Kashmir (where the capital, Srinagar, is located). Another 45 percent of the people were concentrated in Jammu, which leads down the foothill slopes of the Himalayas to the edge of the Punjab. The small remainder of the population was scattered through the mountains, including Pakhtuns in Gilgit and other parts of the northwest. Of these population groups, the people of the mountain-encircled Vale of Kashmir are almost all Muslims, whereas the majority of Jammu's population is Hindu.

But the important feature of the state of Jammu and Kashmir was that its rulers were Hindu, not Muslim, although the overall population was more than 75 percent Muslim. Thus, the rulers were faced with a difficult decision in 1947—to go with Pakistan and thereby exclude themselves from Hindu India, or to go with India and thereby incur the wrath of the majority of the people. Hence, the maharajah of Kashmir sought to remain outside both Pakistan and India and to retain the status of an autonomous separate unit. This decision was followed, after the partitioning of India and Pakistan, by a Muslim uprising against Hindu rule in Kashmir. The maharajah asked for the help of India, and Pakistan's forces came to the aid of the Muslims. After more than a year's fighting and through the intervention of the United Nations, a cease-fire line was established that left Srinagar, the Vale, and most of Jammu and Kashmir, including nearly four-fifths of the territory's population, in Indian hands (Fig. 8–23, inset map). In due course,

tioned opposite the Khyber Pass less than 125 miles (200 kilometers) from the Pakistan border. Kabul's pre-1979 government actively encouraged Pakhtuns living in Pakistan's Northwest Frontier Province to demand their own

state of Pakhtunistan (Pathanistan). Pakistan's moderate response was to hasten the integration of its northwestern areas into the national state through improved communications, education, and other facilities, but Afghan irredentism

this line began to appear on maps as the final boundary settlement, and Indian governments have proposed that it be so recognized.

Why should two countries, whose interests would be served by peaceful cooperation, allow a distant mountainland to trouble their relationship to the point of war? There is no single answer to this question, but there are several areas of concern for both sides. In the first place, Pakistan is wary of any situation whereby India would control vital irrigation waters needed in Pakistan. As the map shows, the Indus River, the country's lifeline, crosses Kashmir. Moreover, other tributary streams of the Indus originate in Kashmir, and it was in the Punjab that Pakistan learned the lessons of dealing with India for water supplies. Second, the situation in Kashmir is analogous to the one that led to the partition of the whole subcontinent: Muslims are under Hindu domination. The majority of Kashmir's people are Muslims, so Pakistan argues that free choice would deliver Kashmir to the Islamic Republic. A free plebiscite is what the Pakistanis have sought—and the Indians have thwarted. Furthermore, Kashmir's connections with Pakistan prior to partition were much stronger than those between Kashmir and India, although India has invested heavily in improving its links to Jammu and Kashmir since the military stalemate. In recent years, it did seem more likely that the cease-fire line would, indeed, become a stable boundary between India and Pakistan. (The incorporation of the state of Jammu and Kashmir into the Indian federal union was accomplished as far back as 1975 when India was able to reach an agreement with the state's chief minister and political leader.)

The drift toward stability, however, was sharply reversed at the end of the 1980s as a new crisis engulfed Kashmir. Extremist Muslim groups, demanding independence, escalated a long-running insurgency into a new phase of open violence in 1988; this brought on a swift crackdown by the Indian military, whose harsh tactics prompted an already disgruntled citizenry to far more strongly support the separatists. By 1990, Kashmir was under tight curfew, the Pakistanis were charging the Indians with widespread human-rights violations, and the Indians were accusing the Pakistanis of inciting secession and supplying arms. With India prepared to go to any length to prevent the loss of Kashmir—whose populace now clearly aspires to break free of Indian control—the specter of the third war with Pakistan over this territory since 1947 loomed ominously.

continued. Islamabad, thus, is situated between pressure areas in both Kashmir *and* Afghanistan.

The Soviet invasion and subsequent occupation of Afghanistan (1979–1989) greatly aggravated these internal regional difficulties,

because over 3.5 million Afghan refugees—about one-third of the pre-invasion population of Afghanistan—crowded into Pakistan's border zone during the 1980s. Although the Soviets withdrew in 1989, guerrilla warfare

continued inside Afghanistan, making it too dangerous for most of these emigrants to return home. Therefore, Pakistan now not only must cope with the world's largest concentration of refugees, but also with intensifying ethnic strife as the huge Pakhtun population becomes an ever-larger presence on the national scene.

SRI LANKA: ISLAND OF THE SOUTH

Sri Lanka (formerly Ceylon), the compact, pear-shaped island located just 22 miles (35 kilometers) across the Palk Strait from the southern tip of the Indian peninsula, is the fourth independent state to have emerged from the British sphere of influence in South Asia (Fig. 8–25). Sovereign since 1948, Sri Lanka has had to cope with political as well as economic problems, some of them quite similar to those facing India and Pakistan and others quite different. There were good reasons to create a separate independence for Sri Lanka. This is neither a Hindu nor a Muslim country; the majority—some 70 percent—of its 17.4 million people are Buddhists. Furthermore, unlike India or Pakistan, Sri Lanka is plantation country (a legacy of the European period), with export agriculture still the mainstay of the external economy.

The majority of Sri Lanka's people are not Dravidian, but are of Aryan origin with a historical link to ancient northern India. After the fifth century B.C., their ancestors began to migrate to Ceylon, a relocation that took several centuries to complete and brought to this southern island the advanced culture of the northwestern

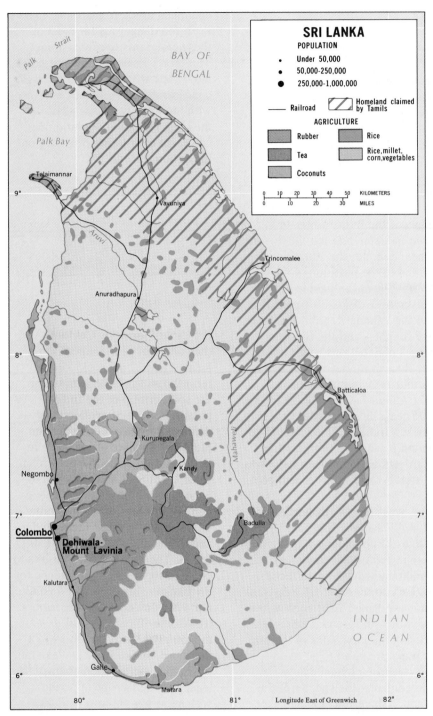

FIGURE 8–25

Tamil language to northern Sri Lanka, and eventually came to constitute a substantial minority (now 18 percent) of the country's population. Their numbers were markedly strengthened during the second half of the nineteenth century when the British brought hundreds of thousands of Tamils from the adjacent mainland to Ceylon to work on the plantations that were being laid out. Sri Lanka has sought the repatriation of this ethnic element in its population, and an agreement to that effect was even signed with India. In 1978, however, Tamil was granted the status of a national language of Sri Lanka.

Sri Lanka is not a large island (25,000 square miles/65,000 square kilometers), but it displays considerable topographic diversity. The upland core lies in the south, where elevations reach over 8000 feet (2500 meters) and sizable areas exceed 5000 feet (1500 meters). Steep, thickly-forested slopes lead down to an encircling lowland, most of which lies below 1000 feet (300 meters). Northern Sri Lanka, capped by the Jaffna Peninsula, is entirely low lying. Rivers, the sources of ricefield irrigation waters, flow radially from the interior highland across this lowland rim.

Since the decline of the Sinhalese Empire, focused on centrally located Anuradhapura, the moist southwest has been the leading zone of productive capacity (the plantations are concentrated here) and the population core. Three plantation crops have been successful (Fig. 8–25): coconuts in the hot lowlands, rubber up to about 2000 feet (600 meters), and tea, the product for which Sri Lanka is most famous, in the highlands above. Sri Lanka is the world's second largest tea exporter, and this commodity accounts for about one-third of the country's annual

portion of the subcontinent. Part of that culture was the Buddhist religion; another component was the knowledge of irrigation techniques. Today, the descendants of these early invaders, the Sinhalese, speak a language (Sinhala)

belonging to the Indo-Aryan linguistic family of northern India.

The darker-skinned Dravidians from southern India never came in sufficient numbers to challenge the Sinhalese. They introduced the Hindu way of life, brought the

exports by value. Whereas Sri Lanka's plantation agriculture is very productive and quite efficient, the same cannot be said for the island's ricelands. As recently as the 1960s, it was necessary to import half the rice consumed in Sri Lanka, a situation that was detrimental to the general economic situation. Accordingly, the government made it a priority to reconstruct plainland irrigation systems, repopulate the lowlands (until its recent eradication, malaria was an obstacle to settlement there), and intensify rice cultivation. The result has been a substantial increase in rice production, and Sri Lanka is approaching self-sufficiency.

In a country so heavily agricultural, it is not surprising to find very little industrial development except for factories that process plantation and farm products. Sri Lanka appears to have very little in the way of mineral resources; graphite is the most valuable mineral export. The industries that have developed, other than those processing foodstuffs, depend on Sri Lanka's relatively small local market. Predictably, these include cement, shoes, textiles, paper, china, glassware, and the like. The majority of these establishments are located in Colombo (1.8 million), the country's capital, largest city, and leading port.

Attention to economic development programs in Sri Lanka has become increasingly difficult since 1983 because of heightened Tamil-Sinhalese ethnic tensions. Tamil grievances at their treatment by the Sinhalese majority mounted steadily during the 1970s, focusing on this minority's alleged inability to achieve equal rights in education, employment, landownership, and linguistic and political representation. By the early 1980s, the perceived failure of the Sri Lankan government to respond adequately

to Tamil demands led to insurrection and to Tamil calls for the creation of an independent homeland through a Cyprus-like partitioning of Sri Lanka. Tamil extremists in 1984 launched a guerrilla war that disrupted much of the country. The Tamils wanted nothing less than their own separate state, which would be called *Eelam* and cover the territory in the north and east shown in stripes in Fig. 8–25.

As the conflict intensified in the mid-1980s, both the Indian and Sri Lankan governments sought to mediate a settlement. India's concern here was practical: well over 50 million Tamils reside in Tamil Nadu State on the mainland adjacent to Sri Lanka, and the notion of expanded Tamil separatism could become a problem for the New Delhi regime as well. Invited by the Colombo government in 1987, Indian troops entered the country to help subdue the Tamil guerrillas. Although the Tamil fighters were soon driven out of their stronghold in northernmost Sri Lanka, the Indians were unable to disarm them in their new jungle sanctuary to the southeast. After two years of stalemate and considerable bloodshed, the Sri Lankan government accepted a guerrilla offer to open peace talks, contingent upon the departure of the Indians. As the 1990s opened, the last Indian troops were withdrawn; the Tamil forces then declared their war at an end, forming a political party in order to vie for control of northeastern Sri Lanka. Negotiations seemed likely to produce something short of partition (even though the Tamils were expected to win political power in their region); but the real hope was that a solution could finally be found that would prevent the resumption of armed conflict—a diminishing prospect as sporadic fighting again broke out in mid-1990.

THE ~~NOT~~ MOUNTAINOUS NORTH

South Asia, as we noted at the beginning of this chapter, is one of the world's most clearly defined geographic realms. Walls of mountains stand between India and China, mountains that defy penetration even in this age of modern highways. As Fig. 8–1 reminds us, those mountains serve as more than barriers: great life-giving rivers rise there, their melting snows sustaining millions in valleys and plains far below. Control over those headwaters has produced centuries of conflict, and the results are etched on the political map.

Across South Asia's North, from Afghanistan in the west through Jammu and Kashmir and Nepal to Bhutan in the east, lies a tier of mountain states and territories; remote, isolated, and vulnerable, these countries are remnants of a frontier history. Their vulnerability is illustrated by the recent misfortunes of one of them, Afghanistan, and the disappearance of another, Sikkim. The latter, wedged between Nepal and Bhutan, was absorbed by India in 1975 and ceased to be a separate country. However, the kingdoms of Nepal and Bhutan retain their independence.

Nepal

Nepal, the size of Illinois and containing a population of just under 20 million, lies directly northeast of India's Hindu coreland. It is a country of three geographic zones: a southern, subtropical, fertile lowland called the Terai; a central belt of Himalayan foothills with swiftly flowing streams and deep

FIGURE 8–26 Against the spectacular backdrop of the main Himalayan range, Nepal's lower mountains are increasingly denuded of their forests as population pressures force farming and fuel-wood extraction toward ever higher elevations.

sands of temples and pagodas ranging from the simple to the ornate grace the cultural landscape, especially in the valley of Kathmandu, the country's core area. Although over a dozen languages are spoken, 90 percent of the people also speak Nepali, a language related to Indian Hindi.

Nepal's problems are those of underdevelopment and centrifugal political forces. Living space is limited, population pressure is high and rising steadily, and environmental degradation is a consequence. Deforestation is particularly severe—over one-third of Nepal's alpine woodlands have been cutover into wastelands since the 1960s (Fig. 8–26)—as the growing population of subsistence farmers is forced to expand into higher-altitude wilderness zones for sufficient crop-raising space (on steep terraces) and to obtain the firewood that supplies most of Nepal's energy needs. But soil quality is poor throughout the uplands, new farms are soon abandoned after a few seasons of declining productivity, and the land denudation process intensifies. Moreover, the steep slopes and the awesome power of the wet monsoon rains accelerate soil erosion in treeless areas, and so much silt is now transported out of the Himalayas that (according to some researchers) river flooding is heightening in the crowded lower Ganges and Brahmaputra basins. With about half its farmland already abandoned to erosion and with 95 percent of its population engaged in subsistence agriculture (rice, corn, wheat, and millet), Nepal today faces a serious ecological crisis.

In terms of political geography, the country suffers from regionalism and from conflict between those who support the monarchy and others who favor a more representative form of government. The southern Terai is a world

valleys; and the spectacular high Himalayas themselves (topped by Mount Everest) in the north. The capital, Kathmandu (375,000), lies in the east-central part of the country in an open valley of the central hill zone. Nepal is a ma-terially poor but culturally rich country. The Nepalese are a people of many sources, including India, Tibet, and interior Asia; about 90 percent are Hindu, but Nepal's Hinduism is a unique blend of Hindu and Buddhist ideals. Thou-

apart from the hills of the central zone, and the peoples of the west have origins and traditions quite different from those in the east. Welding these diverse elements into a unified nation in a land-locked, internally fragmented country is Nepal's greatest challenge.

Bhutan

East of Nepal lies the Kingdom of Bhutan, only one-third as large as Nepal, with just 1.6 million inhabitants. Here Buddhism is the state religion, and the official language is a variant of Tibetan. The capital, Thimphu (20,000), lies in the west-central part of this mountainous state. Bhutan was part of the British colonial sphere but remained a separate entity. When India became independent in 1947, a treaty was drawn up that permitted Bhutan to continue to control its internal affairs yet required its leaders to consult India on matters of foreign relations.

It seemed a prescription for absorption, but Bhutan has survived—in part because it is self-sufficient in food, its terraced hill-slopes producing enough grain to sustain all of its people. Comparatively secure in their mountain fortress, the Bhutanese have changed but slowly. The king retains much of his power; subsistence farming is the mainstay of the economy. Timber exploitation, hydroelectric power generation, and tourism have potential, but development has been minimal. Centuries ago Bhutan's rival princes built forts to control the trade routes between Tibet and points south. That trade has ceased, the kingdom is consolidated, and Bhutan's isolation continues.

PRONUNCIATION GUIDE

Agra (*ahg*-ruh)

Ahmadabad (*ah*-muh-duh-bahd)

Amritsar (um-*rit*-sahr)

Andhra Pradesh (ahn-druh-pruh-*desh*)

Anuradhapura (unna-rahd-uh-*poora*)

Aryans (*ahr*-yunz)

Aśoka (uh-*shoh*-kuh)

Assam (uh-*sahm*)

Aurangzeb (*aw*-rung-zeb)

Badshahi [Mosque] (bud-*shah*-hee [mosk])

Baluchistan (buh-loo-chih-*stahn*)

Bangladesh (bang-gluh-*desh*)

Bengal[i] (beng-*gahl*-[ee])

Bhopal (boh-*pahl*)

Bhutan (boo-*tahn*)

Bihar (bih-*hahr*)

Brahmaputra (brahm-uh-*pooh*-truh)

Buddhism (*bood*-ism)

Caste (*cast*)

Ceylon (seh-*lonn*)

Charisma (kuh-*rizz*-muh)

Chittagong (*chitt*-uh-gahng)

Chota Nagpur (choat-uh-*nahg*-poor)

Colombo (kuh-*lum*-boh)

Cooum (*koo*-um)

Coromandel (kor-uh-*mandle*)

Deccan (*decken*)

Delhi (*delly*)

Dhaka (*dahk*-uh)

Dravidian (druh-*viddy*-un)

Eelam (*ee*-lum)

Gandhi (*gondy*)

Ganges/Gangetic (*gan*-jeez/*gan*-jettick)

Ghats (*gahtss*)

Gilgit (*gill*-gutt)

Goa (*go*-uh)

Godavari (guh-*dah*-vuh-ree)

Golconda (goll-*kon*-duh)

Gujarat (goo-juh-*raht*)

Gurkha (*ghoor*-kuh)

Haryana (hah-ree-*ahna*)

Himalayas (him-*ahl*-yuz/himma-*lay*-uz)

Hindu Kush (hin-dooh-*koosh*)*

Hindustan (hin-dooh-*stahn*)

Hooghly (*hoo*-glee)

Hyderabad (*hide*-uh-ruh-bahd)

*Double ''o'' pronounced as in ''book.''

Igneous (*igg*-nee-uss)

Islamabad (iss-*lahm*-uh-bahd)

Jaffna (*jahf*-nuh)

Jammu (*juh*-mooh)

Jamshedpur (*jahm*-shed-poor)

Jamuna (*jum*-uh-nuh)

Kabul (*kah*-bull)

Kanarese (*kahn*-uh-reece)

Karachi (kuh-*rah*-chee)

Karakoram (kahra-*kor*-rum)

Karnataka (kahr-*naht*-uh-kuh)

Kashmir (*kash*-meer)

Kathmandu (kat-man-*dooh*)

Khalistan (kahl-ee-*stahn*)

Konkan (*kahng*-kun)

Ladakh (luh-*dahk*)

Lahore (luh-*hoar*)

Littoral (*lit*-uh-rul)

Loess (*lerss*)

Madhya Pradesh (mahd-yuh-pruh-*desh*)

Madras (muh-*drahss*)

Maharashtra (mah-huh-*rah*-shtra)

Malabar (*mal*-uh-bahr)

Malayalam (mal-uh-*yah*-lum)

Malaysia (muh-*lay*-zhuh)

Malthus/Malthusian (*mal*-thuss/mal-*thoo*-zee-un)

Mauryan (*maw*-ree-un)
Meghna (*may*-gnuh)
Mizoram (mih-*zor*-rum)
Mohajirs (moh-hah-*jeerz*)
Multan (mool-*tahn*)
Naga[land] (*nahga*-[land])
Nagpur (*nahg*-poor)
Negritos (neh-*gree*-toze)
Nehru (*nay*-rooh)
Nepal (nuh-*pahl*)
Orissa (aw-*rissa*)
Pakhtuns (puck-*toonz*)
Pakistan (*pah*-kih-stahn)
Palk (*pawk*)

Pashtuns [see Pushtuns]
Pathans (puh-*tahnz*)
Peshawar (puh-*shah*-wahr)
Punjab (pun-*jahb*)
Pushtuns (*pah*-shtoonz)
Raj (*rahdge*)
Rawalpindi (rah-wull-*pin*-dee)
Sepoy (*see*-poy)
Shi'ite (*shee*-ite)
Siddhartha (sid-*dahr*-tuh)
Sikh (*seek*)
Sikkim (*sick*-um)
Sinhala (sin-*hahla*)
Sinhalese (sin-hah-*leeze*)

Sri Lanka (sree-*lahng*-kuh)
Srinagar (srih-*nug*-arr)
Sunni (*soo*-nee)
Taj Mahal (*tahj* muh-*hahl*)
Tamil [Nadu] (*tammle* [*nah*-doo])
Tapti (*tahp*-tee)
Telugu (*telloo*-goo)
Terai (teh-*rye*)
Thimphu (thim-*pooh*)
Turkestan (ter-kuh-*stahn*)
Urdu (*oor*-doo)
Uttar Pradesh (ootar-pruh-*desh*)
Varanasi (vuh-*rahn*-uh-see)
Vindhya (*vin*-dyuh)

REFERENCES AND FURTHER READINGS

(BULLETS [•] DENOTE BASIC INTRO-
DUCTORY WORKS ON THE REALM
OR SYSTEMATIC-ESSAY TOPIC)

Bayliss-Smith, Tim P. & Wanmali, Sudhir, eds. *Understanding Green Revolutions: Agrarian Change and Development Planning in South Asia* (London & New York: Cambridge University Press, 1984).

Bhardwaj, Surinder M. *Hindu Places of Pilgrimage in India: A Study in Cultural Geography* (Berkeley & Los Angeles: University of California Press, 1973).

Brush, John E. "Spatial Patterns of Population in Indian Cities," in Dwyer, Denis J., ed., *The City in the Third World* (New York: Barnes & Noble, 1974), pp. 105–132.

Burki, Shahid J. *Pakistan: A Nation in the Making* (Boulder, Colo.: Westview Press, 1986).

• Butzer, Karl W. *Geomorphology From the Earth* (New York: Harper & Row, 1976).

• Chorley, Richard J. et al. *Geomorphology* (London & New York: Methuen, 1985).

Costa, Frank J. et al., eds. *Asian Urbanization: Problems and Processes* (Forestburgh, N.Y.: Lubrecht and Cramer, 1988).

Costa, Frank J. et al., eds. *Urbanization in Asia: Spatial Dimensions and Policy Issues* (Honolulu: University of Hawaii Press, 1989).

Dumont, Louis. *Homo Hierarchus: The Caste System and Its Implications* (Chicago: University of Chicago Press, 1970).

Dutt, Ashok K. "Cities of South Asia," in Brunn, Stanley D. & Williams, Jack F., eds., *Cities of the World: World Regional Urban Development* (New York: Harper & Row, 1983), pp. 324–368.

• Dutt, Ashok K. & Geib, Margaret. *An Atlas of South Asia* (Boulder, Colo.: Westview Press, 1987).

• Dutt, Ashok K. et al. *India: Resources, Potentialities, and Planning* (Dubuque, Iowa: Kendall/Hunt, 1972).

Er-Rashid, Haroun. *Geography of Bangladesh* (Boulder, Colo.: Westview Press, 1977).

Farmer, Bertram H. *An Introduction to South Asia* (London & New York: Methuen, 1984).

Gupte, Pranay. *Vengeance: India After the Assassination of Indira Gandhi* (New York: W.W. Norton, 1985).

Gwatkin, Davidson R. & Brandel, Sarah K. "Life Expectancy and Population Growth in the Third World," *Scientific American*, May 1982, pp. 57–65.

Hennayake, Shantha K. & Duncan, James S. "A Disputed Homeland: Sri Lanka's Civil War," *Focus*, Spring 1987, pp. 20–27.

Huke, Robert E. "The Green Revolution," *Journal of Geography*, 84 (1985): 248–254.

Ives, Jack D. & Messerli, Bruno. *Himalayan Dilemma: Reconciling Development and Conservation* (London & New York: Routledge, 1989).

Johnson, Basil L.C. *Bangladesh* (Totowa, N.J.: Barnes & Noble, 2 rev. ed., 1982).

• Johnson, Basil L.C. *Development in South Asia* (New York: Viking Penguin, 1983).

Johnson, Basil L.C. *India: Resources and Development* (Totowa, N.J.: Barnes & Noble, 1979).

Karan, Pradyumna P. *Bhutan* (Lexington: University Press of Kentucky, 1967).

Karan, Pradyumna P. *Nepal: A Cultural and Physical Geography* (Lexington: University Press of Kentucky, 1960).

Kosinski, Leszek A. & Elahi, K. Maudood, eds. *Population Redistribution and Development in South Asia* (Hingham, Mass.: D. Reidel, 1985).

Lall, Arthur S. *The Emergence of Modern India* (New York: Columbia University Press, 1981).

Lapierre, Dominique. *The City of Joy* [Calcutta] (Garden City, N.Y.: Doubleday, trans. Kathryn Spink, 1985).

Lukacs, John R., ed. *The People of South Asia: The Biological An-*

thropology of India, Pakistan, and Nepal (New York: Plenum Press, 1984).

Manogaran, Chelvadurai. *Ethnic Conflict and Reconciliation in Sri Lanka* (Honolulu: University of Hawaii Press, 1987).

Miller, E. Willard & Miller, Ruby M. *Southern Asia: A Bibliography on the Third World* (Monticello, Ill.: Vance Bibliographies, No. P-921, 1982).

Murton, Brian J. "South Asia," in Klee, Gary A., ed., *World Systems of Traditional Resource Management* (New York: Halsted Press/ V.H. Winston, 1980), pp. 67-99.

Muthiah, S., ed. *A Social and Economic Atlas of India* (New York: Oxford University Press, 1987).

Newman, James L. & Matzke, Gordon E. *Population: Patterns, Dynamics, and Prospects* (Englewood Cliffs, N.J.: Prentice-Hall, 1984).

• Noble, Allen G. & Dutt, Ashok K., eds. *India: Cultural Patterns and Processes* (Boulder, Colo.: Westview Press, 1982).

Robinson, Francis, ed. *The Cambridge Encyclopedia of India, Pakistan, Bangladesh, Sri Lanka, Nepal, Bhutan, and the Maldives* (New York: Cambridge University Press, 1989).

Schwartzberg, Joseph E., ed. *An Historical Atlas of South Asia* (Chicago: University of Chicago Press, 1978).

Siddiqi, A. *Pakistan: Its Resources and Development* (Hong Kong: Asian Research Service, 1985).

Sopher, David E. *Geography of Religions* (Englewood Cliffs, N.J.: Prentice-Hall, 1967).

Sopher, David E., ed. *An Exploration of India: Geographical Perspectives on Society and Culture* (Ithaca, N.Y.: Cornell University Press, 1980).

• Spate, Oskar H.K. & Learmonth, Andrew T.A. *India and Pakistan: A General and Regional Geography* (London: Methuen, 2 vols., 1971).

• Spencer, Joseph E. & Thomas, William L. *Asia, East by South: A Cultural Geography* (New York: John Wiley & Sons, 2 rev. ed., 1971).

Sukhwal, B.L. *India: Economic Resource Base and Contemporary Political Patterns* (New York: Envoy Press, 1987).

Tambiah, Stanley J. *Sri Lanka: Ethnic Fratricide and the Dismantling of Democracy* (Chicago: University of Chicago Press, 1986).

Thornbury, William D. *Principles of Geomorphology* (New York: John Wiley & Sons, 2 rev. ed., 1969).

Weisman, Steven R. "India: Always Inventing Itself," *New York Times Magazine*, December 11, 1988, pp. 50–51, 62, 64, 110–111.

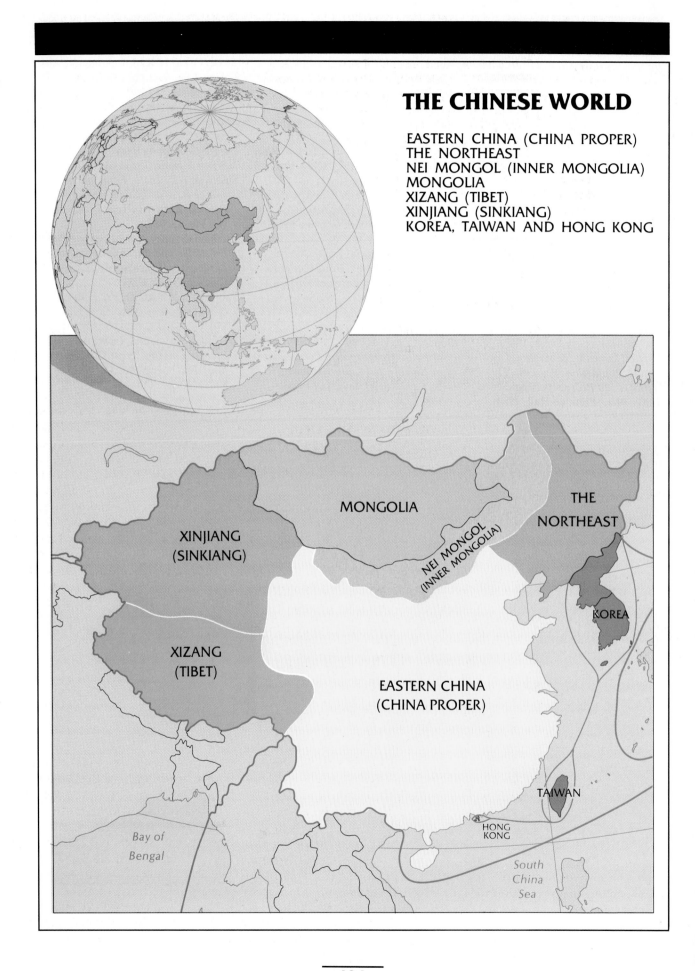

THE CHINESE WORLD

EASTERN CHINA (CHINA PROPER)
THE NORTHEAST
NEI MONGOL (INNER MONGOLIA)
MONGOLIA
XIZANG (TIBET)
XINJIANG (SINKIANG)
KOREA, TAIWAN AND HONG KONG

MONGOLIA

THE
NORTHEAST

XINJIANG
(SINKIANG)

NEI MONGOL
(INNER MONGOLIA)

KOREA

XIZANG
(TIBET)

EASTERN CHINA
(CHINA PROPER)

TAIWAN

HONG
KONG

Bay of
Bengal

South
China
Sea

CHAPTER 9

CHINA'S WORLD OF CONTRADICTIONS

When we in the Western world chronicle the rise of civilization, we tend to focus on the historical geography of Southwest Asia, the Mediterranean, and Western Europe. Ancient Greece and Rome were crucibles of culture; Mediterranean and Atlantic waters were avenues of its diffusion. China lay remote, so we believe, barely connected to this Western realm of achievement and progress. The Chinese themselves take quite a different view. Events on the western margins of the great Eurasian landmass were deemed irrelevant to theirs, the most advanced and refined culture on earth. Roman emperors were rumored to be powerful and Rome to be a great city, but nothing could match the omnipotence of China's rulers, and the city of Xian far eclipsed Rome as a center of sophistication. Chinese civilization existed long before ancient Greece and Rome emerged, and it was still there long after those empires had collapsed. China, the Chinese teach themselves, is eternal. It was, and always will be, the center of the civilized world.

We should remember this when we study China's regional geography, because 4000 years of Chinese culture and perception will not change overnight—not even in a generation. When most of your

parents were college students, China had once more closed its doors to the outside world, and there were just a few dozen foreigners in the entire country, then approaching 1 billion inhabitants. When Chinese leaders decided that an opening to the Western world would be advantageous, President Nixon was invited to Beijing. That historic occasion, in 1972, ended decades of isolation— but on China's terms. Foreign companies were invited to engage in business partnerships and commerce in China, thousands of Chinese students went to the United States and other countries to learn needed skills, and tourists climbed

the Great Wall, descended into the Ming Tombs, and explored the Forbidden City. But in 1989, when China's new economic policies led to popular demands for political reform, China's rulers ordered its armies to suppress this unwanted campaign. Students, workers, and sympathizers were massacred on Beijing's Avenue of Eternal Peace, and many were killed even on Chinese communism's holy ground, Tiananmen Square. The arrests and executions that followed reminded the world that China is a land of contradictions. China's rulers wanted economic change without political reform. They invited foreign involvement but rejected world condemnation.

ETERNAL CHINA

China lies at the heart of the East Asian geographic realm. China's 1.1 billion people constitute the largest national population on earth, but China is not the only substantial country in East Asia. South Korea (44 million) and North Korea (24 million) are neighbors to the east; island Taiwan has more than 20 million people; and Hong Kong, formerly

S Y S T E M A T I C   E S S A Y

GEOGRAPHY OF DEVELOPMENT

Scholars from many academic disciplines study **development**—the economic, social, and institutional growth of national states. We introduced this topic in the opening chapter (see pp. 38–43), where we reported that the widening development gap is deeply rooted in **colonialism**. The rapidly advancing Western countries, the colonizers, gained a decisive head start economically while their dependencies remained suppliers of raw materials and consumers of manufactured goods. Today, over 77 percent of the world's population resides in the underdeveloped countries (UDCs) of the Third World, most of which were formerly colonized. It also was noted that underdevelopment is characterized by certain specific symptoms; viewed spatially, these symptoms produce the *core-periphery contrasts* discussed on pp. 41–42.

Various attempts have been made to improve our understanding of the development process. One global model, still much discussed, was formulated by the economist Walt Rostow in the 1960s; his model suggested that all developing countries follow an essentially similar path through five interrelated growth stages. In the earliest of these stages, a *traditional society* engages mainly in subsistence farming, is locked in a rigid social structure, and resists technological change. When Stage 2—*preconditions for takeoff*—is reached, a progressive group of leaders moves the country toward greater flexibility, openness, and diversification. Old ways are abandoned by substantial numbers of people; birth rates decline; more products than farm-related goods are made and sold; transportation improves. A sense of national unity and purpose may emerge to help nurture this progress. If that happens, Stage 3—*takeoff*—is reached. Now the country experiences something akin to an industrial revolution, and sustained growth takes hold. Urbanization and industrialization proceed, and technological and mass-production breakthroughs occur. If the economy continues to develop, it enters Stage 4—*drive to maturity*. Technologies diffuse nationwide, sophisticated industrial specialization occurs, international trade expands. Some countries reach a still higher Stage 5, that of *high mass consumption*, marked by high incomes, the robust production of consumer goods and services, and a majority of workers in the tertiary and quaternary sectors. Today we could add a further stage to Rostow's model: the *postindustrial* stage of ultrasophisticated high technology.

On today's world map (Fig. 9–1), traditional societies of Stage 1 are heavily concentrated in tropical Africa; the most advanced developed countries (DCs) associated with Stages 4 and 5 lie mainly in North America and Western Europe. Any assessment of the stage of development of a country involves difficult problems. For example, by some measures (such as those used in Fig. 9–1) countries in the "takeoff" stage include Mexico, Brazil, Venezuela, Egypt, Turkey, and China. But while some of these (and other) countries in the takeoff stage seem to be headed for the next upward stage, others are stagnating or sliding backward. Nigeria during its 1970s oil boom seemed poised for a "takeoff"; it now has most of the characteristics of a Stage 2, not a Stage 3, society.

In their book, *China: The Geography of Development and Modernization*, Clifton Pannell and Laurence Ma evaluate China's development problems from a geographic perspective. They cite environmental problems, population pressures, spatial dimensions (the sheer size of the country is an obstacle), inadequate transport networks, regional inequalities, and traditional resistance to change as major impediments to development. Prophetically, they point to the need for political stability as essential to China's eco-

a part of China and due to be returned to Chinese control by the British government in 1997, has a population of 5.8 million. Landlocked Mongolia makes up in size what it lacks in population: with only 2.2 million inhabitants, its territory, wedged between China and the Soviet Union, is three times as large as France.

China's more than 1000 million people are heirs to what may well be the world's oldest continuous national culture and civilization.

The present capital, Beijing (Peking), lies near the nucleus of ancient China. Over 4000 years ago, this source area gave rise to a state that expanded and contracted as its fortunes changed, a state that was sometimes united, sometimes

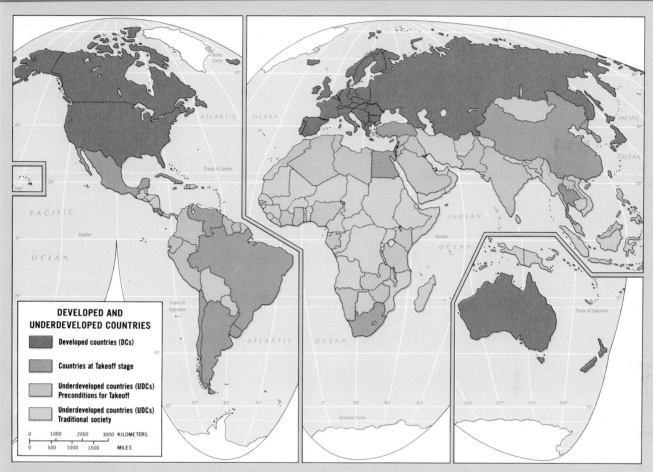

FIGURE 9–1

nomic progress—and remind their readers that political instability, one of development's greatest hindrances, is also one of China's frequent problems.

In the modern world, the prognosis for UDCs is generally not good. The UDCs have been thrust into a global system of exchange and capital flow over which they have little or no control. Unable in many instances to stem the tide of population growth, UDCs see the prices of their products fall while interest rates rise; they must borrow, and are plunged into debt that deepens every year. Thus, the poorest of the world's countries owe billions of dollars to the richest, a situation that reminds the peoples and leaders of the UDCs that the geography of development is the geography of inequality. While UDCs must get their economic and political houses in order (corruption is yet another obstacle to development), the DCs must modify a discriminatory system of international commerce that aborts many a takeoff.

divided by feudal rivalries. But in the heartland there was always a China—crucible of culture, source of innovations, and focus of power. Over more than 40 centuries the Chinese created for themselves a world apart, a society with strong traditions, values, and philosophies that could reject and repel influences from the outside. It was an isolated society, confident of its strength and superiority. Even when that strength was finally broken and its superiority punctured, that confidence remained. China is a world unto itself.

The Chinese themselves deliberately contributed to the isolation of their country from foreign contacts and influences. Tradition and

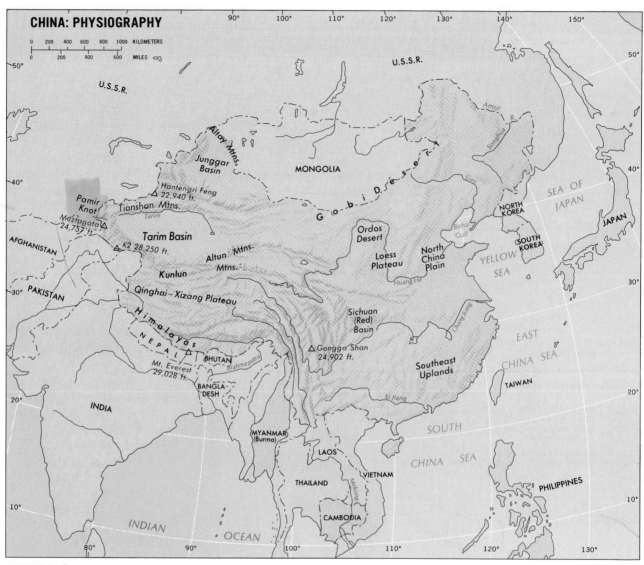

CHINA: PHYSIOGRAPHY

FIGURE 9–2

geography made it possible (Fig. 9–2). To China's north lie the rugged mountain ranges of eastern Siberia and the barren country of Mongolia. To the northwest, beyond Xinjiang (Sinkiang), the mountains open—but onto the vast Kirghiz Steppe. To the west and southwest lie the Tianshan and Pamir Mountains, the forbidding Qinghai-Xizang (Tibetan) Plateau, and the great Himalayan wall. And to the south there are the mountain slopes and tropical rainforests of Southeast Asia.

But more telling even than these imposing physical barriers to contact and interaction is the factor of distance. Until recently, China has always been far from the modern source areas of industrial innovation and technological change. True, China—as the Chinese emphasize—was itself such a hearth, but China's contributions to the outside world were very limited. China did interact to some extent with Korea, Japan, and parts of Southeast Asia—but compare this to the impact of the Arabs, who ranged far and wide and who brought their religion and political influences to areas from Mediterranean Europe to Bangladesh,

East Africa to Indonesia. When Europe became the center of change, China lay farther away by sea than almost any other part of the world, farther even than the coveted East Indies of island Southeast Asia.

Today, modern communications notwithstanding, China still is distant from almost anywhere else on earth. Going by rail from the heartland of China's Eurasian neighbor, the Soviet Union, is a long and tedious journey of several days. Direct overland connections with India are practically nonexistent. Communications with South-

TEN MAJOR GEOGRAPHIC QUALITIES OF CHINA

1. China's population represents over one-fifth of all humanity. Territorially, China ranks third among the world's countries.

2. China is one of the world's oldest continuous civilizations.

3. China's civilization developed over a long period in considerable isolation, protected by physiographic barriers and by sheer distance from other source areas.

4. The Chinese state and national culture evolved from a core area that emerged in the north, near the present capital of Beijing (Peking). China's culture hearth has remained there ever since.

5. Foreign intervention had disastrous impacts on Chinese society, from European colonialism to Japanese imperialism. Intensified regionalism and territorial losses are only two of the many resulting afflictions.

6. China occupies the eastern flank of Eurasia. Its sphere of influence was reduced by Russian expansionism in East Asia.

7. China's enormous population is strongly concentrated in the country's eastern regions. Western zones remain comparatively empty and open, and are also more arid and far less productive.

8. China's communist-designed transformation after 1949 involved unprecedented regimentation and the imposition of effective central authority, with results that are perhaps permanently imprinted on the cultural landscape.

9. China's recent modernizing drive notwithstanding, the country remains a dominantly rural society with limited urbanization and industrialization.

10. Rural China is a land of enduring traditions. Neither the Communist Revolution nor the modernization drive has truly changed the villagers' way of life. Many old values persist, and the teachings of Kongfuzi (Confucius) are still remembered.

east Asia, though improving, are still rather tenuous. Links with Taiwan and the two Koreas would be easier, but here political impediments get in the way. North Korea has not shared China's recent enthusiasm for economic reform, and industrializing South Korea has not had diplomatic or other official links to China. This leaves Japan as China's only well-connected, industrialized trading partner.

CHINA IN THE WORLD TODAY

For all its isolation and remoteness, China has moved to the center stage of world attention in recent years. Napoleon long ago remarked that China was asleep and whoever would awaken the Chinese giant would be sorry. Today, China is wide awake. It was stung by Japanese aggression in the 1930s and 1940s; after the end of World War II, the growing communist tide took power following a bitter civil war in which the United States supported the losing side. European colonialism, Japanese imperialism, and a communist ideology combined to stir China into action. The foreigners were ousted, China's old order was rejected and destroyed, and since 1949 the communist regime has been engaged in a massive effort to remake China in a new image of unity and power.

This effort has taken China from a position of backwardness and weakness to one of considerable strength. China is not yet a superpower capable of contending for global influence, but China is on the move. Economic and political geographers are mindful of the speed with which Japan rose from isolation and stagnation to imperial power, and some believe that China is on the verge of a similar ascent. At a time when the two strongest powers on earth are trying to achieve a world in which they can coexist, China looms as a new threat to this joint monopoly over ultimate power.

China's potential for world influence was underscored by its initial explosion of an atomic bomb in the 1960s and its subsequent development of a nuclear arsenal as well as an expanding capacity to deliver such weapons over a long range. After the communist takeover in 1949, the Chinese were assisted by the Soviets in their industrial and military development, but an ideological quarrel ended Sino-Soviet cooperation, and the Russian technicians were sent home. The Chinese accused the Soviets of ''revisionism''—the dilution of communist ideology with doses of capitalism. So it was that China, awakened to communism by the revolutionary example of Russia, took on the mantle of the "purest" of communist systems and rejected the modern Soviet version. When anti-Soviet feeling in China reached its peak in the late 1960s and Soviet and Chinese military forces exchanged shots across the Amur and Ussuri rivers, it was not difficult to hear the echo of Napoleon's famous words.

Momentous changes came to China during the 1970s. Following China's entry into the United Nations in 1971 (replacing the delegation from the Nationalist holdout island of Taiwan), a U.S. president visited Beijing (Peking) in 1972. Diplomatic relations with Japan were established, and several trade agreements followed; China was an important oil supplier to Japan during the energy crises of the 1970s. In 1976, both Premier Zhou Enlai (Chou En-lai) and the architect of modern Communist China, Mao Zedong (Mao Tse-tung), died, and a power struggle ensued after Mao's death. This struggle involved "moderate" and "radical" factions of the Communist Party and had been building even during Mao's final illness. The basic issues centered on the course of economic development and the future of education in China. The pragmatic moderates wanted to speed economic growth by offering workers some material rewards; they also proposed that the schools spend less time on ideological instruction and more on practical technical training. The radicals feared that such changes would weaken China's "pure" communism and produce new elite people, setting back three decades of class struggle. Eventually, the pragmatists prevailed; by the opening of the 1980s, a new China had begun to emerge. Thousands of American visitors were admitted, and diplomatic relations returned to normal. And most important, China's planners laid out a new course for post-Mao

THE *PINYIN* SYSTEM

If a Western traveler in China were to stop and ask for directions to Peking, the response would quite probably be a shake of the head. The Chinese call their capital *Beijing*, not Peking. For centuries, the romanization of Chinese has led to countless errors of this kind, in part because Chinese dialects are also inconsistent and the Chinese in northern China may pronounce a place name somewhat differently from those living in the south.

In 1958, the Chinese government adopted the *pinyin* system of standard Chinese, not to teach foreigners how to spell and pronounce Chinese names and words, but to establish a standard form of the Chinese language throughout China. The *pinyin* system is based on the pronunciation of Chinese characters in Northern Mandarin, the Chinese spoken in the region of the capital and the north in general.

This new linguistic standard caught on more slowly outside China. However, by 1980 all press reports from China were exclusively using the *pinyin* spelling of Chinese names, and in the United States and in other parts of the world, newspapers and magazines were adopting the system. Atlases, wall maps, and other geographic materials (including textbooks) were revised. In a few instances, the change was resisted, and the old spellings of the predecessor Wade-Giles system persist. But *pinyin* has today become the world standard for romanized Chinese.

Strange as the familiar Chinese place names seem when transcribed into *pinyin*, the new system is not difficult to learn or understand. Most of the sounds are clear to someone who speaks

China—the "Four Modernizations." In industry, defense, science, and agriculture, China launched an era of modernization as rapid as the country could achieve.

While China's economic modernization proceeded, old ideological tensions continued to strain its political leadership. Consumer goods (including transistor radios) diffused across China's markets; tens of thousands of Chinese students were sent to study at American and other foreign universities. Fresh political ideas were published, discussed, and debated, and the new economic climate had its inevitable political consequences. In 1989, a huge throng of

students and workers occupied Beijing's hallowed central plaza, Tiananmen Square, demanding political reforms to go along with the country's economic liberalization. As the students remained encamped on the square, a power struggle developed among the authorities. As the demonstrators called for the democratization of China's political system, China's old authoritarian traditions prevailed at the top, and a brutal crackdown followed (Fig. 9–3). Thousands were killed as the army took control, not only in Beijing but also in other cities where demonstrations had broken out. Later, leaders of the pro-democracy movement were rounded up, im-

English, except for the *q* and the *x*. The *q* sounds like the *ch* in *cheese*; the *x* is pronounced *sh* as in *sheer*. Thus, the Chinese spell China *Xinhua* (which actually means ''new China'') to reflect more closely how the name should sound.

Some familiar place names, spelled the Wade-Giles and the *pinyin* way, are listed below:

Peking	Beijing
Canton	Guangzhou
Tibet	Xizang
Singkiang	Xinjiang
Yangtze Kiang	Chang Jiang (River)
Szechwan	Sichuan
Hwang Ho	Huang He (River)
Tientsin	Tianjin

And some personal names:

Mao Tse-tung	Mao Zedong
Teng Hsiao-ping	Deng Xiaoping

In this chapter both the old Wade-Giles and new *pinyin* spellings are given as needed, so that it will be easy to become familiar with the new names. The maps have been revised to conform to the *pinyin* system. In the text, the *pinyin* version is given first and the old spelling is shown in parentheses.

FIGURE 9–3 One of the most vivid photographic images of the June 4, 1989 crackdown on the Beijing demonstrators: the lone student protester temporarily halting a column of tanks near Tiananmen Square.

prisoned, and, in numerous cases, executed. It was a reminder of that old adage about Chinese indifference to foreign norms or opinions: the repression was carried out before global television networks and in the face of world condemnation.

China's expanding role in the world is a major portent for the twenty-first century. The score may have been temporarily settled at home, but China has an external agenda to be reckoned with. Before the turmoil of 1989, China had settled with the United Kingdom on the return to Beijing's control of the Crown Colony of Hong Kong on December 31, 1997—a date now seen in a very different light by the colony's inhabitants. Already, China has absorbed Tibet (Xizang), and Beijing's leaders regard Taiwan's independent status as a temporary separation.

But the most serious future issues may involve China's southern and northern flanks. Southeast Asia contains large Chinese minorities, and China has a historic interest in that geographic realm. Following the U.S. defeat in Vietnam, the Soviet Union seized the opportunity to establish a sphere of influence in Indochina. This has become a matter of great concern in China, but it is not the only place of potential conflict with its powerful communist neighbor. In the north, Russian imperialism cost China dearly, and China claims large areas of the Soviet Far East on historic grounds (see Fig. 2–24). As China's military capacity grows, such latent conflicts take on new significance. The possibility that two nuclear powers with large arsenals and sophisticated delivery systems will face each other across a single (and disputed) boundary on the same continent may become a dominant menace to the peace and survival of the world.

EVOLUTION OF THE CHINESE STATE

China may have developed in comparative isolation for more than 4000 years, but there is evidence that China's earliest core area, which was positioned around the confluence of the Huang He (Yellow) and the Wei Rivers (Fig. 9–4), received stimuli from other distant and possibly slightly earlier civilizations—the river-based cultures of Southwest and South Asia. From Mesopotamia, the Indus Valley, and the Ganges Delta area, techniques of irrigation and metalworking, innovations in agriculture, and possibly even the practice of writing reached the Huang He Basin, probably overland along the almost endless routes across interior mountains, deserts, and steppes (see Fig.

FIGURE 9–4

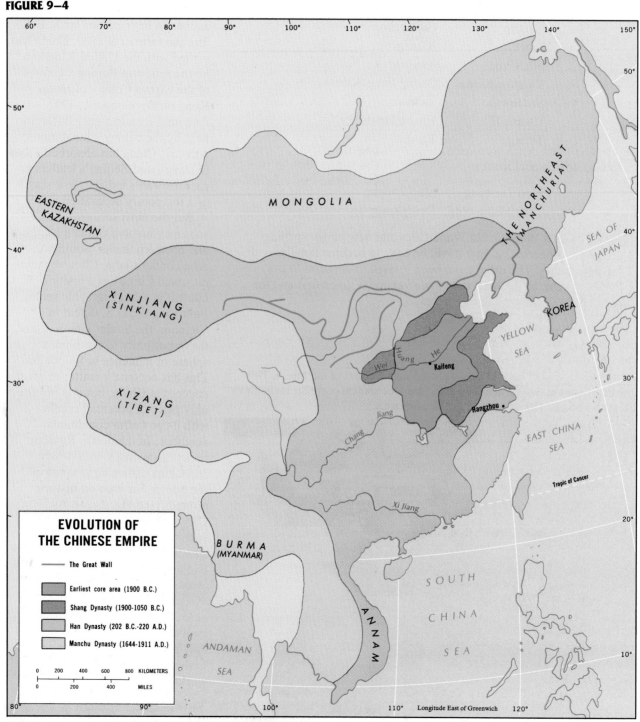

EVOLUTION OF THE CHINESE EMPIRE

— The Great Wall

Earliest core area (1900 B.C.)

Shang Dynasty (1900-1050 B.C.)

Han Dynasty (202 B.C.-220 A.D.)

Manchu Dynasty (1644-1911 A.D.)

6–3). The way these early Chinese grew their rice crops provides evidence that they learned from the Mesopotamians; the water buffalo most likely came from the Indian subcontinent. But quite soon the distinctive Chinese element began to appear. By the time the record becomes reliable and continuous, around 1800 B.C., Chinese cultural individuality was already strongly established.

The oldest dynasty of which much is known is the Shang Dynasty (sometimes called Yin), centered in the Huang/Wei confluence zone from perhaps 1770 B.C. to about 1120 B.C. Walled cities were built during the Shang period, and the Bronze Age commenced during Shang rule. For more than a thousand years after the beginning of the Shang Dynasty, North China was the center of development in this part of Asia. The Zhou (Chou) Dynasty (ca. 1120–221 B.C.) sustained and consolidated what had begun during the Shang period.

Eventually, agricultural techniques and population numbers combined to press settlement in the obvious direction—southward, where the best opportunities for further expansion lay. During the brief Qin (Chin) Dynasty (221–207 B.C.), the lands of the Chang (Yangzi) Jiang (*jiang* means "river") were opened up, and settlement spread as far south as the lower Xi (Hsi) Jiang (West River).

Han China

A pivotal period in the historical geography of China now lay ahead: the Han Dynasty (207 B.C.–A.D. 220). The Han rulers brought unity and stability to China, and they enlarged the Chinese sphere of influence to include Korea, the Northeast, Mongolia, Xinjiang (Sinkiang), and, in Southeast Asia, Annam (located in what is now central Vietnam). Thus, the Chinese rulers established control over the bases of China's constant harassers—the nomads of the surrounding steppes, deserts, and mountains—and protected (in Xinjiang) the main overland avenue of westward contact between China and the rest of Eurasia.

The Han period was a formative one in the evolution of China (Fig. 9–4). Not only was Chinese military power stronger than ever before, but there were also changes in the systems of landownership as the old feudal order broke down and private individual property was recognized, and the silk trade grew into China's first external commerce. To this day, most Chinese, recognizing that much of what is China first came about during this period, still call themselves the "People of Han."

Han China was the Roman Empire of East Asia. Great achievements were made in art, architecture, the sciences, and other spheres. Like the Romans, the Han Chinese had to contend with unintegrated hostile peoples on their empire's margins. In China, these peoples occupied the mountainous country south of the national territory, and the Han rulers built military outposts to keep these frontiers quiet. Again, like the Roman Empire, the China of Han fell into decline and disarray, and more than a dozen successor states arose after A.D. 220 to compete for primacy. Not until the Sui Dynasty (581–618) did consolidation begin again, to be continued during the Tang Dynasty (618–907), another period of national stability and development.

Later Dynasties

Following the Tang period, when China once again was a great national state, the Song Dynasty (960–1279) carried the culture to unprecedented heights despite continuing problems with marauding nomadic peoples. This time the threat came from the north; in 1127, the Song rulers had to abandon their capital at Kaifeng on the Huang He in favor of Hangzhou (Hangchow) in the southeast. Nevertheless, China during the Song Dynasty was in many ways the world's most advanced state. It had several cities of more than 1 million people, paper money was in use, commerce intensified, literature flourished, schools multiplied in number, the arts thrived as never before, and the philosophies of Kongfuzi (Confucius), for many centuries China's guide, were modernized, printed, and mass-distributed for the first time. Eventually Song China fell to conquerors from the outside, the Mongols led by Kublai Khan. The Mongol authority, known as the Yuan Dynasty (1279–1368), made China part of a vast empire that extended all the way across Asia to Eastern Europe. But it lasted less than a century and had little effect on Chinese culture. Instead, it was the Mongols who adopted Chinese civilization!

In 1368, a Chinese local ruler led a rebellion that ousted the Mongols, signaling the ascent of the great indigenous Ming Dynasty (1368–1644). Under the Ming rulers China's greatness was restored, its territory once again consolidated from the Great Wall in the north (Fig. 9–5) to Annam in the south. Finally, in 1644, another northern, foreign nomadic people forced their way to control over China. They came from the far northeast (see *box*), but unlike the Mongols, they sustained and nurtured Chinese traditions of administration, authority, and national culture. The Manchu (or Qing) Dynasty (1644–1911) extended the Chinese sphere of influence to include Mongolia, Xinjiang, Xizang (Tibet), Burma (now Myanmar), Indochina, eastern Kazakhstan, Korea, and a large part of what is today the Soviet Far East (Fig.

9–4). But the Manchus also had to contend with the rising power of Europe and the West, and the encroachment of Russia; large portions of the Chinese spheres in interior and Southeast Asia were lost to China's voracious competitors.

It is often said that the year 1949, when Communist Party Chairman Mao Zedong proclaimed the People's Republic of China, marked the beginning of a new dynasty that is not so different from the old, communist doctrine notwithstanding. Certainly some of China's old traditions continue in

MANCHURIA: THE NORTHEAST

Three provinces, Liaoning, Jilin (Kirin), and Heilongjiang (Heilungkiang), constitute China's Northeast, a region bounded by the Soviet Union on the north and east, Mongolia in the west, and North Korea in the southeast (see Fig. 9–21). This was the home of the conquering Manchus, and was later the scene of foreign domination. Russians and Japanese struggled for control over the area; ultimately, Japan created a dependency here in the 1930s that it called *Manchukuo*. More recently, the region came to called *Manchuria*, but this is not an accepted regional appellation in China. Chinese geographers refer to the region encompassed by these three northeastern provinces as simply the *Northeast*—a practice followed in this chapter.

FIGURE 9–5   The Great Wall of China, completed ca. 200 B.C., was an attempt to separate eastern China's sedentary farmers from the pastoral herders of the Asian interior. Here, the wall snakes its way across the rugged terrain just north of Beijing.

the new communist era, but in many more ways the new China is a totally overhauled society. Benevolent or otherwise, the rulers of old China headed a country in which—for all its splendor, strength, and cultural richness— the fate of the landless and the serf was often indescribably miserable, famine and disease could decimate the population of entire regions, local lords could repress the people with impunity, children were sold, and brides were bought. Contact with Europe, however peripheral, brought ruin to Chinese urban society where slums, starvation, and deprivation were commonplace. Mao's long tenure and apparent omnipotence may remind us of the dynastic rulers' frequent longevity and absolutism, but the new China he left behind is vastly different from the old.

A CENTURY OF CONVULSION

China long withstood the European advent in East Asia with a self-assured superiority based on the strength of its culture and the continuity of the state. There was no market for the British East India Company's rough textiles in a country long used to finely fabricated silks and cottons. There was little interest in the toys and trinkets the Europeans produced in the hope of bartering for Chinese tea and pottery. And even when Europe's sailing ships made way for steam-driven vessels and newer and better factory-made textiles were offered in trade for China's tea and silk, China continued to reject the European imports that were still, initially at least, too expensive and too inferior to compete with China's handmade goods. Long after India

had fallen into the grip of mercantilism and economic imperialism, China was able to maintain its established order. This was no surprise to the Chinese; after all, they had held a position of undisputed superiority among the countries of eastern Asia as long as could be remembered, and they had dealt with overland and seaborne invaders before.

The nineteenth century finally shattered the self-assured isolationism of China as it proved the superiority of the new Europe. On two fronts, the economic and the political, the European powers destroyed China's invincibility. In the economic sphere, they succeeded in lowering the cost and improving the quality of manufactured goods, especially textiles, and the handicraft industry of China began to collapse in the face of unbeatable competition. In the political sphere, the demands of the British merchants and the growing English presence in China led to conflicts. In the early part of the century, the central issue was the importation into China of opium, a dangerous and addictive intoxicant. Opium was destroying the very fabric of Chinese culture, weakening the society, and rendering China an easy prey for colonial profiteers. As the Manchu government moved to stamp out the opium trade in 1839, armed hostilities broke out, and soon the Chinese sustained their first defeats. Between 1839 and 1842, the Chinese fared badly, and the First Opium War signaled the end of Chinese sovereignty.

British forces penetrated up the Chang Jiang (Yangzi) and controlled several areas south of that river (Fig. 9–6); Beijing hurriedly sought a peace treaty. As a result, leases and concessions were granted to foreign merchants. Hong Kong Island was ceded to the British, and five ports, including Guangzhou (Canton) and Shanghai, were opened to foreign com-

merce. No longer did the British have to accept a status that was inferior to the Chinese in order to do business; henceforth, negotiations would be pursued on equal terms. Opium now flooded into China, and its impact on Chinese society became even more devastating. Fifteen years after the First Opium War, the Chinese tried again to stem the disastrous narcotic tide, and again they were bested by the foreigners who had attached themselves to their country. Now the cultivation of the opium poppy in China itself was legalized. Chinese society was disintegrating, and the scourge of this drug abuse was not defeated until the revival of Chinese power early in the twentieth century.

But before China could reassert itself, much of what remained of China's sovereignty was steadily eroded away (see *box*). The Germans obtained a lease on Qingdao (Tsingtao) on the Shandong (Shantung) Peninsula in 1898, and in the same year the French acquired a sphere of influence in the far south at Zhanjiang (Fig. 9–6). The Portuguese took Macao; the Russians obtained a lease on Liaodong (Liaotung) in the Northeast as well as railway concessions there; even Japan got into the act by annexing the Ryukyu Islands and, more importantly, Formosa (Taiwan) in 1895.

After millennia of cultural integrity, economic security, and political continuity, the Chinese world lay open to the aggressions of foreigners whose innovative capacities China had denied to the end. But now ships flying European flags lay in the ports of China's coasts and rivers; the smokestacks of foreign factories rose above the landscapes of its great cities. The Japanese were in Korea, which had nominally been a Chinese vassal; the Russians had entered China's Northeast. The foreign invaders even took to fighting among themselves, as did Japan and Rus-

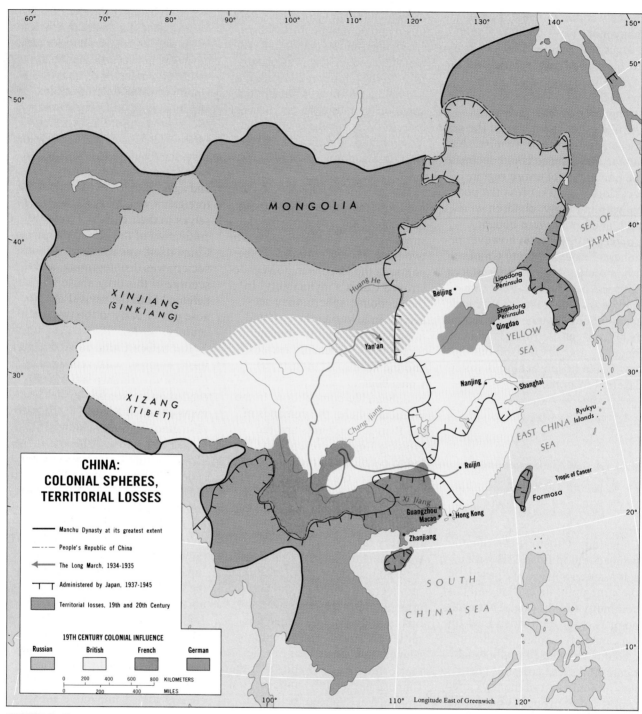

FIGURE 9–6

sia in ''Manchuria'' (as the foreigners now called the Northeast) in 1904.

Rise of a New China

In the meantime, organized opposition to the foreign presence in China was gathering strength, and this century opened with a large-scale revolt against all outside elements. Bands of revolutionaries roamed the cities as well as the countryside, attacking not only the hated foreigners but also Chinese citizens who had adopted Western cultural traits. Known as the Boxer Rebellion (after a loose translation of the Chinese name for these revolutionary groups), the 1900 uprising was put down with much bloodshed. Simultaneously, another revolutionary movement was gaining support, aimed against the Manchu leadership itself. In 1911, the emperor's

EXTRATERRITORIALITY

FIGURE 9–7 Sha Mian Island in Guangzhou (Canton), a colonial-era enclave where foreigners resided beyond the jurisdiction of the Chinese.

A sign of China's weakening during the second half of the nineteenth century was the application in its cities of a European doctrine of international law—**extraterritoriality**. This principle denotes a situation in which foreign states or international organizations and their representatives are immune from the jurisdiction of the country in which they are present. This, of course, constitutes an erosion of the sovereignty of the state hosting these foreign elements, especially when the practice goes beyond the customary immunity of embassies and persons in diplomatic service.

In China, extraterritoriality reached unprecedented proportions. The best residential suburbs of the large cities, for example, were declared to be ''extraterritorial'' parts of foreign countries and were made inaccessible to Chinese citizens. Sha Mian Island in the Pearl River in Guangzhou (Canton) was a favorite extraterritorial enclave of that city (Fig. 9–7). A sign at the only bridge to the island stated ''No Dogs or Chinese'' and still stands today as a reminder of foreign excesses. In this way, the Chinese found themselves unable to enter their own public parks and many buildings without permission from foreigners. Christian missionaries fanned out into China, their bases fortified with extraterritorial security. To the Chinese, this involved a loss of face that contributed to the bitter opposition against the presence of all foreigners, a resentment that finally exploded in the Boxer Rebellion of 1900.

improve the country's overall position. The Japanese captured Germany's holdings on the Shandong Peninsula, including the city of Qingdao, during World War I; when the victorious European powers met at Versailles to divide the territorial spoils, they affirmed Japan's rights in the area. This led to another significant demonstration of Chinese reassertion as nationwide protests and boycotts of Japanese goods were organized in what became known as the May Fourth Movement. One participant in these demonstrations was a charismatic young man named Mao Zedong.

Nonetheless, China after World War I remained a badly divided country. By the early 1920s, there were two governments—one in Beijing and another in the southern city of Guangzhou (Canton) where the famous Chinese revolutionary, Sun Yat-sen, was the central figure. Neither government could pretend to control much of China. The Northeast was in complete chaos, petty states were emerging all over the central part of the country, and the Guangzhou ''parliament'' controlled only a part of Guangdong (Kwangtung) Province (Fig. 9–8). Yet, it was just at this time that the power groups that were ultimately to struggle for supremacy in China were formed. While Sun Yat-sen was trying to form a viable Nationalist government in Guangzhou, the Chinese Communist Party was formed by a group of intellectuals in Shanghai. Several of these intellectuals had been leaders in the May Fourth Movement, and in the early 1920s they received help from the Communist Party of the Soviet Union. Mao Zedong was already a prominent figure in these events.

Initially, there was cooperation between the new Communist Party and the Nationalists led by Sun Yat-sen. The Nationalists were stronger and better organized, and

garrisons were attacked all over China, and in a few months the 267-year-old dynasty was overthrown. Indirectly, it too was yet another casualty of the foreign intrusion, and it left China divided and disorganized.

The end of the Manchu era and the proclamation of a republican government in China did little to

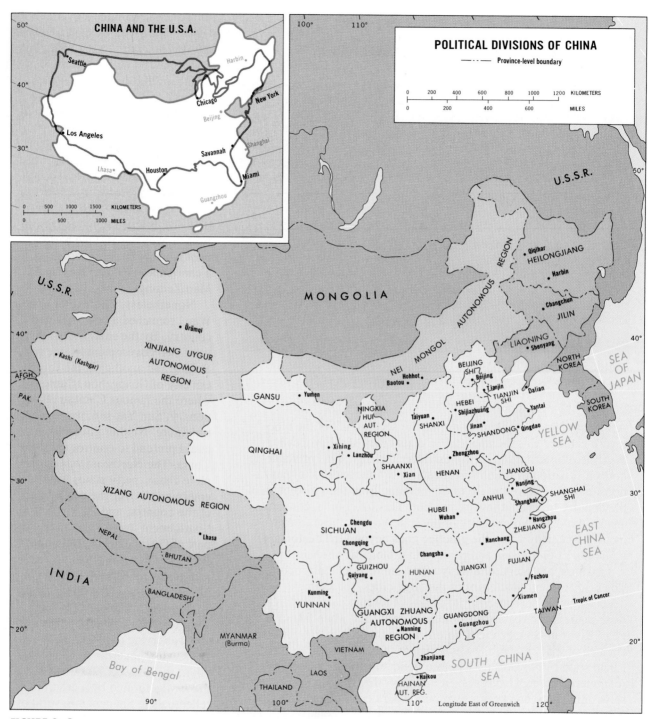

CHINA AND THE U.S.A.

POLITICAL DIVISIONS OF CHINA
- · - · - Province-level boundary

FIGURE 9–8

they hoped to use the communists in their anti-foreign (especially anti-British) campaigns. By 1927, the foreigners were on the run; the Nationalist forces entered cities and looted and robbed at will while aliens were evacuated or, failing that, sometimes killed. As the Nationalists continued their

drive northward and success was clearly in the offing, internal dissension arose. Soon, the Nationalists were as busy purging the communists as they were pursuing foreigners, and the central figure to emerge in this period was Chiang Kai-shek. Sun Yat-sen died in 1925, and when the Nation-

alists established their capital at Nanjing (Nanking) in 1928, Chiang was the country's leader.

Three-Way Struggle

The post-Manchu period of strife and division in China was quite similar to other times when, following a lengthy period of comparative stability under dynastic rule, the country fragmented into rival factions. In the first years of the Nanjing government's **hegemony**, the campaign against the communists intensified and many thousands were killed. Chiang's armies drove them ever deeper into the interior (Mao himself escaped the purges only because he was in a remote rural area at the time); for a while, it seemed that Nanjing's armies would break the back of the communist movement in China.

The Long March

A core area of communist peasant forces survived in the zone where the provinces of Jiangxi (Kiangsi) and Hunan adjoin in southeastern China (Fig. 9–8) and defied Chiang's attempts to destroy them. Their situation grew steadily worse, however, and in 1933 the Nationalist armies were on the verge of encircling this last eastern communist stronghold. The communists decided to avoid inevitable strangulation by leaving. Nearly 100,000 people—armed soldiers, peasants, local leaders—gathered near Ruijin and started to walk westward in 1934. It was a momentous event in modern China, and among the leaders of the column were Mao Zedong and Zhou Enlai. The Nationalists rained attack after attack on the marchers, but they never succeeded in wiping them out completely; as the communists marched, they were joined by new sympathizers.

The Long March (see route in Fig. 9–6), as this drama has come to be called, first took them to Yunnan Province, where they turned north to enter western Sichuan (Szechwan). They then traversed Gansu (Kansu) Province and eventually reached their goal, the mountainous interior near Yanan (Yenan) in Shaanxi (Shensi) Province. The Long March covered nearly 6000 miles (10,000 kilometers) of China's more difficult terrain, and the Nationalists' continuous attacks killed an estimated 75,000 of the original participants. Only about 20,000 survived the epic migration, but among them were Mao and Zhou, who were convinced that a new China would emerge from the peasantry of the rural interior to overcome the urban easterners whose armies could not eliminate them.

The Japanese

While the Nanjing government was pursuing the communists, foreign interests made use of the situation to further their own objectives in China. The Soviet Union held a sphere of influence in Mongolia, and was on the verge of annexing a piece of Xinjiang (Sinkiang). Japan was dominant in the Northeast, where it had control over ports and railroads. The Nanjing government tried to resist the expansion of Japan's sphere of influence. The effort failed, and the Japanese set up a puppet state in the region; they appointed a ruler and called their new dependency Manchukuo.

The inevitable war between the Japanese and Chinese broke out in 1937. There were calls for a suspension of the Nationalist-Communist struggle in the face of the common enemy, but after a brief armistice, the factional conflict arose again while both sides fought the Japanese. China now became divided into three regions: the areas taken by the Japanese during their quick offensive of 1937–1938 (when they took much of eastern China, including the principal ports—see Fig. J–2); the Nationalist zone centered on China's wartime capital of Chongqing (Chungking); and the Communist-held areas of the interior west. The Japanese, by engaging Chiang's forces, gave the Communists an opportunity to build their strength and prestige in China's western regions.

Victory

After Japan was defeated by the U.S.-led Western powers in 1945, the full-scale Nationalist-Communist struggle quickly resumed. The United States, hopeful for a stable and friendly government in China, sought to mediate the conflict, but did so while recognizing the Nationalist faction as the legitimate government. Its chances of mediation were impaired by this position and also by the military aid given to Nationalist forces. By 1948, it was clear that Mao Zedong's armies would defeat the forces of Chiang Kai-shek and that the final victory was only a matter of months. Chiang kept moving his capital—back to Guangzhou (where Sun Yat-sen had built the first Nationalist government), and then to Chongqing (Chungking). But late in 1949, following disastrous defeats in which hundreds of thousands of Nationalist forces were killed, the remnants of Chiang's faction fled to the island of Taiwan, taking control there and proclaiming the Republic of China. Meanwhile, in Beijing, standing in front of the assembled masses at the Gate of Heavenly Peace on Tiananmen Square, Mao Zedong on October 1, 1949 announced the birth of the People's Republic of China.

REGIONS OF CHINA

China is a vast country that contains a wide variety of human and physical landscapes. China's territorial proportions are almost identical to those of the contiguous

United States (although China extends somewhat farther both to the north and to the south, as the inset map in Fig. 9–8 shows). On the other hand, there is nothing in China to compare with California—there is no west coast, not even an easy outlet into central Eurasia or the Soviet Union. China faces the Pacific as the U.S. faces the Atlantic, but China's far west consists of deserts and mountains without a coastline.

The map reveals some other similarities between the United States and China. In both countries, the core area lies in the eastern part of the national territory. In China, however, the degree of eastern concentration is much stronger, because there is no California-like subsidiary to the core area. And China's core region is much more populous, since

FIGURE 9–9

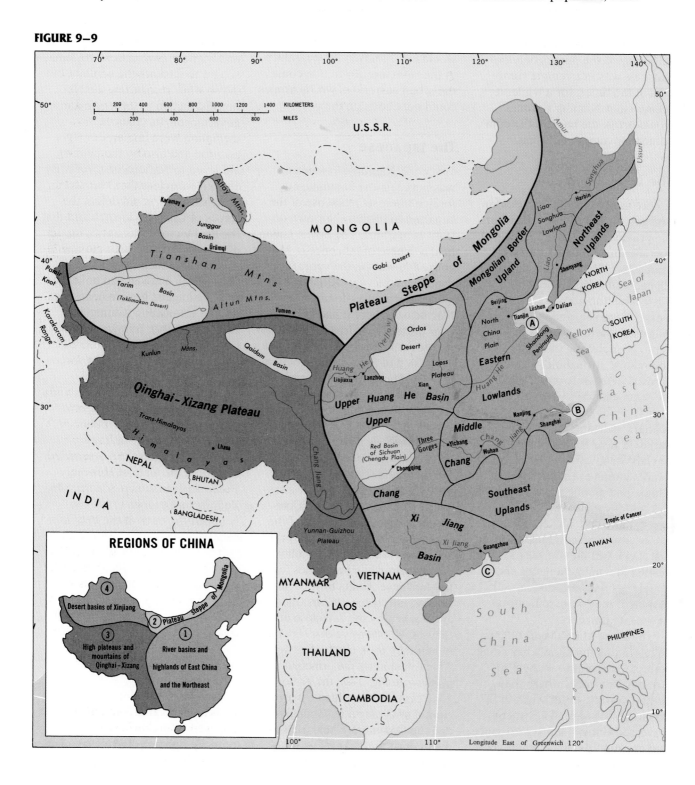

China's total population is more than four times as large as that of the United States. Scrutiny of the political map of China (Fig. 9–8) suggests still another pattern familiar to Americans. China's smallest provinces lie in its eastern areas, and Texas-size units mark the west. This reflects both history and geography; even populous China has remote, isolated, sparsely peopled expanses where small administrative units would make no sense.

Physiographically, China divides into about the same number of regions as the United States, its landscapes ranging from lush sub-tropical lowlands in the southeast to rugged, snowcapped mountains in Xizang (Tibet) and from fertile river plains in its core area to barren deserts in the western interior. To grasp the geographic layout of a country as large and complex as China, it is useful to begin with broad outlines and to focus on smaller areas later. Consider first the inset map in Fig. 9–9. At this small scale, China is divided into four major regions: ① the river basins and highlands of East China and the Northeast; ② the plateau steppe of Mongolia; ③ the high plateaus and mountains of Qinghai-Xizang; and ④ the desert basins of Xinjiang. Strictly speaking, these are not physiographic *provinces*—they are broader divisions. Comparable regions in the U.S. would be, say, the Interior Lowlands, which encompass several different kinds of low-lying landscapes, or the Appalachian Highlands, also diverse enough to be divided into many distinct sub-areas.

Looking now at the main map in Fig. 9–9, it is clear that the four physiographic divisions shown in the inset map each contains a number of well-defined subregions. Eastern China's landscapes (Region ①) consist of the basins of three large rivers (and several

smaller streams) that flow into the Pacific Ocean. As we will note later, these great river systems are quite distinctive. Eastern China also includes the valleys, plains, and uplands of the cold-weather Northeast; the Ordos Desert around which the Huang He (Yellow River) makes its huge northward loop; the famous, teeming Sichuan (Red) Basin; and the Southeast Uplands. Again, the Qinghai-Xizang Plateau incorporates the great Himalaya Mountains in the south; an enormous, snowcapped plateau in the center; and the Kunlun Mountains and Qaidam Basin in the north. And China's northwestern quadrant, called the Desert Basins of Xinjiang, consists of more than a series of arid lowlands: it is bisected by the mountainous Tianshan and flanked by the Altun Mountains to the south and the Altay Mountains to the north. The region gets its name from its two dominant desert basins, the Tarim Basin in the south (containing the Taklimakan Desert) and the significant Junggar Basin in the north.

As it happens, these broad physiographic divisions of China coincide substantially with the human spatial mosaic. The Eastern China of Fig. 9–9 (Region ①) is what is often called *China Proper*, the ''real'' China with its enormous population clusters, intensively-farmed fields, countless villages, and crowded cities. Coupled with the increasingly important Northeast, Eastern China is the national core area, the heart of the People's Republic. Region ② in Fig. 9–9 is a transitional zone from China Proper to Mongolia. Qinghai-Xizang in the southwest is the region (③) of Tibet, where local inhabitants still view China as a foreign invader and imperial power. Xinjiang to the northwest is the region (④) formerly called Sinkiang, a remote and distant frontier facing the Soviet Union.

River Basins and Highlands of Eastern China and the Northeast (Region ①)

Eastern China is a densely peopled land of river valleys separated by highlands, a land of vast wheatfields in the north and rice paddies in the south (Fig. 9–10). This region contains the greater length of China's three most important rivers: the Huang He (Yellow), Chang Jiang (Yangzi), and Xi Jiang (West), marked as ④, ⑧, and ⓒ, respectively, on the larger map in Fig. 9–9. All three rivers rise on eastern slopes of the Qinghai-Xizang/Yunnan-Guizhou plateau complex and flow eastward, the Huang He through the most circuitous and longest course, the Xi through the most direct and shortest. The upper courses of the Huang and Chang rivers lie in two neighboring but distinct physiographic regions; their lower reaches, however, lie in an area that can with justification be identified as a single region, the Eastern Lowlands. In human terms, the Eastern Lowlands region is China's most important one, including as it does the North China Plain and the cities of Beijing and Tianjin (Tientsin), and the productive lower Chang Basin with Nanjing and China's largest city, Shanghai. In every respect, this is China's core area with its greatest population concentration, highest percentage of urbanization (nearing 50 percent, as opposed to 23 percent for the country as a whole), enormous agricultural production, growing industrial complexes, and intensive communications networks.

The Chang Jiang

Of the three great river systems that constitute this heartland of China Proper (Fig. 9–11), the

FIGURE 9—10 The lowlands of eastern China are covered by farm fields that must be worked very intensively to feed the country's huge population. These women are cultivating rice near Guilin, in the northern part of the Xi Basin, against the backdrop of this area's well-known limestone tower-and-dome topography.

Chang Jiang (the middle one) is in almost every respect the most important, because it traverses China's most populous and productive areas. Physiographically, the course of this river flows through three basins, of which the westernmost, the Sichuan (Szechwan) or Red Basin, contains one of China's largest population clusters (about 80 million people). Its middle course begins near the town of Yichang, where the Chang emerges from a series of gorges and where the huge new Gezhouba Dam has been constructed to control floodwaters and generate vast quantities of hydroelectric power. This middle basin is one of China's leading agricultural regions. And just upstream from Nanjing (2.0

million), the Chang enters its lower basin and quickly reaches its delta near Shanghai (6.9 million).

The Chang Jiang is China's most navigable waterway. Ocean-going ships can sail over 600 miles (1000 kilometers) up the river to the Wuhan conurbation (Wuhan [3.2 million] is short for Wuchang, Hanyang, and Hankou); boats of up to 1000 tons can travel twice that distance to Chongqing (2.4 million) in the Sichuan Basin. Several of the Chang's tributaries are also navigable, and a total of more than 18,500 miles (30,000 kilometers) of water transport routes operate in the entire Chang Basin. Therefore, the Chang Jiang is one of China's leading transportation corridors, and with its tributaries

attracts the trade of a vast area, including nearly all of middle China and sizable parts of the adjacent north and south. Funneled down the Chang, most of this enormous volume is transshipped at Shanghai, whose metropolitan size of nearly 7 million reflects the productivity and great population of this hinterland.

Early in Chinese history, when the Chang's basin was being opened up and rice and wheat cultivation began, a canal was built to link this granary to the northern core of old China. Over 1000 miles (1600 kilometers) in length, this was the longest artificial waterway in the world, but during the nineteenth century it fell into disrepair. Known as the Grand Canal

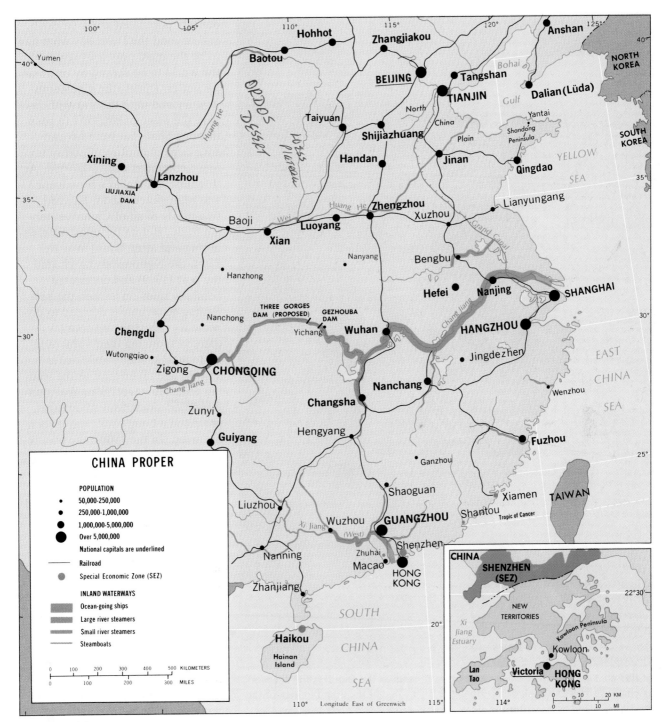

FIGURE 9–11

(Fig. 9–11), it was dredged and rebuilt during the period when the Nationalists held control over eastern China. After 1949, the communist regime continued this restoration effort, and much of the canal is now once again open to barge traffic, supplementing the huge fleet of vessels that hauls domestic interregional trade along the east coast.

The bulk of China's internal trade in agricultural as well as industrial products is either derived from, or distributed to, the Chang region. International trade also goes principally through Shanghai (Fig. 9–12), whose port normally handles half of the country's overseas tonnage; the rest is split among China's other leading ports, including Tianjin (Tientsin) on the Bohai Gulf in the north and Guangzhou (Canton) on the lower

FIGURE 9–12 Shanghai's busy waterfront (known as the Bund) is the heart of its port that lines the bank of the Huangpu River, one of the distributaries of the massive Chang Delta.

densely populated areas on earth, and beyond the Delta lies what has to be the most populous region in the world to be served by one major outlet. During the nineteenth century and until the war with Japan, the principal exports to pass through Shanghai were tea and silk; large quantities of cotton textiles and opium were imported. At that time, Shanghai's prominence was undisputed, and it handled two-thirds of all of China's external trade. But its fortunes suffered during and after World War II. First, the Nationalists blockaded the port (1949) and conducted bombing raids on it; then the new Beijing government decided to disperse its industries up-country, thereby reducing their vulnerability to attack. Meanwhile, safer Tianjin had taken over as the leading port. However, Shanghai's unparalleled situational advantages promised a comeback, and it soon occurred. In the 1960s, the port regained its dominant position, and the industrial complex (textiles, food processing, metals, shipyards, rubber, chemicals) resumed its expansion. Today, Shanghai ranks among the world's largest urban concentrations, and its population is growing five times faster than the national rate of increase.

The vast majority of the Chang Basin's 400-odd million people are farmers on cooperatives and collectives that produce the country's staples and cash crops. Sichuan is typical of the amazing variety of crops that are grown: besides such cereals as rice, wheat, and corn, there are soybeans, tea, sweet potatoes, sugarcane, and a wide range of fruits and vegetables. There, as in the Chang's lower basin, rice predominates (Fig. 9–13). However, a look at Figs. I–9 and I–10 will suggest why the Sichuan-Chang line is close to the northern margin for rice cultivation. To the north, temperatures decline and rainfall diminishes as well; even in

Xi River in the south. Shanghai, just a regional town until the mid-nineteenth century, rose to prominence as a result of its selection as a treaty port by the British. Ever since, its unequaled locational advantages have sustained its position as China's leading city in almost every respect. This metropolis lies on the Huangpu River at one corner of the Chang Delta, an area of about 20,000 square miles (50,000 square kilometers) containing more than 50 million people. Some two-thirds of these are farmers who produce food, silk filaments, and cotton for the city's industries.

Thus, Shanghai has as its immediate hinterland one of the most

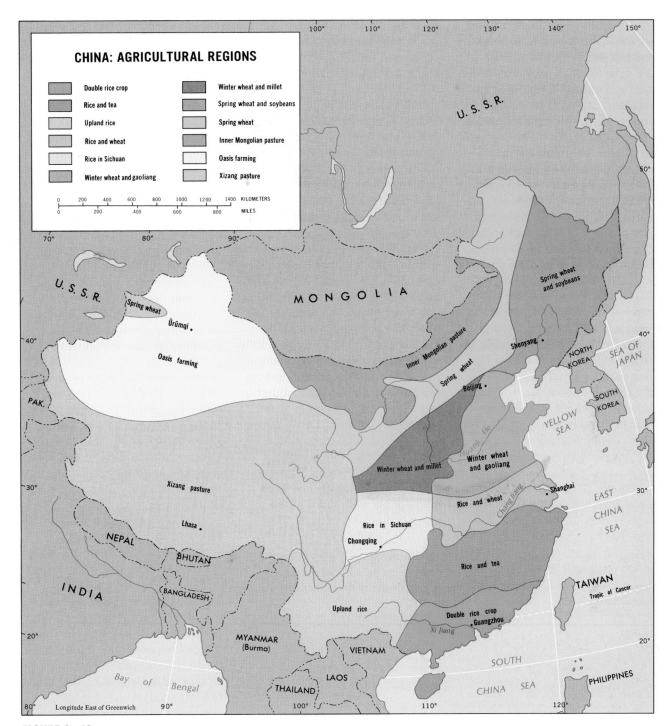

FIGURE 9–13

the irrigated areas, wheat rather than rice is the grain crop. In the Chang Basin, rice rotates with winter wheat. In effect, this is the transition zone; a line drawn from southern Shaanxi Province east to mid-coastal Jiangsu Province (as in Fig. 9–13) approximates the northern limit of rice cultivation.

The Huang He and North China Plain

As Fig. 9–9 shows, the Eastern Lowlands merge northward into the lower basin of the Huang He and the North China Plain. Here (Fig. 9–13), winter wheat and barley are planted to the south of Bei-

jing (spring wheat is raised to the north), and in the spring other crops follow the winter wheat. Millet, sorghum (gaoliang), soybeans, corn, a variety of fruit and vegetables, tobacco, and cotton are also cultivated in this northern zone of China Proper. This part of China was subdivided into very small parcels of land, and the communist regime has effected a major reorganization of landholding. At the same time, an enormous effort has been made to control the flood problem that has bedeviled the Huang He Basin for uncounted centuries, and to expand the land area under irrigation.

The North China Plain is one of the world's most heavily populated agricultural areas, with about 1400 people per square mile (540 people per square kilometer) of cultivated land (Fig. 9–14). Here the ultimate hope of the Beijing government lay less in land redistribution than in raising yields through improved fertilization, expanded irrigation facilities, and the more intensive use of labor. A series of dams on the Huang He now somewhat reduces the flood danger, but outside the irrigated areas the ever-present problem of rainfall variability and drought persists. The North China Plain has not produced any substantial food surplus, even under normal circumstances; so when the weather turns unfavorable, the situation

soon becomes precarious. The specter of famine may have receded, but the food situation is still uncertain in this northern part of China Proper.

The two major cities of the Huang He Basin are the historic capital, Beijing, and the port city of Tianjin on the Bohai Gulf, both positioned near the northern edge of the North China Plain. In common with many of China's harbor sites, that of the river port of Tianjin is not particularly good; but this city is well situated to serve the northern sector of the Eastern Lowlands, the nearby capital, the upper Huang Basin, and Nei Mongol (Inner Mongolia) beyond. Like Shanghai, Tianjin had its modern start as a treaty port, but the city's major growth awaited communist rule. For many decades, it had remained a center for light industry and a flood-prone harbor; after 1949, a new artificial port was constructed and flood canals were dug. More important, Tianjin was chosen as a site for major industrial development, and large investments were made in the

chemical industry (in which Tianjin now leads China), in such basic industries as iron and steel production, in heavy machinery manufacturing, and in the textile industry. Today, with a population of 4.9 million, Tianjin is the center of the country's third-largest industrial complex, after the Northeast's Liaoning Province and its old competitor, Shanghai.

Beijing (5.8 million), on the other hand, has chiefly remained the political, educational, and cultural center of China. Although industrial development also occurred here after the communist takeover, its dimensions have not been comparable to Tianjin. The communist administration did, however, greatly expand the municipal area of Beijing, which is not controlled by the province of Hebei (Hopeh), but is directly under the central government's authority. In one direction, Beijing was enlarged all the way to the Great Wall—30 miles (50 kilometers) to the north—so that the "urban" area includes hundreds of thousands of farmers.

Just as Shanghai symbolizes bustling, crowded, industrial, trade-oriented China, so Beijing represents Chinese history, power, and order. Facing Beijing's main thoroughfare are the Forbidden City—the treasure-laden preserve of the rulers and elite of old—and, directly opposite, the starkly contrasting Tiananmen Square flanked by modern buildings, the (now bloodstained) scene of mass assemblages of millions during the communist era. The great Gate of Heavenly Peace (Fig. 9–15) overlooks both ancient city and modern square, and here in 1949 Mao Zedong proclaimed the birth of the People's Republic of China. Beijing's relative location is also remarkable. Not only does the city lie close to the northern margin of the North China Plain and not far from the Great Wall and the Ming

FIGURE 9–14 A densely packed farm village on the overcrowded North China Plain just south of Beijing. Many of these settlements became individual communes during the reorganization of China's social order in the early years of communism.

FIGURE 9–15 The Gate of Heavenly Peace—scene of Mao's famous 1949 proclamation—on another historic occasion: the raising of the anti-government "Goddess of Liberty" statue on May 31, 1989, five days before the Red Army savagely crushed the student occupation of Tiananmen Square.

crosses the Loess Plateau. This region is named after the powdery, wind-blown deposit that forms its unusual landscape. This material probably had its origin in the rocks pulverized by the Late Cenozoic glaciers that advanced into northern Eurasia as they pushed into North America's Midwest (see pp. 12–13). Persistent winds accumulated a mantle of loess as much as 250 feet (75 meters) thick, burying the pre-existing landscape. Loess is a very fertile material; but unlike ordinary soil, its fertility does not decrease with depth. It also is easily excavated, and people have lived in hollowed-out "caves" for thousands of years. The landscape of the Loess Plateau (Fig. 9–16) reflects intensive cultivation and above-ground as well as underground settlement by a dense population. The latter entails risks: when earthquakes strike the Loess Plateau, the sub-surface dwellings quickly collapse, and over the centuries countless inhabitants have lost their lives this way.

In the past, the Huang River often posed a deadly threat by breaking through its levees and flooding its densely peopled, low-lying alluvial plain, drowning millions. The lower course, with dozens of distributaries, swung wildly across the North China Plain, sometimes draining into the Bohai Gulf, then shifting to an outlet south of the Shandong Peninsula. Massive amounts of yellowish silt pour from the interior (hence *Yellow* River); approaching China's coast by air, one can also see that silt coloring the *Yellow* Sea dozens of miles from shore. Over time, the Chinese have built extensive systems of dikes, dams, and artificial levees, and the threat of flooding is not as great as it was a century ago. Still, "China's Sorrow," as the Huang He is sometimes called, continues to pose a risk to its

Tombs, but China's greatest archaeological discovery, Peking Man, also was made near the city. These fossil human remains prove that communal life, organized hunting, and the use of fire existed at the present site of the city as long as 350,000 years ago.

As Fig. 9–9 shows, the Huang River rises on the Qinghai-Xizang (Tibetan) Plateau and, before reaching the Eastern Lowlands and the North China Plain, forms a huge clockwise bend that almost encircles the Ordos Desert. Downstream from the last, the Huang

FIGURE 9–16 The distinctive agricultural landscape of the highly fertile Loess Plateau, located in the Huang He's semiarid middle basin.

neighbors. As recently as 1983, its undrinkable, unnavigable, unpredictable waters claimed a large number of flood victims.

North China is not all farmland and villages, however. A dense, heavily used railroad network connects market centers, ports, industrial complexes, and raw material sites. Beneath the loess of the Loess Plateau lie sizable coal deposits and oilfields (Fig. 9–17). Taiyuan (2.0 million) in Shanxi (Shansi) Province is the site of a major iron and steel complex, machine manufacturing plants, and chemical industries. The Beijing-Tianjin area, as already noted, is one of China's leading manufacturing clusters. Additionally benefiting from its relative location at the center of China's heartland, North

China is the country's political as well as economic focus.

The Xi Basin and South China

China's southernmost major river is actually called the West (Xi) River, but neither in length nor in terms of the productivity of the region through which it flows is this river comparable to the Huang or the Chang. In this Xi Jiang Basin, there is much less level land. Moreover, in the hills and mountains of South China, there are millions of local indigenous people not yet acculturated to the civilization of the Han Chinese (Fig. 9–18). Hence, the western part of the Xi Basin consists largely of an

area designated for these ethnic minorities, the Guangxi Zhuang Autonomous Region. In the east lies prosperous Guangdong (Kwangtung) Province, the populous delta of the Xi River, and Guangzhou (Canton), the south's largest and most important city (3.4 million).

Between the deltas of the Xi and Chang rivers lie the Southeast Uplands, a region of rugged relief which for a long time in Chinese history remained a refuge for southern tribes against the encroachment of the people of Han from the north. With its steep slopes and narrow valleys, this region has very little agricultural potential, although a massive land development program was begun here in the 1960s. For many years,

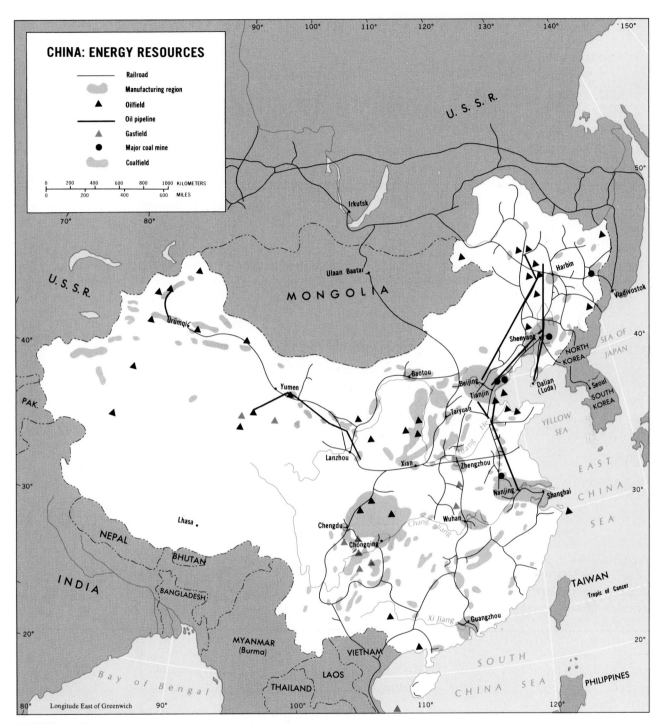

CHINA: ENERGY RESOURCES

- —— Railroad
- Manufacturing region
- ▲ Oilfield
- —— Oil pipeline
- ▲ Gasfield
- ● Major coal mine
- Coalfield

FIGURE 9–17

this has been one of China's most outward-looking regions, with a considerable emigration (via Guangzhou) to the Philippines and Southeast Asia, substantial overseas trade in the leading commercial product (tea), and a seafaring tradition for which the people of Fujian (Fukien) Province have become famous.

The regional geography of southern China stands in sharp contrast against the north. Whereas northern China is the land of the ox and even the camel, southern China uses the water buffalo.

Northern China grows wheat for food and cotton for sale; South China grows rice and tea. North China has a rather dry, continental climate; southern China is subtropical and humid. Northern China has long looked inward to interior Asia, whereas South China

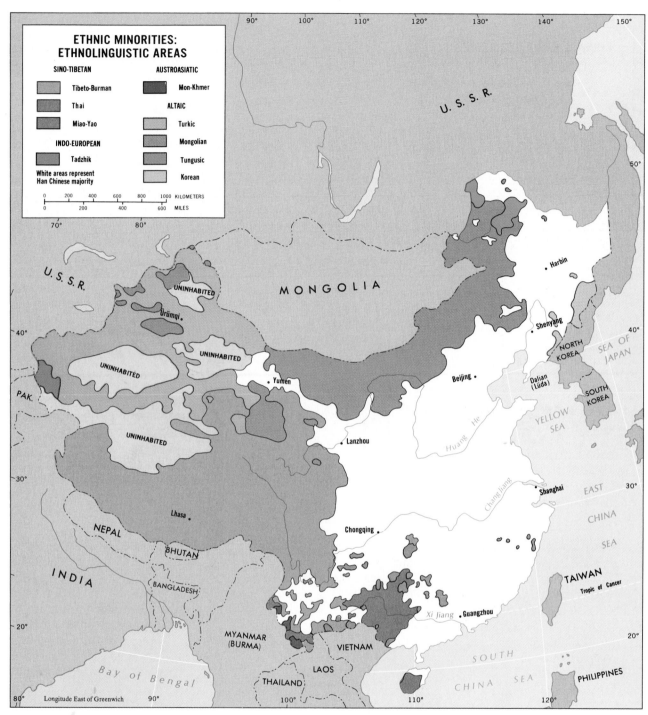

**ETHNIC MINORITIES:
ETHNOLINGUISTIC AREAS**

SINO-TIBETAN

Tibeto-Burman

Thai

Miao-Yao

INDO-EUROPEAN

Tadzhik

White areas represent
Han Chinese majority

AUSTROASIATIC

Mon-Khmer

ALTAIC

Turkic

Mongolian

Tungusic

Korean

FIGURE 9–18

has oriented itself outwardly to the sea and even to lands beyond. With its highly diverse ethnic population and multilingual character, southern China carries strong Southeast Asian imprints (Fig. 9–18).

The physiographic map (Fig. 9–2) suggests that the Xi Jiang is no Chang, and this impression is soon verified. Not only is the Xi much shorter, but for a great part of its course it lies in mountainous or hilly terrain. West of Wuzhou

the upper valley is less than a half-mile wide, and, confined as it is, the river is subject to great fluctuations in level. Except for the delta, whose 3300 square miles (8500 square kilometers) support a population of almost 20 million, there

FIGURE 9–19   The hills of subtropical South China are frequently clothed in human-made terraces that expand the available agricultural space. These stepped ricefields are located in Guizhou Province southeast of Sichuan's Red Basin.

is little here to compare with the lower Chang Basin. Cut by a large number of distributaries and by levee-protected flood canals, the Xi Delta is South China's largest area of flat land and the site of its largest population cluster; here too lies Guangzhou (Canton), the south's leading city.

Subtropical and moist, southern China provides a year-round growing season. In the lower-lying areas, rice is double-cropped. One planting takes place in mid- to late winter, with a harvest in late June or shortly thereafter; a second crop is then planted in the same paddyfield and harvested in mid-autumn. It is even possible to raise some vegetables or root crops between the rice plantings! As in the

uplands surrounding the Sichuan Basin, whole areas of hillside have been transformed into a multitude of artificial terraces (Fig. 9–19); here, too, rice is grown, although the higher areas permit the harvesting of only a single crop each year. But again there is time for a vegetable crop or some other planting before the next rice is sown. Fruits, sugarcane, tea, corn—wherever farming is possible, southern China is tremendously productive, and the range of produce is almost endless. If only there were more level land and fewer people: this is still one of China's food-deficient regions, always requiring imports of grain.

The problems of southern China also relate in part to the distribu-

tion of the region's ethnic minorities (Fig. 9–18). These indigenous population groups are unlike the people of Han, and here in the south China takes on an almost Southeast Asian character. The minority groups are both large and numerous, and they have been somewhat exempt from certain Chinese regulations (for example, those limiting family size). As a result, natural rates of population increase are the highest of all China's regions, land pressure is enormous, and environmental deterioration (especially soil erosion) is a serious problem. Small wonder that the south cannot feed itself.

With its 3.4 million people, Guangzhou (Canton) is the urban focus for the entire South China region. Despite its steadily narrowing valley, the Xi Jiang is navigable for several hundred miles (Fig. 9–11), so that hinterland connections are effective. Hong Kong (see *box*), just over 100 miles (160 kilometers) away on the coast, overshadows Guangzhou in size as well as trade volume. Guangzhou was allowed to fade as a manufacturing center during the first three decades of communism, but the 1980s witnessed a steady revival that resulted in some spectacular growth rates by the final years of the decade (undoubtedly propelled by proximity to the external capitalist influences of Hong Kong). In fact, the entire social atmosphere in Guangzhou and southern China also differs from that of Beijing and the north. Compared to orderly Beijing, Guangzhou is lively, open, even boisterous, reminiscent of the Canton of old; and a myriad of foreign goods can be found on burgeoning black markets from Guangzhou to Kunming (1.7 million).

The Northeast

As Fig. 9–9 shows, China's heartland also extends into the Northeast (which was defined in the *box* on p. 504). Physiographically, the Northeast resembles eastern China in some respects but not in others. The Liao-Songhua Lowland, where most of the people and activities are concentrated, does not compare to the great basins of the Chang, Huang, or Xi; it is essentially an erosional plain rather than a depositional basin. The relief of the adjoining Northeast Uplands, which border North Korea, is quite similar to that of the Southeast Uplands lying between the Chang and Xi rivers. What really makes the Northeast's landscape look different is the colder, more wintry climate of this, China's

HONG KONG

FIGURE 9–20 Panoramic view of crowded Hong Kong, seen from atop Victoria Peak on Hong Kong Island, looking northeast across the harbor toward the burgeoning Kowloon Peninsula.

The mouth of China's southernmost major river, the Xi Jiang, opens into a wide estuary below the city of Guangzhou (Canton). At the head of this estuary, on the northeast side, lies the British crown colony of Hong Kong (Fig. 9–11 inset). Hong Kong consists of three parts: the island of Hong Kong (32 square miles/82 square kilometers), the Kowloon Peninsula on the mainland opposite this island, and, adjacent to the west, the so-called New Territories. The total area of this British dependency is only just over 400 square miles (1000 square kilometers), but the population is very large; in 1991 it had reached 5.8 million, 98 percent Chinese in its ethnic composition. Hong Kong Island and the Kowloon Peninsula were ceded permanently by China to Britain in 1841 and 1860, respectively, but the rest of the New Territories were leased on a 99-year basis in 1898.

With its excellent deep-water harbor, Hong Kong is the major entrepôt of the western Pacific between Shanghai and Singapore. Undoubtedly, the Chinese could have recaptured the territory during the successful campaign of the late 1940s, but both Beijing and London saw advantages in maintaining the status quo. During the period of isolation and communist reorganization, Hong Kong provided the People's Republic of China with a convenient place of contact with the Western world without any need for long-distance entanglement. Even when hundreds of thousands of Chinese from Guangdong and Fujian provinces fled to Hong Kong and the city was a place of rest and recuperation for American forces during the Indochina War, the Chinese government continued to tolerate the

British colonial presence. Indeed, the colony depends on China for vital supplies, including fresh water and food. After Japan, China is Hong Kong's chief source of imports.

The colony is extremely crowded. Hong Kong Island, where the capital (Victoria) is located, is beyond the saturation point, and urban sprawl now extends into the Kowloon Peninsula where high-rise apartments continue to go up (Fig. 9–20). Until the early 1950s, Hong Kong was a trading colony; then the Korean War and the U.N. embargo on trade with China cut the dependency's connections with its hinterland, and an economic reorientation was necessary. This came about in only a few years, as a huge textile industry and many other light manufacturing industries developed. Today, textiles and fabrics make up over one-third of the exports by value, and the volume of electrical equipment and appliances is growing. These consumer goods go to the United States, China, Germany, the United Kingdom, and Japan. In the case of neighboring China, Hong Kong also accounts for about 80 percent of all foreign investment and one-fourth of the total external trade of the People's Republic, making it the major Chinese economic gateway to the outside world.

The crucial politico-geographical issue now involving Hong Kong is the approaching end, in 1997, of British control over the colony. After long and difficult negotiations during the mid-1980s, the British government and China's leaders agreed on a formula for transition to Chinese rule. When this plan was approved, real estate prices declined, Hong Kong's stock market crashed, and its dollar plunged, but confidence was soon restored. Not only did China promise that Hong Kong's way of life would be allowed to continue unchanged for 50 years after 1997, but Hong Kong seemed to realize its own strength once again. After all, the lion's share of all foreign investment in China belongs to Hong Kong, and the colony accounts for a substantial proportion of China's foreign trade. Visions of Hong Kong as the future New York of China—its banking, financial, and trading center—caused a new boom. Property values rose and exceeded pre-crash levels; the stock market rose to record heights; the Hong Kong dollar rose to its highest value ever.

Of course, the Beijing government's crushing of the pro-democracy movement in 1989 triggered another cycle of decline, one that may not abate as the early 1990s unfold. But as 1997 looms ever larger, China's evolving new relationship with Hong Kong may have consequences far beyond the confines of the colony. If China's promise is fulfilled, future leaders of Taiwan may look anew at Chinese offers of reunification under special conditions. It is also worth noting that the Portuguese have followed Britain's lead and negotiated the return of their tiny colony of Macao (located across the Xi estuary from Hong Kong) to the Chinese, scheduled for 1999. As the colonial era comes to a close, the world will be watching closely to see if China's deeds match its words.

northern frontier. The Liao-Songhua Lowland and the Northeast Uplands converge southward on the silt-plagued Liaodong Gulf, where Dalian (Lüda) is the port city near the tip of the adjacent Liaodong Peninsula.

In terms of economic geography, there can be no mistaking the vital national role of the Northeast (*Dongbei* in Chinese). With the elimination of foreign interference here after 1949, massive Chinese immigration into the region and vigorous development of its considerable resources have combined to make this an integral and crucial part of the People's Republic. Today, the approximately 10 percent of China's population that resides in the Northeast produces almost 25 percent of the country's heavy industrial goods and one-sixth of its total manufactures by value.

Northeast China has lengthy, and in places sensitive, foreign boundaries. North Korea lies to the southeast, and the Soviet Union to the east and north; Chinese Nei Mongol (Inner Mongolia) and the **buffer state** of Mongolia lie to the west (Fig. 9–21). As this map shows, the shortest route from Vladivostok to the Soviet heartland is right across this region via Harbin, China's northernmost large city. This is the rail corridor in which the Soviets long had an interest, finally relinquished by treaty in 1950. With the heightened tension between the two countries, the Soviets have come to rely on the route skirting the Chinese border, via Khabarovsk—a trip that is 300 miles (480 kilometers) longer.

Japan also had imperial designs on the Northeast, but the aftermath of Japan's defeat in World War II was the confirmation of Chinese hegemony in the region. The Japanese contributed to the legacy by leaving behind railroads, factories, and other infrastructural elements for the region's development. The Soviets in the late 1940s

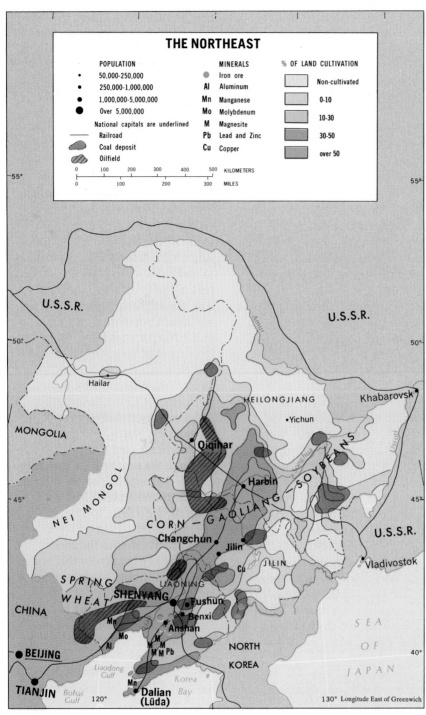

FIGURE 9–21

the core area of China Proper. The lowland axis formed by the basins of the Liao and Songhua rivers extends from the Liaodong Gulf northeastward for over 500 miles (800 kilometers) to the Amur Valley on the Soviet border. Although the growing season here is a great deal shorter than in most of lowland China Proper, it is still long enough in the Liao Basin to permit the cultivation of grains such as spring wheat, soybeans, gaoliang (sorghum), barley, and corn. Although the Northeast has drought problems and severe winters, its southern zone contains some excellent fertile soils. Toward the northern margins, extensive stands of oak and other hardwood forests on higher slopes have great value in timber-poor China.

Han Chinese have been moving into the northeastern provinces for centuries, driven by hunger and the devastating Huang He floods, dislocated by the ravages of war, and attracted by the region's opportunities of land and space. Since 1949, the settlement and development of the Northeast has been promoted by official policy, and, as we noted, this region has become one of China's leading industrial complexes. The largest city, Shenyang, has a population of 4.3 million and has emerged as the Chinese Pittsburgh. All this is based on large iron ore and coal deposits, which—like the best farmland—lie concentrated in the Liao Basin (Fig. 9–21). Shenyang is located centrally amid these resources: there is coal 100 miles (160 kilometers) to the west and less than 30 miles (50 kilometers) to the east of the city (at Fushun), and iron ore is found about 60 miles (100 kilometers) to the southwest near Anshan. Since the ore is not particularly high in iron content, coal is hauled from Fushun (and also from northern China Proper) to the site of the iron reserves; this is the cheapest

removed machinery, railroad rolling stock, and a large quantity of hoarded gold; China controlled an area of only 1.2 million people, but with enormous economic possibilities. Today, however, the three Northeast provinces of Heilongjiang, Jilin, and Liaoning contain just over 100 million people.

The geographic layout of the Northeast is such that the areas of greatest productive capacity and largest population lie in the south, where they form an extension of

way to convert these low-grade ores to finished iron and steel, and Anshan (1.5 million) has now become one of China's leading steel-producing centers. But Shenyang remains the Northeast's biggest and most diversified industrial city (Fig. 9–22). Machine fabrication plants and other engineering works in Shenyang supply the entire country with various types of equipment, including drills, lathes, machine tools, and the like; moreover, as the economic capital of northeastern China's most productive farming area, this metropolis is also an agricultural processing center of national importance.

The southern part of the Northeast, however, is not the only area that has coalfields and iron ores. From near the North Korean border (where the iron ore is of higher quality) to the far northeast, where the coal is of best quality, the region possesses natural resources in abundance: aluminum ore, molybdenum, lead, zinc, and limestone. As Fig. 9–17 shows, the Northeast is also very well endowed with pe-

troleum deposits; the largest reserves are found in the giant Daqing field between Harbin and Qiqihar (Fig. 9–21), which produces nearly one-half of China's entire annual crude oil output. Reflecting the northward march of development in the Northeast, the railway crossroads of Harbin has now become a city of 2.7 million, with large machine manufacturing plants (especially farm equipment), a wide range of agricultural processing factories, and such industries as leather products, nylon, and plastics. Not only does Harbin lie at the convergence of five railroads, but it also lies at the head of navigation on the Songhua River, connecting the city to the towns of the far northeast.

Communist doctrine in this Northeast frontier zone has always been somewhat relaxed. Wages have consistently been higher than in China Proper in order to attract the skills that developing industries needed; in the 1980s, economic reforms in industrial organization were pioneered here. In

terms of planning, the rebuilt industries of Shenyang, Anshan, and Harbin, in some instances, are models of everything the communist regime would like them to be; schools, apartments, recreational facilities, hospitals, and even old-age homes are all part of the huge industrial complex—and having a job means that a worker has access to all of these. But most important, in the northeastern provinces large-scale, efficient agricultural development stands in contrast to the parceled chaos of old China. Here the farms have been laid out more recently; they have always been larger, more effectively collectivized, and more heavily mechanized. In many ways, then, the Northeast is the image in which economic planners would like to remake all China.

Uplands and Basins of Inner Asian China (Regions ②, ③, ④)

Our survey of China's regional geography now turns from the populous, intensively-utilized east to the sparsely settled western and northwestern interior—a frontier land of rugged terrain and deserts, unsettled political boundaries, and spotty and tenuous human occupance. This Inner Asian China covers a huge territory—some 1.3 million square miles (3.3 million square kilometers), an area bigger than all of India—but contains fewer than 50 million of China's 1.1 billion-plus people. As Fig. 9–9 (p. 510) shows, Inner Asian China consists of the country's three remaining physiographic divisions (which also coincide with the human regional pattern): the Plateau Steppe of Mongolia (Region ②); the High Plateaus and Mountains of Qinghai-Xizang (Region ③); and the Desert Basins of Xinjiang (Region ④).

FIGURE 9–22 A few of the more than 10,000 factories that dot Shenyang, the Northeast's leading manufacturing center. This scene is also a classic example of the bleakness and pollution that mark China's contemporary industrial landscapes.

China were different, this program was launched using doctrines applied earlier in the Soviet Union. The idea was to expropriate all farmland from its owners and redistribute it equitably, but far more of China's peasants sold produce on the market than had been the case in the U.S.S.R.; thus, the risk that collectivization would reduce production was greater.

Nevertheless, the reform program was initiated almost at once in every village of Han China (the ethnic minorities were at first excluded) and involved the redistribution of the land of the landlords among all the landless families of the villages. Allocations were based on the number of people per family and was done strictly according to these totals. The landlords themselves were included and received their proportionate share of what had been their own land. Meanwhile, the villagers who were not landlords but who did own the land they worked were allowed to keep their properties. Hence, the reform program did not produce a truly egalitarian situation.

Collectivization

During the early 1950s, therefore, the whole pattern of landownership in China was being radically changed. In response, there was an increase in agricultural output, but it was not as great as the authorities had planned. True, the taxes that the new landowners had to pay on their acquired land were much lower than those previously paid their landlords, so they had some money available to buy fertilizer and improve yields. But in the absence of any prospect of large-scale mechanization of agriculture, the Chinese leaders now began to seek a method whereby higher productivity could be

achieved. They devised a new program based on the countrywide formation of mutual-aid teams, in which all the peasants in a given community were encouraged to render assistance to each other during the planting and harvest seasons.

In 1955, the first stage of **collectivized agriculture** started. All the peasants in a village pooled their land and their labor. Compensation would be in proportion to the size of each share and to the labor one had contributed; the farmers retained the right to withdraw their land from the cooperative if they wished. In the beginning participation was slow, but within months the government pressed for greater cooperation; by coercion and often quite brutal means, compliance was raised to nearly 100 percent by 1956. In that year, the cooperatives were being turned into collective farms on which it was not the share of the land but the amount of the farmer's labor that determined returns. Thus, by 1957 virtually the whole countryside was organized into socialist collectives.

Communization

In 1958, the program of collectivization was carried one step further. In less than a year, over 120 million peasant households, most of them already organized in collectives, were reorganized into about 26,500 People's Communes containing about 20,000 people each. This massive <u>modification</u> of China's entire socioeconomic structure was intended to be the country's "<u>Great Leap Forward</u>" from socialism to true communism. Now the private landholdings of the collective sys.em were abolished. Teams of party organizers traveled throughout China, and opposition was harshly put down while the new system was being

imposed. In essence, the new communes were to be the economic, social, and political units of the Chinese communist state.

But most drastic was the way these communes affected the daily lives of the people. The adults, men and women, were organized into a hierarchy of "production teams" with military designations (sections, companies, battalions, brigades). Communal quarters were built, families were disrupted through distant work assignments, children were put into boarding schools, and households were viewed as things of the past. (Later, the Chinese Communist Party rescinded this order and allowed families to stay together.) The wage system of the collectives was changed in favor of an arrangement whereby the farmer or worker received free food and clothing plus a small salary—another step toward the communist ideal.

The impact of the commune system on the face of China was immediate and far-reaching. Workers by the thousands tackled projects such as irrigation dams and roads; fences and hedges between the lands of former collectives were torn down, and the fields consolidated. Some villages were leveled, others enlarged. New roads were laid out to serve the new system better. Schools to accommodate the children of parents in the workers' brigades were built. Each commune was given the responsibility to maintain its own budget, to make capital investments, and to pay the state a share of its income to replace the taxation formerly levied on the collectives and individual peasants.

Despite opposition, these revolutionary changes were accepted because most Chinese farmers know that some form of communal organization is necessary in their heavily overcrowded country.

Nonetheless, the commune system faltered by the early 1960s. For one thing, it was all done with too much haste, too little planning, and too little preparation for what was truly a profound change from private family living to a communal existence. Even more devastating were the severe droughts and destructive floods that plagued much of China at this time. It was enough to cause a moderation of the communization program and the restoration of such incentives as private peasant plots.

Cultural Revolution

It was natural that China's revolutionary leaders would attempt to transform agriculture before turning their attention to other programs of change. China's soil had to produce far greater yields if China's people were to be adequately fed; from China's land would also have to come the revenues to sustain industrial and technological development; on China's farms there prevailed much of the worst corruption in China's precommunist society. But Mao Zedong and his associates also wanted to rid Chinese society of its Kongzi (Confucian) traditions and prescriptions (see *box*), and to substitute the principles of communism as a philosophical basis for the new Chinese order. Furthermore, the new leaders intended to give China industrial strength based on domestic resources and generated by heavy investment in this sector.

The industrial growth of China since 1949 has been spectacular, helped at first by Soviet technicians and loans, and sustained by new mineral discoveries resulting from intensified exploration. The energy picture (Fig. 9–17) brightened as oil reserves were developed in the Northeast, Sichuan, and Xinjiang; coal reserves extend

from the Northeast to the Huang Basin to Sichuan, and Chinese production in the late 1980s ranked first in the world. In iron ore and other vital raw materials, China's domestic supplies are as good as those of the United States or the U.S.S.R. China's planners have sought to diversify the country's manufacturing base and to disperse factories into the interior away from the eastern industrial heartland; as the map shows, to a certain extent they have succeeded.

While the industrialization effort surged ahead, China experienced major difficulties in the sociopolitical arena when its leaders ordered the reorientation of the population from Kongzi (Confucian) to communist precepts. Mao Zedong, apparently fearing that the China of the 1960s and 1970s might abandon its revolutionary fervor and become "revisionist" like the Soviet Union, initiated the "Great Proletarian Cultural Revolution" in 1966 to rekindle the old enthusiasms and to recoup and conceal losses sustained during the ill-fated Great Leap Forward. The Cultural Revolution centered initially on education, and adult learning programs were launched in the communes. University professors now joined workers on assembly lines and peasants in the fields; the idea was to eliminate the old Kongzi (Confucian) distinction between mental and manual labor, and to prevent the emergence of an educated elite in the urban areas that would live in comfort off the labor of the peasants.

But serious problems arose. China's schools were closed down on the grounds that the entry system was unfair to the mass of the students and that teachers perpetuated "bourgeois" principles. In this way, millions of young people found themselves at loose ends—ready to be recruited into

the Red Guards, the paramilitary organization that was to become the heart of the renewed revolution. But just as the Great Leap Forward simply did not mobilize enough of the needed effort in China, so the Cultural Revolution failed to get the necessary commitment from the people. The Red Guards were soon fighting cadres organized to support and protect local party leaders. Additional factions were created, even the army was threatened, and the Red Guards wound up doing battle with rival pro-Mao groups. Once again, the country was badly dislocated by a wholesale revolutionary program.

Between 1967 and 1979, the Communist Party as well as the country was at times badly divided; suspicions and accusations were brought to the surface and could not easily be submerged again. After Mao's death in 1976, these divisions led to a power struggle that nearly plunged China into civil war. But worst of all, China had once again been set back by the instability the whole drawn-out affair had caused. That China was able to absorb the collectivization drive, the Great Leap Forward, and the lingering effects of the Cultural Revolution—and could still demonstrate substantial progress in agriculture, industry, military technology, and several other spheres—is testimony to the amazing capacity of this huge country and its resilient people to overcome enormous odds.

The Four Modernizations

Historical geographers of the future will undoubtedly point to the period following Mao's death in 1976 as a crucial time in the development of the contemporary Chinese state. With the power struggle resolved in favor of the pragmatic moderates, China's new rulers were able to turn their attention to the country's urgent development needs in several spheres. Premier Hua Guofeng decided to include Deng Xiaoping, previously dismissed as too "moderate," as the country's deputy premier. Deng's pragmatic approach to China's future soon made its mark on government policies, and by 1982 he became the strongest member of the leadership group.

In a country where exhortations, slogans, and other calls for action are commonplace, Deng Xiaoping introduced his plea for the "Four Modernizations." (Although Deng is often credited with this concept, it was Zhou Enlai who introduced it as long ago as 1964—only to see it engulfed by the Cultural Revolution.) This program involved (1) the rapid modernization and mechanization of agriculture, (2) the immediate upgrading of defense forces, (3) the modernization and expansion of industry, and (4) the development of science, technology, and medicine. Importantly, Deng proposed that Mao Zedong's long-prevailing policy of self-reliance be abandoned through the expansion of foreign trade, the purchase of foreign technology and machinery, and the use of foreign scientists to help China in its modernization drive. Moreover, capitalist-type incentives to spur production would be used, and education would return to practices where success in examinations rather than political attitudes and associations counted most.

It was not difficult to see in this new determination to modernize the reflection of Japan in the late nineteenth century (Japan, too, broke its long-term isolation and embarked on an intensive development drive). China's initial target year is 2000. By then, its planners hope, agriculture will be substan-

KONGFUZI (CONFUCIUS)

Confucius (*Kongfuzi* or *Kongzi* in *pinyin* spelling) was China's most influential philosopher and teacher, whose ideas dominated Chinese life and thought for over 20 centuries. But the leaders who took control in 1949 considered Kongzi ideals incompatible with communist doctrine, and the elimination of Kongzi principles was one of Mao Zedong's primary objectives. Kongfuzi left his followers a wealth of "sayings," many of which were frequently quoted as a part of daily life in China; Mao's "Red Book" of quotations was part of the campaign to erase that tradition.

Kongfuzi was born in 551 B.C. and died in 479 B.C. He was one of many philosophers who lived and wrote during China's classical (Zhou dynastic) period. The Kongzi school of thought was one of several to arise during this era; the philosophies of Daoism (Taoism) also emerged at this time. Kongfuzi was appalled at the suffering of the ordinary people in China, the political conflicts, and the harsh rule by feudal lords. In his teaching, he urged that the poor assert themselves and demand the reasons for their treatment at the hands of their rulers (thereby undermining the absolutism of government in China); he also tutored the indigent as well as the privileged, giving the poor an education that had hitherto been denied them and ending the aristocracy's exclusive access to the knowledge that constituted power.

Kongfuzi, therefore, was a revolutionary in his time—but he was no prophet. Indeed, he had an aversion to supernatural mysticism and argued that human virtues and abilities should determine a person's position and responsibilities in society. In those days, it was believed that China's aristocratic rulers had divine ancestors and governed in accordance with the wishes of those godly connections. Kongfuzi proposed that the dynastic rulers give the reins of state to ministers chosen for their competence and merit. This was another Kongzi heresy, but in time the idea came to be accepted and practiced.

His earthly philosophies notwithstanding, Kongfuzi took on the mantle of a spiritual leader after his death. His ideas spread to Korea, Japan, and Southeast Asia, and temples were built in his honor all over China. As so often happens, Kongfuzi was a leader whose teachings were far ahead of his time. His thoughts emerged

tially mechanized and self-sufficiency even in years of drought assured; the gap between China and the Soviet Union in military capacity will no longer exist; industrialization will rank China as one of the world's leading manufacturers; and the initial dependence on foreign technology and science will have lessened. These are optimis-

tic goals, but throughout the 1980s the groundwork was being laid. Foreign technicians were in China to develop mineral resources and plan new industrial projects; *Special Economic Zones* (*SEZs*) were preparing to accommodate a planned infusion of outside investment (see *box*); the need for foreign revenue was among the

from the mass of philosophical writing of his day later to become guiding principles during the formative Han Dynasty. Kongfuzi had written that the state should not exist for the pleasure and power of the aristocratic elite; it should be a cooperative system, and its principal goal should be the well-being and happiness of the people. As time went on, a mass of Kongzi writings evolved; many of these Kongfuzi never wrote, but they were attributed to him nonetheless.

At the heart of this body of literature lie the *Confucian Classics*, 13 texts that became the focus of education in China for 2000 years. In the fields of government, law, literature, religion, morality, and in every conceivable way the Classics were the Chinese civilization's guide. The whole national system of education (including the state examinations through which everyone, poor or privileged, could achieve entry into the arena of political power) was based on the Classics. Kongfuzi was a champion of the family as the foundation of Chinese culture, and the Classics prescribe a respect for parents and the aged that was a hallmark of Chinese society. It has been said that to be Chinese—whether Buddhist, Christian, or even communist—one would have to be a follower of Kongfuzi; hardly any conversation of substance could be held without reference to some of his principles.

When the Western powers penetrated China, Kongzi philosophy came face to face with practical Western education. For the first time, a segment of China's people (initially small) began to call for reform and modernization, especially in teaching. Kongzi principles could guide an isolated China, but they were found wanting in the new age of competition. The Manchus resisted change, and during their brief tenure the Nationalists under Sun Yat-sen tried to combine Kongzi and Western knowledge into a neo-Kongzi philosophy. But it was left to the communists, beginning in 1949, to attempt to substitute an entirely new set of principles to guide Chinese society. Kongzi thought was attacked on all fronts, the Classics were abandoned, ideological indoctrination pervaded the new education, and even the family was assaulted during the early years of communization. However, it is extremely difficult to eradicate two millennia of cultural conditioning in a few decades. Thus, the fading spirit of Kongfuzi will haunt physical and mental landscapes in China for years to come.

factors that led China to open its doors to hundreds of thousands of tourists. The decade, of course, ended with the political crackdown, and a major result was the slowing of the development thrust that carried beyond the opening of the 1990s.

As we have observed previously for Third World countries, high rates of population growth inhibit national economic development. China's leaders clearly recognize this relationship: a centerpiece of their plans to quadruple economic output between 1980 and 2000 is the sharp curtailment of the annual rate of natural increase. This population growth rate did come down during the 1970s—from over 3 percent early in the decade to about 1.7 percent in 1979—but the pace of reduction was insufficient to match post-1980 modernization plans. The new goals were announced in 1979: China's target population for 2000 would be 1.2 billion (the 1991 total was 1.136 billion), mandating an increase rate until then of only about 1 percent yearly. In order to meet that figure, this and the next generation of Chinese would have to restrict the sizes of their families to a single child. In 1980, a massive publicity campaign was launched (Fig. 9–25), and the new policy swiftly gained popular acceptance.

The initial results were impressive, given China's demographics and the challenge of instituting such widespread and immediate change in attitudes regarding family formation—the 1980–1985 annual increase rate was only 1.2 percent. During the second half of the 1980s, some relaxations in population policy entered the picture: certain farming and fishing families were allowed at least a second child as essential labor; rural families could seek permission for a second child if the first baby was a girl (a measure to halt the rise of female infanticide); ethnic minority groups of under 10 million were exempted. At the same time, however, evidence mounted that more and more couples were voluntarily complying with the one-child policy (better housing is available to these people as well as a second income for the working mother). Nonetheless, China is no longer on track toward meeting the 1.2 billion total at the end of the century. Planners now speak optimistically of a 2000 population totaling 1.270 billion; U.S. demographers offer a more realistic projection of 1.292 billion.

China's cultural landscapes reflect the successes as well as the failures of the modernizing drive.

FIGURE 9–25 Billboards and posters are often part of the Chinese public landscape, especially in the South. Here, in the city of Chengdu in central Sichuan Province, the message concerning official population policy is clear enough—in English as well as Chinese.

In much of rural China, to be sure, the evidence of three turbulent decades is still visible—traditional dwellings in disrepair, brick buildings of the commune period rising above the old village roofs, fading slogans on their walls. But the big story in the countryside today is the progress that China has made during the past decade, which Sylvan Wittwer and his associates (in the book *Feeding a Billion*) conclude amounts to one of the most striking examples of agricultural development in human history. By diversifying their crops and applying all sorts of new technologies, Chinese farmers increased the country's total food production by an amazing *60 percent* between 1979 and 1986. Significantly, this breakthrough could not have been accomplished without a healthy dose of capitalism: peasants contract to deliver a given amount of a crop to the state but are then free to produce as much more as they wish for sale on the open market. Moreover, although the state still owns all farmland, the communes have nearly disappeared in favor of a system of small plots tended by millions of individual farm families. In view of this remarkable productivity, China is not only achieving agricultural self-sufficiency, but may soon become a food exporter to help earn badly needed foreign revenues.

Urban China, now home to over 250 million people, is more likely to present scenes that resemble those of nineteenth-century England during the early decades of the Industrial Revolution. Pollution-belching smokestacks rise as a forest from city blocks; factories large and small vie for space with apartment buildings, schools, and architectural treasures of the past (Fig. 9–22). There seem to be no zoning regulations, nor any protection for the environment. Turn a street corner, and the deafening roar of machines drowns out everything else; walk along the street and as the roar fades another earsplitting sound replaces it. Across the street is a school, and one wonders how students can learn under such conditions day after day. Gaze upward and the sun is shrouded by an ever-present cloud of pollution so thick that the air is dangerous to breathe. A bridge ahead brings another vivid portrait of the cost of China's industrial march: the water below is little more than an industrial sewer, filled with waste and colored gray by tons of toxic chemicals.

All this must be seen against the reality of a China that in 1990 was still far from being a developed

SPECIAL ECONOMIC ZONES

In conjunction with the Four Modernizations program, the Chinese government in 1979 initiated a new "open-door" policy in order to attract technology transfers and substantial investments from abroad. A key element in this policy was the creation of **Special Economic Zones (SEZs)**, manufacturing and export centers that would lure foreign capital and factories through special tax benefits and other economic incentives. Four SEZs were initially established, all on the southeastern coast: Shenzhen, Zhuhai, Shantou, and Xiamen (Fig. 9–11). To date, only Shenzhen—which adjoins Hong Kong—has elicited a major response, but developments there point the way to future progress.

Shenzhen's success and vigorous recent growth as a Special Economic Zone is undoubtedly due to its location just across the Chinese border from Hong Kong (Fig. 9–11, inset map). A sleepy fishing village of 20,000 in 1980, Shenzhen a decade later is a burgeoning modern metropolis of more than 1.1 million (perhaps the world's fastest-growing), with a rapidly expanding skyline that includes China's tallest building. More than 3000 agreements have been signed with foreign businesses since 1980, generating hundreds of millions of dollars (U.S.) in investments—about one-third of the total foreign investment that flowed into the People's Republic between 1980 and 1989. At least 800 companies have begun production, mainly of labor-intensive industrial goods such as electronic appliances and toys, capitalizing on significantly lower Chinese wage rates. Although these manufactures have not yet generated much export income, Shenzhen is nonetheless thriving, thanks to its booming retailing activities and tourism. Today, despite the 1989 political crackdown and its aftermath, Shenzhen's planners are optimistic about the zone's continuing growth and are supervising the development of hotels, roads, telecommunications, banks, and other facilities that will make this ultramodern showcase city more attractive to the international business community and tourists.

But the SEZ incentive has not had all the results the Chinese government had hoped for. Of the more than 20 sites chosen for this kind of development, all along the eastern seaboard, few have thus far shown signs of takeoff. To spur the flagging program along, Beijing's leaders in 1988 declared the entire near-tropical island of Hainan (Fig. 9–11) the newest SEZ, revealing ambitious plans for creating a free-market economy, a huge new city, a high-tech industrial complex, a duty-free port, and world-class resort facilities—in effect, a second Hong Kong. After a promising start, however, the political shift of mid-1989 thwarted development as both the government and foreign investors began to have second thoughts. Therefore, through the early 1990s, Shenzhen is the only SEZ performing up to original expectations, obviously owing its success to its adjacency to China's major economic gateway from the non-communist world.

country. Annual income per person in recent years was about U.S. $300—lower than the level of poverty-stricken Haiti. Urbanization hovered around 22 percent; more than 75 percent of the massive labor force worked in agriculture. Such statistics, of course, underscore the magnitude of the effort on which the new China is now embarked. And overshadowing all is the political redirection that took hold in 1989, when the hardliners prevailed and the pro-democracy movement was crushed. If the goals of the Four Modernizations program are to be reached around the turn of the century, that status quo cannot persist very far into the 1990s.

TAIWAN NOT

Less than 125 miles (200 kilometers) off the southeastern coast of China lies Taiwan, the island where the defeated Chiang Kai-shek and the remainder of his Nationalist faction took refuge in 1949. In all, about 2 million refugees reached Taiwan between the end of World War II and the mainland revolution four years later. Assisted by the United States, Taiwan has survived as "Nationalist China" on the very doorstep of the People's Republic.

Mountainous Taiwan is one island in a huge arc of islands and archipelagoes that stretches along mainland Asia's eastern coast from north of Japan to Indonesia in the far south. Unlike many of the other islands, Formosa (as the Portuguese called this island when they tried to colonize it) is quite compact with rather smooth coastlines and few good natural harbors. In common with several of its northern and southern neighbors, the island's topography has a linear, north-south orientation and a mountainous spine. In Taiwan, this backbone lies in the eastern half of the island (Fig. 9–26), and elevations in places exceed 10,000 feet (3000 meters). Eastward from this forested mountain spine, the land surface drops rapidly to the coast and there is very little space for settlement and agriculture; but westward there is an adjacent belt of lower hills and, facing the Taiwan Strait, a substantial coastal plain. The overwhelming majority of Taiwan's 20.4 million inhabitants live in this western zone, near the northern end of which is the capital, Taipei (6.7 million).

There have been times when Taiwan was under the control of rulers based on the mainland, but history shows that its present separate status is really nothing new. Known to the Chinese since at least the Sui Dynasty, the island was not settled by mainlanders until the 1400s. For a time there was intermittent Chinese interest in Taiwan, but then the Portuguese arrived. During the 1600s, there was conflict here among the Europeans, including the Dutch and the Spaniards; however, in 1661 a Chinese general landed with his mainland army and ousted the foreign invaders. This was during the rise of the Manchus, and, much like the Nationalists of the 1940s, thousands of Chinese refugees fled to Taiwan rather than face Manchu domination. But before 1700, the island fell to the Manchu victors; by then, the indigenous population of Malayan stock was already far outnumbered and in

FIGURE 9–26

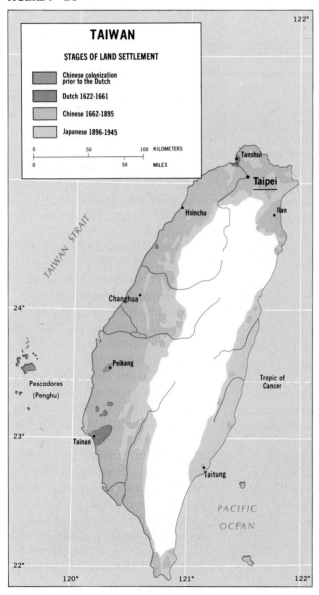

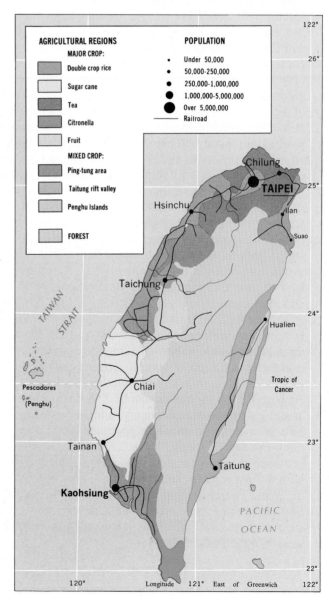

the process of retreating into the highlands.

Taiwan did not escape the fate of China itself during the second half of the nineteenth century. In 1895 it fell to Japan, and for the next half-century it was under foreign rule. The Japanese saw more of the same possibilities in Formosa that they saw in Northeast China: the island could be a source of food and raw materials as well as a market for Japanese products. To ensure the latter, Japan launched a prodigious development program in Taiwan, involving road and railroad construction, irrigation projects, hydroelectric schemes, mines (mainly for coal), and factories. The whole island was transformed. Crop yields rose rapidly as the area of cultivated land was expanded and better farming methods were introduced.

Japan's rule ended with its collapse at the end of World War II. After 1945, the island was briefly returned to Chinese control, but before long it became the last stronghold of the Nationalists and again its mainland connections were severed. A large influx of refugee Nationalists arrived during the late 1940s, and they were fortunate; here was one of the few parts of China where they could have found a well-functioning economy, productive farmlands, the beginnings of industry, good communications—and the capacity to absorb a large immigrant population.

The year 1949 brought a new beginning with the installation of Chiang Kai-shek's regime in Taipei and the arrival of the mainland immigrant horde, which constituted one-fifth of the total population in 1950. In some ways, the problems faced by Chiang's regime were similar to those of the new mainland government. The Japanese had achieved much, but the war had brought destruction; and, as on the nearby continent, there

was a need for land reform and increased agricultural yields. While U.S. assistance helped reconstruct Taiwan's transport network and industrial facilities, the Taipei government initiated a highly successful program of agricultural reform. Through economic incentives, seed improvement, expanded irrigation, more fertilizers, and new double-crop rotations, it has been possible to increase the per acre yields of Taiwan's major crops by a factor of more than two since 1950. Rice is the leading staple, and about two-thirds of the harvest is from double-cropped land (Taiwan is at the same subtropical latitude as the mainland's Xi Basin); wheat and sweet potatoes are also important staples, and sugarcane in the lower areas and tea in the uplands are grown for cash (Fig. 9–26, right map).

In contrast to heavily rural China, more than 70 percent of Taiwan's inhabitants live in urban areas. The Japanese began an industrialization program that temporarily faltered during the Nationalist takeover, but the 1960s

witnessed the island's industrial resurgence. By the mid-1980s, agriculture accounted for only 20 percent of Taiwan's labor force and per capita income was over U.S. $3000 per year, statistics no other large Asian country could match (except, of course, Taiwan's present-day model— Japan). All this has been accomplished despite the limitations of Taiwan's raw material base. Energy sources, however, are ample. Along the western flank of the mountain backbone there are coal deposits, and numerous streams on this well-watered isle provide good opportunities for hydroelectric power development; there even is some oil and natural gas. But the textile industry, one of Taiwan's major foreign revenue earners, depends on imported raw cotton; the aluminum industry gets its bauxite from Indonesia. Nevertheless, since 1970 Taiwan's planners have been successfully moving toward the establishment of a domestic iron and steel industry, nuclear power plants, shipyards, a larger chemical industry, and

FIGURE 9–27   A busy street in downtown Taipei. Taiwan's capitalism has produced a cityscape that differs radically from that of the communist People's Republic.

an electrified and more comprehensive railroad network.

In the 1990s, Taiwan's urban-commercial economy is advancing steadily. Taipei, the capital city, contrasts strongly with cities of similar size in the People's Republic; busy traffic, abundant consumer goods, and many other aspects of Taiwan's capitalist economy produce a strikingly different atmosphere and urban landscape (Fig. 9–27). The newest effort is to shift the Taiwanese economy away from labor-intensive manufacturing toward today's high-technology industries, particularly personal computers, telecommunications, and precision electronic instruments. To that end, such projects as the Science-Based Industrial Park in the city of Hsinchu (Taiwan's Silicon Valley, an hour's drive from Taipei) are being given the highest development priority. In fact, the western Pacific rim is now often described as the domain of the **Four Tigers of the Orient**—Taiwan, Hong Kong, South Korea, and Singapore—as these burgeoning beehive countries successfully follow the Japanese example with impressive annual size increases in their economies that have recently ranged up to 10 percent.

In the international arena, Taiwan's position has become more difficult as a result of China's emergence from its isolation. Taiwan was expelled from the United Nations to make way for the People's Republic; the United States, Taiwan's vital ally, closed its embassy, and since 1979 has been under pressure to lower its commitments to Taipei in favor of a more productive relationship with Beijing. In 1976, the Canadian government refused to permit Taiwan to participate in the Montreal Olympic Games. But Taiwan remains, in effect, a U.S. protectorate, a bastion of support for the American presence in the

western Pacific. At any rate, mainland China has opposed a policy wherein the United States would recognize both Beijing and Taipei, so that a two-China policy is no solution to this persistent problem.

KOREA

The peninsula of Korea extends from the East Asian mainland toward Japan (Fig. 9–28) and, not surprisingly, has experienced a

FIGURE 9–28

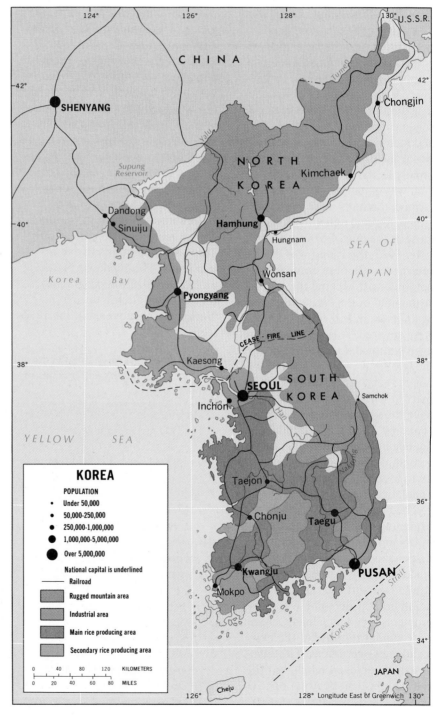

turbulent history. For uncounted centuries it has been a marchland, a pawn in the struggles among the three powerful countries that surround it. Korea has been a dependency of China and a colony of Japan; when it was freed of Japan's oppressive rule in 1945, it was divided for administrative purposes by the victorious allied powers. This division gave North Korea above the 38th parallel to the forces of the Soviet Union, and South Korea to those of the United States. In effect, Korea traded one master for two new ones. The country was never reunited because North Korea immediately fell under the communist ideological sphere, and South Korea, with massive American aid, became part of East Asia's non-communist perimeter (of which Japan and Taiwan are also parts). Once again, it was the will of external powers that prevailed over the desires of the Korean people.

In 1950, North Korea sought to reunite the country by force and invaded South Korea across the 38th parallel, an attack that drew a U.N. military response led by the United States. This was the beginning of the devastating Korean War (1950–1953) in which North Korea's forces pushed far to the south, only to be driven back into their own half of Korea almost to the Chinese border, whereupon the Red Chinese army entered the war to drive the U.N. forces southward again. Eventually, a cease-fire was arranged in 1953, but not before the people and the land had been ravaged in a way that was unprecedented even in Korea's violent past. The heavily guarded boundary between North and South, demarcated in 1953, remains unchanged today, an ever-present symbol of divided Korea.

The partitioning of Korea into two political units happens to coincide closely with the regional division any physical geographer might suggest as an initial breakdown of the country. As Fig. I–9 shows, South Korea is moister than the North; Fig. I–10 emphasizes the temperate maritime climatic conditions that prevail over most of the South as opposed to the more extreme continental character of the North. Figure I–13 shows much of the Korean peninsula to lie under mountain soils, but the most productive soils are found in a zone that lies largely in South Korea. And Fig. I–14 shows that most of Korea's 68 million people are concentrated in the western part of the country, an area that widens southward so that the majority of the population is in South Korea, roughly within a triangle to the west of a line drawn from Seoul to Pusan.

Thus, a great number of contrasts exist between North and South Korea. The North is continental, the South peninsular; the North is more mountainous than the South. The North can grow only one crop annually and depends on wheat and millet, whereas in much of South Korea multiple cropping is possible and the staple is rice. The North has significantly fewer people than the South (23.6 million against 44.2 million), but the North has a food deficit; the South comes close to feeding itself, and even enjoys occasional surpluses. Another contrast lies in the distribution of Korea's raw materials for industry: North Korea has always produced more coal and iron ore than the South, and the majority of all other mineral resources also comes from the North. Similarly, North Korea has maintained a great lead in hydroelectric power development, an advantage initiated by the Japanese that has been sustained. In recent years, several discoveries of coal and iron have been made in South Korea, but the overall balance in bases for heavy industry remains strongly in favor of North Korea.

Although it is practically impossible to obtain data concerning North Korea's external trade, one thing is certain: the superimposed political boundary between the two Koreas has been virtually airtight, and no trade has passed across it. North Korea's trading connections have been mainly with China and the Soviet Union, and those of South Korea with the United States, Japan, and Western Europe. Yet it must be clear from what has just been said that North Korea, with greater advantages for heavy industry on the peninsula, could transact much business with less favored South Korea. The two Koreas are potentially interdependent in so many ways; even within the industrial sector itself there are **regional complementarities**. Although North Korea specializes in heavy manufacturing, lighter industries such as cotton textiles and food processing still prevail in South Korea—but heavy industries are now developing (especially around Pusan) since the discovery of iron ores and Samchok's coal (Fig. 9–28). North Korea's chemical industries produce fertilizers; South Korean farms need them. North Korea has much electric power; the transmission lines that used to carry it to the South were severed right after the postwar division of the country.

South Korea also has the largest part of the domestic market as well as the biggest cities, the old capital of Seoul (at 16.8 million, the fourth largest city in the world) and the southern metropolises of Pusan (5.0 million) and Taegu (2.7 million). (Pyongyang, the North Korean headquarters [Fig. 9–29], is just approaching 2 million.) Seoul is the country's primate city (Fig. 9–30), whose location in the waist of the peninsula, midway between the industrialized northern and agriculturally productive

FIGURE 9–29 Speckles of colored balloons cannot mask the drabness of Pyongyang, dominated by Kim Il Sung's Tower of the "Juche Idea" (rear).

FIGURE 9–30 The strikingly Westernized landscape of central Seoul. This city, which hosted the 1988 Olympics, is only 125 miles but a world apart from gloomy communist Pyongyang.

southern regions of Korea, has been an advantage. Pusan is nearest to Japan and grew in phases during this century, first under Japanese stimulus and later as the chief U.S. entry point. Pyongyang was also developed by the Japanese, and it lies at the center of a major manufacturing region, which is merging with the mining-industrial area of the northwest.

The Koreans, North and South, are a single people with common ways of life, religious beliefs, historic and emotional ties, and a common language. Throughout the twentieth century, the Koreans have barely known what self-determination in their own country would be like, but they have not lost sight of their aspirations. After seemingly endless suffering through invasions and wars, most of them precipitated by the "national" interests of other states, Korea today, ideologically and politically divided as it is, still retains the ingredients of a cohesive national entity. In the 1970s, the first efforts were begun to thaw the hostilities between North and South Korea through the reopening of trade connections, but this attempt has not resulted in a breakthrough.

Serious efforts at reconciliation will almost certainly have to await the passing of the present North Korean regime, led by the aged Kim Il Sung—a long-time leader installed by Stalin. The juxtaposition of the cityscapes of Pyongyang and Seoul (Figs. 9–29 and 9–30)— only 125 miles (200 kilometers) apart—underscores the current ideological polarization that keeps the two Korean entities apart. Perhaps in the near future, with great-power "spheres of influence" in a state of flux, it might become more feasible for a united Korea to function as a cultural and economic whole. The potential advantages for both North and South will certainly continue to exist until that day.

PRONUNCIATION GUIDE

Users are advised to review the box on the *pinyin* system on pp. 500–501.

Altay (*al*-tye)
Altun (ahl-*tun*)*
Amur (uh-*moor*)
Annam (uh-*nahm*)
Anshan (ahn-*shahn*)
Beijing (bay-*zhing*)
Bohai (bwoh-*hye*)
Cadre (*kah*-dray)
Canton (kan-*tonn*)
Chang Jiang (chung-jee-*ahng*)
Chengdu (chung-*doo*)
Chiang Kai-shek (jee-*ahng* kye-*sheck*)
Chongqing (chong-*ching*)
Confucius (kun-*few*-shuss)
Dalai Lama (dah-lye-*lahma*)
Dalian (dah-lee-*enn*)
Daoism (*dau*-ism)
Daqing (dah-*ching*)
Deng Xiaoping (*dung* shau-*ping*)
Dongbei (dung-*bay*)
Dzungaria (joong-*gah*-ree-uh)
Fujian (foo-jee-*enn*)
Fushun (foo-*shun*)*
Gansu (gahn-*soo*)
Gaoliang (gow-lee-*ahng*)
Gezhouba (guh-*joh*-bah)
Gobi (*goh*-bee)
Guangdong (gwahng-*dung*)
Guangxi Zhuang (gwahng-shee-*jwahng*)
Guangzhou (gwahng-*joh*)
Guilin (gway-*lin*)
Guizhou (gway-*joh*)
Han (*hahn*)
Hangzhou (hahng-*joh*)
Hankou (hahn-*koh*)
Hanyang (hahn-*yahng*)
Hebei (huh-*bay*)
Heilongjiang (hay-long-jee-*ahng*)

Hsinchu (hsin-*choo*)
Hua Guofeng (*hwah* gwoh-*fung*)
Huang He (*hwahng-huh*)
Huangpu (*hwahng-poo*)
Hunan (hoo-*nahn*)
Jiangsu (jee-ahng-*soo*)
Jiangxi (jee-ahng-*shee*)
Jilin (jee-*linn*)
Juche (*joo*-chay)
Junggar (*joong*-gahr)
Kaifeng (kye-*fung*)
Karamay (kah-*rah*-may)
Kazakh[stan] (*kuzz*-uck [stahn])
Khabarovsk (kab-uh-*roffsk*)
Kim Il Sung (kim-ill *soong*)
Kirghiz (*keer*-geeze)
Kongfuzi (kung-*foodzee*)
Kongzi (*kung*-dzee)
Korea (*career*)
Kublai Khan (koob-lye-*kahn*)
Kunlun (*koon*-loon)
Kunming (koon-*ming*)
Lanzhou (lahn-*joh*)
Lhasa (*lah*-suh)
Li (*lee*)
Liao (lee-*au*)
Liaodong (lee-au-*dung*)
Liaoning (lee-au-*ning*)
Loess (*lerss*)
Lüda (*loo*-dah)
Macao (muh-*kau*)
Manchu (man-*choo*)
Manchukuo (mahn-*joh*-kwoh)
Manchuria (man-*choory*-uh)
Mao Zedong (*mau* zee-*dung*)
Molybdenum (muh-*libb*-dun-um)
Mongol (*mung*-goal)
Mongolia (mung-*goh*-lee-uh)
Myanmar (mee-ahn-*mah*)
Nanjing (nahn-*zhing*)
Nei Mongol (nay-*mung*-goal)
Ordos (*ord*-uss)
Peking (pea-*king*)
Pinyin (pin-*yin*)

Potala (poh-*tah*-lah)
Pusan (*poo*-sahn)
Pyongyang (pea-*awhng*-yahng)
Qaidam (*chye*-dahm)
Qanats (*kah*-nahts)
Qin (*chin*)
Qing (*ching*)
Qingdao (ching-*dau*)
Qinghai (ching-*hye*)
Qiqihar (chee-*chee*-har)
Rostow [Walt] (*russ*-stoff)
Ruijin (rway-*jeen*)
Ryukyu (ree-*yoo*-kyoo)
Samchok (sahm-*jook*)
Seoul (*soal*)
Shaanxi (shahn-*shee*)
Sha Mian (shah-mee-*ahn*)
Shandong (shahn-*dung*)
Shanghai (shang-*hye*)
Shantou (shahn-*toh*)
Shanxi (shahn-*shee*)
Shenyang (shun-*yahng*)
Shenzhen (shun-*zhen*)
Sichuan (zeh-*chwahn*)
Sinicization (sine-ih-sye-*zay*-shun)
Song (*sung*)
Songhua (*sung*-hwah)
Sui (*sway*)
Sun Yat-sen (*soon* yaht-*senn*)
Taegu (*tag*-oo)
Taipei (tye-*bay*)
Taiwan (tye-*wahn*)
Taiyuan (tye-*yoo*-ahn)
Taklimakan (tahk-luh-muh-*kahn*)
Tarim (*tah*-reem)
Tiananmen (*tyahn*-un-men)
Tianjin (tyahn-*jeen*)
Tianshan (tyahn-*shahn*)
Tibet (tuh-*bett*)
Ulaan Baatar (oo-lahn-*bah*-tor)
Ürümqi (oo-*room*-chee)
Ussuri (woo-*soo*-lee)
Uygur (Uighur) (*wee*-ghoor)

*"U" or final "u" pronounced as in "put."

Versailles (vair-*sye*)	Xian (shee-*ahn*)	Yuan (*yoo*-ahn)
Vietnam (vee-et-*nahm*)	Xi Jiang (shee-jee-*ahng*)	Yumen (*yoo*-mun)
Vladivostok (fluddy-*foss*-stock)	Xinhua (*sheen*-hwah)	Yunnan (yoon-*nahn*)
Wei (*way*)	Xinjiang (shin-jee-*ahng*)	Zhanjiang (jahn-jee-*ahng*)
Wuchang (woo-*chahng*)	Xizang (sheedz-*ahng*)	Zhou (*joh*)
Wuhan (woo-*hahn*)	Yangzi (*yang*-dzee)	Zhou Enlai (*joh* en-lye)
Wuzhou (woo-*joh*)	Yenan (yen-*ahn*)	Zhuhai (joo-*hye*)
Xiamen (shah-*men*)	Yichang (yee-*chahng*)	

REFERENCES AND FURTHER READINGS

(BULLETS [●] DENOTE BASIC INTRODUCTORY WORKS ON THE REALM OR SYSTEMATIC-ESSAY TOPIC)

Amsden, Alice. *Asia's Next Giant: Late Industrialization in South Korea* (New York: Oxford University Press, 1989).

● Buchanan, Keith et al. *China: The Land and People* (New York: Crown, 1981).

● Cannon, Terry & Jenkins, Alan, eds. *The Geography of Contemporary China: The Impact of Deng Xiaoping's Decade* (London & New York: Routledge, 1990).

Cheng, Chu-Yuan. *Behind the Tiananmen Square Massacre: Social, Political, and Economic Ferment in China* (Boulder, Colo.: Westview Press, 1990).

Chisholm, Michael. *Modern World Development* (Totowa, N.J.: Barnes & Noble, 1982).

Chiu, T.N. & So, C.L., eds. *A Geography of Hong Kong* (New York: Oxford University Press, 2 rev. ed., 1987).

● Cressey, George B. *Asia's Lands and Peoples: A Geography of One-Third of the Earth and Two-Thirds of Its People* (New York: McGraw-Hill, 3 rev. ed., 1963).

● de Souza, Anthony R. & Porter, Philip W. *The Underdevelopment and Modernization of the Third World* (Washington, D.C.: Association of American Geographers, Commission on College Geography, Resource Paper No. 28, 1974).

● Dickenson, John P. et al. *A Geography of the Third World* (London & New York: Methuen, 1983).

Fairbank, John King. *China's Revolution from 1800 to the Present* (New York: Harper & Row, 1986).

Geelan, P.J.M. & Twitchett, D.C., eds. *The Times Atlas of China* (New York: D. Van Nostrand, 2 rev. ed., 1984).

● Ginsburg, Norton S., ed. *The Pattern of Asia* (Englewood Cliffs, N.J.: Prentice Hall, 1958).

● Ginsburg, Norton S. & Lalor, Bernard A., eds. *China: The 80s Era* (Boulder, Colo.: Westview Press, 1984).

Goodman, David, ed. *China's Regional Development* (London & New York: Routledge, 1989).

Ho, Samuel P.S. *Economic Development of Taiwan* (New Haven, Conn.: Yale University Press, 1978).

Hoare, James & Pares, Susan. *Korea: An Introduction* (London & New York: Routledge, 1988).

● Hook, Brian, ed. *The Cambridge Encyclopedia of China* (New York: Cambridge University Press, 1982).

Information China (Elmsford, N.Y.: Pergamon Press, 3 vols., 1988).

Jao, Y.C. & Leung, Chi-Keung, eds. *China's Special Economic Zones: Policies, Problems and Prospects* (New York: Oxford University Press, 1986).

● Jingzhi, Sun, ed. *The Economic Geography of China* (New York: Oxford University Press, 1988).

Kelly, Ian. *Hong Kong: A Political-Geographic Analysis* (Honolulu: University Press of Hawaii, 1987).

● Kirkby, Richard J.R. *Urbanization in China: Town and Country in a Developing Economy, 1949–2000 A.D.* (New York: Columbia University Press, 1985).

Lattimore, Owen. *Inner Asian Frontiers of China* (New York: American Geographical Society, 1940).

Linder, Staffan B. *The Pacific Century: Economic and Political Consequences of Asian-Pacific Dynamism* (Stanford, Calif.: Stanford University Press, 1986).

Ma, Laurence J.C. & Hanten, Edward W., eds. *Urban Development in Modern China* (Boulder, Colo.: Westview Press, 1981).

Mabogunje, Akinlawon L. *The Development Process: A Spatial Perspective* (New York: Holmes & Meier, 1981).

Miller, E. Willard & Miller, Ruby M. *The Far East: A Bibliography on the Third World* (Monticello, Ill.: Vance Bibliographies, No. P-918, 1982).

Morris, Jan. *Hong Kong* (New York: Random House, 1988).

Murphey, Rhoads. *The Fading of the Maoist Vision: City and Country in China's Development* (London & New York: Methuen, 1980).

Murphey, Rhoads et al., eds. *The Chinese: Adapting the Past, Building the Future* (Ann Arbor: University of Michigan, Center for Chinese Studies, 1986).

Pannell, Clifton W. "Regional Shifts in China's Industrial Output," *The Professional Geographer*, 40 (February 1988): 19–31.

● Pannell, Clifton W., ed. *East Asia: Geographical and Historical Approaches to Foreign Area Studies* (Dubuque, Iowa: Kendall/Hunt, 1983).

● Pannell, Clifton W. & Ma, Laurence J.C. *China: The Geography of Development and Modernization* (New York: Halsted Press/V.H. Winston, 1983).

Rostow, Walt W. *The Stages of Economic Growth* (London & New York: Cambridge University Press, 2 rev. ed., 1971).

Salisbury, Harrison E. *The Long March: The Untold Story* (New York: Harper & Row, 1985).

Schinz, Alfred. *Cities in China* (Berlin: Gebrüder Borntraeger, 1989).

Shabad, Theodore. *China's Changing Map: National and Regional Development, 1949–1971* (New York: Praeger, 1972).

Sit, Victor F.S., ed. *Chinese Cities: The Growth of the Metropolis Since 1949* (New York: Oxford University Press, 1985).

Smil, Vaclav. *The Bad Earth: Environmental Degradation in China* (Armonk, N.Y.: M.E. Sharpe, 1984).

Smil, Vaclav. "China's Food," *Scientific American*, December 1985, pp. 116–124.

• Smith, Christopher J. *China: People and Places in the Land of One Billion* (Boulder, Colo.: Westview Press, 1990).

Smith, Michael et al. *Asia's New Industrial World* (London & New York: Methuen, 1985).

Spence, Jonathan D. *The Search for Modern China* (New York: W.W. Norton, 1990).

• Spencer, Joseph E. & Thomas, William L. *Asia, East by South: A Cultural Geography* (New York: John Wiley & Sons, 2 rev. ed., 1971).

Theroux, Paul. *Riding the Iron Rooster: By Train Through China* (New York: G.P. Putnam's Sons, 1988).

• Tregear, Thomas R. *China: A Geographical Survey* (New York: John Wiley & Sons/Halsted Press, 1980).

Vining, Daniel R. "The Growth of Core Regions in the Third World," *Scientific American*, April 1985, pp. 42–49.

Wittwer, Sylvan et al. *Feeding a Billion: Frontiers of Chinese Agriculture* (East Lansing, Mich.: Michigan State University Press, 1987).

Zhao, Songqiao. *Physical Geography of China* (New York & Beijing: John Wiley & Sons/Science Press, 1986).

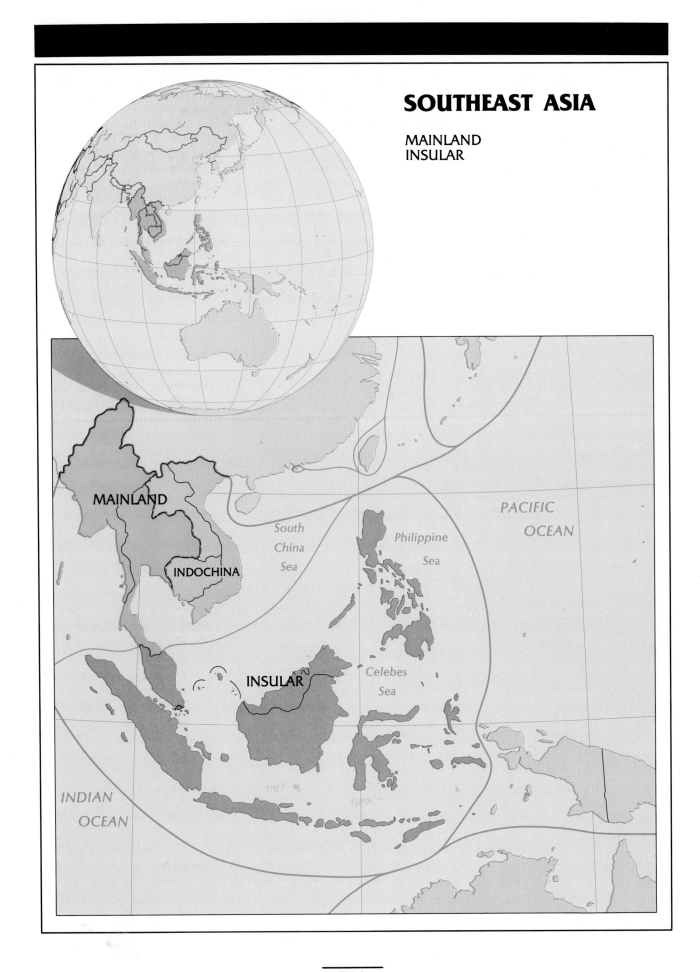

SOUTHEAST ASIA

MAINLAND
INSULAR

MAINLAND

INDOCHINA

INSULAR

South
China
Sea

Philippine
Sea

PACIFIC
OCEAN

Celebes
Sea

INDIAN
OCEAN

CHAPTER 10

SOUTHEAST ASIA: BETWEEN THE GIANTS

Southeast Asia, the largest continent's southeastern corner of peninsulas and islands, is bounded by India on the northwest and by China on the north; its western coasts are washed by the Indian Ocean, and to the east stretches the vast Pacific. From all these directions, Southeast Asia has been penetrated by outside forces. From India came traders; from China came settlers. From across the Indian Ocean came the Arabs to engage in commerce and the Europeans in pursuit of empires. And from across the Pacific came the Americans. Southeast Asia has been the scene of countless contests for power and primacy—the competitors have come from near and far. Like Eastern Europe, Southeast Asia is a world region of great cultural diversity, and it is also a shatter belt.

The concepts of **shatter belt** and **buffer zone** go hand in hand, and the map of Southeast Asia will remind you somewhat of Eastern Europe: a mosaic of smaller countries on the periphery of one of the world's largest states. Just as pressures and stresses from without and within fractured the political geography of Eastern Europe, Southeast Asia's limited space also shows the effects of divisive forces. As in Eastern Europe, the map has changed even in recent

IDEAS AND CONCEPTS

shatter belt (2)

buffer zone

political geography

maritime boundaries

territorial morphology

genetic boundary classification

refugee flows

southeast asia city model

exclusive economic zone

world-lake concept

times. In 1965, Singapore broke away from Malaysia and became the realm's smallest political entity territorially, a modern city-state. Even more recently, a boundary disappeared in 1976 when North and South Vietnam were united. Again, as in Eastern Europe, boundary disputes continue; Vietnam and Cambodia disagree on the position of long stretches of their common border.

Since the politico-geographical map (Fig. 10–1) is so complicated, it should be studied attentively. One good way to strengthen your mental map of this realm is to follow the mainland coastline from

west to east. The westernmost state in the realm is Myanmar (called Burma before 1989), the only country that borders both India and China; Myanmar also shares the "neck" of the Malay Peninsula with Thailand, heart of the mainland region. The south of the peninsula is part of Malaysia—except for Singapore, at the very tip of it. Facing the Gulf of Thailand is Cambodia. Still moving generally eastward, we reach Vietnam, a strip of land that extends all the way to the Chinese border. And surrounded by its neighbors is landlocked Laos, remote and isolated. That leaves the islands that constitute insular Southeast Asia: the Philippines in the north and Indonesia in the south, and between them the offshore portion of Malaysia, situated on the largely Indonesian island of Borneo. Completing the map is Brunei, smallest country in the realm in terms of population but, as we will see, important in the regional picture.

These are the countries of a realm in which there is no dominant state—no Brazil, no China, no India. Southeast Asia is a realm of mountain barriers, unproductive uplands, rugged coastlines, and far-flung islands—as well as fertile valleys and deltas, rich volcanic soils, and productive plains (Fig.

10–3). No single dominant core of indigenous development emerged here. Cultural diversity prevails: the realm is a mosaic of ethnic and linguistic groups, of various and different religions, of contrasting ways of life. Its historical geography, still etched on the map, is one of foreign intervention and competition. Less than a half-century ago, at the end of World War II, only Thailand was an independent state, itself a buffer between French colonialism to the east and British power to the west. The colonial era has ended, but its aftermath still hangs heavily over fragmented Southeast Asia.

POPULATION PATTERNS

Compared to the huge population numbers and densities in the habitable regions of South Asia and China, demographic totals for the countries of Southeast Asia seem to pale by comparison. Laos, territorially quite a large country, has a population of just over 4 million and an average population density that is half of desert-dominated Iran's. Indeed, much of interior mainland Southeast Asia has population densities similar to those of savanna Africa, the margins of South America's Amazon Basin, and the Soviet Muslim South (Fig. I–14). Even the more densely inhabited coastal areas of Southeast Asia have fewer people and smaller agglomerations than elsewhere in southern and eastern Asia. There is nothing in this realm to compare to the immense scale of human clusters in India's Ganges Lowland or the North China Plain. The whole pattern is different: Southeast Asia's fewer dense population clusters are rela-

---

TEN MAJOR GEOGRAPHIC QUALITIES OF SOUTHEAST ASIA

1. The Southeast Asian realm is fragmented into numerous peninsulas and islands.
2. Southeast Asia, like Eastern Europe, exhibits shatter-belt characteristics. Pressures on this realm from external sources have always been strong.
3. Southeast Asia exhibits intense cultural fragmentation, reflected by complex linguistic and religious geographies.
4. The legacies of powerful foreign influences (Asian as well as non-Asian) continue to mark the cultural landscapes of Southeast Asia.
5. Southeast Asia's politico-geographical traditions involve frequent balkanization, instability, and conflict.
6. Population in Southeast Asia tends to be strongly clustered, even in rural areas.
7. Compared to neighboring regions, mainland Southeast Asia's physiologic population densities remain relatively low.
8. Rapid population growth has prevailed in the island regions of Southeast Asia, notably in the Philippines, during much of the twentieth century.
9. Intraregional communications in Southeast Asia remain poor. External connections are often more effective than internal linkages.
10. The boundaries of the Southeast Asian realm are problematic. Transitions occur into the adjoining South Asian, Chinese, and Pacific realms.

---

tively small and lie separated from one another by areas of much sparser human settlement. Nonetheless, when everything is added up, this realm does exhibit some fairly large numbers: the 1991 Southeast Asian population total of 464 million is about 100 million higher than North Africa/Southwest Asia, 89 percent of Subsaharan Africa's total, and 93 percent of Europe's total.

Why do Southeast Asia's population patterns differ from those of its giant neighbors to the north and west? When there is such population pressure and such land shortage in adjacent realms, why has Southeast Asia not been flooded

by waves of immigrants? Several factors have combined to inhibit large-scale invasions. In the first place, obstacles hinder travel along the overland routes into Southeast Asia. In discussing the Indian subcontinent, we noted the barrier effect of the densely forested hills and mountains that lie along the border between northeastern India and northwestern Myanmar (Burma). North of Myanmar lies forbidding Xizang (Tibet) and northeast of Myanmar and north of Laos is the high Yunnan Plateau. Transit is easier into northern Vietnam via southeastern China, and along this avenue considerable contact and migration

FIGURE 10–1

have indeed occurred. Moreover, contact within Southeast Asia itself is not enhanced by the rugged, somewhat parallel ridges that hinder east-west communications between the fertile valleys of the realm (Fig. 10–3). Second—and the population map reflects this—

there are limits to the agricultural opportunities in Southeast Asia. Much of the realm is covered by dense tropical rainforests, part of an ecological complex whose effect on human settlement we have previously observed in other low-latitude zones of the world. Ex-

cept in certain locales, the soils of mainland Southeast Asia are excessively leached (diluted of chemical nutrients) by generally heavy rains. In the areas of monsoonal rainfall regimes and savanna climate (the latter prevailing across much of the mainland interior),

S Y S T E M A T I C   E S S A Y

POLITICAL GEOGRAPHY

Southeast Asia is a laboratory for the study of **political geography**. This systematic field is one of the oldest in geography and focuses on the spatial expressions of political behavior. Boundaries on land and on the oceans, the roles of capital cities, power relationships among states, administrative systems, voter behavior, conflicts over resources, and even matters involving outer space have politico-geographical dimensions. We have already been introduced to some political geography when we studied the approaching unification of Western Europe (*supranationalism*), the historic crushing of Eastern Europe (*balkanization*), the support given by Pakistan to Muslims in Kashmir (*irredentism*), and the competition for hegemony over Eurasia (*geopolitics*).

The field of political geography grew from geographers' interest in the spatial nature of the national state. What are the ingredients of a nation-state? Why do some states survive over many centuries, while others (such as the Ottoman Empire, the Austro-Hungarian Empire, and now perhaps the Soviet Empire) collapse? About a century ago, Friedrich Ratzel proposed a theory that likened the nation-state to a biological organism (see pp. 73–76). Just as an organism is born, grows, matures, and eventually dies, Ratzel argued, states go through stages of birth (around a culture hearth or core area), expansion (perhaps by colonization), maturity (stability), and eventual collapse. Only the sporadic absorption of new land and people, he suggested, could stave off the state's decline. It was a blueprint for imperialism!

Later, political geographers

realized that every state has ties that bind it (*centripetal forces*) and stresses that tend to break it apart (*centrifugal forces*). When the centripetal or unifying forces are much stronger, the state succeeds; when the centrifugal or divisive forces prevail, the state fails. In Japan today the centripetal forces are very strong; but in Lebanon centrifugal forces dominate. Identifying and measuring these forces, obvious as they may be, is one of the challenges in political geography.

We sometimes call countries *nations*, but many countries contain more than one nation, which is why it is better to call them *states*. It might be appropriate to call Finland a nation, but Nigeria (as we noted in Chapter 7) contains three nations and many minority peoples. States that are also pluralistic societies, such as South Africa, exhibit powerful centrifugal forces that must be overcome. Even Canada confronts cultural division, spatially expressed, that could lead to its fragmentation; thus, Canada is not yet a nation.

The most basic device in political geography is the world political map (Fig. I–17). This map reveals the enormous range in the sizes of states; some are microstates (or ministates), while others are giants. The map also suggests why the United Nations officially recognizes a group of Geographically Disadvantaged States (GDS): more than two dozen states have no **maritime boundaries** and are landlocked. As we noted in the case of Bolivia, this can have a disastrous effect on the fortunes of a country. Another aspect of the world map lies in the **territorial**

morphology, or physical shape, of states. In this chapter we examine in some detail the shapes of Southeast Asian states, and discuss what effect their morphology may have had on their development.

States have capitals, core areas, administrative divisions, and boundaries. Boundaries are sensitive parts of the anatomy of a state: just as people are territorial about their individual properties, so nations and states are sensitive about their territories and limits. Boundaries, in effect, are contracts between neighboring states. The *definition* of a boundary is likely to be found in an elaborate treaty that verbally describes its precise location. Cartographers then perform the *delimitation* (official mapping) of what the treaty stipulates. Certain boundaries are actually placed on the ground as fences, walls, or other artificial barriers; this represents the *demarcation* of the boundary. The world map shows that some boundaries have a sinuous form, while others are straight lines. Boundaries can therefore be classified as *geometric* (straight-line or curved), *physiographic* (coinciding with rivers or mountain crests), or *anthropogeographic* (marking breaks or transitions in the cultural landscape).

Another way to view boundaries has to do with their evolution or genesis. This **genetic boundary classification** was established by Richard Hartshorne, a leading American political geographer. Hartshorne reasoned that certain boundaries were defined and delimited before the present-day human landscape developed. In Fig. 10–2 (upper-left map), the United States–Canada boundary is an

GENETIC POLITICAL BOUNDARY TYPES

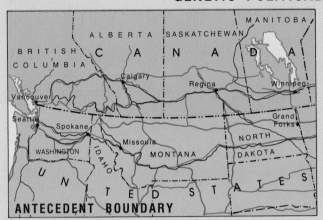

ANTECEDENT BOUNDARY

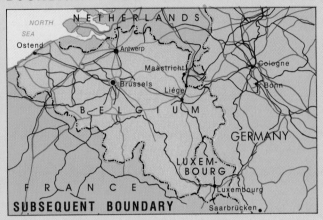

SUBSEQUENT BOUNDARY

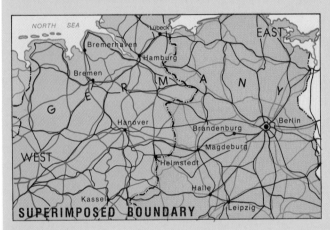

SUPERIMPOSED BOUNDARY

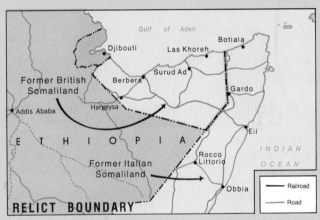

RELICT BOUNDARY

FIGURE 10–2

example of this *antecedent* type. Westward expansion in the United States and Canada, on opposite sides of the 49th parallel, created two corridors of settlement, transport routes, and other features of the cultural landscape. Also note the contrasting administrative divisions of the two countries as shown on this map: states in the U.S., provinces in Canada.

Other boundaries evolved as the cultural landscape of an area took shape. These *subsequent* boundaries are exemplified by the map in the upper right of Fig. 10–2, which shows Belgium and its neighbors. A long-term process of adjustment produced what is on the map today, with every mile of boundary codified in international treaties.

But the map of Belgium does not display a boundary that is nevertheless a reality—the internal border between the Flemish and Walloon communities that make Belgium one of those countries that cannot qualify as a nation-state. Brussels, the capital, lies near this transition from Flemish (Dutch) to Walloon (French) Belgium. Belgium's cultural division is a pervasive fact in its daily life.

Some boundaries are forcibly drawn across a unified cultural landscape. After World War II, defeated Germany was divided into two countries by such a *superimposed* boundary (lower-left map of Fig. 10–2). That boundary also was strongly demarcated (as part of the Iron Curtain) to prevent an

exodus from East Germany to West Germany. The reunification of Germany eliminated the function of this boundary.

The boundary between East and West Germany will now become a *relict* boundary. Relict boundaries are those that have ceased to function, but whose imprints are still evident on the cultural landscape. The lower-right map in Fig. 10–2 shows one such example: the former boundary between the two colonies of Italian and British Somaliland. When the colonial powers departed and Somalia became a unified state, that dividing line became a relict boundary; but evidence of it persists to this day, even on the map in the contrasting names of villages and towns.

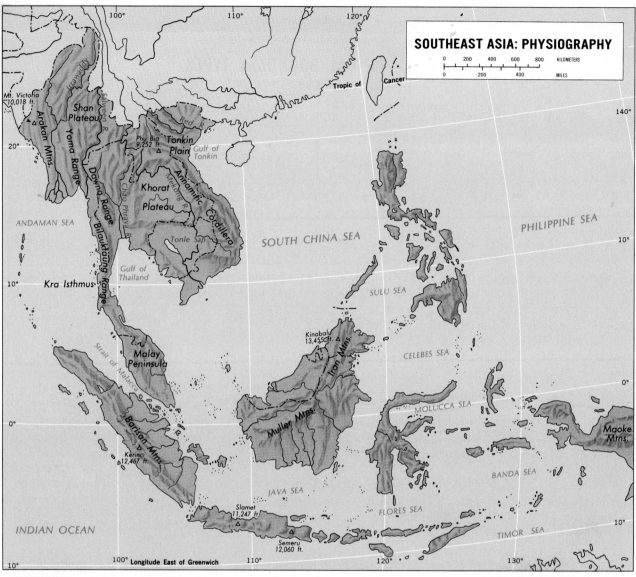

FIGURE 10–3

there is a dry season. But the limitations imposed by savanna conditions on agriculture are all too familiar: high evapotranspiration rates, long droughts, hard-baked soils, meager fertility, high runoff, and erosional problems add up to anything but a peasant farmer's paradise in these parts of Southeast Asia.

The Clusters

Population in Southeast Asia has become concentrated in three kinds of natural environments. First, there are the valleys and deltas of Southeast Asia's major rivers, where alluvial soils have been formed. Four major rivers stand out. In Myanmar, the Irrawaddy rises near the border with Xizang (Tibet) and creates a delta as it empties into the Andaman Sea. In Thailand, the Chao Phraya traverses the length of the country and flows into the Gulf of Thailand. In southern Vietnam, there is the extensive delta of the great Mekong River, which rises in the high mountains on the Qinghai-Xizang border in interior China and

crosses the entire Indochinese peninsula. And in northern Vietnam lies the Red River, whose lowland (the Tonkin Plain) is probably the most densely settled area in mainland Southeast Asia. Each of these four river basins contains one of the realm's major mainland population clusters (Fig. I–14).

Second, Southeast Asia is known for its volcanic mountains—at least in the island archipelagoes. In certain parts of these island chains the conditions are right for the formation of deep, dark, and rich volcanic soils, especially across much of Jawa

(Java).* The population map indicates the significance of this fertility: the island of Jawa is one of the world's most densely populated and intensively cultivated areas. On Jawa's productive land live about 120 million people, approximately two-thirds of the inhabitants of all the islands of Indonesia and almost a quarter of the population of the entire realm.

Another look at Fig. I–14 indicates that one additional area remains to be accounted for: the belt of comparatively dense population that extends along the western coast of the lower Malay Peninsula, apparently unrelated to either alluvial or volcanic soils. This represents the third basis for population agglomeration in Southeast Asia—the plantation economy. Actually, plantations were introduced by the European colonizers throughout most of insular (as well as parts of coastal mainland) Southeast Asia, but nowhere did they so thoroughly transform the economic geography as in Malaya. Rubber trees were planted on tens of thousands of acres and colonial Malaya became the world's leading exporter of this product. Undoubtedly, modern Malaysia would not have developed so populous a core area without its plantation economy (which is also marked by recent increases in palm oil production).

Where there are no alluvial soils, no volcanic soils, and no productive plantations, as in the rainforested areas and the steep-sloped uplands where there is little

* As in Africa, names and spellings have changed with independence. In this chapter, the contemporary spellings will be used, except when reference is made to the colonial period. Thus, Indonesia's four major islands are Jawa, Sumatera, Kalimantan (the Indonesian part of Borneo), and Sulawesi. The Dutch called them Java, Sumatra, Dutch Borneo, and Celebes.

level land, far fewer people manage to make a living. Here the practice of shifting subsistence cultivation prevails (see Fig. 5–2), sometimes augmented by hunting, fishing, and the gathering of wild nuts, berries, and the like. This practice is similar to what we observed in tropical-interior South America and Equatorial Africa; because land that was once cleared for cultivation must be left alone for years in order to regenerate, rather sparse populations are all that can be sustained. Here in the forests and uplands of Southeast Asia, these nonsedentary cultivators still live in considerable isolation from the peoples of the core areas. The forests are dense and difficult to penetrate, distances are great, and steep slopes only add to the obstacles that inhibit efficient surface communications.

Over a major part of their combined length, the political boundaries of continental Southeast Asia traverse these rather thinly peopled inland areas. But here are the roots of some of the region's political troubles: these unstable interior zones have never been effectively integrated into the states of which they are part. The people who live in the forested hills may not be well disposed toward those who occupy the dominant core areas. Therefore, these frontier-like inner reaches of Southeast Asia—with their protective isolation and distance from the seats of political power—are fertile ground for revolutionary activities if not for agriculture (which can turn illicit, too, as in the case of the huge opium poppy harvests of the notorious ''Golden Triangle'' where the borders of Myanmar, Laos, and Thailand converge). Thus, the integration of highland and lowland economies, the improvement of connections between national cores and peripheries, is a very important task for Southeast Asian governments.

INDOCHINA

The French colonialists called their Southeast Asian possessions *Indochina*, and the name is appropriate for the bulk of the mainland region, as it suggests the two leading Asian influences that have affected the realm for the past 2000 years. Overland immigration was mainly southward from southern China, resulting from the expansion of the Chinese Empire. The Indians came from the west by way of the seas, as their trading ships plied the coasts and settlers from India founded colonies on Southeast Asian shores in the Malay Peninsula, the lower Mekong Plain, and on Jawa and Borneo.

With the migrants from the Indian subcontinent came their faiths: first Hinduism and Buddhism, later Islam. The Muslim religion was also promoted by the growing number of Arab traders who appeared on the scene, and Islam became the dominant religion in Indonesia where nearly 90 percent of the population adheres to it today. But in Myanmar, Thailand, and Cambodia, Buddhism remained supreme, and in all three countries, the overwhelming majority of the people are now adherents. In culturally diverse Malaysia, the Malays are Muslims (to be a Malay is to be a Muslim), and almost all Chinese are Buddhists; but most Malaysians of Indian ancestry continue to adhere to the Hindu way of life. Although Southeast Asia has generated its own local cultural expressions, most of what remains in tangible form has resulted from the infusion of foreign elements. For instance, the main temple at Angkor Wat, constructed in Cambodia during the twelfth century, remains a monument to Indian architecture of that time (Fig. 10–4).

The *Indo* part of Indochina, then, refers to the cultural im-

FIGURE 10—4 The realm's most famous relic, the main temple at Cambodia's Angkor Wat, survived that country's recent wars with only minor damage, but vandalism and neglect remain problems.

prints from South Asia: the Hindu presence, the importance of the Buddhist faith (which came to Southeast Asia via Sri Lanka [Ceylon] and its seafaring merchants), the influences of Indian architecture and art (especially sculpture), writing and literature, and social structures and patterns.

The Chinese role in Southeast Asia has been substantial as well. Chinese emperors coveted Southeast Asian lands, and China's power at times reached deeply into the realm. Social and political upheavals in China sent millions of Sinicized people southward. Chinese traders, pilgrims, sailors, fishermen, and others sailed from southeastern China to the coasts of Southeast Asia and established settlements there. Over time, those settlements attracted more Chinese emigrants, and Chinese

influence in the realm grew. Not surprisingly, relations between the Chinese settlers and the earlier inhabitants of Southeast Asia have at times been strained, even violent. The Chinese presence in Southeast Asia is long-term, but the invasion has continued into modern times. The economic power of Chinese minorities and their role in the political life of the area have led to conflicts.

The Chinese initially profited from the arrival of the Europeans, who stimulated the growth of agriculture, trade, and industries; here they found opportunities they did not have at home. The Chinese established rubber holdings, found jobs on the docks and in the mines, cleared the bush, and transported goods in their sampans. They brought with them skills that proved to be very useful, and as

tailors, shoemakers, blacksmiths, and fishermen they prospered. The Chinese also proved to be astute in business; soon, they not only dominated the region's retail trade, but also held prominent positions in banking, industry, and shipping. Thus, their importance has always been far out of proportion to their modest numbers in Southeast Asia. The Europeans used them for their own designs but found the Chinese to be stubborn competitors at times—so much so that eventually they tried to impose restrictions on Chinese immigration. Previously, the United States, when it took control of the Philippines, had also sought to stop the influx of Chinese into those islands.

When the European colonial powers withdrew and Southeast Asia's independent states emerged, Chinese population sectors ranged from nearly 50 percent of the total in Malaysia (1963) to barely over 1 percent in Myanmar (Fig. 10–5). In Singapore, Chinese today constitute 77 percent of the population of 2.7 million; when Singapore seceded from Malaysia in 1965, the Chinese component in the latter was reduced to about 35 percent. In Indonesia, the percentage of Chinese in the total population is not high (an estimated 3 percent), but the Indonesian population is so large that even this small percentage indicates a Chinese sector of more than 5 million. In Thailand, on the other hand, many Chinese have married Thais, and the Chinese minority of 14 percent has become a cornerstone of Thai society, dominant in trade and commerce.

In general, Southeast Asia's Chinese communities remained quite aloof and formed their own separate societies in the cities and towns. They kept their culture and language alive by maintaining social clubs, schools, and even residential suburbs that, in practice if not by law, were Chinese in character. There was a time when they were in the middle between the Europeans and Southeast Asians, and when the hostility of the local people was directed toward white people as well as toward the Chinese. But since the withdrawal of the Europeans, the Chinese have become the main target of this antagonism, which remained strong because of the Chinese involvement in money-lending, banking, and trade monopolies. Moreover, there is the specter of an imagined or real Chinese political imperialism along Southeast Asia's northern flanks.

The *china* in Indochina, therefore, represents a wide range of penetrations. The source of most of the old invasions was in southern China, and Chinese territorial consolidation provided the im-

FIGURE 10–5

petus for successive immigrations. Mongoloid racial features carried southward from East Asia mixed with the preexisting Malay stock to produce a transition from Chinese-like people in the northern mainland to dark-skinned Malay types in the distant Indonesian east. Although Indian cultural influences remained strong, Chinese modes of dress, plastic arts, types of houses and boats, and other cultural attributes were adopted throughout Southeast Asia. During the past century, especially during the last half-century, renewed Chinese immigration brought Chinese skills and energies that propelled these minorities to positions of comparative wealth and power in this realm.

THE ETHNIC MOSAIC

It is quite revealing that much of the mainland region of Southeast Asia is often referred to as Indochina, a name that reflects external influences but fails to reveal indigenous cultural-geographical roots. Perhaps this is so because alien forces have been so powerful in the realm. But despite the cultural impacts from India and China and notwithstanding the European colonial era, the great majority of the realm's peoples have regional identities. When you study Fig. 10–6, view it as you saw Fig. 1–12, the language map of Europe.

but which came into use in the 1950s). And finally, there are countries that completely surround others and are therefore *perforated*. Southeast Asia does not have any examples of perforated states (South Africa, enclosing Lesotho, is the most prominent case; see inset map, Fig. 10–8), but all the other morphological types are well represented.

Compact Cambodia

It is appropriate to begin with Cambodia, because this country on the Gulf of Thailand contains one of the cradles of regional civilization. The city of Angkor was the capital of the Khmer Empire—the root of modern Cambodia—from the ninth to the fifteenth century. One of the Khmer kings, Suryavarman II, supervised the construction of the realm's most magnificent temples, the complex known as Angkor Wat (Fig. 10–4). These temples were built to symbolize the universe in accordance with the precepts of Hindu cosmology; Hinduism was the powerful cultural force in mainland Southeast Asia during that time.

The Khmer Empire was a strong element in the historical geography of the realm, but its decline was well under way when the French colonizers came to take control of it in the 1860s. Thailand and Vietnam, Cambodia's neighbors, had invaded the country, and the city of Angkor with its great shrines lay severely damaged and abandoned. France established its Cambodian protectorate's boundaries with Thailand to the west and north, and also defined a boundary to the east between Cambodia and (also French-ruled) Vietnam. The border with Laos, in the northeast, also was an intra-French administrative colonial boundary. In a

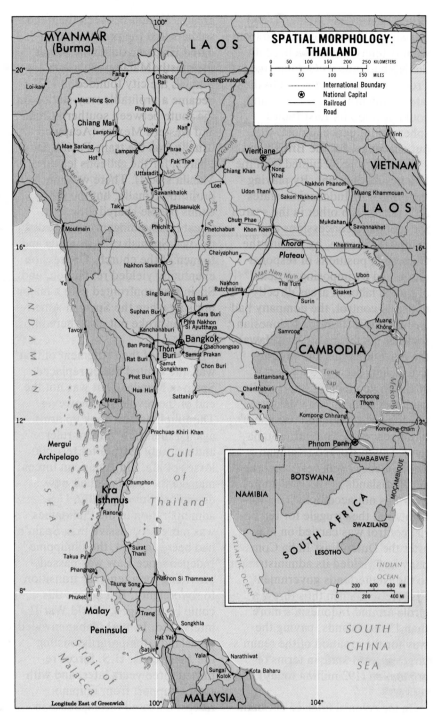

FIGURE 10–8

way the French thus salvaged the Khmer Empire and the Cambodian state, but as a protectorate the country existed in France's colonial background, ultimately under Paris's rule, but really governed by a dynasty of kings.

Geographically, Cambodia enjoyed several advantages, particularly its compact shape. *Compact* states have boundaries on which all points lie about the same distance from the geometric center (in Cambodia's case, Kompong

Thom on the Sen River). This means that such countries enclose a maximum of territory within a minimum of boundary, which can be advantageous in politically sensitive areas. Compact countries such as Cambodia, moreover, do not face the problems confronting some other underdeveloped states of integrating faraway islands, lengthy peninsulas, or other distant territorial extensions into the national framework. Effective political control is most easily established in a compact territory, and for centuries the Cambodians came closer to national integration than perhaps any other Southeast Asian state. Cambodia also had the advantage of remarkable ethnic and cultural homogeneity in so complex a realm: 90 percent of its people are still Khmers, with the rest divided about equally between Vietnamese and Chinese origin.

As the map shows, Southeast Asia's greatest river, the Mekong, enters Cambodia's territory from Laos and crosses it from north to south, creating a great bend before flowing into southern Vietnam and its great delta there. Phnom Penh (1.1 million), the country's modern capital, lies on the river's right bank where it makes that large bend. The ancient capital, Angkor, lies in the northwest, not far from Tonle Sap, a major interior lake that drains into the Mekong.

Cambodia's compactness and comparative isolation, however, could not protect the country against the impact of the Vietnam War (1964–1975). In 1970, the last king was deposed by military rulers; in 1975, that government was overthrown by communist revolutionaries, the so-called Khmer Rouge. These new rulers embarked on a course of terror and destruction in order to reconstruct "Kampuchea" (as they called Cambodia) as a rural society. They

drove townspeople into the countryside where they had no place to live or work, emptied hospitals and sent the sick and dying into the streets, outlawed religion and family life, and killed as many as 2 million Cambodians (out of a total of 7.5 million) in the process. In the late 1970s, Vietnam, victorious in its own war, invaded Cambodia to drive the Khmer Rouge away. But this led to new terror, and a stream of refugees crossed the border into Thailand. Once a self-sufficient country that could feed others, Cambodia now must import food. The economy is dominantly agricultural, and most people grow rice and beans for subsistence. Its population, severely reduced by war, has again surpassed 7 million in the early 1990s. But the country's future remains uncertain with powerful adversaries on its borders.

Laos

North of Cambodia lies Southeast Asia's only landlocked country, Laos. Interior and isolated, Laos changed little during 60 years of French colonial administration (1893–1953). Then, along with other French-ruled areas, Laos became an independent state. Soon its well-entrenched, traditional kingdom fell victim to rivalries between traditionalists and communists, and the old order collapsed. Laos has no fewer than five neighbors, one of which is the East Asian giant, China. A long stretch of its western boundary is formed by the Mekong River, and the important sensitive border with Vietnam to the east lies in mountainous terrain. With 4.1 million people (about half of them ethnic Lao, related to the Thai of Thailand), Laos lies surrounded by comparatively powerful states. The country has no railroads, just a few miles of paved roads, and

very little industry; it is just 17 percent urbanized (the capital, Vientiane, contains only about 450,000 people). Laos remains the region's poorest and most vulnerable entity.

Prorupt Neighbors

Quite a different spatial form is represented by the configurations of Myanmar and Thailand. The main territories of these two states, which contain their respective core areas, are essentially compact—but to the south they share sections of the slender Malay Peninsula. These peninsular portions are long and even narrower, and states with such extensions leading away from the main body of territory are referred to as *prorupt* states. Obviously, Thailand is the best example: its proruption extends nearly 600 miles (1000 kilometers) southward from just west of the capital, Bangkok (Fig. 10–8). Where the Thailand-Myanmar boundary (terminating at the Kra Isthmus) runs along the peninsula, the Thai proruption is in places less than 20 miles (32 kilometers) wide. Naturally, such proruptions can be troublesome, especially when they are as lengthy as this. In the whole state of Thailand, no area lies farther from the core or from Bangkok than the southern extremity of its very tenuous proruption. But at least Thailand's railroad network extends all the way to the Malaysian border; in the case of Myanmar, not only does the railway terminate more than 300 miles (480 kilometers) short of the end of the proruption, but there is not even a permanent road over its southernmost 150 miles (240 kilometers) (Fig. 10–8).

The territorial shape of Thailand and Myanmar may be similar, but their internal geography is quite different. To a large degree, this is the result of differences in relative location. Thailand, as the map shows, occupies the heart of Southeast Asia's mainland region. Historically, it formed a buffer state between French colonial holdings to the east and the British sphere to the west. Indochina to the east became a complex of French dependencies; Myanmar to the west fell under British sway (Fig. 10–7).

Thailand

Thailand's morphology should be viewed in relation to its latitudinal extent. There are humid, nearly equatorial conditions on its southern proruption and more marginally tropical, savanna-like environments over much of the mainland to the north (Fig. I–10). The heart of Thailand is a great low-lying plain, with extensive alluvial flatlands watered by the many tributaries of the Chao Phraya River. Irrigation systems guide the waters to paddies (ricefields) not reached by those streams. In the east lies the Khorat Plateau, where the soils are poorer, moisture is less available, and population is much less dense. Thailand's 57-plus million people are about 75 percent Thai (or Siamese—the country used to be known as Siam). Nearly 15 percent are Chinese, and the remainder are various minorities living in the northern mountains, in the eastern borderlands, and on the southern peninsula. The Thais are concentrated in the country's core area and in the capital (Fig. 10–1). The Chinese are heavily clustered in the urban areas, especially in and around the capital city, Bangkok, where they dominate trade and commerce. Thailand remains a bastion of Buddhism; over 95 percent of the population adheres to this faith (Fig. 10–9).

Thailand lay sufficiently far

FIGURE 10–9 As the map of world religions (Fig. 6–2) indicates, the heart of mainland Southeast Asia is dominated by Buddhism. The splendor of this Buddhist temple complex in Bangkok attests to the continuing appeal of this influential faith despite the pressures that emanate from the region's persistent conflicts.

from the Vietnam War arena to escape its ravages, and the economy grew so rapidly during the 1980s that there was much talk about the country's emergence as Asia's Fifth Tiger (alongside Taiwan, South Korea, Hong Kong, and Singapore). Not only does the country export a great deal of rice (especially to Japan), but another farm product, tapioca roots (exported to Europe as animal feed), has recently earned the Thais even more income. Corn, grown in the drier savannalands, also is a valu-

able export. Teak from northern forests, rubber from trees on the peninsula, and tin from mines in the west are further sources of foreign revenue. The discovery of oil and natural gas in the Gulf of Thailand has helped Thailand's international balance of payments.

But the political geography of Thailand is troubled. Although it remains officially known as the Kingdom of Thailand, a military group took control in 1958, and this has been followed by additional coups. Opposition to the military government has developed among the minorities in the remote hills and on the peninsula. A communist rebellion has caused difficulties in the north, notably in the hinterland of the important northwestern city of Chiang Mai (260,000). And in the east, thousands of refugees have entered Thailand from Cambodia.

All these problems seem far removed from Bangkok (6.0 million), the thriving capital, main port, chief industrial center, and focus of the entire state. This great metropolis lies at or below sea level (land subsidence is a severe problem), and it is riddled by a network of canals whose boat traffic is as busy as rush-hour jams on a Western highway (Fig. 10–10). Bang-

kok is graced by numerous beautiful palaces, monasteries, and shrines. Tourism is another major source of income, not only from Europe and America but increasingly from prospering Asian countries, especially Japan. Thailand's relative location at the heart of mainland Southeast Asia symbolizes its proud heritage of independence, royal succession, culture, and tradition.

Myanmar (Burma)

Myanmar stands in contrast to Thailand in many ways. Not only is it Thailand's western neighbor, but Myanmar also is the realm's northernmost country and possesses the longest of all borders

with China. Here, too, the spatial morphology of the state is further complicated by a shift in the Burmese core area that took place during colonial times. Prior to the colonial period, the focus of the embryonic state was in the so-called dry zone, between the Arakan Mountains and the Shan Plateau (Fig. 10–3). Mandalay in Upper Myanmar, today containing 650,000 people, was the main urban node. Then the British developed the rice potential of the Irrawaddy Delta, and Rangoon (now called Yangon, with 2.9 million residents), a less important city occupied earlier by the British, became the new capital. The old and new core areas are connected by the Irrawaddy in its

FIGURE 10–10 Bangkok, often called "Venice of the East," is crisscrossed by *klongs* or canals that are filled with boats of every stripe. Floating markets such as these remind us that the busy waterways are also an important setting for the city's commercial life.

function as a water route, but the center of gravity in modern Myanmar now lies in the south.

The lowland core area is the domain of the majority Burman population, but the surrounding areas are occupied by 11 other peoples (Fig. 10–6), all of whom have come under the domination of the majority since the British unification of colonial Burma. The Karens live to the southeast of the Irrawaddy Delta, on Myanmar's proruption and in adjacent Thailand. The Shans mostly inhabit the eastern plateau that borders northwestern Thailand, the Kachins live near the Chinese boundary in the far north, and the Chins reside in the highlands along the Indian border in the extreme west.

Other groups also form part of the country's complex population of just under 43 million, and this heterogeneity was further intensified during the period of Britain's India-connected administration (1886–1937) when more than one million Indians entered the country. These Indians came as shopkeepers, money-lenders, and commercial agents, and their presence heightened Burmese resentment against British policies. During World War II, the national division of Myanmar was brought sharply into focus when the lowland Burmans welcomed the Japanese intrusion, whereas the peoples of the surrounding hill country, who had seen less of the British maladministration and who had little sympathy for their Burman compatriots anyway, generally remained pro-England.

The political geography of Myanmar constitutes a particularly good example of the role and effect of territorial shape and internal state structure. Not only is this a prorupt state, with its southern extension into the upper Malay Peninsula; Myanmar's core area is surrounded on the west, north, and east by a horseshoe of great

mountains—mountains where many of the country's 11 minority peoples had their homelands before British occupation and subsequent Burman control. In 1976, nine of these indigenous peoples, opposed to the Burman government, formed a union representing about 8 million people in the country. What these peoples demand is the right of self-determination in their own homelands. The Karens, for instance, a nation of 3 million (about 7 percent of Myanmar's population), have proclaimed that they wish to create an autonomous territory for themselves within a federal Myanmar. The Shans of the far north, a mountain people totaling nearly 4 million, demand similar rights. These centrifugal forces bedevil the central government (even though rebellious minorities have suffered a string of setbacks since 1985), which constantly seeks to establish tighter authority over its outlying areas.

The economic geography of Myanmar is potentially more encouraging. The country can still feed itself from its Irrawaddy paddies (although isolationist trade policies have ended its days as a leading rice exporter); the Salween River, in a parallel valley east of the Irrawaddy, has fertile and productive ricefields as well. With its warm monsoonal climate, Myanmar's fine alluvial soils also yield harvests of sugar cane, beans and other vegetables, and, in drier areas, cotton and peanuts. Known mineral deposits have only begun to be exploited; tin has been mined for many years, and there is petroleum in the lower Irrawaddy Valley not far from Yangon. More than anything else, however, the repressive national policies of a brutal military government have all but halted Myanmar's development in recent years. The state needs internal consensus and liberalization—and far better political leadership than that of the en-

trenched present regime, which has brought impoverished and exhausted Myanmar to the ranks of the world's poorest countries.

Elongated Vietnam

The third category of territorial shape represented in Southeast Asia is the *elongated* or attenuated state. By this is meant that a state is at least six times as long as its average width. Familiar examples on the world map are Chile in South America, Norway and Italy in Europe, and Malawi in Africa. Elongation presents obvious and recurrent problems; even if the core area lies in the middle of the state, as in Chile, the distant areas at either end may not be effectively connected and integrated within the state system. In Norway, the core area lies near one end of the country, with the result that the opposite extremity takes on the characteristics of a remote frontier. In Italy, as we saw in Chapter 1, contrasts between north and south pervade all aspects of national life, and they reflect the respective exposures of these two areas to different mainstreams of European-Mediterranean change. If a state is elongated and at the same time possesses more than one core area, strong centrifugal stresses will arise. This has been the classic case in Vietnam, one of the three political entities into which former French Indochina was divided.

French Indochina incorporated three major ethnic groups: the Laotians, the Cambodians, and the Vietnamese. Even before independence, the Laotians and Cambodians possessed their own political areas (which later were to become the states of Laos and Cambodia), and this left a 1200-mile (2000-kilometer) belt of territory to the east, averaging under 150 miles (240 kilometers) in width, facing the South China Sea. This domain of the Annamites extended all the way from the Chinese border to the Mekong Delta. Administratively, the French divided this elongated stretch of land into three units: Cochin China in the south, Annam in the middle, and Tonkin in the north (Fig. 10–7). The capitals of these areas became familiar during the Vietnam War (1964–1975): Saigon (now Ho Chi Minh City) was the focus of Cochin China and the headquarters of all Indochina, the ancient city of Hué was the center of Annam, and Hanoi was the capital of Tonkin. Cochin China and Tonkin were both incipient core areas, for they lay astride the populous and productive deltas of the Mekong and Red Rivers, respectively (Fig. 10–1).

In 1940–1941, the Japanese forcibly penetrated Indochina, and during their nearly four years of occupation the concept of a united Vietnam emerged. The Japanese did not discourage rising Vietnamese nationalism, especially when it became apparent that the tide of war was turning against them; Japan would rather have an Annamese government in Vietnam than a French one. So, during the period of Japanese authority, a strong coalition of pro-independence movements arose in Indochina—the Viet Minh League. In 1945, when the Japanese surrendered, this organization seized control and proclaimed the Republic of Vietnam a sovereign state under the leadership of Ho Chi Minh (a communist). For some time after the Japanese defeat, the Chi-

nese actually occupied Tonkin and northern Annam, but theirs was a sympathetic involvement as well, and the Viet Minh grew in strength during that time.

Meanwhile, the French had proposed that Vietnam, Laos, and Cambodia should join as associated states in the fourth-republic concept of the French union, but this plan was rejected throughout Indochina. As the French reestablished themselves where they had always been strongest—in Saigon and the Mekong Delta—they started negotiations with the Viet Minh. These talks soon broke down, and in 1946 a full-scale war broke out between the French and the Viet Minh. France's base lay in Cochin China, and the Viet Minh had their greatest strength in Tonkin a thousand miles to the

FIGURE 10–11   This recent photograph of Ho Chi Minh City shows it still bears the imprints of its pre-1975 capitalist days as Saigon (which many still call the city). It has been estimated that the infrastructure of southern Vietnam is at least 30 years ahead of the north—a gap that is not closing in the 1990s.

north. Vietnam's pronounced elongation favored the Viet Minh and their numerous nationalist, communist, and revolutionary sympathizers. After an enormously costly eight-year war, the crucial battle was fought in the north (at Dien Bien Phu in 1954) not far from the Chinese border, and the French lost. In the aftermath, the country was partitioned into communist-led North Vietnam and non-communist South Vietnam.

The developments in Vietnam during the 1960s and 1970s are still fresh in memory. Opposition to the Saigon government in South Vietnam turned into armed insurrection, with North Vietnam providing arms and personnel to aid the Viet Cong insurgents; U.S. advisors and armed forces in support of the shaky Saigon regime (sent in to halt the wider spread of communism) at one time exceeded a half million in number. Despite the severe bombing of North Vietnam and its capital, Hanoi, and extensive defoliation efforts in the protective forests of the South, what had begun as a scattered rebellion ended in 1975 in the defeat of the national government of South Vietnam and the ouster of its ally, the United States. This sequence of events closely conformed to the *insurgent state* model that was outlined in Chapter 4 (pp. 275–276).

The unification of North and South Vietnam in 1976 produced one of the world's most decidedly elongated states and, with a population of 70.3 million (1991), Southeast Asia's second largest. Not surprisingly, the country's attenuation continues to dominate its regional geography. A totalitarian central government notwithstanding, the old divisions between north, center, and south still prevail. Saigon may now be Ho Chi Minh City (3.7 million), but it remains a world apart from

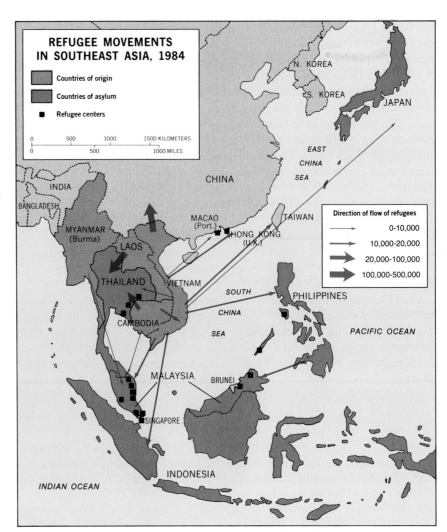

FIGURE 10–12

Hanoi (1.7 million) and the distant north (Fig. 10–11). Since unification, Vietnam has been involved in an intermittent armed conflict with China along its northern boundary. In these efforts, Vietnam functions as a client state of the Soviet Union, which provides crucial economic aid. In return, the Soviets use Vietnam as a vital military base for their warplanes and navy—the only such outpost of the U.S.S.R. in this part of the world.

One of the by-products of successful insurgency movements is the human upheaval that occurs when these types of states become fully established. Large numbers of those who opposed the rise of

the new order, fearing for their lives, feel compelled to flee the country and thereby join the growing world population of refugees (estimated to be 15 million in 1990). We have described many such involuntary migrations in earlier chapters, but since 1975 none have been more tragic than the **refugee flows** generated by the turbulence in Indochina (Fig.10–12). One major exodus involved those who desperately sailed from Vietnam in small, barely seaworthy boats. Of the nearly 2 million of these "Boat People" estimated to have set sail on the wide South China Sea since the communist takeover in the mid-1970s, more than half perished at the hands of storms, pirates, leaky vessels, exposure, and starvation; this emigration continued through the 1980s (as the map shows), and thousands of new "Boat People" per month were still coming out of Vietnam in 1990. Figure 10–12 also indicates a second refugee flow of substantial proportions—the 250,000-plus "Land People" who staggered into Thailand across Cambodia's western border, escaping from the genocide perpetrated by the brutal Khmer Rouge regime and, later, the Vietnamese invasion that endured throughout the 1980s.

Because the non-communist countries of Southeast Asia did not permit most of these migrants to settle permanently within their borders, the refugees were primarily resettled elsewhere in the world. Of the 1-million-plus such surviving refugees who fled the various communist purges in Vietnam, Cambodia, and Laos between 1975 and 1990, approximately 900,000 were resettled in the United States, 85,000 in both Canada and France, 60,000 in Australia, 40,000 in the United Kingdom (including Hong Kong), 20,000 in West Germany, and 40,000 in Switzerland and other smaller European countries; in addition, China has accepted well over 300,000 refugees.

Challenges of Fragmentation

On the peninsulas and islands of Southeast Asia's southern and eastern periphery lie five of the realm's ten states (Fig. 10–1). Few regions in the world contain so diverse a set of countries. Malaysia, the former British colony, consists of two major areas separated by hundreds of miles of South China Sea. The realm's southernmost state, Indonesia, sprawls across

thousands of islands from Sumatera in the west to New Guinea in the east. North of the Indonesian archipelago lies the Philippines, a nation that once was a U.S. colony. These are three of the most severely *fragmented* states on earth, and each has faced the challenges that such politico-spatial division brings.* This insular region of Southeast Asia also contains two small but important sovereign entities: a city-state and a sultanate. The city-state is Singapore, once a part of Malaysia (and one instance in which internal centrifugal forces were too great to be overcome). The sultanate is Brunei, an oil-rich Muslim territory on the island of Borneo that seems transplanted from the Persian Gulf. Few parts of the world are more varied or more interesting geographically.

---

*The inset map in Fig. 10–8 shows the fifth of the spatial forms of states, the only case not represented in Southeast Asia—the *perforated* state. South Africa is shown, perforated by the state of Lesotho.

Divided Malaysia

Malaysia's ethnic and cultural divisions are etched into its landscapes. The Malays are traditionally a rural people. They originated in this region and displaced earlier aboriginal peoples, now no longer significant numerically. The Malays, who constitute over half of the Malaysian population of 18.3 million, possess a strong cultural unity expressed in a common language, adherence to the Muslim faith, and a sense of territoriality that arises from their Malayan origins and their collective view of Chinese, South Asian, and other foreign intruders. Although they have held control over the government, the Malays often express a fear of the more aggressive, commercially oriented, and urbanized Chinese minority (who comprise 32 percent of the population).

Malay-Chinese differences worsened during World War II when the Japanese (who occupied the area) elevated the Malays into positions of authority but ruthlessly persecuted the Chinese, driving

FIGURE 10–13 The Malaysian capital of Kuala Lumpur, whose cityscape is an amalgam of multiple cultural influences set against the skyscrapers of the modern CBD.

many of them into the forested interior where they founded a communist-inspired resistance movement that long continued to destabilize the region. The British returned, then yielded after a system of interracial cooperation had been achieved. But social tensions continued, and in 1969 resulted in racial clashes in Kuala Lumpur. The landscape of this capital city (of 1.8 million) reflects Malaysia's cultural mosaic (Fig. 10–13); but trade and commerce are in the hands of the Chinese, around whose shops and businesses the city's life revolves. Such contrasts, often reinforced by the visible remnants of colonialism, are typical of the Southeast Asian urban experience (see *box*).

The map is essential to any appraisal of Malaysia as a politico-geographical entity. The Malay Peninsula was Britain's most important colonial possession in Southeast Asia; by comparison, Britain's holdings on the Indonesian archipelago (Fig. 10–7) were quite neglected. The British focused their attention on Malaya and created a substantial economy there. The Strait of Malacca, between the Malay Peninsula and Sumatera, became one of the world's busiest and most strategic waterways, and Singapore, at the southern end of it, a prized possession. The map confirms Singapore's locational advantage at the entrance to the strait and near the southern end of the South China Sea (see *box*).

The core area of Malaysia lies on the western side of the peninsula (Fig. 10–1). Here are the capital, the best surface communications, the plantations, the industries, and most of the mines. The Malays of this area, moreover, have the political decision-making power by virtue of their majority over the Chinese. But there is no strong alliance between the Malays of the mainland pen-

THE SOUTHEAST ASIAN CITY

In their survey of urbanization trends in this realm, Thomas Leinbach and Richard Ulack offered some noteworthy generalizations about the larger cities of post-colonial Southeast Asia. First, they are all experiencing rapid growth; between 1950 and 1990, the region's urban population has almost doubled in relative size (from 15 to 28 percent) and quintupled in absolute numbers (26 to 130 million). Second, despite a waning overall presence, foreigners still play a decisive role in the commercial lives of cities, with the Japanese influence particularly prominent today. Third, the most recent episode of urban agglomeration has been heavily concentrated in the large coastal cities, reinforcing their old colonial-era dominance through renewal of their functions as collection-distribution nodes for interior hinterlands as well as leading ports for external trade and shipping. And fourth, they exhibit similar internal land-use patterns, a spatial structuring worth examining in some detail.

The general intraurban pattern of residential and nonresidential activities is summarized in the **Southeast Asia city model** developed by Terence McGee in his book, *The Southeast Asian City* (Fig. 10–14). The old colonial port zone, its functions renewed in the post-colonial period, is the city's focus together with the largely commercial district that surrounds it. Although no formal central business district is evident, its elements are present as separate clusters within the land-use belt beyond the port: the government zone, the Western commercial zone (a colonialist remnant, which is practically a CBD by itself), the alien commercial zone—usually dominated by Chinese merchants whose residences are attached to their places of business—and the mixed land-use zone that contains miscellaneous economic activities, including light industry. The other nonresidential areas are the market-gardening zone at the urban periphery and, still farther from the city, an industrial park or "estate" of recent vintage. The residential zones in McGee's construct are quite reminiscent of the Griffin-Ford model of the Latin American city (Fig. 5–10, p. 300) which, as we saw, could be extended to the Third World city in general.

Among the similarities between the two are: the hybrid sector/ring framework; an elite residential sector that includes new sub-

insula and the majority population in the territories of Sarawak and Sabah, which constitute the offshore component of Malaysia on the island of Borneo. Thus, the balance of power among Malaysia's culturally and geographically separated sectors is fragile.

Nonetheless, Malaysia has prospered. Effective economic leadership has yielded returns in

the form of steady development, even in the face of risks: the decline of natural rubber and variations in the price of tin on world markets were offset in the 1980s by the expansion of palm oil production on plantations (which produce an inexpensive shortening demanded by the manufacturers of processed foods in developed countries). Sabah and Sarawak

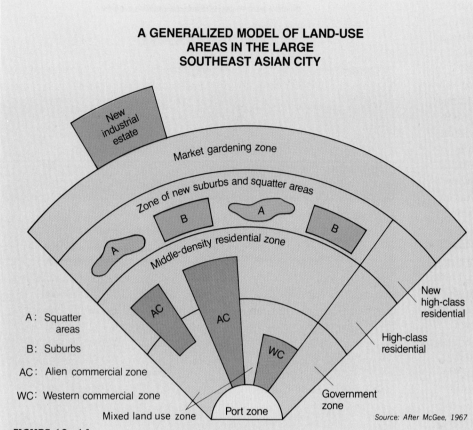

**A GENERALIZED MODEL OF LAND-USE
AREAS IN THE LARGE
SOUTHEAST ASIAN CITY**

New industrial estate

Market gardening zone

Zone of new suburbs and squatter areas

Middle-density residential zone

A

B

A

B

New high-class residential

High-class residential

AC

AC

WC

Government zone

Port zone

Mixed land use zone

A: Squatter areas

B: Suburbs

AC: Alien commercial zone

WC: Western commercial zone

Source: After McGee, 1967

FIGURE 10–14

urbanization; an inner-city zone of comfortable middle-income housing (with new suburban offshoots in the McGee schema); and peripheral concentrations of low-income squatter settlement. The differences are relatively minor and can partly be accounted for by local cultural and historical variations. If the Griffin-Ford model can be viewed as a generalization of the spatial organization of the Third World city, then the McGee model illustrates the departures that occur in coastal cities that were laid out as major colonial ports but continue their development within a now independent country.

have been yielding increasing amounts of petroleum, and economic diversification has brought greater security. Foreign manufacturers, attracted by the skills and wages of Malaysia's work force, have located hundreds of factories there, notably in the electronics industry. Divided Malaysia has to date surmounted its geographic obstacles.

Archipelagic Indonesia

The very term *archipelago* denotes fragmentation, and Indonesia is the world's most expansive archipelagic state. Spread across some 13,000 islands, Indonesia's 192 million inhabitants live separated and clustered—separated by water, and clustered on islands large and

small. The map of Indonesia (Fig. 10–1) requires some attention. There are five large islands, of which one, easternmost New Guinea, is shared with the independent state of Papua New Guinea (to be discussed in the vignette following this chapter). The other four major islands are collectively known as the Greater Sunda Islands: Jawa (Java), the most populous and important; Sumatera (Sumatra) in the west; Kalimantan (the Indonesian portion of Borneo) in the center; and wishbone-shaped Sulawesi (Celebes) to the east. Extending eastward from Jawa are the Lesser Sunda Islands, including Bali and, near the eastern end, Timor. Another important island chain within Indonesia is constituted by the Moluccan Islands, which lie between Sulawesi and New Guinea. The central water body of Indonesia is the Java Sea.

When the Dutch arrived to colonize this archipelago, it was by no means a unified country. There had already been a Hindu invasion, and Buddhism also made inroads among the local peoples. But in the sixteenth century, Islam arrived to become entrenched as the dominant religion, spreading rapidly before the seventeenth-century arrival of the Dutch colonizers (Indonesia today remains the world's largest Muslim country). Not only was there religious division, but there also was no political unity. Thus, the Dutch could exploit the rivalries among local rulers and did so with the help of Chinese, whom they brought to the ''Netherlands East Indies'' to assist in the colonial administration.

This administration was designed to exploit the ''spice islands'' as much as possible. The forced-crop and forced-labor practices the Dutch installed under what they called the *Cultuur Stelsel* (Culture System) were extremely harsh. The system did what it was designed to do: sugar, coffee, indigo, tea, tobacco, tapioca, and other products streamed through the ports of the islands on their way to European consumers. But word of its cruelties reached the Netherlands, and there was an outcry against it. By that time, however, the diverse Indonesians

SINGAPORE

FIGURE 10–15 The ultramodern skyline of the beehive city-state of Singapore, one of the Four Tigers of the East Asian/Western Pacific rim.

Singapore, in 1965, became Southeast Asia's smallest independent state in terms of territory (just over 240 square miles/600 square kilometers) as well as in population (2.7 million in 1991). Situated at the southern tip of the Malay Peninsula where the Straits of

had a common unifying cause: opposition to the Dutch and a desire for independence.

The Culture System was officially abandoned in 1870 (although the forced growing of coffee continued until 1917), and a new and more liberal colonial policy evolved. But throughout the first half of this century the Indonesian

drive for self-determination intensified, and following World War II independence was proclaimed in 1945. The Dutch then fought a losing battle to regain their East Indies, finally yielding in 1949. The easternmost territory of Irian Jaya (West Irian) on New Guinea was awarded to Indonesia as recently as 1969; and formerly

Malacca open into the South China Sea and the waters of Indonesia (Fig. 10–1, inset map), this port city had been a part of Britain's Southeast Asian empire and (briefly) a member state in the Malaysian federation. It had grown to become the world's fourth-largest port by number of ships served, and it was always a distinct and individual entity—physically (the Johor Strait separates the island from Malaysia's mainland) and culturally (the population is 77 percent Chinese, 15 percent Malay, and 6 percent South Asian). When Singapore seceded from Malaysia in 1965, a major conflict between city and federation was averted.

This city-state has thrived since independence as Southeast Asia's only developed country (Fig. 10–15), capitalizing on its relative location and its function as an entrepôt between the Malay Peninsula, Southeast Asia, Japan, and the other industrialized nations. Crude oil from the Middle East is unloaded and refined at Singapore, then shipped to Asian destinations. Raw rubber from the adjacent peninsula and Sumatera is shipped to Japan, the United States, China, and other countries. Timber from Malaysia, rice, and spices are also processed and forwarded via Singapore; in turn, automobiles, machinery, and equipment are imported for transshipment to Malaysia and other countries in the realm.

Significantly, Singapore's manufacturing sector is expanding rapidly and has now overtaken the entrepôt function in terms of its contribution to the national income. Taking a page from Hong Kong's book and aware of world price fluctuations and their effects on international trade, Singapore's planners have encouraged the diversification of industries. Foreign investment is attracted and given very advantageous terms. In addition to the refineries, Singapore now has shipbuilding and repair yards, food-processing plants, sawmills, and a growing number of small industries serving the local market. High technology is particularly stressed in the 1990s as the city-state relishes its role as one of the Four Tigers of the western Pacific (see p. 538). In this regard, Singapore is most actively developing its services and consultancy functions, based on its growing technological expertise and increasingly well-educated labor force. Agrotechnology is a leading concern today, and several new industrial parks specialize in food research that can be quickly applied to improve crop yields in the realm's less developed countries.

Portuguese Timor became Indonesia's 27th province in 1976, only after Indonesian armed forces had invaded the area.

The Dutch had chosen Jawa as their colonial headquarters, and established their capital city (Batavia, now renamed Jakarta [9.9 million]) there. Today, Jawa is still the undisputed core area of Indonesia; with about 120 million inhabitants, Jawa is one of the world's most densely peopled places and one of the most agriculturally productive (Fig. 10–16). In no other way—population, urbanization, communications, productivity—can any of the other islands compare. Sumatera, much larger in size, has under 40 million inhabitants; this island also ranks second in economic terms, with large rubber plantations on its eastern coastal lowland. Kalimantan (the Indonesian name for their portion of the island of Borneo) is huge but sparsely populated by only about 9 million residents; Indonesia shares Borneo not only with Malaysia, but also with the Islamic Sultanate of Brunei (see *box*). And Sulawesi, east of Borneo, has an estimated 13 million inhabitants, barely more than one-tenth of Jawa. Thus Jawa dominates, a situation that in a fragmented domain can produce problems. Distance and water inhibit circulation and contact, and Indonesia has coped with secessionist uprisings on several of its islands. Old differences and disagreements tend to emerge again after the common enemy has been defeated.

Viewed from this perspective, the persistence of Indonesia as a unified state is another politico-geographical wonder on a par with post-colonial India. Wide waters and high mountains have helped perpetuate cultural distinctions and differences; centrifugal political forces have been powerful and, on more than one occasion, nearly pulled the country apart. But Indonesia's national integration appears to have strengthened, just as the country's motto—*Unity in Diversity*—underscores. With more than 300 discrete ethnic clusters, over 250 individual languages, and just about every religion practiced in the world, Indonesian nationalism has faced enormous odds—overcome to some extent by development based on the archipelago's considerable resource base. This includes sizable petroleum reserves, large rubber and palm oil plantations (Sumatera shares the Malay Peninsula's environments), extensive lumber resources, major tin deposits offshore from eastern Sumatera, and soils that produce tea, coffee, and other cash crops. However, Indonesia's large population continues to grow at the annual rate of 2.0 percent, a long-term threat to the country's future—and rice and wheat already figure prominently among annual imports.

What Indonesia has achieved is etched against the country's continuing cultural complexity. There are dozens of distinct aboriginal cultures; virtually every coastal community has its own roots and traditions. And the majority, the rice-growing Indonesians, include not only the numerous Jawanese—who are Muslims largely in name only and have their own cultural identity—but also the Sundanese (who constitute 14 percent of Indonesia's population), the Madurese (8 percent), and others. Perhaps the best impression of the cultural mosaic comes from the string of islands that extends eastward from Jawa to Timor (Fig. 10–1). The rice-growers of Bali are

BRUNEI

Brunei is an anomaly in Southeast Asia—an oil-exporting Islamic sultanate far from the Persian Gulf. Located on the north coast of Borneo, sandwiched between Malaysian Sarawak and Sabah, the Brunei sultanate is a former British-protected remnant of a much larger Islamic kingdom that once controlled all of Borneo and areas beyond. Brunei achieved full independence in 1984. With a mere 2225 square miles (5700 square kilometers)—slightly larger than Delaware—and only 285,000 people, Brunei is dwarfed by the other political entities of Southeast Asia. But the discovery of oil in 1929 (and natural gas in 1965) heralded a new age for this remote territory. Today, Brunei is one of the largest oil producers in the British Commonwealth, and new offshore discoveries suggest that production will increase. As a result, the population is growing rapidly by immigration (64 percent of Brunei's residents are Malay, 20 percent Chinese), and the sultanate enjoys one of the highest standards of living in Southeast Asia (per capita income in 1987 was U.S. $20,000). Most of the people live near the oilfields in the western corner of the country and in the capital in the east (Bandar Seri Begawan). The evidence of a development boom can be seen in modern apartment houses, shopping centers, and hotels—a sharp contrast to many other towns on Borneo. There are some marked internal contrasts as well; Brunei's interior still remains an area of subsistence agriculture and rural isolation, virtually untouched by the modernization of the coastal zone.

FIGURE 10–16 The relentless advance of Jawa's paddyfields is occurring in order to keep pace with the rapidly growing population. As in China, the terracing of steep slopes is increasingly necessary, but the rich volcanic soils permit upland rates of agricultural productivity that can be as high as the output of the valleys and coastal plains.

mainly Hindus; the population of Lombok is mainly Muslim, with some Balinese Hindu immigrants. Sumbawa is a Muslim community, but the next island, Flores, is mostly Roman Catholic. On Timor Protestant groups predominate, and this island remains marked by its long-time division into a Dutch-controlled and a Portuguese-owned sector (Fig. 10–7).

An independence movement on former Portuguese Timor was subdued by invading Indonesian forces, a campaign that was followed by severe dislocation and famine. Nor is Indonesia's relationship with its easternmost island sector of New Guinea yet stable. The people of New Guinea are Papuans, not Indonesians (Fig. 10–6). They traded one foreign master (the Dutch) for another, and there has been resistance to Indonesia's administration. With the Papuan state of Papua New Guinea across the border on the eastern half of the island of New Guinea (see map p. 584), Indonesia—now itself branded as a colonial power—faces yet another challenge.

Philippine Fragmentation

After Indonesia and Vietnam, the Philippines, with 69 million people, is Southeast Asia's next most populous state (Fig. 10–17). However, few of the generalizations that can be made about the realm would apply without qualification to this island-chain country, and the Philippines' location relative to the mainstreams of change in this part of the world has had much to do with this. The islands, inhabited by peoples of Malay ancestry with Indonesian strains, shared with much of the rest of Southeast Asia an early period of Indian cultural influence, strongest in the south and southwest and diminishing northward. Next came a Chinese invasion, felt more strongly on the largest island of

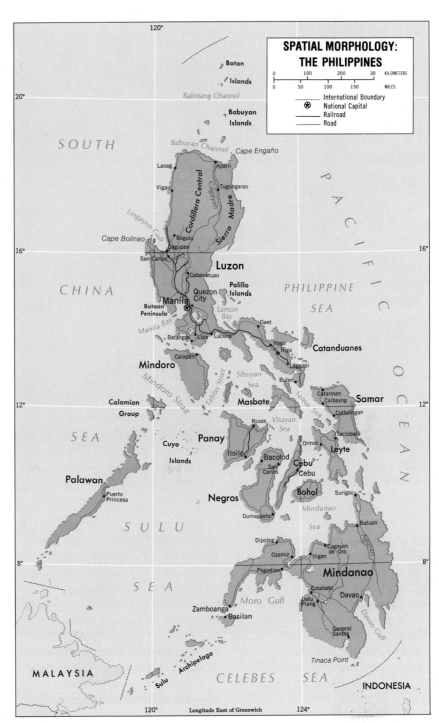

FIGURE 10–17

Luzon in the northern part of the Philippine archipelago. Islam's arrival was delayed somewhat by the position of the Philippines well to the east of the mainland and to the north of the Indonesian islands. The southern Muslim beachheads, however, were soon overwhelmed by the Spanish invasion of the six-teenth century; today, the Philippines, adjacent to the world's largest Muslim state (Indonesia), is 83 percent Roman Catholic, 9 percent Protestant, and only 5 percent Muslim.

Out of the Philippine melting pot, where Mongoloid-Malay, Arab, Chinese, Japanese, Spanish,

FIGURE 10–18   An aerial view of the heart of Manila. The Philippine capital is bisected by the Pasig River, and its center is as densely built up as any Western CBD.

and American elements have met and mixed, has emerged the distinctive culture of the Filipino. It is not a homogeneous or a unified culture, but in Southeast Asia it is in many ways unique. One example of its absorptive qualities is demonstrated by the way the Chinese infusion has been accommodated: although the "pure" Chinese minority numbers a mere 1.5 percent of the population (far less than in most Southeast Asian countries), a much larger portion of the Philippine population carries a decidedly Chinese ethnic imprint. What has happened is that the Chinese have intermarried, producing a sort of Chinese-mestizo element that constitutes more than 10 percent of the total population. In another cultural sphere, the country's ethnic mixture and variety are paralleled by its great linguistic diversity. Nearly 90

Malay languages, major and minor, are spoken by the approximately 70 million people of the Philippines; only about 1 percent still use Spanish. Visayan is the language most commonly spoken, and about 50 percent of the population is able to use English. At independence in 1946, the largest of the Malay languages, Tagalog or Pilipino, was adopted as the country's official language, and its general use is strongly promoted through the educational system. English is learned as a subsidiary language and remains the chief *lingua franca*; an English-Tagalog hybrid ("Taglish") is increasingly heard today, remarkably cutting across all levels of society.

The widespread use of English in the Philippines, of course, results from a half-century of American rule and influence, beginning in 1898 when the islands were ceded

to the United States by Spain under the terms of the treaty that followed the Spanish-American War. The United States took over a country in open revolt against its former colonial master and proceeded to destroy the Filipino independence struggle, now directed against the new foreign rulers. It is a measure of the subsequent success of U.S. administration in the Philippines that this was the only dependency in Southeast Asia that during World War II sided against the Japanese in favor of the colonial power. United States rule had its good and bad features, but the Americans did initiate reforms that were long overdue, and they were already in the process of negotiating a future independence for the Philippines when the war intervened in 1941.

The reforms begun by the United States in the Philippines and

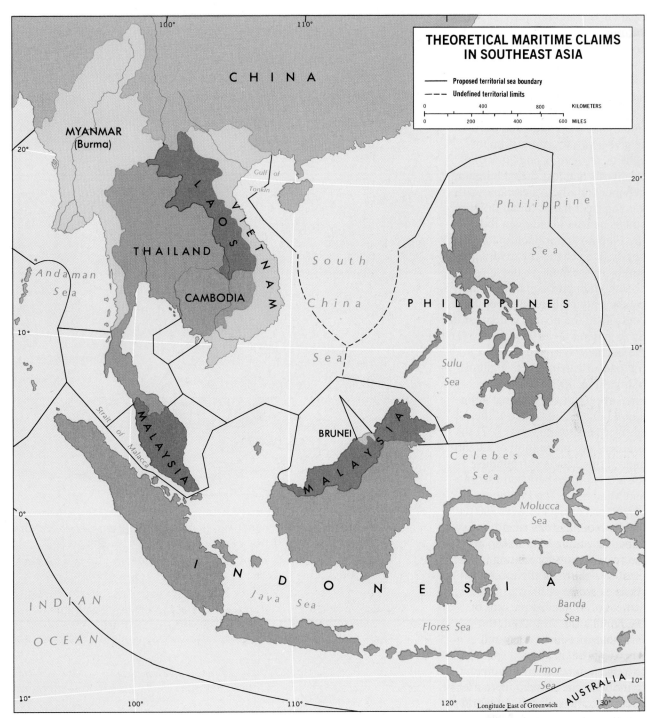

THEORETICAL MARITIME CLAIMS
IN SOUTHEAST ASIA

—— Proposed territorial sea boundary
- - - Undefined territorial limits

FIGURE 10–19

continued by the Filipinos themselves after postwar independence in 1946 were designed to
eliminate the worst aspects of
Spanish rule. Spanish provincial control had been facilitated
through allocations of good farmland as rewards to loyal representatives of church and state. The

quick acceptance of Catholicism
and its diffusion throughout the islands helped consolidate this system; exploitation of land and labor
(often by force) were the joint objectives of the priests and political
rulers alike. When, after three centuries of such exploitation, Spanish colonial policy showed signs of

relaxation during the nineteenth
century, it was too late. Crops
from the Americas were introduced (tobacco became one lucrative
product), and the belated effort
was made to integrate the Philippine economy with that of Spanish-
influenced America. But what
the Philippines needed most the

Spaniards could not provide: land reform. As everywhere else in the colonial world, the main issue between the colonizers and the colonized in the Philippines was land—agricultural land. And, as elsewhere, the Spanish colonizers found that what had been easy to give away was almost impossible to retrieve. Long after the Spaniards had lost their Philippine dependency, the Americans, with a much freer hand, found the same still to be true—and even today the Filipino government, after nearly five decades of sovereignty, still faces the same issue.

The Philippines's population, concentrated where the good farmlands lie in the plains, is densest in three general areas (Fig. I–14): (1) the northwestern and south-central part of Luzon (metropolitan Manila [Fig. 10–18], with over one-seventh of the national population [10.2 million], lies at the southern end of this zone); (2) the southeastern proruption of Luzon; and (3) the islands surrounding the Visayan Sea between Luzon and Mindanao. The Philippine archipelago consists of over 7000 mostly mountainous islands, of which Luzon and Mindanao are the two largest (accounting for almost two-thirds of the total area). About a dozen of these islands, however, contain 95 percent of the population. In Luzon, the farmlands producing rice and sugarcane lie on alluvial soils; in extreme southeastern Luzon and in the Visayan Islands, there are good volcanic soils. When world market prices are high, sugar is the most valuable export of the agriculture-dominated Philippines; timber, copra, and coconut oil are also major exports, but most Filipino farmers are busy raising the subsistence crops, rice and corn. As in the other Southeast Asian countries, there is a considerable range of supporting food crops.

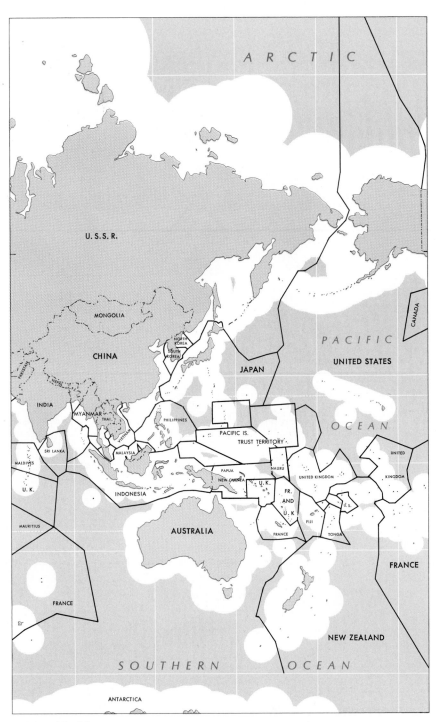

FIGURE 10–20

Unlike the other countries in this realm, the Philippine state now faces a major overpopulation problem. Its annual rate of natural increase currently stands at 2.8 percent, which yields a doubling time of only 25 years. Undoubtedly, a contributing factor in recent population growth is the influence of the Roman Catholic church, one of the Third World's most conservative on the issue of birth control and family planning. A disturbing parallel trend today is the proliferation of poverty and malnutrition in many parts of the

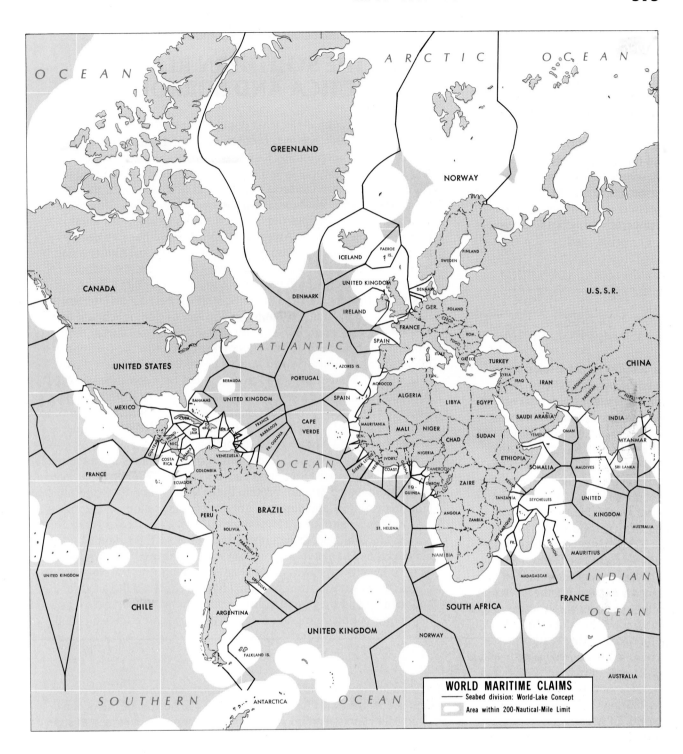

WORLD MARITIME CLAIMS
— Seabed division: World-Lake Concept
☐ Area within 200-Nautical-Mile Limit

Philippines, which portends disaster if the 69 million Filipinos of 1991 are really to become 138 million over the next quarter-century.

Thus, the Philippine state exhibits both rule and exception in Southeast Asia. Culturally and demographically, it is in several ways unique; economically, it shares with the rest of Southeast Asia the liability of unbalanced and painfully slow development. The Philippines' lumber and agricultural exports are supplemented by copper and iron ores; revenues must be spent on the purchase of machinery, various kinds of equipment, fuels, and even some foods. All this exemplifies the country's persistent underdevelopment, and regional frustrations are in the process of deepening a political crisis that has been building for many years. That crisis came to a head in 1986 with the ouster of the long-entrenched regime of Ferdinand Marcos. He was replaced by Corazon Aquino (widow of the assassinated opposition leader, Benigno Aquino), whose government has been bedeviled by numerous internal difficulties, including a troublesome commuist-led insurgency; in the difficult effort to restore political and economic stability, democracy struggles to survive.

NOT

LAND AND SEA

In a realm of peninsulas and islands such as Southeast Asia, the surrounding and intervening waters are of extraordinary significance. They may afford more effective means of contact than the land itself; as trade and migration routes, they sustain internal as well as external circulation. On the other hand, the waters between the individual islands of a fragmented state also function as a divisive force. There are literally tens of thousands of islands in insular Southeast Asia; although some are productive and in effective maritime contact with other parts of the realm and beyond, many of these islands and islets are comparatively isolated. In this respect, of course, Southeast Asia is not unique in the world. In previous chapters, we discussed the peninsular character of Western Europe and the insular nature of Caribbean America.

In all these regions, peoples and governments possess a keen awareness of the historic role of the seas. States with coastlines have extended their sovereignty over some of their adjacent waters. This is not a new principle; in Europe, the concept that a

THE SOUTHERN REALM: ANTARCTICA AND SURROUNDINGS

The Southern realm is the least populous and most remote of the world's geographic regions. It consists of the Antarctic continent and the waters surrounding it—the Southern Ocean. The Antarctic landmass is almost totally covered by an enormous and thick ice-sheet; the Southern Ocean is a giant swirl of frigid water, moving in an easterly (clockwise) direction around Antarctica. This is not an attractive picture, but the Southern realm has long attracted pioneers and explorers. Antarctica's coasts have been visited by navigators from various countries, by whale and seal hunters, and by explorers who established temporary stations on the margins of the landmass and planted the flags of their nations there. Between 1895 and 1914, the journey to the South Pole became an international obsession; Roald Amundsen, the Norwegian, reached it first in 1911. All this led to the creation of national claims during the ensuing interwar period. The geographic effect was the partitioning of Antarctica into pie-shaped sectors centered on the South Pole (Fig. 10–21). One of the areas of contention between states was the Antarctic Peninsula (facing South America), where British, Argentinean, and Chilean claims overlapped—a situation that remains unresolved. Only a single Antarctic sector is free of such claims: Marie Byrd Land (shown in white on Fig. 10–21).

Why should states be interested in territorial claims in so remote and difficult an area? Both land and sea contain resources that may some day become crucial: protein in the waters, and fuels and minerals beneath the land surface. Antarctica (5.5 million square miles/ 14.2 million square kilometers) is almost twice as large as Australia, and the Southern Ocean is nearly as large as the Atlantic. However distant actual exploitation may be, several countries want to keep their stakes in the Southern realm. But the *claimant* states (those with territorial claims) have recognized the need for cooperation and the potential for conflict. In the 1950s, there was a major international program of geophysical research, the so-called

state should own some offshore waters is centuries old. Initially, such coastal claims—on a world scale—were quite modest. But as time went on, some states began to extend their territorial waters farther and farther out, and other countries followed suit.

These widening maritime claims soon involved not only the ocean itself, but also the seabed below, the floor of the ocean and the rocks beneath. Technological advances in recent decades have

made it possible to explore and mine the bottom of the ocean to ever greater depths, and soon only the deepest trenches may lie beyond the grasp of modern machines. Already the *continental shelves* (submerged zones adjoining continents to a depth averaging 660 feet [200 meters]) are intensively exploited, and more than 25 percent of all the oil brought to the surface each year now comes from offshore wells. Minerals are also being mined, even from the deep

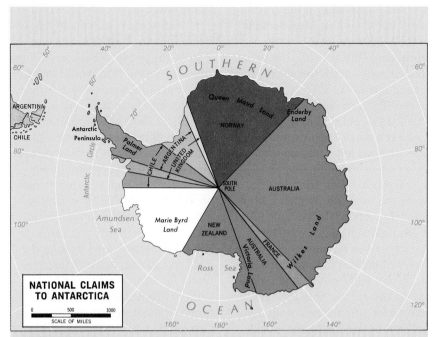

FIGURE 10–21

International Geophysical Year. The spirit of cooperation that made this program possible extended to the political sphere, and led to the 1961 signing of the Antarctic Treaty (to which 39 countries now subscribe); this agreement ensures continued scientific cooperation, prohibits military activities, safeguards the environment, and holds national claims in abeyance. But the Antarctic Treaty required review in 1991, and there has been a growing concern that it might be weakened. In particular, it does not settle the question of resource exploitation; in an age of growing national self-interest and increasing resource needs, there is the possibility that international rivalry in the Southern realm will intensify and produce a confrontation the treaty cannot prevent. The Southern realm is truly the globe's last frontier, and the partition of its lands and waters is a process fraught with dangers.

World War II ended, as many as 40 countries claimed a mere 3 nautical miles (1 nautical mile equals 1.15 statute miles or /1.85 kilometers) of territorial waters, and only 9 countries had a *territorial sea* wider than 6 nautical miles. But by 1990, only 12 countries still restricted their claim to 3 miles, whereas 99 countries claimed a 12-mile territorial sea; moreover, another 11 countries demanded sovereignty over 200 miles of territorial waters!

In Southeast Asia, Indonesia announced in 1957 that it would claim as national territory all waters within 12 nautical miles of its outer islands *and* all waters between all the islands of its far-flung archipelago (Fig. 10–19). This had the effect of making the entire Java Sea, Flores Sea, Banda Sea, and most of the Celebes Sea territorial waters. A similar claim by Malaysia, following establishment of its federation with Sarawak and Sabah, appropriated much of the southern South China Sea. These claims to territorial waters in Southeast Asia, as elsewhere, also involve the underlying continental shelf—here known to contain abundant petroleum reserves but still only partially explored.

Efforts to contain the scramble for the oceans began in the 1920s with a conference convened under the auspices of the League of Nations. In the 1950s, the United Nations organized the first U.N. Conference on the Law of the Sea (UNCLOS). This led to a series of meetings of which the most recent, UNCLOS III, lasted nearly a decade. It ended in 1982 with the signing of a convention that allowed states to claim not only a 12-mile territorial sea, but also an adjacent 12-mile contiguous zone—a zone where the coastal state may exercise certain controls (for example, over pollution). Moreover, the convention permits states to claim a so-called

floors of abyssal plains in the central areas of ocean basins. Manganese nodules, potato-sized concentrations of valuable industrial minerals, are being drawn to the surface through pipes lowered more than 11,500 feet (3500 meters) to the ocean floor; besides containing up to 40 percent manganese (an important ferroalloy in the manufacture of steel), these nodules also yield copper, nickel, and cobalt.

These developments place the

underdeveloped countries in a disadvantageous position, for they do not possess the capital and technology to extract the resources contained within their own offshore zones. They do, however, have collective influence in the United Nations, and their response has been to extend their claims to territorial waters far beyond prevailing limits. This process has become known as the "scramble for the oceans," and it is still in progress today. When

Exclusive Economic Zone (EEZ), extending up to 200 nautical miles from the coast (or 188 miles from the outer edge of a 12-mile territorial sea). In this wide ocean space, coastal states can control fishing by other countries' fleets, mineral exploration, and other activity. And the coastal state may sell leases to corporations that want to explore, and perhaps exploit, resources beneath the continental shelf.

But many countries face waters that are much less than 400 miles wide, and the effect of the EEZ allocation was to reduce greatly the remaining oceanic high seas. As the map of Southeast Asia (Fig. 10–19) shows, waters between Malaysia and its neighbors, when divided by median lines (lines drawn midway between two states' coastlines), leave no free and open sea in that area. So it

is in the North Sea, in the Caribbean, in the Gulf of Mexico, and in the Mediterranean. On the world map (Fig. 10–20), the white-shaded zones mark the approximate limits of the EEZ as defined by the 1982 convention, if claimed by all states entitled to it. Note how even small Pacific archipelagoes become huge marine territories when 200-mile arcs are drawn around their islands. (Because the map is drawn on a Mercator projection, the oceanic areas in the high latitudes are proportionally not as large as they seem.)

When viewed in the perspective of the past half-century, the scramble for the oceans now appears headed for still another stage. Already, a notion called the **world-lake concept** has made its appearance. This would involve the extension of EEZ principles not just to a 200-mile limit, but to median

lines drawn across all the world's seas and oceans, as has already been done in the North Sea (see Fig. 1–16) and the Caribbean. The resulting world map would make neighbors of Canada and Portugal, the United States and Japan. Figure 10–20 shows the approximate location of median-line boundaries drawn according to the world-lake concept. The final elimination of the last of the high seas may seem an inconceivable development at present, but it seemed equally inconceivable 50 years ago that a 200-mile priority zone such as the EEZ would ever achieve international sanction. The oceans—together with Antarctica, where a threat of territorial competition also looms (see *box*)—are this planet's last frontiers of international territorial competition and expansion.

PRONUNCIATION GUIDE

Abyssal (uh-*bissle*)

Amundsen, Roald (*ah*-moon-sun, *roh*-ahl)

Andaman (*ann*-duh-mun)

Angkor [Wat] (*ang*-kor [*wot*])

Annam (uh-*nahm*)

Annam[ese]/[ite] (anna-[*meeze*]/[*mite*])

Antarctica (ant-*ark*-tick-uh)

Aquino, Corazon/Benigno (uh-*keenoh*, *kor*-ruh-zoan/beh-*neen*-yoh)

Arakan (ahr-uh-*kahn*)

Archipelago (ark-uh-*pell*-uh-goh)

Bali[nese] (*bah*-lee [*neeze*])

Banda (*bahn*-duh)

Bandar Seri Begawan (*bun*-dahr *serry* buh-*gah*-wun)

Bataks (buh-*tahks*)

Batavia (buh-*tay*-vee-uh)

Borneo (*boar*-nee-oh)

Brunei (broo-*nye*)

Burmans (*berr*-munz)

Cambodia (kam-*boh*-dee-uh)

Celebes (*sell*-uh-beeze)

Chao Phraya (*chow* pruh-yah)

Chiang Mai (chee-*ahng*-mye)

Cochin (*koh*-chin)

Copra (*koh*-pruh)

Cultuur Stelsel (kuhl-*toor*-stell-sul)

Dien Bien Phu (dyen-byen-*fooh*)

Filipinos (filla-*pea*-noze)

Flores (*flaw*-rihss)

Hanoi (han-*noy*)

Hartshorne (*hartss*-horn)

Ho Chi Minh [see Minh, Ho Chi]

Hué (*hway*)

Irian Jaya (ih-ree-ahn *jye*-uh)

Irrawaddy (ih-ruh-*woddy*)

Jakarta (juh-*kahr*-tuh)

Java (*jah*-vuh)

Jawa (*jah*-vuh)

Johor (juh-*hoar*)

Kachins (kuh-*chinz*)

Kalimantan (kalla-*man*-tan)

Kampuchea (kahm-pooh-*chee*-uh)

Karens (kuh-*renz*)

Khmer [Rouge] (kuh-*merr* [*roozh*])

Khorat (koh-*raht*)

Kompong Thom (kahm-pong-*tom*)

Kra Isthmus (*krah iss*-muss)

Kuala Lumpur (*kwahl*-uh loom-poor)*

Lao (*lau*)

Laos (*lauss*)

Laotian (lay-*oh*-shun)

Lesotho (leh-*soo*-too)

*Each "u" in Lumpur pronounced as in "look."

Lombok   (*lahm*-bahk)
Luzon   (loo-*zahn*)
Madurese   (muh-dooh-*reece*)
Malacca   (muh-*lahk*-uh)
Malawi   (muh-*lah*-wee)
Malay   (muh-*lay*)
Malaya   (muh-*lay*-uh)
Malaysia   (muh-*lay*-zhuh)
Mekong   (*may*-kong)
Mercator   (mer-*cater*)
Mindanao   (min-duh-*nau*)
Minh, Ho Chi   (*minn*, ho-chee)
Moluccans   (muh-*luck*-unz)
Montagnards   (*mon*-tun-yardz)
Myanmar   (mee-ahn-*mah*)
New Guinea   (noo-*ghinny*)
Papua[ns]   (pahp-*oo*-uh[nz])

Philippines   (*fill*-uh-peenz)
Phnom Penh   (puh-*nom pen*)
Pilipino   (pilla-*pea*-noh)
Prorupt   (pro-*ruppt*)
Qinghai   (ching-*hye*)
Sabah   (*sahb*-ah)
Saigon   (sye-*gahn*)
Salween   (*sal*-ween)
Sarawak   (suh-*rah*-wahk)
Shan   (*shahn*)
Siam   (sye-*amm*)
Singapore   (*sing*-uh-poar)
Sinicized   (*sye*-nuh-sized)
Sri Lanka   (sree-*lahng*-kuh)
Sulawesi   (soo-luh-*way*-see)
Sultanate   (*sull*-tun-ut)
Sumatera   (suh-*mah*-tuh-ruh)

Sumatra   (suh-*mah*-truh)
Sumbawa   (soom-*bah*-wuh)
Sunda[nese]   (*soon*-duh [*neeze*])
Suryavarman   (soory-*ahva*-mun)
Tagalog   (tah-*gah*-log)
Tamil   (*tammle*)
Thailand   (*tye*-land)
Timor   (*tee*-more)
Tonkin   (*tahn-kin*)
Tonle Sap   (tahn-lay-*sap*)
Vientiane   (vyen-*tyahn*)
Viet Minh   (vee-et-*minn*)
Vietnam   (vee-et-*nahm*)
Visayan   (vuh-*sye*-un)
Xizang   (sheedz-*ahng*)
Yangon   (yahn-*koh*)
Yunnan   (yoon-*nahn*)

REFERENCES AND FURTHER READINGS

(BULLETS [•] DENOTE BASIC INTRO-
DUCTORY WORKS ON THE REALM
OR SYSTEMATIC-ESSAY TOPIC)

"Antarctica: Is Any Place Safe from Mankind?" *Time*, Cover Story, January 15, 1990, pp. 56–62.

Blake, Gerald H., ed. *Maritime Boundaries and Ocean Resources* (Totowa, N.J.: Rowman & Littlefield, 1988).

Broek, Jan O.M. "Diversity and Unity in Southeast Asia," *Geographical Review*, 34 (1944): 175–195.

Burling, Robbins. *Hill Farms and Padi Fields: Life in Mainland Southeast Asia* (Englewood Cliffs, N.J.: Prentice-Hall, 1965).

Costa, Frank J. et al., eds. *Urbanization in Asia: Spatial Dimensions and Policy Issues* (Honolulu: University of Hawaii Press, 1989).

Couper, Alastair D. "Who Owns the Oceans?" *Geographical Magazine*, September 1983, pp. 450–457.

Couper, Alastair D., ed. *The Times Atlas of the Oceans* (New York: Van Nostrand Reinhold, 1983).

de Blij, Harm J. "A Regional Geography of Antarctica and the Southern Ocean," *University of Miami Law Review*, 33 (1978): 299–314.

• Dobby, Ernest H.G. *Southeast Asia* (London: London University Press, 11 rev. ed., 1973).

Drake, Christine. *National Integration in Indonesia: Patterns and Policies* (Honolulu: University of Hawaii Press, 1989).

• Dutt, Ashok K., ed. *Southeast Asia: Realm of Contrasts* (Boulder, Colo.: Westview Press, 3 rev. ed., 1985).

• Dwyer, Denis J., ed. *South East Asian Development* (New York: Wiley/Longman, 1990).

• Fisher, Charles A. *Southeast Asia: A Social, Economic and Political Geography* (New York: E.P. Dutton, 2 rev. ed., 1966).

• Fryer, Donald W. *Emerging South-East Asia: A Study in Growth and Stagnation* (New York: John Wiley & Sons, 2 rev. ed., 1979).

Glassner, Martin I. *Neptune's Domain: A Political Geography of the Sea* (Winchester, Mass.: Unwin Hyman, 1990).

• Glassner, Martin I. & de Blij, Harm J. *Systematic Political Geography* (New York: John Wiley & Sons, 4 rev. ed., 1989).

Hartshorne, Richard. "Suggestions on the Terminology of Political Boundaries," *Annals of the Association of American Geographers*, 26 (1936): 56–57.

• Hill, Ronald D., ed. *South-East Asia: A Systematic Geography* (New York: Oxford University Press, 1979).

• Hill, Ronald D. & Bray, Jennifer M., eds. *Geography and the Environment in Southeast Asia* (Hong Kong: Hong Kong University Press, 1978).

Johnston, Douglas M. & Saunders, Phillip M. *Ocean Boundary-Making: Regional Issues and Developments* (London & New York: Routledge, 1988).

Karnow, Stanley. *In Our Image: America's Empire in the Philippines* (New York: Random House, 1989).

• Kasperson, Roger E. & Minghi, Julian V. *The Structure of Political Geography* (Chicago: Aldine, 1969).

Kumar, Raj. *The Forest Resources of Malaysia: Their Economics and Development* (Singapore: Oxford University Press Singapore, 1986).

Leinbach, Thomas R. & Sien, Chia Lin, eds. *South-East Asian Transport* (New York: Oxford University Press, 1989).

Leinbach, Thomas R. & Ulack, Richard. "Cities of Southeast

THE PACIFIC WORLD AND ITS ISLAND REGIONS

Between the Americas to the east and Asia and Australia to the west lies the vast Pacific Ocean, larger than all the world's land areas combined. In this great ocean lie tens of thousands of islands, some large (New Guinea is by far the biggest), most small (many are uninhabited). This fragmented, culturally complex realm, despite the preponderance of water, does possess regional identities. It includes the Hawaiian Islands, Tahiti, Fiji, Tonga, Samoa—fabled names in a world apart.

Indonesia and the Philippines are not part of the Pacific realm; neither are Australia and New Zealand. Before the European invasion, Australia and New Zealand would have been included— Australia as a discrete Pacific region on the basis of its indigenous black population, and New Zealand because its Maori population has Polynesian affinities. But black Australians and Maori New Zealanders have been engulfed by the Europeanization of their coun-

TEN MAJOR GEOGRAPHIC QUALITIES OF THE PACIFIC REALM

1. The Pacific realm's total area is the largest of all geographic realms. Its land area, however, is among the smallest.
2. The bulk of the land area of the Pacific realm lies on the island of New Guinea.
3. Papua New Guinea, with an estimated population of 4.1 million, alone contains over three-fifths of the Pacific realm's population.
4. The Pacific realm consists of three regions: Melanesia (including New Guinea), Micronesia, and Polynesia.
5. The Pacific realm is the most markedly fragmented of all world realms.
6. The Pacific realm's islands and cultures may be divided into volcanic "high-island" cultures and coral "low-island" cultures.
7. The Hawaiian Islands, the 50th state of the United States, lie in the northern sector of Polynesia. As in New Zealand, indigenous culture has been submerged under Westernization.
8. In Polynesia, local culture is nearly everywhere under severe strain in the face of external influences.
9. Indigenous Polynesian culture has a remarkable consistency and uniformity throughout the Polynesian region, its enormous dimensions and dispersal notwithstanding.
10. The Pacific realm is in politico-geographical transition as islands attain independence or redirect their political associations.

tries, and the regional geography of Australia and New Zealand today is dominantly Western, not Pacific. Only on the island of New

Guinea do Pacific peoples remain the dominant cultural element. Although the realm's contents are fragmented, scattered, and remote, the Pacific World does possess three distinct regions: Melanesia, Micronesia, and Polynesia.

MELANESIA

New Guinea lies at the western end of a Pacific region that extends eastward to Fiji and includes the Solomon Islands, Vanuatu, and New Caledonia (Fig. P–1). These islands are inhabited by Melanesian peoples who have very dark skins and dark hair (*melas* means black); the region as a whole is called Melanesia. Some cultural geographers include the Papuan peoples of New Guinea in the Melanesian race, but others suggest that the Papuans are more closely related to the aboriginal (indigenous) Australians. In any case, Melanesia is by far the most populous Pacific region (ca. 5.6 million in 1991). New Guinea alone has a population of more than 7 million (although statistics are unreliable), but this island is divided into two halves by a geometric boundary that separates non-Melanesian Irian Jaya (West Irian), now an Indonesian province, from independent Papua New Guinea (P.N.G.).

With an estimated 4.1 million inhabitants today, P.N.G. became a sovereign state in 1975 after nearly a century of British and Australian administration. It is one of the world's poorest and least developed countries, with much of the mountainous interior—where the Papuan population is clustered in tiny villages (Fig. P–2)—hardly touched by the changes that transformed neighboring Australia. The largest town and capital, Port Moresby, has more than 300,000 residents; only about 15 percent of the people of P.N.G. live in urban areas, but the current rate of urbanization is phenomenal. Although English is used by the educated minority, about 55 percent of the population remains illiterate and over 700 languages are spoken by the Papuan and Melanesian communities. The Melanesians are concentrated in the northern and eastern coastal areas of the country, and here, as in the other islands of this region, they grow root crops and bananas for subsistence. However, recent discoveries of major mineral deposits (copper, gold, oil) point to the country's bright development potential; there is also enormous scope for increasing the output of such profitable export crops as palm oil, coffee, and cocoa.

MICRONESIA

North of Melanesia and east of the Philippines lie the islands that comprise the region known as Micronesia (*micro* means small). In this case, the name refers to the size of the islands, not the physical appearance of the population. The 2000-plus islands of Micronesia are not only small (many of them no larger than one square mile), but they are also much lower-lying, on an average, than those of Melanesia. There are some volcanic islands (''high islands,'' as the people call them), but they are outnumbered by islands composed of coral, the ''low islands'' that barely lie above sea level. Guam, with 210 square miles (550 square kilometers), is Micronesia's largest island, and no island elevation anywhere in Micronesia reaches 3300 feet (1000 meters).

The region until 1986 was largely a U.S. Trust Territory (the last of the post–World War II trusteeships supervised by the United Nations), but that status has now changed. The former U.S.-administered territory is today divided into four island groups: the Northern Mariana Islands, the Republic of Palau, the Federated States of Micronesia, and the Republic of the Marshall Islands (Fig. P–1). The last two political units have achieved limited sovereignty

FIGURE P–2 A typical village landscape in underdeveloped, interior Papua New Guinea. ''P.N.G.'' contains more than one-half of the Pacific realm's total population.

FIGURE P—1

under the continued guidance and financial support of the United States. The Northern Marianas chose to become a U.S. commonwealth, along the lines of Puerto Rico. But Palau continues as the last of the trust territories, scheduled to become independent in "free association" with the United States when 75 percent of its citizens agree to the terms of a U.S. sovereignty compact; in referenda held thus far, Palauan voters have not yet given their approval.

The Micronesians are not nearly as numerous as the Melanesians (totaling only about 400,000), but they nevertheless comprise a distinct racial group in the Pacific realm. Culturally, it is useful to distinguish their communities as **high-island cultures** based on the better-watered volcanic islands where agriculture is the mainstay, or **low-island cultures** on the sometimes drought-plagued coral islands where fishing is the chief

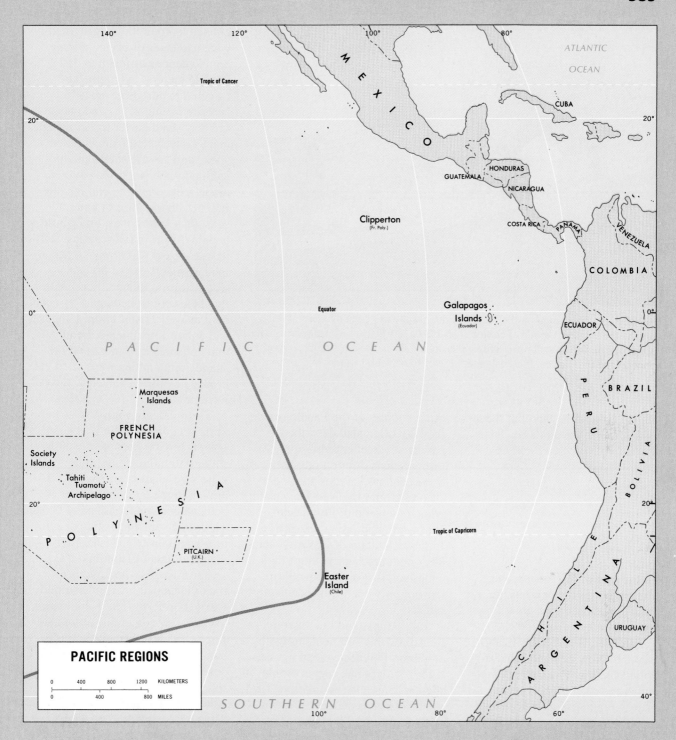

PACIFIC REGIONS

mode of subsistence. These numerous Micronesian communities have developed a large number of locally spoken languages, many of them mutually unintelligible.

But these islands do not exist in isolation. There is a certain complementarity between the islands of farmers and the islands of fishing people, and Micronesians—especially the low-islanders—are skilled seafarers. The trade for food and basic needs encourages circulation; sometimes the threat of a devastating typhoon (hurricane) compels the low-islanders to seek the safety of higher ground elsewhere. Thus, movement in the Pacific realm has always been by water. Even today many islanders are expert boaters, often choosing a water route when a road is also available. In Fig. P–3, a canoe loaded with people and goods sets off from Suva (the capital of Fiji) to a small town a few miles away; the trip could also be made by land

FIGURE P–3   Movement in the Pacific realm has always been by water, which remains the transport mode of choice. This canoe has just set off from Suva, the capital of Fiji, for a small town a few miles away. Even though a modern road connects the two places, these islanders prefer to travel by water.

vehicle around the bay, but these Fijians prefer the more familiar alternative.

POLYNESIA

To the east of Micronesia and Melanesia lies the heart of the Pacific, enclosed by a great triangle stretching from the Hawaiian Islands to Chile's Easter Island to New Zealand. This is Polynesia (Fig. P–1), a region of numerous islands (*poly* means many), ranging from volcanic mountains rising above the Pacific's waters (Mauna Kea on Hawaii reaches nearly 13,800 feet/ over 4200 meters), clothed by luxuriant tropical forests and drenched by well over 100 inches of rainfall each year, to low coral atolls where a few palm trees form the only vegetation and where drought is a persistent and recurrent problem. The Polynesians have somewhat lighter-colored skin and wavier hair than do the

other peoples of the Pacific realm; they are often also described as having an excellent physique. Anthropologists differentiate between these original Polynesians and a second group, the Neo-Hawaiians, who are a blend of Polynesian, European, and Asian ancestries. In the U.S. state of Hawaii—actually an archipelago of more than 130 islands—Polynesian culture has not only been Europeanized, but also Orientalized.

Its vastness and the diversity of its natural environments notwithstanding, Polynesia clearly constitutes a geographic region within the Pacific realm (population: 1.8 million). Polynesian culture, spatially fragmented though it is, exhibits a remarkable consistency and uniformity from one island to the next, from one end of this widely dispersed region to the other; this consistency is particularly expressed in vocabularies, technologies, housing, and art forms. The Polynesians are uniquely adapted to their **maritime environment**, and long before European sailing ships began to arrive in their waters, Polynesian seafar-

ers had learned to navigate their wide expanses of ocean in huge double canoes as much as 150 feet (45 meters) in length. They traveled hundreds of miles to favorite fishing zones and engaged in inter-island barter trade, using maps constructed from bamboo sticks and cowrie shells and navigating by the stars. However, modern descriptions of a Pacific Polynesian paradise of emerald seas, lush landscapes, and gentle people distort harsh realities. Polynesian society was forced to accommodate much loss of life at sea when storms claimed their boats, families were ripped apart by accident as well as migration, hunger and starvation afflicted the inhabitants of smaller islands, and the island communities were often embroiled in violent conflicts and cruel retributions.

The political geography of Polynesia is complex. The Hawaiian Islands in 1959 became the 50th state to join the United States. The state's population surpassed 1.1 million in the late 1980s, over 80 percent living on the island of Oahu. There, the superimposition of cultures is exquisitely illustrated by the panorama of Honolulu's skyscrapers against the famous extinct volcano at nearby Diamond Head (Fig. P–4). The Kingdom of Tonga, now containing a population of 105,000, became an independent country in 1970 after seven decades as a British protectorate. The British-administered Ellice Islands were renamed Tuvalu; along with the Gilbert Islands to the north (now renamed Kiribati), they received independence from Britain in 1978. Other islands continued under French control (including the Marquesas Islands and Tahiti), under New Zealand's administration (Rarotonga), and under British, U.S., and Chilean flags.

In the process of politico-geographical fragmentation, Polynesian culture has been dealt some

FIGURE P–4   The circle closes: our world regional survey returns us to the developed world of familiar Waikiki Beach and Honolulu, capital of the fiftieth U.S. state.

severe blows. Land developers, hotel builders, and tourist dollars have set Tahiti on a course along which Hawaii has already traveled far. The Americanization of eastern Samoa has created a new society quite different from the old. Polynesia has lost much of its ancient cultural consistency; today, the region is a patchwork of new and old—the new often bleak and barren with the old under intensifying pressure.

In the early 1990s, the diverse island states and dependencies of the Pacific realm are beginning to cooperate in the search for solutions to their many shared problems. A key development is the recent emergence of the South Pacific Forum, whose 13 members include Australia, the Cook Islands, Fiji, Kiribati, Nauru, New Zealand, Niue, Papua New Guinea, the Solomon Islands, Tonga, Tuvalu, Vanuatu, and Western Samoa. The Forum's agencies have helped foster regional cooperation on fisheries, tourism, and air and sea transportation. But today this organization is moving actively toward discussion and confrontation of the realm's major political issues—increasing Western aid to accelerate economic development, expediting the termination of colonialism in French New Caledonia and the American trust territories (almost completed in Micronesia, as we saw), and the banning of nuclear weapons testing, particularly by France.

The heightening competition of the superpowers is another concern, propelled by the recently enlarged presence of the Soviet Union, which has secured fishing rights to the waters of Kiribati and established diplomatic relations with Vanuatu, Fiji, and a number of other island countries. The United States is carefully watching these Soviet moves because Moscow's new landing rights and access to port facilities here could lead to an attempt to alter the existing military and geopolitical balance in the U.S.-dominated Pacific Basin. The South Pacific Forum is headquartered in the modern, centrally-located Fijian capital of Suva and has so far proven to be an effective institution in bringing together Melanesians, Micronesians, Polynesians, and their transplanted European counterparts from Australia and New Zealand to define their mutual problems, discuss their common causes, and debate the future course of their unique oceanic realm.

PRONUNCIATION GUIDE

Atoll (*ay*-tole)
Ellice (*ell*-uss)
Fiji (*fee*-jee)
Fijian (fuh-*jee*-un)
Guam (*gwahm*)
Irian Jaya (*ih*-ree-ahn *jye*-uh)
Kiribati (*kih*-ruh-bahss)
Maori (*mah*-aw-ree/*mau*-ree)
Mariana (marry-*anna*)
Marquesas (mahr-*kay*-suz)
Mauna Kea (mau-nuh-*kay*-uh)

Melanesia (mella-*nee*-zhuh)
Micronesia (mye-kroh-*nee*-zhuh)
Nauru (nah-*oo*-roo)
New Caledonia (noo-kalla-*doan*-yuh)
New Guinea (noo-*ghinny*)
Niue (nee-*oo*-ay)
Oahu (uh-*wah*-hoo)
Palau (puh-*lau*)
Papua New Guinea (pahp-*oo*-uh noo-*ghinny*)

Papuans (pahp-*oo*-unz)
Polynesia (polla-*nee*-zhuh)
Port Moresby (port *morz*-bee)
Rarotonga (rarra-*tahng*-guh)
Samoa (suh-*moh*-uh)
Suva (*soo*-vuh)
Tahiti (tuh-*heet*-ee)
Tonga (*tahng*-guh)
Tuvalu (too-*vahl*-oo)
Vanuatu (vahn-uh-*wah*-too)
Waikiki (*wye*-kuh-*kee*)

REFERENCES AND FURTHER READINGS

(BULLETS [•] DENOTE BASIC INTRO-DUCTORY WORKS)

Atlas of Hawaii (Honolulu: University Press of Hawaii, 2 rev. ed., 1983).

• Brookfield, Harold C., ed. *The Pacific in Transition: Geographical Perspectives on Adaptation and Change* (New York: St. Martin's Press, 1973).

• Brookfield, Harold C. & Hart, Doreen. *Melanesia: A Geographical Interpretation of an Island World* (New York: Barnes & Noble, 1971).

Bunge, Frederica M. & Cooke, Melinda W., eds. *Oceania: A Regional Study* (Washington, D.C.: U.S. Government Printing Office, 1984).

Campbell, Ian C. *A History of the Pacific Islands* (Berkeley: University of California Press, 1990).

Carter, John, ed. *Pacific Islands Yearbook* (Sydney: Pacific Publications Pty., annual).

Chapman, Murray & Prothero, R. M., eds. *Circulation in Population Movement: Substance and Concepts from the Melanesian Case* (Boston: Routledge & Kegan Paul, 1985).

Couper, Alastair D., ed. *Development and Social Change in the Pacific Islands* (London & New York: Routledge, 1989).

• Freeman, Otis W., ed. *Geography of the Pacific* (New York: John Wiley & Sons, 1951).

Friis, Herman R., ed. *The Pacific Basin: A History of Its Geographical Exploration* (New York: American Geographical Society, Special Publication No. 38, 1967).

Gourevitch, Peter A., guest ed. "The Pacific Region: Challenges to Policy and Theory," *Annals of the American Academy of Political and Social Science*, 505 (September 1989).

Grossman, Lawrence S. *Peasants, Subsistence Ecology, and Development in the Highlands of Papua New Guinea* (Princeton, N.J.: Princeton University Press, 1984).

Howard, A., ed. *Polynesia: Readings on a Culture Area* (Scranton, Pa.: Chandler, 1971).

Howells, W. *The Pacific Islanders* (New York: Scribner's, 1973).

• Howlett, Diana. *Papua New Guinea: Geography and Change* (Melbourne, Australia: Thomas Nelson, 1973).

Kissling, Christopher C., ed. *Transport and Communications for Pacific Microstates: Issues in Organization and Management* (Suva, Fiji: University of the South Pacific, Institute of Pacific Studies, 1984).

Langdon, Frank C. & Ross, Douglas A. *Superpower Maritime Strategy in the Pacific* (London & New York: Routledge, 1990).

Levison, Michael et al. *The Settlement of Polynesia: A Computer Simulation* (Minneapolis: University of Minnesota Press, 1973).

• "Mobility and Identity in the Island Pacific," Special Issue, *Pacific Viewpoint*, 26, No. 1 (1985).

Morgan, Joseph R., ed. *Hawaii* (Boulder, Colo.: Westview Press, 1983).

Oliver, Douglas L. *The Pacific Islands* (Honolulu: University Press of Hawaii, reprinted ed., 1975).

Sager, Robert J. "The Pacific Islands: A New Geography," *Focus*, Summer 1988, pp. 10–14.

Segal, Gerald. *Rethinking the Pacific* (New York: Oxford University Press, 1990).

Spate, Oskar H.K. *The Pacific Since Magellan: Monopolists and Freebooters* (Minneapolis: University of Minnesota Press, 1983).

Spate, Oskar H.K. *The Spanish Lake: A History of the Pacific Since Magellan* (Beckenham, U.K.: Croom Helm, 1979).

• Vayda, Andrew P., ed. *Peoples and Cultures of the Pacific* (New York: Natural History Press, 1968).

• Ward, R. Gerard, ed. *Man in the Pacific Islands: Essays on Geographical Change in the Pacific* (New York: Oxford University Press, 1972).

AREA AND DEMOGRAPHIC DATA FOR THE WORLD'S STATES

(SMALLEST MICROSTATES OMITTED)

	Area		Population (millions)			Annual Natural Increase (%)	Doubling Time (years)	Life Expectancy at Birth (years)	1991 Population Density	
	1000 Miles²	1000 Km²	1980	1991	2000				Per Mile²	Per Km²
WORLD	51,529.5	133,461.4	4,485.9	5,422.0	6,322.7	1.8	39	63	94	41
Developed Realms	21,206.0	54,923.5	1,135.3	1,219.6	1,267.1	0.6	122	73	58	22
Underdeveloped Realms	30,323.5	78,537.9	3,350.6	4,197.7	5,049.6	2.1	32	60	138	53
Europe	1,879.9	4,868.9	483.5	500.8	506.4	0.3	269	74	266	103
Albania	11.1	28.7	2.7	3.3	3.8	2.0	35	70	299	115
Austria	32.4	83.9	7.6	7.6	7.6	0.1	1155	75	235	91
Belgium	11.8	30.6	9.8	9.9	9.7	0.1	577	75	840	324
Bulgaria	42.8	110.9	8.9	9.0	9.0	0.1	770	71	210	81
Czechoslovakia	49.4	127.9	15.3	15.7	16.2	0.2	301	71	318	123
Denmark	16.6	43.0	5.1	5.1	5.2	−0.0	(−)	75	309	119
Finland	130.2	337.2	4.8	5.0	5.0	0.3	277	75	38	15
France	211.2	547.0	53.8	56.6	57.9	0.4	161	76	268	103
Germany, East	41.8	108.3	16.7	16.6	16.5	0.1	990	73	397	153
Germany, West	96.0	248.6	61.6	61.4	59.6	−0.1	(−)	75	639	247
Greece	51.0	132.1	9.6	10.1	10.2	0.1	630	74	197	76
Hungary	35.9	93.0	10.7	10.5	10.5	−0.2	(−)	70	294	113
Iceland	39.8	103.1	0.2	0.3	0.3	1.0	69	78	6	2
Ireland	27.1	70.2	3.4	3.6	3.7	0.8	89	73	133	51
Italy	116.3	301.2	56.2	57.6	58.2	0.0	2310	74	495	191
Luxembourg	1.0	2.6	0.4	0.4	0.4	0.1	1155	74	376	145
Malta	0.3	0.8	0.3	0.4	0.4	0.7	99	75	1,183	457
Netherlands	15.9	41.2	14.1	15.0	15.2	0.4	158	77	942	364
Norway	125.1	324.0	4.1	4.2	4.3	0.2	315	76	34	13
Poland	120.7	312.6	35.6	38.6	39.9	0.6	114	71	320	124
Portugal	34.3	88.8	9.9	10.5	10.5	0.3	257	74	305	118
Romania	91.7	237.5	22.2	23.4	24.4	0.5	141	70	255	98
Spain	194.9	504.8	37.5	39.5	40.8	0.3	210	76	202	78
Sweden	173.7	449.9	8.3	8.5	8.3	0.2	367	77	49	19
Switzerland	15.9	41.2	6.4	6.6	6.6	0.3	267	77	418	161
United Kingdom	94.2	244.0	56.0	57.5	57.3	0.2	289	75	611	236
Yugoslavia	98.8	255.9	22.3	24.0	25.0	0.6	114	71	243	94
Australia– New Zealand	3,070.1	7,951.6	17.7	20.5	22.4	0.8	87	76	7	3
Australia	2,966.2	7,682.5	14.6	17.1	18.7	0.8	88	76	6	2
New Zealand	103.9	269.1	3.1	3.4	3.6	0.8	83	74	33	13

	Area		Population (millions)			Annual Natural Increase (%)	Doubling Time (years)	Life Expectancy at Birth (years)	1991 Population Density	
	1000 Miles^2	1000 Km^2	1980	1991	2000				Per Mile^2	Per Km^2
Soviet Union	8,600.4	22,275.0	265.5	294.5	311.6	1.0	70	69	34	13
North America	7,509.9	19,450.6	251.8	279.3	296.7	0.7	97	75	37	14
Canada	3,831.0	9,922.3	24.1	26.8	28.4	0.8	92	76	7	3
United States	3,678.9	9,528.4	227.7	252.5	268.3	0.7	98	75	69	26
Japan	145.7	377.4	116.8	124.4	130.0	0.5	141	78	854	330
Middle America	1,054.7	2,731.7	121.5	154.2	181.8	2.3	30	67	146	56
Bahamas	5.4	14.0	0.2	0.3	0.3	1.2	60	71	47	18
Barbados	0.2	0.5	0.2	0.3	0.3	0.7	94	75	1,314	507
Belize	8.9	23.1	0.1	0.2	0.2	3.0	23	69	21	8
Costa Rica	19.7	51.0	2.4	3.1	3.8	2.5	28	76	158	61
Cuba	44.2	114.5	9.7	10.7	11.4	1.1	62	74	242	93
Dominican Rep.	18.7	48.4	5.8	7.4	8.6	2.5	28	65	394	152
El Salvador	8.7	22.5	4.7	5.4	6.3	2.7	26	63	621	240
Guadeloupe	0.7	1.8	0.3	0.4	0.4	1.4	51	73	501	193
Guatemala	42.0	108.8	7.1	9.5	11.8	3.1	23	62	226	87
Haiti	10.7	27.7	5.4	6.7	7.8	2.2	32	55	622	240
Honduras	43.3	112.1	3.8	5.3	6.8	3.2	22	64	122	47
Jamaica	4.2	10.9	2.2	2.6	2.9	1.7	41	74	612	236
Martinique	0.4	1.0	0.3	0.3	0.4	1.3	55	74	851	329
Mexico	761.6	1,972.5	70.1	91.0	107.2	2.4	29	68	119	46
Neth. Antilles	0.4	1.0	0.3	0.2	0.2	1.4	51	76	470	182
Nicaragua	50.2	130.0	2.5	3.8	5.1	3.5	20	62	75	29
Panama	29.8	77.2	1.9	2.5	2.9	2.2	31	72	83	32
Puerto Rico	3.4	8.8	3.2	3.4	3.4	1.2	56	74	995	384
Saint Lucia	0.2	0.5	0.1	0.2	0.2	2.2	31	71	783	302
Trinidad & Tobago	2.0	5.2	1.2	1.3	1.5	2.0	34	70	648	250
South America	6,874.7	17,805.5	242.5	301.8	352.6	2.0	35	67	44	17
Argentina	1,068.3	2,766.9	28.2	32.8	36.2	1.4	51	70	31	12
Bolivia	424.2	1,098.7	5.5	7.5	9.3	2.6	27	53	18	7
Brazil	3,286.5	8,512.0	122.4	153.5	179.5	2.0	34	65	47	18
Chile	292.1	756.5	11.0	13.4	15.3	1.6	43	71	46	18
Colombia	439.7	1,138.8	26.1	32.5	38.0	2.0	34	64	74	29
Ecuador	109.5	283.6	8.0	11.0	13.6	2.6	27	65	100	39
French Guiana	35.1	90.9	0.1	0.1	0.1	2.7	26	66	3	1
Guyana	83.0	215.0	0.8	0.8	0.8	1.9	36	66	10	4
Paraguay	157.0	406.6	3.2	4.4	5.5	2.9	24	67	28	11
Peru	496.2	1,285.2	17.6	22.4	26.4	2.1	33	61	45	17
Suriname	63.0	163.2	0.4	0.4	0.5	2.1	33	69	7	3
Uruguay	68.0	176.1	2.9	3.0	3.2	0.8	83	71	45	17
Venezuela	352.1	911.9	16.3	20.1	24.1	2.4	29	70	57	22
North Africa/ Southwest Asia	6,177.0	15,998.4	268.6	367.6	473.4	2.8	25	61	60	23
Afghanistan	250.0	647.5	15.2	15.6	26.6	2.6	26	42	62	24
Algeria	919.6	2,381.8	18.8	26.6	33.7	3.2	22	62	29	11
Bahrain	0.3	0.8	0.3	0.5	0.7	2.8	25	67	1,644	635
Cyprus	3.6	9.3	0.6	0.7	0.8	1.0	71	76	197	76
Egypt	386.9	1,002.1	42.1	57.9	71.2	2.8	24	59	150	58
Iran	636.3	1,648.0	38.8	57.6	75.7	3.6	20	62	91	35
Iraq	167.9	434.9	13.1	19.5	27.2	3.8	18	66	116	45
Israel	7.8	20.2	3.8	4.8	6.4	1.6	43	75	593	238

	Area		Population (millions)			Annual Natural Increase (%)	Doubling Time (years)	Life Expectancy at Birth (years)	1991 Population Density	
	1000 Miles²	1000 Km²	1980	1991	2000				Per Mile²	Per Km²
Jordan	37.7	97.6	3.1	4.3	5.7	3.5	20	69	113	44
Kuwait	6.9	17.9	1.4	2.2	2.8	2.8	25	73	320	123
Lebanon	4.0	10.4	2.7	3.4	4.1	2.1	33	68	860	332
Libya	679.4	1,759.6	3.0	4.3	5.6	3.1	22	65	6	2
Morocco	275.1	712.5	21.1	27.1	33.3	2.6	27	61	99	38
Oman	82.0	212.4	0.9	1.5	2.1	3.3	21	55	18	7
Qatar	4.2	10.9	0.2	0.5	0.6	2.7	25	69	110	42
Saudi Arabia	830.0	2,149.7	9.4	15.8	22.4	3.4	20	63	19	7
Somalia	246.0	637.1	5.9	8.7	10.4	2.6	26	44	35	14
Sudan	967.5	2,505.8	18.7	25.9	33.6	2.8	24	49	27	10
Syria	71.5	185.2	8.8	13.0	17.6	3.8	18	65	182	70
Tunisia	63.2	163.7	6.5	8.3	9.9	2.4	29	64	131	51
Turkey	300.9	779.3	46.0	57.8	68.6	2.2	32	64	192	74
Un. Arab Emirates	32.3	83.7	1.0	1.8	2.1	2.3	30	69	55	21
Yemen	203.9	528.1	7.2	10.1	13.6	3.4	20	49	50	19
Subsaharan Africa	8,163.8	21,144.2	365.9	524.4	699.1	2.9	24	50	64	25
Angola	481.4	1,246.8	7.0	8.9	11.5	2.6	27	43	19	7
Benin	43.5	112.7	3.5	5.0	7.1	3.0	23	45	115	44
Botswana	231.8	600.4	0.9	1.3	1.8	3.4	20	57	6	2
Burkina Faso	105.9	274.3	6.1	9.2	11.8	2.8	25	46	87	34
Burundi	10.7	27.7	4.2	5.9	7.8	3.3	21	52	549	212
Cameroon	183.6	475.5	8.6	11.4	14.5	2.6	26	50	62	24
Cent. African Rep.	240.5	622.9	2.3	2.9	3.7	2.5	28	46	12	5
Chad	495.8	1,284.1	4.4	5.1	6.3	2.0	35	39	10	4
Comoro Islands	0.8	2.1	0.4	0.4	0.7	3.3	21	55	534	206
Congo	132.0	341.9	1.6	2.4	3.2	3.4	21	56	18	7
Djibouti	8.9	23.1	0.4	0.4	0.6	3.0	23	46	48	18
Equatorial Guinea	10.8	28.0	0.3	0.4	0.4	1.9	37	45	38	15
Ethiopia	472.4	1,223.5	39.8	51.9	71.1	2.1	33	41	110	42
Gabon	103.3	267.5	0.8	1.1	1.6	1.9	36	50	11	4
Gambia	4.4	11.4	0.6	0.9	1.1	2.5	28	42	199	77
Ghana	92.1	238.5	12.1	15.5	20.4	3.1	22	53	168	65
Guinea	94.9	245.8	5.0	7.4	9.2	2.4	29	41	78	30
Guinea-Bissau	13.9	36.0	0.8	1.0	1.2	1.9	37	44	72	28
Ivory Coast	123.8	320.6	8.1	13.0	18.5	3.6	19	52	105	41
Kenya	225.0	582.8	16.4	26.1	37.6	4.1	17	57	116	45
Lesotho	11.7	30.3	1.3	1.8	2.4	2.8	25	55	156	60
Liberia	43.0	111.4	1.9	2.6	3.5	3.2	22	54	61	24
Madagascar	226.7	587.2	8.7	12.3	16.6	3.1	22	53	54	21
Malawi	45.7	118.4	6.0	9.3	11.6	3.3	21	48	204	79
Mali	478.8	1,240.1	6.9	9.4	12.3	2.9	24	43	20	8
Mauritania	398.0	1,030.8	1.5	2.1	2.7	2.7	26	45	5	2
Mauritius	0.8	2.1	1.0	1.1	1.3	1.3	54	68	1,437	555
Moçambique	302.3	783.0	12.1	16.1	20.4	2.6	27	46	53	21
Namibia	318.3	824.4	1.0	1.9	2.6	3.1	22	55	6	2
Niger	489.2	1,267.0	5.5	7.9	10.6	2.9	24	44	16	6
Nigeria	356.7	923.9	77.1	122.1	160.9	2.9	24	47	342	132
Rwanda	10.2	26.4	5.1	7.5	10.4	3.4	20	48	733	283
Senegal	76.0	196.8	5.8	7.6	9.7	2.6	27	45	99	38
Sierra Leone	27.9	72.3	3.4	4.3	5.4	2.4	29	40	153	59
South Africa	471.4	1,220.9	28.7	40.5	51.5	2.6	27	63	86	33
Swaziland	6.7	17.4	0.6	0.8	1.1	3.4	21	54	122	47
Tanzania	364.9	945.1	18.6	28.3	39.6	3.6	19	52	77	30

	Area		Population (millions)			Annual Natural Increase (%)	Doubling Time (years)	Life Expectancy at Birth (years)	1991 Population Density	
	1000 Miles^2	1000 Km^2	1980	1991	2000				Per Mile^2	Per Km^2
Subsaharan Africa (Cont.)										
Togo	21.9	56.7	2.6	3.7	4.9	3.4	20	54	168	65
Uganda	91.1	235.9	12.8	18.2	24.7	3.4	20	50	200	77
Zaïre	905.6	2,345.5	28.6	37.0	49.3	3.1	23	51	41	16
Zambia	290.6	752.7	5.8	8.8	12.2	3.7	19	52	30	12
Zimbabwe	150.8	390.6	7.6	10.9	15.2	3.6	19	57	72	28
South Asia	1,710.3	4,429.7	890.2	1,148.3	1,387.2	2.3	30	57	671	259
Bangladesh	55.6	144.0	88.1	121.3	153.4	2.8	25	52	2,181	842
Bhutan	18.1	46.9	1.3	1.6	1.9	2.1	32	48	88	34
India	1,237.1	3,204.1	685.1	871.3	1,042.5	2.2	32	58	704	272
Maldives	0.1	0.3	0.2	0.2	0.3	3.7	19	60	2,270	877
Nepal	54.4	140.9	15.0	19.6	24.4	2.5	28	52	361	139
Pakistan	319.9	828.5	85.7	116.8	145.3	2.9	24	54	365	141
Sri Lanka	25.1	65.0	14.8	17.4	19.4	1.6	43	70	695	268
China & Its Sphere	4,394.6	11,382.5	1,095.6	1,231.8	1,400.3	1.4	50	67	280	108
China	3,691.5	9,561.0	1,013.6	1,135.5	1,291.6	1.4	49	66	308	119
Hong Kong	0.4	1.0	5.0	5.8	6.3	0.8	91	76	14,490	5,595
Korea, North	46.5	120.4	17.9	23.6	28.4	2.4	29	70	508	196
Korea, South	38.0	98.4	39.6	44.2	48.8	1.3	53	68	1,164	449
Mongolia	604.3	1,565.1	1.7	2.2	2.8	2.9	24	65	4	1
Taiwan	13.9	36.0	17.8	20.4	22.4	1.1	62	73	1,469	567
Southeast Asia	1,735.4	4,494.7	361.9	463.5	547.9	2.2	32	61	267	103
Brunei	2.2	5.7	0.2	0.3	0.3	2.7	25	71	123	48
Cambodia	69.9	181.0	5.7	7.2	8.5	2.3	31	48	102	39
Indonesia	741.0	1,919.2	151.2	192.0	222.0	2.0	35	58	259	100
Laos	91.4	236.7	3.5	4.1	5.0	2.5	28	49	45	17
Malaysia	128.4	332.6	14.0	18.3	20.9	2.5	28	67	142	55
Myanmar (Burma)	261.2	676.5	34.4	42.7	51.1	2.3	30	58	164	63
Philippines	115.8	299.9	49.3	68.6	85.5	2.8	25	66	592	229
Singapore	0.2	0.5	2.4	2.7	2.9	1.2	59	73	13,723	5,299
Thailand	198.1	513.1	47.7	57.4	65.5	1.7	41	65	290	112
Vietnam	127.2	329.4	53.5	70.3	86.0	2.6	27	64	553	213
Pacific	213.0	551.7	4.4	6.0	7.3	2.7	26	56	28	11
Fiji	7.1	18.4	0.6	0.8	0.8	2.3	31	67	111	43
French Polynesia	1.5	3.9	0.2	0.2	0.3	2.4	29	71	137	53
New Caledonia	7.4	19.2	0.1	0.2	0.2	1.8	39	69	21	8
Papua New Guinea	178.7	462.8	3.0	4.1	5.1	2.7	26	54	23	9
Solomon Islands	11.5	29.8	0.2	0.3	0.5	3.6	19	69	30	12
Vanuatu	5.7	14.8	0.1	0.2	0.2	3.2	21	69	30	12
Western Samoa	1.1	2.8	0.2	0.2	0.2	2.8	25	66	175	67

OPPORTUNITIES IN GEOGRAPHY

The final section of the introductory chapter (pp. 50–53) reviewed the recent development of North American geography and the growing place of regional studies. The chapters that followed provided an idea of the wide range of topics that geographers pursue, particularly in our many discussions of concepts and the overviews of geography's subfields in the Systematic Essays. But there are specializations within each of those topics and, in an introductory book such as this, there was not enough space to discuss all of them. This appendix is designed to help you should you decide to major or minor in geography and/or to consider it as a career option.

AREAS OF SPECIALIZATION

As in all disciplines, areas of concentration or specialization change over time. As we noted for North American geography in the twentieth century, there was a period when most geographers were physical geographers and the physical landscape was the main objective of geographic analysis. Then the pendulum swung toward human (cultural) geography, and

students everywhere focused on the imprints made by human activity on the surface of the earth. Still later, the spatial-organization tradition became a major area of interest. In the meantime, geography's attraction for some students lay in technical areas: in cartography, in remote sensing, in computer-assisted data analysis, and in geographic information systems.

All this meant that geography posed—and continues to pose—a challenge to its professionals. New developments require that we keep up to date; but also, we must continue to build on established foundations.

Regional Geography

One of these established foundations, of course, is regional geography, which encompasses a large group of specializations. Some geographers specialize in the theory of regions: how they should be defined, how they are structured, how their internal components work. This led in the direction of *regional science*, and some geographers preferred to call themselves regional scientists. But make no mistake: regional science is regional geography.

Another, older approach to regional geography involves specialization in an area of the world, ranging in size from a geographic realm to a single region or even a state or part of a state. There was a time when regional geographers, because of the interdisciplinary nature of their knowledge, were sought after by government agencies. Courses in regional geography abounded in universities' geography departments. Regional geographers played key roles in international studies and research programs. But then the drive to make geography a more rigorous science and to search for universal (rather than regional) truths contributed to a decline in regional geography. The results were not long in coming, and lately you have probably seen the issue of "geographic illiteracy" discussed in newspapers and magazines. Now the pendulum that has affected geography throughout its existence is swinging back again, and regional geography and regional specialization are reviving. It is a propitious time to consider regional geography as a professional field.

Your Personal Interests

Geography, we have pointed out, is united by several bonds, of which regional geography is but one. Regional geography exemplifies the spatial view that all geographers hold. The spatial approach to study and research binds physical and human geographers, regionalists, and topical specialists. Another unifying theme is our interest in the relationships between human societies and natural environments. We have referred to that topic frequently in this book; as an area of specialization it has gone through difficult times. Perhaps more than anything, geography remains a field of *synthesis,* of understanding interrelationships.

Geography also is a field science, using "field" in another context. In the past, almost all major geography departments required a student's participation in a "field camp" as part of a master's degree program; thus, many undergraduate programs included field experience. It was one of those bonding practices where students and faculty with diverse interests met, worked together, and learned from one another. Today, few field camps of this sort are offered but that does not change what geography is all about. If you see an opportunity for field experience with professional geographers—even just a one-day reconnaissance—take it. But realize this: a few days in the field with geographic instruction may hook you for life.

Some geographers, in fact, are far better field-data gatherers than analysts or writers. In this respect they are not alone: this also happens in archaeology, in geology, and in biology (among other field disciplines). This does not mean that these field workers do not contribute significantly to knowledge. Often, in a research team, some of the members are better in the field and others excel in subsequent analysis. From all points of view, fieldwork is important.

Geography, then, is practiced in the field and in the office, in physical and human contexts, in generality and detail. Small wonder that so many areas of specialization have developed. If you check the undergraduate catalogue of your college or university, you will see some of these specializations listed as semester-length courses. But no geography department, no matter how large, could offer them all.

How does an area of specialization develop, and how can one become a part of it? The way in which geographic specializations have developed tells us much about the entire discipline. Some major areas, now old and established, began as research and theory-building by one scholar and his or her students. These graduate students dispersed to the faculties of other universities and began teaching what they had learned. Thus, for example, did Carl Sauer's cultural geography spread from the University of California, Berkeley.

It is one of the joys of geography that the basics and methods, once learned, are applicable to so many features of the human and physical world. Geographers have specialized in areas as disparate as shopping-center location and glacier movement, tourism and coastal erosion, real estate and wildlife, retirement communities and sports. Many of these specializations started with the interests and energies of one scholar. Some thrived and grew into major geographic pursuits; others remained one-person shows, but with potential. When you discuss your own interest with a faculty advisor, you may refer to a university where you would like to do graduate work. "Oh yes," the answer may be, "Professor X is in their geography department, working on just that." Or perhaps your advisor will suggest another university, where a member of the faculty is known to be working on the topic in which you are interested. That is the time to write a letter of inquiry. What is the professor working on now? Are graduate students involved? Are there research funds available? What are the career prospects after graduation?

The Association of American Geographers or **AAG** (1710 16th Street, N.W., Washington, D.C. 20009-3198 [(202) 234-1450]) recognizes more than three dozen so-called Specialty Groups. In academic year 1990–1991, the Specialty Group list included the following branches of Geography:

Africa
Aging and the Aged
American Indians
Applied
Asia
Bible
Biogeography
Canadian Studies
Cartography
China
Climate
Coastal and Marine
Contemporary Agriculture and
 Rural Land Use
Cultural
Cultural Ecology
Energy and Environment
Environmental Perception and
 Behavioral Geography
Geographic Information Systems
Geographic Perspectives on
 Women
Geography in Higher Education
Geomorphology
Hazards
Historical
Industrial
Latin America
Mathematical Models and
 Quantitative Methods
Medical
Microcomputers
Political
Population
Recreation, Tourism, and Sport
Regional Development and
 Planning
Remote Sensing
Rural Development
Socialist Geography

Soviet Union and Eastern
  Europe
Transportation
Urban
Water Resources

If you write or call the AAG at the address or telephone number given above, they will be glad to send you the newest listing of these groups which also includes the address of the current chairperson of each Specialty Group. We encourage you to directly contact the person who chairs the group(s) you are interested in. You may be assured that these elected leaders are ready, enthusiastic, and willing to provide you with the information you are looking for.

All this may seem far in the future. Still, the time to start planning for graduate school is now. Applications for admission and financial assistance must be made shortly after the *beginning* of your senior year! That makes your junior year a year of decision.

AN UNDERGRADUATE PROGRAM

The most important concern for any geography major or minor, however, is basic education and training in the field. An undergraduate curriculum contains all or several of the following courses (titles may vary):

1. Introduction to Physical Geography I (Natural landscapes, landforms, soils, elementary biogeography)
2. Introduction to Physical Geography II (Climatology, elementary oceanography)
3. Introduction to Human Geography (Principles and topics of cultural and economic geography)
4. World Regional Geography (Major world realms)

These beginning courses are followed by more specialized courses, including both substantive and methodological ones:

5. Introduction to Quantitative Methods of Analysis
6. Introductory Cartography
7. Analysis of Remotely Sensed Data
8. Cultural Geography
9. Political Geography
10. Urban Geography
11. Economic Geography
12. Historical Geography
13. Geomorphology
14. Geography of United States-Canada, Europe, and/or other major world realms

You can see how Physical Geography I would be followed by Geomorphology, and how Human Geography now divides into such areas as intermediate and/or advanced cultural and economic geography. As you progress, the focus becomes even more specialized. Thus, Economic Geography may be followed by:

15. Industrial Geography
16. Transportation Geography
17. Agricultural Geography

At the same time, regional concentrations may come into sharper focus:

18. Geography of Western Europe (or other regions)

In these more advanced courses, you will use the technical know-how from courses numbered **5, 6,** and **7** (and perhaps others). Now you can avail yourself of the opportunity to develop these skills further. Many departments offer such courses as:

19. Advanced Quantitative Methods (involving numerous computer applications)
20. Advanced Cartography (usually involving Geographic Information Systems)
21. Advanced Satellite Imagery Interpretation

From this list (which represents only a part of a comprehensive curriculum), it is evident that you cannot, even in four years of undergraduate study, register for all courses. The geography major in many universities requires a minimum of only 30 (semester-hour) credits—just 10 courses, fewer than half those listed here. This is another reason to begin thinking about specialization at an early stage.

Because of the number and variety of possible geography courses, most departments require that their majors complete a core program that includes courses in substantive areas as well as theory and methods. That core program is important, and you should not be tempted to put these courses off until your last semesters. What you learn in the core program will make what follows (or should follow) much more meaningful.

You should also be aware of the flexibility of many undergraduate programs, something that can be especially important to us geographers. Imagine that you are majoring in geography and develop an interest in Southeast Asia. But the regional specialization of the geographers in your department may be focused somewhere else— say Africa. However, courses on Southeast Asia are indeed offered by other departments such as anthropology, history, or political science. If you are going to be a regional specialist, then those courses will be very useful and should be part of your curriculum. But you are not able to receive geography credit for them. After taking these courses successfully, however, you may be able to register for an independent study or reading course in the Geography of Southeast Asia if a faculty member is willing to guide you. Always discuss such matters with the undergraduate advisor or chairperson of your geography department.

LOOKING AHEAD

By now, as a geography major, you will be thinking of the future—either in terms of graduate school or a salaried job. In this connection, if there is one important lesson to keep in mind, it is to *plan ahead* (a redundancy for emphasis). The choice of a graduate school is one of the most important you will make in your life. The professional preparation you acquire as a graduate student will affect your competitiveness on the job market for years to come.

Choosing a Graduate School

Your choice of a graduate school hinges on several factors, and the geography program it offers is one of them. Possibly, you are constrained by residency factors and your choice involves the schools of only one state. Your undergraduate record and grade point average affect the options. As a geographer, you may have strong feelings in favor of—or against—particular parts of the country (see Fig. 3–21!). And although some schools may offer you financial support, others may not.

Certainly the programs and

specializations of the prospective graduate department are extremely important. If you have settled on your own area of interest, it is wise to find a department that offers opportunities in that direction. If you have yet to decide, it is best to select a large department with several options. Some students are so impressed by the work and writings of a particular geographer that they go to his or her university solely to learn from, and work with, that scholar.

In every case, information and preparation are crucial. Many a prospective graduate student has arrived on campus eager to begin work with a favorite professor—only to find that the professor is away on a sabbatical leave!

Fortunately, information can be acquired with little difficulty. One of the most useful publications of the AAG is the *Guide to Departments of Geography*, published each fall. A copy of this annual directory should be available in the office of your geography departments, but if you plan to enter graduate school, a personal copy would be an asset (the AAG charged $25 for nonmembers in 1990). Not only does the *Guide* describe the programs, requirements, financial aid, and other aspects of geography departments in the United States and Canada; it also lists all faculty members and their current research and teaching specializations. You will find the *Guide* indispensable in your decision-making.

You may discover one particular department that stands out as the most interesting, most appropriate to your plans. But do not limit yourself to one school. After careful investigation, it is best to rank a half-dozen schools (or more), write for admission application forms to all of them, and apply to several. Multiple applications are costly, but the investment is worth it.

Assistantships and Scholarships

One reason to apply at several universities has to do with financial (and other) support for which you may be eligible. If you have a reasonably well-rounded undergraduate program behind you and a good record of achievement, you are eligible for a position known as TA (teaching assistant) in a graduate department. Such assistantships usually offer full or partial tuition plus a monthly stipend (during the nine-month academic year). Conditions vary, but they may make it possible to attend a university that would otherwise be out of reach. A tuition waiver alone can be worth more than $10,000 annually. Application for an assistantship is made directly to the department. Write to the department contact listed in the latest AAG *Guide*, who will either respond directly to you or forward your inquiry to the departmental committee that evaluates applications.

What does a TA do? The responsibilities vary, but often teaching assistants are expected to lead discussion sections (of a larger class taught by a professor), laboratories, or other classes. They prepare and grade examinations and help undergraduates tackle problems arising from their courses. It is an excellent way to determine your own ability and interest in a teaching career.

In some instances, especially in larger geography departments, RAs (research assistantships) are available. When a member of the faculty is awarded a large grant for some research project (e.g., by the National Science Foundation), that grant may make possible the appointment of one or more research assistants. These RAs perform tasks generated by the project and are rewarded by a modest salary (usually comparable to a TA's). Normally, RAs do not receive tuition waivers, but sometimes the department and the graduate school can arrange a waiver to make the research assistantship more attractive to better students. Usually, RAs are chosen from among graduate students already on campus, students who have proven their interest and ability. But sometimes an incoming student is appointed. Always ask about opportunities.

Moreover, geography students are among those eligible for many scholarships and fellowships offered by universities and off-campus organizations. When you write your introductory letter, be sure to inquire about other forms of financial aid.

JOBS FOR GEOGRAPHERS

You may decide to take a job upon completing the bachelor's degree rather than going on to graduate school. Again, this is a decision best made early in the junior year for two main reasons: (1) to tailor your curriculum toward a vocational objective, and (2) to start searching for a job well before graduation.

Internships

A very good way to enter the job market—to become familiar with the working environment—is by taking an internship in an agency, office, or firm. Many organizations find it useful to take interns. It helps train beginning professionals; it gives the company an opportunity to observe the performance of trainees. Many an intern has ul-

timately been employed by his or her company. Some employers have even suggested what courses the intern should take in the next academic year to improve future performance. For example, an urban- or regional-planning office that employs an intern might suggest that the intern add an advanced cartography course or an urban planning course to his or her program of study.

Some offices will appoint interns on a continuing basis, say two afternoons a week around the year; others make full-time, summer-only appointments available. Occasionally, the internship can be linked to your departmental curriculum, yielding academic credit as well as vocational experience. Your department undergraduate advisor or the chairperson will be the best source of assistance.

One of the most interesting internship programs is offered by the National Geographic Society in Washington, D.C. Every year, the Society invites three groups of about eight interns each to work with the permanent staff of its various departments. Application forms are available in every geography department in the United States, and competition is strong. The application itself is a useful exercise since it tells you what the Society (and other offices) look for in your qualifications.

Professional Opportunities

A term you will sometimes see in connection with jobs is *Applied Geography*. This, one supposes, is to distinguish geography teaching from practical geography. But, in fact, all professional geography in education, business, government, and elsewhere, is "applied." In the past, a large majority of geog-

raphy graduates became teachers—in elementary and high schools, colleges and universities. More recently, geographers have entered other areas in increasing numbers. In part, this is related to the decline in geographic education in the schools, but it also reflects the growing recognition of geographic skills by employers in business and government.

Nevertheless, what you as a geographer can contribute to a corporation is not yet as clear to the managers of many companies as it should be. The anybody-can-do-geography attitude is a form of ignorance you will undoubtedly confront. (This is so even in pre-collegiate education, where geography was submerged in social studies—and taught by teachers many of whom never took a course in the discipline!)

So, once employed, you may have to prove not only yourself, but also the usefulness of the skills and capacities you bring to your job. There is a positive side to this. Many employers, once dubious about hiring a geographer, learn how geography can contribute—and become enthusiastic users of geographic talent. Where one geographer is employed, whether in a travel firm, publishing house, or planning office, you will soon find more.

Geographers are employed in every area: in business, government, and education. Planning also is a profession that employs many geographers. Employment in business has grown in recent years. Government at national, state, and local levels always has been a major employer of geographers. And education, where a decline was long suffered, is now reviving, and the demand for geography teachers will grow again.

To secure more detailed information about employment, write for the inexpensive booklet entitled *Careers in Geography* from the AAG (the address was given earlier). Or simply check with your geography department, whose office is very likely to have a copy for you to peruse.

Business

With their education in global and international affairs, their knowledge of specialized areas of interest to business, and their training in cartography, methods of quantitative analysis, and good writing, geographers with a bachelor's degree are attractive business-employment prospects. Some undergraduate students already have chosen the business they will enter, and for them there are departments that offer concentrations in their area. Several geography departments, for instance, offer a curriculum that concentrates on tourism and travel—a business in which geographic skills are especially useful.

These days, many companies want graduates who have a strong knowledge of international affairs and fluency in at least one foreign language in addition to their skills in other areas. The business world is quite different from the academic, and the transition is not always easy. Your employer will want to use your abilities to enhance income and profit. You may, at first, be placed in a job where your geographic skills are not immediately applicable, and it will be up to you to look for opportunities to do so.

One of our students was in such a situation some years ago. She was one of more than a dozen new employees doing what was, essentially, clerical work. (Some companies will use this procedure to determine a new employee's punctuality, work habits, adaptability, and productivity.) One day she heard news that the company was considering establishing a branch in East Africa. She had a regional interest in East Africa as an undergraduate and had even taken a year of Swahili-language training. On her own time, she wrote a carefully documented memorandum to the company's president and vice presidents, describing factors that should be taken into consideration in the projected expansion. That report evinced her regional skills; locational insights; knowledge of the local market and transport problems; the probable cultural reaction to the company's product; the country's political circumstances; and her capacity to present such issues effectively. She supported her report with good maps and several illustrations. Soon she received a special assignment to participate in the planning process, and her rise in company ranks had begun. She had seized the opportunity and demonstrated the utility of her geographic skills.

Geography graduates have established themselves in businesses of all kinds: banking, international trade, manufacturing, retail, and many more. Should you join a large firm, you may be pleasantly surprised by the number of other geographers who hold jobs there— not under the title of geographer but under countless other titles, ranging from analyst and cartographer to market researcher and program manager. These are positions for which the appointees have competed with other graduates, including business graduates. As we noted earlier, once an employer sees the assets a geographer can bring, the role of geographers in the company is assured.

Government

Government has long been a major employer of geographers at the national and state as well as the local level. In *Careers in Geography,* the AAG booklet, it is estimated that at least 2500 geographers are working for the government, about half of them for the federal government. In the U.S. State Department, for example, there is an Office of the Geographer staffed by professional geographers. Other agencies where geographers are employed include the Defense Mapping Agency, the Bureau of the Census, the U.S. Geological Survey, the Central Intelligence Agency, and the Army Corps of Engineers. Other employers are the Library of Congress, the National Science Foundation, and the Smithsonian Institution. Many other branches of government also have positions for which geographers are eligible.

Opportunities also exist at state and local levels. Several states have their own Office of the State Geographer; all states have agencies engaged in planning, resource analysis, environmental protection, and transportation policy-making. All these agencies need geographers who have skills in cartography, remote sensing, database analysis, and the operation of geographic information systems.

Securing a position in government requires early action. If you want a job with the federal government, start at the beginning of your senior year. Every state capital and many other large cities have a Federal Job Information Center (FJIC). The Washington office is at 1900 E Street, N.W., Washington, D.C. 20415. You may request information about a particular agency and its job opportunities; it also is appropriate to write directly to the Personnel Office of the agency or agencies in which you are interested. Also write to the Office of Personnel Management (OPM), Washington, D.C. 20415; OPM also has offices in most large cities.

Planning

Planning has become one of geography's allied professions. The planning process is a complex one in which people trained in many fields participate. Geographers, with their cartographic, locational, regional, and analytical skills, are sought after by planning agencies. Many an undergraduate student gets that first professional opportunity as an intern in a planning office.

Planning is done by many agencies and offices at levels ranging from the federal to the municipal. Cities have planning offices, as do regional authorities. Working in such an office can be a rewarding experience because it involves the solving of social and economic problems for the future, the conservation and protection of the environment, the weighing of diverse and often conflicting arguments and viewpoints, and much interaction with workers trained in other fields. Planning is a superb learning experience.

A career in planning can be much enhanced by a background in geography, but you will have to adjust your undergraduate curriculum to include courses in such areas as public administration, public financing, and other related fields. Thus, a career in planning requires early planning on your part. At many universities, the geography department is closely associated with the planning department, and your faculty advisor can inform you about course requirements. But if you have your eye on a particular office or agency, you should also request information from its director about desired and required skills.

Planning is by no means a monopoly of government. Government-related organizations such as the Agency for International Development (AID), the World Bank, and the International Monetary Fund (IMF) have planning offices, as do nongovernmental organizations such as banks, airline companies, industrial firms, and multinational corporations. In the private sector, the opportunities for planners have been expanding and you may wish to explore these possibilities. An important address is the planners' equivalent of the geographers' AAG, the American Planning Association, 1776 Massachusetts Avenue, N.W., Washington, D.C. 20036. The AAG's *Careers in Geography* provides additional information on this expanding field and where you might go to study for it.

Teaching

If you are presently a freshman or sophomore, your graduation may coincide with the end of a long decline in the geography-teaching profession. Just 25 years ago, teaching geography in elementary or high school was the goal of thousands of undergraduate geography majors. But then began the submergence of geography into the hybrid field called "social studies," and prospective teachers no longer needed to have any training in geography in many states. In Florida, for example, teachers were required to take courses in regional geography and conservation (as taught in geography departments), but in the 1970s those requirements were dropped. Education planners fell victim to the myth that the teaching of geography does not require any training. What was left of geography often was taught by teachers whose own fields were history, civics, or basketball coaching!

If you have read the papers, you have seen reports of the predictable results. "Geographic illiteracy" has become a common complaint (often by the same education planners who pushed geography into the social-studies program and eliminated teacher-education requirements). Now the pendulum is swinging the other way again. States are returning to the education requirement so that teachers will learn some geography. And geography is returning to elementary and high school curricula—assisted by the National Geographic Society, which supports state-level Geographic Alliances of educators and academic geographers. The need for geography teachers will soon be on the upswing again.

So this may be a good time to consider teaching geography as a career. But you should do some research because states vary in their progressiveness in this area. Opportunities will become more widespread as the 1990s proceed. You should visit the School or College of Education in your college or university and ask questions about this. Also write not only the AAG, but also the National Council for Geographic Education (NCGE) at Indiana University of Pennsylvania, Indiana, Penna. 15705. Your state geographic society or Geographic Alliance also may be helpful; ask your department advisor or chairperson for details.

The opportunities in geography are many, but often they are not as obvious as those in other fields. You will find that good and timely preparation produces results and, frequently, unexpected rewards. The discussion in this appendix provides a comprehensive answer to the oft-asked question, "What can one do with Geography?" If you wish to explore these issues further, besides the AAG's *Careers in Geography,* we recommend a new book entitled *On Becoming a Professional Geographer,* edited by Martin S. Kenzer (Merrill, 1989).

We wish you every success in all your future endeavors. And speaking for the entire community of professional geographers, we would be delighted to have you join our ranks should you choose a geography-related career.

GLOSSARY

Absolute location The position or place of a certain item on the surface of the earth as expressed in degrees, minutes, and seconds of **latitude**,* 0° to 90° north or south of the equator, and **longitude**, 0° to 180° east or west of the *prime meridian* passing through Greenwich, England (a suburb of London).

Accessibility The degree of ease with which it is possible to reach a certain location from other locations. Accessibility varies from place to place and can be measured.

Acculturation Cultural modification resulting from intercultural borrowing. In cultural geography, the term is used to designate the change that occurs in the culture of indigenous peoples when contact is made with a society that is technologically more advanced.

Acid rain A growing environmental peril whereby acidified rainwater severely damages plant and animal life. Caused by the oxides of sulfur and nitrogen that are released into the atmosphere when coal, oil, and natural gas are burned, especially in major manufacturing zones.

Agglomerated (nucleated) settlement A compact, closely packed settlement (usually a hamlet or larger village) sharply demarcated from adjoining farmlands.

Agglomeration Process involving the clustering of people or activities. Often refers to manufacturing plants and

* Words in boldface type are defined elsewhere in this Glossary.

businesses that benefit from close proximity because they share skilled-labor pools and technological and financial amenities.

Agrarian Relating to the use of land in rural communities, or to agricultural societies in general.

Agriculture The purposeful tending of crops and livestock in order to produce food and fiber.

Alluvial Refers to the mud, silt, and sand (*alluvium*) deposited by rivers and streams. *Alluvial plains* adjoin many larger rivers; they consist of such renewable deposits that are laid down during floods, creating fertile and productive soils. Alluvial **deltas** mark the mouths of rivers such as the Mississippi and the Nile.

Altiplano High-elevation plateau, basin, or valley between even higher mountain ranges. In the Andes Mountains of South America, altiplanos lie at 10,000 feet (3000 meters) and even higher.

Altitudinal zonation Vertical regions defined by physical-environmental zones at various elevations, particularly in the highlands of South and Middle America. See *puna, tierra caliente, tierra fría, tierra helada,* and *tierra templada.*

Antecedent Antecedent, like **subsequent** and **superimposed**, is a term used in human as well as physical geography. Antecedent is something that goes before. In physical geography, a river that is antecedent is one that is older

than the landscape through which it flows. In political geography, an *antecedent boundary* is one that was there before the cultural landscape emerged and stayed in place while people moved in to occupy the surrounding area. An example is the 49th parallel boundary, dividing the United States and Canada between the Pacific Ocean and Lake of the Woods in northernmost Minnesota.

Anthracite coal Highest carbon content coal (therefore of the highest quality), that was formed under conditions of high pressure and temperature that eliminated most impurities. Anthracite burns almost without smoke and produces high heat.

Apartheid Literally, "apartness." The Afrikaans term given to South Africa's policies of racial separation, and the highly segregated socio-geographical patterns they have produced— a system now being dismantled.

Aquaculture The use of a river segment or an artificial body of water such as a pond for the raising and harvesting of food products, including fish, shellfish, and even seaweed. Japan is among the world's leaders in aquaculture.

Aquifer An underground reservoir of water contained within a porous, water-bearing rock layer.

Arable Literally, cultivable. Land fit for cultivation by one farming method or another.

Archipelago A set of islands grouped closely together, usually elongated into a chain.

Area A term that refers to a part of the earth's surface with less specificity than **region**. For example, *urban area* alludes very generally to a place where urban development has taken place, whereas *urban region* requires certain specific criteria upon which a delimitation is based (e.g., the spatial extent of commuting or the built townscape).

Areal interdependence A term related to **functional specialization**. When one area produces certain goods or has certain raw materials or resources and another area has a different set of resources and produces different goods, their needs may be *complementary*; by exchanging raw materials and products, they can satisfy each other's requirements. The concepts of areal interdependence and **complementarity** are related: both have to do with exchange opportunities between regions.

Arithmetic density A country's population, expressed as an average per unit area (square mile or square kilometer), without regard for its distribution or the limits of **arable** land—see also **physiologic density**.

Aryan From the Sanskrit *Arya* (''noble''), a name applied to an ancient people who spoke an Indo-European language and who moved into northern India from the northwest. Although properly a language-related term, Aryan has assumed additional meanings, especially racial ones.

Atmosphere The earth's envelope of gases that rests on the oceans and land surface and penetrates the open spaces within soils. This layer of nitrogen (78 percent), oxygen (21 percent), and traces of other gases is densest at the earth's surface and thins with altitude. It is held against the planet by the force of gravity.

Autocratic An autocratic government holds absolute power; rule is often by one person or a small group of persons who control the country by despotic means.

Balkanization The fragmentation of a region into smaller, often hostile political units.

Barrio Term meaning ''neighborhood'' in Spanish. Usually refers to an urban community in a Middle or South American city; also applied to low-income, inner-city concentrations of Hispanics in such southwestern U.S. cities as Los Angeles.

Bauxite Aluminum ore; an earthy, reddish-colored material that usually contains some iron as well. Soil-forming processes such as leaching and redeposition of aluminum and iron compounds contribute to bauxite formation, and many deposits exist at shallow depths in the wet tropics.

Biome A macroscale plant community occupying a large geographical area, marked by similarity in vegetation structure and/or appearance.

Birth rate The *crude birth rate* is expressed as the annual number of births per 1000 individuals within a given population.

Bituminous coal Softer coal of lesser quality than **anthracite** (more impurities remain) but of higher grade than **lignite**. Usually found in relatively undisturbed, extensive horizontal layers, often close enough to the surface to permit strip-mining. When heated and converted to coking coal or *coke,* it is used to make iron and steel.

Break-of-bulk point A location along a transport route where goods must be transferred from one carrier to another. In a port, the cargoes of oceangoing ships are unloaded and put on trains, trucks, or perhaps smaller river boats for inland distribution.

Buffer zone A set of countries separating ideological or political adversaries. In southern Asia, Afghanistan, Nepal, and Bhutan were parts of a buffer zone between British and Russian-Chinese imperial spheres. Thailand was a *buffer state* between British and French colonial domains in mainland Southeast Asia.

Caliente See *tierra caliente.*

Cartel An international syndicate formed to promote common interests in some economic sphere through the formulation of joint pricing policies and the limitation of market options for consumers. The Organization of Petroleum Exporting Countries (OPEC) is a classic example.

Cartography The art and science of making maps, including data compilation, layout, and design. Also concerned with the interpretation of mapped patterns.

Caste system The strict social segregation of people—specifically in India's Hindu society—on the basis of ancestry and occupation.

Cay Pronounced *kee.* A low-lying small island usually composed of coral and sand. Often part of an island chain such as the Florida Keys or the Bahamas archipelago.

Central business district (CBD) The downtown heart of a city, the CBD is marked by high land values, a concentration of business and commerce, and the clustering of the tallest buildings.

Centrality The strength of an urban center in its capacity to attract producers and consumers to its facilities; a city's ''reach'' into the surrounding region.

Centrifugal forces A term employed to designate forces that tend to divide a country—such as internal religious, linguistic, ethnic, or ideological differences.

Centripetal forces Forces that unite and bind a country together—such as a strong national culture, shared ideological objectives, and a common faith.

Charismatic Personal qualities of certain leaders that enable them to capture and hold the popular imagination, to secure the allegiance and even the devotion of the masses. Gandhi, Mao Zedong, and Franklin D. Roosevelt are good examples in this century.

City-state An independent political entity consisting of a single city with (and sometimes without) an immediate **hinterland**. The ancient city-states of Greece have their modern equivalent in Singapore.

Climate A term used to convey a generalization of all the recorded weather observations over time at a certain place or in a given area. It represents an "average" of all the weather that has occurred there. In general, a tropical location such as the Amazon Basin has a much less variable climate than areas located, say, midway between the equator and the pole. In low-lying tropical areas the *weather* changes little; the climate is rather like the weather on any given day. But in the middle latitudes, there may be summer days to rival those in the tropics and winter days so cold that they resemble polar conditions.

Climax vegetation   The final, stable vegetation that has developed at the end of a succession under a particular set of environmental conditions. This vegetative community is in dynamic equilibrium with its environment.

Coal   See **anthracite** and **bituminous coal.**

Collectivization   The reorganization of a country's agriculture that involves the expropriation of private holdings and their incorporation into relatively large-scale units, which are farmed and administered cooperatively by those who live there. This system has transformed Soviet agriculture and went beyond the Soviet model in China's program of communization.

Colonialism   See **imperialism.**

Common Market   Name given to a group of 12 European countries that belong to a **supranational** association to promote their economic interests (see Fig. 1–32A). Official name is European Economic Community (EEC), nowadays shortened to European Community (EC).

Compact state   A politico-geographical term to describe a state that possesses a roughly circular, oval, or rectangular territory in which the distance from the geometric center to any point on the boundary exhibits little variance. Cambodia, Uruguay, and Poland are examples of this shape category.

Complementarity   Regional complementarity exists when two regions, through an exchange of raw materials and/or finished products, can specifically satisfy each other's demands.

Concentric zone model   A geographical model of the American central city that suggests the existence of five concentric rings arranged around a common center (see Fig. 3–1A).

Condominium   In political geography, this denotes the shared administration of a territory by two governments.

Coniferous forest   A forest of cone-bearing, needleleaf evergreen trees with straight trunks and short branches, including spruce, fir, and pine.

Contagious diffusion   The distance-controlled spreading of an idea, innovation, or some other item through a local population by contact from person to person—analogous to the communication of a contagious illness.

Contiguous   A word of some importance to geographers that means, literally, to be in contact with, adjoining, or adjacent. Sometimes we hear the continental (*conterminous*) United States minus Alaska referred to as contiguous. Alaska is not contiguous to these "lower 48" states because Canada lies in between; neither is Hawaii, separated by over 2000 miles of ocean.

Continental drift   The slow movement of continents controlled by the processes associated with **plate tectonics.**

Continental shelf   Beyond the coastlines of the continents the surface beneath the water, in many offshore areas, declines very gently until the depth of about 660 feet (200 meters). Beyond the 660-foot line the sea bottom usually drops off sharply, along the *continental slope*, toward the much deeper mid-oceanic basin. The submerged continental margin is called the continental shelf, and it extends from the shoreline to the upper edge of the continental slope.

Conurbation   General term used to identify large multi-metropolitan complexes formed by the coalescence of two or more major urban areas. The Boston-Washington **Megalopolis** along the U.S northeastern seaboard is an outstanding example.

Copra   Meat of the coconut; fruit of the coconut palm.

Cordillera   Mountain chain consisting of sets of parallel ranges, especially the Andes in northwestern South America.

Core area   In geography, a term with several connotations. *Core* refers to the center, heart, or focus. The core area of a **nation-state** is constituted by the national heartland—the largest population cluster, the most productive region, the area with greatest **centrality** and **accessibility**, probably containing the capital city as well.

Core-periphery relationships   The contrasting spatial characteristics of, and linkages between, the *have* (core) and *have-not* (periphery) components of a national or regional system; elaborated in the Opening Essay (pp. 41–42).

Corridor   In general, refers to a spatial entity in which human activity is organized in a linear manner, as along a major transport route or in a valley confined by highlands. Specific meaning in politico-geographical context is a land extension that connects an otherwise **landlocked** state to the ocean. History has seen several such corridors come and go. Poland once had a corridor (it now has a lengthy coastline); Bolivia lost a corridor to the Pacific Ocean between Peru and Chile.

Cultural diffusion   The process of spreading and adoption of a cultural element, from its place of origin across a wider area.

Cultural ecology   The multiple interactions and relationships between a culture and its natural environment.

Cultural landscape   The forms and artifacts sequentially placed on the physical landscape by the activities of various human occupants. By this progressive imprinting of the human presence, the physical landscape is modified into the cultural landscape, forming an interacting unity between the two.

Cultural pluralism   A society in which two or more population groups, each practicing its own **culture**, live adjacent to one another without mixing inside a single **state.**

Culture See pp. 6–9, especially the *box* on p. 7.

Culture area A distinct, culturally discrete spatial unit; a region within which certain cultural norms prevail.

Culture complex A related set of **culture traits** such as prevailing dress codes, cooking, and eating utensils.

Culture hearth Heartland, source area, innovation center; place of origin of a major culture.

Culture realm A cluster of regions in which related culture systems prevail. In North America, the United States and Canada form a culture realm, but Mexico belongs to a different one.

Culture trait A single element of normal practice in a culture—such as the wearing of a turban.

Cyclical movement Movement (for example, **nomadic** migration) that has a closed route repeated annually or seasonally.

Death rate The *crude death rate* is expressed as the annual number of deaths per 1000 individuals within a given population.

Deciduous A deciduous tree loses its leaves at the beginning of winter or the start of the dry season.

Definition In political geography, the written legal description (in a treaty-like document) of a boundary between two countries or territories—see **delimitation**.

Delimitation In political geography, the translation of the written terms of a boundary treaty (the **definition**) into an official cartographic representation.

Delta **Alluvial** lowland at the mouth of a river, formed when the river deposits its alluvial load on reaching the sea. Often triangular in shape—hence the use of the Greek letter whose symbol is Δ.

Demarcation In political geography, the actual placing of a political boundary on the landscape by means of barriers, fences, walls, or other markers.

Demographic transition model Three-stage model, based on Western Europe's experience, of changes in population growth exhibited by countries undergoing industrialization. High **birth rates** and **death rates** are followed by plunging death rates, producing a huge net population gain; this is followed by the convergence of birth and death rates at a low overall level.

Demographic variables Births (fertility), deaths (mortality), and migration (population redistribution) are the three basic demographic variables.

Demography The interdisciplinary study of population—especially birth and death rates, growth patterns, longevity, and related characteristics.

Density of population The number of people per unit area. Also see **arithmetic density** and **physiologic density** measures.

Desert An arid area supporting very sparse vegetation, receiving less than 10 inches (25 centimeters) of precipitation per year. Usually exhibits extremes of heat and cold because the moderating influence of moisture is absent.

Desertification The encroachment of **desert** conditions on moister zones along the desert margins. Here plant cover and soils are threatened by desiccation, in part through overuse by humans and their domestic animals and, possibly, also because of inexorable shifts in the earth's environmental zones.

Determinism See **environmental determinism**.

Development The economic, social, and institutional growth of **states**.

Devolution In political geography, the disintegration of a **nation-state** as the result of emerging or reviving regionalism.

Dhows Wooden boats with characteristic triangular sail, plying the seas between Arabian and East African coasts.

Diffusion The spatial spreading or dissemination of a culture element (such as a technological innovation) or some other phenomenon (e.g., a disease outbreak). See also **contagious**, **expansion**, **hierarchical**, and **relocation diffusion**.

Dispersed settlement In contrast to **agglomerated** or **nucleated** settlement, dispersed settlement is characterized by the wide spacing of individual homesteads. This lower-density pattern is characteristic of rural North America.

Distance decay The various degenerative effects of distance on human spatial structures. The degree of spatial interaction diminishes as distance increases; therefore, people and activities try to arrange themselves in geographic space to minimize the "friction" effects of overcoming distance, which involves the costs of time as well as travel.

Diurnal Daily.

Divided capital In political geography, a country whose administrative functions are carried on in more than one city is said to have divided capitals.

Domestication The transformation of a wild animal or wild plant into a domesticated animal or a cultivated crop to gain control over food production. A necessary evolutionary step in the development of humankind—the invention of **agriculture**.

Double cropping The planting, cultivation, and harvesting of two crops successively within a single year on the same plot of farmland.

Doubling time The time required for a population to double in size.

Ecological trilogy The economic and communal ties that bind urbanites (who dominate), villagers, and nomads in Iran and other traditional societies of North Africa and Southwest Asia. See quotation on pp. 382–383.

Ecology Strictly speaking, this refers to the study of the many interrelationships between all forms of life and the natural environments in which they have evolved and continue to develop. The study of *ecosystems* focuses on the interactions between specific organisms and their environments. See also **cultural ecology**.

Economies of scale The savings that accrue from large-scale production whereby the unit cost of manufacturing decreases as the level of operation enlarges. Supermarkets operate on this principle and are able to charge lower prices than small grocery stores.

Ecumene The habitable portions of the earth's surface where permanent human settlements have arisen.

Elite A small but influential upper-echelon social class whose power and privilege give it control over a country's political, economic, and cultural life.

Elongated state A state whose territory is decidedly long and narrow in that its length is at least six times greater than its average width. Chile and Vietnam are two classic examples on the world political map.

Emigrant A person migrating away from a country or area; an out-migrant.

Empirical Relating to the real world, as opposed to theoretical abstraction.

Enclave A piece of territory that is surrounded by another political unit of which it is not a part.

Entrepôt A place, usually a port city, where goods are imported, stored, and transshipped. Thus, an entrepôt is a **break-of-bulk point**.

Environmental determinism The view that the natural environment has a controlling influence over various aspects of human life, including cultural development. Also referred to as *environmentalism*.

Erosion A combination of gradational forces that shape the earth's surface landforms. Running water, wind action, and the force of moving ice combine to wear away soil and rock. Human activities often speed erosional processes, such as through the destruction of natural vegetation, careless farming practices, and overgrazing by livestock.

Escarpment A cliff or steep slope; frequently marks the edge of a plateau.

Estuary The widening mouth of a river as it reaches the sea. An estuary forms when the margin of the land has subsided somewhat (or as ocean levels rise following glaciation periods) and seawater has invaded the river's lowest portion.

Evapotranspiration The loss of moisture to the **atmosphere** through the combined processes of evaporation from the soil and transpiration by plants.

Exclave A bounded (non-island) piece of territory that is part of a particular **state** but lies separated from it by the territory of another state.

Exclusive Economic Zone (EEZ) An oceanic zone extending up to 200 nautical miles from a shoreline, within which the coastal **state** can control fishing, mineral exploration, and additional activities by all other countries.

Expansion diffusion The spreading of an innovation or an idea through a fixed population in such a way that the number of those adopting grows continuously larger, resulting in an expanding area of dissemination.

Extraterritoriality Politico-geographical concept suggesting that the property of one **state** lying within the boundaries of another actually forms an extension of the first state.

Favela **Shantytown** on the outskirts or even well within an urban area in Brazil.

Fazenda Coffee plantation in Brazil.

Federal state A political framework wherein a central government represents the various entities within a **nation-state** where they have common interests—defense, foreign affairs, and the like—yet allows these various entities to retain their own identities and to have their own laws, policies, and customs in certain spheres.

Federation The association and cooperation of two or more **nation-states** or territories to promote common interests and objectives.

Ferroalloy A metallic mineral smelted with iron to produce steel of a particular quality. Manganese, for example, provides steel with tensile strength and the ability to withstand abrasives. Other ferroalloys are nickel, chromium, cobalt, molybdenum, and tungsten.

Fertile Crescent Crescent-shaped zone of productive lands extending from near the southeast Mediterranean coast through Lebanon and Syria to the **alluvial** lowlands of Mesopotamia. Once more fertile than today, this is one of the world's great source areas of agricultural innovations.

Feudalism Prevailing politico-geographical system in Europe during the Middle Ages when land was owned by the nobility and was worked by peasants and serfs. Feudalism also existed in other parts of the world, and the system persisted into this century in Ethiopia and Iran, among other places.

Fjord Narrow, steep-sided, elongated, and inundated coastal valley deepened by glacier ice that has since melted away, leaving the sea to penetrate. Norway's coasts are marked by fjords, as are certain coasts in Canada, Greenland, and Chile.

Floodplain Low-lying area adjacent to a mature river, often covered by **alluvial** deposits and subject to the river's floods.

Forced migration Human migration flows in which the movers have no choice but to relocate.

Formal region A type of region marked by a certain degree of homogeneity in one or more phenomena; also called *uniform* region or *homogeneous* region.

Forward capital Capital city positioned in actually or potentially contested territory, usually near an international border; it confirms the state's determination to maintain its presence in the region in contention.

Four Tigers of the Orient Taiwan, South Korea, Hong Kong, and Singapore. These burgeoning beehive countries of the western Pacific rim, using postwar Japan as a model, have experienced significant modernization, industrialization, and Western-style economic growth since the 1970s.

Fragmented state A state whose territory consists of several separated parts, not a **contiguous** whole. The individual parts may be isolated from each other by the land area of other states or by international waters.

Francophone Describes a country or region where other languages are also spoken, but where French is the *lingua franca* or the language of the **elite**. Quebec is Francophone Canada.

Fría See *tierra fria*.

Frontier Zone of advance penetration, of contention; an area not yet fully integrated into a national state.

Functional region A region marked less by its sameness than its dynamic internal structure; because it usually focuses on a central node, also called *nodal* or *focal* region.

Functional specialization The production of particular goods or services as a dominant activity in a particular location. Certain cities specialize in producing automobiles, computers, or steel; others mainly serve tourists.

Geographic realm The basic spatial unit in our world regional classification. Each realm is defined in terms of a synthesis of its total human geography—a composite of its leading cultural, economic, historical, political, and appropriate environmental features.

Geometric boundaries Political boundaries **defined** and **delimited** (and occasionally **demarcated**) as straight lines or arcs.

Geomorphology The geographic study of the configuration of the earth's solid surface—the world's landscapes and their constituent landforms.

Geopolitik **(Geopolitics)** A school of political geography that involved the use of quasi-academic research to encourage and justify a national policy of expansionism and imperialism.

Reached its height in pre-World War II Germany.

Ghetto An intraurban region marked by a particular ethnic character. Often an inner-city poverty zone, such as the *black ghetto* in the American central city. Ghetto residents are involuntarily segregated from other income and racial groups.

Glaciation See **Pleistocene Epoch**.

Green Revolution The successful recent development of higher yield, fast-growing varieties of rice and other cereals in certain Third World countries. This has led to increased production per unit area and a temporary narrowing of the gap between population growth and food needs.

Gross national product (GNP) The total value of all goods and services produced in a country during a given year. In some **underdeveloped countries (UDCs)**, where a substantial number of people practice **subsistence** and where the collection of information is difficult, GNP figures may be unreliable.

Growing season The number of days between the last frost in the spring and the first frost of the fall.

Growth pole An urban center with certain attributes that, if augmented by a measure of investment support, will stimulate regional economic development in its **hinterland**.

Hacienda Literally, a large estate in a Spanish-speaking country. Sometimes equated with **plantation**, but there are important differences between these two types of agricultural enterprise (see pp. 259–261).

Heartland theory The hypothesis, proposed by British geographer Halford Mackinder during the first two decades of this century, that any political power based in the heart of Eurasia (see Fig. 2–23) could gain sufficient strength to eventually dominate the world. Further, since Eastern Europe controlled access to the Eurasian interior, its ruler would command the vast "heartland" to the east.

Hegemony The political dominance of a country (or even a region) by another country. The Soviet Union's postwar grip on Eastern Europe, which lasted from 1945 to 1990, was a classic example.

Helada. See *tierra helada*.

Hierarchical diffusion A form of diffusion in which an idea or innovation spreads by "trickling down" from larger to smaller adoption units. An urban **hierarchy** is usually involved, encouraging the leapfrogging of innovations over wide areas, with geographic distance a less important influence.

Hierarchy An order or gradation of phenomena, with each level or rank subordinate to the one above it and superior to the one below. The levels in a national urban hierarchy are constituted by hamlets, villages, towns, cities, and (frequently) the **primate city**.

High seas Areas of the oceans away from land, beyond national jurisdiction, open and free for all to use.

Highveld A term used in Southern Africa to identify the high, grass-covered plateau that dominates much of the region. **Veld** means "grassland" in Dutch and Afrikaans. The lowest-lying areas in South Africa are called *lowveld*; areas that lie at intermediate elevations are the *middleveld*.

Hinterland Literally, "country behind," a term that applies to a surrounding area served by an urban center. That center is the focus of goods and services produced for its hinterland and is its dominant urban influence as well. In the case of a port city, the hinterland also includes the inland area whose trade flows through that port.

Humus Dark-colored upper layer of a soil that consists of decomposed and decaying organic matter such as leaves and branches, nutrient-rich and giving the soil a high fertility.

Iconography The identity of a region as expressed through its cherished symbols; its particular cultural landscape and personality.

Immigrant A person migrating into a particular country or area; an in-migrant.

Imperialism The drive toward the creation and expansion of a colonial empire and, once established, its perpetuation.

Indentured workers Contract laborers who sell their services for a stipulated period of time.

Infrastructure The foundations of a society: urban centers, communications, farms, factories, mines, and such facilities as schools, hospitals, postal services, and police and armed forces.

Insular Having the qualities and properties of an island. Real islands are not alone in possessing such properties of **isolation**: an **oasis** in the middle of a desert also has qualities of insularity.

Insurgent state Territorial embodiment of a successful guerrilla movement. The anti-government insurgents establishing a territorial base in which they exercise full control; thus, a state within a state.

Intermontane Literally, "between mountains." The location can bestow certain qualities of natural protection or **isolation** to a community.

Internal migration Migration flow within a **nation-state**, such as ongoing westward and southward movements in the United States and eastward movement in the Soviet Union.

International migration Migration flow involving movement across international boundaries.

Intervening opportunity The presence of a nearer opportunity that diminishes the attractiveness of sites farther away.

Irredentism A policy of cultural extension and potential political expansion aimed at a national group living in a neighboring country.

Irrigation The artificial watering of croplands. In Egypt's Nile Valley, *ba-sin* irrigation is an ancient method that involved the use of floodwaters that were trapped in basins on the floodplain and released in stages to augment rainfall. Today's *perennial* irrigation requires the construction of dams and irrigation canals for year-round water supply.

Isobar A line connecting points of equal atmospheric pressure.

Isohyet A line connecting points of equal rainfall total.

Isolation The condition of being geographically cut off or far removed from mainstreams of thought and action. It also denotes a lack of receptivity to outside influences, caused at least partially by inaccessibility.

Isoline A line connecting points of equal value.

Isotherm A line connecting points of equal temperature.

Isthmus A **land bridge**; a comparatively narrow link between larger bodies of land. Central America forms such a link between North and South America.

Juxtaposition Contrasting places in close proximity to one another.

Land alienation One society or culture group taking land from another. In Subsaharan Africa, for example, European colonialists took land from indigenous Africans and put it to new uses, fencing it off and restricting settlement.

Land bridge A narrow **isthmian** link between two large landmasses. They are temporary features—at least in terms of geologic time—subject to appearance and disappearance as the land or sea-level rises and falls.

Landlocked An interior country or **state** that is surrounded by land. Without coasts, a landlocked state is at a disadvantage in a number of ways—in terms of access to international trade routes, and in the scramble for possession of areas of the **continental shelf** and control of the **exclusive economic zone** beyond.

Land reform The spatial reorganization of agriculture through the allocation of farmland (often expropriated from landlords) to peasants and tenants who never owned land. Also, the consolidation of excessively fragmented farmland into more productive, perhaps cooperatively-run farm units.

Late Cenozoic Ice Age The latest in a series of ice ages that mark the earth's environmental history, lasting from ca. 3.5 million to 10,000 years ago. As many as 24 separate advances and retreats of continental icesheets took place, the height of activity occurring during the **Pleistocene Epoch** (ca. 2 million to 10,000 years ago).

Latitude Lines of latitude are **parallels** that are aligned east-west across the globe, from 0° latitude at the equator to 90° north and south latitude at the poles. Areas of low latitude, therefore, lie near the equator in the tropics; high latitudes are those in the north polar (Arctic) and south polar (Antarctic) regions.

Lignite Also called brown coal, a low-grade variety of coal somewhat higher in fuel content than *peat* but not nearly as good as the next higher grade, **bituminous coal**. Lignite cannot be used for most industrial processes, but it is important as a residential fuel in certain parts of the world.

Lingua franca The term derives from "Frankish language," and applied to a tongue spoken in ancient Mediterranean ports that consisted of a mixture of Italian, French, Greek, Spanish, and even some Arabic. Today it refers to a "common language," a second language that can be spoken and understood by many peoples, although they speak other languages at home.

Littoral Coastal; along the shore.

Llanos Name given to the savanna-like grasslands of the Orinoco River's wide basin in the interior parts of Colombia and Venezuela.

Location theory A logical attempt to explain the locational pattern of an economic activity and the manner in which its producing areas are interrelated. The agricultural location theory contained in the **Von Thünen model** is a leading example.

Loess Deposit of very fine silt or dust that is laid down after having been windborne for a considerable distance. Loess is notable for its fertility under irrigation and its ability to stand in steep vertical walls when eroded by a river or (as in China's Loess Plateau) excavated for cave-type human dwellings.

Longitude Angular distance (0° to 180°) east or west as measured from the prime **meridian** (0°) that passes through the Greenwich Observatory in suburban London, England. For much of its length in the mid-Pacific Ocean, the 180th meridian functions as the *international date line*.

Maghreb Westernmost segment of the North Africa/Southwest Asia realm, consisting of the countries of Morocco, Algeria, and Tunisia.

Malthusian Designates the early-nineteenth-century viewpoint of Thomas Malthus, who argued that population growth was outrunning the earth's capacity to produce sufficient food. *Neo-Malthusian* refers to those who subscribe to such positions in modern contexts.

Maquiladora The term given to new industrial plants in Mexico's northern (U.S.) border zone. These foreign-owned factories assemble imported components and/or raw materials, and then export finished manufactures, mainly to the United States. Most import duties are minimized, bringing jobs to Mexico and the advantages of low wage rates to the foreign entrepreneurs.

Marchland A **frontier** or area of uncertain boundaries that is subject to various national claims and an unstable political history. The term refers to the movement of various national armies across such zones.

Megalopolis Term used to designate large coalescing supercities that are forming in diverse parts of the world;

was used specifically to refer to the Boston-Washington multi-metropolitan corridor in the northeastern U.S., but the term is now used generically with a lower-case *m* as a synonym for **conurbation**.

Mental map The structured spatial information an individual acquires in his or her perception of the surrounding environment and more distant places. A *designative* mental map involves the objective recording of geographical information received (e.g., Florida is a narrow peninsular state at the southeastern corner of the United States). An *appraisive* mental map entails the subjective processing of spatial information according to a person's biases (e.g., Florida is a nice place to live because it has a balmy climate, opportunities for water recreation, and a growing postindustrial economy).

Mercantilism Protectionist policy of European **states** during the sixteenth to the eighteenth centuries that promoted a state's economic position in the contest with other countries. The acquisition of gold and silver and the maintenance of a favorable trade balance (more exports than imports) were central to the policy.

Meridian Line of **longitude**, aligned north-south across the globe, that together with **parallels** of **latitude** forms the global grid system. All meridians converge at both poles and are at their maximum distances from each other at the equator.

Mestizo The root of this word is the Latin for *mixed*; it means a person of mixed white and American Indian ancestry.

Metropolitan area See **urban (metropolitan) area**.

Mexica Empire The name the ancient Aztecs gave to the domain over which they held **hegemony** on the north-central mainland of Middle America.

Migration A change in residence intended to be permanent. See also **forced**, **internal**, **international**, and **voluntary migration**.

Migratory movement Human relocation movement from a source to a des-

tination without a return journey, as opposed to **cyclical movement**.

Model An idealized representation of reality created to demonstrate certain of its properties. A **spatial** model focuses on a geographical dimension of the real world.

Monotheism The belief in, and worship of, a single god.

Monsoon Refers to the seasonal reversal of wind (and moisture) flows in certain parts of the subtropics and lower-middle latitudes. The *dry monsoon* occurs during the cool season when dry offshore winds prevail. The *wet monsoon* occurs in the hot summer months, which produce onshore winds that bring large amounts of rainfall. The air-pressure differential over land and sea is the triggering mechanism. Monsoons make their greatest regional impact in the coastal and near-coastal zones of South Asia, Southeast Asia, and China.

Mulatto A person of mixed African (black) and European (white) ancestry.

Multinationals Internationally active corporations that strongly influence the economic and political affairs of many countries.

Multiple nuclei model The Harris-Ullman model that showed the mid-twentieth-century American central city to consist of several zones arranged around nuclear growth points (see Fig. 3–1C).

Nation Legally a term encompassing all the citizens of a **state**, it has also taken on other connotations. Most definitions now tend to refer to a group of tightly-knit people possessing bonds of language, ethnicity, religion, and other shared cultural attributes. Such homogeneity actually prevails within very few states.

Nation-state A country whose population possesses a substantial degree of cultural homogeneity and unity. The ideal form to which most **nations** and **states** aspire—a political unit wherein the territorial state coincides with the area settled by a certain national group or people.

Natural increase rate Population growth measured as the excess of live births over deaths per 1000 individuals per year. Natural increase of a population does not reflect either **emigrant** or **immigrant** movements.

Natural resource As used in this book, any valued element (or means to an end) of the environment, including minerals, water, vegetation, and soil.

Nautical mile By international agreement, the nautical mile—the standard measure at sea—is 6076.12 feet in length, equivalent to approximately 1.15 statute miles (1.85 kilometers).

Network (transport) The entire regional system of transportation connections and nodes through which movement can occur.

Nomadism **Cyclical movement** among a definite set of places. Nomadic peoples are mostly **pastoralists**.

Norden A regional appellation for Northern Europe's three Scandinavian countries (Denmark, Norway, and Sweden), Finland, and Iceland.

Nucleated settlement See **agglomerated settlement**.

Nucleation Cluster; **agglomeration**.

Oasis An area, small or large, where the supply of water permits the transformation of the surrounding desert into a green cropland; the most important focus of human activity for miles around. It may also encompass a densely populated **corridor** along a major river where irrigation projects stabilize the water supply—as along the Nile in Egypt, which can be viewed as an elongated set of oases.

Occidental Western. See **Oriental**.

Oligarchy Political system involving rule by a small minority, an often corrupt **elite**.

Organic theory Concept that suggests that the **state** is in some ways analogous to a biological organism, with a life cycle that can be sustained through cultural and territorial expansion.

Oriental The root of the word *oriental* is from the Latin for *rise*. Thus, it has to do with the direction in which one sees the sun "rise"—the east; *oriental* therefore means Eastern. *Occidental* originates from the Latin for *fall*, or the "setting" of the sun in the west; *occidental* means Western.

Orographic precipitation Mountain-induced precipitation, especially where air masses are forced over topographic barriers. Downwind areas beyond such a mountain range experience the relative dryness known as the **rain shadow effect**.

Pacific ring of fire Zone of crustal instability along tectonic **plate** boundaries, marked by earthquakes and volcanic activity, that rings the Pacific Ocean basin.

Paddies (paddyfields) Ricefields.

Pangaea A vast, singular landmass consisting of most of the areas of the present continents (including the Americas, Eurasia, Africa, Australia, and Antarctica), which existed until near the end of the Mesozoic era when plate divergence and **continental drift** broke it apart. The "northern" segment of this Pangaean supercontinent is called *Laurasia*, the "southern" part *Gondwana* (see Fig. 7–7).

Parallel An east-west line of **latitude** that is intersected at right angles by **meridians** of **longitude**.

Páramos See **puna**.

Pastoralism A form of agricultural activity that involves the raising of livestock. Many peoples described as herders actually pursue mixed agriculture, in that they may also fish, hunt, or even grow a few crops. But pastoral peoples' lives do revolve around their animals.

Peasants In a **stratified** society, peasants are the lowest class of people who depend on agriculture for a living. But they often own no land at all and must survive as tenants or day workers.

Peninsula A comparatively narrow, finger-like stretch of land extending from the main landmass into the sea. Florida and Korea are examples.

Peon (*peone*) Term used in Middle and South America to identify people who often live in serfdom to a wealthy landowner; landless **peasants** in continuous indebtedness.

Per capita *Capita* means individual. Income, production, or some other measure is often given per individual.

Perforated state A state whose territory completely surrounds that of another state. South Africa, which encloses Lesotho and is perforated by it, is a classic example (see inset map, Fig. 10–8).

Periodic market Village market that opens every third or fourth day or at some other regular interval. Part of a regional network of similar markets in a preindustrial, rural setting where goods are brought to market on foot (or perhaps by bicycle) and barter remains a major mode of exchange.

Permafrost Permanently frozen water in the soil and bedrock, as much as 1000 feet (300 meters) in depth, producing the effect of completely frozen ground. Can thaw near surface during brief warm season.

Physiographic political boundaries Political boundaries that coincide with prominent physical breaks in the natural landscape—such as rivers or the crest ridges of mountain ranges.

Physiographic region (province) A region within which there prevails substantial natural-landscape homogeneity, expressed by a certain degree of uniformity in surface **relief**, climate, vegetation, and soils.

Physiologic density The number of people per unit area of **arable** land.

Pidgin A language that consists of words borrowed and adapted from other languages. Originally developed from commerce among peoples speaking different languages.

Pilgrimage A journey to a place of great religious significance by an individual or by a group of people (such as a pilgrimage to Mecca for Muslims).

Plantation A large estate owned by an individual, family, or corporation and organized to produce a cash crop. Almost all plantations were established within the tropics; in recent decades, many have been divided into smaller holdings or reorganized as cooperatives.

Plate tectonics Bonded portions of the earth's mantle and crust, averaging 60 miles (100 kilometers) in thickness. More than a dozen such plates exist (see Fig. I–6), most of continental proportions, and they are in motion. Where they meet one slides under the other, crumpling the surface crust and producing significant volcanic and earthquake activity. A major mountain-building force.

Pleistocene Epoch Recent period of geologic time that spans the rise of humanity, beginning about 2 million years ago. Marked by *glaciations* (repeated advances of continental icesheets) and milder *interglaciations* (icesheet retreats). Although the last 10,000 years are known as the *Recent* Epoch, Pleistocene-like conditions seem to be continuing and the present is probably another Pleistocene interglaciation; the glaciers likely will return. See also **Late Cenozoic Ice Age**.

Plural society See **cultural pluralism**.

Polder Land reclaimed from the sea adjacent to shore by constructing dikes and pumping out the water. Technique used widely along the coast of the Netherlands since the Middle Ages to add badly needed additional space for settlements and farms.

Pollution The release of a substance, through human activity, which chemically, physically, or biologically alters the air or water it is discharged into. Such a discharge negatively impacts the environment, with possible harmful effects on living organisms—including humans.

Population explosion The rapid growth of the world's human population during the past century, attended by ever-shorter **doubling times** and accelerating *rates* of increase.

Population structure Graphic representation of a population by sex and age, as in Fig. 1–2 (p. 61).

Postindustrial economy Emerging economy, in the United States and a handful of other highly advanced countries, as traditional industry is overshadowed by a high-technology productive complex dominated by services, information-related, and managerial activities.

Primary economic activity Activities engaged in the direct extraction of **natural resources** from the environment—such as mining, fishing, lumbering, and especially **agriculture**.

Primate city A country's largest city—ranking atop the urban **hierarchy**—most expressive of the national culture and usually (but not always) the capital city as well.

Process Causal force that shapes spatial pattern or structure as it unfolds over time.

Proletariat Lower-income working class in a community or society. People who own no capital or means of production and who live by selling their labor.

Prorupt A type of **state** territorial shape that exhibits a narrow, elongated land extension leading away from the main body of territory.

Protectorate In Britain's system of colonial administration, the protectorate was a designation that involved the guarantee of certain rights (such as the restriction of European settlement and **land alienation**) to peoples who had been placed under the control of the Crown.

Puna In Andean South America, the highest-lying habitable altitudinal zone—ca. 12,000 to 15,000 feet (3600 to 4500 meters)—between the tree line (upper limit of the *tierra fría*) and the snow line (lower limit of the *tierra helada*). Too cold and barren to support anything but the grazing of sheep and other hardy livestock. Also known as the *páramos*.

Push-pull concept The idea that **migration** flows are simultaneously stimulated by conditions in the source area, which tend to drive people away, and by the perceived attractiveness of the destination.

Qanat In desert zones, particularly in Iran and western China, an underground tunnel built to carry **irrigation** water from nearby mountains (where **orographic precipitation** occurs) to the arid flatlands below.

Quaternary economic activity Activities engaged in the collection, processing, and manipulation of *information*.

Rain shadow effect The relative dryness in areas downwind of or beyond mountain ranges caused by **orographic precipitation**, whereby moist air masses are forced to deposit most of their water content in the highlands.

Region A commonly used term and a geographic concept of central importance. An **area** on the earth's surface marked by certain properties.

Regional complementarity See **complementarity**.

Relative location The regional position or **situation** of a place relative to the position of other places. Distance, **accessibility**, and connectivity affect relative location.

Relict boundary A political boundary that has ceased to function, but the imprint of which can still be detected on the cultural landscape.

Relief Vertical difference between the highest and lowest elevations within a particular area.

Relocation diffusion Sequential **diffusion** process in which the items being diffused are transmitted by their carrier agents as they evacuate the old areas and relocate to new ones. A disease can move from one population cluster to another in this manner, running its course in one area before fully invading the next. The most common form of relocation diffusion involves the spreading of innovations by a **migrating** population.

Remote sensing The acquisition of geographic information from great distances and over broad areas, using instruments mounted on high-altitude aircraft or orbiting satellites. Unlike camera photography, remote sensors make use of the entire spectrum of visible and nonvisible light (see Fig. 6–17).

Rural density A measure that indicates the number of persons per unit area living in the rural areas of a country, outside of the urban concentrations.

Sahel Semiarid zone extending across most of Africa between the southern margins of the arid Sahara and the moister tropical savanna and forest zone to the south. Chronic drought, **desertification**, and overgrazing have contributed to severe famines in this area since 1970.

Scale Representation of a real-world phenomenon at a certain level of reduction or generalization. In **cartography**, the ratio of map distance to ground distance; indicated on a map as bar graph, representative fraction, and/or verbal statement.

Scale economies See **economies of scale.**

Secondary economic activity Activities that process raw materials and transform them into finished industrial products. The *manufacturing* sector.

Sector model A structural model of the American central city that suggests land-use areas conform to a wedge-shaped pattern focused on the downtown core (see Fig. 3–1B).

Sedentary Permanently attached to a particular area, a population fixed in its location. The opposite of **nomadic.**

Selva Tropical rainforest.

Separate development The spatial expression of South Africa's "grand" **apartheid** scheme, wherein nonwhite groups were required to settle in segregated "homelands." During the 1980s, the implementation of this plan collapsed, and by 1990 its dismantling—together with the system of apartheid—was well under way.

Sequent occupance The notion that successive societies leave their cultural imprints on a place, each contributing to the cumulative cultural landscape.

Shantytown Unplanned slum development on the margins of Third World cities, dominated by crude dwellings and shelters mostly made of scrap wood, iron, and even pieces of cardboard.

Sharecropping Relationship between a large landowner and farmers on the land wherein the farmers pay rent for the land they farm by giving the landlord a share of the annual harvest.

Shatter belt **Region** caught between stronger, colliding external cultural-political forces, under persistent stress and often fragmented by aggressive rivals. Eastern Europe and Southeast Asia are classic examples.

Shifting agriculture Cultivation of crops in recently cut and burned tropical forest clearings, soon to be abandoned in favor of newly cleared nearby forest land. Also known as *slash-and-burn agriculture.*

Sinicization Giving a Chinese cultural imprint; Chinese **acculturation.**

Site The internal locational attributes of a place, including its local spatial organization and physical setting.

Situation The external locational attributes of a place; its **relative location** or regional position with reference to other nonlocal places.

Slash-and-burn agriculture See **shifting agriculture.**

Spatial Pertaining to space on the earth's surface. Synonym for *geographic(al).*

Spatial diffusion See **diffusion.**

Spatial interaction See **complementarity, transferability,** and **intervening opportunity.**

Spatial model See **model.**

Squatter settlement See **shantytown.**

State A politically organized territory that is administered by a sovereign government and is recognized by a significant portion of the international community. A state must also contain a permanent resident population, an organized economy, and a functioning internal circulation system.

Stratification (social) In a layered or stratified society, the population is divided into a **hierarchy** of social classes. In an industrialized society, the **proletariat** is at the lower end; **elites** that possess capital and control the means of production are at the upper level. In the traditional **caste system** of Hindu India, the "untouchables" form the lowest class or caste, whereas the still-wealthy remnants of the princely class are at the top.

Subsequent boundary A political boundary that developed contemporaneously with the evolution of the major elements of the cultural landscape through which it lies.

Subsistence Existing on the minimum necessities to sustain life; spending most of one's time in pursuit of survival.

Suburban downtown Significant concentration of diversified economic activities around a highly **accessible** suburban location, including retailing, light industry, and a variety of major corporate and commercial operations. Late-twentieth-century coequal to the American central city's **central business district (CBD)** (see Fig. 3–15).

Superimposed boundary A political boundary emplaced by powerful outsiders on a developed human landscape. Usually ignores pre-existing cultural-spatial patterns, such as the border that now divides North and South Korea.

Supranational A venture involving three or more national states—political, economic, or cultural cooperation to promote shared objectives. Europe's Economic Community or **Common Market** is such an organization.

System Any group of objects or institutions and their mutual interactions. Geography treats systems that are expressed spatially, such as **regions.**

Systematic geography Topical geography: cultural, political, economic geography, and the like.

Takeoff Economic concept to identify a stage in a country's development when conditions are set for a domestic Industrial Revolution, which occurred in Britain in the late eighteenth century and in Japan in the late nineteenth following the Meiji Restoration.

Tectonics   See **plate tectonics**.

Tell   The lower slopes and narrow coastal plains along the Atlas Mountains in northwesternmost Africa, where the majority of the **Maghreb** region's population is clustered.

Templada   See *tierra templada*.

Terracing   The transformation of a hillside or mountain slope into a step-like sequence of horizontal fields for intensive cultivation (see Figs. J–7 and 9–19).

Territoriality   A country's or more local community's sense of property and attachment toward its territory, as expressed by its determination to keep it inviolable and strongly defended.

Territorial morphology   A **state's** geographical shape, which can have a decisive impact on its spatial cohesion and political viability. A **compact** shape is most desirable; among the less efficient shapes are those exhibited by **elongated, fragmented, perforated**, and **prorupt** states.

Territorial sea   Zone of seawater adjacent to a country's coast, held to be part of the national territory and treated as a segment of the sovereign **state**.

Tertiary economic activity   Activities that engage in *services*—such as transportation, banking, retailing, education, and routine office-based jobs.

Theocracy   State whose government is under the control of a ruler who is deemed to be divinely guided or under the control of a group of religious leaders, as in post-Khomeini Iran. The opposite of the theocratic state is the *secular* state.

Tierra caliente   The lowest of the altitudinal zones into which the human settlement of Middle and South America is classified according to elevation. The *caliente* is the hot humid coastal plain and adjacent slopes up to 2500 feet (750 meters) above sea level.

The natural vegetation is the dense and luxuriant tropical rainforest; the crops are sugar, bananas, cacao, and rice in the lower areas, and coffee, tobacco, and corn along the higher slopes.

Tierra fría   The cold, high-lying zone of settlement in Andean South America, extending from about 6000 feet (1850 meters) in elevation up to nearly 12,000 feet (3600 meters). Coniferous trees stand here; upward they change into scrub and grassland. There are also important pastures within the *fría*, and wheat can be cultivated. Several major population clusters in western South America lie at these altitudes.

Tierra helada   The highest and coldest altitudinal zone in Andean South America (lying above 15,000 feet [4500 meters]), an uninhabitable environment of permanent snow and ice that extends upward to the Andes' highest peaks of more than 20,000 feet (6000 meters).

Tierra templada   The intermediate altitudinal zone of settlement in Middle and South America, lying between 2500 feet (750 meters) and 6000 feet (1850 meters) in elevation. This is the "temperate" zone, with moderate temperatures compared to the *tierra caliente* below. Crops include coffee, tobacco, corn, and some wheat.

Time-space convergence   The increasing nearness of places that occurs as modern transportation breakthroughs progressively reduce the time-distance between them. The trip by boat from New York to San Francisco before the Civil War took months. After 1870, the transcontinental railroad cut the travel time to less than two weeks; by 1930, trains made the journey in three days. After 1945, propeller planes made the trip in about 12 hours; and by 1960, non-stop jet planes achieved today's travel time of five hours.

Totalitarian   A government whose leaders rule by absolute control, tolerating no differences of political opinion.

Transculturation   Cultural borrowing that occurs when different cultures of approximately equal complexity and technological level come into close

contact. In **acculturation**, by contrast, an indigenous society's culture is modified by contact with a technologically superior society.

Transferability   The capacity to move a good from one place to another at a bearable cost; the ease with which a commodity may be transported.

Underdeveloped countries (UDCs) Countries that, by various measures, suffer seriously from negative economic and social conditions, including low **per capita** incomes, poor nutrition, inadequate health, and related disadvantaged circumstances.

Unitary state   A **nation-state** that has a centralized government and administration that exercises power equally over all parts of the state.

Urbanization   A term with several connotations. The proportion of a country's population living in urban places is its level of urbanization. The **process** of urbanization involves the movement to, and the clustering of, people in towns and cities—a major force in every geographic realm today. Another kind of urbanization occurs when an expanding city absorbs rural countryside and transforms it into suburbs; in the case of the Third World city, this also generates peripheral **shantytowns**.

Urban (metropolitan) area   The entire built-up, non-rural area and its population, including the most recently constructed suburban appendages. Provides a better picture of the dimensions and population of such an area than the delimited municipality (central city) that forms its heart.

Urban realms model   A spatial generalization of the large, late-twentieth-century American city. It is shown to be a widely dispersed, multicentered metropolis consisting of increasingly independent zones or *realms*, each focused on its own **suburban downtown**; the only exception is the shrunken central realm, which is focused on the **central business district** (see Fig. 3–17).

Veld Open grassland on the South African plateau, becoming mixed with scrub at lower elevations where it is called *bushveld*. As in Middle and South America, there is an altitudinal zonation into *highveld*, *middleveld*, and *lowveld*.

Voluntary migration Population movement in which people relocate in response to perceived opportunity, not because they are forced to move.

Von Thünen model Explains the location of agricultural activities in a commercial, profit-making economy. A **process** of spatial competition allocates various farming activities into concentric rings around a central market city, with profit-earning capability the determining force in how far a crop locates from the market. The original (1826) *Isolated State* model now applies to the continental scale (Fig. 1–11); in certain Third World areas, however, transportation conditions still shape an application reminiscent of the original (Fig. 6–30).

Water table When precipitation falls on the soil, some of the water is drawn downward through the pores in the soil and rock under the force of gravity. Below the surface it reaches a level where it can go no further; there it joins water that already saturates the rock completely. This water that ''stands'' underground is *groundwater*, and the upper level of the zone of saturation is the *water table*.

World geographic realm See **geographic realm**.

World-lake concept The extension of **Exclusive Economic Zone** principles beyond the 200-mile limit, so that all the seas and oceans would be carved up by the claims of the world's coastal countries. The black median-line boundaries in Fig. 10–20 show what this hypothetical global maritime partitioning would look like.

PHOTO CREDITS

INTRODUCTION

Fig. I–1: Robert Caputo.
Fig. I–2: Craig Aurness/Westlight.
Fig. I–4: Mark & Evelyne Bernheim/Woodfin Camp & Associates.
Fig. I–7: V. Shone/Gamma-Liaison.
Fig. I–18: Michael DeCamp/The Image Bank.
Fig. I–20: Brian Seed.
Fig. I–21: Jean Gaumy/Magnum.

CHAPTER 1

Fig. 1–6: Peter Menzel.
Fig. 1–8: Brian Brake/Photo Researchers.
Fig. 1–13: Rapho Sarval/Photo Researchers.
Fig. 1–14: Rozbroj/Explorer/Photo Researchers.
Fig. 1–15: Jean Paul Nacivet/Leo de Wys.
Fig. 1–18: Malanca/Sipa Press.
Fig. 1–19: Peter Turnley/Black Star.
Fig. 1–22: David Warren/Shostal Associates.
Fig. 1–23: Michael & Barbara Reed/Earth Scenes.
Fig. 1–26: Harvey Lloyd/The Stock Market.
Fig. 1–28: Dino Fracchia/Agencia Contrasto.
Fig. 1–30: Porterfield/Chickering/Photo Researchers.
Fig. 1–31: Luc Delahaye/Sipa-Press.

AUSTRALIA VIGNETTE

A–1: Courtesy Harm de Blij.
A–4: Courtesy Harm de Blij.
A–7: David Moore/Black Star.
A–9: G. R. Roberts/Documentary Photos.

CHAPTER 2

Fig. 2–3: Nik Wheeler/Black Star.
Fig. 2–4: David Turnley/Black Star.
Fig. 2–6: Tass/Sovfoto.
Fig. 2–10: The Stock Market.
Fig. 2–13: Paolo Koch/Photo Researchers.
Fig. 2–14: Jim Brandenburg.
Fig. 2–15: Laski/Sipa Press.
Fig. 2–18: Fred McConnaughey/Photo Researchers.
Fig. 2–19: Howard Sochureck/Woodfin Camp & Associates.
Fig. 2–20: George Holton/Photo Researchers.
Fig. 2–21: Sovfoto/Eastfoto.
Fig. 2–22: Sovfoto/Eastfoto.

CHAPTER 3

Fig. 3–2: Ted Spiegel/Black Star.
Fig. 3–5: F. Gohier/Photo Researchers.
Fig. 3–7: See attached.
Fig. 3–10: See attached.
Fig. 3–14: See attached.
Fig. 3–15: Joe Rodriguez/Black Star.
Fig. 3–23: Larry Lefever/Grant Heilman Photography.
Fig. 3–25: Jeffrey Smith/Woodfin Camp & Associates.
Fig. 3–26: North Star Steel Co./Christopher Studio, Inc.
Fig. 3–28: James Blank.
Fig. 3–29: Shostal Associates.
Fig. 3–31: Allen Green/Photo Researchers.
Fig. 3–32: James Blank.

JAPAN VIGNETTE

J–4: D. Bartruff/FPG International.
J–5: Courtesy Harm de Blij.
J–6: Ben Simmons/The Stock Market.
J–7: Robert Perron/f/Stop Pictures.
J–8: Ken Straiton/The Stock Market.
J–9: John Burbank/The Image Works.
J–10: Courtesy Harm de Blij.

CHAPTER 4

Fig. 4–4: Alon Reininger/Woodfin Camp & Associates.
Fig. 4–5: J. D. Cunningham.
Fig. 4–9: Porterfield/ Chickering/Photo Researchers.
Fig. 4–10: James P. Blair/National Geographic Society.
Fig. 4–11: Courtesy Jim Forbes.
Fig. 4–12: Courtesy Harm de Blij.
Fig. 4–14: Guillereno Aldana.
Fig. 4–16: Courtesy Harm de Blij.
Fig. 4–17: Peter C. Poulides.
Fig. 4–19: Gamma-Liaison.

CHAPTER 5

Fig. 5–4: Suzanne L. Murphy/FPG International.
Fig. 5–7: Renato de Sousa/Abril Imagens.
Fig. 5–11: Courtesy Harm de Blij.
Fig. 5–12: Nicholas Devore III/Photographers Aspen.
Fig. 5–14: Luis Villota/The Stock Market.
Fig. 5–16: Victor Englebert/Black Star.
Fig. 5–18: Courtesy Harm de Blij.
Fig. 5–19: Courtesy Harm de Blij.

BRAZIL VIGNETTE

B–1: Claudio Versiani/Abril Imagens.
B–3: Courtesy Harm de Blij.
B–4: Ellis Herwig/The Picture Cube.
B–5: Nicholas Devore III/Bruce Coleman, Inc.
B–6: S. Jorge/The Image Bank.
B–7: Wil Wilkins/Photo Researchers.
B–8: Loren McIntyre/Woodfin Camp & Associates.

CHAPTER 6

Fig. 6–4: Robert Azzi/Woodfin Camp & Associates.
Fig. 6–6: Lu Ming-Shen/Tom Stack & Associates.
Fig. 6–11: FPG International.
Fig. 6–15: Stern/Black Star.
Fig. 6–17: Courtesy NASA.
Fig. 6–18: M. Timothy O'Keefe/Bruce Coleman, Inc.
Fig. 6–21: Chip Hires/Gamma-Liaison.
Fig. 6–22: Dennis Stock/Magnum.
Fig. 6–23: Bill Foley/Woodfin Camp & Associates.
Fig. 6–24: Tom McHugh/Photo Researchers.
Fig. 6–26: Courtesy Harm de Blij.
Fig. 6–28: Leo de Wys, Inc.
Fig. 6–29: Tony Suau/Black Star.
Fig. 6–31: Steve McCurry/Magnum.

CHAPTER 7

Fig. 7–3: Courtesy Harm de Blij.
Fig. 7–5: Nicholas Devore III/Photographers Aspen.
Fig. 7–6: George Holton/Photo Researchers.
Fig. 7–8: Owen Franken/Stock, Boston.
Fig. 7–11: Ray Wilkinson/Woodfin Camp & Associates.
Fig. 7–12: Stuart Cohen/Comstock.
Fig. 7–13: Mark & Evelyne Bernheim/Woodfin Camp & Associates.
Fig. 7–14: Mark & Evelyne Bernheim/Woodfin Camp & Associates.
Fig. 7–16: Richard & Mary Magruder/The Image Bank.
Fig. 7–21: Mark & Evelyne Bernheim/Woodfin Camp & Associates.
Fig. 7–22: Agence France-Presse.
Fig. 7–23: P. Robert/Sygma.
Fig. 7–24: Jason Lauré.
Fig. 7–26: Courtesy Harm de Blij.
Fig. 7–27: Stuart Cohen.
Fig. 7–29: Thomas W. Friedman/Photo Researchers.
Fig. 7–31: Thomas W. Friedman/Photo Researchers.

SOUTH AFRICA VIGNETTE

SA–1: Ian Berry/Magnum.
SA–3: Courtesy Harm de Blij.
SA–4: Courtesy Harm de Blij.
SA–5: Nicholas Devore III/Photographers Aspen.
SA–6: G. Mendel/Magnum.

CHAPTER 8

Fig. 8–2: William Thompson.
Fig. 8–6: Roland & Sabrina Michaud/Woodfin Camp & Associates.
Fig. 8–7: Robert Weinreb/Bruce Coleman, Inc.
Fig. 8–9: UPI/Bettman.
Fig. 8–11: Kate Bader.
Fig. 8–12: Henebry Photography.
Fig. 8–13: Jacques Jangoux/Peter Arnold, Inc.
Fig. 8–15: Steve McCurry/Magnum.
Fig. 8–17: Erja Lahdenpera/Lehpikuva/Woodfin Camp & Associates.
Fig. 8–20: Paolo Koch/Photo Researchers.
Fig. 8–22: Chip Hires/Gamma-Liaison.
Fig. 8–24: Franke Keating/Photo Researchers.
Fig. 8–26: Bill Wassman/The Stock Market.

CHAPTER 9

Fig. 9–3: Charles Cole/Sipa Press.
Fig. 9–5: Georg Gerster/Comstock.
Fig. 9–7: Courtesy Harm de Blij.
Fig. 9–10: Hiroji Kubota/Magnum.
Fig. 9–12: M. Timothy O'Keefe/Bruce Coleman, Inc.
Fig. 9–14: Courtesy Harm de Blij.
Fig. 9–15: Peter Charlesworth/JB Pictures.
Fig. 9–16: Peter Charlesworth/JB Pictures.
Fig. 9–19: Georg Gerster/Comstock.
Fig. 9–20: Sheryl McNee/Southern Stock Photos.
Fig. 9–22: Gary I. Rothstein.
Fig. 9–24: Reinhold Messner.
Fig. 9–25: Courtesy Harm de Blij.
Fig. 9–27: Bruce Stromberg/The Stock Market.
Fig. 9–29: Hiroji Kubota/Magnum.
Fig. 9–30: FPG International.

CHAPTER 10

Fig. 10–4: Christophe Loviny/Gamma-Liaison.
Fig. 10–9: FPG International.
Fig. 10–10: Paul Chesley/Photographers Aspen.
Fig. 10–11: Larry Downing/Woodfin Camp & Associates.
Fig. 10–13: Porterfield/Chickering/Photo Researchers.
Fig. 10–15: Jean Kugler/FPG International.
Fig. 10–16: Garry Milburn/Tom Stack & Associates.
Fig. 10–18: Georg Gerster/Comstock.

PACIFIC VIGNETTE

P–2: Thomas Hopker/Woodfin Camp & Associates.
P–3: Courtesy Harm de Blij.
P–4: Paul Chesley/Photographers Aspen.

LIST OF MAPS

GEOGRAPHICAL INDEX (Gazetteer)

ALL ENTRIES REFER TO MAP CONTENTS

INDEX